T0178369

Lecture Notes in Artificial Intelligence 11440

Subseries of Lecture Notes in Computer Science

More information about this series at http://www.springer.com/series/1244

Qiang Yang · Zhi-Hua Zhou ·
Zhiguo Gong · Min-Ling Zhang ·
Sheng-Jun Huang (Eds.)

Advances in Knowledge Discovery and Data Mining

23rd Pacific-Asia Conference, PAKDD 2019
Macau, China, April 14–17, 2019
Proceedings, Part II

Editors
Qiang Yang
Hong Kong University of Science and Technology
Hong Kong, China

Zhiguo Gong
University of Macau
Taipa, Macau, China

Sheng-Jun Huang
Nanjing University of Aeronautics and Astronautics
Nanjing, China

Zhi-Hua Zhou
Nanjing University
Nanjing, China

Min-Ling Zhang
Southeast University
Nanjing, China

ISSN 0302-9743 ISSN 1611-3349 (electronic)
Lecture Notes in Artificial Intelligence
ISBN 978-3-030-16144-6 ISBN 978-3-030-16145-3 (eBook)
https://doi.org/10.1007/978-3-030-16145-3

Library of Congress Control Number: 2019934768

LNCS Sublibrary: SL7 – Artificial Intelligence

This Springer imprint is published by the registered company Springer Nature Switzerland AG
The registered company address is: Gewerbestrasse 11, 6330 Cham, Switzerland

PC Chairs' Preface

It is our great pleasure to introduce the proceedings of the 23rd Pacific-Asia Conference on Knowledge Discovery and Data Mining (PAKDD 2019). The conference provides an international forum for researchers and industry practitioners to share their new ideas, original research results, and practical development experiences from all KDD-related areas, including data mining, data warehousing, machine learning, artificial intelligence, databases, statistics, knowledge engineering, visualization, decision-making systems, and the emerging applications.

We received 567 submissions to PAKDD 2019 from 46 countries and regions all over the world, noticeably with submissions from North America, South America, Europe, and Africa. The large number of submissions and high diversity of submission demographics witness the significant influence and reputation of PAKDD. A rigorous double-blind reviewing procedure was ensured via the joint efforts of the entire Program Committee consisting of 55 Senior Program Committee (SPC) members and 379 Program Committee (PC) members.

The PC Co-Chairs performed an initial screening of all the submissions, among which 25 submissions were desk rejected due to the violation of submission guidelines. For submissions entering the double-blind review process, each one received at least three quality reviews from PC members or in a few cases from external reviewers (with 78.5% of them receiving four or more reviews). Furthermore, each valid submission received one meta-review from the assigned SPC member who also led the discussion with the PC members. The PC Co-Chairs then considered the recommendations and meta-reviews from SPC members, and looked into each submission as well as its reviews and PC discussions to make the final decision. For borderline papers, additional reviews were further requested and thorough discussions were conducted before final decisions.

As a result, 137 out of 567 submissions were accepted, yielding an acceptance rate of 24.1%. We aim to be strict with the acceptance rate, and all the accepted papers are presented in a total of 20 technical sessions. Each paper was allocated 15 minutes for oral presentation and 2 minutes for Q/A. The conference program also featured three keynote speeches from distinguished data mining researchers, five cutting-edge workshops, six comprehensive tutorials, and one dedicated data mining contest session.

We wish to sincerely thank all SPC members, PC members and externel reviewers for their invaluable efforts in ensuring a timely, fair, and highly effective paper review and selection procedure. We hope that readers of the proceedings will find that the PAKDD 2019 technical program was both interesting and rewarding.

February 2019

Zhiguo Gong
Min-Ling Zhang

General Chairs' Preface

On behalf of the Organizing Committee, it is our great pleasure to welcome you to Macau, China for the 23rd Pacific-Asia Conference on Knowledge Discovery and Data Mining (PAKDD 2019). Since its first edition in 1997, PAKDD has well established as one of the leading international conferences in the areas of data mining and knowledge discovery. This year, after its four previous editions in Beijing (1999), Hong Kong (2001), Nanjing (2007), and Shenzhen (2011), PAKDD was held in China for the fifth time in the fascinating city of Macau, during April 14–17, 2019.

First of all, we are very grateful to the many authors who submitted their work to the PAKDD 2019 main conference, satellite workshops, and data mining contest. We were delighted to feature three outstanding keynote speakers: Dr. Jennifer Neville from Purdue University, Professor Hui Xiong from Baidu Inc., and Professor Josep Domingo-Ferrer from Universitat Rovira i Virgili. The conference program was further enriched with six high-quality tutorials, five workshops on cutting-edge topics, and one data mining contest on AutoML for lifelong machine learning.

We would like to express our gratitude to the contributions of the SPC members, PC members, and external reviewers, led by the PC Co-Chairs, Zhiguo Gong and Min-Ling Zhang. We are also very thankful to the other Organizing Committee members: Workshop Co-Chairs, Hady W. Lauw and Leong Hou U, Tutorial Co-Chairs, Bob Durrant and Yang Yu, Contest Co-Chairs, Hugo Jair Escalante and Wei-Wei Tu, Publicity Co-Chairs, Yi Cai, Xiangnan Kong, Gang Li, and Yasuo Tabei, Proceedings Chair, Sheng-Jun Huang, and Local Arrangements Chair, Andrew Jiang. We wish to extend our special thanks to Honorary Co-Chairs, Hiroshi Motoda and Lionel M. Ni, for their enlightening support and advice throughout the conference organization.

We appreciate the hosting organization University of Macau, and our sponsors Macao Convention & Exhibition Association, Intel, Baidu, for their institutional and financial support of PAKDD 2019. We also appreciate the Fourth Paradigm Inc., ChaLearn, Microsoft, and Amazon for sponsoring the PAKDD 2019 data mining contest. We feel indebted to the PAKDD Steering Committee for its continuing guidance and sponsorship of the paper award and student travel awards.

Last but not least, our sincere thanks go to all the participants and volunteers of PAKDD 2019—there would be no conference without you. We hope you enjoy PAKDD 2019 and your time in Macau, China.

February 2019

Qiang Yang
Zhi-Hua Zhou

Organization

Organizing Committee

Honorary Co-chairs

Hiroshi Motoda	Osaka University, Japan
Lionel M. Ni	University of Macau, SAR China

General Co-chairs

Qiang Yang	Hong Kong University of Science and Technology, SAR China
Zhi-Hua Zhou	Nanjing University, China

Program Committee Co-chairs

Zhiguo Gong	University of Macau, China
Min-Ling Zhang	Southeast University, China

Workshop Co-chairs

Hady W. Lauw	Singapore Management University, Singapore
Leong Hou U	University of Macau, China

Tutorial Co-chairs

Bob Durrant	University of Waikato, New Zealand
Yang Yu	Nanjing University, China

Contest Co-chairs

Hugo Jair Escalante	INAOE, Mexico
Wei-Wei Tu	The Fourth Paradigm Inc., China

Publicity Co-chairs

Yi Cai	South China University of Technology, China
Xiangnan Kong	Worcester Polytechnic Institute, USA
Gang Li	Deakin University, Australia
Yasuo Tabei	RIKEN, Japan

Proceedings Chair

Sheng-Jun Huang	Nanjing University of Aeronautics and Astronautics, China

Local Arrangements Chair

Andrew Jiang	Macao Convention & Exhibition Association, China

Steering Committee

Co-chairs

Ee-Peng Lim	Singapore Management University, Singapore
Takashi Washio	Institute of Scientific and Industrial Research, Osaka University, Japan

Treasurer

Longbing Cao	Advanced Analytics Institute, University of Technology, Sydney, Australia

Members

Dinh Phung	Monash University, Australia (Member since 2018)
Geoff Webb	Monash University, Australia (Member since 2018)
Jae-Gil Lee	Korea Advanced Institute of Science & Technology, Korea (Member since 2018)
Longbing Cao	Advanced Analytics Institute, University of Technology, Sydney, Australia (Member since 2013, Treasurer since 2018)
Jian Pei	School of Computing Science, Simon Fraser University (Member since 2013)
Vincent S. Tseng	National Cheng Kung University, Taiwan (Member since 2014)
Gill Dobbie	University of Auckland, New Zealand (Member since 2016)
Kyuseok Shim	Seoul National University, Korea (Member since 2017)

Life Members

P. Krishna Reddy	International Institute of Information Technology, Hyderabad (IIIT-H), India (Member since 2010, Life Member since 2018)
Joshua Z. Huang	Shenzhen University, China (Member since 2011, Life Member since 2018)
Ee-Peng Lim	Singapore Management University, Singapore (Member since 2006, Life Member since 2014, Co-chair 2015–2017, Chair 2018–2020)
Hiroshi Motoda	AFOSR/AOARD and Osaka University, Japan (Member since 1997, Co-chair 2001–2003, Chair 2004–2006, Life Member since 2006)

Rao Kotagiri	University of Melbourne, Australia (Member since 1997, Co-chair 2006–2008, Chair 2009–2011, Life Member since 2007, Co-sign since 2006)
Huan Liu	Arizona State University, USA (Member since 1998, Treasurer 1998–2000, Life Member since 2012)
Ning Zhong	Maebashi Institute of Technology, Japan (Member since 1999, Life Member since 2008)
Masaru Kitsuregawa	Tokyo University, Japan (Member since 2000, Life Member since 2008)
David Cheung	University of Hong Kong, SAR China (Member since 2001, Treasurer 2005–2006, Chair 2006–2008, Life Member since 2009)
Graham Williams	Australian National University, Australia (Member since 2001, Treasurer 2006–2017, Co-sign since 2006, Co-chair 2009–2011, Chair 2012–2014, Life Member since 2009)
Ming-Syan Chen	National Taiwan University, Taiwan (Member since 2002, Life Member since 2010)
Kyu-Young Whang	Korea Advanced Institute of Science & Technology, Korea (Member since 2003, Life Member since 2011)
Chengqi Zhang	University of Technology Sydney, Australia (Member since 2004, Life Member since 2012)
Tu Bao Ho	Japan Advanced Institute of Science and Technology, Japan (Member since 2005, Co-chair 2012–2014, Chair 2015–2017, Life Member since 2013)
Zhi-Hua Zhou	Nanjing University, China (Member since 2007, Life Member since 2015)
Jaideep Srivastava	University of Minnesota, USA (Member since 2006, Life Member since 2015)
Takashi Washio	Institute of Scientific and Industrial Research, Osaka University (Member since 2008, Life Member since 2016, Co-chair 2018–2020)
Thanaruk Theeramunkong	Thammasat University, Thailand (Member since 2009, Life Member since 2017)

Past Members

Hongjun Lu	Hong Kong University of Science and Technology, SAR China (Member 1997–2005)
Arbee L. P. Chen	National Chengchi University, Taiwan (Member 2002–2009)
Takao Terano	Tokyo Institute of Technology, Japan (Member 2000–2009)

Tru Hoang Cao	Ho Chi Minh City University of Technology, Vietnam (Member 2015–2017)
Myra Spiliopoulou	Information Systems, Otto-von-Guericke-University Magdeburg (Member 2013–2019)

Senior Program Committee

James Bailey	University of Melbourne, Australia
Albert Bifet	Telecom ParisTech, France
Longbin Cao	University of Technology Sydney, Australia
Tru Cao	Ho Chi Minh City University of Technology, Vietnam
Peter Christen	Australian National University, Australia
Peng Cui	Tsinghua University, China
Guozhu Dong	Wright State University, USA
Benjamin C. M. Fung	McGill University, Canada
Bart Goethals	University of Antwerp, Belgium
Geoff Holmes	University of Waikato, New Zealand
Qinghua Hu	Tianjin University, China
Xia Hu	Texas A&M University, USA
Sheng-Jun Huang	Nanjing University of Aeronautics and Astronautics, China
Shuiwang Ji	Texas A&M University, USA
Kamalakar Karlapalem	IIIT Hyderabad, India
George Karypis	University of Minnesota, USA
Latifur Khan	University of Texas at Dallas, USA
Byung S. Lee	University of Vermont, USA
Jae-Gil Lee	KAIST, Korea
Gang Li	Deakin University, Australia
Jiuyong Li	University of South Australia, Australia
Ming Li	Nanjing University, China
Yu-Feng Li	Nanjing University, China
Shou-De Lin	National Taiwan University, Taiwan
Qi Liu	University of Science and Technology of China, China
Weiwei Liu	University of New South Wales, Australia
Nikos Mamoulis	University of Ioannina, Greece
Wee Keong Ng	Nanyang Technological University, Singapore
Sinno Pan	Nanyang Technological University, Singapore
Jian Pei	Simon Fraser University, Canada
Wen-Chih Peng	National Chiao Tung University, Taiwan
Rajeev Raman	University of Leicester, UK
Chandan K. Reddy	Virginia Tech, USA
Krishna P. Reddy	IIIT Hyderabad, India
Kyuseok Shim	Seoul National University, Korea
Myra Spiliopoulou	Otto-von-Guericke-University Magdeburg, Germany
Masashi Sugiyama	RIKEN/The University of Tokyo, Japan
Jiliang Tang	Michigan State University, USA

Kai Ming Ting	Federation University, Australia
Hanghang Tong	Arizona State University, USA
Vincent S. Tseng	National Chiao Tung University, Taiwan
Fei Wang	Cornell University, USA
Jianyong Wang	Tsinghua University, China
Jie Wang	University of Science and Technology of China, China
Wei Wang	University of California at Los Angeles, USA
Takashi Washio	Osaka University, Japan
Jia Wu	Macquarie University, Australia
Xindong Wu	Mininglamp Software Systems, China
Xintao Wu	University of Arkansas, USA
Xing Xie	Microsoft Research Asia, China
Jeffrey Xu Yu	Chinese University of Hong Kong, SAR China
Osmar R. Zaiane	University of Alberta, Canada
Zhao Zhang	Soochow University, China
Feida Zhu	Singapore Management University, Singapore
Fuzhen Zhuang	Institute of Computing Technology, CAS, China

Program Committee

Saurav Acharya	University of Vermont, USA
Swati Agarwal	BITS Pilani Goa, India
David Albrecht	Monash University, Australia
David Anastasiu	San Jose State University, USA
Luiza Antonie	University of Guelph, Canada
Xiang Ao	Institute of Computing Technology, CAS, China
Sunil Aryal	Deakin University, Australia
Elena Baralis	Politecnico di Torino, Italy
Jean Paul Barddal	Pontifícia Universidade Católica do Paraná, Brazil
Arnab Basu	Indian Institute of Management Bangalore, India
Gustavo Batista	Universidade de São Paulo, Brazil
Bettina Berendt	KU Leuven, Belgium
Raj K. Bhatnagar	University of Cincinnati, USA
Arnab Bhattacharya	Indian Institute of Technology, Kanpur, India
Kevin Bouchard	Université du Quebec a Chicoutimi, Canada
Krisztian Buza	Eotvos Lorand University, Hungary
Lei Cai	Washington State University, USA
Rui Camacho	Universidade do Porto, Portugal
K. Selcuk Candan	Arizona State University, USA
Tanmoy Chakraborty	Indraprastha Institute of Information Technology Delhi, India
Shama Chakravarthy	University of Texas at Arlington, USA
Keith Chan	Hong Kong Polytechnic University, SAR China
Chia Hui Chang	National Central University, Taiwan
Bo Chen	Monash University, Australia
Chun-Hao Chen	Tamkang University, Taiwan

Lei Chen	Nanjing University of Posts and Telecommunications, China
Meng Chang Chen	Academia Sinica, Taiwan
Rui Chen	Samsung Research America, USA
Shu-Ching Chen	Florida International University, USA
Songcan Chen	Nanjing University of Aeronautics and Astronautics, China
Yi-Ping Phoebe Chen	La Trobe University, Australia
Yi-Shin Chen	National Tsing Hua University, Taiwan
Zhiyuan Chen	University of Maryland Baltimore County, USA
Jiefeng Cheng	Tencent Cloud Security Lab, China
Yiu-ming Cheung	Hong Kong Baptist University, SAR China
Silvia Chiusano	Politecnico di Torino, Italy
Jaegul Choo	Korea University, Korea
Kun-Ta Chuang	National Cheng Kung University, Taiwan
Bruno Cremilleux	Université de Caen Normandie, France
Chaoran Cui	Shandong University of Finance and Economics, China
Lin Cui	Nanjing University of Aeronautics and Astronautics, China
Boris Cule	University of Antwerp, Belgium
Bing Tian Dai	Singapore Management University, Singapore
Dao-Qing Dai	Sun Yat-Sen University, China
Wang-Zhou Dai	Nanjing University, China
Xuan-Hong Dang	IBM T.J. Watson Research Center, USA
Jeremiah Deng	University of Otago, New Zealand
Zhaohong Deng	Jiangnan University, China
Lipika Dey	Tata Consultancy Services, India
Bolin Ding	Data Analytics and Intelligence Lab, Alibaba Group, China
Steven H. H. Ding	McGill University, Canada
Trong Dinh Thac Do	University of Technology Sydney, Australia
Gillian Dobbie	University of Auckland, New Zealand
Xiangjun Dong	Qilu University of Technology, China
Dejing Dou	University of Oregon, USA
Bo Du	Wuhan University, China
Boxin Du	Arizona State University, USA
Lei Duan	Sichuan University, China
Sarah Erfani	University of Melbourne, Australia
Vladimir Estivill-Castro	Griffith University, Australia
Xuhui Fan	University of Technology Sydney, Australia
Rizal Fathony	University of Illinois at Chicago, USA
Philippe Fournier-Viger	Harbin Institute of Technology (Shenzhen), China
Yanjie Fu	Missouri University of Science and Technology, USA
Dragan Gamberger	Rudjer Boskovic Institute, Croatia
Niloy Ganguly	Indian Institute of Technology Kharagpur, India
Junbin Gao	University of Sydney, Australia

Wei Gao	Nanjing University, China
Xiaoying Gao	Victoria University of Wellington, New Zealand
Angelo Genovese	Università degli Studi di Milano, Italy
Arnaud Giacometti	University Francois Rabelais of Tours, France
Heitor M. Gomes	Telecom ParisTech, France
Chen Gong	Nanjing University of Science and Technology, China
Maciej Grzenda	Warsaw University of Technology, Poland
Lei Gu	Nanjing University of Posts and Telecommunications, China
Yong Guan	Iowa State University, USA
Himanshu Gupta	IBM Research, India
Sunil Gupta	Deakin University, Australia
Michael Hahsler	Southern Methodist University, USA
Yahong Han	Tianjin University, China
Satoshi Hara	Osaka University, Japan
Choochart Haruechaiyasak	National Electronics and Computer Technology Center, Thailand
Jingrui He	Arizona State University, USA
Shoji Hirano	Shimane University, Japan
Tuan-Anh Hoang	Leibniz University of Hanover, Germany
Jaakko Hollmén	Aalto University, Finland
Tzung-Pei Hong	National University of Kaohsiung, Taiwan
Chenping Hou	National University of Defense Technology, China
Michael E. Houle	National Institute of Informatics, Japan
Hsun-Ping Hsieh	National Cheng Kung University, Taiwan
En-Liang Hu	Yunnan Normal University, China
Juhua Hu	University of Washington Tacoma, USA
Liang Hu	University of Technology Sydney, Australia
Wenbin Hu	Wuhan University, China
Chao Huang	University of Notre Dame, USA
David Tse Jung Huang	University of Auckland, New Zealand
Jen-Wei Huang	National Cheng Kung University, Taiwan
Nam Huynh	Japan Advanced Institute of Science and Technology, Japan
Akihiro Inokuchi	Kwansei Gakuin University, Japan
Divyesh Jadav	IBM Research, USA
Sanjay Jain	National University of Singapore, Singapore
Szymon Jaroszewicz	Polish Academy of Sciences, Poland
Songlei Jian	University of Technology Sydney, Australia
Meng Jiang	University of Notre Dame, USA
Bo Jin	Dalian University of Technology, China
Toshihiro Kamishima	National Institute of Advanced Industrial Science and Technology, Japan
Wei Kang	University of South Australia, Australia
Murat Kantarcioglu	University of Texas at Dallas, USA
Hung-Yu Kao	National Cheng Kung University, Taiwan

Shanika Karunasekera	University of Melbourne, Australia
Makoto P. Kato	Kyoto University, Japan
Chulyun Kim	Sookmyung Women University, Korea
Jungeun Kim	Korea Advanced Institute of Science and Technology, Korea
Kyoung-Sook Kim	Artificial Intelligence Research Center, Japan
Yun Sing Koh	University of Auckland, New Zealand
Xiangnan Kong	Worcester Polytechnic Institute, USA
Irena Koprinska	University of Sydney, Australia
Ravi Kothari	Ashoka University, India
P. Radha Krishna	National Institute of Technology, Warangal, India
Raghu Krishnapuram	Indian Institute of Science Bangalore, India
Marzena Kryszkiewicz	Warsaw University of Technology, Poland
Chao Lan	University of Wyoming, USA
Hady Lauw	Singapore Management University, Singapore
Thuc Duy Le	University of South Australia, Australia
Ickjai J. Lee	James Cook University, Australia
Jongwuk Lee	Sungkyunkwan University, Korea
Ki Yong Lee	Sookmyung Women's University, Korea
Ki-Hoon Lee	Kwangwoon University, Korea
Sael Lee	Seoul National University, Korea
Sangkeun Lee	Korea University, Korea
Sunhwan Lee	IBM Research, USA
Vincent C. S. Lee	Monash University, Australia
Wang-Chien Lee	Pennsylvania State University, USA
Yue-Shi Lee	Ming Chuan University, Taiwan
Zhang Lei	Anhui University, China
Carson K. Leung	University of Manitoba, Canada
Bohan Li	Nanjing University of Aeronautics and Astronautics, China
Jianmin Li	Tsinghua University, China
Jianxin Li	Deakin University, Australia
Jundong Li	Arizona State University, USA
Nan Li	Alibaba, China
Peipei Li	Hefei University of Technology, China
Qian Li	University of Technology Sydney, Australia
Rong-Hua Li	Beijing Institute of Technology, China
Shao-Yuan Li	Nanjing University, China
Sheng Li	University of Georgia, USA
Wenyuan Li	University of California, Los Angeles, USA
Wu-Jun Li	Nanjing University, China
Xiaoli Li	Institute for Infocomm Research, A*STAR, Singapore
Xue Li	University of Queensland, Australia
Yidong Li	Beijing Jiaotong University, China
Zhixu Li	Soochow University, China

Defu Lian	University of Electronic Science and Technology of China, China
Sungsu Lim	Chungnam National University, Korea
Chunbin Lin	Amazon AWS, USA
Hsuan-Tien Lin	National Taiwan University, Taiwan
Jerry Chun-Wei Lin	Western Norway University of Applied Sciences, Norway
Anqi Liu	California Institute of Technology, USA
Bin Liu	IBM Research, USA
Jiajun Liu	Renmin University of China, China
Jiamou Liu	University of Auckland, New Zealand
Jie Liu	Nankai University, China
Lin Liu	University of South Australia, Australia
Liping Liu	Tufts University, USA
Shaowu Liu	University of Technology Sydney, Australia
Zheng Liu	Nanjing University of Posts and Telecommunications, China
Wenpeng Lu	Qilu University of Technology, China
Jun Luo	Machine Intelligence Lab, Lenovo Group Limited, China
Wei Luo	Deakin University, Australia
Huifang Ma	Northwest Normal University, China
Marco Maggini	University of Siena, Italy
Giuseppe Manco	ICAR-CNR, Italy
Silviu Maniu	Universite Paris-Sud, France
Naresh Manwani	International Institute of Information Technology, Hyderabad, India
Florent Masseglia	Inria, France
Tomoko Matsui	Institute of Statistical Mathematics, Japan
Michael Mayo	The University of Waikato, New Zealand
Stephen McCloskey	The University of Sydney, Australia
Ernestina Menasalvas	Universidad Politécnica de Madrid, Spain
Xiangfu Meng	Liaoning Technical University, China
Xiaofeng Meng	Renmin University of China, China
Jun-Ki Min	Korea University of Technology and Education, Korea
Nguyen Le Minh	Japan Advanced Institute of Science and Technology, Japan
Leandro Minku	The University of Birmingham, UK
Pabitra Mitra	Indian Institute of Technology Kharagpur, India
Anirban Mondal	Ashoka University, India
Taesup Moon	Sungkyunkwan University, Korea
Yang-Sae Moon	Kangwon National University, Korea
Yasuhiko Morimoto	Hiroshima University, Japan
Animesh Mukherjee	Indian Institute of Technology Kharagpur, India
Miyuki Nakano	Advanced Institute of Industrial Technology, Japan
Mirco Nanni	ISTI-CNR, Italy

Richi Nayak	Queensland University of Technology, Australia
Raymond Ng	University of British Columbia, Canada
Wilfred Ng	Hong Kong University of Science and Technology, SAR China
Cam-Tu Nguyen	Nanjing University, China
Hao Canh Nguyen	Kyoto University, Japan
Ngoc-Thanh Nguyen	Wroclaw University of Science and Technology, Poland
Quoc Viet Hung Nguyen	Griffith University, Australia
Arun Reddy Nelakurthi	Arizona State University, USA
Thanh Nguyen	Deakin University, Australia
Thin Nguyen	Deakin University, Australia
Athanasios Nikolakopoulos	University of Minnesota, USA
Tadashi Nomoto	National Institute of Japanese Literature, Japan
Eirini Ntoutsi	Leibniz University of Hanover, Germany
Kouzou Ohara	Aoyama Gakuin University, Japan
Kok-Leong Ong	La Trobe University, Australia
Shirui Pan	University of Technology Sydney, Australia
Yuangang Pan	University of Technology Sydney, Australia
Guansong Pang	University of Adelaide, Australia
Dhaval Patel	IBM T.J. Watson Research Center, USA
Francois Petitjean	Monash University, Australia
Hai Nhat Phan	New Jersey Institute of Technology, USA
Xuan-Hieu Phan	University of Engineering and Technology, VNUHN, Vietnam
Vincenzo Piuri	Università degli Studi di Milano, Italy
Vikram Pudi	International Institute of Information Technology, Hyderabad, India
Chao Qian	University of Science and Technology of China, China
Qi Qian	Alibaba Group, China
Tang Qiang	Luxembourg Institute of Science and Technology, Luxembourg
Biao Qin	Renmin University of China, China
Jie Qin	Eidgenössische Technische Hochschule Zürich, Switzerland
Tho Quan	Ho Chi Minh City University of Technology, Vietnam
Uday Kiran Rage	University of Tokyo, Japan
Chedy Raissi	Inria, France
Vaibhav Rajan	National University of Singapore, Singapore
Santu Rana	Deakin University, Australia
Thilina N. Ranbaduge	Australian National University, Australia
Patricia Riddle	University of Auckland, New Zealand
Hiroshi Sakamoto	Kyushu Institute of Technology, Japan
Yücel Saygin	Sabanci University, Turkey
Mohit Sharma	Walmart Labs, USA
Hong Shen	Adelaide University, Australia

Wei Shen	Nankai University, China
Xiaobo Shen	Nanjing University of Science and Technology, China
Victor S. Sheng	University of Central Arkansas, USA
Chuan Shi	Beijing University of Posts and Telecommunications, China
Motoki Shiga	Gifu University, Japan
Hiroaki Shiokawa	University of Tsukuba, Japan
Moumita Sinha	Adobe, USA
Andrzej Skowron	University of Warsaw, Poland
Yang Song	University of New South Wales, Australia
Arnaud Soulet	University of Tours, France
Srinath Srinivasa	International Institute of Information Technology, Bangalore, India
Fabio Stella	University of Milan-Bicocca, Italy
Paul Suganthan	University of Wisconsin-Madison, USA
Mahito Sugiyama	National Institute of Informatics, Japan
Guangzhong Sun	University of Science and Technology of China, China
Yuqing Sun	Shandong University, China
Ichigaku Takigawa	Hokkaido University, Japan
Mingkui Tan	South China University of Technology, China
Ming Tang	Institute of Automation, CAS, China
Qiang Tang	Luxembourg Institute of Science and Technology, Luxembourg
David Taniar	Monash University, Australia
Xiaohui (Daniel) Tao	University of Southern Queensland, Australia
Vahid Taslimitehrani	PhysioSigns Inc., USA
Maguelonne Teisseire	Irstea, France
Khoat Than	Hanoi University of Science and Technology, Vietnam
Lini Thomas	International Institute of Information Technology, Hyderabad, India
Hiroyuki Toda	NTT Corporation, Japan
Son Tran	New Mexico State University, USA
Allan Tucker	Brunel University London, UK
Jeffrey Ullman	Stanford University, USA
Dinusha Vatsalan	Data61, CSIRO, Australia
Ranga Vatsavai	North Carolina State University, USA
Joao Vinagre	LIAAD—INESC TEC, Portugal
Bay Vo	Ho Chi Minh City University of Technology, Vietnam
Kitsana Waiyamai	Kasetsart University, Thailand
Can Wang	Griffith University, Australia
Chih-Yu Wang	Academia Sinica, Taiwan
Hongtao Wang	North China Electric Power University, China
Jason T. L. Wang	New Jersey Institute of Technology, USA
Lizhen Wang	Yunnan University, China
Peng Wang	Southeast University, China
Qing Wang	Australian National University, Australia

Peixiang Zhao	Florida State University, USA
Pengpeng Zhao	Soochow University, China
Yanchang Zhao	CSIRO, Australia
Zhongying Zhao	Shandong University of Science and Technology, China
Zhou Zhao	Zhejiang University, China
Huiyu Zhou	University of Leicester, UK
Shuigeng Zhou	Fudan University, China
Xiangmin Zhou	RMIT University, Australia
Yao Zhou	Arizona State University, USA
Chengzhang Zhu	University of Technology Sydney, Australia
Huafei Zhu	Nanyang Technological University, Singapore
Pengfei Zhu	Tianjin University, China
Tianqing Zhu	University of Technology Sydney, Australia
Xingquan Zhu	Florida Atlantic University, USA
Ye Zhu	Deakin University, Australia
Yuanyuan Zhu	Wuhan University, China
Arthur Zimek	University of Southern Denmark, Denmark
Albrecht Zimmermann	Université de Caen Normandie, France

External Reviewers

Ji Feng
Xuan Huo
Bin-Bin Jia
Zhi-Yu Shen
Yanping Sun
Xuan Wu
Zheng-Fan Wu
Yafu Xiao
Yang Yang
Meimei Yang
Han-Jia Ye
Peng Zhao

Sponsoring Organizations

University of Macau

Macao Convention & Exhibition Association

Intel

Baidu Inc.

Contents – Part II

Sequential Pattern Mining

Weakly Supervised Learning

Recommender System

Deep Learning Models and Applications

Semi-interactive Attention Network for Answer Understanding in Reverse-QA

Qing Yin[1], Guan Luo[2], Xiaodong Zhu[3], Qinghua Hu[1], and Ou Wu[1](✉)

[1] Tianjin University, Tianjin 300110, China
{qingyin,huqinghua,wuou}@tju.edu.cn

[2] NLPR, Chinese Academy of Sciences, Beijing, China
gluo@nlpr.ia.ac.cn

[3] University of Shanghai for Science, Shanghai, China
zhuxd81@gmail.com

Abstract. Question answering (QA) is an important natural language processing (NLP) task and has received much attention in academic research and industry communities. Existing QA studies assume that questions are raised by humans and answers are generated by machines. Nevertheless, in many real applications, machines are also required to determine human needs or perceive human states. In such scenarios, machines may proactively raise questions and humans supply answers. Subsequently, machines should attempt to understand the true meaning of these answers. This new QA approach is called reverse-QA (rQA) throughout this paper. In this work, the human answer understanding problem is investigated and solved by classifying the answers into predefined answer-label categories (e.g., *True*, *False*, *Uncertain*). To explore the relationships between questions and answers, we use the interactive attention network (IAN) model and propose an improved structure called semi-interactive attention network (Semi-IAN). Two Chinese data sets for rQA are compiled. We evaluate several conventional text classification models for comparison, and experimental results indicate the promising performance of our proposed models.

Keywords: Question answering · Reverse-QA · Attention · LSTM

1 Introduction

Question answering (QA) is applied in many real applications, such as robots and intelligent customer service. The goal of QA is to provide a satisfactory answer depending on users' question [1]. QA can provide a more natural way for humans to acquire information than traditional search engines [2].

In nearly all existing QA studies and applications, the questions are raised by humans and the answers are generated by machines. In other words, in existing QA, humans are the questioners and machines are the answerers. Therefore, selecting a satisfactory answer from candidate answer corpora, which are also

Q. Yang et al. (Eds.): PAKDD 2019, LNAI 11440, pp. 3–15, 2019.
https://doi.org/10.1007/978-3-030-16145-3_1

known as answer selection, is the key problem in QA [3]. In addition to meeting users' information requirements, machines in some real applications, such as telephone survey [15], are also required to actively acquire the exact needs or feedbacks of users. Accordingly, machines may choose to proactively raise questions to users and then analyze their answers. In other words, machines are the questioners and humans are the answerers. This process is a reverse of the conventional QA process and is called reverse-QA (rQA) in this paper. Figure 1 shows the conventional QA and rQA processes.

In conventional QA, the key problem is understanding users' questions. On the contrary, the key problem in rQA is understanding users' answers. In the present study, two types of machine-launched questions are considered, namely, true-or-false (T/F) and multiple-choice (MC) questions. Table 1 shows two illustrative examples for T/F and MC questions. Some human answers in Table 1 are easy to analyze. For example, the "Yes" and "No" answers to the first question can clearly distinguish the category. However, other answers are vague and difficult to process for their exact meanings. For example, the answer "I was a teacher last year" for the second question is a 'false' response, but it is easily classified as a 'true' response. Hence, understanding users' answers is not a trivial task.

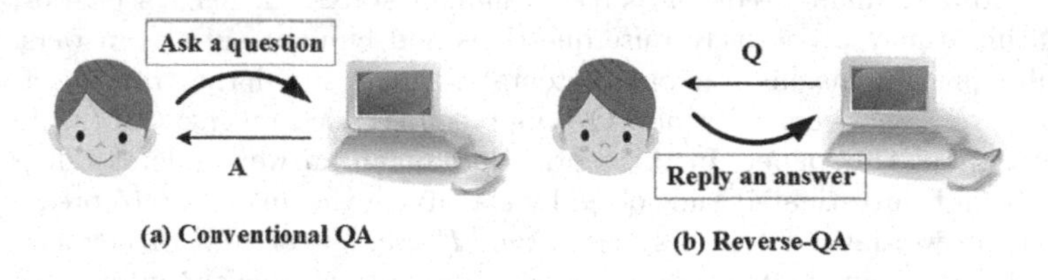

Fig. 1. Difference between conventional QA (a) and reverse-QA (b).

Table 1. Illustrative examples of the T/F and the MC questions.

Type	Question	(Possible) answers by human
T/F	Do you like running?	Yes/A little/No/Sometimes
T/F	Are you a teacher?	Sometimes/I'm not sure/You guess/ I was a teacher last year
MC	Would you like coffee or tea?	Coffee/Tea/No, thanks/Either is ok
MC	Are you usually walking, cycling, or driving to work?	Walk/Except cycling/I lost my job/By train/It all depends

As far as we know, it is the first to focus on the rQA procedure and corresponding answer understanding. Considering that no public data set is available for this work, two data sets[1] are compiled to construct and test the models for

[1] These two data sets have been uploaded to Github and the URL is provided after anonymous review.

rQA. We simply take the answer understanding in rQA as a classification task and use several common text classification techniques. These classification algorithms ignore the relationships between a (machine) question and an (human) answer. To this end, two new models based on deep neural network (DNN) are proposed. The first model is based on the interactive attention network (IAN) [18]. The second model is a simplified but a more effective version of IAN and is called semi-interactive attention network (Semi-IAN). The experimental results suggest the potential of the proposed models. The contributions of this work are summarized as follows:

- We investigate a new QA procedure called rQA. Moreover, a new problem called answer understanding for rQA is proposed. Two data sets are collected and labeled depending on two common question types, namely, T/F and MC. The two data sets can be used to construct and evaluate new models.
- Two new models are proposed to capture the semantic relationships between questions and answers. The proposed IAN model is based on the raw IAN [18], which is initially designed for opinion mining. The proposed Semi-IAN model, which considers questions as background, achieves highest accuracy throughout the experiments.

2 Related Work

2.1 QA

QA is a crucial NLP task that depends on natural language understanding and domain knowledge [4]. Given a question from users, QA returns an answer via answer selection or generation based on a knowledge base. In most existing QA studies, the answer selection is implemented by the matching between a question and candidate answers or documents. The answer that has the highest match score is usually selected and returned to users. According to the matching procedure, most existing QA methods can be divided into two categories:

- Hard-crafted feature-based methods. This category of methods extracts lexical features to represent questions and candidate answers [5,6]. Chen et al. [7] proposed a feature fusion strategy for various features, including carefully crafted lexical, syntactic, and word order features.
- Deep feature-based methods. This category of methods extracts deep features via a CNN or long-short time memory network (LSTM) [8]. Kadlec et al. [9] presented a pointer-style attention mechanism for text feature representation in QA.

Some other works have focused on questions [10] and visual QA [11] .

2.2 Text Classification

Text classification aims to predict the category of an input text sample. The category can be semantic (e.g., political and economic) or sentimental (e.g., positive and negative) [19]. Besides some rule-based methods [12], most existing methods are based on machine learning theories. Nearly all classical (shallow) classifiers have been used in text classification, such as support vector machine (SVM) [13], KNN [14], and logistic regression (LR), etc.

In recent years, the emergence of deep learning as a powerful technique for nearly all NLP tasks has facilitated the adoption of classical DNNs (e.g., CNN [16] and Recurrent Neural Network (RNN) [17]) for these tasks.

3 Methodology

Answer understanding in rQA can be formulated into a text classification problem as follows. By considering a machine-question and human-answer pair (q, s) and a predefined answer-label A, we aim to predict the category c $(c \in A)$ for s.

In this study, two common types of questions are considered, namely, T/F and MC. The two types of questions correspond to two scenarios, namely, T/F and MC rQA. The primary difference between T/F and MC rQA lies in the definitions of the answer-label set A.

In the T/F rQA, the answer-label set A can be set as $\{True, False, Uncertain\}$ regardless of the question. In the MC rQA, we assume that the option set is I, and the answer-label set A is the union of all the subsets of I plus the $Uncertain$ element. For example, if I is $\{opt1, opt2, opt3\}$, then the set A is $\{\{opt1\}, \{opt2\}, \{opt3\}, \{opt1, opt2\}, \{opt2, opt3\}, \{opt1, opt3\}, \{opt1, opt2, opt3\}, Null, Uncertain\}$.

The following part introduces how the answer category c is inferred in the two rQA scenarios above.

3.1 Answer Understanding for T/F rQA

In T/F rQA, we aim to classify the human answer s into one element (category) of the predefined answer-label set $\{True, False, Uncertain\}$. Intuitively, most existing classification methods can be used.

3.1.1 Text Classification-Based Methods

If s or the simple concatenation of q and s is taken as a piece of input texts, then three typical methods are obtained and listed below:

- **Rule-based method:** This method relies on some key words, such as '*ok*', '*yes*', '*not*'. These key words directly indicate a 'true' or 'false' answer.
- **Bag-of-words:** This method firstly extracts a bag-of-word (BOW) feature vector for the input text and then classifies the texts using conventional shallow classifiers such as SVM [13], LR.
- **DNN-based method:** This method first extracts a deep feature vector for the input text and then classifies it on the basis of the softmax layer of the involved DNN.

3.1.2 The Proposed Models

The above-mentioned text classification-based methods independently extract the feature vectors of the question and the answer or extract one feature vector from the texts by simply concatenating the question and the answer. These two strategies simply ignore the semantic relationship between the question and the answer. Intuitively, the question and answer texts can facilitate the analysis of their counterpart. Alternatively, their feature extraction procedures should not be independent.

In opinion mining, Ma et al. [18] proposed an IAN to extract features for target and contextual texts. In IAN, the target information is used in feature extraction for contextual texts, and the latter is also used in feature extraction for the former. This network is used with a slight modification to utilize the relationship between a pair of question and answer. The overall architecture is shown in Fig. 2(a).

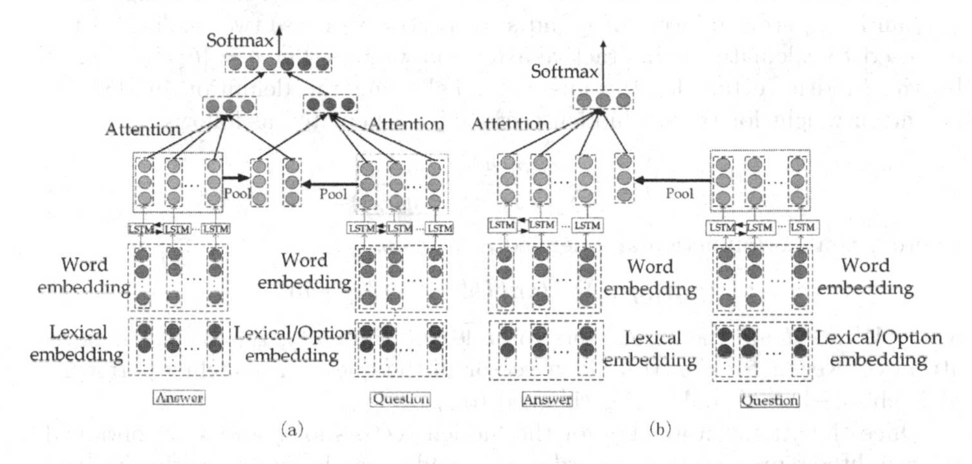

Fig. 2. (a) IAN for answer understanding in rQA. (b) Structure of the Semi-IAN. (In question modeling part of MC, the additional embedding is option embedding.)

We let w_q^t and w_s^t denote the tth words in the question q and the answer s, respectively. The embeddings of w_q^t and w_s^t consist of two parts. The first part is word embedding, and the second part is lexical embedding. The word embedding is implemented by the standard word2vec algorithm. The lexical embedding relies on a ρ-hot encoding [22] on a pre-compiled dictionary of several key words. In this study, our dictionary contains six classes of key words: affirmative, privative, suspicious, positive, negative, supposed.

The two input embedding sequences are then fed into a bi-directional LSTM to infer the hidden representation of each word in a sentence. In our model, the forward LSTM at the tth input word is as follows:

$$\begin{aligned}
i_t &= \sigma(W_i[c_{t-1}, h_{t-1}, x_t, l_t] + b_i) \\
f_t &= \sigma(W_f[c_{t-1}, h_{t-1}, x_t, l_t] + b_f) \\
o_t &= \sigma(W_o[c_t, h_{t-1}, x_t, l_t] + b_o) \\
d_t &= \sigma(W_d[c_{t-1}, h_{t-1}, x_t, l_t] + b_d) \\
c_t &= i_t \otimes d_t + f_t \otimes d_{t-1} \\
h_t &= o_t \otimes \tanh(c_t)
\end{aligned} \tag{1}$$

where x_t and l_t are the word and lexical embedding for the tth word, respectively; i_t and d_t are the input vectors of the input unit and the input gate, respectively, for the tth word; o_t and h_t are the output and hidden vectors, respectively; f_t is the output of the forget gate; c_t is the internal state of the memory cell in a LSTM unit; σ is the sigmoid active function. The backward LSTM is very similar to the forward one except that the input sequence is fed in a reversed way. The output of the bi-LSTM and the lexical embedding of words are concentrated.

After the hidden vectors for each input word are obtained, two pooling vectors q_{avg} and s_{avg} are produced for q and s, respectively. These two pooling vectors are used to calculate the interactive attention weights. Let $h = [h_s^1, h_s^2, \cdots, h_s^n]$ be the hidden vectors for the answer s. Following the definition in [18], the attention weight for the tth hidden vector h_s^t is calculated as follows:

$$\alpha_t = \frac{exp(\gamma(h_s^t, q_{avg}))}{\sum_k exp(\gamma(h_s^k, q_{avg}))} \tag{2}$$

where γ is a score function and defined as follows:

$$\gamma(h_s^t, q_{avg}) = tanh(h_s^t \cdot W \cdot q_{avg}^T + b) \tag{3}$$

where W and b are the parameters to be learnt. Similarly, Eq. (2) indicates the attention weight for the tth hidden vector in the question modeling part when h_s^t is changed to h_q^t and q_{avg} is changed to s_{avg}.

Once the attention weights for the hidden vectors for q and s are obtained, the weighted representations recorded as q_r and s_r can be subsequently obtained. The final feature vector q_s is the concatenation of q_r and s_r, i.e., $q_s = [q_r^T, s_r^T]^T$.

The proposed model is based the original IAN proposed by Ma et al. [18] with slight modifications. Intuitively, the answer is the focus and the question is the background. Nevertheless, the answer and the question are symmetric and equal in the model shown in Fig. 2. To this end, an improved model is proposed and shown in Fig. 2(b). Because this model only used partial interactive information, it is called semi-interactive attention network (Semi-IAN).

In the Fig. 2(b) model, the final feature vector q_s does not contain the feature representation q_r of the question part. Alternatively, the question is only used as a background text in answer understanding. The experimental results show that this Semi-IAN model outperforms the raw IAN.

3.2 Answer Understanding for MC rQA

In MC rQA, the size of the answer-label set A depends on the number of candidate options for selection. In other words, the number of categories varies depending on the concrete question. Conventional classification technique is inappropriate for the scenario with varied number of categories. Consequently, the original answer classification in MC rQA should be transformed into a new classification problem with a fixed number of categories. In this work, the transformation is implemented as follows. Without loss of generality, let the option set I of one MC question be $\{option1,\ option2,\ option3\}$. The raw answer classification is transformed into three new classification subtasks. The first sub-task is about *option*1; the second is about *option*2; the third is about *option*3. Each subtask infers an answer category from the set *True*, *False*, *Uncertain* for the corresponding option. With the above transformation, the new classification problem is with a fixed number (i.e., three) categories.

The model for the transformed MC question is same as the model introduced in Sect. 3.1.2 with only one difference that lexical embedding in question modeling part is replaced with option embedding. The option embedding is also based on the ρ-hot encoding to indicate the current option to be considered, as shown in Fig. 2. Therefore, if k options exist, the model should be run k times to infer the category for each option depending on the input question and human answers. For example, if the predicted categories for the three options are $Ture$, $False$, $True$, then the final output category c is $\{option1,\ option3\}$.

4 Experimental Data Construction

Existing QA and text classification benchmark data sets are inappropriate for training and evaluating rQA models. Therefore, two data sets are compiled with a standard labeling process. The two data sets are named rQAData1 and rQAdata2 for the T/F rQA and MC rQA, respectively. The type of MC questions we studied is limited in the type that the options appear in the question, which we call option-contained MC questions.

For the two data sets, the questions are constructed as follows. First, seven domains are selected, namely, encyclopedia, insurance, personal, purchases, leisure interests, medical health, and exercise. Ten graduate students, specifically six males and four females, were invited to participate in the data compiling using Email advertising from our experimental laboratory. All the participants are Chinese and in the age of [22], [30]. Considering that the question and answer generations are not difficult to understand, we did not give special instructions to the participants. Each participant was allowed to construct 50 to 60 questions. We obtain 20 insurance questions, 30 encyclopedia questions, and 40 questions in each of the remaining areas for T/F rQA; for MC rQA, we obtain 20 questions in the insurance field and 40 questions in each of the remaining categories. Finally, 503 questions are obtained after deleting some invalid questions. Among that, the numbers of questions in rQAData1 and rQAData2 are 250 and 253, respectively.

The answers are constructed as follows. The 503 questions are equally assigned to the 10 participants, and each question is given 18 to 22 answers. The participants also labeled their answers.

For the rQAData1, the types of answers are roughly divided into affirmative, negative, uncertain, and unrelated. Given that the uncertain and unrelated answers are similar in function to the next question, we classify them as the same class. In this way, each answer is tagged with '1' for true, '0' for false, and '2' for uncertain or unrelated. Each sample consists of three components: question, answer, and label. The numbers of training and testing samples are 4,080 and 1,000, respectively.

For the rQAData2, the number of options for each MC question are different and cannot be categorized uniformly. Thus, we add the option information to the MC questions and get a series of transformed MC questions as described in Sect. 3.2. Therefore, the same answer to the same question will have different labels for dissimilar options. Similarly, '1' indicates that the answer is an 'true' answer to the current option, '0' implies that the answer is a 'false' answer to the current option, and '2' denotes that the answer is 'uncertain' answers to the current option or the answer is meaningless to this question. Each sample consists of four components: question, option, answer, and label. There are 12,923 transformed MC questions, and the numbers of training and testing samples are 9,074 and 3,876, respectively. Table 2 presents the brief summary.

Table 2. Statistics of our rQA datasets.

Data sets	Train samples	Test samples	*False/True/Uncertain*
rQAData1	4,047	1,000	2,153/2,266/628
rQAData2	9,047	3,876	6,257/4,872/1,814

5 Experiment

5.1 Comparative Methods

To demonstrate the validity of our models, the following methods are considered in the experimental comparison.

- **Rule-based method:** This method is introduced in Sect. 3.1.1.
- **BOW+:** The main idea of BOW is to extract features with a BOW model and send them to a classifier [20] In the subsequent experiments, we use LR and SVM as the classification algorithms.
- **CNN/LSTM/Bi-LSTM (A):** The CNN/LSTM/Bi-LSTM (A) network is used to extract deep features from answers only. The features are then fed to the softmax classification layer to obtain the possible results of the category.
- **CNN/LSTM/Bi-LSTM (A+Q):** Unlike CNN/LSTM/Bi-LSTM (A), these methods extract deep features from the concatenation of answers and questions.

Table 3. Classification accuracies on different training set settings.

Model	rQAData1(A)	rQAData1(A+Q)	rQAData2(A)	rQAData2(A+Q)
Ruled-Based	0.438	/	0.298	/
BOW+LR	0.486	0.473	0.318	0.314
BOW+SVM	0.649	0.612	0.503	0.532
CNN	0.673	0.615	0.521	0.530
LSTM	0.685	0.652	0.532	0.534
Bi-LSTM	0.708	0.669	0.534	0.530
IAN$^+$	/	0.720	/	0.578
Semi-IAN	/	**0.735**	/	**0.585**

Our proposed methods are listed as follows:

- **IAN$^+$**: This method is introduced in Sect. 3.1.2. It is similar to the raw IAN model with a slight modification.
- **Semi-IAN:** The network structure of this method is shown in Fig. 2(b).

5.2 Training Settings

All the DNN models are trained by applying Keras that is equipped with Tensorflow. Both our data sets use accuracy as the metric. The division of training and test data is shown in Table 2. The specific training settings used in our experiment are listed as follows:

- In BOW, we put words that appear more than twice into the dictionary.
- For SVM, parameters C and g are searched via five-fold cross validations from {0.1, 1, 5, 10, 100} and {0.01, 0.1, 1, 5, 10}, respectively. For LR, the codes in MATLAB are used, and all the parameters are set to default.
- For deep models, the word embedding dimension is set to 300 by GloVe [21].
- In both IAN$^+$ and Semi-IAN, the ρ-hot encoding [22] is used in the lexical and option embeddings. In ρ-hot encoding, the size k is searched in [1, 2, 4, $\cdots$, 16]; the parameter ρ is searched in [0.1, 0.2, $\cdots$, 1].

5.3 Overall Competing Results

Table 3 shows the classification accuracies for all competing methods. Among these methods, the rule-based method has the worst effect. The two BOW methods achieve better accuracies than the rule-based method. Although the SVM-based method is 0.2% higher than some deep network methods, the overall method based on deep learning has a better effect. On both data sets, the accuracies of the LSTM-based method are considerably higher than those of the CNN-based method. This finding illustrates that LSTM is more suitable for the problem investigated in this study than CNN. The underlying reason is that

LSTM can effectively extract the semantic expression of text information with complex time correlation and different lengths.

On rQAData1, Bi-LSTM outperforms LSTM over 1% and 3%. On rQAData2, the accuracies of Bi-LSTM-based and LSTM-based approaches are same, basically.

In general, the performance of most CNN/LSTM/Bi-LSTM (A+Q) is worse than that of CNN/LSTM/Bi-LSTM (A) on our data sets. The reason is that although the adding of the questions is definitely useful, simple concatenation of answers and questions is not conducive to the judgment of answers. It is necessary to propose new methods of introducing questions.

Compared with the Bi-LSTM (A+Q) model, IAN^+ improves the performance by approximately 5.1% and 4.8% on the rQAData1 and rQAData2, respectively. We can see that Semi-IAN achieves the best performance. The main reason may be that the IAN^+ and Semi-IAN provide a more effective way to combine the answer and question texts than the simple way that discards question texts or directly concatenates answer and question texts. Furthermore, the Semi-IAN model highlights the answer texts compared with IAN^+.

In Table 3, the accuracies on rQAData2 are lower than those on rQAData1. The classification problem investigated on rQAData2 contains additional information, that is, the options. Additional information also brings more challenges, so the classification for rQAData2 is more difficult than that for rQAData1. The main reason is the understanding for MC answers requires more domain knowledge. In our feature work, we will introduce knowledge graph for the involved domains into our models.

5.4 Discussion on the Key Modules in Our Models

The input embedding and attention modules are crucial for a deep model. The lexical and option embeddings used in our input layer incorporate domain knowledge into the network. We first evaluate this embedding strategy.

Table 4 shows the results of the Bi-LSTM, IAN^+ and Semi-IAN with or without additional embedding containing lexical and option embeddings on the two data sets. In the six groups of comparison, four groups showed the effectiveness of additional coding. In particular, there was 1% improvement in our semi-INA model, indicating that the proposed lexical and option embedding are useful.

Three important parameters are involved in our lexical and option, namely, size k, ρ in lexical embedding, and ρ in option embedding. We record the classification performance with different values of these parameters in the experiments. The green curves in Fig. 3(a) and (b) show the accuracy variations in terms of different ρ values in lexical embedding. The blue curve in Fig. 3(b) shows the accuracy variations in terms of different ρ values in option embedding. The three curves indicate that the tuning of the ρ value in lexical embedding is useful. In Fig. 3(c) and (d), the accuracies when $k = 14$ for rQData1 and $k = 8$ for rQData2 are larger than those when $k = 1$ for both sets. The tuning for k is also useful.

Table 4. Results of with or without lexical and option embedding.

Model	W/O	rQAData1	rQAData2
Bi-LSTM (A+Q)	W	0.665	0.495
Bi-LSTM (A+Q)	O	0.669	0.491
IAN$^+$	W	0.720	0.578
IAN$^+$	O	0.728	0.546
Semi-IAN	W	**0.735**	**0.585**
Semi-IAN	O	0.726	0.575

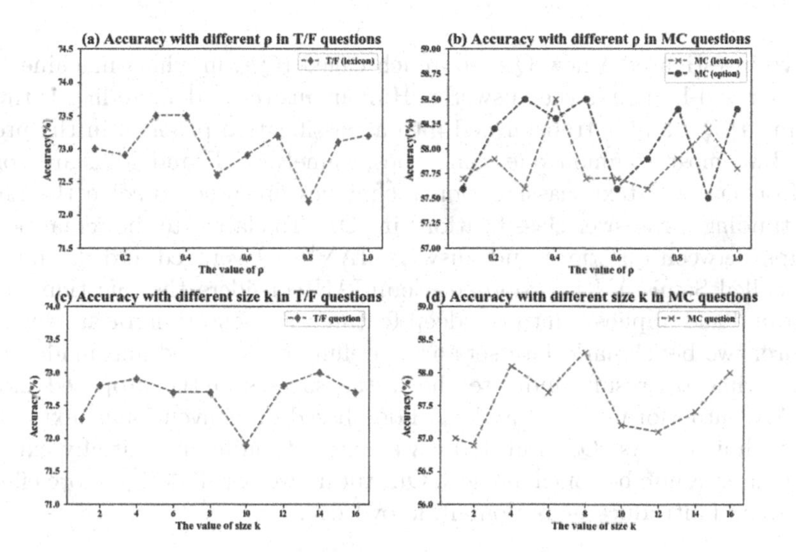

Fig. 3. Results of accuracy under different parameter values. (Color figure online)

Next, we investigate the effectiveness of a new attention mechanism, namely, CRF attention [23], which is proven useful in text sentiment analysis. Table 5 shows the results of IAN$^+$ and Semi-IAN with and without the CRF attention layer. On both data sets, the accuracies of Semi-IAN with CRF attention mechanism have different degrees of reduction. Although the IAN$^+$ with CRF attention exhibits a certain improvement compared with the model without additional attention, its accuracy remains lower than that of our Semi-IAN. In summary, the CRF attention does not exhibit an outstanding performance in the experiment. The partial reason may lie in that the lengths of answers are usually short, whereas CRF attention is suitable for texts with moderate lengths. In our future work, we will investigate more attention mechanisms to solve this problem.

Table 5. Results of models with or without the CRF attention layer.

Model	W/O CRF attention	rQAData1	rQAData2
IAN$^+$	W	0.722	0.548
IAN$^+$	O	0.720	0.578
Semi-IAN	W	0.716	0.583
Semi-IAN	O	0.735	0.585

6 Conclusion

We have investigated a new QA approach called rQA, in which machine is the questioner and human is the answerer. Human answer understanding is the key problem in rQA and is transformed into a classification problem in the present study. Two most common question types, namely, T/F and MC, are considered. Conventional text classification techniques are used to solve the answer understanding (or answer classification) in rQA. To elaborate the semantic relationships between questions and answers, IAN$^+$ is leveraged and an improved model called Semi-IAN is proposed. Semi-IAN considers the questions as the background and applies it into the deep feature representation for answers. Furthermore, two benchmark data sets are carefully constructed and made public. The experimental results indicate the initial success of the proposed models. Semi-IAN outperforms IAN$^+$ and methods based on conventional text classification techniques. As rQA and its answer understanding are initially explored, there remains a number of challenges. Our future work will design more effective networks and introduce more domain knowledge.

Acknowledgments. This work is partially supported by NSFC (61673377 and 61732011), and Tianjin AI Funding (17ZXRGGX00150).

References

1. Kumar, A., et al.: Ask me anything: dynamic memory networks for natural language processing. In: ICML, pp. 1378–1387 (2016)
2. Hixon, B., Clark, P., Hajishirzi, H.: Learning knowledge graphs for question answering through conversational dialog. In: NAACL-HLT 2015, pp. 851–861 (2015)
3. Tan, M., Santos, C., Xiang, B., Zhou, B.: LSTM-based deep learning models for non-factoid answer selection. arXiv preprint arXiv:1511.04108 (2015)
4. Xiong, C., Zhong, V., Socher, R.: Dynamic coattention networks for question answering. In: ICLR (2017)
5. Richardson, M., Burges, C.J.C., Renshaw, E.: MCTest: a challenge dataset for the open-domain machine comprehension of text. In: EMNLP, pp. 1532–1543 (2014)
6. Wang, H., Bansal, M., Gimpel, K., Mcallester, D.: Machine comprehension with syntax, frames, and semantics. In: ACL& IJNLP, pp. 700–706 (2015)
7. Chen, D., Bolton, J., Manning, C.D.: A thorough examination of the CNN/Daily mail reading comprehension task. In: ACL (2016)

8. Hill, F., Bordes, A., Chopra, S., Weston, J.: The Goldilocks principle: reading children's books with explicit memory representations. In: ICLR (2016)
9. Kadlec, R., Schmid, M., Bajgar, O., Kleindienst, J.: Text understanding with the attention sum reader network. arXiv preprint arXiv:1603.01547 (2016)
10. Bao, J., Duan, N., Yan, Z., Zhou, M., Zhao, T.: Constraint-based question answering with knowledge graph. In: COLING, pp. 2503–2514 (2016)
11. Malinowski, M., Rohrbach, M., Fritz, M.: Ask your neurons: a neural-based approach to answering questions about images. In: ICCV, pp. 1–9 (2015)
12. Sasaki, M., Kita, K.: Rule-based text categorization using hierarchical categories. In: IEEE International Conference on SMC, pp. 2827–2830 (1998)
13. Kiritchenko, S., Zhu, X., Cherry, C., Mohammad, S.M.: NRC-Canada-2014: detecting aspects and sentiment in customer reviews. In: SemEval, pp. 437–442 (2014)
14. Deng, Z., Zhu, X., Cheng, D., Zong, M., Zhang, S.: Efficient kNN classification algorithm for big data. Neurocomputing **195**(26), 143–148 (2016)
15. Lipps, O., Pekari, N., Roberts, C.: Undercoverage and nonresponse in a list-sampled telephone election survey. J. Eur. Surv. Res. Assoc. **9**(2), 71–82 (2015)
16. Zhang, X., Zhao, J., Cun, Y.L.: Character-level convolutional networks for text classification. In: NIPS, pp. 649–657 (2015)
17. Tang, D., Qin, B., Liu, T.: Document modeling with gated recurrent neural network for sentiment classication. In: EMNLP 2015, pp. 1422–1432 (2015)
18. Ma, D., Li, S., Zhang, X., Wang, H.: Interactive attention networks for aspect-level sentiment classification. In: IJCAI, pp. 4068–4074 (2017)
19. Zhang, L., Wang, S, Liu, B.: Deep learning for sentiment analysis: a survey. WIREs: Data Min. Knowl. Disc., 25 (2018)
20. Mullen, T., Collier, N.: Sentiment analysis using support vector machines with diverse information sources. In: EMNLP, pp. 412–418 (2004)
21. Pennington, J., Socher, R., Manning, C.D.: Glove: global vectors for word representation. In: EMNLP, pp. 1532–1543 (2014)
22. Wu, O., Yang, T., Yang, M., Li, M.: ρ-hot lexical embedding-based two-level LSTM for sentiment analysis. arXiv preprint arXiv: 1803.07771 (2018)
23. Wang, B., Lu, W.: Learning latent opinions for aspect-level sentiment classification. In: AAAI (2018)

Neural Network Based Popularity Prediction by Linking Online Content with Knowledge Bases

Wayne Xin Zhao[1(✉)], Hongjian Dou[1], Yuanpei Zhao[1], Daxiang Dong[2], and Ji-Rong Wen[1]

[1] School of Information, Renmin University of China, Beijing, China
batmanfly@gmail.com, {hongjiandou,YuanpeiZhao,jrwen}@ruc.edu.cn
[2] Baidu Inc., Beijing, China
dongdaxiang@baidu.com

Abstract. Predicting the popularity of online items has been an important task to understand and model online popularity dynamics. Feature-based methods are one of the mainstream approaches to tackle this task. However, most of the existing studies focus on some specific kind of auxiliary data, which is usually platform- or domain- dependent. In existing works, the incorporation of auxiliary data has put limits on the applicability of the prediction model itself. These methods may not be applicable to multiple domains or platforms. To address these issues, we propose to link online items with existing knowledge base (KB) entities, and leverage KB information as the context for improving popularity prediction. We represent the KB entity by a latent vector, encoding the related KB information in a compact way. We further propose a novel prediction model based on LSTM networks, adaptively incorporating KB embedding of the target entity and popularity dynamics from items with similar entity information. Extensive experiments on three real-world datasets demonstrate the effectiveness of the proposed model.

Keywords: Deep learning · Popularity prediction · Knowledge base

1 Introduction

With the rapid development of Web platforms, various online items (*a.k.a.*, online content), such as Amazon e-books and YouTube videos, are available to users. The increasing of online items has intensified the competition for users' attention [24], since only a small number of items become popular. In order to better understand and model online popularity dynamics, the task of predicting the popularity of web content [16,23] has become very important and attracted much attention from the research community.

H. Dou—Co-first author.

Q. Yang et al. (Eds.): PAKDD 2019, LNAI 11440, pp. 16–28, 2019.
https://doi.org/10.1007/978-3-030-16145-3_2

Traditional methods try to build prediction models (*e.g.*, regression models) on time series data of historical popularity statistics [23]. Since web content is usually associated with rich auxiliary data, many studies further leverage different kinds of feature information for improving the prediction performance, including content features [19], user features [27] and spatial features [8]. These feature-based approaches utilize both time series and auxiliary data in order to learn a better prediction model. However, most of the existing studies focus on some specific kind of auxiliary data, which is usually platform- or domain-dependent. In existing works, the incorporation of auxiliary data has put limits on the applicability of the prediction model itself. These methods may not be applicable to multiple domains or platforms. For example, it is difficult to directly apply an image-based prediction model to a prediction task which takes text data as auxiliary input. There is a need to develop a more general way to characterize and utilize auxiliary data for the task of popularity prediction.

To address these difficulties, in this paper, we propose to leverage knowledge bases (KBs) for improving the prediction of the popularity of online items. KBs store entity information in triples of the form (HEAD ENTITY, RELATION, TAIL ENTITY), typically corresponding to attribute information of entities. Compared with other less-structured data forms, KBs provide a general way to flexibly characterize context information of entities from various domains, and emphasize the interconnection of data. Many large-scale KBs have been released for public usage, such as FREEBASE [7] and YAGO [22]. For utilizing the KB information, we use a heuristic data linkage method to associate FREEBASE entities with online contents. With such a linkage, we are able to utilize rich KB information of online items from a variety of domains. We hypothesize KBs are of useful information to improve popularity prediction of online content. Following [2], we propose to learn vectorized representations (*a.k.a.*, embedding) for entities and relations in a latent space. In this way, we encode related KB information of an online item into a compact embedding vector. Inspired by recent works on deep learning for popularity prediction [12,14], we propose a novel LSTM-based neural network for using KB data for popularity prediction. The proposed model not only utilizes the associated KB information of an online item, but also improves the performance using related KB entities, called *KB neighbors*. It is able to adaptively utilize the KB information and automatically extract useful data characteristics.

To evaluate our model, we construct extensive experiments on three real-world datasets, and the results demonstrate the effectiveness of the proposed model compared with several competitive baselines. To our knowledge, it is the first time KB data is utilized in popularity prediction, and our approach provides a general way to utilize useful auxiliary data for different domains.

2 Related Work

A classic approach to popularity prediction is to build regression or classification prediction models [16,23] by taking as input the previous popularity statistics.

They make the predictions by characterizing temporal dependence or correlation patterns in the time series data. Since simple prediction models may not be effective to capture complex temporal characteristics, follow-up studies have introduced a series of more powerful prediction models, such as reinforced Poisson process [20], multi-dimensional time-series model [17], lifetime-aware regression model [15] and transfer autoregressive model [4]. With rich context data on the Web, many studies propose to leverage these auxiliary features for improving popularity prediction [5], including content features [19], user features [27], structural features [11] and spatial features [8].

Recently, deep learning has become a popular technique to address various complicated tasks. A typical deep learning approach to popularity prediction is to utilize Recurrent Neural Networks (RNN) to capture temporal dependencies and build better predictors [12,18,21,25,28]. They mainly rely on the excellence of RNN in modeling sequence data. Furthermore, several studies also adopt neural networks as a mapping mechanism to leverage various feature information for popularity prediction, including event signal [6], cascade [3,14] and multi-modality data [26].

Our work is closely related to the above works but has a different focus, *i.e.*, how to leverage KB information for improving popularity prediction. Although auxiliary data has been explored to some extent, to our knowledge, no work has utilized KB data for popularity prediction. As will be shown in the model and experiment parts, it is not trivial to integrate and model KB information into the prediction model. We have made the initiative attempt on this direction.

3 Problem Definition

Let $\mathcal{I}$ denote a set of items on an online platform, *e.g.*, an AMAZON ebook or a LAST.FM music. Assume that an observation window $[1, n]$ of n time steps (*a.k.a.*, intervals) is given[1]. At the t-th time step, each individual item i receives a value measuring its popularity within the current step, denoted by v_t^i. Popularity values reflect the received online attention for an item, *e.g.*, the number of reviews or clicks. By sorting these values by time ascendingly, we can form a time series of popularity values for item i, namely $\{v_1^i, \cdots v_t^i, \cdots v_n^i\}$, called *popularity time series*. We are often interested in future popularity. Let $v_{n,m}^i$ denote the incremental popularity in the m steps after time n, so we have $v_{n,m}^i = \sum_{t=n+1}^{n+m} v_t^i$.

Besides popularity time series, we assume that a knowledge base (KB) is also available as the input. A KB is defined over an entity set $\mathcal{V}$ and a relation set $\mathcal{R}$, containing a set of KB triples. A KB triple $\langle e_1, r, e_2 \rangle$ denotes there exists relation r from $\mathcal{R}$ between two entities e_1 and e_2 from $\mathcal{V}$, stating a fact stored in KB. For example, a KB triple (CHINA, HASCAPITALCITY, BEIJING) describes that *Beijing* is the capital city of *China*. Since we assume it is possible to link online items with KB entities, item set $\mathcal{I}$ can be considered as a subset of KB

[1] Note we use 1 to n to indicate a relative time span. Not all the items share the same absolute time span (*i.e.*, lifespan).

entity set $\mathcal{V}$, so we have $\mathcal{I} \subset \mathcal{V}$. By linking an online item with a KB entity, we can obtain all its related KB information.

Knowledge-based popularity prediction is to predict the incremental popularity value $v^i_{n,m}$ for an item i after m time steps given previous n popularity values and its KB information. Following [3], we predict the incremental popularity to avoid data dependency. Our definition is general in that we parameterize the task setting with two numbers n and m. When $m = 1$, the task becomes the next-step popularity prediction; when $m = +\infty$, the task becomes final popularity prediction. Also, the granularity of time steps (*e.g.,* day, month or year) and the scale of popularity values (*e.g.,* absolute or normalized values) can be set accordingly for different tasks. We will specify the details in Sect. 5.

4 The Proposed Model

In this section, we present the proposed model for the task of knowledge-based popularity prediction. We start with a base model which adopts the standard LSTM architecture, and then extend the model by incorporating KB information in two aspects, namely KB embedding and KB neighbors.

4.1 A LSTM-Based Popularity Prediction Model

Recurrent Neural Networks (RNN) have been shown effective to capture the temporal dependencies in sequence data, especially the Long Short Term Memory (LSTM) networks [9]. Similar to RNN, the LSTM network generates the current hidden state vector $\boldsymbol{h}_t$ conditioned on previous hidden state vector $\boldsymbol{h}_{t-1}$ and current input vector $\boldsymbol{x}_t$, so we have $\boldsymbol{h}_t = \text{LSTM}(\boldsymbol{h}_{t-1}, \boldsymbol{x}_t; \Theta)$, where LSTM$(\cdot)$ is the LSTM unit and Θ denotes all the related parameters. We adopt the LSTM network as the main architecture to build the prediction model.

For our task, the input at each time t is the observed popularity value v^i_t. In this case, $\boldsymbol{x}_t$ degenerates into a scalar value. Our task is specified by two numbers n and m. For item i, when LSTM receives n input values, it makes the prediction of m-step incremental value $\hat{v}^i_{n,m}$ using a function $g(\cdot)$ conditioned on the n-th hidden state vector $\boldsymbol{h}^i_n \in \mathbb{R}^L$ of item i. Formally, we have $\hat{v}^i_{n,m} = g(\boldsymbol{h}^i_n)$, where the superscript of i indicates item i and $g(\cdot)$ is set to a linear function.

4.2 Enhancing the Prediction with KB Embeddings

The above prediction model mainly captures the temporal correlation or dependence in time series data. In our setting, we also have KB data available, which contains potentially useful information for popularity prediction. Next, we study how to integrate KB information into the prediction model.

Knowledge Base Embedding. Given an online item i, let e_i denote its corresponding entity in KB. Since KB is originally framed as a set of triples, we can obtain a set of related triples where e_i plays the head or tail entity. Using the related triples, the first solution is to represent each e_i by a one-hot relation-based vector. However, such a feature vector has a large dimension size and is usually sparse. For effectively encoding KB information for e_i, we propose to learn a distributed vector $\boldsymbol{e}_i \in \mathbb{R}^D$. To learn KB embedding, we use the commonly used model TransE [2] to minimize the loss of the triples $\sum_{\{\langle e_1,r,e_2\rangle\}} \parallel \boldsymbol{e}_1 + \boldsymbol{r} - \boldsymbol{e}_2 \parallel$. We train the TransE model using all the triples in KB instead of using only those related to linked entities. The learned KB embedding provides a general and compact representation for KB information, which is more flexible to use and integrate.

Adaptive Integration of KB Embeddings. Now, we study how to integrate KB embedding into the LSTM-based prediction model. For popularity prediction, KB embedding is likely to contain both useful and irrelevant information, even noise. It may not work well to directly incorporate the KB embedding into the prediction model. To leverage KB embedding $\boldsymbol{e}_i$, we first transform it into a vector that is more suitable for the current task

$$\tilde{\boldsymbol{e}}_i = \text{MLP}(\boldsymbol{e}_i), \tag{1}$$

where MLP$(\cdot)$ is a standard Multi-layer Perceptron containing two hidden layers and using *relu* as the activation function in our work.

For item i, we have both the hidden state vector $\boldsymbol{h}_n^i$ learned from time series data and the transformed embedding $\tilde{\boldsymbol{e}}_i$ learned from KB data. We need to consider how to effectively combine these two vectors. Instead of setting a fixed weight, the model should be able to adaptively tune the combination weight based on the current state. To achieve this, we adopt the gate mechanism to combine the transformed KB embedding $\tilde{\boldsymbol{e}}_i$ and hidden state vector $\boldsymbol{h}_n^i$

$$z_n^i = \text{sigmoid}(\boldsymbol{W}^E \tilde{\boldsymbol{e}}_i + \boldsymbol{U}^E \boldsymbol{h}_n^i), \tag{2}$$

$$\tilde{\boldsymbol{h}}_n^i = z_n^i \cdot \tilde{\boldsymbol{e}}_i + (1 - z_n^i) \cdot \boldsymbol{h}_n^i, \tag{3}$$

where $z_n^i \in (0, 1)$ is the adaptive combination weight, $\boldsymbol{W}^E$ and $\boldsymbol{U}^E$ are parameter matrices, and $\tilde{\boldsymbol{h}}_n^i$ is the KB-enhanced representation of item i at the n-th time step. In our model, we first adopt nonlinear transformation to learn suitable representations of KB embedding for popularity prediction. Then, the gate mechanism tries to balance the two factors conditioned on the current hidden state. A benefit of the gate-based combination method is that even for the same item we can have different combination weights at varying time steps, adaptively integrating KB information.

4.3 Enhancing the Prediction with KB Neighbors

Two items with similar KB information are likely to have similar or correlated popularity dynamics. Hence, we further incorporate the popularity dynamics of related items with similar entity information to improve popularity prediction.

For convenience, we call the two items in the same domain with similar KB information *KB neighbors*. Now, our problems become how to identify KB neighbors and integrate the information of KB neighbors for popularity prediction.

KB Neighbor Identification. To measure the relatedness (or similarity) between two items using KB data, an intuitive idea is to compute the path reachability over the KB graph. However, the KB graph is usually very huge, and it is item-consuming to run graph search algorithms for each individual entity. Based on the learned KB embedding, we propose to compute the distance between entity embeddings for measuring item relatedness. Formally, given two entities e_1 and e_2, we compute the KB embedding distance via a distance function $f(\boldsymbol{e}_1, \boldsymbol{e}_2)$, where $f(\cdot)$ can be flexibly set to any distance function for vectors, *e.g.*, cosine and L_1 norm. In this way, we can rank the candidate items ascendingly by their KB embedding distance with the target item. Our idea is similar to that in [16]. They select the neighbors based on historical popularity trends, highly relying on the training set; while we select the neighbors using KB information, independent of historical popularity trends. In order to reduce data dependency, we further remove all the candidate items which has a prolonging lifetime with the target entity. For efficiency consideration, we keep top K related entities (with the same entity type) as KB neighbors.

Attentive Integration of KB Neighbors. With the identified KB neighbors, we next describe how to utilize the information of KB neighbors for improving the prediction performance. Given a target item, for each neighbor k, we still use the LSTM network to encode their popularity dynamics up to the n-th time step into a hidden vector

$$\boldsymbol{h}_n^k = LSTM(\{v_1^k, \cdots, v_n^k\}; \Theta'), \tag{4}$$

where we use a different configuration Θ' for the LSTM network compared with the one for encoding the target item, because $\boldsymbol{h}_n^k$s are mainly used to improve the prediction for item i instead of item k itself. To integrate multiple hidden vectors of KB neighbors, we adopt the attention mechanism [1] to set the summation weights $\{\alpha_k^i\}$ conditioned on item i. Formally, α_k^i is defined as follows

$$\alpha_k^i = \frac{\exp(w(\boldsymbol{h}_n^i, \tilde{\boldsymbol{e}}_k))}{\sum_{k'=1}^{K}(\exp(w(\boldsymbol{h}_n^i, \tilde{\boldsymbol{e}}_{k'}))}, \tag{5}$$

where $\boldsymbol{h}_n^i$ is the derived hidden state vector of item i using only time series data, $\tilde{\boldsymbol{e}}_k$ is the transformed KB embedding of item k, and $w(\boldsymbol{h}_n^i, \tilde{\boldsymbol{e}_k})$ is set using the following function

$$w(\boldsymbol{h}_n^i, \tilde{\boldsymbol{e}}_k) = \boldsymbol{a}^\top \tanh(\boldsymbol{W}^N \boldsymbol{h}_n^i + \boldsymbol{U}^N \tilde{\boldsymbol{e}}_k), \tag{6}$$

where $\boldsymbol{W}^N$ and $\boldsymbol{U}^N$ are the parameter matrices, and $\boldsymbol{a}$ is the parameter vector. With the obtained attentive weights, we can encode the information from the K KB neighbors into a unique vector $\tilde{\boldsymbol{h}}_n^i$:

$$\check{\boldsymbol{h}}_n^i = \sum_{k=1}^{K} \alpha_k^i \cdot \boldsymbol{h}_n^k. \tag{7}$$

Similar to the gate mechanism in Sect. 4.2, our model adaptively sets the attention weights conditioned on the current hidden state. It is able to alleviate the problem that two items have different correlation patterns at varying time steps. The attention mechanism can be viewed as a key-value retrieval procedure, where the query $\boldsymbol{h}_n^i$ is the time series representation of the target item, and the keys $\{\tilde{\boldsymbol{e}}_k\}_{k=1}^K$ are the transformed KB embeddings of KB neighbors. The derived result is the attentive combination of time series representations of KB neighbors $\{\boldsymbol{h}_n^k\}_{k=1}^K$. Finally, our item representation $\boldsymbol{s}_n^i$ for popularity prediction is a vector concatenation of $\tilde{\boldsymbol{h}}_n^i$ and $\check{\boldsymbol{h}}_n^i$,

$$\boldsymbol{s}_n^i = \tilde{\boldsymbol{h}}_n^i \oplus \check{\boldsymbol{h}}_n^i, \tag{8}$$

where $\tilde{\boldsymbol{h}}_n^i$ is the representation learned using only the information from the item itself (including both time series and KB data) defined in Eq. 3 and $\check{\boldsymbol{h}}_n^i$ is the representation learned using K KB neighbors defined in Eq. 7. We define the loss over the training set as follows

$$L = \sum_{i \in \mathcal{D}} \sum_{t \in [1,n]} \ell(v_{t,m}^i, \hat{v}_{t,m}^t), \tag{9}$$

where $\mathcal{D}$ is the item set, $v_{t,m}^i$ and $\hat{v}_{t,m}^t$ are the ground-truth or predicted incremental popularity value for item i in the time span $(t, t+m)$, and $\ell(\cdot)$ is the loss function, which is set to *Mean Absolution Error*. We learn our model parameters by using mini-batch gradient descent with the Adam optimizer.

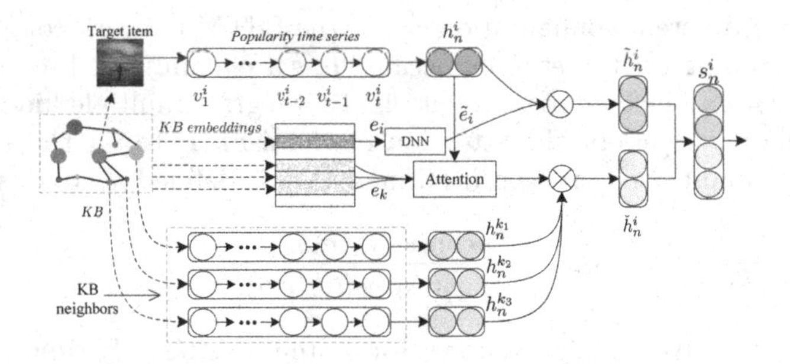

Fig. 1. The overall schematic diagram of the proposed model.

We present the overall schematic diagram of the proposed model in Fig. 1. It is clear to see that the model consists of two parts: one utilizes the information of the target item itself, and the other utilizes the information of its KB neighbors. KB information is used in both aspects. First, it is transformed as a direct signal to enhance the prediction; second, it is used as the keys of the attention module.

By using non-linear query-key matching mechanism in Eq. 6, our model is more capable of inferring the usefulness of each KB neighbor for the target item. Since we filter out neighbors with a prolonging lifetime, our model will not use any information after the observed window. After obtaining s_n^i, we still adopt the linear function $g(s_n^i)$ to generate the final prediction. We call the proposed model *KB-enhanced Popularity Prediction Network* (**KB-PPN**).

5 Experiments and Analysis

This section present the experiment setup and result analysis.

5.1 Experimental Setup

Construction of the Datasets. In our task, we need to prepare both KB and popularity time series data. We adopt the linked KB4Rec dataset shared in [10, 29] for our evaluation. KB4Rec dataset has linked the public KB FREEBASE with three popular recommender system datasets in three domains, namely music, movie and book. The detailed linkage process and statistics can be found in [29]. For each linked item, we can obtain both its popularity and KB information. To measure the popularity value, for the music dataset, we use the listening count, while for the other two datasets, we use the number of received ratings. It is uninteresting to predict the popularity of items with either a small popularity value or a short lifespan. We rank the items by its total popularity, and then select top items covering at least 40% of the entire time span of the dataset. To train TRANSE, we start with linked entities as seeds and expand the graph with one-step search. In order to exclude temporal evidence from KB, we remove all the triples related to a temporal relation (*e.g.,* RELEASEDATE) together with the entities in the triples. For the music dataset, we use a month as a time step; while the other datasets are much more sparse, we use a year as a time step. Following [16], we construct the ten-fold cross validation for evaluation. The final results are averaged from ten runs. We summarize the detailed statistics of the three linked datasets in Table 1.

Table 1. Statistics of our datasets after the preprocessing procedure. APPS denotes the average popularity value per step, and #extended and #relations denote the numbers of entities and relations in the extended graph for training TRANSE.

Datasets	#selected	#linked	#extended	#relations	Time span (yr.)	APPS
Music	37,000	23,120	214,524	19	2006–2014	577
Movie	12,000	9,260	1,125,100	91	1995–2014	277
Book	4,000	2,228	313,956	49	1997–2015	41

Evaluation Metrics. Following [16], we adopt three standard measurements as evaluation metrics: (1) Mean Absolute Percentage Error (*MAPE*) measures the average derivation between the predicted and observed popularity, defined as $MAPE = \frac{1}{N}\sum_{i=1}^{N}|\frac{\hat{v}-v}{v}|$; (2) Accuracy ($ACC$) measures the fraction of items correctly predicted for a given error tolerance ϵ, defined as $ACC = \frac{1}{N}|\{\forall i : |\frac{\hat{v}-v}{v}| < \epsilon\}|$, where the threshold ϵ is set to 0.15 in this paper; and (3) Mean Relative Squared Error ($MRSE$) measures the relative error between the predicted and observed popularity, defined as $MRSE = \frac{1}{N}\sum_{i=1}^{N}\left(\frac{\hat{v}}{v} - 1\right)^2$.

Comparison Methods. The comparison methods are as follows.

- Multivariate Linear Regression (MLR) [16]: it predicts the popularity of an item using a linear combination of previous popularity values.
- MRBF [16]: it is an extension of the basic MLR model by considering the similarity between the item and known examples from training set.
- Support Vector Regression (SVR) [13]: Khosla *et al.* adopt SVR model using linear kernel to predict popularity with time series data as features.
- Random Forest (RF): We use the tree-based ensemble approach to capturing complex data characteristics for popularity prediction.
- Long Short-Term Memory (LSTM): LSTM improves RNN with a better capacity of encoding long sequences.
- The State Frequency Memory (SFM) [28]: it is the extension of the basic LSTM model by capturing multi-frequency time series patterns, which is a recently published work on popularity prediction.
- Our model: we prepare three variants of our model: (1) only using KB embedding in Eq. 3, denoted by KB-PPN$_{+E}$, (2) using both KB embedding in Eq. 3 and KB neighbors in Eq. 4 without attention, denoted by KB-PPN$_{+E+N}$, and (3) our full model in Eq. 8, denoted by KB-PPN$_{full}$.

For MLR, SVR and LSTM, we also implement the corresponding variants by directly integrating KB embedding as features, denoted by MLR$_{+E}$, SVR$_{+E}$ and LSTM$_{+E}$ (using simple concatenation).

Parameter Setting. All the models have some parameters to tune. We either follow the reported optimal parameter settings or optimize each model separately using 10-fold cross validation. For all the neural network models, following [14], we vary the hidden layer size L in $\{32, 64, \cdots, 512\}$, the number of hidden layers in $\{1, 2, 3\}$, the activation function in {relu, tanh, sigmoid}, the batch size in $\{64, 128, \cdots, 1024\}$, and the initial learning rate in $\{0.02, 0.01, \cdots, 10^{-4}\}$. The embedding size of TransE D is selected from $\{50, 100, \cdots, 300\}$, and the number of neighbors K is selected from $\{1, 2, 3, 4, 5\}$. In order to avoid over-fitting, the dropout rate is chosen from $\{0.2, 0.3, ..., 0.8\}$.

5.2 Results and Analysis

We present the main comparison results of different methods for popularity prediction in Table 2. Since different datasets have varying time spans and popularity scales, we set different values for n and m, where n is the number of previous steps (seen) and m is the number of future steps (predicted). For ease

Table 2. Performance comparisons of different methods on popularity prediction. "↓ / ↑" indicate smaller is better or worse. ▲% denotes the improvement of our model over the best performance of all the baselines.

Datasets	Music ($n = 3, m = 9$)			Movie ($n = 2, m = 4$)			Book ($n = 2, m = 4$)		
Models	MAPE (↓)	ACC (↑)	MRSE (↓)	MAPE (↓)	ACC (↑)	MRSE (↓)	MAPE (↓)	ACC (↑)	MRSE (↓)
MLR	0.212	0.450	0.079	0.222	0.427	0.080	0.288	0.324	0.130
MLR_{+E}	0.211	0.456	0.078	0.222	0.427	0.080	0.273	0.344	0.119
MRBF	0.210	0.461	0.079	0.212	0.452	0.075	0.272	0.341	0.115
SVR	0.209	0.461	0.078	0.206	0.460	0.070	0.269	0.337	0.111
SVR_{+E}	0.204	0.471	0.074	0.204	0.463	0.068	0.257	0.361	0.105
RF	0.206	0.466	0.076	0.211	0.460	0.078	0.261	0.367	0.112
RF_{+E}	0.206	0.468	0.075	0.209	0.460	0.077	0.257	0.360	0.106
LSTM	0.208	0.458	0.076	0.196	0.488	0.064	0.261	0.348	0.104
SFM	0.206	0.469	0.075	0.195	0.482	0.063	0.256	0.355	0.101
LSTM_{+E}	0.205	0.469	0.075	0.192	0.495	0.062	0.246	0.374	0.096
KB-PPN_{+E}	0.196	0.488	0.068	0.188	0.499	0.058	0.242	0.386	0.094
KB-PPN_{+E+N}	0.193	0.494	0.066	0.189	0.491	0.059	0.238	0.391	0.090
KB-PPN_{full}	**0.189**	**0.501**	**0.062**	**0.182**	**0.505**	**0.054**	**0.232**	**0.397**	**0.085**
▲%	7.35%	6.37%	16.22%	5.21%	2.02%	12.90%	5.59%	6.15%	11.46%

of result analysis, we categorize the comparison methods into two groups, namely traditional methods and neural network methods.

Among all the traditional baselines, SVR and RF perform better than the others, since they adopt more powerful modeling mechanisms (*i.e.*, margin-based optimization or tree-based non-linear transformation) and are likely to yield better performance. Another interesting observation is that the improvement of KB embedding using MLR is smaller than that of using SVR and RF. It indicates that the learned embedding may not be directly useful using linear models.

For neural network models, SFM achieves a better performance than LSTM, since it is able to capture multi-frequency time series patterns. Next, we examine the effect of different ways to integrate KB embeddings of the target item. It is clear to see that KB-PPN_{+E} is substantially better than LSTM_{+E}. The major difference is that our model KB-PPN_{+E} adopts an adaptive way to integrate KB embedding, while LSTM_{+E} simply adopts a vector concatenation, which is less effective to utilize the information of KB embedding. By additionally integrating the information of KB neighbors, KB-PPN_{+E+N} and KB-PPN_{full} perform better than all the other comparison methods, which indicates the usefulness of KB neighbors. While, KB-PPN_{full} further improves over KB-PPN_{+E+N} due to the use of attention mechanisms. The above findings show that the KB information is useful to improve popularity prediction. In particular, the integration way of these information is important to the final performance.

Our task is parameterized with two setting parameters of n and m. A good prediction model should be able to work well in various cases of n and m. To examine the performance stability of our model, we vary n and m alternatively, and compare our method with baselines. As shown in Fig. 2, we can see that our model is consistently better than the selected baselines in various cases, indicating the robustness of the proposed model. Our model KB-PPN_{full} has

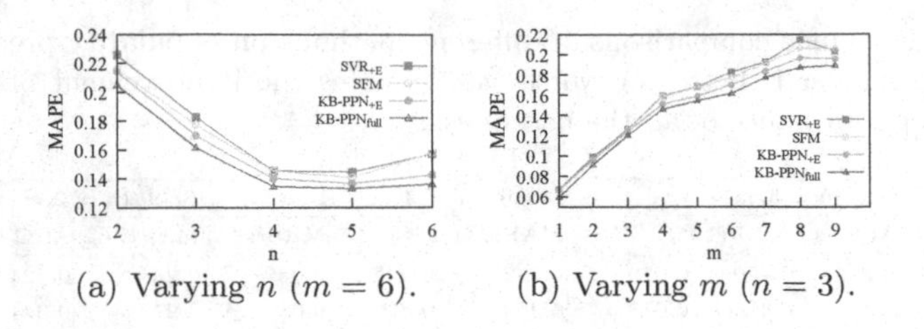

(a) Varying n ($m = 6$). (b) Varying m ($n = 3$).

Fig. 2. Varying the number of previous steps n and the number of future steps m.

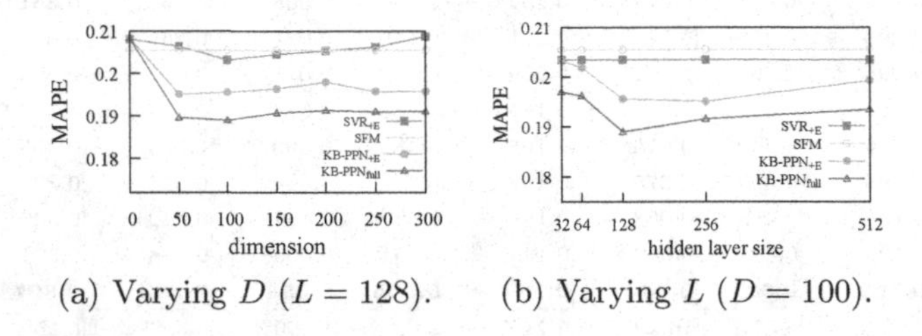

(a) Varying D ($L = 128$). (b) Varying L ($D = 100$).

Fig. 3. Performance tuning for the KB embedding size D and the hidden layer size L.

several parameters to tune, including the number of KB neighbors K, the KB embedding size D and the hidden layer size L. We find the number of KB neighbors should be set to a small value. In our experiments $K = 3$ yields the best performance. We next tune the parameters of D and L. As Fig. 3 shows, $D = 100$ and $L = 128$ gives the best performance. Overall, the performance of our model is relatively stable and consistently better than the baselines.

6 Conclusion

In this paper, we proposed to heuristically link online items with existing KB entities, and leverage KB data for improving popularity prediction. Experiment results showed that both KB embedding of the target item and popularity dynamics of its KB neighbors are useful for our task. As future work, we will test the proposed approach in more domains. Since not all the entities can find corresponding KB entries, it will be interesting to study how to enhance the prediction performance of non-linked items with KBs. We will also consider using more complicated sequence neural networks such as SFM [28].

Acknowledgements. This work was partially supported by National Natural Science Foundation of China under the grant numbers 61872369 and 61832017, and the Science and Technology Project of Beijing (Z181100003518001).

References

1. Bahdanau, D., Cho, K., Bengio, Y.: Neural machine translation by jointly learning to align and translate. Computer Science (2014)
2. Bordes, A., Usunier, N., García-Durán, A., Weston, J., Yakhnenko, O.: Translating embeddings for modeling multi-relational data. In: NIPS, pp. 2787–2795 (2013)
3. Cao, Q., Shen, H., Cen, K., Ouyang, W., Cheng, X.: Deephawkes: bridging the gap between prediction and understanding of information cascades. In: CIKM, pp. 1149–1158 (2017)
4. Chang, B., Zhu, H., Ge, Y., Chen, E., Xiong, H., Tan, C.: Predicting the popularity of online serials with autoregressive models. In: CIKM, pp. 1339–1348 (2014)
5. Cheng, J., Adamic, L., Dow, P.A., Kleinberg, J.M., Leskovec, J.: Can cascades be predicted? In: WWW, pp. 925–936 (2014)
6. Ding, X., Zhang, Y., Liu, T., Duan, J.: Deep learning for event-driven stock prediction. In: IJCAI, pp. 2327–2333 (2015)
7. Google: Freebase data dumps (2016). https://developers.google.com/freebase/data
8. Grover, A., Kapoor, A., Horvitz, E.: A deep hybrid model for weather forecasting. In: SIGKDD, pp. 379–386 (2015)
9. Hochreiter, S., Schmidhuber, J.: Long short-term memory. Neural Comput. **9**(8), 1735–1780 (1997)
10. Huang, J., Zhao, W.X., Dou, H., Wen, J.R., Chang, E.Y.: Improving sequential recommendation with knowledge-enhanced memory networks. In: The 41st International ACM SIGIR Conference on Research & Development in Information Retrieval, pp. 505–514. ACM (2018)
11. Imamori, D., Tajima, K.: Predicting popularity of twitter accounts through the discovery of link-propagating early adopters. In: CIKM, pp. 639–648 (2016)
12. Jia, X., et al.: Incremental dual-memory LSTM in land cover prediction. In: SIGKDD (2017)
13. Khosla, A., Sarma, A.D., Hamid, R.: What makes an image popular? In: WWW, pp. 867–876 (2014)
14. Li, C., Ma, J., Guo, X., Mei, Q.: Deepcas: an end-to-end predictor of information cascades. In: WWW, pp. 577–586 (2017)
15. Ma, C., Yan, Z., Chen, C.W.: LARM: a lifetime aware regression model for predicting youtube video popularity. In: CIKM, pp. 467–476 (2017)
16. Pinto, H., Almeida, J.M.: Using early view patterns to predict the popularity of youtube videos. In: WSDM, pp. 365–374 (2013)
17. Proskurnia, J., Grabowicz, P.A., Kobayashi, R., Castillo, C., Cudré-Mauroux, P., Aberer, K.: Predicting the success of online petitions leveraging multidimensional time-series. In: WWW, pp. 755–764 (2017)
18. Qin, Y., Song, D., Chen, H., Cheng, W., Jiang, G., Cottrell, G.: A dual-stage attention-based recurrent neural network for time series prediction. In: IJCAI, pp. 2627–2633 (2017)
19. Roy, S.D., Mei, T., Zeng, W., Li, S.: Towards cross-domain learning for social video popularity prediction. IEEE Trans. Multimed. **15**(6), 1255–1267 (2013)
20. Shen, H.W., Wang, D., Song, C., Barabsi, A.: Modeling and predicting popularity dynamics via reinforced poisson processes. In: AAAI, pp. 291–297 (2014)
21. Shi, X., et al.: Deep learning for precipitation nowcasting: a benchmark and a new model (2017)
22. Suchanek, F.M., Kasneci, G., Weikum, G.: YAGO: a core of semantic knowledge. In: WWW, pp. 697–706 (2007)

23. Szabo, G., Huberman, B.A.: Predicting the popularity of online content. Commun. ACM **53**(8), 80–88 (2008)
24. Tatar, A., de Amorim, M.D., Fdida, S., Antoniadis, P.: A survey on predicting the popularity of web content. J. Internet Serv. Appl. **5**(1), 8:1–8:20 (2014)
25. Wang, Y., Liu, S., Shen, H., Gao, J., Cheng, X.: Marked temporal dynamics modeling based on recurrent neural network. In: Kim, J., Shim, K., Cao, L., Lee, J.-G., Lin, X., Moon, Y.-S. (eds.) PAKDD 2017. LNCS (LNAI), vol. 10234, pp. 786–798. Springer, Cham (2017). https://doi.org/10.1007/978-3-319-57454-7_61
26. Wu, B., Cheng, W., Zhang, Y., Huang, Q., Li, J., Mei, T.: Sequential prediction of social media popularity with deep temporal context networks. In: IJCAI, pp. 3062–3068 (2017)
27. Wu, B., Mei, T., Cheng, W.H., Zhang, Y.: Unfolding temporal dynamics: predicting social media popularity using multi-scale temporal decomposition. In: AAAI, pp. 272–278 (2016)
28. Zhang, L., Aggarwal, C., Qi, G.J.: Stock price prediction via discovering multi-frequency trading patterns. In: SIGKDD (2017)
29. Zhao, W.X., He, G., Dou, H., Huang, J., Ouyang, S., Wen, J.R.: Kb4rec: a dataset for linking knowledge bases with recommender systems. arXiv preprint arXiv:1807.11141 (2018)

Passenger Demand Forecasting with Multi-Task Convolutional Recurrent Neural Networks

Lei Bai[1](✉), Lina Yao[1], Salil S. Kanhere[1], Zheng Yang[2], Jing Chu[2], and Xianzhi Wang[3]

[1] School of Computer Science and Engineering, University of New South Wales, Sydney, Australia
baisanshi@gmail.com, {lina.yao,salil.kanhere}@unsw.edu.au
[2] School of Software, Tsinghua University, Beijing, China
hmilyyz@gmail.com, j-zhu16@mails.tsinghua.edu.au
[3] School of Software, University of Technology Sydney, Sydney, Australia
sandyawang@gmail.com

Abstract. Accurate prediction of passenger demands for taxis is vital for reducing the waiting time of passengers and drivers in large cities as we move towards smart transportation systems. However, existing works are limited in fully utilizing multi-modal features. First, these models either include excessive data from weakly correlated regions or neglect the correlations with similar but spatially distant regions. Second, they incorporate the influence of external factors (e.g., weather, holidays) in a simplistic manner by directly mapping external features to demands through fully-connected layers and thus result in substantial bias as the influence of external factors is not unified. To tackle these problems, we propose an end-to-end multi-task deep learning model for passenger demand prediction. First, we select similar regions for each target region based on their Point-of-Interest (PoI) information or historical demand and utilize Convolutional Neural Networks (CNN) to extract their spatial correlations. Second, we map external factors to future demand levels as part of the multi-task learning framework to further boost prediction accuracy. We conduct experiments on a large-scale real-world dataset collected from a city in China with a population of 1.5 million. The results demonstrate that our model significantly outperforms the state-of-the-art and a set of baseline methods.

Keywords: Demand prediction · Muti-task learning · Spatial-temporal correlations · Convolutional recurrent neural networks

1 Introduction

Taxis are an integral mode of transportation in cities and serve a large number of passengers on a daily basis. However, traditional taxi services are slow in

Q. Yang et al. (Eds.): PAKDD 2019, LNAI 11440, pp. 29–42, 2019.
https://doi.org/10.1007/978-3-030-16145-3_3

adopting Information and Communication technologies (ICT) to improve their efficiency and provide better services to commuters. In recent years, many online peer-to-peer ridesharing services such as Uber and Didi have successfully filled this void. These services allow customers to book a ride through their mobile apps. Drivers are then matched with customers based on their proximity. These services often employ dynamic pricing models and have significantly impacted the taxi markets in most countries. Despite the sophisticated ICT technologies adopted, these ride-sharing solutions have still not cracked the code on the relationship between passenger demand and ride supply. On the one hand, drivers often have to drive a long way before they can find passengers due to low demand volumes in their locations [5]; on the other hand, passengers may experience long delays in obtaining a ride due to the high demand in their locations. This imbalance between the demand and supply incurs excessive delays and energy consumption, thus calling for an effective passenger demand prediction method for efficient scheduling of taxis and shared cars.

A variety of techniques have been proposed to address this problem in the literature. Traditional methods [6,7] utilize time series models such as Auto-Regressive Integrated Moving Average (ARIMA) and its variants to predict traffic. These methods only consider temporal correlation. However, recent studies [2,3,8] have revealed that a region's passenger demand is also related to other regions demand and thus utilizing the spatial relationship between regions could positively help predict future passenger demand. There are two ways of utilizing the spatial relationships in literature: (1) Treating the whole city as an image (a two-dimensional matrix) and applying CNN [2,12] or Convolutional Long-Short Term Memory (ConvLSTM) [1,13] directly to this image to capture relationships among all regions. Although this method can find all possible relationships, it may also introduce weak or negative correlations. As a result, this method may adversely impact the prediction outcomes. Also, processing the data in this manner for a large city requires several CNN layers, which consumes significant resources. (2) The second focuses on discovering local relationships [3]. This method treats the target region, and it's surrounding regions as an image. The foundation of this method is that "near things are more related than distant things". However, it neglects the fact that remote regions could also share strong similarities in passenger demand patterns if they have similar properties (for example they have similar PoIs such as schools or hospitals). Besides, both of these two approaches can only be applied if the city is partitioned by a grid-based method. Furthermore, the integration of ubiquitous technologies makes it possible to collect a vast amount of multi-modal data from urban spaces (e.g., historical crowd flow, weather, holidays), many of which may influence taxi demand and thus promote better prediction. However, previous works either don't take these external features into account [6,8] or directly map external features to future passenger demand [1,4,13], which can lead to large biases because the influence of external factors is not uniform to all regions.

In this work, we propose an end-to-end unified deep learning framework to predict passenger demand. Our model readily scales to large urban areas and also incorporates insights from urban data sources of different types. More specifically, the model takes historical passenger demand, historical crowd outflow data, PoI information, weather data, air quality data and time meta (time of day, day of week, and holidays) as inputs for predicting the passenger demand in future for all regions. We first select similar regions for each target region by their PoI information or historical demand, then utilize CNN and LSTM to extract their spatial-temporal relationships. Our method is more flexible than the two approaches listed above which either consider all regions or co-located regions, as it can both filter weakly-related adjacent regions and find similar remote regions. Moreover, our method operates without knowledge of how the city is partitioned, i.e., either using road networks or relying on grids. Besides, to better utilize external features, we design an auxiliary task under the multi-task learning framework which predicts the demand level (e.g., high, medium or low) for each region in next time interval to further improve the passenger demand prediction task. Besides, our framework does not need hand-crafted features and is extensible to other datasets. The contributions of our work include the following:

- We propose a multi-task deep learning based framework for passenger demand prediction to solve the supply-demand imbalance problem in urban transportation systems. The framework takes features from multiple urban datasets into consideration and incorporates their joint influence on future passenger demands.
- We propose a similarity-based CNN model to capture the spatial similarity exclusively with similar regions. Our model emphasizes highly correlated regions, while simultaneously filtering out the influence of weakly related regions.
- We propose to predict and classify future passenger demand level as an auxiliary task under the multi-task learning framework to better utilize the power of external features and enhance the prediction accuracy of passenger demand value. None of the above aspects have been examined thoroughly in previous works.
- We conduct extensive experiments on a large-scale real-world dataset collected from a major city in China covering 1.5 million people, and demonstrate that our method outperforms a series of baselines and state-of-the-art methods.

2 Proposed Approach

2.1 Problem Formulation

In this section, we will first give some notations and definitions used to formalize the passenger demand prediction problem.

Notation 1: Region. We utilize the road networks based partition [4] to divide the entire city into blocks as it is more flexible and can integrate semantic meanings into regions. The entire city is divided into N regions, represented as a set: $\{r_1, r_2, ..., r_i, ...r_N\}$.

Notation 2: External Features. These represent the following information: weather data, air quality data, PoI information and time meta (e.g., time of day, day of week, holidays).

Definition 1: Passenger Demand. The passenger demand of region $r_i (i \in [1, N])$ in a given period t is defined as the number of taxi requests originating in this region during this time period, which can be represented as $D_t(r_i)$.

Definition 2: Crowd Outflow [11]. We use $P_T(r_i)$ to denote the set of people in region r_i at time T. The crowd outflow of region r_i during time interval t can be defined as $C_t(r_i) = P_T(r_i) \setminus P_{T+\Delta T}(r_i)$.

Passenger Demand Prediction. Let $S_t(r_i)$ denotes all the historically observed data (passenger demand, crowd outflow) for region r_i in time period t, $E_{t+1}(r_i)$ denotes all external features in time interval $t+1$ (since the weather in time interval $t+1$ is unknown, we can use the predicted weather or the weather in time t), passenger demand prediction aims to predict:

$$D_{t+1}(r_i) = \mathcal{F}(S_t(r_i), S_{t-1}(r_i), S_{t-2}(r_i), ..., S_{t-h}(r_i), E_{t+1}(r_i))$$

where $D_{t+1}(r_i)$ is the passenger demand for region r_i in time interval $t+1$, h is the historical window of time that is used for prediction. We define our prediction function $\mathcal{F}(\cdot)$ on all regions and previous time period up to $t-h$.

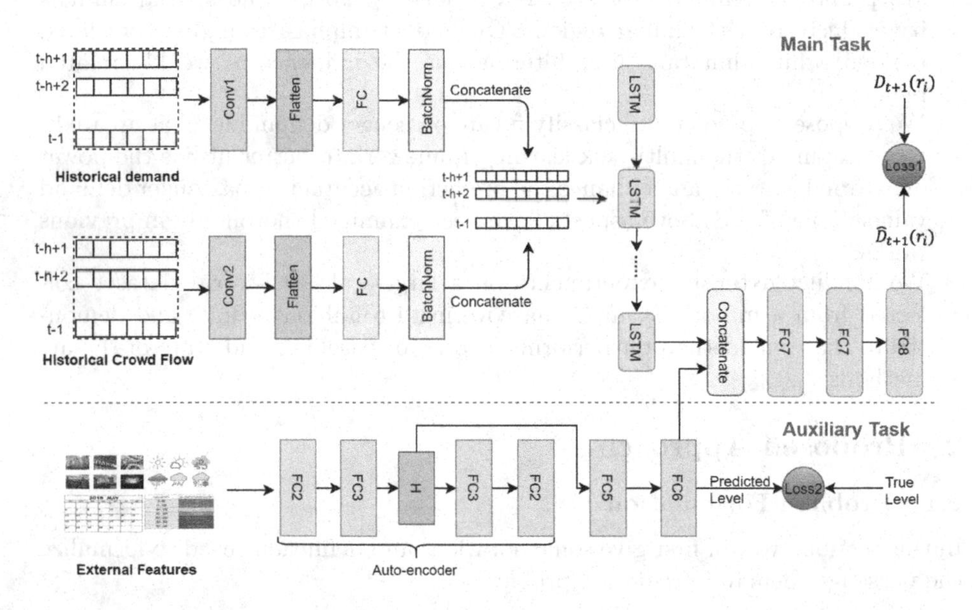

Fig. 1. Multi-Task Convolutional Recurrent Neural Networks (MT-CRNN)

2.2 Multi-Task CRNN (MT-CRNN) Framework

We design a multi-task deep learning framework (shown in Fig. 1) that contains two tasks: (1) The main task involves predicting the precise passenger demand in the next time interval by capturing the spatial-temporal relationships within historical observed data from selected regions; (2) The auxiliary task is to predict and classify the level of passenger demand (e.g., whether it is high, low or medium in the next time interval) to get a better representation of external features for the main task.

Auxiliary Task: Predict Future Passenger Demand Level. For passenger demand prediction and related tasks (such as crowd prediction, air quality prediction, rainfall prediction and so on), it is not apparent how external features can be used. External features from different domains may have different attributes, i.e., they may vary from dynamic to static and from continuous to categorical. Misusing external features may adversely impact the accuracy of the final prediction. The state-of-the-art methods [1,4,13] usually map external features to the value of the passenger demand directly, which can result in significant errors and thus doesn't make the best use of external features. Our intuition is that external features have a closer relationship with a more granular measure of the passenger demand rather than the precise value. Mapping external features to passenger demand level could lead to a better representation of external features. Here, we use categorical values to denote levels. For example, level 0, 1 and 2 map to low, medium and high passenger demand respectively.

Table 1. Passenger demand level generation

Level	Condition	Label
Extreme	$D_{t+1}(r_i) \geq 3 * Ar_i$	3
High	$2 * Ar_i \leq D_{t+1}(r_i) < 3 * Ar_i$	2
Medium	$1 * Ar_i \leq D_{t+1}(r_i) < 2 * Ar_i$	1
Low	$D_{t+1}(r_i) < Ar_i$	0

Label Generation. Predicting passenger demand level of the next time interval is a classification task. Given the external features as input, the classifier outputs the label of the corresponding passenger demand level. To train a classifier, we organize external features as a vector. We use $E_{t+1}(r_i)$ to represent the vector of region r_i in time interval $t+1$ and then label $E_{t+1}(r_i)$ by comparing $D_{t+1}(r_i)$ with the average passenger demand of region r_i in a day. In this paper, labels are generated according to Table 1, where A_{r_i} is the average passenger demand of region r_i in the corresponding day. Considering the high imbalance of passenger demand in large cities, region-specific average demands are more meaningful than a comprehensive average demand for the entire city. We emphasize that predicting demand level would not involve unavailable data because we only need to generate demand level when training the model. We do not need to generate demand level in the predicting phase. While some external features (such as weather) of the next time interval are also not obtainable in testing, we can use the predicted value or the value in the last period.

Classification. As shown in Fig. 1, the classification portion of the auxiliary task is composed of an auto-encoder and two fully-connected layers. $E_{t+1}(r_i)$ is fed into an auto-encoder at first to fuse features from different domains together while keeping most of the useful information. The encoding and decoding processes are implemented with two-layer fully-connected neural networks: $H_{t+1}(r_i) = encoder(E_{t+1}(r_i))$ and $\hat{E}_{t+1}(r_i) = decoder(H_{t+1}(r_i))$. Then the hidden representation $H_{t+1}(r_i)$ is fed into two fully connected layers for classification.

Main Task: Predicting Future Passenger Demand. As introduced in Sect. 1, previous works are inapplicable to non-grid based city partition datasets. When applied to grid-based datasets, they either introduce excessive data from weakly correlated regions or miss out on exploiting correlations from spatially distant but similar regions. To overcome these problems, we propose to extract spatial correlations only from regions that are similar to the target region. We propose two strategies to measure similarities between different regions.

Measuring by Historical Order Sequence. A direct way to measure the similarity between different regions is to calculate the correlations (e.g., Pearson Coefficient) with historical demand. Let $D_{0\sim t}(r_i)$ represent historical order sequence of region r_i from time 0 to t in the training data. Then the similarity of region r_i and r_j can be defined as:

$$Similarity_{r_i,r_j} = Pearson(D_{0\sim t}(r_i), D_{0\sim t}(r_j)) \tag{1}$$

Measuring by PoI Similarity. PoI information can also be used to measure the region similarity. Our motivation is that the existence of certain PoIs in a region can directly influence the passenger demand patterns in that region. For example, if a region has many shopping malls, then passenger demand of that region would significantly increase on weekends and holidays. PoI data can thus be used to characterize a region, analyze the region's passenger demand patterns and find regions that have similar characteristics. Consequently, regions with similar categories of PoI are likely to share similar patterns of passenger demand. Considering that different regions are of different size, we normalize the PoI information of each region with the area of that region. For region r_i and region r_j, similarity between r_i and r_j is:

$$Similarity_{r_i,r_j} = \|\frac{poi_{r_i}}{Area_{r_i}} - \frac{poi_{r_j}}{Area_{r_j}}\|_1 \tag{2}$$

where $Area_{r_i}$ and $Area_{r_j}$ represent the area of region r_i and region r_j respectively, $\|\cdot\|_1$ represents the L1 norm.

After obtaining the pairwise similarities between all regions, we select the m most similar regions for region r_i, represented as $r_{i\_s1}, r_{i\_s2}, ..., r_{i\_sm}$. We organize their passenger demand and crowd outflow data of the same time interval as a vector separately. The main task (predicting precise passenger demand) treats the passenger demand and crowd outflow data in the previous h time intervals of the target region and m similar regions as input to model the spatial and temporal correlations.

Convolutional Neural Networks. As shown in Fig. 1, historical passenger demand and crowd outflow are fed into two CNN networks separately to extract spatial relationships. In the following, we will omit the descriptions for crowd outflow data transformation as they are the same with processing passenger demand data. For each time interval in the previous h time intervals, we only use demand data during this period. Consider region r_i as the target region, we have the most similar m regions $r_{i\_s1}, r_{i\_s2}, ..., r_{i\_sm}$ of r_i. At time interval t, we treat and mix these $1 + m$ region's passenger demand as a $1 \times (2 * m)$ image respectively, represented as:

$$D_t(r_i, r_{i\_s1}, r_{i\_s2}, ..., r_{i\_sm}) = (D_t(r_i), D_t(r_{i\_s1}), D_t(r_i), D_t(r_{i\_s2}), \\ ..., D_t(r_i), D_t(r_{i\_sm})) \tag{3}$$

For time interval t in the previous h time intervals, the CNN takes $D_t(r_i, r_{i\_s1}, r_{i\_s2}, ..., r_{i\_s3})$ as input and feeds it into a convolutional layer. After the convolutional operation, a flatten layer is used to transfer the output of the convolutional layer into a vector. Next we use a fully-connected layer to lower the dimension and get $R_d^t(r_i)$. Using the same approach, we can also get $R_c^t(r_i)$ for crowd outflow in time t of region r_i. Before feeding these two representations into the LSTM layer, we concatenate them together:

$$R_{dc}^t(r_i) = R_d^t(r_i) + R_c^t(r_i) \tag{4}$$

LSTM Layer. The representations extracted from the CNN are fed into an LSTM layer to capture the temporal relationships between future passenger demand and previous h time interval's passenger demand and crowd outflow. Notice that we use previous h time intervals' passenger demand and crowd outflow as input to CNN and extract representations for each time interval separately, so we get h representations. We only save the output of the last LSTM cell for further processing:

$$Q_{t-h+1}^t(r_i) = lstm(R_{dc}^{t-h+1}(r_i), R_{dc}^{t-h+2}(r_i), ..., R_{dc}^t(r_i)) \tag{5}$$

where $lstm$ represents the transformation of all cells in LSTM layer, $Q_{t-h+1}^t(r_i)$ is the output of the last LSTM cell, it represents the captured spatial-temporal information of region r_i and corresponding top m most similar regions from time interval $t - h + 1$ to t.

Combination and Prediction. To fuse the information from spatial, temporal and external part together, we concatenate $Q_{t-h+1}^{t}(r_i)$ with $\hat{H}_{t+1}(r_i)$ together to form $U_{t+1}(r_i)$:

$$U_{t+1}(r_i) = Q_{t-h+1}^{t}(r_i) + \hat{H}_{t+1}(r_i) \tag{6}$$

Finally $U_{t+1}(r_i)$ is fed into three fully-connected layers to get the final predicted passenger demand $\hat{D}_{t+1}(r_i)$. Up to this point, the objective function of the proposed network is composed of three parts: (1) Constraint of auto-encoder in auxiliary task part $\mathcal{L}_1$; (2) Loss of passenger demand level prediction in auxiliary task part $\mathcal{L}_2$; (3) Loss of final passenger demand prediction in main task part $\mathcal{L}_3$:

$$\mathcal{L}_1 = MSE(\hat{E}_{t+1}(r_i) - E_{t+1}(r_i)) \tag{7}$$

$$\mathcal{L}_2 = Cross\_entropy(\hat{L}_{t+1}(r_i) - L_{t+1}(r_i)) \tag{8}$$

$$\mathcal{L}_3 = MSE(\hat{D}_{t+1}(r_i) - D_{t+1}(r_i)) \tag{9}$$

where MSE is the mean square error, $Cross\_entropy$ is the cross entropy loss, and $L_{t+1}(r_i)$ is the true label of passenger demand level for region r_i in $t+1$. Then the overall loss is:

$$\mathcal{L}(\theta) = \mathcal{L}_1 + \mathcal{L}_2 + \mathcal{L}_3 \tag{10}$$

where θ represents all learnable parameters in the network. It is obtained via back-propagation and Adadelta optimizer.

3 Experiments

3.1 Dataset

We use real-world collected datasets to evaluate our method. There are five datasets collected from Dec 5th, 2016 to Feb 4th, 2017 in Shenyang, a big city in China [11]:

- **Passenger Demand Data**: This dataset contains taxi request data of Didi Chuxing. Each item contains the time and location (latitude and longitude) of a request. We pre-process this dataset to map requests to related regions and time intervals. In our experiment, we set the time interval to 1 h.
- **Crowd Outflow Data**: This data is extracted from the cellular networks of the same city which covers more than 1.5×10^6 mobile users. They are also mapped to related regions and time intervals, with the time interval set to 1 h.
- **Meteorological Data**: The meteorological dataset contains information about weather and air quality, including temperature, wind speed, visibility, weather, and air quality level. Temperature, wind speed, and visibility readings are continuous and updated every one hour. Weather and air quality level are categorical data.

- **PoI data**: We collected PoI data of 12 categories, including offices, entertainment facilities, hotels, shopping malls, residences (i.e., apartments), schools, banks, restaurants, government facilities, bus stations, tourist attractions, and hospitals. Each PoI item contains name and location (latitude and longitude). We pre-process this dataset to map PoI data to related regions.
- **Time Meta**: Time meta includes hour of day, day of week, and holiday information.

3.2 Experiment Settings

The model is implemented in TensorFlow 1.8. Due to the page limitations, we have excluded the discussion on parameter tuning. In our experiments, the length of the time interval used is 1 h. We set the number of the most similar regions m to 3 and the historical time window h to 8, which means previous 8 h passenger demand and crowd outflow of the most similar three regions are used to predict the passenger demand in next hour. We use 32 kernels with size 1×3 in CNN, and the stride is 1×1. The output dimension of CNN is re-scaled to 32 by FC layer. The hidden layer of auto-encoder is 24 dimensions. We set the learning rate to 0.02, batch size to 140, and use previous 80% of the data for training and the rest 20% for testing. To evaluate the model, we use Root Mean Square Error $RMSE = \sqrt{\frac{1}{\epsilon}\sum_i(\hat{D}_{t+1}(r_i) - D_{t+1}(r_i))^2}$ and Mean Absolute Error $MAE = \frac{1}{\epsilon}\sum_i \|\hat{D}_{t+1}(r_i) - D_{t+1}(r_i)\|$ of **all regions** to evaluate our model, where ϵ is the number of total time intervals in testing data.

3.3 Experimental Results

Overall Comparison. To validate our model, we compare it with the following methods.

- HA: The historical average model predicts future passenger demand by calculating the average value of previous passenger demand in the same related time interval in the same region.
- ARIMA: The Auto-Regressive Integrated Moving Average model is a widely used time series prediction model which is a generalization of Auto-Regressive Moving Average (ARMA) model.
- SARIMA: The Seasonal Auto-Regressive Integrated Moving Average model is a variance of the ARIMA model, which can capture the seasonality in a time series data.
- OLSR: The Ordinary Least Square Regression model is a kind of linear regression model, it can estimate the relationship between multiple variables.
- MLP: The Multiple Layer perceptron is a typical class of feed-forward neural network. It has multiple layers and non-linear activation function.
- LSTM: As introduced in Sect. 4, LSTM is a variation of recurrent neural networks, which is prominent in sequence data processing.

- XGBoost [14]: XGBoost is a boosting tree-based machine learning method, which is used to achieve state-of-the-art results on many data mining challenges.
- DMVST-Net [2]: DMVST-Net is a state-of-the-art method for predict passenger demand. It is a deep learning based method which considers both spatial and temporal correlations.

Table 2. Overall comparison with different methods

Index	Method	RMSE	MAE
1	HA	25.028	95.573
2	ARIMA	23.702	93.829
3	SARIMA	23.293	91.682
4	OSLR	22.003	87.348
5	MLP	21.889	85.265
6	LSTM	21.799	88.049
7	XGBoost [14]	20.497	79.489
8	DMVST-Net [2]	20.231	80.753
9	**MT-CRNN (PoI)**	19.602	76.469
10	**MT-CRNN (Order)**	19.467	74.438

HA only considers the historical demand as input, while all other aforementioned models employ all features to predict future passenger demand. MLP contains four fully connected layers, while LSTM only has one layer. As described in Sect. 1, DMVST-Net [2] can only be used when the city is partitioned to grids. In order to compare with DMVST-Net, we fed inputs of our model to DMVST-Net.

We show the experimental results in Table 2. From the table, we can observe that the performance of simple neural networks such as MLP and LSTM is not good. They don't show much improvement over traditional methods such as SARIMA and OSLR. In contrast, state-of-the-art methods XGBoost and DMVST-Net achieve 18.17% and 19.57% improvement in RMSE over HA, respectively. However, our model produces the lowest RMSE (19.602 when measuring similarity by PoI and 19.467 when measuring similarity by historical order sequence) among all the methods. Furthermore, our method achieves 3.80% (RMSE) and 7.82% (MAE) relative improvement over DMVST-Net.

Component Analysis. We also evaluated some variations of our MT-CRNN model to study the effect of different components and our auxiliary task setting, including:

- ST-D: This model only performs the main task of the MT-CRNN model and uses historical demand data as input. Similar region selection, CNN and LSTM layers are the same with MT-CRNN. The loss function is $\mathcal{L}_1$.

- ST-DC: Similar to ST-D, ST-DC further integrates crowd flow data as additional input.
- ST-DE: ST-DE is a single task model with the same design as MT-CRNN, but it doesn't take crowd flow data as input. The loss function is $\mathcal{L}_1 + \mathcal{L}_3$.
- ST-DCE: In this model, the final loss function is $\mathcal{L}_1 + \mathcal{L}_3$, which transforms MT-CRNN to a single task model.
- MT-DE: MT-DE is a multi-task model with the same design as MT-CRNN, but it doesn't utilize crowd outflow data.
- MT-FC: In this model, the final loss function is $\mathcal{L}_2 + \mathcal{L}_3$, which means the decoder part of the auto-encoder is not trained. Thus, the auto-encoder is transformed into a two-layer fully-connected neural network.
- MT-GA: Instead of labeling the passenger demand level by the target region's average demand, this model labels the passenger demand with the entire region's average demand.

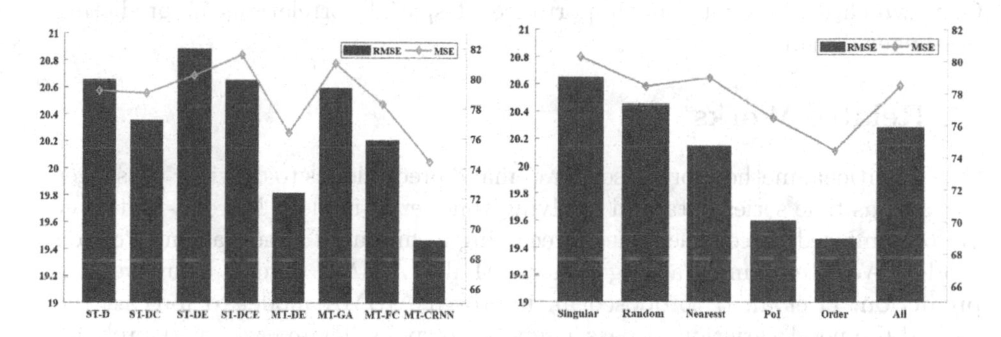

Fig. 2. Component analysis

Fig. 3. Prediction with different correlated regions

From Fig. 2, we can observe that MT-CRNN outperforms ST-DCE and MT-DE outperforms ST-DE, which justify the importance of our multi-task setting. Secondly, ST-DC outperforms ST-D, ST-DCE outperforms ST-DE and MT-CRNN outperforms MT-DE, which shows that prediction accuracy is better when crowd outflow data is included in addition to historical passenger demand data. Besides, ST-DCE performs worse than ST-DC and ST-DE performs worse than ST-D, which demonstrates that the improper use of external features adversely impacts prediction accuracy. Thirdly, the results for MT-FC shows that an auto-encoder is better than fully-connected neural networks in extracting hidden features from external data. Finally, by comparing MT-GA with MT-CRNN, we can show that considering region-specific average demand is better than average demand over the entire city.

Spatial Correlation Analysis. We also evaluated the performance of our model with data from different regions to analyze spatial correlations. As shown in Fig. 3, we included the following:

- Singular: Only consider the target region's historical data;
- Random: Randomly select correlated regions for the target region. We randomly select regions five times and present the average prediction results.
- Nearest: Similar with DMVST-Net, it only considers spatially nearby regions to capture their spatial correlations;
- PoI: Select correlated regions by PoI similarity;
- Order: Select correlated regions by historical demand series similarity;
- All: Similar with DeepST [12], it captures the spatial correlations within the whole city.

We can observe that predicting only with selected similar region's data is better than all other strategies, which shows the advantages of our similarity-based CNN in capturing spatial correlations. Moreover, all strategies are better than One, which demonstrate the importance of spatial correlations in predicting passenger demand.

4 Related Works

One traditional method for passenger demand prediction is to consider passenger demand as time series data and applying time series models. Moreira-Matias et al. [6] combined three time-series forecasting techniques (Time-Varying Poisson Model, Weighted Time-Varying Poisson Model, ARIMA model) to arrive at a prediction. Li et al. [7] proposed an improved ARIMA model to forecast the spatial-temporal variation of passengers in hotspots. These early works rely on GPS trajectories data from a subset of the entire taxis, which may not necessarily reveal the actual passenger demand. In recent years, some researchers have applied deep learning methods in smart transportation systems [10]. Yu et al. [9] proposed to use Long-Short Term Memory (LSTM) network to capture the temporal relationship in historical observations and used auto-encoder to process static features. However, they didn't consider spatial correlations. Wang et al. [5] presented a neural network framework based on fully-connected layers and residual network to predict the gap between passenger demand and supply. Their approach cannot accurately capture the sequential relationship. Another way to capture the spatial correlation is treating the city as an image (a two-dimensional matrix) and applying CNN to it. Zhang et al. [3] propose a spatial-temporal model to predict citywide crowd flow. They represent city-wide crowd flow as a multi-dimensional image and use CNN and residual network to extract spatial relationships. Yao et al. [2] further designed "local CNN" to extract spatial relationship within surrounding regions and construct a weighted graph to represent similarity among regions. A positive aspect is that all these deep learning based methods take external features (weather, holiday, time meta) into consideration. However, they transform external features using fully-connected layers or auto-encoder, which are incapable of fully realising their potential.

5 Conclusions

In this paper, we proposed a Multi-Task Convolutional Recurrent Neural Network (MT-CRNN) framework to forecast the passenger demand with multiple features from different domains. We captured the spatial-temporal correlations of historical passenger demand by the convolutional recurrent neural network based on the historical demand of selected similar regions. To better utilize external features, we designed an auxiliary task for predicting passenger demand level under the guideline of multi-task learning. Experimental results show that our model significantly outperforms a series of baselines and gains 3.8% improvement (RMSE) over state-of-the-art methods and that the auxiliary task can improve the final passenger demand prediction accuracy.

References

1. Ke, J., et al.: Short-term forecasting of passenger demand under on-demand ride services: a spatio-temporal deep learning approach. Transp. Res. Part C: Emerg. Technol. **85**, 591–608 (2017)
2. Yao, H., et al.: Deep multi-view spatial-temporal network for taxi demand prediction. In: AAAI (2018)
3. Zhang, J., et al.: Deep spatio-temporal residual networks for citywide crowd flows prediction. In: AAAI (2017)
4. Deng, D., et al.: Latent space model for road networks to predict time-varying traffic. In: Proceedings of the 22nd ACM SIGKDD International Conference on Knowledge Discovery and Data Mining. ACM (2016)
5. Wang, D., et al.: DeepSD: supply-demand prediction for online car-hailing services using deep neural networks. In: 2017 IEEE 33rd International Conference on Data Engineering (ICDE). IEEE (2017)
6. Moreira-Matias, L., et al.: Predicting taxi-passenger demand using streaming data. IEEE Trans. Intell. Transp. Syst. **14**(3), 1393–1402 (2013)
7. Li, X., et al.: Prediction of urban human mobility using large-scale taxi traces and its applications. Front. Comput. Sci. **6**(1), 111–121 (2012)
8. Li, Y., et al.: Taxi booking mobile app order demand prediction based on short-term traffic forecasting. Transp. Res. Rec.: J. Transp. Res. Board **2634**, 57–68 (2017)
9. Yu, R., et al.: Deep learning: a generic approach for extreme condition traffic forecasting. In: Proceedings of the 2017 SIAM International Conference on Data Mining. Society for Industrial and Applied Mathematics (2017)
10. Zheng, Y., et al.: Urban computing: concepts, methodologies, and applications. ACM Trans. Intell. Syst. Technol. (TIST) **5**(3), 38 (2014)
11. Chu, J., et al.: Passenger demand prediction with cellular footprints. In: 2018 15th Annual IEEE International Conference on Sensing, Communication, and Networking (SECON). IEEE (2018)
12. Zhang, J., et al.: DNN-based prediction model for spatio-temporal data. In: Proceedings of the 24th ACM SIGSPATIAL International Conference on Advances in Geographic Information Systems. ACM (2016)

13. Xingjian, S.H.I., et al.: Convolutional LSTM network: a machine learning approach for precipitation nowcasting. In: Advances in Neural Information Processing Systems (2015)
14. Chen, T., Guestrin, C.: XGBoost: a scalable tree boosting system. In: Proceedings of the 22nd ACM SIGKDD International Conference on Knowledge Discovery and Data Mining. ACM (2016)

Accurate Identification of Electrical Equipment from Power Load Profiles

Ziyi Wang, Chun Li, and Lin Shang(✉)

National Key Laboratory for Novel Software Technology,
Nanjing University, Nanjing 210023, China
{zywang,lichun}@smail.nju.edu.cn, shanglin@nju.edu.cn

Abstract. It is essential for the power industries to identify the running electrical equipment automatically. For power monitoring, the load profile data vary with the equipment's types. Proceeding from the fundamental features of load time series, we propose a method to identify electrical equipment from power load profiles accurately. Aiming to improve the classification accuracy and generalization performance of convolutional neural network (CNN), we combine the training process of generative adversarial networks (GANs) with CNN, which employs the generated samples to enhance the classification accuracy. The CNN and discriminator in our approach share the first convolution layer for extracting richer features. We evaluate our method on UCR data sets comparing with 12 existing methods. Furthermore, we compare our model with LSTM, GRU and CNN on the electrical equipment load data, which is from industries in certain area. The final results show that our model has a higher equipment identification accuracy than other deep learning models.

Keywords: Power load profiles · CNN · GAN · Time series

1 Introduction

In power industries, equipment monitoring system adopts a variety of technologies to analyze the data generated from electrical equipments or power load data. Most electrical equipments are sophisticated and expensive, and the cost is high to maintain them. With the load shifting or electrical equipment variation, it is necessary to ensure the safety of the equipment. Apart from inspecting these electrical equipments manually, there are some common methods, such as parameter determination and image recognition [1,2]. But in some cases, the exception occurred will only lead to the load shift, with the normal equipment state, which will be potentially dangerous. If the machine learning methods can be used to identify equipment from the load profile, it can carry out more analysis and management.

The load profile is essentially a class of time series data. Most existing Time Series Classification (TSC) approaches fall into two categories [3]: distance-based

Q. Yang et al. (Eds.): PAKDD 2019, LNAI 11440, pp. 43–55, 2019.
https://doi.org/10.1007/978-3-030-16145-3_4

methods and feature-based methods. For distance-based methods, the key idea is to compute the similarity between any given two time series. Such as K-nearest neighbors (KNN) or support vector machines (SVMs). The most remarkable distance-based method is dynamic time warping (DTW). For feature-based methods, the key part is to get a good representations of time series. Shapelet [4] is a relatively successful feature-based method, which are time series subsequences in some sense maximally representative of a class. Also a new symbolic representation for time series was proposed in [5]. In addition to these traditional methods, CNN model has recently been introduced into time series classification. Although the CNN model has been prevailing in the field of time series classification, the powerful learning ability of CNN could lead to over-fitting in insufficient and constrained data. We must carefully set the parameters to achieve a better generalization performance. There exist some very effective tricks hacks to prevent CNN from over-fitting. However the most effective way to prevent over-fitting is to increase the amount of training data.

Recently, unsupervised learning has attracted significant interest in the field of machine learning, among which generative modelling is significantly. The most prominent generative models are the variational autoencoder (VAE) [6] and the generative adversarial network (GAN) [7]. More and more GAN researchers aim to learn the distribution of raw data. When GAN is trained well, generator can generate new samples of high quality [8].

In the industrial work, it is constrained to obtain the required data and the CNN may be over-fitting in such training set. To improve the accuracy of electrical equipment identification, in this paper we propose a method of electrical equipment identification by combining the training process of GAN and CNN. We add the CNN into the original discriminative model for the classification, so that CNN and discriminator will share the first convolution layer. In our model, generator will continually generate 'fake samples' during training process. These 'fake samples' are also used for features extraction in the first convolution layer. If the generated 'fake samples' are of high quality, which means the generated samples are realistic, then the first convolution layer can extract richer features. As is known, the discriminator updates the parameters according to *d_loss*. In this paper, we set a factor for *d_loss* and we define it as d_loss_factor ($0 \leq d\_loss\_factor \leq 1$). The d_loss_factor is used to control the effect of *d_loss* on the first convolution layer.

The main contribution of our paper is to propose a new deep learning model, which can not only improve the accuracy of equipment identification, but also can generate some samples of high quality. We also conduct comprehensive experiments and compare with 12 existing time series classification methods. The results show that our model achieve the highest accuracy on 8 UCR datasets. Moreover, our model has higher equipment identification accuracy than the CNN and the generalization error of our model is smaller than CNN.

2 Our Approach and Model Architecture

In this section, we will introduce our approach and model architecture. Figure 1 shows the framewok of our model.

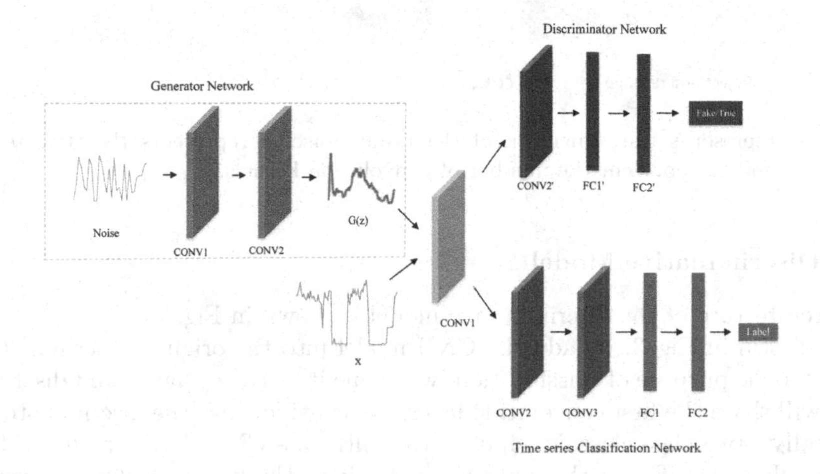

Fig. 1. The framework of our model

As shown in Fig. 1, our model is composed of generator network and discriminator network. When the generator generates some 'fake samples', we send these samples to the discriminator network together with the raw samples for feature extraction in the first convolution layer. 'Fake samples' do not enter CNN model because they have not class labels. CNN and discriminator update their network parameters individually through loss function, but update the parameters of the first convolution layer at the same time.

GANs have been known to be unstable to train, often resulting in generators that produce nonsensical outputs. In this paper, we design our experiments based on existing techniques introduced by DCGAN [9].

2.1 Generative Model

The architecture of the generator is shown in Fig. 2.

The generative model learn to capture the data distribution [7]. A n dimensional uniform distribution noise z is projected to a convolutional representation with a features. A series convolutions (since we did not use pooling in the discriminator, we did not use the fractionally-strided convolutions in the generator.) then convert the features into the generated samples. The generator updates the parameters according to the loss function of generator.

In most cases, the length of the time series data is different, so the dimension of the input noise is not fixed.

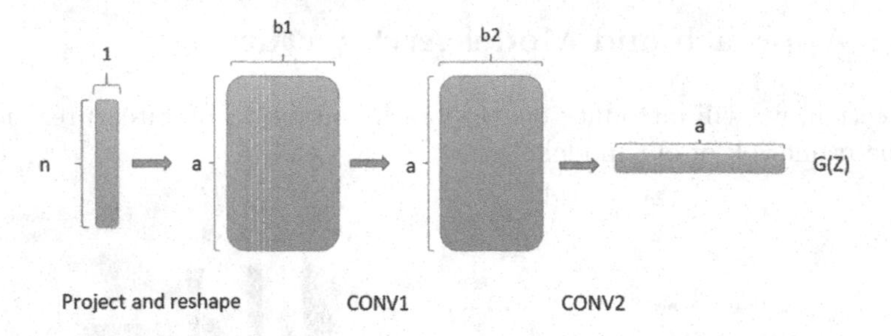

Fig. 2. n represents the dimension of the input noise, a represents the time series length, $b1$ and $b2$ represent the number of convolution kernels.

2.2 Discriminative Model

The architecture of the discriminative model is shown in Fig. 3.

As shown in Fig. 3, we add the CNN model into the original discriminative model for the purpose of classification, we define it as C, so that C and discriminator will share the first convolution layer. We consider that the discriminator is essentially a two-class classifier, and C is a multi-class CNN classifier. Regardless the two-class classifier or the multi-class classifier, the first convolution layer is used to extract the basic features. Not only the 'raw samples' from the training set, but also the 'fake samples' generated by the generator are included in the samples through the discriminator. When the generated 'fake samples' are of high quality, the first convolution layer can extract richer features. In this case, the CNN will have a better generalization performance. When the generated 'fake samples' are not of high quality, we inevitably introduce a lot of noise. In this case, the CNN will have a worse generalization performance. So we set a factor for d_loss, defined as d_loss_factor ($0 \leq d\_loss\_factor \leq 1$). The d_loss_factor is used to control the effect of d_loss on the first convolution layer, and the d_loss_factor is a hyper parameter in our model and we give a better initialization value for d_loss_factor in Sect. 3.3.

Our model can be considered as a kind of multitask learning [10] model, which improve generalization by pooling the examples arising out of two tasks. In the same way that additional training examples put more pressure on the parameters of the model toward values that generalize well, when part of a model is shared across tasks, that part of the model is more constrained toward good values (assuming the sharing is justified), often yielding better generalization [11].

2.3 Loss Function

We define the loss function of generator as G_loss:

$$G\_loss = E_{z \sim p_z(z)} log[1 - D(G(z^{(i)}))] \tag{1}$$

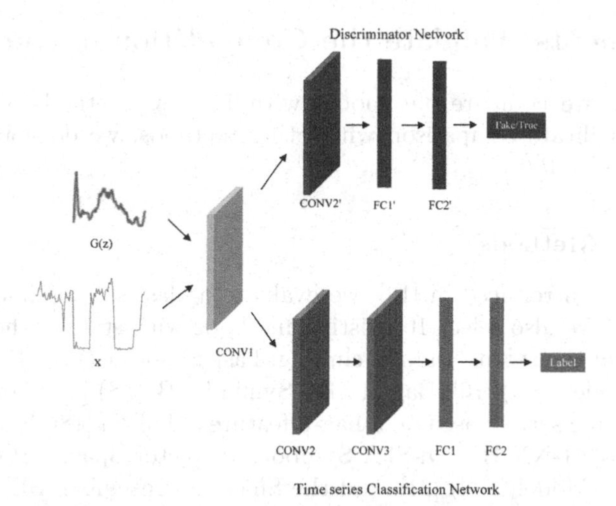

Fig. 3. The CONV1 represents the public part, the upper part is the discriminator network, the lower part is the time series classification network.

We train generative model to minimize $E_{z \sim p_z(z)} log[1 - D(G(z^{(i)}))]$, that means we hope that the discriminative model can discriminate the generated samples as real samples as far as possible.

The loss function of the discriminative model consists of two components, one is the loss of discriminator, another is the loss of CNN. So we define the loss function of discriminative model as D_loss:

$$\begin{aligned} D\_loss = E_{z \sim p_z(z),(x,y) \sim p_{data}(x,y)}[[d\_loss\_factor(\\ -log(D(x^{(i)})) - log(1 - D(G(z^{(i)}))))] + C(x^{(i)}, y^{(i)})] \end{aligned} \tag{2}$$

$$C = -\frac{1}{N}[\sum_{i=1}^{N}\sum_{j=1}^{k} I(y^{(i)} = j) log p_j^{(i)}] \tag{3}$$

C represents the loss function of the CNN model, N represents the number of samples and k represents that the samples have k different classes.

We train discriminative model to minimize the $-log(D(x^{(i)})) - log(1 - D(G(z^{(i)})))$ of assigning the correct label to both training examples and samples from generative model. Meanwhile, we minimize the $C(x^{(i)}, y^{(i)})$ so that the CNN model has a higher classification accuracy.

The d_loss_factor ($0 \leq d\_loss\_factor \leq 1$) is used to control the effect of d_loss on the first convolution layer. When $d\_loss\_factor = 0$, $D\_loss = E_{z \sim p_z(z),(x,y) \sim p_{data}(x,y)}$ $[C(x^{(i)}, y^{(i)})]$. In this case, the D_loss represents the loss function of CNN model and the accuracy of our model is equal to the CNN model. When d_loss_factor is set to a small value (i.e. $d\_loss\_factor = 0.1$), the discriminator loss has a weak influence on the first convolution layer. In the experiment, we can adjust the value of d_loss_factor according to the quality of the generated samples.

3 Experiments: Validate the Competition of Our Model

In this section, we compare our model with 12 TSC methods on UCR [12] datasets. To facilitate comparison with other methods, we do not list all the data sets in the UCR.

3.1 Baseline Methods

As the baselines introduced in [13], we evaluate a classical baseline method 1-NN DTW [14]. We also select 10 existing methods with state-of-the-art results published within the recent years, including: Fast Shapelet (FS) [15], SAX with vector space model (SV) [16], Bag-of-SFA-Symbols (BOSS) [5], Shotgun Classifier (SC) [17], time series based on a bag-offeatures (TSBF) [18], Elastic Ensemble (PROP) [19], 1-NN Bag-Of-SFA-Symbols in Vector Space (BOSSVS) [20], Learn Shapelets Model(LTS) [21], and the Shapelet Ensemble (SE) model [22]. For reference, we also list the results of flat-COTE (COTE), an ensemble model proposed by Bagnall et al. [22], which uses the weighted votes over 35 different classifiers.

In addition to the above methods, we also compare with MCNN and GAN-CNN. MCNN is a successful deep learning model currently for time series classification. GAN-CNN trains GAN using different labels of data to generate 'fake samples' of the corresponding label and then put the 'fake samples' in the raw data set. GAN-CNN uses this new data set to train CNN. GAN-CNN model is simple, but the training cost of this model is too large. If a training data set has k different classes, we need to train k different GAN models, and then use them to generate the samples of the corresponding labels. GAN-CNN is essentially a serial training of GAN and CNN. Our model can train GAN and CNN at the same time.

3.2 Datasets

We evaluate all methods thoroughly on the UCR [12] datasets. The UCR datasets contain some of the time series data extracted from various real-world domains. Although most of the datasets in the UCR archive are not large enough, we use these training sets directly to train our model, but do not use window slicing to increasing the size of the training size. We selected 23 datasets from more than 80 UCR datasets to verify the competitiveness of our model. We did not experiment on all more than 80 data sets because our model is not a general time series classification model. When the dimensions of the time series are too high and the number of training samples are limited, which will lead to GAN to be hard to train.

3.3 Hyper Parameters

In order to compare the generalization performance of our model with CNN and GAN-CNN, we test the CNN and GAN-CNN with the same architecture and number of parameters as our model. The hyper parameters of the CNN model include the number of kernels in each convolution layer and the number of channels in fully connected layer. For the data set in the experiment, we randomly select 20% training samples as the validation set to select the hyper parameters. Our model and GAN-CNN use the same hyper parameters as CNN. Because the samples generated by the generator may contain a lot of noise, so we set a d_loss_factor to control the effect of the generated samples on the CNN training process. By default, we set $d\_loss\_factor = 0.1$.

3.4 Experimental Results on UCR Datasets

We show the testing errors in the Table 1. The experimental results of the first 12 baselines come from [13]. All data sets of UCR are divided into training sets and testing sets, so these results can be directly compared. We can see that our model is very competitive, achieving the highest accuracy on 8 datasets as well as our model has a better generalization performance than CNN. The BOSS is a ensemble classifier and also achieves the highest accuracy on 8 datasets, but the average rank of Boss is 6.7, the average rank of our model is 5.08.

The GAN-CNN also has a good performance and achieves the highest accuracy on 6 datasets. But as mentioned before, GAN-CNN has an unacceptable time complexity and it cannot control the effect of the generated samples on the CNN training process. Once the generated samples contain a lot of noise, the final performance of GAN-CNN is badly than CNN.

In the experiment, we have observed the effect of d_loss_factor on the classification accuracy of our model, and we suggest that d_loss_factor be set as a relatively small value. In general, when $d\_loss\_factor = 0.1$, our model has a better generalization performance.

4 Electrical Equipment Identification from Power Load Profiles

In this section, we conduct the identification procedure on five different scale load training sets and test on the same testing set. We compare our model ($d\_loss\_factor = 0.1$) with CNN, Long Short-Term Memory (LSTM), and Gated Recurrent Units (GRUs). LSTM is well-suited to classify, process and predict time series given time lags of unknown size and duration between important events. Compared with LSTM, GRUs have been shown to exhibit better performance on smaller datasets [23].

4.1 Dataset

Dataset is obtained and randomly selected from one district load power of one province, which contains the load data for ten different electrical equipments. Each power load profile contains the 96-load values and we use the pre- and post-value to fill the missing values. All data have been normalized before classification.

Table 1. Testing error for 23 UCR time series datasets

Dataset	DTW	SV	BOSS	SE1	TSBF	TSF	BOSSVS	PROP	LS	SE	COTE	MCNN	CNN	GAN-CNN	Our Model
Adiac	0.396	0.417	0.22	0.373	0.245	0.261	0.302	0.353	0.437	0.435	0.233	0.231	0.225	0.3	**0.2148**
Beef	0.367	0.467	0.2	0.133	0.287	0.3	0.267	0.367	0.24	0.167	0.133	0.367	**0.0677**	**0.0677**	**0.0677**
CBF	0.003	0.007	**0**	0.01	0.009	0.039	0.001	0.002	0.006	0.003	0.001	0.002	0.112	0.069	0.033
ChlorineCon	0.352	0.334	0.34	0.312	0.336	0.26	0.345	0.36	0.349	0.3	0.314	0.203	0.17	**0.15**	0.17
CinCECGTorso	0.349	0.344	0.125	**0.021**	0.262	0.069	0.13	0.062	0.167	0.154	0.064	0.058	0.069	0.039	0.0348
Coffee	**0**	**0**	**0**	**0**	0.004	0.071	0.036	**0**	**0**	**0**	**0**	0.036	**0**	**0**	**0**
CricketX	0.246	0.308	0.259	0.297	0.278	0.287	0.346	0.203	0.209	0.218	0.154	0.182	**0**	**0**	**0**
ECGFiveDays	0.232	0.003	**0**	0.055	0.183	0.07	**0**	0.178	**0**	0.001	**0**	**0**	0.055	0.033	0.042
FaceAll	0.192	0.244	0.21	0.247	0.234	0.231	0.241	0.152	0.217	0.263	0.105	0.235	0.084	0.128	**0.0835**
FaceFour	0.17	0.114	**0**	0.034	0.051	0.034	0.034	0.091	0.048	0.057	0.091	**0**	**0**	0.091	**0**
GunPoint	0.093	0.013	**0**	0.06	0.011	0.047	**0**	0.007	**0**	0.02	0.007	**0**	0.047	0.047	0.0267
ItalyPower	0.05	0.089	0.053	0.053	0.096	0.033	0.086	0.039	0.03	0.048	0.036	0.03	0.04	0.027	**0.0253**
Lighting2	0.131	0.23	0.148	**0.098**	0.257	0.18	0.262	0.115	0.177	0.344	0.164	0.164	0.279	0.246	0.246
Lighting7	0.274	0.342	0.342	0.274	0.262	0.263	0.288	0.233	**0.197**	0.26	0.247	0.219	0.3	0.3	0.3
MedicalImage	0.263	0.516	0.288	0.305	0.269	**0.232**	0.474	0.245	0.27	0.396	0.258	0.26	0.295	0.28	0.234
MoteStrain	0.165	0.117	**0.073**	0.113	0.135	0.118	0.115	0.114	0.087	0.109	0.085	0.079	0.143	0.128	0.13
OliveOil	0.167	0.133	0.1	0.133	0.09	0.1	0.133	0.133	0.56	0.1	0.1	0.133	**0.067**	**0.067**	**0.067**
SonyAIBORobot	0.275	0.306	0.321	0.238	0.175	0.235	0.265	0.293	0.103	0.067	0.146	0.23	0.14	**0.065**	0.099
SonyAIBORobotII	0.169	0.126	0.098	**0.066**	0.196	0.177	0.188	0.124	0.082	0.115	0.076	0.07	0.175	0.121	0.142
SyntheticControl	0.007	0.013	0.03	0.033	0.008	0.023	0.04	0.01	0.007	0.017	**0**	0.003	0.003	0.013	0.003
Trace	**0**	**0**	**0**	0.05	0.02	**0**	**0**	0.01	**0**	0.02	0.01	**0**	0.02	0.02	0.01
TwoLeadECG	**0**	0.004	0.004	0.029	0.001	0.112	0.015	**0**	0.003	0.004	0.015	0.001	0.076	0.082	0.0088
wafer	0.02	0.002	**0.001**	0.002	0.004	0.047	**0.001**	0.003	0.004	0.002	**0.001**	0.002	0.003	0.003	0.003
#best	3	2	8	4	0	2	4	2	5	1	4	4	5	6	8

Table 2. Load profile of three types of electrical equipment

Equipment name	Equipment	Power Load Profile
Recycling air blower motor		
Sanders		
Refined rubber machine		

Table 2 shows examples for three different electrical equipments and their power load profiles We can see that the load curves of some electrical equipments have obvious characteristics, for example the Sanders, which has an obvious low load Valley. Our model can well identify the abnormal power load profiles. The load curves of some electrical equipments have not regular pattern, such as Recycling air blower motor and Refined rubber machine, which are worth our attention when they are maintained.

4.2 Experimental Results

Table 3 shows the testing accuracy for load profiles of 5 different training sets. For example, the 500 means the training set contain about 500 samples. We can see that, the LSTM and GRU have not achieved a good performance on this data set. As is known to all, LSTM or GRU is widely used in Natural Language Processing (NLP), because there is a strong dependency between words. However, there is no obvious dependency between the previous moment value and the next moment value in the power load profiles mentioned in this paper, although it belongs to time series data. The value of each moment largely depends on the external factors, such as the current working state of each electrical equipment. In addition, the dimensions of some kind of time series data are often too high (i.e. 1000+), which will significantly cause the vanishing gradient problem or the exploding gradient problem. The test results show that our model achieves highly competitive performance.

Table 3. Testing accuracy for load profiles of 5 different training sets

NetWork	500	1000	1500	2000	2500
CNN	61.3	64.2	67	81.3	83.7
LSTM	47.6	47.7	48.1	53.3	57.9
GRU	40.9	44.6	46.4	47.2	50.0
Our model	**63.2**	**66.15**	**69.42**	**83.7**	**84.9**

Figure 4 shows the trend of accuracy of CNN and our model on testing set when the model trained on five training sets of different size. We can see that when the training set is small, the accuracy of CNN and our model are not high. With the increase of the number of training samples, the generalization performance will be greatly improved. When the number of training samples is increased to a certain extent, the speed of generalization performance improvement will be slowed down.

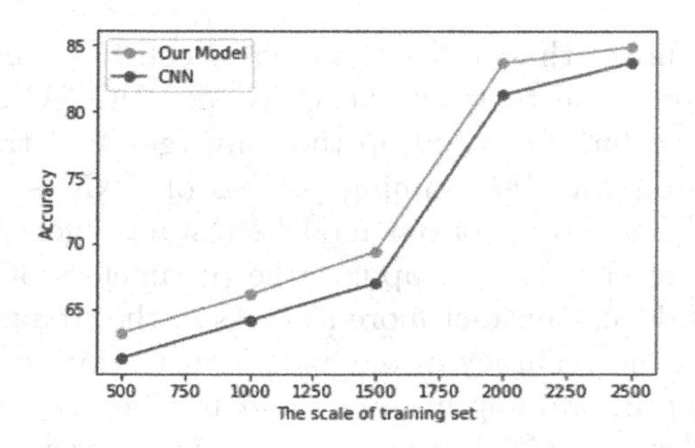

Fig. 4. The trend of the accuracy on testing set

We use our model to identify electrical equipment on the testing set. Figure 5 shows the identification accuracy of ten types of electrical equipments when the size of the training set is 2500. We can see that some electrical equipments, such as 1 and 2, have higher identification accuracy. We believe that these electrical equipments have stable load characteristic. Once the load profiles of these equipments are mistakenly identified, we are confident that the equipment may be out of order or there exist some potentially dangerous. Some equipments, which have an unstable load profile. It is not easy to estimate the operation state through the load profiles and we must check the equipment carefully. So if we have trained such a classifier, we can monitor the operation of the electrical equipments in real time and give the corresponding warning so that the cost of manual maintenance is greatly reduced.

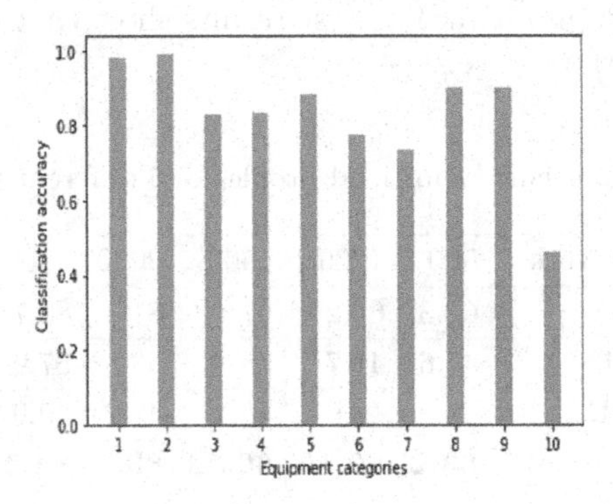

Fig. 5. Identification accuracy of ten types of electrical equipments.(equipment categories 1: '4524RTC', 2: 'CT3 High-pressure oxygen generation', 3: 'Sanders', 4: 'Molding equipment II', 5: 'Mixer III', 6: 'CT3-EB1 and EB2', 7: 'Recycling air blower motor', 8: '4443RTC', 9: '4435RTC', 10: 'Refined rubber machine')

4.3 Training Process Analysis

In this section, we demonstrate the training process of CNN and our model on the load data.

From Fig. 6 we can see that, CNN can quickly converge on the training set, our model takes more time to train the CNN and DCGAN. In general, setting a small *d_loss_factor* can speed up the convergence of training process of our model. The reason why the training process of CNN is more stable than our model is that the generator of our model constantly generate new samples and these samples are also used to update the parameters of first convolution layer. Due to our model can extract more features in the training process, which results in sometimes the accuracy of our model on testing set exceeds that on training set. The gap of training accuracy and testing accuracy of our model is significantly smaller than CNN under the same hyper-parameters that means our model has high potential for improvement.

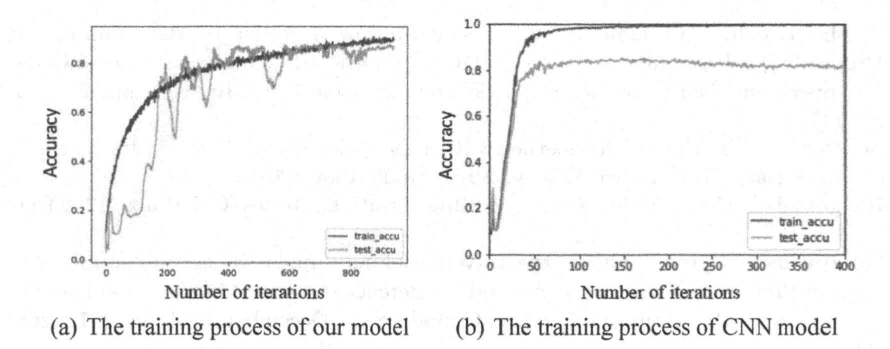

(a) The training process of our model (b) The training process of CNN model

Fig. 6. The training process of two models

5 Discussion and Conclusions

For electrical equipment identification from power data profiles, we propose a new deep learning model, which can improve the generalization performance and significantly reduce the generalization error of CNN model through the new samples generated by GAN. It is not acceptable on time complexity to use different labels of data to train GAN to generate samples of corresponding labels. In our model, CNN and the discriminator share the first convolution layer and both the CNN and GAN are trained at the same time. We test the validity of our model on 23 UCR datasets compared with 12 existing methods and achieve the highest accuracy on 8 datasets. Note that when the training data is too small, DCGAN cannot learn the true distribution of the data and the generated samples lack diversity, we use d_loss_factor to control the effect of d_loss on the first convolution layer. We compare our model with CNN, LSTM and GRU on power load data, the final results show that our model has a higher equipment identification accuracy than other deep learning models.

Acknowledgment. We would like to thank Keith for his help and suggestions in writing this paper. This work is supported by the National Natural Science Foundation of China (No. 61672276) and Natural Science Foundation of Jiangsu, China (BK20161406).

References

1. Qiu, J., Wang, H., Lin, D., He, B.: Nonparametric regression-based failure rate model for electric power equipment using lifecycle data. IEEE Trans. Smart Grid **6**(2), 955–964 (2015)
2. Warnier, M., Dulman, S., Koç, Y., Pauwels, E.: Distributed monitoring for the prevention of cascading failures in operational power grids. Int. J. Crit. Infrastruct. Prot. **17**(3), 245–251 (2015)
3. Xing, Z., Pei, J., Keogh, E.J.: A brief survey on sequence classification. SIGKDD Explor. **12**(1), 40–48 (2010)

4. Ye, L., Keogh, E.J.: Time series shapelets: a new primitive for data mining. In: Proceedings of the 15th ACM SIGKDD International Conference on Knowledge Discovery and Data Mining, Paris, France, 28 June - 1 July 2009, pp. 947–956 (2009)
5. Schäfer, P.: The BOSS is concerned with time series classification in the presence of noise. Data Min. Knowl. Discov. **29**(6), 1505–1530 (2015)
6. Kingma, D.P., Welling, M.: Auto-encoding variational bayes. CoRR abs/1312.6114 (2013)
7. Goodfellow, I.J., et al.: Generative adversarial nets. In: Advances in Neural Information Processing Systems 27: Annual Conference on Neural Information Processing Systems 2014, Montreal, Quebec, Canada, 8–13 December 2014, pp. 2672–2680 (2014)
8. Denton, E.L., Chintala, S., Szlam, A., Fergus, R.: Deep generative image models using a Laplacian pyramid of adversarial networks. In: Advances in Neural Information Processing Systems 28: Annual Conference on Neural Information Processing Systems 2015, Montreal, Quebec, Canada, 7–12 December 2015, pp. 1486–1494 (2015)
9. Radford, A., Metz, L., Chintala, S.: Unsupervised representation learning with deep convolutional generative adversarial networks. CoRR abs/1511.06434 (2015)
10. Caruana, R.A.: Multitask connectionist learning. In: Connectionist Models Summer School, pp. 372–379 (1993)
11. Goodfellow, I., Bengio, Y., Courville, A.: Deep Learning. MIT Press (2016). http://www.deeplearningbook.org
12. Chen, Y., et al.: The UCR time series classification archive, July 2015. www.cs.ucr.edu/~eamonn/time_series_data/
13. Cui, Z., Chen, W., Chen, Y.: Multi-scale convolutional neural networks for time series classification. CoRR abs/1603.06995 (2016)
14. Berndt, D.J., Clifford, J.: Using dynamic time warping to find patterns in time series. In: Knowledge Discovery in Databases: Papers from the 1994 AAAI Workshop, Seattle, Washington, July 1994. Technical report WS-94-03, pp. 359–370 (1994)
15. Keogh, E.J., Rakthanmanon, T.: Fast shapelets: a scalable algorithm for discovering time series shapelets. In: Proceedings of the 13th SIAM International Conference on Data Mining, Austin, Texas, USA, 2–4 May 2013, pp. 668–676 (2013)
16. Senin, P., Malinchik, S.: SAX-VSM: interpretable time series classification using SAX and vector space model. In: 2013 IEEE 13th International Conference on Data Mining, Dallas, TX, USA, 7–10 December 2013, pp. 1175–1180 (2013)
17. Schäfer, P.: Towards time series classification without human preprocessing. In: Perner, P. (ed.) MLDM 2014. LNCS (LNAI), vol. 8556, pp. 228–242. Springer, Cham (2014). https://doi.org/10.1007/978-3-319-08979-9_18
18. Baydogan, M.G., Runger, G.C., Tuv, E.: A bag-of-features framework to classify time series. IEEE Trans. Pattern Anal. Mach. Intell. **35**(11), 2796–2802 (2013)
19. Lines, J., Bagnall, A.: Ensembles of elastic distance measures for time series classification. In: Proceedings of the 2014 SIAM International Conference on Data Mining, Philadelphia, Pennsylvania, USA, 24–26 April 2014, pp. 524–532 (2014)
20. Schäfer, P.: Scalable time series classification. Data Min. Knowl. Discov. **30**(5), 1273–1298 (2016)
21. Grabocka, J., Schilling, N., Wistuba, M., Schmidt-Thieme, L.: Learning time-series shapelets. In: The 20th ACM SIGKDD International Conference on Knowledge Discovery and Data Mining, KDD 2014, New York, NY, USA, 24–27 August 2014, pp. 392–401 (2014)

22. Bagnall, A., Lines, J., Hills, J., Bostrom, A.: Time-series classification with COTE: the collective of transformation-based ensembles. In: 32nd IEEE International Conference on Data Engineering, ICDE 2016, Helsinki, Finland, 16–20 May 2016, pp. 1548–1549 (2016)
23. Chung, J., Gülçehre, Ç., Cho, K., Bengio, Y.: Empirical evaluation of gated recurrent neural networks on sequence modeling. CoRR abs/1412.3555 (2014)

Similarity-Aware Deep Attentive Model for Clickbait Detection

Manqing Dong[1(✉)], Lina Yao[1], Xianzhi Wang[2], Boualem Benatallah[1], and Chaoran Huang[1]

[1] Department of Computer Science, University of New South Wales, Sydney, Australia
manqing.dong@unsw.edu.au

[2] School of Software, University of Technology Sydney, Sydney, Australia

Abstract. Clickbait is a type of web content advertisements designed to entice readers into clicking accompanying links. Usually, such links will lead to articles that are either misleading or non-informative, making the detection of clickbait essential for our daily lives. Automated clickbait detection is a relatively new research topic. Most recent work handles the clickbait detection problem with deep learning approaches to extract features from the meta-data of content. However, little attention has been paid to the relationship between the misleading titles and the target content, which we found to be an important clue for enhancing clickbait detection. In this work, we propose a deep similarity-aware attentive model to capture and represent such similarities with better expressiveness. In particular, we present the ways of either using similarity only or integrating it with other available quality features for the clickbait detection. We evaluate our model on two benchmark datasets, and the experimental results demonstrate the effectiveness of our approach by outperforming a series of competitive state-of-the-arts and baseline methods.

1 Introduction

Clickbait is a type of web links designed to entice users to enter specific web-pages or videos[1]. Clickbait titles are generally written in an exaggerated or ambiguous way to attract curious readers to the hyper-linked content. For example, "*You will never believe what happened when...*" and "*This is the biggest mistake you can make...*" are two representative titles of clickbait[2]. Most clickbaits are created for financial purposes. For example, Web publishers regard clickbait as a useful tool to draw attention to their websites and make money from advertisements. However, clickbaits are often malicious to the readers despite the potential benefit to the advertiser, as they are mostly misleading or meaningless

[1] https://en.wikipedia.org/wiki/Clickbait.

[2] https://www.thedailybeast.com/saving-us-from-ourselves-the-anti-clickbait-movement.

Q. Yang et al. (Eds.): PAKDD 2019, LNAI 11440, pp. 56–69, 2019.
https://doi.org/10.1007/978-3-030-16145-3_5

articles. For most of the time, the content of such articles is not even related to the title, making the detection of clickbait not only necessary but also highly significant.

Research on clickbait detection has been active in recent years. Potthast [12] made one of the first few early attempts. They consider the features from both titles and the linked web page, including linguistic information (e.g., the mean word length and sentiment polarity) and side information (e.g., the writer of the titles). They feed the features into traditional classifiers such as logistic regression, Naive Bayes, and random forest [10] and attain the accuracy of around 80%. Later, deep learning methods [5,20] are increasingly studied owing to their advantages in dealing with high-dimensional data and extracting non-linear relationship among features [8]. Most of the top teams in 2017's Clickbait Challenge[3] use deep learning based methods. Zhou's classifier [20], which won the first place, is a is a Recurrent Neural Network [8] based framework that considers the context of words, more specifically, a hybrid of bidirectional Gated Recurrent Unit (GRU) and attention model [18]. Another work worth mentioning is conducted by Maria et al. [5], which takes both image and the text representations into consideration and different deep learning methods, like Convolutional Neural Network (CNN) and Long Short Term Memory (LSTM) [8], are tried for the prediction.

Until now, few works have investigated the similarities between the misleading titles and the linked web contents for clickbait detection. Clickbaits are not necessarily spams, instead, they may actually contain genuine information but with rather low quality (e.g., unmatched contents and titles). This makes it possible to improve the performance of clickbait detection. In a recent work, Biyani et al. [2] utilize similarities between the title and the top five sentences in the bodies as features blended with traditional texture information for detecting clickbaits. Another work by Kumal et al. [7] used Siamese Networks for measuring the text and visual similarities and combined the similarities as the input of several fully connected layers. Yet existing studies have several limitations: (i) they consider the similarity as features in a linear manner and therefore lack the expressiveness when compared to non-linear methods; (ii) current efforts on leveraging such similarities typically use the partial/local information, such as quantifying the similarity between the titles and the top five sentences of content; on the other hand, they overlooking the hidden global information in the entire content. To overcome these challenges, we propose a deep attentive similarity model for capturing the discriminative information from local and global similarities. This way, we provide a way of untangling the non-linear connections between content and titles for the further prediction. The global similarity in this work measures the similarities over the pair of inputs. Specially, to alleviate the impact of noise, we propose using attentive local similarities to select the most useful similarity information for the final prediction. In a nutshell, we make the following contributions:

[3] https://www.clickbait-challenge.org/.

- We propose a deep attentive similarity model which is capable of capturing both global and local similarities of the pair of inputs. The model represents local similarities as vectors to combine them with other features for the future prediction easily.
- We introduce the ways of either using only similarity information or combining the similarity with other features to detect clickbait. We further employ an attention-based bidirectional Gated Recurrent Units (GRU) model to obtain robust representations of textual inputs.
- We evaluate our framework on two benchmark datasets of clickbait detection. The experimental results demonstrate its effectiveness in detecting clickbait and its competitive performance against the baselines.

The rest of this paper is organized as follows. Section 2 describes the related work of deep semantic similarity model. Section 3 defines the relevant operators, the target problem, and the framework of our model. Section 4 describes the real-world dataset and corresponding experimental results. Section 5 gives the concluding remarks.

2 Related Work

2.1 Clickbait Detection

As an arguably new research topic, first attempts on this problem extract latent features [12,17]. For example, Chen et al. [3] considered both content cues and non-text cues. Specifically, they extract features from lexical and semantic levels for content cues and features like user behavior, information about figures for non-text cues. Those features are then used over various classifiers (e.g. Naive Bayes classifier, SVM classifier) for the prediction tasks.

Instead of extracting features manually, some recent works utilize word vectors [13] for representing the textual information in order to take advantages of deep learning methods [4]. For example, Zheng et al. [19] transformed the titles into word embeddings and then used text-Convolutional Neural Networks as classifier. Also, Recurrent Neural Network (RNN) based methods are widely used in detecting the clickbaits, due to the efficiency in dealing with sequential data. In fact, RNN was used by all the top five teams in the aforementioned Clickbait Challenge. On the basis of RNN, Glenski et al. [5] used LSTM, and Zhou [20] used attentive bi-GRU for learning the textual inputs.

However, limited works exploited the similarity information for detecting the clickbait, although this can directly indicate the matching level between titles and contents. In early works Biyani et al. [2] made the few attempts that used the similarities information with several features, including n-grams and metrics for evaluating the informality. They then fed those features into a gradient boost decision tree (GBDT) classifier. Kumal et al. [7] used Siamese Networks for measuring the text similarity between titles and bodies, and image similarity between figures and descriptions. They then concatenated the similarities for the final prediction.

2.2 Deep Semantic Similarity Model

The Deep Semantic Similarity Model (DSSM) [6] was originally designed for web search ranking, which is a latent semantic model with a deep structure that projects queries Q and documents D into a common low-dimensional space to calculate the semantic similarity. DSSM differs from the traditional latent semantic models in the use of deep neural networks that learn the latent representations. In particular, DSSM first maps the input features x to the latent semantic space l by:

$$layer_1 = W_1 x \tag{1}$$

$$layer_i = f(W_i layer_{i-1} + b_i), i = 2, \ldots, N-1 \tag{2}$$

$$l = f(W_N layer_{N-1} + b_N) \tag{3}$$

where $layer_i$ is the ith intermediate hidden layer, W_i is the ith weight matrix, b_i is the ith bias matrix, and f is the activation function, e.g., sigmoid function. Then, the semantic relevance score between a query Q and a document D is measured:

$$R(Q,D) = cosine(l_Q, l_D) = \frac{l_Q^T l_D}{\| l_Q \| \| l_D \|} \tag{4}$$

Learning the DSSM is equivalent to maximizing the fraction of the similarity between queries Q and matching documents D_+ in the entire collection of either matching and mismatching documents. A typical improvement of DSSM is to change the way of learning the latent representations, e.g., by changing the deep neural networks with convolutional neural networks (CDSSM) [14] or with long short term memory (LSTM-DSSM) [11]. In this work, we follow the idea of DSSM to calculate the similarities in the latent space, but the latent space is produced in different ways for different types of inputs. Since the similarity in DSSM is a constant, we regard this similarity value as the global similarity and learn a local similarities vector from it for the further prediction. The next section gives the details.

3 Methodology

In this paper, we define a piece of information as clickbait when *the title does not match the content.* Given a set of titles $H = \{h_1, h_2, \ldots, h_N\}$, and their bodies $B = \{b_1, b_2, \ldots, b_N\}$, the goal is to predict a label $Y = \{y_1, y_2, \ldots, y_N\}$ of these pairs, where $y_i = 1$ if headline i is a clickbait. Our framework includes three parts: learning latent representations, learning the similarities, and using the similarity for the further predictions. Figure 1 illustrates the last two parts.

3.1 Learn Latent Representations

Here we consider transform the titles H and bodies B into the latent representations: L_H and L_B, where $L_H, L_B \in \mathbb{R}^M$. We first preprocess the text information,

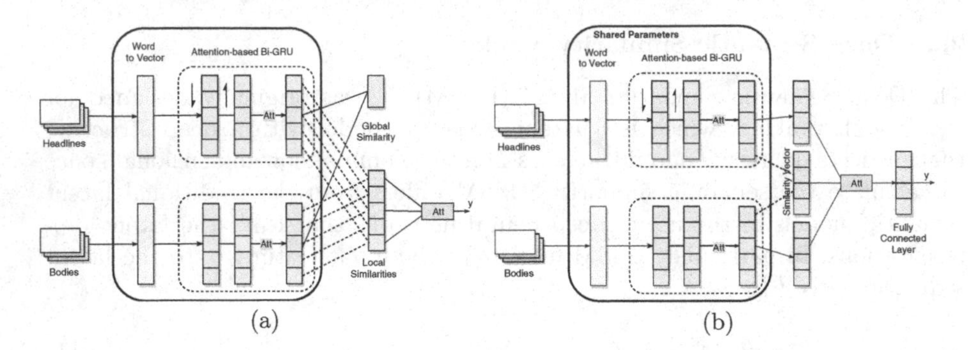

Fig. 1. The illustration of our proposed model. The left (a) shows the learning of the global similarity and local similarities. The words will first be transformed to vectors and go through the attention based bidirectional GRU models. The global similarity is then learned from the label which minimize the distance between matching titles and bodies. And the local similarity is calculated accordingly. And the right one (b) shows the combined method for doing the further prediction.

where we remove all the punctuation and stop words, make the sentence in a lower form, and do word lemmatization [15]. And we then transform the cleaned inputs as word vectors [13].

We apply the attention-based bidirectional GRU [18], one of the most popular RNN based models, to obtain hidden representations, which have shown effectiveness in dealing with natural languages tasks in recent years, by using a gating mechanism to track the state of sequences without using separate memory cells [1]. Given a $b_i, i \in [1, N]$, we first get a set of word embedding vectors $w_{i,t}$, where $t \in [1, T_i]$, T_i is the number of words in body i. Then, we use the bidirectional GRU to get annotations of words by summarizing information from both directions of a word. The bidirectional GRU contains a forward $\overrightarrow{GRU}$, which reads the sentence from w_{i1} to w_{iT_i}, and a backward $\overleftarrow{GRU}$, which reads the sentence from w_{iT_i} to w_{i1}:

$$\overrightarrow{w'_{it}} = \overrightarrow{GRU}(w_{it}), t \in [1, T_i] \tag{5}$$

$$\overleftarrow{w'_{it}} = \overleftarrow{GRU}(w_{it}), t \in [T_i, 1] \tag{6}$$

Then, we get the hidden representation w'_{it} by concatenating the forward hidden state $\overrightarrow{w'_{it}}$ and backward hidden state $\overleftarrow{w'_{it}}$: $w'_{it} = [\overrightarrow{w'_{it}}, \overleftarrow{w'_{it}}]$. And an attention mechanism is used to extract words that are important to the sentence and aggregate the representation of those words to get the latent representation L_{b_i}:

$$u_t = tanh(W_w w'_{it} + b_w) \tag{7}$$

$$a_t = \frac{exp(u_t^T u_w)}{\Sigma_t exp(u_t^T u_w)} \tag{8}$$

$$L_{b_i} = \Sigma_t a_t w'_{it} \tag{9}$$

Finally, we get the latent representation of the bodies L_B. In a similar way, we can obtain the hidden representation L_H.

3.2 Learn the Similarities

For getting the global similarity, similar to DSSM, we calculate it as the cosine similarity between the L_H and L_B:

$$r(H,B) = cosine(L_H, L_B) = \frac{L_H^T L_B}{\| L_H \| \| L_B \|} \tag{10}$$

This similarity $r(H,B)$ is a constant within $[0,1]$ (if the input space is in a positive space), a higher value of which stands for a higher level of consistency between the titles and bodies. Intuitively, we want to maximize this similarity score between the matching titles and minimize the similarity score between the mismatching pairs. For using only global similarity to predict the clickbait, we use $R(H,B) = softmax[r(H,B),(1-r(H,B))]$ as the balance value of the global similarities. Thus, the prediction for the matching is $\hat{y} = \text{argmax}_y(P(y|h,b))$, where $P(Y|H,B) = R(H,B)$. We use cross entropy for measuring the loss, which is:

$$\mathcal{L} = -\Sigma_{Y=0,1} Y \log P(Y|H,B) \tag{11}$$

Then the optimization goal is to minimize this loss:

$$\underset{\Theta}{\text{argmin}}\ \mathcal{L} + \lambda \| \Theta \|_2 \tag{12}$$

where the $\mathcal{L}$ is the loss function of Eq. 11 and the right norm is for the regularization of the parameters, λ is the hyperparameter. And we take Adam as our optimization method [8]. This way, maximizing the global similarity between matching titles and bodies also helps us update the matching latent representations accordingly. That means the corresponding latent representations for titles and bodies will be as close as possible. Plus the global matching similarity is usually sensitive to some noise like partial occlusion [16]. We learn the local similarities for a better matching representation. Recall that we have latent representations $L_H, L_B \in \mathbb{R}^M$, we set the local block size as $\mu, \mu < M$, and we move from left to right with $\nu, \nu < (M-\mu)$ strides to the next local block. Then we have $K = [\frac{M-\mu}{\nu}]$ local blocks, that is, the latent L_H can be represented as $L_H = [L_{H,1}^T, L_{H,2}^T, \ldots, L_{H,K}^T]^T$ and so as the L_B. Thus, the local similarities are then calculated by

$$LS(H,B) = (r(L_{H,1}, L_{B,1}), ..., r(L_{H,K}, L_{B,K}))^T \tag{13}$$

We use an attentive mechanism to select the most useful similarities, i.e., the local similarities $LS(H,B)$, for the final prediction. More specifically, we apply the self-attention mechanism [9] for getting the attention values (which serve as self-learned weight values),

$$A = softmax(V_a \tanh(W_a LS(H,B)^T)) \tag{14}$$

where $W_a \in \mathbb{R}^K$ and $V_a \in \mathbb{R}^{K\times K}$ are two weight matrices, and A is the attention matrix for the local similarity. Then let

$$P = softmax(W_P(A \times LS(H,B)) + b_P) \tag{15}$$

we get the prediction for the clickbait as $\hat{y} = argmax_y P$, where W_P and b_P are weights and biases. Similarity, for using only local similarities to predict the clickbaits, we choose the combination of cross entropy and regularization as the loss function and optimize it using Adam optimization method.

3.3 Learn for Prediction

So far, we introduced how global similarities and local similarities are learned and how to utilize only attentive local similarities to the detection. Here, we will introduce the classification method which combines the features with the similarities.

Learning from raw textual information could help mine clickbait indicators such as the writing style and quality, and learning the similarity can lead to the matching degrees. To combine these two useful clues, we adopt an attentive way for the final prediction, which is shown in Fig. 1(b). We first use fully connected layers to map the hidden representations L_H and L_B into layers with K dimension.

$$L'_H = f(W_H L_H + b_H) \tag{16}$$

$$L'_B = f(W_B L_B + b_B) \tag{17}$$

Denote $LS(H,B)$ as L'_{LS}, then we get a concatenation layer $L' = [L'_H, L'_{LS}, L'_B]$. Similar to Eqs. 14 to 15, we calculate the self attention values $A_{L'}$ and use them get the combination layer L''. The combination layer is then fed into multilayer perceptrons and we get the $P - softmax(W_P L'' + b_P)$. Then the prediction is $\hat{y} = argmax_y P$. Similarly, we set the loss as the combination of cross entropy and L2-norm of the parameters, and we learn the parameters with Adam optimization.

4 Experiments

In this section, we will test our model on two benchmark datasets. We will first give some details about these two datasets, and then present the comparison results of our method and several related works. Furthermore, we conduct the sensitivity analysis of the proposed method with different parameter settings.

4.1 Dataset Description

Here we use two datasets for evaluating the model.

- **Clickbait Challenge**[4] is a benchmark dataset for the clickbait detection that released in 2017. The dataset contains over 20,000 labelled pairs of posts for training and validation. There are five judges, each giving a clickbait score (from 0 to 1) to label the post. And a higher score stands for the higher probability of a post being clickbait. Then we regard the post with the mean score over 0.5 as clickbait.
- **FNC dataset**[5] is from the Fake News Challenge in 2017. The data describe pairs of titles and bodies and are labeled as 'agree', 'disagree', 'discuss' and 'unrelated'. We regard data with label 'unrelated' as clickbait. The dataset contains 49,972 pairs of titles and bodies for training and 25,413 pairs for the testing.

As mentioned in the latent representation learning part, we first preprocess the texts by removing the stop words and lemmatization. The processed Clickbait Challenge dataset has an average of 10 words in the titles and 50 words in the bodies, while the FNC dataset has an average of 8 and 200 words accordingly. We further vectorize the data using word-embedding techniques [13], which can be conducted unsupervised. Given that some titles only contain one word and this word is unique among the corpus, we train the word vectors with the "Min Word Count" set to 1.

4.2 Comparison Methods

We have compared our LSDA model with a series of baseline models and state-of-the-arts. Where the first two are latent semantic similarity based models and the other four are the most current works for clickbait detection.

- Huang et al. [6]: propose a deep semantic similarity model (DSSM) that uses deep neural networks, to get the latent representations of the inputs and calculate the similarity in the latent representation space. They then use the calculated similarities as introduced in learning global similarity for the prediction. The difference is they use N-gram to preprocess the textual features.[6]
- Shen et al. [14]: propose a similar structure to DSSM, yet they use convolutional neural networks for latent representations.
- Kumar et al. [7]: propose a hybrid method for detecting the clickbait. They first use attentive bidirectional RNN based methods for learning the inputs, and then concatenate the latent inputs with the relationship information that learned with Siamese Net for the final prediction.

---

4 https://www.clickbait-challenge.org/.
5 http://www.fakenewschallenge.org/.
6 https://en.wikipedia.org/wiki/N-gram.

- Zheng et al. [19]: only consider using characteristics from titles to detect the clickbait. They first transform the titles into word vectors and then use text-CNN for predicting the labels.
- Glenski et al. [5]: consider information from both titles and bodies. They also learn from the textual information by firstly vectorizing them and learn for the predictions with using LSTM networks.
- Zhou et al. [20]: use attentive bi-GRU model for learning the hidden representations of titles and bodies. Then, the two learned hidden representations are concatenated and fed into fully connected layers for the prediction.

Different settings are considered in terms of evaluation. We denote the combination of the local similarities and the raw input features as LSDA, and the variant of our method that considers deep local similarity but not the attention by LSD, as shown in Fig. 1(b). For our experimental setting, we initialize the weight and bias parameters with random variables. Besides, we set the word embedding dimension as 100 and the hidden size M as 100. The comparison results are shown in Table 1.

Table 1. Comparison results

Methods	Clickbait Challenge				FNC dataset			
	Accuracy	Precision	Recall	F1-score	Accuracy	Precision	Recall	F1-score
Huang et al. [6]	0.817	0.655	0.661	0.658	0.747	0.894	0.740	0.811
Shen et al. [14]	0.833	0.683	0.643	0.662	0.756	0.959	0.762	0.853
Kumar et al. [7]	0.826	0.699	0.474	0.565	0.859	0.920	0.877	0.907
Zheng et al. [19]	0.844	0.654	0.653	0.653	0.789	0.852	0.845	0.857
Glenski et al. [5]	0.827	0.642	0.621	0.631	0.868	0.925	0.884	0.913
Zhou et al. [20]	0.856	0.719	0.650	0.683	0.879	0.924	0.897	0.919
LSD	0.847	0.697	0.675	0.686	0.885	0.928	0.901	0.923
LSDA	0.860	0.722	0.699	0.710	0.894	0.933	0.912	0.928

The evaluation is conducted with four commonly used metrics: accuracy, recall, precision, and F1 score[7]. Generally, the comparison results show the effectiveness of our proposed method for detecting the clickbait detection. We observed that both the CNN- and RNN-based models perform better than traditional deep neural networks. This may largely resort to the capability of CNN and RNN in capturing the location information. The attention-based bidirectional GRU shows superior performance in dealing with textual information. Besides the superiority of the bi-AttGRU itself, which we use for learning latent representations, both the similarity information and the attention mechanism help with the final prediction.

[7] https://en.wikipedia.org/wiki/Precision_and_recall.

4.3 Sensitivity Analysis

In this part, we test the model's sensitivity to different parameter settings on Clickbait Challenge dataset. Similar results can also be found using the FNC dataset. As mentioned above, we mainly have parameters related to the latent representation learning part and the similarity learning part, as well as some hyper-parameters to indicate the learning rate. As the default setting, we separate the training dataset with a ratio of 80%, set the dimensions of word vectors to 50, and pad each sentence to the same length for the input of attentive Bi-GRU model. We also set size of hidden units in Bi-GRU to 50, the local similarity block size to 50 and the default learning rate for Adam optimizer to 0.001.

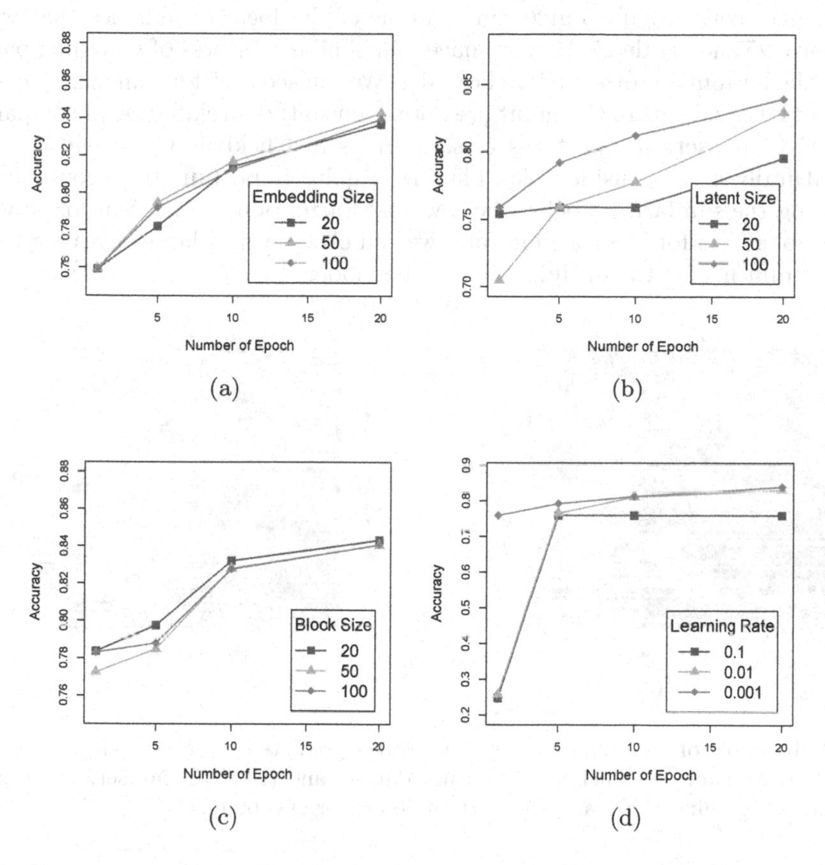

Fig. 2. Sensitivity towards different parameter settings: (a) word embedding size, (b) dimension of latent representations, (c) local similarity block size, and (d) learning rate.

In particular, we compare the following parameters settings: the word embedding size, the latent representation size, the local similarity block size, and the optimizer learning rate. Figure 2 presents the comparison results, where the horizontal axis shows the learning epochs and the vertical axis stands for accuracy.

We can tell that model with smaller word embedding size have lower accuracy in predicting the clickbait, which can be the result of the inadequate grasping of the content information in low dimensional word space. On the other hand, the model with large word embedding size requires more time to learn for a decent result. And it can be observed that the dimension of the latent feature representations do affect the results, where larger latent size helps with higher predicting performance. Noted that the average length of bodies of the FNC dataset is 200, thus for training a dataset with word embedding 100, each sample will be sized 200 × 100, which ends up with latent representations with higher dimensions, and contain relatively more information. For the block size of the local similarities, it can be claimed that a smaller block size performs better than bigger ones. Figure 3 gives an example of the local similarities that with block size 25 and stride 25, thus we have four similarity scores of matching pairs, where the bottom 50 rows are for clickbaits. We can see that for some instances of clickbaits, the subsets of the input are not significantly unrelated. And compared with FNC dataset, the patterns of similarities in Clickbait Challenge dataset are naturally more considerable. Thus it is quite important to automatically weighting the similarity blocks where we used attention mechanism for solving this problem. As for the learning rate, we can observe that larger learning rates make it difficult for the optimization of the models.

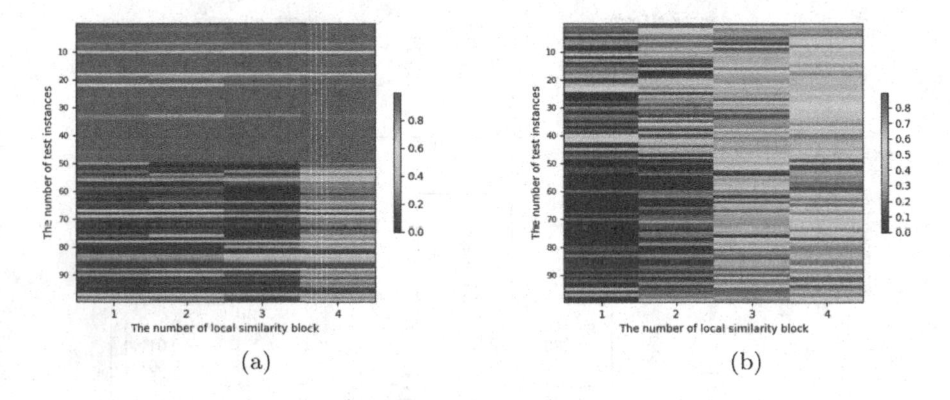

Fig. 3. Example of local similarities of correctly predicted instances with block size 25 and stride 25 on (a) Clickbait Challenge dataset and (b) FNC dataset. The top 50 rows are for genuine clicks and the bottom 50 are for clickbaits.

We also consider the impact of different settings for the model to find which part help more with the final prediction. In particular, we consider the following: whether local similarity performs better than global ones; whether adding the latent representations is helpful; and the efficiency of adding the attention. We give five variants of the models, and Fig. 4 shows the results. GS represents the model that considers using only the global similarity for the prediction; LS stands for the model using only local similarities, and similarly, LSA stands

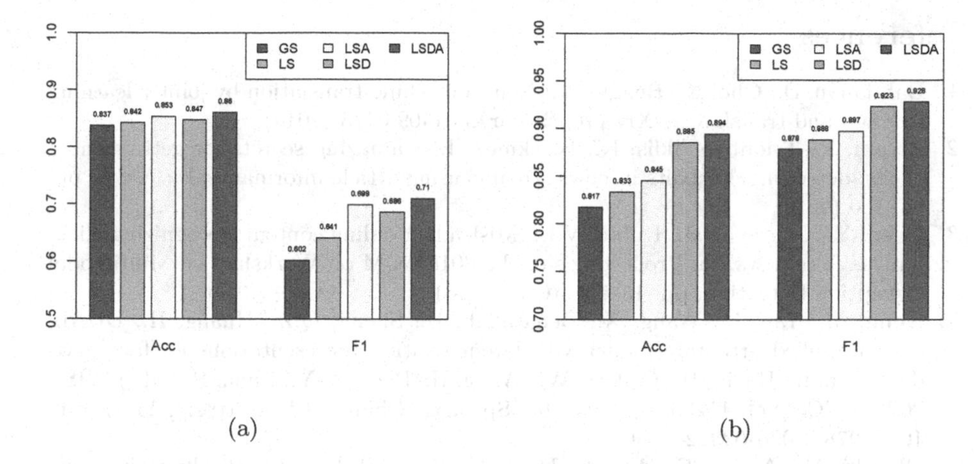

Fig. 4. Ablation studies on variants of models on (a) Clickbait Challenge dataset and (b) FNC dataset.

for the model that adds attentions to local similarities, LSD combining latent representations with local similarities, and LSDA is for a combination method shown in Fig. 1(b).

First, we can see that models with concerning local similarities perform better than a model that only considers the global similarity, which might reasons from the sensitiveness of the global similarity to the data noises. Besides, we can observe that concatenating the raw features with the similarity information do help with the prediction, especially on the FNC dataset. While raw features can be used to extract some text patterns like content quality, we can say that these patterns are helpful for detecting the clickbaits. And comparing LSA with LS, and LSDA with LSD, it can be observed that adding the attention mechanism helps to improve the results. Generally, those results indicate the model is effective for extracting essential information in different settings, and further demonstrate the superiority of the attentive similarity in encoding the features.

5 Conclusions

In this paper, we solve the problem of clickbait detection from the similarity perspective, as opposed to the traditional feature engineering which lack the properties in representing the matching information between titles and targeted bodies. We have presented a local similarity-aware deep attentive model that learns both local similarities and raw input features to make predictions in an attentive manner. To the best of our knowledge, the model is novel in the area of clickbait detection and yields competitive results among a series of baseline and state-of-the-arts methods on two real world datasets. Noted that we have not considered other features like image information in this work, which may also be found on those clickbait web-pages. This will be included in our future investigations.

References

1. Bahdanau, D., Cho, K., Bengio, Y.: Neural machine translation by jointly learning to align and translate. arXiv preprint arXiv:1409.0473 (2014)
2. Biyani, P., Tsioutsiouliklis, K., Blackmer, J.: 8 amazing secrets for getting more clicks: detecting clickbaits in news streams using article informality. In: AAAI, pp. 94–100 (2016)
3. Chen, Y., Conroy, N.J., Rubin, V.L.: Misleading online content: recognizing clickbait as false news. In: Proceedings of the 2015 ACM on Workshop on Multimodal Deception Detection, pp. 15–19. ACM (2015)
4. Dong, M., Yao, L., Wang, X., Benatallah, B., Sheng, Q.Z., Huang, H.: DUAL: a deep unified attention model with latent relation representations for fake news detection. In: Hacid, H., Cellary, W., Wang, H., Paik, H.-Y., Zhou, R. (eds.) WISE 2018. LNCS, vol. 11233, pp. 199–209. Springer, Cham (2018). https://doi.org/10.1007/978-3-030-02922-7_14
5. Glenski, M., Ayton, E., Arendt, D., Volkova, S.: Fishing for clickbaits in social images and texts with linguistically-infused neural network models. arXiv preprint arXiv:1710.06390 (2017)
6. Huang, P.S., He, X., Gao, J., Deng, L., Acero, A., Heck, L.: Learning deep structured semantic models for web search using clickthrough data. In: International Conference on Information & Knowledge Management, pp. 2333–2338. ACM (2013)
7. Kumar, V., Khattar, D., Gairola, S., Kumar Lal, Y., Varma, V.: Identifying clickbait: a multi-strategy approach using neural networks. In: The 41st International ACM SIGIR Conference on Research & Development in Information Retrieval, pp. 1225–1228. ACM (2018)
8. LeCun, Y., Bengio, Y., Hinton, G.: Deep learning. Nature **521**(7553), 436 (2015)
9. Lin, Z., et al.: A structured self-attentive sentence embedding. arXiv preprint arXiv:1703.03130 (2017)
10. Nasrabadi, N.M.: Pattern recognition and machine learning. J. Electron. Imaging **16**(4), 049901 (2007)
11. Palangi, H., et al.: Deep sentence embedding using long short-term memory networks: analysis and application to information retrieval. IEEE/ACM Trans. Audio Speech Lang. Process. (TASLP) **24**(4), 694–707 (2016)
12. Potthast, M., Köpsel, S., Stein, B., Hagen, M.: Clickbait detection. In: Ferro, N., et al. (eds.) ECIR 2016. LNCS, vol. 9626, pp. 810–817. Springer, Cham (2016). https://doi.org/10.1007/978-3-319-30671-1_72
13. Řehůřek, R., Sojka, P.: Software framework for topic modelling with large corpora. In: Proceedings of the LREC 2010 Workshop on New Challenges for NLP Frameworks, pp. 45–50. ELRA, Valletta, May 2010
14. Shen, Y., He, X., Gao, J., Deng, L., Mesnil, G.: A latent semantic model with convolutional-pooling structure for information retrieval. In: ACM International Conference on Conference on Information and Knowledge Management, pp. 101–110. ACM (2014)
15. Turney, P.D., Pantel, P.: From frequency to meaning: vector space models of semantics. J. Artif. Intell. Res. **37**, 141–188 (2010)
16. Wang, D., Lu, H., Bo, C.: Visual tracking via weighted local cosine similarity. IEEE Trans. Cybern. **45**(9), 1838–1850 (2015)
17. Wang, X., et al.: Truth discovery via exploiting implications from multi-source data. In: Conference on Information and Knowledge Management, pp. 861–870. ACM (2016)

18. Yang, Z., Yang, D., Dyer, C., He, X., Smola, A., Hovy, E.: Hierarchical attention networks for document classification. In: Proceedings of the 2016 Conference of the North American Chapter of the Association for Computational Linguistics: Human Language Technologies, pp. 1480–1489 (2016)
19. Zheng, H.T., Chen, J.Y., Yao, X., Sangaiah, A.K., Jiang, Y., Zhao, C.Z.: Clickbait convolutional neural network. Symmetry **10**(5), 138 (2018)
20. Zhou, Y.: Clickbait detection in tweets using self-attentive network. arXiv preprint arXiv:1710.05364 (2017)

Topic Attentional Neural Network for Abstractive Document Summarization

Hao Liu, Hai-Tao Zheng[✉], and Wei Wang

Tsinghua-Southampton Web Science Laboratory, Graduate School at Shenzhen, Tsinghua University, Shenzhen, China
{liuhao17,w-w16}@mails.tsinghua.edu.cn, zheng.haitao@sz.tsinghua.edu.cn

Abstract. Abstractive summarization is a renewed and challenging task of document summarization. Recently, neural networks, especially attentional encoder-docoder architecture, have achieved impressive progress in abstractive document summarization. However, the saliency of summary, which is one of the key factors for document summarization, still needs improvement. In this paper, we propose Topic Attentional Neural Network (TANN) which incorporates topic information into neural networks to tackle this issue. Our model is based on attentional sequence-to-sequence structure but has paired encoders and paired attention mechanisms to deal with original document and topic information in parallel. Moreover, we propose a novel selection method called *topic selection*. This method uses topic information to improve the standard selection method of beam search and chooses a better candidate as the final summary. We conduct experiments on the CNN/Daily Mail dataset. The results show our model obtains higher ROUGE scores and achieves a competitive performance compared with the state-of-the-art abstractive and extractive models. Human evaluation also demonstrates our model is capable of generating summaries with more informativeness and readability.

Keywords: Abstractive summarization · Neural network · Topic information · Attention mechanism

1 Introduction

Document summarization is a task to produce a concise and condensed summary which covers the core information of the original document. Automatic summarization models can be divided into two categories: extractive and abstractive. Extractive approaches select important segments from the original document and rearrange them to construct a summary. Totally different from extractive, abstractive approaches potentially generate new phrases or sentences. Requiring deeper understanding of natural language, abstractive approaches are more difficult and face great challenges, such as saliency, coherence and readability.

Recently, the models based on attentional sequence-to-sequence (seq2seq) framework have demonstrated great advantages for abstractive document summarization [12,15,17]. However, the saliency of summary, which plays a vital role

Q. Yang et al. (Eds.): PAKDD 2019, LNAI 11440, pp. 70–81, 2019.
https://doi.org/10.1007/978-3-030-16145-3_6

in document summarization, is still not satisfactory and needs improvement. Meanwhile, topic information, which is one of the most important features of original document, can also help to identify the key information but has not been paid enough attention yet.

In this paper, to increase the saliency of summaries and make generated summaries cover more core information, we propose Topic Attentional Neural Network (TANN) for abstractive document summarization. Our model expands attentional seq2seq structure by using paired encoders and paired attention mechanisms to deal with original document and topic information in parallel. To obtain topic information, we employ two classic topic-extracted methods: Latent Dirichlet Allocation (LDA) [1] and TF-IDF. Then, we use these two types of topic information to train our model respectively.

Moreover, we also utilize topic information to improve the selection method of beam search. Beam search algorithm is widely used for generating outputs in neural networks [3,12,17]. It can generate several sequences as candidates and chooses one of them as the final output. In summarization task, the standard selection method chooses the candidate with the highest conditional probability as final summary. However, the conditional probability of language model is based on the whole training dataset, which take no specific features of source document into account. To solve the problem, we propose a selection method called *topic selection*. Our method utilizes topic information to calculate a score for each candidate and choose the one with the highest score. The experimental results show that *topic selection* help to produce more informative summaries. Our main contributions can be listed as follows:

- TANN introduces topic information into neural networks to produce summaries with more salient information. As far as we know, it is the first time that topic information is used to improve document-level abstractive summarization.
- We propose a novel selection method *topic selection* which further utilizes topic information to improve the standard selection method of beam search. With the help of topic information, our selection method helps to improve the informativeness of summary.
- Experiment on the CNN/Daily Mail dataset demonstrate that our model achieves competitive results compared with state-of-the-art abstractive and extractive models. Human evaluation also demonstrates our model produces informative summaries with high readability.

This paper is organized as follows. Section 2 introduces related work about abstractive summarization. Section 3 describes our model. Section 4 describes the experiment and gives discussion. In Sect. 5, we conclude this paper.

2 Related Work

While a large number of past works for document summarization are extractive approaches [2,10,13], abstractive approaches generate summaries by understanding the source document, which are closer to the way human writes summaries.

Recently, neural networks applied in abstractive approaches have been intensively studied. Rush et al. are the first to introduce neural network for abstractive text summarization [15]. Their model is based on convolutional encoder-decoder architecture and shows a promising path of applying seq2seq in abstractive summarization. Chopra et al. [3] and Nallapati et al. [12] extend this work by using RNN in place of CNN. However, because of the fixed vocabulary of these models, the generated summaries tend to cause the out-of-vocabulary (OOV) problem. To overcome this issue, Gu et al. propose CopyNet [6] and Gulcehre et al. propose pointer network [7], which both extend seq2seq structure by copying OOV words directly from the source text.

Due to the lack of large document-level dataset, most past works are sentence-level summarization models [3,15], which summarize a document to one sentence. Nallapati et al. address this issue by introducing the CNN/Daily Mail dataset which consists of news from CNN and Daily Mail website [12]. Then, Paulus et al. propose the intra-attention networks with reinforcement learning [14]. See et al. introduce the coverage mechanism into summarization system to address the repetition on the CNN/Daily Mail dataset [17].

So far, few works have considered about using external information to improve abstractive summarization. Nallapati et al. use feature-rich word embedding as the input of model [12]. This embedding expands original word embedding with some linguistic information such as POS tags and named-entities. Li et al. use keyword representation as the extra input for attention to guide the summarization [9]. In this paper, we use topic information as one of the external information and utilizes it to produce more salient summaries.

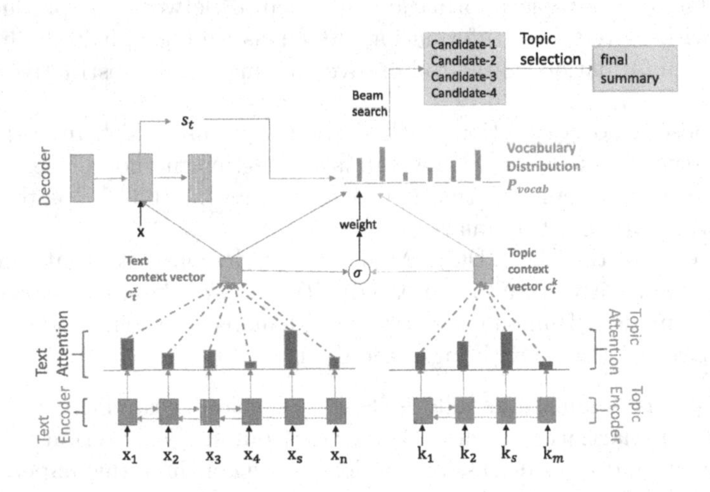

Fig. 1. TANN model with paired encoders and paired attention mechanisms.

3 Our Model

3.1 Overview

We build our model based on attentional encoder-decoder framework. In particular, our model has paired encoder and paired attention mechanisms. We use copying mechanism and coverage mechanism to avoid the OOV and repetition. We show the structure of our model in Fig. 1. In encoding step, the inputs are original document $x = \{x_1, x_2, ..., x_n\}$ and pre-extracted topic information $k = \{k_1, k_2, ..., k_m\}$. Then, the text context vector c_t^x and topic context vector c_t^k are calculated by the respective attention mechanisms. Next, these two context vectors are both used to compute vocabulary distribution in the decoder. Specially, we use a learnable parameter *weight* to make model learn the importance of topic information. Finally, we use beam search algorithm to generate candidate sequences and apply *topic selection* method to choose the final summary.

3.2 Paired Encoder

The input of our model consists of original document text x and topic words k. The text encoder and topic encoder both use a bidirectional LSTM [16]. For the text encoder, it uses the source word x as input and produces two LSTM states as forword state and backward state:

$$\overrightarrow{h_i^x}, \overleftarrow{h_i^x} = biLSTM\left(x\right) \tag{1}$$

Then, we combine these two states into the final hidden state h_i^x as:

$$h_i^x = relu\left(W\left[\overrightarrow{h_i^x}, \overleftarrow{h_i^x}\right] + b\right) \tag{2}$$

where W and b are learnable. Using all the words $x = \{x_1, x_2, ..., x_n\}$ as input, text encoder produces a sequence of hidden states $\{h_1^x, h_2^x, ..., h_n^x\}$. Similarly, the topic encoder reads m topic words $k = \{k_1, k_2, ..., k_m\}$ and builds $\{h_1^k, h_2^k, ..., h_m^k\}$ as the representation of topic information.

3.3 Paired-Attentional Decoder

We adopt the decoder with paired attention mechanisms to deal with text information and topic information respectively. For text attention, the text context vector c_t^x at step t is calculated based on the encoder hidden state h_i^x and the decoder hidden state s_t:

$$e_{t,i}^x = v^T \tanh\left(linear\left(h_i^x, s_t\right)\right) \tag{3}$$

$$a_{t,i}^x = softmax\left(e_t^x\right) \tag{4}$$

$$c_t^x = \sum_i a_{t,i}^x h_i^x \tag{5}$$

where v is learned vector. Likewise, topic attention mechanism also computes the topic context vector c_t^k in the similar way.

Specially, a learnable parameter *weight* is computed as:

$$weight = \sigma\left(W_t c_t^x + W_k c_t^k + b\right) \tag{6}$$

where W_t, W_k, b are learnable. The *weight* is regarded as a parameter that indicates how much topic context vector is involved in the generation of words. Then, the vocabulary distribution is calculated as:

$$P_{vocab} = softmax\left(V'\left(V\left[s_t, c_t^x, weight \cdot c_t^k\right] + b\right) + b'\right) \tag{7}$$

where V', V, b, b' are learnable, s_t is decoder hidden state. Finally, the training loss is defined as:

$$loss_m = -\frac{1}{T}\sum_{t=0}^{T} logP_{vocab}\left(y_t\right) \tag{8}$$

3.4 Copying and Coverage

Copying Mechanism. A number of works for sequence generation tasks have solved the problem of OOV words by copying corresponding words directly from original document [6,17,18]. Following these works, we employ copying mechanism in our model. We use a variable *gate* as a switch to choose whether generating a word from the fixed vocabulary ($gate = 1$), or copying from the source ($gate = 0$). The *gate* at step t is computed as:

$$gate = \sigma\left(linear\left(c_t^x, s_t, x_t\right)\right) \tag{9}$$

Then, the next word y_t is predicted by:

$$P\left(y_t\right) = \begin{cases} P_{vocab}\left(y_t\right), & gate = 1 \\ \sum_{i:x_i=y_t} a_{i,t}^x, & gate = 0 \end{cases} \tag{10}$$

where $a_{i,t}^x$ is the text attention distribution of x_i.

Coverage Mechanism. As for the task of producing long summary, repetition is a common problem and impairs the quality of summaries. Following See et al. [17], we use the coverage mechanism to address this issue. Specially, at each step t, our model keep a coverage vector which sums the text attention distribution $a_{*,i}^x$ before t:

$$cov^t = \sum_{j=0}^{t-1} a_{j,i}^x \tag{11}$$

The vector cov^t indicates how much attention has the model paid for the input word x_i before step t. Then, We use coverage vector as an additional input for text attention mechanism and change the calculation of $e_{t,i}^x$ in Eq. (3) as:

$$e_{t,i}^x = v^T \tanh\left(linear\left(h_i^x, s_t, cov^t\right)\right) \tag{12}$$

We define the loss of coverage mechanism as:

$$loss_{cov} = \sum_{i=1}^{t} min\left(a_i^x, cov^t\right) \tag{13}$$

Then, the final loss of our model is defined as:

$$loss_{final} = loss_m + loss_{cov} \tag{14}$$

3.5 Topic Selection

Beam search algorithm, which is used for generating sequence, produces words step by step in decoder. For each decoding step, the beam search keeps top K sequences with high conditional probability, where K is a hyper-parameter called beam size. Therefore, at the end of beam search, the algorithm produces K sequences which are viewed as candidates for summary. The standard selection method selects the one with the highest conditional probability as the summary. Taking the concrete topic information into account, we propose a brand selection method *topic selection*. We firstly obtain the candidate sequences by using beam search algorithm and sort them by conditional probability. Then, for each of the sequence, we calculate a feature value for every word x_i as:

$$feature\left(x_i\right) = \begin{cases} \sum_t a_{t,i}^k, & x_i \in \ topic\ word \\ 0, & x_i \notin \ topic\ word \end{cases} \tag{15}$$

where $a_{t,i}^k$ is the topic attention distribution at timestep t. This distribution vector can stand for the degree of participation of each topic word, and it can be used for measuring the importance of the word. Next, we calculate the score of each candidate sequence s by suming the feature value of every word:

$$Score\left(s\right) = \sum_{x_i \in s} feature\left(x_i\right) \tag{16}$$

Finally, we resort the candidates and choose the one with the highest *Score* as final summary. If the *Score* of candidates are equal, we choose the one with higher conditional probability, which makes our method consider not only conditional probability but topic information.

4 Experiments

4.1 Dataset

We conducted experiments on the CNN/Daily Mail dataset[1] which consists of news stories in CNN and Daily Mail website [8,12]. The corpora has 312,085 articles paired with human-written multi-sentence summaries. This dataset has two version: non-anonymized and anonymized. Following See et al. [17], we obtain the non-anonymized version by same processing steps[2], which divide the dataset into 287,226 training pairs, 13,368 validation pairs and 11,490 test pairs.

[1] https://cs.nyu.edu/~kcho/DMQA/.

[2] https://github.com/abisee/cnn-dailymail.

4.2 Topic Information Acquisition

We extract topic words of each document by using two classic topic models: TF-IDF and LDA. These two models are both trained on the full CNN/Daily Mail dataset. For TF-IDF, we firstly build individual vocabulary for each document and calculate TF and IDF for each word. Next, we select top 50 percent of words in the document vocabulary as the topic words. Finally, we rearrange topic words following the order of their appearance in the original document. As for LDA, we obtain topic information by GibbsLDA++[3] which using Gibbs sampling for parameter estimation and inference. We set the hyperparameters of LDA as $\alpha = 0.04$ and $\beta = 0.01$. For each document, we pick top 25 topics and each topic has 16 topic words as the final topic information.

4.3 Implementation

We train our model with 128-dimensional word embeddings. In particular, we train word embeddings directly from scratch. We use bi-LSTM for both text encoder and topic encoder. For each encoder, we use the hidden state dimension as 256. We select 50k most frequently used words from both source documents and human-written summaries, then put them together as the vocabulary. Our model is trained using Adagrad optimizer [4] with learning rate 0.15 and the initial accumulator value is set to 0.1. We use mini-batches of size 16 and the encoder size is set to 400. For decoding time, we set the decoder size as 100 for training and 120 for testing. Beam size is fixed as 4 during beam search.

4.4 Results and Discussion

In this section, we firstly report and analysis the results of ROUGE scores. Then, we give discussion on the performance of two types of topic information (TF-IDF and LDA) and *topic selection*. Finally, we report the result of human evaluation.

Quantitative Analysis. We evaluate our model with ROUGE [5] scores which are widely used in summarization task. We compare our models with some state-of-the-art extractive models (lead-3 [12] and SummaRuNNer [11]) and abstractive models. The overall performance evaluation is demonstrated in Table 1. Due to the different topic-words acquisition methods, we use several notations (**m1** to **m8**) to represent our models. From the Table 1, we can see that our basic model **m1** and **m5** both outperform baseline PG model (the sixth row in Table 1) on all ROUGE scores, which shows the effectiveness of topic attention. After using *topic selection* to choose the final summaries, **m2** and **m6** obtain ROUGE scores improvement, indicating *topic selection* method helps to differentiate the candidate sequences. With coverage mechanism employed, **m7** model has shown a competitive performance in abstractive models. Moreover, it is observed that **m8** achieves best performance on non-anonyized dataset and even exceeds strong

[3] http://gibbslda.sourceforge.net/.

Table 1. The results is full-length F1 scores for ROUGE-1, ROUGE-2 and ROUGE-L on the CNN/daily mail test set. All ROUGE scores have a 95% confidence interval of at most ±0.25. Models with subscript * were trained and tested on the anonymized version dataset. Best results on non-anonymized dataset are bolded.

Model	ROUGE-1	ROUGE-2	ROUGE-L
Lead-3 [12]	40.34	17.70	36.57
SummaRuNNer [11]*	39.60	16.20	35.30
Abstractive model [12]*	35.46	13.30	32.65
DeepRL, ML [14]*	39.87	15.82	36.90
DeepRL, Intra-attn [14]*	41.16	15.75	39.08
PG [17]	36.44	15.66	33.42
PG, coverage [17]	39.53	17.28	36.38
TANN, LDA (**m1**)	37.48	16.46	35.13
TANN, LDA, *topic selection* (**m2**)	38.32	16.96	35.45
TANN, LDA, coverage (**m3**)	39.75	17.45	36.62
TANN, LDA, coverage, *topic selection* (**m4**)	40.09	17.78	36.92
TANN, TF-IDF, (**m5**)	37.78	16.56	35.34
TANN, TF-IDF, *topic selection* (**m6**)	38.82	17.16	35.48
TANN, TF-IDF, coverage (**m7**)	40.29	17.89	36.92
TANN, TF-IDF, coverage, *topic selection* (**m8**)	**40.56**	**18.01**	**37.15**

extractive baseline lead-3. Besides, though different version of dataset may cause some deviations in the comparison of ROUGE, **m8** still achieves best ROUGE-2 score.

To illustrate the effectiveness of our models, we show an example in Fig. 2. Compared to human summary, it can be seen that with the help of topic information (green font), **m4** and **m8** both capture the important information (bold font) which Pointer-generator misses. Moreover, our model may identify the core information accurately and generate less unnecessary information (red font).

LDA or TF-IDF. It can be observed that our models with TF-IDF yield higher ROUGE scores than the models with LDA (such as the result of **m8** and **m4**). Meanwhile, the final value of parameter *weight* also supports this result. We show the learning curve of *weight* in Fig. 3. As we can see, the initial value of *weight* is about 0.5 for both **m4** and **m8**. After the training, the *weight* of **m8** stabilizes around to 0.65 and that of **m4** finally drops to about 0.45. It indicates that the topic information acquired by TF-IDF may be more involved in the generation process. This result in part because TF-IDF method extracts topic words directly from original document which may represent more important details. Moreover, we observe that these two learning curves both quickly increase

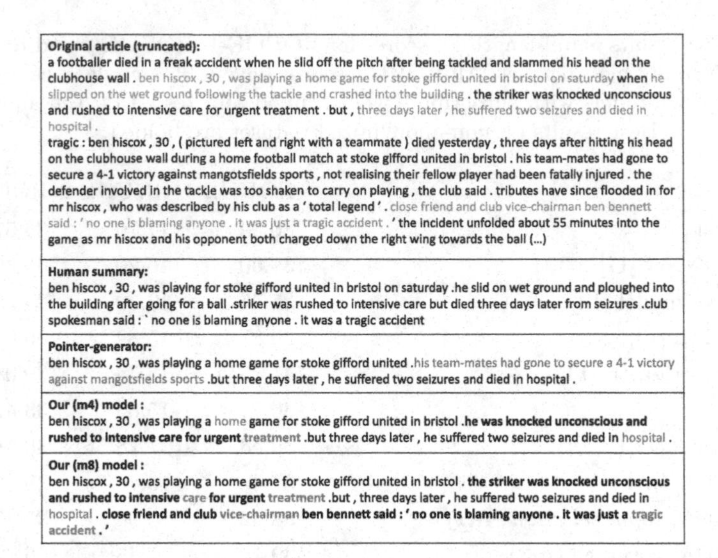

Original article (truncated):
a footballer died in a freak accident when he slid off the pitch after being tackled and slammed his head on the clubhouse wall . ben hiscox , 30 , was playing a home game for stoke gifford united in bristol on saturday when he slipped on the wet ground following the tackle and crashed into the building . the striker was knocked unconscious and rushed to intensive care for urgent treatment . but , three days later , he suffered two seizures and died in hospital .
tragic : ben hiscox , 30 , (pictured left and right with a teammate) died yesterday , three days after hitting his head on the clubhouse wall during a home football match at stoke gifford united in bristol . his team-mates had gone to secure a 4-1 victory against mangotsfields sports , not realising their fellow player had been fatally injured . the defender involved in the tackle was too shaken to carry on playing , the club said . tributes have since flooded in for mr hiscox , who was described by his club as a ' total legend ' . close friend and club vice-chairman ben bennett said : ' no one is blaming anyone . it was just a tragic accident . ' the incident unfolded about 55 minutes into the game as mr hiscox and his opponent both charged down the right wing towards the ball (...)

Human summary:
ben hiscox , 30 , was playing for stoke gifford united in bristol on saturday .he slid on wet ground and ploughed into the building after going for a ball .striker was rushed to intensive care but died three days later from seizures .club spokesman said : ` no one is blaming anyone . it was a tragic accident

Pointer-generator:
ben hiscox , 30 , was playing a home game for stoke gifford united .his team-mates had gone to secure a 4-1 victory against mangotsfields sports .but three days later , he suffered two seizures and died in hospital .

Our (m4) model :
ben hiscox , 30 , was playing a home game for stoke gifford united in bristol **.he was knocked unconscious and rushed to intensive care for urgent** treatment .but three days later , he suffered two seizures and died in hospital .

Our (m8) model :
ben hiscox , 30 , was playing a home game for stoke gifford united in bristol . **the striker was knocked unconscious and rushed to intensive** care **for urgent** treatment .but , three days later , he suffered two seizures and died in hospital . **close friend and club** vice-chairman **ben bennett said : ' no one is blaming anyone . it was just a** tragic accident . '

Fig. 2. Typical comparison. **m4** and **m8** both capture the most important information (bold font) which Pointer-generator misses. With the guidance of topics (green font), our models tend to generate less unnecessary information (red font). (Color figure online)

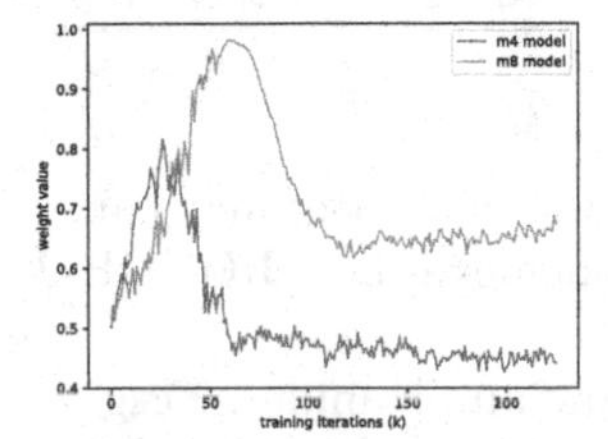

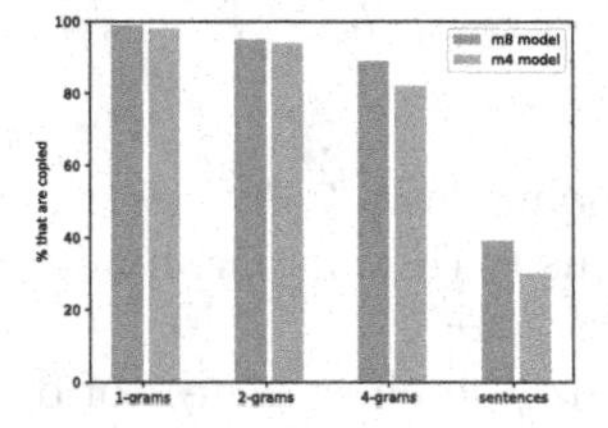

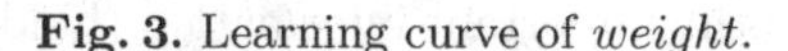

Fig. 3. Learning curve of *weight*.

Fig. 4. Copying rate of n-grams.

and peak at early iterations of training. It may because without pre-trained word-embedding, topic context vector is more informative than text context vector at the beginning of training.

We also study whether it has a link between topic-extracted methods and how abstractive the models are. For our two final models **m4** and **m8**, we compare the copying rate of n-grams which computes n-grams that both appear in the source document and generated summary. The result is shown in Fig. 4. For 1-gram and 2-grams, our two models have similar performance. However, for sentence-level output, **m4** has lower copying rate (about 33%) than that of **m8** (about 41%), showing **m4** is more abstractive and tends to generate diverse summaries. The reason for this result may be that LDA method can generate some novel words, which enrich the topic information.

Topic Selection. The results show that *topic selection* method helps the model to achieve better performance on all ROUGE scores, indicating that topic words directly help to identify the key information. As an example shown in Fig. 5, the candidate sequences are similar but still contain different information (bold font). With the guidance of topic words *"fitness-enthusiast"* and *"sign"*, our *topic selection* select the best candidate sequence Candidate-2 as the final summary, even though Candidate-1 has the highest conditional probability and is selected by standard selection method.

Candidate-1: nbc journalist jeff rossen filmed an airport worker at laguardia airport in new york doing push-ups on the runway after loading up a plane .footage shows the employee wearing his high-visibility jacket and gloves while performing the aerobic stunt .
Candidate-2: nbc journalist jeff rossen filmed an airport worker at laguardia airport in new york doing push-ups on the runway after loading a plane .footage shows the employee wearing his high-visibility jacket and gloves while performing the aerobic stunt **.even after his 12th push-up , the fitness-enthusiast shows no sign of slowing down .**
Candidate-3: nbc journalist jeff rossen filmed an airport worker at laguardia airport in new york doing push-ups on the runway .footage shows the employee wearing his high-visibility jacket and gloves while performing the aerobic stunt .
Candidate-4: nbc journalist jeff rossen filmed an airport worker at laguardia airport in new york doing push-ups on the runway after loading a plane .footage shows the employee wearing his high-visibility jacket and gloves while performing the aerobic stunt .
Human Summary: nbc journalist jeff rossen filmed an airport worker at laguardia airport in new york doing push-ups on the runway after loading up a plane .footage shows the employee wearing his high-visibility jacket and gloves while performing the aerobic stunt .even after his 12th push-up , the fitness-enthusiast shows no sign of slowing down .
Standard selection method: nbc journalist jeff rossen filmed an airport worker at laguardia airport in new york doing push-ups on the runway after loading up a plane .footage shows the employee wearing his high-visibility jacket and gloves while performing the aerobic stunt .
Topic selection method: nbc journalist jeff rossen filmed an airport worker at laguardia airport in new york doing push-ups on the runway after loading a plane .footage shows the employee wearing his high-visibility jacket and gloves while performing the aerobic stunt **.even after his 12th push-up , the fitness-enthusiast shows no sign of slowing down .**

Fig. 5. An example to show effectiveness of *topic selection*. All generated summaries is produced by **m8**. The topic words is green font. The important information missed by standard selection method is bold font. (Color figure online)

Human Evaluation. We also perform human evaluation to further evaluate our model. We use the following as evaluation criteria: (1) *Informativeness*, the main ideas and important details of article are shared; (2) *Coherence*, ideas are expressed clearly without repetition; (3) *Readability*, the generated summaries are fluent and grammatical.

We compare two final models **m4** and **m8** with lead-3 baseline [17] and PG with coverage [17]. For the process of human evaluation, we randomly pick 100 different samples from the test set. We show the original articles and four generated summaries to the human judges. The judges evaluate each summary by scoring 1–5 point according to each criterion described above. The 5-point means "best", while 1-point means "worst". Each sample is evaluated by 3 judges. The score of each criterion is averaged across all human judges.

Table 2. Human evaluation result. Best results are bolded

Model	Informativeness	Coherence	Readability
Lead-3	3.45	3.40	3.46
PGC	3.35	3.48	3.38
m4	3.52	**3.56**	3.52
m8	**3.56**	3.50	**3.60**

We invite 10 graduate students as our judges. The results are shown in Table 2. Both **m4** and **m8** outperform state-of-the-art abstractive pointer-generator and lead-3 extractive baseline. Compared to pointer-generator, it is observed that our models show competitive performance on informativeness and readability, indicating our models may provide more key information. Comparing the results of **m4** and **m8**, while being inferior to the other two criteria, **m4** shows advantage on Coherence. It indicates the richer topic information produced by LDA can help to improve the diversity of summaries in some extent.

5 Conclusion

In this paper, we propose topic attentional neural network (TANN) to utilize topic information for abstractive document summarization. We also propose a novel selection method named *topic selection* to improve the selection method of beam search. Experiments on the CNN/Daily Mail dataset demonstrate that, with the help of topic information, our model achieves a competitive performance with state-of-the-art abstractive and extractive methods and is able to produce summaries with more salient information. Human evaluation also demonstrates our model generates summaries with high informativeness and readability. In the future, we plan to extend our model with Generative Adversarial Network to generate more diverse summaries.

Acknowledgements. This research is supported by National Natural Science Foundation of China (Grant No. 61773229), Basic Scientific Research Program of Shenzhen City (Grant No. JCYJ20160331184440545), and Overseas Cooperation Research Fund of Graduate School at Shenzhen, Tsinghua University (Grant No. HW2018002).

References

1. Blei, D.M., Ng, A.Y., Jordan, M.I.: Latent Dirichlet allocation. J. Mach. Learn. Res. Arch. **3**, 993–1022 (2003)
2. Cheng, J., Lapata, M.: Neural summarization by extracting sentences and words. In: Meeting of the Association for Computational Linguistics, pp. 484–494 (2016)
3. Chopra, S., Auli, M., Rush, A.M.: Abstractive sentence summarization with attentive recurrent neural networks. In: Conference of the North American Chapter of the Association for Computational Linguistics: Human Language Technologies, pp. 93–98 (2016)

4. Duchi, J.C., Hazan, E., Singer, Y.: Adaptive subgradient methods for online learning and stochastic optimization. J. Mach. Learn. Res. **12**, 2121–2159 (2011)
5. Flick, C.: Rouge: a package for automatic evaluation of summaries. In: The Workshop on Text Summarization Branches Out, p. 10 (2004)
6. Gu, J., Lu, Z., Li, H., Li, V.O.K.: Incorporating copying mechanism in sequence-to-sequence learning. Meeting of the Association for Computational Linguistics, pp. 1631–1640 (2016)
7. Gulcehre, C., Ahn, S., Nallapati, R., Zhou, B., Bengio, Y.: Pointing the unknown words. Meeting of the Association for Computational Linguistics, pp. 140–149 (2016)
8. Hermann, K.M., et al.: Teaching machines to read and comprehend. In: Neural Information Processing Systems, pp. 1693–1701 (2015)
9. Li, C., Xu, W., Li, S., Gao, S.: Guiding generation for abstractive text summarization based on key information guide network. In: Proceedings of the 2018 Conference of the North American Chapter of the Association for Computational Linguistics: Human Language Technologies, Volume 2 (Short Papers), pp. 55–60. Association for Computational Linguistics (2018). http://aclweb.org/anthology/N18-2009
10. McDonald, R.: A study of global inference algorithms in multi-document summarization. In: Amati, G., Carpineto, C., Romano, G. (eds.) ECIR 2007. LNCS, vol. 4425, pp. 557–564. Springer, Heidelberg (2007). https://doi.org/10.1007/978-3-540-71496-5_51
11. Nallapati, R., Zhai, F., Zhou, B.: SummaRuNNer: a recurrent neural network based sequence model for extractive summarization of documents. In: National Conference on Artificial Intelligence, pp. 3075–3081 (2017)
12. Nallapati, R., Zhou, B., Santos, C.N.D., Gulcehre, C., Xiang, B.: Abstractive text summarization using sequence-to-sequence RNNs and beyond. In: Conference on Computational Natural Language Learning, pp. 280–290 (2016)
13. Nishikawa, H., Arita, K., Tanaka, K., Hirao, T., Makino, T., Matsuo, Y.: Learning to generate coherent summary with discriminative hidden semi-Markov model. In: Proceedings of COLING 2014, the 25th International Conference on Computational Linguistics: Technical Papers, pp. 1648–1659. Dublin City University and Association for Computational Linguistics (2014). http://www.aclweb.org/anthology/C14-1156
14. Romain Paulus, C.X., Socher, R.: A deep reinforced model for abstractive summarization. In: The 2018 International Conference on Learning Representations (Submitted for Publication)
15. Rush, A.M., Chopra, S., Weston, J.: A neural attention model for abstractive sentence summarization. Empirical Methods in Natural Language Processing, pp. 379–389 (2015)
16. Hochreiter, S., Schmidhuber, J.: Long short-term memory. Neural Comput. **9**(8), 1735–1780 (1997)
17. See, A., Liu, P.J., Manning, C.D.: Get to the point: summarization with pointer-generator networks. In: Proceedings of the 55th Annual Meeting of the Association for Computational Linguistics (Volume 1: Long Papers), pp. 1073–1083. Association for Computational Linguistics (2017). https://doi.org/10.18653/v1/P17-1099, http://www.aclweb.org/anthology/P17-1099
18. Vinyals, O., Fortunato, M., Jaitly, N.: Pointer networks. Neural Information Processing Systems, pp. 2692–2700 (2015)

Parameter Transfer Unit for Deep Neural Networks

Yinghua Zhang(✉), Yu Zhang, and Qiang Yang

Department of Computer Science and Engineering,
Hong Kong University of Science and Technology, Kowloon, Hong Kong
{yzhangdx,yuzhangcse,qyang}@cse.ust.hk

Abstract. Parameters in deep neural networks which are trained on large-scale databases can generalize across multiple domains, which is referred as "transferability". Unfortunately, the transferability is usually defined as discrete states and it differs with domains and network architectures. Existing works usually heuristically apply parameter-sharing or fine-tuning, and there is no principled approach to learn a parameter transfer strategy. To address the gap, a Parameter Transfer Unit (PTU) is proposed in this paper. PTU learns a fine-grained nonlinear combination of activations from both the source domain network and the target domain network, and subsumes hand-crafted discrete transfer states. In the PTU, the transferability is controlled by two gates which are artificial neurons and can be learned from data. The PTU is a general and flexible module which can be used in both CNNs and RNNs. It can be also integrated with other transfer learning methods in a plug-and-play manner. Experiments are conducted with various network architectures and multiple transfer domain pairs. Results demonstrate the effectiveness of the PTU as it outperforms heuristic parameter-sharing and fine-tuning in most settings.

Keywords: Transfer learning · Deep neural networks

1 Introduction

Deep Neural Networks (DNNs) are able to model complex functional mappings between inputs and outputs, and they produce competitive results in a wide range of areas, including speech recognition, computer vision, natural language processing, etc. Yet most successful DNNs belong to the supervised learning paradigm, and they require large-scale labeled data for training. Otherwise, they are likely to suffer from over-fitting. The data-hungry nature makes it prohibitive to use DNNs in low-resource domains where labeled data are scarce. There is a gap between the lack of training data in real-world scenarios and the data-hungry nature of DNNs.

Q. Yang et al. (Eds.): PAKDD 2019, LNAI 11440, pp. 82–95, 2019.
https://doi.org/10.1007/978-3-030-16145-3_7

The aforementioned dilemma can be addressed by *transfer learning*, which boosts learning in a low-resource *target* domain by leveraging one or more data-abundant *source* domain(s) [12,18]. It is found that parameters in a DNN are *transferable*, i.e., they are general and suitable for multiple domains [10,20,21]. The generalization ability of parameters is referred as "transferability". Two popular parameter-based transfer learning methods are *parameter-sharing* and *fine-tuning*. Parameter-sharing assumes that the parameters are highly transferable. The parameters in the source domain network are directly copied to the target domain network and they are kept "frozen". The fine-tuning method assumes that the parameters in the source domain network are useful, but they need to be trained with target domain data to better adapt to the target domain. One typical example is Domain Adaptation Network (DAN) [8] which applies both techniques. The first three convolutional layers in the DAN are shared, and the next two layers are fine-tuned.

Though parameter-based transfer learning is effective, it suffers from two limitations. Firstly, the parameter transferability is manually defined as discrete states, usually "random", "fine-tune", and "frozen" [10,20,21]. But the transferability at a fine-grained scale has not been considered. A block of parameters, for example, all the filters of a convolutional layer, are treated as a whole. If they are regarded as transferable, all the parameters are retained, though some of them are irrelevant or even introduce noises to the target domain; if they are considered as not transferable, they are completely discarded and the baby is thrown out with the bathwater. The second limitation is that the parameter transferability differs with domains and network architectures and there is no principled approach to learn the optimal transfer strategy. A parameter transfer strategy is obtained by assigning transfer states to different blocks of parameters. To find an optimal strategy, one straightforward solution is the hold-out method, where a part of the training data is reserved as a validation set and the network is decomposed into multiple parts and each part is assigned a transfer state. In the hold-out method, the optimal strategy can be found by choosing the one with the smallest validation error. By denoting by M and L the number of transfer states and the number of parts in the network, the number of possible strategies is M^L. The hold-out method is rather inefficient because it involves long training time and tremendous computational costs.

To tackle the two limitations of existing parameter-based transfer methods, we propose a Parameter Transfer Unit (PTU). Transfer learning with PTUs involves an already trained source domain network and a target domain network, and the two networks are connected by the PTU(s). A PTU produces a weighted sum of the activations from both networks. There are two gates in a PTU, a fine-tune gate and an update gate. The fine-tune gate adapts source domain activations to the target domain, and the update gate decides whether to transfer from the source domain. The two gates control the parameter transferability at a fine-grained scale and can be learned from data.

The contributions of the proposed method are two folds.

1. A principled parameter transfer method. We propose a novel parameter transfer unit, which subsumes hand-crafted discrete transfer states and allows parameter transfer at a fine-grained scale. The PTU is learned in an end-to-end approach.
2. Plug-and-play usage. The PTU can be used in both Convolutional Neural Networks (CNNs) and Recurrent Neural Networks (RNNs). It is a general and flexible transfer method as it can be easily integrated with almost all the existing models which intuitively apply the parameter-sharing or fine-tuning techniques. Experimental results show that transfer learning with the PTU outperforms the heuristic parameter transfer methods.

2 Related Works

Though deep learning models have been extensively studied, there are limited research works addressing transfer learning for DNNs. The most popular transfer method for DNNs is parameter-based transfer. It is shown that parameters of the low-level layers in a CNN are transferable [20]. For natural language processing tasks, Zoph *et al.* and Mou *et al.* study the parameter transferability in machine translation [21] and sentence classification tasks [10]. These works define parameter transferability as discrete states and conduct empirical studies. However, the conclusions drawn from these studies can hardly generalize to a new domain or a new network architecture. On the contrary, the proposed PTU defines the transferability at a fine-grained scale and learns the transferability in a principled approach.

Another line of research works use parameter-sharing and fine-tuning in joint with other transfer learning methods, for example, feature-based transfer learning [2,8,17]. These works heuristically apply the conclusions from the empirical studies, and proposing principled parameter-based transfer learning methods is not their main focus. We show that integrating the PTU with these models can further improve their performance.

The most relevant work to the proposed method is the cross-stitch network [9]. Instead of assigning blocks of parameters as "transferable" or"not transferable", a soft parameter-based transfer method is adopted. Knowledge sharing between networks in two tasks is achieved with a "cross-stitch" unit. It learns a linear combination of activations from different networks. The proposed PTU is different from the cross-stitch network in the following three aspects. First, the transferability is controlled by linear combination coefficients in the cross-stitch unit, while the PTU learns a non-linear combination which is more expressive. Secondly, the cross-stitch unit is proposed for multi-task learning while the PTU is designed for transfer learning where the target domain performance is the main focus. If the cross-stitch network is applied in the transfer learning setting, it becomes a degenerated case of the PTU. Thirdly, the PTU is applied and evaluated with both CNNs and RNNs while the cross-stitch unit is only evaluated with CNNs.

3 Parameter Transfer Unit (PTU)

In this section, we present the proposed PTU. First, three hand-crafted discrete transfer states are introduced as background knowledge. Then we introduce the use of the proposed PTU in CNNs and RNNs. Finally, we discuss several extensions of PTU to handle the scalability issue.

Fig. 1. Three hand-crafted discrete transfer states

3.1 Three Transfer States

There are usually three states for parameter transfer, sorted in an ascending order of the transferability, as shown in Fig. 1.

1. Random: the parameters are randomly initialized and learned with the target domain data only;
2. Fine-tune: the parameters are initialized with those from the source domain network, and then fine-tuned with the target domain data;
3. Frozen: the parameters are initialized with those from the source domain network, and keep unchanged during the training process in the target domain. When parameter-sharing is applied to a convolution layer (or a RNN cell), the parameters of that layer are frozen.

3.2 PTU for CNNs

An overview of transfer learning with PTUs in a CNN is shown in Fig. 2. The whole network, denoted by PTU-CNN, is composed of three parts, a source domain network, a target domain network and a few PTUs, which are denoted by blue blocks, green blocks and red circles in Fig. 2. We focus on the setting where there are limited labeled samples in the target domain. A labeled target domain data sample is denoted by $(\mathbf{x}^{\mathcal{T}}, y^{\mathcal{T}})$.

Let L denote the number of layers in the target domain network, and the target domain network shares an identical architecture to the source domain network from the first layer to the $(L-1)$-th layer. This allows parameter transfer between different tasks or heterogeneous domains where label spaces differ. The parameters in the source domain network are frozen, and the parameters in the target domain network are randomly initialized and learned with target domain data only. PTUs are placed between the two networks in a layer-wise manner to combine activations from both domains. In the training phase, only the target

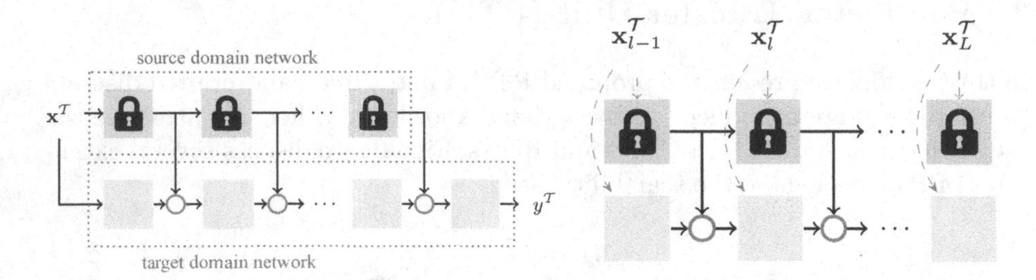

Fig. 2. An overview of the PTU-CNN. In the training phase, the source domain network is frozen, and the target domain network together with the PTUs are optimized. (Color figure online)

Fig. 3. Unrolled PTU-RNN. The source/target domain RNN cell is denoted by blue/green blocks. The connections from the inputs to the RNN cells are denoted by dashed gray arrows. (Color figure online)

domain network and the PTUs are optimized. Domain-specific knowledge can be encoded by the target domain network, and the PTUs learn how to transfer from the source domain network. In the inference phase, a target domain sample is fed into both networks, following the flows shown by the arrows in Fig. 2, and finally a predicted label is produced by the output layer of the target domain network.

Let l denote the l-th layer in the target domain network ($l = 1, \ldots, L-1$), and $\mathbf{h}_l^{\mathcal{S}}/\mathbf{h}_l^{\mathcal{T}}$ denote the output of the l-th layer in the source/target domain network, respectively. Given $\mathbf{h}_l^{\mathcal{S}}$ and $\mathbf{h}_l^{\mathcal{T}}$, a PTU learns a nonlinear combination, denoted by $\tilde{\mathbf{h}}_l^{\mathcal{T}}$, and feeds $\tilde{\mathbf{h}}_l^{\mathcal{T}}$ to the $(l+1)$-th layer of the target domain network.

There are two gates in a PTU, a fine-tune gate $\mathbf{r}_l$ and an update gate $\mathbf{z}_l$, as defined in Eq. (1):

$$\mathbf{r}_l = \sigma(\mathbf{W}_l^r[\mathbf{h}_l^{\mathcal{S}}, \mathbf{h}_l^{\mathcal{T}}]), \quad \mathbf{z}_l = \sigma(\mathbf{W}_l^z[\mathbf{h}_l^{\mathcal{S}}, \mathbf{h}_l^{\mathcal{T}}]), \tag{1}$$

where $[\cdot]$ denotes a concatenation operation and σ denotes the sigmoid function. The gates are artificial neurons whose parameters are denoted by $\mathbf{W}_l^r$ and $\mathbf{W}_l^z$, respectively. They take the activations $\mathbf{h}_l^{\mathcal{S}}$ and $\mathbf{h}_l^{\mathcal{T}}$ as inputs, and output a value between 0 and 1 for each element in the activations. Then the outputs of the gates mask the hidden activations and yield the combined activation $\tilde{\mathbf{h}}_l^{\mathcal{T}}$, as defined in Eq. (2):

$$\begin{aligned} \mathbf{h}_l^f &= (1 - \mathbf{r}_l) * \mathbf{h}_l^{\mathcal{S}} + \mathbf{r}_l * \phi(\mathbf{W}_l^h \mathbf{h}_l^{\mathcal{S}}), \\ \tilde{\mathbf{h}}_l^{\mathcal{T}} &= (1 - \mathbf{z}_l) * \mathbf{h}_l^{\mathcal{T}} + \mathbf{z}_l * \mathbf{h}_l^f, \end{aligned} \tag{2}$$

where $*$ denotes the Hadamard product, ϕ denotes an activation function, usually the hyperbolic tangent function or the Rectified Linear Unit (ReLU), and there is a linear transformation characterized by $\mathbf{W}_l^h$, which adapts the source domain activations to the target domain. The nonlinear transformation $\phi(\mathbf{W}_l^h \mathbf{h}_l^{\mathcal{S}})$ is equivalent to fine-tuning. The fine-tune gate produces a weighted sum of the source domain activations with and without fine-tuning, denoted by $\mathbf{h}_l^f$. The update gate determines how to combine the target domain activations with the transformed source domain activations. Details of the PTU are shown in Fig. 4.

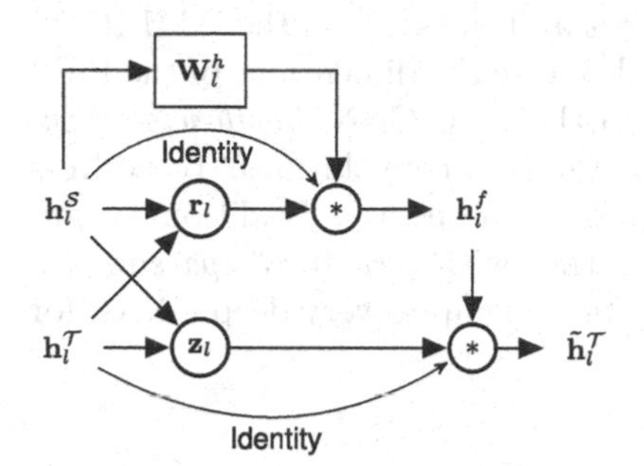

The gates look at the activations, $\mathbf{h}_l^S$ and $\mathbf{h}_l^T$, from both networks. The fine-tune gate $\mathbf{r}_l$ decides how to adapt the source domain activations, and the update gate $\mathbf{z}_l$ determines how to combine the target domain activations $\mathbf{h}_l^T$ with the transformed source domain activations $\mathbf{h}_l^f$.

Fig. 4. Details of the PTU

Relationship with the Hand-Crafted Discrete Transfer States. In extreme cases, the PTU degenerates to the hand-crafted discrete transfer states. That is, when the update gate $\mathbf{z}_l$ equals 0, the fine-tune gate $\mathbf{r}_l$ is ignored and the activations completely come from the target domain; otherwise, it takes source domain information into consideration. When the fine-tune gate $\mathbf{r}_l$ equals 0, the source domain activations $\mathbf{h}_l^S$ are highly transferable and they can be directly copied to the target domain. Otherwise, transformed source domain activations are used. Thus the PTU subsumes the three discrete transfer states. In most cases, the output of the PTU is a fine-grained combination of the activations from both networks.

Relationship with the Cross-Stitch Network. If the cross-stitch network is applied in the transfer learning setting, it becomes a degenerated case of the PTU-CNN. This is because the information flow from the target domain network to the source domain network is blocked, and the non-linear transformation of the PTU subsumes the linear combination in the cross-stitch network.

3.3 PTU for RNNs

We mainly focus on the PTU in CNNs so far, and here we extend the PTU for RNNs, denoted by PTU-RNN. As shown in Fig. 3, a sequence $\mathbf{x}^T$ with L steps is inputted where $\mathbf{x}^T = \{\mathbf{x}_1^T, \ldots, \mathbf{x}_L^T\}$. A RNN can be unrolled into a full network where the parameters in the RNN cell are shared across all time steps. Thus the RNN is able to tackle sequences of arbitrary lengths. The time step l in a RNN can be regarded as the l-th layer in a CNN. Similarly, $\mathbf{h}_l^S/\mathbf{h}_l^T$ denotes the internal hidden state at the l-th time step in the source/target RNN cell, respectively. By building the relationships of the notations in CNNs and RNNs, the PTU can be readily extended to RNNs.

3.4 Scalability

As each PTU introduces three parameters $\mathbf{W}^r$, $\mathbf{W}^z$ and $\mathbf{W}^h$, scalability becomes a challenge. This is because additional parameters take up more computational resources, e.g., GPU memory. In addition, more free parameters require more training data, otherwise, over-fitting is likely to occur. To reduce the computational cost, $\mathbf{W}^r$, $\mathbf{W}^z$ and $\mathbf{W}^h$ are shared across all time steps in the PTU-RNN. But this cannot be applied in the PTU-CNN because the dimensions of the PTU parameters in different layers do not agree. In the PTU-CNN, *depth-wise separable convolutions* [4,15] are used instead of standard convolutions. To address the over-fitting issue, regularization techniques are necessary. Traditional regularizers, such as ℓ_1 and ℓ_2 regularization, together with *structured sparsity* [19] are applied. These techniques allow the PTU to scale up to very deep CNNs, for example, the 28-layer MobileNets [4].

Depth-Wise Separable Convolution. The depth-wise separable convolution is initially proposed in [15], and it can greatly reduce computational costs with a slightly degraded performance [4]. The depth-wise separable convolution factorizes a standard convolution into two steps, a depth-wise convolution and a 1×1 point-wise convolution. In the depth-wise convolution step, a single filter is shared across all the channels. And then the point-wise convolution applies a 1×1 convolution to the output of the depth-wise convolution. For a convolution layer with N filters whose filter size is $K \times K$, the reduction in computational cost is $\frac{1}{N} + \frac{1}{K^2}$.

Structured Sparsity. As $\mathbf{W}^r$, $\mathbf{W}^z$ and $\mathbf{W}^h$ are high-dimensional tensors, structured sparsity learning is imposed to penalizing unimportant weights and improve computation efficiency [19]. Filter-wise and channel-wise *group Lasso* regularization are applied to the parameters in the PTU.

4 Experimental Results

We evaluate the PTU with both CNNs and RNNs on classification tasks. Classification accuracy is adopted as the evaluation metric. All the neural networks are implemented with Tensorflow [1]. Hyper-parameters are selected via the hold-out method.

4.1 Experiments on CNNs

We first describe the experimental setup, and then report numerical results. We also provide an interpretation of the output values of the gates in the PTU.

Experimental Setup. Various network architectures are evaluated on multiple transfer domain pairs. Three transfer settings are composed from four natural image classification datasets and three network architectures, as described below:

1. S1: CIFAR-10 → CIFAR-100 [5] where a LeNet-like 5-layer network is employed.
2. S2: ILSVRC-2012 [14] → Caltech-256 [3] where a VGG-16 network [16] is employed.
3. S3: ILSVRC-2012 → Caltech-256 where a MobilenetV1 network [4] is employed.

Two baseline models are considered:

1. No transfer (NoTL). The parameters are learned from scratch in the target domain.
2. Layer-wise fine-tuning (FT). For a CNN with L layers, if a layer has two possible transfer state, "fine-tune" and"frozen", there are 2^L transfer strategies, which generates prohibitive computation costs. To improve the efficiency, we adopt a strategy that layers are incrementally frozen as the parameter transferability drops when moving from low-level layers to high-level layers in a CNN [20]. That is, there are L fine-tuning strategies, and FT-l denotes a strategy that freezes the first l layers and fine-tunes the remaining layers.

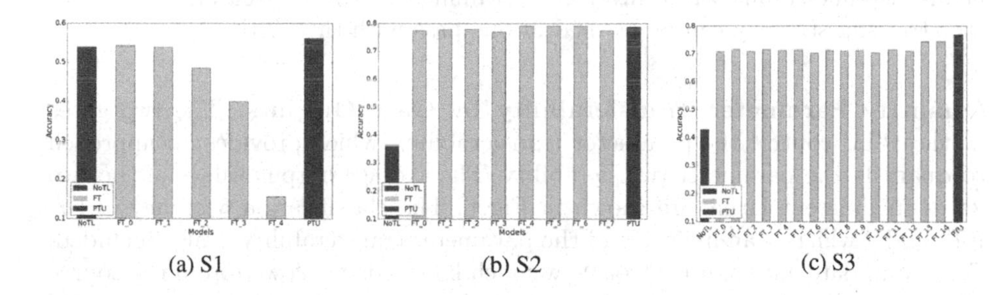

(a) S1 (b) S2 (c) S3

Fig. 5. Image classification accuracy of CNN models

Results. The results of the three methods are depicted in Fig. 5. As there are L fine-tuning models, only the highest test accuracy is summarized in Table 1. The optimal test accuracy of a setting is highlighted with bold face. Δ denotes the relative improvement of the PTU model over the FT model.

As shown in Fig. 5, the parameter transferability differs with domains and architectures. For S1, the parameters in low-level layers are more transferable than those in high-level layers, and the parameter transferability decreases monotonously, which is generally consistent with the conclusions in [20]. But the conclusion does not generalize to S2 and S3. For example, in S3, freezing the first

Table 1. Classification accuracy on CNNs

Models	Settings		
	S1	S2	S3
NoTL	53.92	36.21	42.90
FT	54.28	**79.09**	74.50
PTU	**56.12**	78.68	**76.86**
Δ(%)	3.39	−0.52	3.17

Table 2. Classification accuracy of MNIST → Omniglot

Models	Domains				
	G	L	K	JP-H	JP-K
K-NN	40.28	38.46	32.50	44.23	34.04
NoTL	38.89	41.03	31.67	35.90	26.24
FT	45.83	56.41	48.33	46.79	34.04
PTU	**51.39**	**67.95**	**55.00**	**50.00**	**46.10**
Δ (%)	12.13	20.46	13.80	6.86	35.43

13 layers achieves a higher accuracy than fine-tuning the whole network. These results indicate that the heuristic layer-wise fine-tuning method might not yield the optimal parameter transfer strategy.

Low-resource target domains benefit from parameter-based transfer learning, as both the FT model and the PTU model outperform the NoTL model. Furthermore, the PTU model achieves a comparable performance with the FT model in S2, and obtains the optimal test accuracies in the other 2 settings with a relative improvement around 3%. These results demonstrate the effectiveness of the PTU in various domains and with different network architectures. Unlike the layer-wise fine-tuning which involves L training processes, a reasonable parameter transfer strategy can be learned in one pass with the PTU.

Quantify Parameter Transferability by Gate Outputs. The two gates in the PTU control the parameter transferability, which provides an approach to quantify the parameter transferability. The average output values of the two gates in different layers are shown in Fig. 6. The classification accuracy of the FT model which is an indicator of the parameter transferability is also included. For the update gate $\mathbf{z}$, it controls how much knowledge is flowed from the source domain to the target domain. A larger $\mathbf{z}$ indicates more knowledge transfer. For example, in Fig. 6(a), the FT-1 achieves the optimal test accuracy, and z_1 is also the highest value among all the 4 values. In addition to the update gate, the fine-tune gate $\mathbf{r}$ characterizes how many activations are copied from the source domain, and how many activations need to be transformed before applying to the target domain, which has not been considered by existing works.

Here we quantify the parameter transferability as the average output values of gates which are scalars. A more fine-grained visualization analysis can be performed. For example, a large z value of a filter might help us identify important patterns that are shared between domains. This might demystify DNNs which are considered as a black-box process. Since understanding the neural network via visualization analysis is not the main focus of this paper, it will be left as future works.

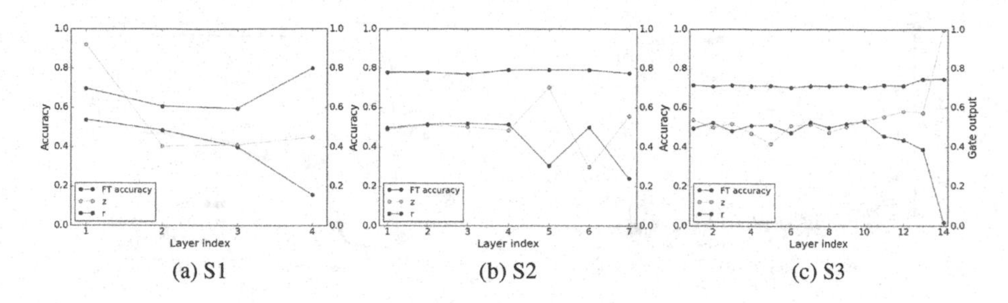

Fig. 6. Parameter transferability of different layers in CNNs

4.2 Experiments on RNNs

The experimental setup is first introduced, and then numerical results are presented.

Experimental Setup. The PTU for RNNs is evaluated with two handwritten character recognition datasets, MNIST [7] as the source domain and Omniglot [6] as the target domain. There are 50 alphabets in the Omniglot dataset, and 5 alphabets are randomly selected and used as target domains. The 5 alphabets are Greek (G), Latin (L), Korean (K), Japanese hiragana (JP-H) and Japanese katakana (JP-K). At each time step, a row of an image is fed into the RNN. A single-layer RNN with 128 hidden units is used as a feature extractor. The hidden state of the last time step is used as the feature and it is fed to a fully-connected layer for classification. The network achieves a classification accuracy at 96% in the source domain. Since the label spaces of the two domains do not agree, the only transferable parameters are those in the RNN cell, and hence there is only one FT strategy. In addition to the NoTL and FT models, a K-Nearest Neighbor (K-NN) classifier are included as well. The K-NN classifier is implemented with scikit-learn [13].

Results. The classification accuracies are listed in Table 2. Since there are only around 1,000 labeled training data in each target domain, the NoTL model performs even worse than a simple K-NN classifier in 4 out of 5 domains. Similar conclusions to the CNN experiments can be drawn. The classification accuracy is improved when a parameter-based transfer learning method is applied, and the proposed PTU further improves over the FT model with a large margin where the relative improvement Δ ranges from 6.86% to 35.43%. The reason that the PTU outperforms heuristic parameter-sharing and fine-tuning might be two folds.

1. The PTU subsumes hand-crafted transfer states by introducing learnable gates. It is more expressive in terms of model capacity.

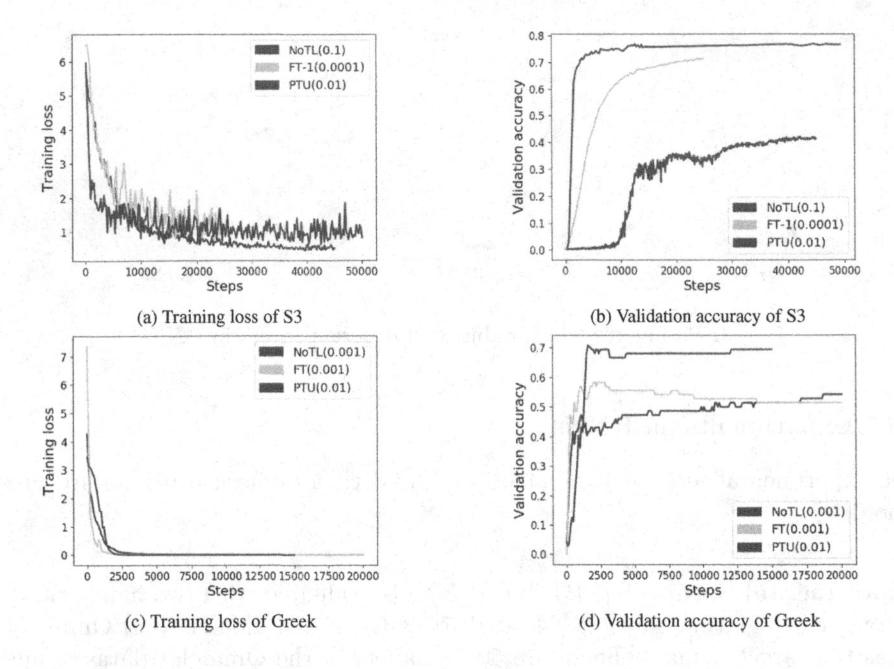

(a) Training loss of S3

(b) Validation accuracy of S3

(c) Training loss of Greek

(d) Validation accuracy of Greek

Fig. 7. Learning curves of two transfer settings

2. In PTU, source domain knowledge is retained as frozen parameters, and the domain-specific knowledge is encoded in the target domain network. On the other hand, the parameters in the FT model are changed during training in the target domain, which might impair the useful knowledge from the source domain.

4.3 Convergence Performance

We investigate the convergence performance of different models by learning curves. The learning curves of two transfer settings, S3 in the CNN experiments and the Greek alphabet as the target domain in the RNN experiments, are shown in Fig. 7. The training loss and validation accuracy as a function of the number of steps are plotted for each setting with the learning rate that yields the optimal test classification accuracy. At each step, the model is optimized with a mini-batch.

For the S3 setting, the NoTL model converges slowly though a large learning rate is used. The validation accuracy is almost 0 in the first 10, 000 steps. The optimization efficiency is significantly improved by parameter transfer. The FT model converges at a similar rate to the NoTL model while its learning rate is 100-times smaller. For the PTU model, the training loss drops quickly, and the validation accuracy saturates. The PTU is rather resistant to over-fitting since its validation accuracy does not deteriorate as the training process continues.

For the Greek setting, the NoTL model gets stuck with a bad local optimum as the training loss decreases while the validation accuracy converges to around 0.5. The FT model starts with the largest training loss and converges the fastest while it suffers from over-fitting. The PTU model uses a learning rate that is 10 times larger than the other two baseline models, as it introduces additional parameters. The PTU model has a better generalization ability as it achieves the highest accuracy on the validation set.

4.4 Integrate PTU with Feature-Based Transfer Learning Method

As a general and flexible module for transfer learning, the PTU can be easily integrated with other transfer learning methods and further improves their performance. The performance of the PTU with a feature-based transfer learning method is evaluated in this section.

The experiment is conducted on an unsupervised domain adaptation task, MNIST → MNIST-M. The source domain is labeled while the target domain is unlabeled. The target domain, MNIST-M, is a variation of MNIST [2]. Domain-invariant representations can learned by minimizing the Maximum Mean Discrepancy (MMD) [8,11,17]. Similar to the experimental setup in Sect. 4.2, a single-layer RNN with 128 hidden units is used as the feature extractor, and then follows a fully-connected classification layer. Three models are evaluated:

1. Source-only (S-only). The model is trained with the labeled source data only, and is directly evaluated on the target domain without any adaptation.
2. MMD with fine-tuning (FT-MMD). The model initializes the parameters with those from the S-only model and fine-tunes them by minimizing the MMD. A linear combination of 19 RBF kernels is used for computing the MMD.
3. MMD with PTU (PTU-MMD). The model replaces the heuristic fine tuning in the FT-MMD with the PTU.

Table 3. Classification accuracy of MNIST → MNIST-M

S-only	FT-MMD	PTU-MMD	Δ (%)
45.83	49.46	**51.29**	3.68

The results are summarized in Table 3. Both feature-based transfer learning models, FT-MMD and PTU-MMD, outperform the S-only model, which demonstrates the efficacy of feature-based transfer learning. Moreover, a relative improvement of 3.68% is achieved by replacing heuristic parameter fine-tuning with the PTU. The results show that the PTU can be successfully applied together with other transfer learning methods to further boost their performance.

5 Conclusion

A principled approach to learn parameter transfer strategy is proposed in this paper. A novel parameter transfer unit (PTU) is designed. The parameter transferability is controlled at a fine-grained scale by two gates in the PTU which can be learned from data. Experimental results demonstrate the effectiveness of the PTU with both CNNs and RNNs in multiple transfer settings where it outperforms heuristic parameter-sharing and fine-tuning. In the future, we will apply the PTU to more challenging settings, for example, image captioning which involves multi-modality data.

References

1. Abadi, M., et al.: TensorFlow: a system for large-scale machine learning. OSDI **16**, 265–283 (2016)
2. Ganin, Y., et al.: Domain-adversarial training of neural networks. J. Mach. Learn. Res. **17**, 591–5935 (2016)
3. Griffin, G., Holub, A., Perona, P.: Caltech-256 object category dataset (2007)
4. Howard, A.G., et al.: MobileNets: efficient convolutional neural networks for mobile vision applications. arXiv preprint arXiv:1704.04861 (2017)
5. Krizhevsky, A., Hinton, G.: Learning multiple layers of features from tiny images (2009)
6. Lake, B.M., Salakhutdinov, R., Tenenbaum, J.B.: Human-level concept learning through probabilistic program induction. Science **350**(6266), 1332–1338 (2015)
7. LeCun, Y., Bottou, L., Bengio, Y., Haffner, P.: Gradient-based learning applied to document recognition. Proc. IEEE **86**(11), 2278–2324 (1998)
8. Long, M., Cao, Y., Wang, J., Jordan, M.: Learning transferable features with deep adaptation networks. In: International Conference on Machine Learning, pp. 97–105 (2015)
9. Misra, I., Shrivastava, A., Gupta, A., Hebert, M.: Cross-stitch networks for multi-task learning. In: Proceedings of the IEEE Conference on Computer Vision and Pattern Recognition, pp. 3994–4003 (2016)
10. Mou, L., et al.: How transferable are neural networks in NLP applications? In: Proceedings of the 2016 Conference on Empirical Methods in Natural Language Processing, pp. 479–489 (2016)
11. Pan, S.J., Tsang, I.W., Kwok, J.T., Yang, Q.: Domain adaptation via transfer component analysis. IEEE Trans. Neural Netw. **22**(2), 199–210 (2011)
12. Pan, S.J., Yang, Q.: A survey on transfer learning. IEEE Trans. Knowl. Data Eng. **22**(10), 1345–1359 (2010)
13. Pedregosa, F., et al.: Scikit-learn: machine learning in Python. J. Mach. Learn. Res. **12**, 2825–2830 (2011)
14. Russakovsky, O., et al.: ImageNet large scale visual recognition challenge. Int. J. Comput. Vis. **115**(3), 211–252 (2015)
15. Sifre, L., Mallat, P.: Rigid-motion scattering for image classification. Ph.D. thesis, Citeseer (2014)
16. Simonyan, K., Zisserman, A.: Very deep convolutional networks for large-scale image recognition. arXiv preprint arXiv:1409.1556 (2014)
17. Tzeng, E., Hoffman, J., Zhang, N., Saenko, K., Darrell, T.: Deep domain confusion: maximizing for domain invariance. arXiv preprint arXiv:1412.3474 (2014)

18. Weiss, K., Khoshgoftaar, T.M., Wang, D.: A survey of transfer learning. J. Big Data **3**(1), 9 (2016)
19. Wen, W., Wu, C., Wang, Y., Chen, Y., Li, H.: Learning structured sparsity in deep neural networks. In: Advances in Neural Information Processing Systems, pp. 2074–2082 (2016)
20. Yosinski, J., Clune, J., Bengio, Y., Lipson, H.: How transferable are features in deep neural networks? In: Advances in Neural Information Processing Systems, pp. 3320–3328 (2014)
21. Zoph, B., Yuret, D., May, J., Knight, K.: Transfer learning for low-resource neural machine translation. In: Proceedings of the 2016 Conference on Empirical Methods in Natural Language Processing, pp. 1568–1575 (2016)

EFCNN: A Restricted Convolutional Neural Network for Expert Finding

Yifeng Zhao(✉), Jie Tang, and Zhengxiao Du

Department of Computer Science and Technology, Tsinghua University, Beijing, China
{zhao-yf16,duzx16}@mails.tsinghua.edu.cn, jietang@tsinghua.edu.cn

Abstract. Expert finding, aiming at identifying experts for given topics (queries) from expert-related corpora, has been widely studied in different contexts, but still heavily suffers from low matching quality due to inefficient representations for experts and topics (queries). In this paper, we present an interesting model, referred to as EFCNN, based on restricted convolution to address the problem. Different from traditional models for expert finding, EFCNN offers an end-to-end solution to estimate the similarity score between experts and queries. A similarity matrix is constructed using experts' document and the query. However, such a matrix ignores word specificity, consists of detached areas, and is very sparse. In EFCNN, term weighting is naturally incorporated into the similarity matrix for word specificity and a restricted convolution is proposed to ease the sparsity. We compare EFCNN with a number of baseline models for expert finding including the traditional model and the neural model. Our EFCNN clearly achieves better performance than the comparison methods on three datasets.

Keywords: Expert finding · Convolution neural network · Similarity matrix

1 Introduction

Online question-and-answer (QA) has become a more popular way for users to share their experiences and to ask questions on the Internet. For example, Quora.com and Zhihu.com, the most popular websites for sharing and acquiring knowledge, attract users to answer millions of questions per day; Toutiao QA, an up-and-coming mobile social platform, has accumulated 580 million Toutiao users and 300 thousand professional writers (authors). The competitive advantage of the online QA platforms is that they provide high-quality answers for users and offers a new direction for professional knowledge sharing. However, at the same time, it also poses new challenges. One central challenge is finding a way to assign those new questions (queries) to potential experts, referred to as expert finding.

Expert finding has been studied by researchers from different communities. Several different methods have been proposed. These include keyword-based

Q. Yang et al. (Eds.): PAKDD 2019, LNAI 11440, pp. 96–107, 2019.
https://doi.org/10.1007/978-3-030-16145-3_8

modeling [2], language modeling [1,3,18], latent semantic indexing [6], and topic modeling [11,15]. Most of these methods represent every expert as a document and cast the problem as a document matching problem. In language models, each document is represented by words with their term frequency-inverse document frequency (TF-IDF) [20] score. Latent semantic indexing learns a low-dimensional representation by decomposing the word feature space, and topic models such as Latent Dirichlet Allocation probabilistically group similar words into topics, and represent documents as distributions over these topics. Obviously, these existing methods mainly represent documents by the frequency or co-frequency of words, but ignore semantic information at the phrase and sentence level. Thus, how to capture and utilize semantic information at the word, phrase, and sentence level of documents remains a challenging problem.

Given a query paper and a candidate expert pool, we represent each expert as a set of documents he/she has written. As a result, how to estimate the similarity score between these documents and the query paper becomes the central issue. Inspired by the success of convolutional neural networks (CNN) [13] in image recognition, we cast this task as an "image recognition" and present a method based on restricted convolution. Specifically, the similarity between any word pair of two documents is calculated to generate a similarity matrix. However, this similarity matrix not only ignores the word specificity, but is also very sparse and position-related. Therefore we introduce IDF into the similarity matrix for word specificity and propose a restricted convolution layer to ease the problem of the sparse and position-related matrix. The experiments on three datasets show that our work performs better than the baselines.

Contributions. In this paper, we define the problem of expert finding and propose a framework based on a restricted convolutional neural network. Based on the similarity matrix, we propose restricted convolution. Compared to a standard convolution, restricted convolution considers the importance of position, and penalizes for similarity far from the center of filters. For taking word specificity into accounts, we further construct a new similarity matrix by combining original similarity matrix and IDF. We prove that the proposed framework can capture and utilize semantic information from word-level to document-level. We compare our framework with several state-of-the-art approaches on three different datasets and experimental results show the effectiveness of our framework.

2 Problem Formulation

Let $S = \{(v_i, d_i)\}_{i=1}^{N}$ denote the set of experts and his/her documents, where v_i is a candidate expert, d_i is the set of support documents authored by (or associated with) expert v_i and N is the expert size. The input of our problem also includes a query d_q, which can be also viewed as a document. There can be various kinds of documents in different applications. For example, in an academic network, the documents could be papers published by researchers, while in a Quora-like website, the documents could be the questions (or answers) that users have asked (or answered). Given this, we can formally define the expert finding problem as follows (Fig. 1):

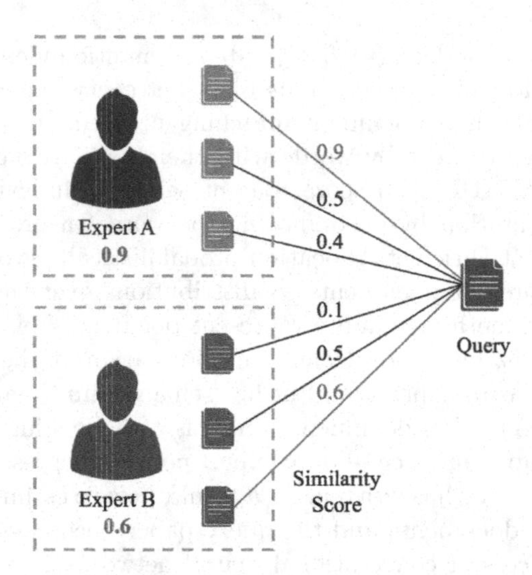

Fig. 1. An example of expert finding: each expert has a set of documents; each document has a similarity score with respect to query. Intuitively, expert A is more relevant than expert B regarding the query.

Definition 1 ***Expert Finding.*** *Given a set S and a query d_q, the objective here is to learn a function f using documents d_i in each expert v_i and the query d_q, in order to predict a ranked list $R(v_i, d_i) \subset S$ with $|R| = k$, which is the top-k relevant experts in S with respect to d_q.*

One challenge here is that experts and query documents are two different kinds of entities. This means that they cannot be represented in a common space and the relevance between an expert and query documents cannot be measured directly. An alternative method is to measure the relevance based on expert v_i's documents d_i, where experts with a more relevant document to the query should be ranked higher. The central problem is how to model the representations for documents and queries so that the similarity score can be easily estimated. In this paper, we propose a model based on the similarity matrix and restricted convolution to address this problem.

3 Our Model

The basic idea of our model is to cast this problem as an "image recognition" problem. Our model first constructs a similarity matrix using the embedding of words contained in the document and the query. Viewing the similarity matrix as an image, a restricted convolutional neural network is employed to learn the representations and also predict the relevance score of a candidate to the query.

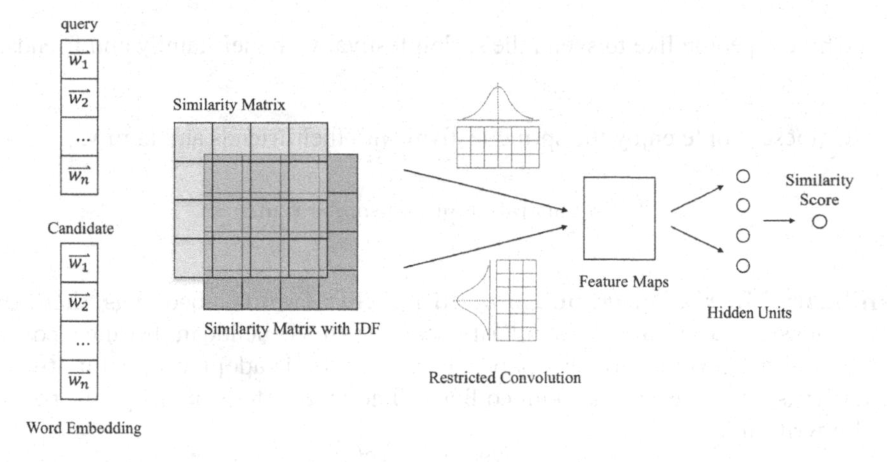

Fig. 2. The overall architecture of EFCNN

3.1 Word Embedding and Similarity Matrix

Word Embedding. It is easy to understand that both the query d_q and candidate document d_i can be represented as word sequences. In order to construct a similarity matrix, a variety of methods can be used to compute the similarity between words. For example, the similarity can be simply defined as 1 or 0 to indicate whether two words are identical; however, that ignores the semantic information between two similar words. Considering the semantic information, we use the word embedding technology, i.e., the Word2Vec model [14], to represent each word as a multi-dimensional vector, and then compute the similarity.

For completeness, we give a brief introduction to the Word2Vec model. Word2Vec employs a neural network to learn word embedding for each word. The neural network architecture (the skip-gram model) consists of an input layer, a projection layer, and an output layer. The objective is to maximize the probability of surrounding words for an input word in the corpus. Therefore, the objective can be written as:

$$\frac{1}{T}\sum_{t=1}^{T}\sum_{w_j\in nb(w_t)}\log p(w_j|w_t)$$

where T is the size of the corpus, $nb(w_t)$ is the set of surrounding words of w_t, and $|nb(w_t)|$ is determined by the window size in training. The probability $p(w_j|w_t)$ is the hierarchical softmax of the word embedding of w_j and w_t. The authors demonstrate that semantic relationships are often preserved in vector operations on word embeddings, e.g., vec("*King*") − vec("*Man*") + vec("*Woman*") results in a vector that is closest to the vector representation of the word "*Queen.*" Due to its high quality and low computational cost, we use Word2Vec embedding as our preferred embedding (Fig. 3).

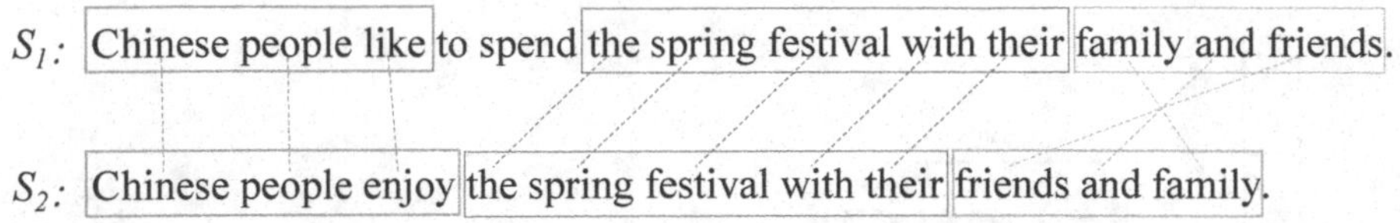

S_1: Chinese people like to spend the spring festival with their family and friends.

S_2: Chinese people enjoy the spring festival with their friends and family.

Fig. 3. An illustration of two similar sentences.

Similarity Matrix Based on Embedding. Given word embeddings, there are many measures to obtain the similarity score, such as euclidean distance, cosine metric and dot product. In this paper, cosine metric is adopted to compute the similarity score between two embeddings. Therefore, the similarity matrix M can be written as:

$$M_{i,j} = \frac{\text{vec}(w_i)^T \text{vec}(w'_j)}{\|\text{vec}(w_i)\| \cdot \|\text{vec}(w'_j)\|} \tag{1}$$

where w_i is the i-th word in query d_q, w'_j is the j-th word in document d_i, $\text{vec}(w)$ is the Word2Vec embedding of the word w, and $\| \cdot \|$ is the norm of Word2Vec embedding.

In this way, similarity matrix M can provide meaningful matching information between query and document at word, phrase, and sentence level. Take two sentences in Fig. 2 as an example, we find that these two sentences are similar at all three mentioned levels. At the word level, these sentences not only have identical word pairs, e.g., "*Chinese-Chinese,*" but also have similar word pairs, e.g., "*like-enjoy.*" At the phrase level, sentences can be broken down into three matching phrase pairs, e.g., "*(Chinese people like)-(Chinese people enjoy).*" These three mentioned phrase pairs roughly construct sentences, which indicates the similarity at the sentence level.

Similarity Matrix with IDF. Similarity matrix mainly focuses on the word similarity of two documents, ignoring how specific and distinctive a word is. Take two sentences in Fig. 2 as an example again, "chinese" and "festival" are always more specific than "people" and "enjoy" and should be given more attention. TF-IDF is the most common measurement for scoring word specificity. In this paper, similarity matrix already contains the whole words in the document, so only the IDF needs to be taken into account. IDF is the logarithmically scaled inverse fraction of the documents that contains the word. There are a whole family of inverse functions, and here we choose the smooth IDF:

$$IDF(w) = log(1 + \frac{N}{n_w})$$

where n_w is the number of documents containing the word w and N is the total number of documents.

Similarity matrix with IDF can be regarded as the supplementary of similarity matrix for word-specific information, but it cannot completely replace the similarity matrix, which is discussed in Sect. 4.2. Formally, it can be written as:

$$M_{i,j}^{IDF} = M_{i,j} * IDF(w_i) * IDF(w'_j)$$

where the smooth IDF will not change the sign of word similarity.

3.2 EFCNN: Expert Finding with Restricted CNN

Restricted Convolution. Inspired by the great success of the CNN in image recognition, [17] views the similarity matrices as images, which can be the input for CNN. However, as shown in Fig. 4, the similarity matrix is slightly different from the traditional image, with its sparse value and regional discontinuity.

To ease this problem, we adopt an intuitive method named restricted convolution to edit the convolution structure to produce position-based filters for each layer. Specifically, the closer a position is to the central axis, the higher its weight is. In this paper, we use two decay functions, including linear and exponential decay.

Linear decay is a common decay function. If the weight attenuation follows linear decay, the weights $w^{(1,k)} \in R^{m \times n}$ of k-th filter in the l-th restricted convolutional layer can be written as:

$$w_{i,j}^{(l,k)} = (\alpha + \frac{||\lfloor \frac{n}{2} \rfloor - j|}{\lfloor \frac{n}{2} \rfloor} * (1 - \alpha)) * w_{i,\lfloor \frac{n}{2} \rfloor}^{(l,k)} \tag{2}$$

where α is the decency coefficient of linear decay.

Exponential decay considers the smooth of decay. If the weight attenuation follows exponential decay, the weights are computed as following:

$$w_{i,j}^{(l,k)} = e^{-\beta |j - \lfloor \frac{n}{2} \rfloor|} * w_{i,\lfloor \frac{n}{2} \rfloor}^{(l,k)} \tag{3}$$

where β is the decency coefficient of exponential decay. Obviously, comparing to standard convolutional layer, each filter in restricted convolution only has one column variable in central axis, which will accelerate the training process. In addition, the restricted convolution can also be transposed as shown in Fig. 2.

Forward Network. Same with standard convolution, the k-th filter in restricted convolutional layer is used to compute dot product between its weights $w^{(l,k)}$ and regions in the input $z^{(l-1)}$. An element-wise activation function δ is applied to obtain a non-linear feature map $z^{(l,k)}$. Formally, we have:

$$\begin{aligned} z^{(0)} &= M \oplus M^{IDF} \\ z_{x,y}^{(l,k)} &= \delta(\sum_{t=0}^{c^{(l-1)}-1} \sum_{i=0}^{m_t - 1} \sum_{j=0}^{n_t - 1} w_{i,j}^{(l,k)} \cdot z_{x+i,y+j}^{(l-1,k)} + b^{(l,k)}) \end{aligned} \tag{4}$$

where m_t, n_t denotes the size of t-th filter, $c^{(l)}$ denotes the number of filters in the l-th layer and $b^{(l,k)}$ is a bias term.

In addition to reducing the spatial size of the feature maps, max-pooling layers also operate independently on every output of convolutional layers and resize them spatially. Therefore, the output $z^{(l,k)}$ of the max-pooling layer can be written as:

$$z_{x,y}^{(l,k)} = \max_{0 \leq i < r_k} \max_{0 \leq j < r_k} z_{x \cdot r_k + i, y \cdot r_k + j}^{(l-1,k)} \tag{5}$$

where r_k denotes the size of the k-th pooling filter, which is set to 2 in our model.

The final feature maps are then turned into a vector and passed through an MLP with several hidden layers. In this paper, we use only two fully-connected layers. For the final output, a single unit is connected to all units of the last hidden layer:

$$s = W_2 \delta(W_1 \cdot z + b_1) + b_2 \tag{6}$$

where W_1 and W_2 are the weights of fully-connected layers with b_1 and b_2 are the bias terms.

Optimization and Model Training. As the task is formalized as a ranking problem, we can utilize pairwise ranking loss such as hinge loss for training. Given a triple (d_q, d_+, d_-), where document d_+ is ranked higher than document d_- with respect to query document d_q, the loss function is defined as:

$$Loss(d_q, d_+, d_-) = max(0, 1 - s(d_q, d_+) + s(d_q, d_-))$$

where $s(d_q, d_+)$ and $s(d_q, d_-)$ are the corresponding predicted similarity scores.

Since the size of expert-finding datasets we use is relatively small, for experiments on these datasets we train our model on a task called citation prediction. Given the abstracts of three documents, the model needs to give higher rank to the document that has citation relationship with the query. Obviously, to complete the task, the model also needs to compute the relevance of two documents. The training dataset of our model is collected from an academic search system Aminer [21].

Training is done through stochastic gradient descent over mini-batches, with the Adagrad update rule [5]. It achieves good performance with a learning rate of 0.001. For regularization, we employ dropout [7] on the penultimate layer, which prevents co-adaptation of hidden units by randomly dropping out, i.e., set to zero. To avoid over-fitting, we apply an early-stop strategy [19].

4 Experiments

4.1 Experimental Setup

To evaluate the proposed model, we conduct the experiments of expert finding problem on three datasets.

Datasets. As the paper-reviewer assignment is private and interest-related, there is no publicly labeled dataset. It is difficult to create one as well. Therefore, for the purpose of evaluation, we collect three datasets from an online system and human judgments.

Paper-Reviewer: This dataset comes from an online system which connects journal editors with qualified journal reviewers. The system recommends journal submissions that are posted by journal editors to qualified reviewers who are willing to review. It includes 540 papers submitted to ten journals and 2,359 experts' invitation responses. Among these responses, 953 are "agree", while the rest are viewed as "decline" (including "unavailable" and "no response"). Basically, we consider "agree" as relevant and "decline" as irrelevant.

Topic-Expert: This dataset is based on papers from Aminer [21]. It consists of 86 papers with 20 candidate experts for each query. In this dataset, we follow a traditional expertise matching setting such as [4,22].

Patent-Relevance: This dataset is based on documents of patents, which comes from the Patent Full-Text Datasets of the United States Patent and Trademark Office. It consists of 67 patent queries with 20 candidate patents for each query.

In Topic-Expert and Patent-Relevance, we gather relevance judgments from college students and experts on patent analysis as the ground truth. The relevance is simply expressed as binary: relevant or irrelevant. In these three datasets, only the first 64 words are chosen for the abstracts of the papers or the patent documents.

Comparison Methods. We compare the following methods in the experiment:

- **BM25:** The relevance score between query q and document d is measured by the BM25 score, where each word in query q is considered as a keyword. The relevance score is defined as:

$$\text{BM25}(q,d) = \sum_{w \in q} \text{IDF}(w) \cdot \frac{\mathcal{N}_d^w \cdot (k_1 + 1)}{\mathcal{N}_d^w + k_1 \cdot (1 - b + b \cdot \frac{\mathcal{N}_d}{\lambda})}$$

  where IDF($\cdot$) is inverse document frequency, $\mathcal{N}_d^w$ is word w's frequency in document d, and $\mathcal{N}_d$ is the length of d. We set $k = 2$, $b = 0.75$, and λ as the average document length;
- **MixMod** [22]**:** While another setting is the same as BM25, the relevance score between query q and expert e is defined as:

$$P(q|e) = \sum_{d_j \in D_i} \sum_{m=1}^{k} \prod_{t_i \in q} P(t_i|\theta_m) P(\theta_m|d_j) P(d_j|e)$$

  where $P(t_i|\theta_m)$ denotes the probability of generating a term given a theme θ_m and $P(\theta_m|d_j)$ denotes the probability of generating a theme given a document d_j;
- **Doc2Vec** [12]**:** We represent each document via Paragraph Vector model. The similarity score between two documents is produced by the cosine metric of their representations;
- **WMD** [10]**:** We apply the Word Mover's Distance (WMD) to measure the similarity between two documents. The WMD is the minimum distance required to transport the words from one document to another based on word embeddings;

Table 1. Results of relevance assignment(%). NG is the simplify of NDCG.

	Paper-reviewer			Topic-expert			Patent-relevance		
Method	NG@1	NG@3	NG@5	NG@3	NG@5	NG@10	NG@3	NG@5	NG@10
BM25	36.9	40.4	41.6	64.2	63.7	66.2	49.8	57.7	62.4
MixMod	35.3	39.9	42.5	57.6	58.2	58.0	45.6	51.3	56.5
Doc2Vec	34.5	41.3	43.4	60.9	63.8	66.2	44.2	48.0	54.7
WMD	41.2	47.7	49.8	62.5	64.5	66.8	57.4	58.5	61.9
LSTM-RNN	34.1	38.6	42.1	58.6	60.5	63.7	58.3	58.1	65.0
MatchPyramid	40.0	47.8	48.8	66.3	66.0	68.7	57.6	59.4	62.9
EFCNN	**43.4**	**49.6**	**52.3**	**67.7**	**67.1**	**70.8**	**59.8**	**61.4**	**65.8**

- **LSTM-RNN** [16]**:** [16] adopts an LSTM to construct sentence representations and uses cosine similarity to output the similarity score;
- **MatchPyramid** [17]**:** The MatchPyramid is a standard CNN built on the standard similarity matrix to get the similarity score.

As for neural models, we can see that Doc2Vec and LSTM-RNN are all sentence representation models, while WMD, MatchPyramid and EFCNN are the interaction-based model.

Parameter Settings. In our model, there are two restricted convolutional layers, both having 64 filters. All filters are set to 3×7 and Batch Normalization [8] is adding to all restricted convolutional layers. The number of hidden units of the fully-connected layer is set to 256. And the hyperparameters α and β are set to 0.2 and 2.0 respectively, which is discussed in Sect. 4.2.

The Word2Vec embedding is learned on AMiner data [21]. The embedding is trained using the Skip-gram architecture [14]. For a fair comparison, word embedding of all comparison models is the same as that of the proposed model and the dimension number of word embedding is set to 150.

Evaluation Metric. Formalized as a ranking problem, the output is a ranked list of experts, where the order depends on the maximum similarity score of their documents regarding the query. The goal is to rank the positive one higher than the negative ones. Therefore, we use NDCG@n [9] as an evaluation metric. Formally, we have:

$$\text{NDCG@}n = \frac{\sum_{i=1}^{n} \frac{2^{r_i}-1}{\log_2(i+1)}}{\sum_{i=1}^{|R|} \frac{2^{R_i}-1}{\log_2(i+1)}}$$

where r_i is the relevance of i-th expert in the output and R represents the list of experts (ordered by their relevant) in the length of n.

4.2 Results and Discussion

Performance Analysis. We compare the performance of all methods on three datasets. Table 1 shows the ranking accuracy of different methods in terms of NDCG, where measures are averaged for all queries on each dataset. Roughly speaking, the neural methods (such as WMD and EFCNN) outperform the traditional methods in most cases. Only taking the exact word matching into account, traditional methods will lose important information easily, while neural methods based on word embedding can learn better representations and deal with the mismatch problem effectively. As for neural models, we can see that interaction based models, such as WMD and MatchPyramid, perform better than representation based models. This is mainly because these model can capture more detailed information from the interaction of documents.

On Paper-Review and Topic-Expert, we also see that our model achieves significant improvement compared to all the baselines. On Paper-Reviewer, EFCNN clearly outperforms the comparison methods in all by 2.2% and 1.9% (p−value $\ll 0.01$ in both cases by t-test) in terms of NDCG@1 and NDCG@5 respectively. It indicates that restricted convolution deals with sparse but position-depended signals effectively.

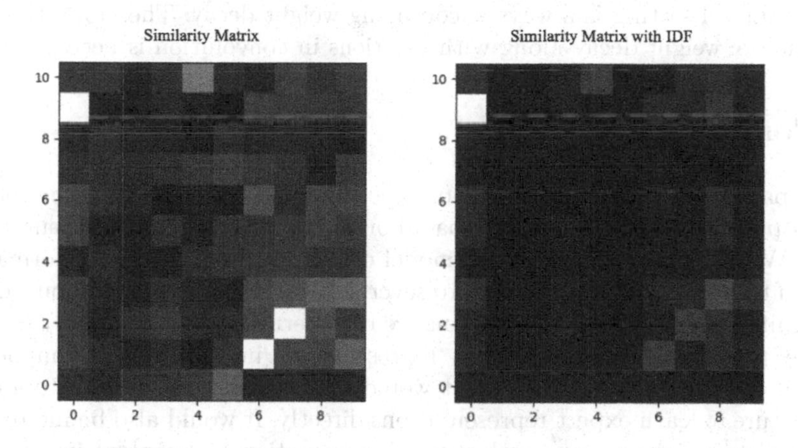

Fig. 4. The visualization result of two matrices based on a document pair. The brighter the pixel is, the larger value it has. The document pair is as follows: D1: Privacy is an enormous problem in online social networking sites. D2: While online social networks encourage sharing information, they raise privacy issues.

How the Similarity Matrix with IDF Works. To have a better understanding of how the similarity matrix with IDF works, we show the pixel images of matrices without/with IDF in Fig. 4. From the pixel image, we can see that similarity matrix focuses more on word similarity. While the similarity matrix with IDF focuses more on details, it will only show the significant result when two words are similar and when both of them are important to the document.

These two matrices support each other, and we will lose some information if we drop any one of them.

Table 2. The effect of hyperparameters α and β on Topic-Expert.

	NG@3	NG@10
EFCNN-Lin($\alpha = 0.2$)	67.7	70.8
EFCNN-Lin($\alpha = 0.5$)	66.2	68.7
EFCNN-Lin($\alpha = 1.0$)	64.5	66.4
EFCNN-Exp($\beta = 1.0$)	66.3	68.7
EFCNN-Exp($\beta = 1.5$)	66.5	68.0
EFCNN-Exp($\beta = 2.0$)	67.3	69.5

Sensitivity Analysis of Hyperparameters. Since there are two different decay functions, our model has two versions, denoted as EFCNN-Lin and EFCNN-Exp. There are two hyperparameters α and β in EFCNN-Lin and EFCNN-Exp, respectively. We further study the effect of different choices of α and β. The experimental result is listed in Table 2. The results indicate that the best model setting is always encouraging weight decay. Therefore, the consideration of weight decay along with positions in convolution is necessary.

5 Conclusions

In this paper, we study the problem of expert finding. We formalize the problem and propose a deep learning model based on restricted convolutional neural networks. We prove that the proposed model can capture the relevant information between two documents. Compared to several state-of-the-art models, our model can significantly improve the performance of expert finding.

The problem of expert finding represents an interesting and important research direction. In future work, it would be intriguing to investigate a deep architecture to learn expert representations directly. It would also be interesting to study how to incorporate both network information and content information together to better learn the expert representations.

References

1. Balog, K., Azzopardi, L., De Rijke, M.: Formal models for expert finding in enterprise corpora. In: Proceedings of the 29th Annual International ACM SIGIR Conference on Research and Development in Information Retrieval, pp. 43–50. ACM (2006)
2. Basu, C., Hirsh, H., Cohen, W.W., Nevill-Manning, C.: Recommending papers by mining the web (1999)

3. Cao, Y., Liu, J., Bao, S., Li, H.: Research on expert search at enterprise track of TREC 2005. In: TREC (2005)
4. Deng, H., King, I., Lyu, M.R.: Formal models for expert finding on DBLP bibliography data. In: ICDM 2008, pp. 163–172 (2008)
5. Duchi, J.C., Hazan, E., Singer, Y.: Adaptive subgradient methods for online learning and stochastic optimization. J. Mach. Learn. Res. **12**, 2121–2159 (2011)
6. Dumais, S.T., Nielsen, J.: Automating the assignment of submitted manuscripts to reviewers. In: Proceedings of the 15th Annual International ACM SIGIR Conference on Research and Development in Information Retrieval, pp. 233–244. ACM (1992)
7. Hinton, G.E., Srivastava, N., Krizhevsky, A., Sutskever, I., Salakhutdinov, R.R.: Improving neural networks by preventing co-adaptation of feature detectors. arXiv preprint arXiv:1207.0580 (2012)
8. Ioffe, S., Szegedy, C.: Batch normalization: accelerating deep network training by reducing internal covariate shift. arXiv preprint arXiv:1502.03167 (2015)
9. Järvelin, K., Kekäläinen, J.: Cumulated gain-based evaluation of IR techniques. ACM Trans. Inf. Syst. (TOIS) **20**(4), 422–446 (2002)
10. J. Kusner, M., Sun, Y., I.Kolkin, N., Q.Weinberger, K.: From word embeddings to document distances. In: International Conference on Machine Learning, vol. 15, pp. 957–966 (2015)
11. Karimzadehgan, M., Zhai, C., Belford, G.: Multi-aspect expertise matching for review assignment. In: Proceedings of the 17th ACM Conference on Information and Knowledge Management, pp. 1113–1122. ACM (2008)
12. Le, Q., Mikolov, T.: Distributed representations of sentences and documents. In: International Conference on Machine Learning, pp. 1188–1196 (2014)
13. LeCun, Y., Bottou, L., Bengio, Y., Haffner, P.: Gradient-based learning applied to document recognition. Proc. IEEE **86**(11), 2278–2324 (1998)
14. Mikolov, T., Chen, K., Corrado, G., Dean, J.: Efficient estimation of word representations in vector space. arXiv preprint arXiv:1301.3781 (2013)
15. Mimno, D., McCallum, A.: Expertise modeling for matching papers with reviewers. In: Proceedings of the 13th ACM SIGKDD International Conference on Knowledge Discovery and Data Mining, pp. 500–509. ACM (2007)
16. Palangi, H., et al.: Deep sentence embedding using long short-term memory networks: analysis and application to information retrieval. IEEE/ACM Trans. Audio Speech Lang. Process. (TASLP) **24**(4), 694–707 (2016)
17. Pang, L., Lan, Y., Guo, J., Xu, J., Wan, S., Cheng, X.: Text matching as image recognition. In: AAAI, pp. 2793–2799 (2016)
18. Petkova, D., Croft, W.B.: Hierarchical language models for expert finding in enterprise corpora. Int. J. Artif. Intell. Tools **17**(01), 5–18 (2008)
19. Prechelt, L.: Automatic early stopping using cross validation: quantifying the criteria. Neural Netw. **11**(4), 761–767 (1998)
20. Salton, G., Buckley, C.: Term-weighting approaches in automatic text retrieval. Inf. Process. Manag. **24**(5), 513–523 (1988)
21. Tang, J., Zhang, J., Yao, L., Li, J., Zhang, L., Su, Z.: ArnetMiner: extraction and mining of academic social networks. In: Proceedings of the 14th ACM SIGKDD International Conference on Knowledge Discovery and Data Mining, pp. 990–998. ACM (2008)
22. Zhang, J., Tang, J., Liu, L., Li, J.: A mixture model for expert finding. In: PAKDD 2008, pp. 466–478 (2008)

CRESA: A Deep Learning Approach to Competing Risks, Recurrent Event Survival Analysis

Garima Gupta(✉), Vishal Sunder, Ranjitha Prasad, and Gautam Shroff

TCS Research Lab, New Delhi, India
{gupta.garima1,s.vishal3,ranjitha.prasad,gautam.shroff}@tcs.com

Abstract. Survival analysis refers to a gamut of statistical techniques developed to infer the *survival time* from time-to-event data. In particular, we are interested in *recurrent* event survival analysis in the presence of one or more *competing risks* in each recurrent time-step, in order to obtain the probabilistic relationship between the input covariates and the distribution of event times. Since traditional survival analysis techniques suffer from drawbacks due to strong parametric model constraints and constant hazard based assumptions, we propose a modern deep learning based flexible probabilistic framework for cause-specific recurrent survival analysis. In single-risk scenarios, we propose an LSTM-based model where the time-steps represent the recurrent events for each participant whose covariates may be static or time-varying. To cater to multi-risk scenarios, we build on the single-risk LSTM model and introduce a cumulative incidence curve approach to handle the multiple competing risks using a joint distribution over the event times and each of the competing risks over multiple time-steps and term the proposed novel architecture as *CRESA*. We use the concordance index per risk and the maximum absolute error in every time-step as the metrics of performance. We demonstrate a superior predictive performance of the proposed approach (single and multiple risk scenarios) as compared to traditional model-based approaches, and deep learning based approaches for synthetic and state-of-the-art public datasets.

Keywords: Recurrent neural networks · Competing risks · Hazard function · Deep learning · LSTM · Cox models · Frailty

1 Introduction

In the broad field of study of temporal data, survival analysis is a well-known statistical technique for the study of temporal events. Classic applications of survival analysis has been in the field of reliability engineering especially for equipments under stress, where accurately measuring the uncertainty associated with events related to the critical parameters of an individual or equipment is paramount. With the advent of data collection technologies, survival analysis has

Q. Yang et al. (Eds.): PAKDD 2019, LNAI 11440, pp. 108–122, 2019.
https://doi.org/10.1007/978-3-030-16145-3_9

been widely used in the field of medical statistics for patient monitoring, treatment plans etc. Typically in survival analysis, time-to-an-event data is modeled using a parametric probabilistic function of some observed, or partially observed covariates. The fundamental issue with most time-to-event data is the presence of *censored* observations, i.e., the observations pertaining to those participants whose event of interest is not observed as it was the end of observation period or due to participants dropping out in the observation period. Note that neglecting censored data can introduce a bias in the inference process, and hence, analyzing such time-to-event data necessitates significantly different statistical and machine learning techniques. One of the popular naive non-parametric technique for obtaining the empirical estimate of the survival function is by using the Kaplan-Meier (KM) method [11]. The drawback of this technique is that it does not incorporate the available covariates in the data. The most popular semi-parametric technique for survival analysis is the Cox proportional hazards (CPH) model [3], which incorporates the available covariates using an exponential probabilistic framework along with strong assumptions regarding the proportionality of hazards in the underlying stochastic process. Several other semi-parametric models explore specific forms of the underlying stochastic process such as a Weiner or a Markov process [4,15,17].

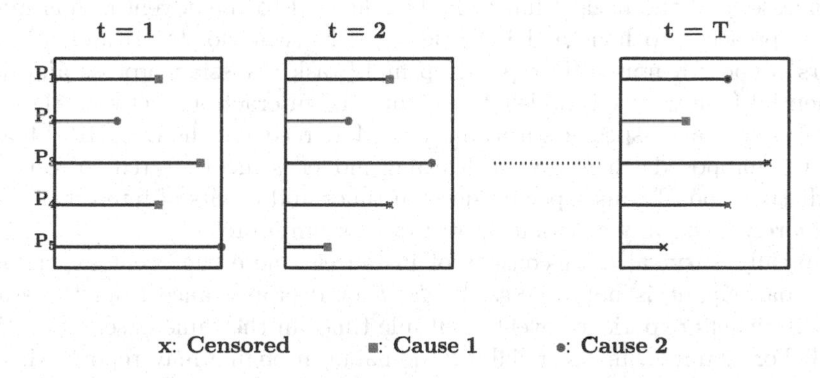

Fig. 1. Typical scenarios encountered in cause-specific recurrent survival analysis depicted on 5 participants, $P_1, \ldots, P_5$. P_1 and P_2 report events in all the time-steps, however the cause of event is not necessarily same in all time-steps. P_3, P_4 and P_5 are censored after a few time-steps.

Recent advances in deep learning have transformed clinical practices especially in the field of machine-learning based diagnostic applications, health-care applications, etc. Although the first paper incorporating neural networks for survival data modeling was two decades ago [5], a renewed interest in survival analysis has emerged since the advent of such modern approaches. Typically, a deep learning architecture is employed to learn the parameters of a general CPH model. For instance, [27] replaces the exponential component of a CPH model with a deep convolutional network. In [12] and [19], the authors train the neural network based on a loss function pertaining to a CPH model to learn the first hitting-time of the event of interest. Further, in [16], the authors propose an LSTM-based feature extractor in conjunction with a conventional CPH model to predict assets' health. In spite of incorporating powerful deep learning based techniques, such approaches necessitate the assumption of a constant hazard rate.

Often, survival analysis data comprises of *competing risks*, where there are at least two possible causes for an event, but only one such event type can actually occur at the time of the event. A typical *cause-specific* approach for analyzing such data is to infer for each event type separately, considering the data reporting alternate events as censored observations. However such approaches may be biased and inaccurate since these competing risks may not be independent. Furthermore, fitting the KM-based survival curve obtained for each event type is also inaccurate [13]. An alternative to the cause censoring approach is the *Cumulative Incidence Curve (CIC)* approach, which estimates the marginal probability of an event. In [7] and [6], the authors model the hazard function by using the cumulative incidence function (CIF). In [20], the authors proposed to fit a single CPH model rather than separate models for each event-type, thus eliminating the need for censoring different event causes. However, all the above models make strong assumptions about the structure of the underlying stochastic processes and the hazard function. This has led to the advent of non-model based approaches to handle the competing risks scenario. In [1] and [22], the authors propose a non-CPH type, deep multi-task Gaussian process and deep exponential family based models to capture the interactions between the data covariates and cause-specific survival times. More recently, the DeepHit [14] approach was proposed where a deep learning model is incorporated to learn the joint distribution of cause-specific survival times and events of interest directly, without relying on any constant-hazard rate assumption.

Typically, survival data consists of instances where the event experienced by the participant is not necessarily *death* or disappearance from the study, and participants experience events multiple times in the same observation time period. For example, in the reliability domain, an engine may report a failure due to valve-related issues, and another failure due to piston-related issues in an observation window, as depicted in Fig. 1. Such a scenario where, for a given participant, cause-specific events are reported more than once are referred to as *recurrent events with competing risks* [13]. KM-type non-parametric estimators have been proposed for non-cause specific recurrent event survival analysis [24]. Typically, to handle scenarios where the recurring events for each participant are identical, the semi-parametric counting process algorithm [2] has been employed, where multiple time-intervals are considered as independent and a semi-parametric stratified Cox based approach has been used [13,25,26]. Note that very few papers handle scenarios which consider cause-specific recurring events in survival analysis. Multistage models with competing risks have been known to handle such scenarios with limited success [21].

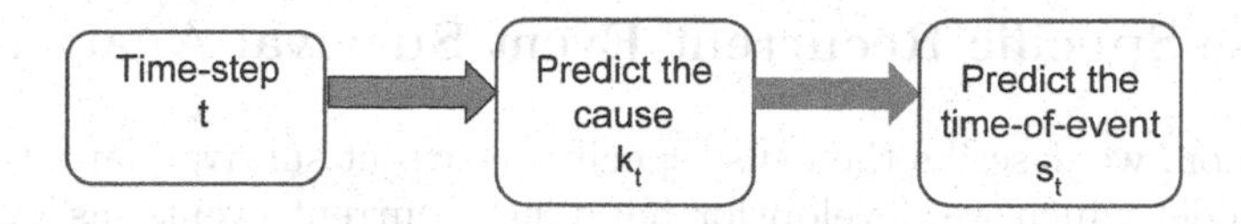

Fig. 2. Prediction hierarchy: at each time-step t, predict the cause of the event (k_t) and the cause-specific time-to-event(s_t) for $t = 1, \dots T$.

Contributions: In this work, we propose a general deep learning architecture which we call as *CRESA* for handling recurrence in the context of cause-specific survival analysis. Our approach consists of designing a deep neural network that directly learns the distributions of time-to-event and competing events. Our proposed technique does not suffer from the drawbacks related to conventional CPH models since we do not assume any underlying stochastic process or proportionality of hazards. Our loss-function can be split into two components, where the first component caters to the log-likelihood of the cause-specific recurrent event survival data, taking into account the censoring of the data and the second component considers the cause-specific ranking loss based on concordance [14]. The features of CRESA are as follows:

- In scenarios consisting of a single-risk ($C = 1$), CRESA is an LSTM-based deep neural network for recurrent event prediction for T time-steps, for scenarios where the covariates remain static or vary across time steps.
- In scenarios consisting of multiple risks ($C > 1$), CRESA is an LSTM-based deep neural network based architecture consisting of C cause specific subnetworks in order to handle C competing risks at each time-step. Hence, our neural network architecture is capable of predicting the time of the event and the cause of event at every time-step, as depicted in Fig. 2.

To the best of authors' knowledge, this is the first instance where recurrent neural network based cause-specific models are proposed in the context of survival analysis. Since no assumptions are made on the underlying time-varying stochastic process governing the covariates and hitting times, this is one of the most flexible frameworks that handles most of the complex events that occur in a typical survival-analysis scenario. We use the concordance index (CI) per risk, and mean absolute error (MAE) at every time-step as the metric of performance. We demonstrate the efficacy of the proposed approach by comparing its predictive performance with traditional CPH model, Random survival forest (RSF) [9] and frailty based recurrent approaches [23]. We also compare the proposed approach with the more recent deep learning based DeepHit [14] and DeepSurv [12] approaches. We test the different approaches on two real-life datasets and a synthetic dataset and demonstrate that the proposed technique performs better than traditional and modern approaches.

2 Cause-Specific Recurrent Event Survival Analysis

In this section, we describe the cause-specific recurrent survival data, the neural network model which we develop for handling recurrent events, as well as the loss function that we use to train and test the model.

2.1 Survival Data

Typically, survival data for each participant is characterized by the individuals' covariates, time of event and censoring information. However, in the context of cause-specific recurrent event survival data, each participant is characterized by the following:

- Observed covariates in the given time-step.
- Time elapsed since the previous event or start of time.
- A label indicating the type of event in the given time-step.
- A label indicating the cause of the event in the given time-step.

We assume that the survival time is discrete and finite, i.e., the survival time takes values in a set $\mathcal{M} = \{1, \ldots, M\}$, defined for a maximum time horizon M. We assume that a participant may experience more than one type of competing risk, such that the corresponding label takes values from $\mathcal{C} = \{1, \ldots, C\}$. The fundamental difference between other temporal data and recurrent event survival data is the presence of *censored observations* in some or all time-steps. For example, in the context of reliability data, it is possible that a failure (considered as an event) occurs twice in the observation time period, leading to two same or different recurrent events experienced by the same participant. However, the participant may still be in the observation time period and report no further failures after the second time-step, which implies that the instance is censored beyond the second time-step. Handling such time-dependent and cause-specific censored data is a crucial aspect of this work.

Each instance is therefore a triple $(\mathbf{x}_t, s_t, d_t, k_t)$ where $\mathbf{x}_t \in \mathbf{X}_t$, denotes the set of covariates in the time-step t. Here, $s_t \in \mathcal{M}$ the time at which the event has occurred, $d_t \in \{0, 1\}$ denotes whether the i-th participant is censored or not, and $k_t \in \mathcal{C}$ denotes the label on the cause due to which an event is reported at time step t. We are given a dataset $D = (\mathbf{x}_t^{(i)}, s_t^{(i)}, d_t^{(i)}, k_t^{(i)})$, where $i = 1, \ldots, N$ denotes the number of participants. In the sequel, we describe the neural network model and the loss function we use to train the model using the above described dataset.

2.2 Model Description

Conventional approaches to survival analysis such as KM plots and the vanilla Cox models fail to provide meaningful insights into the cause-specific survival analysis based prediction tasks. This has led to alternative approaches such as the CIC approach which uses the marginal probabilities of an event in the presence of competing events, and does not require the assumption that these risks are independent. The corresponding cause-specific cumulative incidence function is given by [6]

$$F_k(s^*|\mathbf{x}^*) = P(S \leq s^*, K = k^*|\mathbf{X} = \mathbf{x}^*) = \sum_{s=0}^{s^*} P(S = s, K = k^*|\mathbf{X} = \mathbf{x}^*), \quad (1)$$

i.e., the CIF gives the probability that a particular cause k^* occurs on or before time s^* given the covariates $\mathbf{x}^*$. However, the scenarios that we consider in this paper involves predicting the *recurrent* hitting times, and hence we define the *recurrent* CIF (RCIF), which is the probability that a given event occurs on or before time s_t^* for different time steps t, given the time-dependent covariates $\mathbf{x}_t^*$, given as

$$F_k(s_t^*|\mathbf{x}_t^*) = P(S_t \leq s_t^*, K_t = k_t^*|S_{t-1} = s_{t-1}^*, \ldots, S_1 = s_1^*, K_{t-1} = k_{t-1}^*, \ldots, K_1 = k_1^*, \mathbf{X}_t = \mathbf{x}_t^*)$$
$$= \sum_{s_t=0}^{s_t^*} P(S_t = s_t, K_t = k_t^*|S_{t-1} = s_{t-1}^*, \ldots, S_1 = s_1^*, K_{t-1} = k_{t-1}^*, \ldots, K_1 = k_1^*, \mathbf{X}_t = \mathbf{x}_t^*). \quad (2)$$

Here, the function $P(X = x)$ represents the pdf of the random variable X evaluated at its realization x. Note that the true RCIF is not known, and it is not possible to directly compute it from the dataset. Hence, in order to quantitatively measure how models discriminate across cause-specific risks among participants, it is essential to design the neural network model that computes an approximate RCIF, which is subsequently incorporated into the loss function.

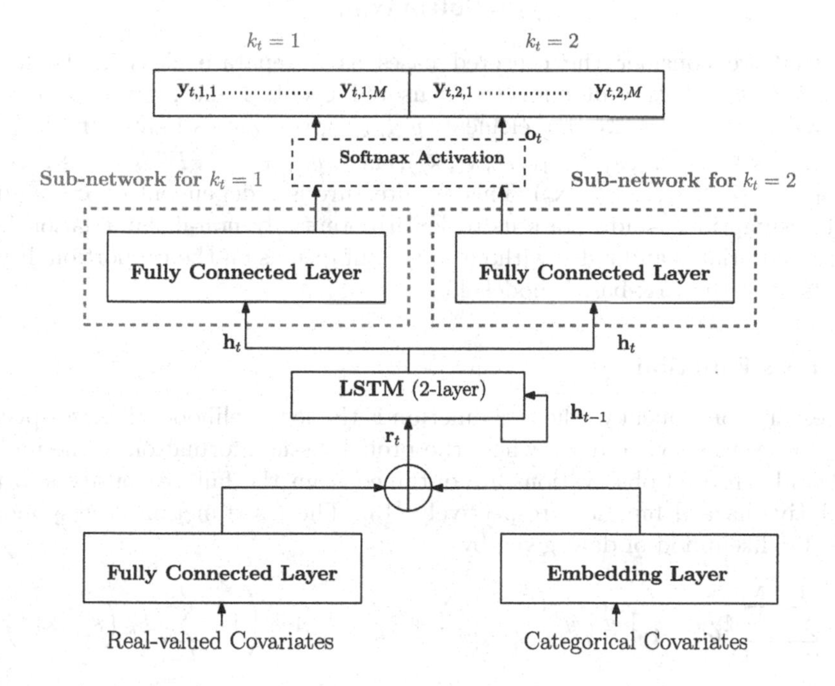

Fig. 3. CRESA consists of LSTM layer, cause-specific sub-networks and a single softmax activation layer.

Accordingly, in order to compute the estimates of the RCIF, we propose the following model, as illustrated in Fig. 3. We encode all real-valued covariates, ${\mathbf{x}_t}^{\text{real}}$ using a single fully connected layer with ReLU activation, denoted as MLP$(\cdot)$, and the categorical covariates, ${\mathbf{x}_t}^{\text{cat}}$ using an embedding lookup, denoted as Lookup$(\cdot)$. Using the concatenation operation denoted as $\bigoplus$, the dense representation, $\mathbf{r}_t$ of the covariates is computed as follows:

$$\mathbf{r}_t = \text{MLP}\left({\mathbf{x}_t}^{\text{real}}\right) \bigoplus \text{Lookup}\left({\mathbf{x}_t}^{\text{cat}}\right). \tag{3}$$

At each time step t, $\mathbf{r}_t$ is given as an input to the 2-layer LSTM [8] in order to obtain the hidden representation $\mathbf{h}_t \in \mathbb{R}^n$, where n is the number of hidden units in the LSTM, as follows:

$$\mathbf{h}_t = \text{LSTM}(\mathbf{r}_t, \mathbf{h}_{t-1}). \tag{4}$$

The hidden state $\mathbf{h}_t$ is used to obtain C cause-specific representations by incorporating C single layer MLPs, represented as MLP$_k(\cdot)$. For brevity and purposes of this work we limit ourselves to cases where $C = 2$ but the model is easily generalizable. Thus, we obtain a final representation $\mathbf{o}_t$ as

$$\mathbf{o}_t = \text{MLP}_1(\mathbf{h}_t) \bigoplus \text{MLP}_2(\mathbf{h}_t). \tag{5}$$

Finally, $\mathbf{o}_t$ undergoes a softmax transform, which results in the final probability distribution function $\mathbf{y}_t = [y_{t,1,1}, ..., y_{t,1,M}, y_{t,2,1}, ..., y_{t,2,M}]$ is given by

$$\mathbf{y}_t = \mathbf{Softmax}(\mathbf{o}_t). \tag{6}$$

Note that we consider the censored cases as a separate class in itself, i.e, the last class M accounts for the censored event. Hence, given a participant with covariates $\mathbf{x}_t$, an element y_{t,s_t,k_t} gives an estimate of $P(S_t = s_t, K_t = k_t | s_{t-1}, \ldots, s_1, k_{t-1}, \ldots, k_1, \mathbf{x}_t)$, i.e., $y_{t,s_t,k_t} = \hat{P}(S_t = s_t, K_t = k_t | s_{t-1}, \ldots, s_1, k_{t-1}, \ldots, k_1, \mathbf{x}_t)$. This architecture is independent of any model-based assumptions, and allows us to learn potentially non-linear relationships between covariates and risks, without any assumptions on the proportionality of hazards as in the Cox-based models [3].

2.3 Loss Function

A necessary component of the loss function is the log-likelihood of cause-specific recurrent event survival data, where the probabilistic information of the uncensored and censored observations are obtained from the failure density and the cumulative hazard function, respectively [13]. The loss function corresponding to the log-likelihood of data given by:

$$\mathcal{L}_1 = \sum_{t=1}^{T} \sum_{i=1}^{N} \mathbb{1}_{(d_t^{(i)}=1)} \log\left(y^{(i)}_{t^{(i)}, s_t^{(i)}, k_t^{(i)}}\right) + \mathbb{1}_{(d_t^{(i)}=0)} \log\left(1 - \sum_{k=1}^{K} \hat{F}_k\left(s_t^{(i)} | \mathbf{x}_t^{(i)}\right)\right). \tag{7}$$

Here $\hat{F}_k\left(s_t^{(i)}|\mathbf{x}_t^{(i)}\right)$ represents the estimate of the risk, computed using (2) by substituting the true RCIF by their estimates. In addition to the log-likelihood, it has been observed that it is essential to penalize the cost function based on relative risks of uncensored participants. The central idea is from concordance [14] which states that at a given time-step t, a participant that experiences a time-to-event s_t should have a higher risk at s_t than a patient who survived longer than s_t. The risk-based penalty denoted by $\mathcal{L}_2$, is as follows:

$$\mathcal{L}_2 = \sum_{t=1}^{T} \alpha_{k,t} \sum_{i \neq j} A_t^{(k,i,j)} \eta\left[\hat{F}_k\left(s_t^{(i)}|\mathbf{x}_t^{(i)}\right), \hat{F}_k\left(s_t^{(i)}|\mathbf{x}_t^{(j)}\right)\right], \tag{8}$$

where $\alpha_{k,t}$ is a parameter that trades off between the log-likelihood and the concordance based loss for the k-th cause in the t-th time-step. Further, $\eta[x, y]$ is any convex loss function. Note that, $A_t^{(k,i,j)}$ indicates if participant i and j are both uncensored, with i experiencing the cause specific risk k, and hence can be compared for relative risks in a given time-step, i.e.,

$$A_t^{(k,i,j)} = \mathbb{1}\left(k_t^{(i)} = k_t, s_t^{(i)} < s_t^{(j)}\right). \tag{9}$$

Hence, the overall cost function is given by $\mathcal{L} = \mathcal{L}_1 + \mathcal{L}_2$. Time complexity of $\mathcal{L}_1$ is a linear function of N while time complexity for $\mathcal{L}_2$ is $\mathcal{O}(N^2T)$.

3 Experiments

In this section, we first describe the datasets we employed, followed by a description of the baselines we used for comparison. We then elaborate on the training details followed by the results and discussion. We provide results for both single-risk and multi-risk cases.

3.1 Dataset I: MIMIC III Clinical Dataset

MIMIC-III (Medical Information Mart for Intensive Care) is a large, freely available clinical dataset developed by the MIT Lab for Computational Physiology [10]. It comprises of more than 40,000 patient instances with information relating to patients admitted to critical care units at a hospital. This dataset contains multiple instances of the same patient being admitted due to a cause-specific risk, and discharged from ICU. We are interested in predicting the length of stay at ICU for recurrent admissions due to same or different competing risk. In a survival framework, we consider discharge of a patient from the ICU as the event of interest, and the observations are censored in the event of patients' death. We consider time-varying real-valued features ($\mathbf{x}_t{}^{\text{real}}$) such as blood pressure, glucose, heart rate, oxygen saturation, respiratory rate, temperature, pH, and categorical features ($\mathbf{x}_t{}^{\text{cat}}$) such as weight, height, ethnicity, gender, Glascow coma scale eye opening, Glascow coma scale motor response, Glascow coma scale verbal response as input covariates for LSTM. We use this dataset for single-risk and multi-risk survival recurrent analysis.

Single-risk Dataset: Single-risk dataset consists of 13021 instances of patients who are admitted into ICU recurrently, due to a single risk. These patients are admitted to the ICU recurrently for a maximum of $T = 5$ number of time-steps, and the time-to-event is one of $M = 101$ target classes.

Multi-risk Dataset: We consider $C = 2$ competing risks for predicting $M = 43$ number of classes (time-of-event occurrences) for each risk, recurring for $T = 5$ number of time steps.

3.2 Dataset II: Engine Failures Dataset

The engine failures dataset is a proprietary dataset which consists of instances of engines whose failures and months in service have been recorded for insurance related investigations. In this dataset, engine failure refers to the failure of a specific part of the engine, and the engine does not stop running after a failure. Cause-specific recurrent failures of the different parts have been reported which allows us to perform multi and single-risk, recurrent event survival analysis on this dataset.

We use real-valued features ($\mathbf{x}_t{}^{\text{real}}$) such as the number of miles covered, time duration since the build of engine and time since the engine is in-service, and categorical features ($\mathbf{x}_t{}^{\text{cat}}$) such as engine type, horse power, rotations per minute, application of engine, design configuration, equipment manufacturer, model name, etc. For this dataset the covariates do not change across time steps, and we expect that the hidden state of the LSTM will be the key differentiator while making predictions for different time steps.

Single-risk Dataset: Single-risk dataset consists of 18221 instances of engines which fail recurrently due to single risk over time. These engines experience a maximum of $T = 3$ recurrences (time-steps) of the risk, and their time-to-event is one of $M = 27$ classes.

Multi-risk Dataset: Multi-risk dataset consists of 31948 instances of engines with $C = 2$. The number of time steps is chosen as $T = 3$ and $M = 27$ represents the number of classes per part failure.

Engines which report failure for less than $T = 3$ are considered as censored for remaining time-steps.

3.3 Dataset III: Synthetic Dataset

We create a synthetic dataset with $C = 2$, $T = 7$, and $M = 7$ per risk, in order to demonstrate the time-tracking efficacy of the proposed framework. We constructed two stochastic processes over T time-steps with parameters and the hitting times inspired by the synthetic dataset proposed in [14]. The covariates for the participant i at time-step t are sampled as $\mathbf{x}_{1,\,t}^{(i)}, \mathbf{x}_{2,\,t}^{(i)}, \mathbf{x}_{3,\,t}^{(i)} \sim \mathcal{N}(0, \mathbf{I}_4)$. Further, the hitting times for a given time-step t are obtained as

$$s_{1,\,t}^{(i)} \sim \exp((\gamma_3^T \mathbf{x}_{3,\,t}^{(i)})^2 + (\gamma_1^T \mathbf{x}_{1,\,t}^{(i)})), \quad s_{2,\,t}^{(i)} \sim \exp((\gamma_3^T \mathbf{x}_{3,\,t}^{(i)})^2 + (\gamma_2^T \mathbf{x}_{2,\,t}^{(i)})). \quad (10)$$

In order to account for the recurrence of events, the hitting-times are varied over the time-steps using an autoregressive model given by

$$s_{k,\,t}^{(i)} = \rho_k s_{k,(t-1)}^{(i)} + \sqrt{1-\rho_k^2}\, z_{k,\,t}^{(i)}, \tag{11}$$

where $z_{k,t}^{(i)} \sim \mathcal{N}(0, \mathbf{I})$ and the correlation co-efficient is chosen as $\rho_1 = \rho_2 = 0.6$. For convenience, we set $\gamma_1 = \gamma_2 = \gamma_3 = 0.4$, $s_t^{(i)} = \min[s_{1,\,t}^{(i)}, s_{2,\,t}^{(i)}]$ and participants with $s_t^{(i)} \geq 21$ are considered as censored events.

All datasets are divided into training, validation and test sets in the ratio 3 : 1 : 1. Statistics of censored and uncensored participants across time steps is mentioned in Table 1.

Table 1. Number of censored and uncensored participants across time steps for each competing risks in 3 datasets

			t_1	t_2	t_3	t_4	t_5	t_6	t_7
Synthetic	Risk-1	Censored	597	1093	1562	2000	2452	2914	3280
		Uncensored	24322	23874	23443	23000	22508	22199	21635
	Risk-2	Censored	639	1082	1525	1999	2440	2850	3282
		Uncensored	24442	23951	23470	23001	22600	22037	21803
MIMIC	Risk-1	Censored	0	297	371	389	402	-	-
		Uncensored	2259	2045	1986	1973	1962	-	-
	Risk-2	Censored	0	888	970	989	990	-	-
		Uncensored	9408	1016	224	87	36	-	-
Engine	Risk-1	Censored	0	11455	14321	-	-	-	-
		Uncensored	15376	3443	887	-	-	-	-
	Risk-2	Censored	0	13099	15622	-	-	-	-
		Uncensored	16572	3951	1118	-	-	-	-

3.4 Training Details

The model described in Sect. 2.2 is trained via backpropagation using an SGD optimizer with momentum and Nesterov. We also use cosine annealing of the learning rate with warm restarts as in [18] using $T_0 = 1$ and $T_{mul} = 2$, with minimum and maximum learning rates of 0.005 and 0.5 respectively. Embedding sizes and all neural network hidden states have dimension 64 and dropout is used for regularization. We consider time independent value for $\alpha_{k,t}$, i.e., $\alpha_k = \alpha_{k,t}$, for $t = 1, \ldots, T$. The values of α_1, α_2 and dropout are determined using grid-search. Similar to [14], for $\mathcal{L}_2$, we use $\eta[x, y] = \exp(-(x-y)/\sigma)$ is where σ is set to 1.

3.5 Baselines

We use the following baselines for the **single-risk** scenario:

1. **DeepSurv**: DeepSurv model[1] is proposed in [12]. Post training, we run this model separately for T time steps and report results for each.
2. **DeepHit**: We simulate T DeepHit units [14] with $C = 1$. The hyperparameter setting is the same as in [14].
3. **Shared Frailty model**: We fit a shared Gamma frailty model [13,23] and predict the marginal probability for the T time steps for each participant.

We use the following baselines for the **multiple risk** scenario:

1. **CRESA**$_{\alpha}$ ($\alpha_1 = 0$, $\alpha_2 = 0$): We fix the values of α_1 and α_2 as 0 for our proposed model and compare results to investigate the effect that $\mathcal{L}_2$ has on the results.
2. **DeepHit**: We simulate T DeepHit units [14] to obtain different distributions for T time steps, with $C > 1$. The hyperparameter setting is the same as in [14].
3. **Random Survival Forest (RSF)**: RSF [9] is trained for all time-steps using competing risks data, and post-training it is tested separately on each time-step.

3.6 Performance Metrics and Results

In this subsection we describe the performance metric used to evaluate the neural network model proposed in Sect. 2.2. We use concordance index per time-step as one of the performance metrics given by:

$$\begin{aligned} CI_t &= P\left(\hat{F}\left(s_t^{(i)}|\mathbf{x}_t^{(i)}\right) > \hat{F}\left(s_t^{(i)}|\mathbf{x}_t^{(j)}\right)|s_t^{(i)} < s_t^{(j)}\right) \\ &\approx \frac{\sum_{i\neq j} A_t^{(k,i,j)} \mathbb{1}\left(\hat{F}\left(s_t^{(i)}|\mathbf{x}_t^{(i)}\right) > \hat{F}\left(s_t^{(i)}|\mathbf{x}_t^{(j)}\right)\right)}{\sum_{i\neq j} A_t^{(k,i,j)}} \end{aligned} \tag{12}$$

The other performance metric that we use is MAE, which is defined as follows:

$$MAE_t = \frac{1}{N}\sum_{i=1}^{N} d_t^{(i)}|y_t^{(i)} - s_t^{(i)}|, \tag{13}$$

where $d_t^{(i)}$ is zero if instance i is censored, and hence, the absolute error between the predicted label class is compared to the true label class only for instances which experience events.

We perform comprehensive evaluation of CRESA, and compare it with deep and non-deep baselines mentioned in the previous subsection using concordance index per risk per time-step, and MAE per time-step as performance metrics.

[1] https://github.com/jaredleekatzman/DeepSurv.

Table 2. Concordance indices for single-risk, recurrent survival analysis

Model used	CI_1	CI_2	CI_3	CI_4	CI_5
MIMIC III clinical dataset					
CRESA	0.671	**0.910**	**0.896**	**0.878**	**0.866**
DeepHit	**0.678**	0.808	0.723	0.665	0.633
DeepSurv	0.586	0.775	0.715	0.681	0.660
Shared frailty	0.3386	0.489	0.534	0.5105	0.521
Engine failures dataset					
CRESA	**0.792**	**0.931**	**0.959**	-	-
DeepHit	0.765	0.759	0.849	-	-
DeepSurv	0.509	0.503	0.560	-	-
Shared frailty	0.459	0.482	0.496	-	-

The concordance index results for the single-risk CRESA are given in Table 2. For the MIMIC III dataset, we observe that DeepHit does slightly better than CRESA for the first time-step. For the subsequent time-steps, the CRESA outperforms both baselines by a huge margin. However, CRESA does better for all the time-steps in the case of engine failures dataset. The results clearly show that the temporal information is an important aspect for recurrent event survival analysis which is captured in our model by the LSTM, and hence, the LSTM-based CRESA model performs better than DeepHit which, by its nature, does not exploit any temporal information. The performance of DeepHit and CRESA is similar for the first time-step since CRESA does not have access to any time dependent *hidden state*. As expected, the proposed approach performs better as compared to traditional, model-based shared frailty approach by a huge margin in all the time-steps.

The concordance index results for the multi-risk CRESA are given in Table 3. First, we note that CRESA performs better compared to CRESA_α across datasets, from which we infer that $\mathcal{L}_2$ indeed has an impact on the final values of concordance index. Even in the presence of multiple risks, CRESA performs better compared to the baseline schemes such as DeepHit and RSF except in the first time-step and hence, CRESA continues to gain from the backbone LSTM architecture in the time-steps other than the first time-step, as in the single-risk scenario. From the results pertaining to synthetic dataset, we infer that the advantages that we obtain by using LSTM based architecture carries over to several time-steps.

In Table 4, we present the MAE performance of CRESA as compared to DeepHit. First, we notice that the MAE performance of both the schemes are poor in the first time-step as compared to the subsequent time-steps, in the case of real-life datasets. However, the MAE results are uniform across all time-steps for the synthetic dataset. Note that real-life datasets have large number of classes as compared to the synthetic counterpart. In the first time-step, the true label

Table 3. Concordance indices for multi-risk, recurrent survival analysis

	Risk-1							Risk-2						
	CI_1	CI_2	CI_3	CI_4	CI_5	CI_6	CI_7	CI_1	CI_2	CI_3	CI_4	CI_5	CI_6	CI_7
MIMIC III														
CRESA	0.680	**0.844**	**0.779**	**0.759**	**0.735**	-	-	0.646	**0.855**	**0.802**	**0.780**	**0.768**	-	-
CRESA$_\alpha$	**0.704**	0.761	0.667	0.631	0.602	-	-	0.650	0.778	0.684	0.654	0.632	-	-
DeepHit	0.658	0.686	0.620	0.595	0.578	-	-	0.629	0.773	0.673	0.643	0.621	-	-
RSF	0.474	0.494	0.518	0.521	0.546	-	-	**0.77**	0.74	0.77	0.77	0.746	-	-
Engine failures dataset														
CRESA	0.801	**0.874**	**0.907**	-	-	-	-	**0.864**	**0.912**	0.909	-	-	-	-
CRESA$_\alpha$	0.643	0.692	0.633	-	-	-	-	0.750	0.735	0.650	-	-	-	-
DeepHit	0.726	0.827	0.890	-	-	-	-	0.813	0.882	**0.944**	-	-	-	-
RSF	**0.81**	0.77	0.78	-	-	-	-	0.81	0.792	0.797	-	-	-	-
Synthetic dataset														
CRESA	**0.762**	**0.754**	0.747	**0.729**	**0.728**	0.715	**0.715**	**0.774**	0.759	**0.737**	**0.736**	**0.73**	**0.715**	**0.72**
CRESA$_\alpha$	0.704	0.698	**0.751**	0.720	0.712	**0.736**	0.701	0.761	**0.761**	0.710	0.713	0.723	0.712	0.69
DeepHit	0.579	0.583	0.568	0.582	0.562	0.554	0.558	0.587	0.581	0.571	0.563	0.56	0.554	0.56
RSF	0.509	0.495	0.517	0.502	0.509	0.498	0.512	0.514	0.517	0.505	0.497	0.513	0.51	0.51

Table 4. Mean absolute error (MAE) for multi-risk, recurrent survival analysis

Model used	MAE_1	MAE_2	MAE_3	MAE_4	MAE_5	MAE_6	MAE_7
MIMIC III clinical dataset							
CRESA	**0.7152**	**0.3475**	**0.3049**	**0.307**	**0.3042**	-	-
DeepHit	0.734	0.3544	0.3136	0.3099	0.3066	-	-
Engine failures dataset							
CRESA	**0.8998**	**0.441**	**0.3526**	-	-	-	-
DeepHit	0.989	0.5275	0.4161	-	-	-	-
Synthetic dataset							
CRESA	**0.5833**	**0.5987**	**0.6078**	**0.6146**	**0.6396**	**0.6503**	**0.658**
DeepHit	0.6598	0.6708	0.6832	0.6927	0.7141	0.718	0.723

distribution across the classes are close to being uniform in the real-life datasets, and hence, CRESA faces difficulty in learning the true distribution. However, in the subsequent time-steps, this distribution across the classes is skewed, and hence learning becomes easier. We suspect that a larger dataset and a more complex model will improve the MAE results. Although synthetic datasets have fewer classes, the true label distribution is uniform across time-steps, and hence, uniform trend the MAE results is observed over all the time-steps.

4 Conclusions

In this paper, we proposed CRESA, a novel deep learning architecture for competing risk, recurrent event survival analysis. CRESA employed an LSTM based backbone recurrent neural network to estimate the RCIF, which is the joint distribution of the time-to-event and competing risks, as a function of the covariates in the data. For the single-risk scenario, the CRESA architecture reduces to an LSTM-based deep neural network for recurrent event prediction for T time-steps. We used a loss function that exploited the CIC-based probabilistic information from uncensored and right-censored participants and also penalized incorrect ordering of relative risks in every time-step. We compared the performance of CRESA with the performance of traditional approaches such as frailty-based CPH model and RSF, and modern deep learning approaches such as DeepHit and DeepSurv. We demonstrated that CRESA has a superior performance in terms of both, concordance index and MAE. We also noted that MAE has a strong dependence on the nature of the dataset, such as the number of classes and the distribution of true labels across the classes. Overall, CRESA lends itself as a flexible, non-model based approach to survival analysis for datasets that involve complex events such as recurrence of events and competing risks. In future, we would extend architecture and loss function of CRESA to handle interval and left censoring events as well.

References

1. Alaa, A.M., van der Schaar, M.: Deep multi-task Gaussian processes for survival analysis with competing risks. In: 30th Conference on Neural Information Processing Systems (2017)
2. Andersen, P.K., Gill, R.D.: Cox's regression model for counting processes: a large sample study. Ann. Stat. **10**, 1100–1120 (1982)
3. Cox, D.R.: Analysis of Survival Data. Routledge, London (2018)
4. Doksum, K.A., Hbyland, A.: Models for variable-stress accelerated life testing experiments based on wener processes and the inverse gaussian distribution. Technometrics **34**(1), 74–82 (1992)
5. Faraggi, D., Simon, R.: A neural network model for survival data. Stat. Med. **14**(1), 73–82 (1995)
6. Fine, J.P., Gray, R.J.: A proportional hazards model for the subdistribution of a competing risk. J. Am. Stat. Assoc. **94**(446), 496–509 (1999)
7. Gray, R.J.: A class of K-sample tests for comparing the cumulative incidence of a competing risk. Ann. Stat. **16**, 1141–1154 (1988)
8. Hochreiter, S., Schmidhuber, J.: Long short-term memory. Neural Comput. **9**(8), 1735–1780 (1997)
9. Ishwaran, H., Kogalur, U.B., Blackstone, E.H., Lauer, M.S., et al.: Random survival forests. Ann. Appl. Stat. **2**(3), 841–860 (2008)
10. Johnson, A.E., et al.: MIMIC-III, a freely accessible critical care database. Sci. Data **3**, 160035 (2016)
11. Kaplan, E.L., Meier, P.: Nonparametric estimation from incomplete observations. J. Am. Stat. Assoc. **53**, 457–481 (1958)
12. Katzman, J.L., Shaham, U., Cloninger, A., et al.: DeepSurv: personalized treatment recommender system using a cox proportional hazards deep neural network. BMC Med. Res. Methodol. **18**(1), 24 (2018)
13. Kleinbaum, D.G., Klein, M.: Survival Analysis, vol. 3. Springer, New York (2010). https://doi.org/10.1007/978-1-4419-6646-9

14. Lee, C., Zame, W.R., Yoon, J., van der Schaar, M.: Deephit: a deep learning approach to survival analysis with competing risks (2018)
15. Lee, M.L.T., Whitmore, G.: Proportional hazards and threshold regression: their theoretical and practical connections. Lifetime Data Anal. **16**(2), 196–214 (2010)
16. Liao, L., Ahn, H.i.: Combining deep learning and survival analysis for asset health management. Int. J. Prognostics Health Manage. (2016)
17. Longini, I.M., Clark, W.S., Byers, R.H., Ward, J.W., Darrow, W.W., et al.: Statistical analysis of the stages of HIV infection using a Markov model. Stat. Med. **8**, 831–843 (1989)
18. Loshchilov, I., Hutter, F.: SGDR: stochastic gradient descent with warm restarts (2016)
19. Luck, M., Sylvain, T., Cardinal, H., Lodi, A., Bengio, Y.: Deep learning for patient-specific kidney graft survival analysis. arXiv preprint arXiv:1705.10245 (2017)
20. Lunn, M., McNeil, D.: Applying cox regression to competing risks. Biometrics **51**, 524–532 (1995)
21. Meira-Machado, L., de Uña-Álvarez, J., Cadarso-Suárez, C., Andersen, P.K.: Multi-state models for the analysis of time-to-event data. Stat. Methods Med. Res. **18**(2), 195–222 (2009)
22. Ranganath, R., Perotte, A., Elhadad, N., Blei, D.: Deep survival analysis. arXiv preprint arXiv:1608.02158 (2016)
23. Rondeau, V., Mazroui, Y., Gonzalez, J.R.: Frailtypack: an R package for the analysis of correlated survival data with frailty models using penalized likelihood estimation or parametrical estimation. J. Stat. Softw. **47**(4), 1–28 (2012)
24. Wang, M.C., Chang, S.H.: Nonparametric estimation of a recurrent survival function. J. Am. Stat. Assoc. **94**(445), 146–153 (1999)
25. Wang, P., Li, Y., Reddy, C.K.: Machine learning for survival analysis: a survey. arXiv preprint arXiv:1708.04649 (2017)
26. Wei, L.J., Lin, D.Y., Weissfeld, L.: Regression analysis of multivariate incomplete failure time data by modeling marginal distributions. J. Am. Stat. Assoc. **84**, 1065–1073 (1989)
27. Zhu, X., Yao, J., Huang, J.: Deep convolutional neural network for survival analysis with pathological images. In: Bioinformatics and Biomedicine (BIBM), pp. 544–547. IEEE (2016)

Long-Term Traffic Time Prediction Using Deep Learning with Integration of Weather Effect

Chih-Hsin Chou, Yu Huang, Chian-Yun Huang, and Vincent S. Tseng(✉)

Department of Computer Science, National Chiao Tung University, Hsinchu, Taiwan, Republic of China
vtseng@cs.nctu.edu.tw

Abstract. Traffic time prediction is a classical problem in intelligent transportation domain, which has attracted lots of attention from the research community in last three decades. The existing relevant works have been focused on how to predict the short-term traffic time for paths and roads. In fact, users may have the demand to know the future traffic time in advance as for making personal or commercial schedule. Long-term traffic time prediction is thus an emerging challenging task as there exist many complicated factors that may affect traffic situations, such as weather and congestion conditions. In this paper, we propose a novel deep learning-based framework named Deep Ensemble Stacked Long Short Term Memory (DE-SLSTM), which aims to solve the prediction bias during traffic congestion. To improve the model performance, we integrate the weather effect into the DE-SLSTM for predicting the long-term traffic time. Through a series of experiments, the proposed DE-SLSTM framework is verified to demonstrate excellent performance in terms of effectiveness. To the best of our knowledge, this is the first work on long-term traffic time prediction that considers deep learning techniques.

Keywords: Long-term prediction · Traffic time prediction · Weather effect · Ensemble model

1 Introduction

In the era of population explosion, the number of vehicles around the world increases and the traffic congestion happens more frequently in urban cities. With advanced technology, government is able to collect real-time dynamic traffic data, such as traffic speed and traffic time, and implement policies to alleviate traffic. Hence, a well-designed system called intelligent transportation system (ITS) containing these information has been developed. However, for the most drivers, arriving at the destinations on time is the most important demand. The ability to predict a credible future traffic time is helpful for drivers to plan the trips conveniently in advance. Besides, it also brings benefits for logistics services and provides assistance to the government in controlling the traffic congestion and improving the quality of road traffic. As a result, predicting the future traffic time given these historical road dynamic data is increasingly in the spotlight.

Q. Yang et al. (Eds.): PAKDD 2019, LNAI 11440, pp. 123–135, 2019.
https://doi.org/10.1007/978-3-030-16145-3_10

However, most of the existing works that address the traffic time estimation problem are limited by short-term or real-time prediction. This is an emerging challenging task as there are many complex factors that affect the traffic situations when conducting long-term traffic time prediction. In real-world situations, users may need to get the future traffic time; for example, an user want to get to the other city in near future, then he/she could plan when to depart and how to go if the future traffic time is predicted. Moreover, foul weather will cause traffic congestion and the traffic time will be affected. Hence, considering weather effect in long-term traffic prediction is a requirement of real-world applications.

Predicting the long-term traffic time is a challenging task. In long-term traffic prediction problem, the future traffic time not only depends on the current traffic situation but also has some relationship with the historically periodical traffic patterns. Then, how to catch these useful patterns and combine to the current traffic situation is a critical issue. Moreover, in our daily life, the occurrence of rush hour is extremely less than the normal traffic. It is difficult to leverage the predicting traffic time between both situations. To conquer the above challenges, in this paper, we propose a deep learning-based framework with integration of the weather effect for predicting the long-term traffic time, called deep ensemble stacked Long Short Term Memory (DE-SLSTM).

The main contributions of this paper can be summarized as follows:

- We propose a framework, called deep ensemble stacked Long Short Term Memory (DE-SLSTM). To best of our knowledge, this is the first deep learning framework on long-term traffic time prediction problem.
- we integrate the weather effect into our DE-SLSTM framework and make a comparison between the results of our method with weather effect and the results of our method without weather effect to demonstrate its effectiveness.
- For the difficulty of predicting the traffic time during the congestion, we adopts the concept of cost sensitive into our proposed framework to improve the predicting accuracy when rush hour occurs.

2 Related Work

Traffic time prediction has been widely studied in the past three decades; it can be divided into three categories in terms of predicted time points, which are real-time prediction, short-term prediction and long-term prediction [13].

2.1 Real-Time and Short-Term Traffic Time Prediction

For the real-time and short-term traffic time prediction, several studies had focused on this topic. Among these studies, we can classify them into two categories, which are shallow learning based method and deep learning based method.

In 2003, Wu et al. [15] used the support vector regression (SVR) model to predict the traffic time, but it did not consider the historically periodical traffic patterns. Billings and Yang [2] used historical traffic time to fit the autoregressive integrated moving average (ARIMA) model and performed one-step-ahead traffic time prediction. K-nearest neighbors (KNN) model is one of the most widely used nonparametric models, which can get great result if there are sufficient and creditable historical data [8]. Bajwa [1], Ul et al. [12] presented a KNN prediction model and a genetic algorithm generating adaptive parameters. Furthermore, Qiao et al. [9] incorporated the trend adjustment feature into the traditional KNN model called KNN-T.

Deep learning based method has shown its power in cases of time series prediction. Duan et al. [3] first explored a deep learning model, the Long Short-Term Memory (LSTM) neural network model, for traffic time prediction, leading the topic on traffic time prediction to a new aspect. Siripanpornchana et al. [10] proposed a deep learning architecture based on a concept of Deep Belief Networks (DBN) [4] which utilizes a stack of Restricted Boltzmann Machines (RBM) to automatically learn generic traffic features. Wang et al. [14] further combined convolutional neural network (CNN) and LSTM to capture spatial and temporal dependencies of each local path and takes some factors affecting the traffic time into consideration. Zhang et al. [18] used bidirectional LSTM (BiLSTM) to capture more features for each grids in the path and design a special dual interval loss to enhance the strength of BiLSTM.

2.2 Long-Term Traffic Time Prediction

Recently, deep learning technique has been adopted to many urban computing problems, such as air quality prediction [17], crime prediction [5], and crowd flow prediction [16]. However, long-term traffic time prediction is a topic with less researchers studying on. Klunder et al. [6] adopted KNN combining with weather effect to performing the long-term traffic time predicting. Li et al. [7] used Evolving Fuzzy Neural Network (EFuNN) to predict long-term traffic time. EFuNN perform the fuzzification of the input variables at first. Second, it trains lots of fuzzy rules for the fuzzed input vectors to transform to fuzzed outputs, and the outputs get through the defuzzification layer to get the final predicting results. It encodes the weather effect into the model, but it does not consider the effect in traffic conditions as well.

3 Proposed Framework

Before introducing our proposed framework, we first address the problem definition. Given a set of historical traffic data, road network and the corresponding weather data, the aim is to predict long-term traffic time of a road segment $r_i \in R, i \in [1, 2, ..., n]$. Specifically, given the historical data at time $t-1, t-2, ..., t$, and weather data at time $t+k$; the goal is to predict the traffic time of long-term target at time $t+k$, where $k \geq 6$.

Figure 1 shows our proposed framework, called Deep Ensemble Stacked Long Short Term Memory (DE-SLSTM). The inputs include road network, weather and traffic data. As for the process part, it consists of two main components:

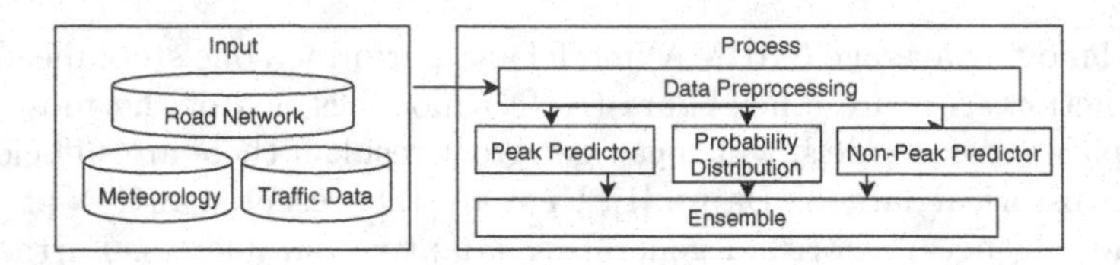

Fig. 1. Our proposed framework

the data pre-processing part and the training part. The output is the predicted traffic time of the road segment with a specific time point.

3.1 Dataset Pre-processing

Missing Data Filling. For filling the missing data, we define the threshold ϵ, and filter the data that the length of the time interval between two adjacent data points is less than the threshold ϵ. Further, We fill the missing data by doing the interpolation of the two continuous data; Otherwise, we search the value of the same time point on the last week or next week to fill the missing data. Finally, we use the average value of the whole data to fill the remaining missing data.

Temporal Attribute Extraction. More extra information may help the predictor to get the accurate result. Table 1 shows the number of dimension and description of the temporal attributes. Five temporal attributes, including month, day of week, time slot, holiday, and peak, are extracted.

Table 1. Temporal attributes

Attribute	# of dimension	Description
Month	12	January to december
Day of week	7	Monday to sunday
Holiday	3	Weekday, weekend, national holiday
Time slot	288	288 time slots in one day (one time slot = five minutes)
Peak	4	Non-peak, morning peak, noon peak, night peak

Data Combination. After filling the missing data and extracting the temporal attributes, we have to combine the traffic data with temporal attributes and weather data at each time point so that it can be fed into deep learning model.

3.2 Model Training

Before training our deep learning model, we first define what is peak traffic time and non-peak traffic time. In fact, peak traffic and non-peak traffic show different features, and the peak traffic time is hard to predict. First of all, We set a threshold (δ) to decide whether the moment of specific road is peak traffic time or non-peak traffic time. For example, if the least limit speed of expressway is 60 km/hr, the threshold would set to be 60 km/hr. Then, we design two predictors in our framework called non-peak predictor and peak predictor to separately deal with the non-peak traffic time and peak traffic time. Finally, We combine the results from both of them to make the final prediction.

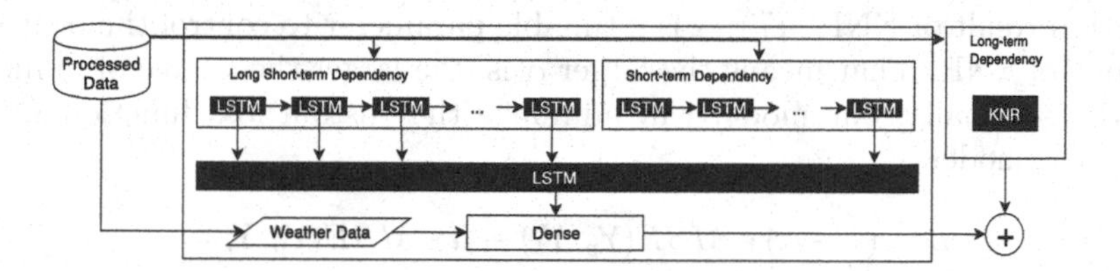

Fig. 2. Traffic time predictor

Non-peak Predictor. Figure 2 shows the architecture of the non-peak traffic time predictor. Processed Data means the output of data from the pre-processing module. To fully utilize the historical data, we consider not only short-term dependency and long short-term dependency but also long-term dependency and weather data with temporal attributes at time point of prediction. Note that the long-term dependency here means the historical dependency of all the input instances. We adopt Long Short Term Memory Network (LSTM) as the core algorithm for catching the short-term dependency and long short-term dependency. For the long-term dependency, we use the K-Nearest Neighbor Regression (KNR) as the core, as it can search the whole historical data.

In detail, we show an input example of non-peak predictor in Fig. 3. For the short-term dependency (green part), we extract all the traffic data with temporal attributes from an hour ago to current time and feed them into a LSTM. As for the long short-term dependency (orange parts), we search the same predicted time point of previous 7 days. Next, we extract all the traffic data with temporal attributes from an hour before the predicted time point to the predicted time point and feed them into a LSTM for each day.

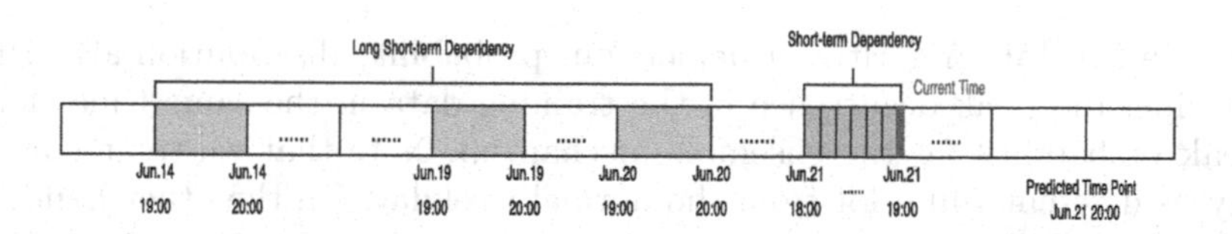

Fig. 3. Input example of traffic time predictor (Color figure online)

To get the prediction result, we concatenate the output of LSTM for short-term dependency with the output of LSTM for long short-term dependency and feed them into another LSTM, which aims to learn higher level temporal features. Last, we concatenate the output of stacked LSTM with the weather data and feed them into fully connected layers to output the final prediction result (Y_p).

For updating the weights in stacked LSTM and optimizing the predictor, We define a custom loss function as shown in Eq. 1. We consider not only the mean squared error (MSE) between the final prediction result (Y_p) and true value (Y_t), but also the error between the final prediction result (Y_p) and the prediction result of KNR (Y_k). α is a tunable parameter to control the degree of influence of KNR. That means the larger α is, the larger the degree of influence of KNR is. Finally, our model will minimize the custom loss function L and update the model.

$$L = (1 - \alpha) * MSE(Y_p, Y_t) + \alpha * MSE(Y_p, Y_k) \tag{1}$$

$$MSE(Y_a, Y_b) = \frac{1}{n}\sum_{i=1}^{n}(Y_{ai} - Y_{bi})^2 \tag{2}$$

Peak Predictor. Indeed, the non-peak predictor still cannot predict well while there is a peak traffic time, as the occurrence of peak traffic is extremely less than the normal traffic. Hence, we adopt the a mechanism called cost sensitive to deal with this issue.

Cost sensitive is one of the effective methods for solving classification of imbalanced data problem [11]. Generally speaking, cost sensitive means to assign a higher weight to multiply by loss function for the minority class. In the traffic time prediction problem, we view the peak traffic time as minority class and give a higher weight multiplied by loss function.

Recall to the Fig. 2, the work-flow of peak predictor is almost the same as the framework of non-peak predictor. The only difference of them is adding a cost sensitive weight function into our custom loss function, which shows in Eq. 3. This design is in order to enhance peak predictor's prediction ability of peak traffic time; On the contrary, it will reduce peak predictor's prediction ability of non-peak traffic time.

$$L = CS((1 - \alpha) * MSE(Y_p, Y_t) + \alpha * MSE(Y_p, Y_k)) \tag{3}$$

Ensemble Model. Algorithm 1 depicts our probability distribution algorithm. We consider the peak occurrence in the training data in the same time slot as the peak probability for the ensemble mechanism. Note that we treat national holiday as different time slot from the normal weekday. On the other hand, the non-peak probability is computed by subtracting peak probability from 1.

We combine the predictors by the probability distribution as an ensemble model. Equation 4 shows the output of ensemble model (Y_p), which is also the output of our model. Y_{npp} means the prediction result of non-peak predictor and Y_{pp} means the prediction result of peak predictor. P_{np}, which is called non-peak probability, is the probability of non-peak probability and P_p, which is called

Algorithm 1. ProbabilityDistribution

```
Input  : D: Training data, R: road section of the prediction, T: time point of the prediction
Output: Probability distribution of R at T
1  Initialize nh_day = check_nh_day(T), speeds ← Ø;
2  if nh_day(T) then
3  |  foreach d in D do
4  |  |  if check_nh_day(d.time_point) & get_time(d.time_point) == get_time(T)  then
5  |  |  |  speeds ← speeds ∪ d.speed;
6  |  |  end
7  |  end
8  else
9  |  foreach d in D do
10 |  |  if get_weekday(d.time_point) == get_weekday(T) &
11 |  |    get_time(d.time_point) == get_time(T) then
12 |  |  |  speeds ← speeds ∪ d.speed;
13 |  |  end
14 |  end
15 end
16 number_peak = get_number_peak(speeds);
17 peak_probability = number_peak/len(speeds);
18 nonpeak_probability = 1 - peak_probability;
19 return peak_probability, nonpeak_probability
```

peak probability, is the probability of traffic peak time. Therefore, the next step is to design a method to generate the probability distribution P_{np} and P_p.

$$Y_p = Y_{npp} * P_{np} + Y_{pp} * P_p \tag{4}$$

4 Experiment Evaluation

4.1 Data Description

Taiwan expressway dataset and Taiwan weather dataset are used in our experiments. Taiwan expressway dataset contains time-stamp, traffic time data, and traffic speed data in Taiwan, while Taiwan weather dataset contains time-stamp, pressure, temperature, relative humidity, wind speed, and precipitation for each weather station in Taiwan. There are totally 322 road segments in Taiwan expressway dataset and 29 weather stations in Taiwan weather dataset. The time interval of Taiwan expressway dataset is 5 min while the time interval of Taiwan weather dataset is 1 h. As for the experiment, we use one-year period from October 2016 to October 2017 as our training data and five-month period from December 2017 to April 2018 as our testing data.

4.2 Evaluation Metrics

In order to assess the performance of our DE-SLSTM framework from different aspects, we use three kinds of evaluation metrics, mean absolute percentage error (MAPE), mean absolute error (MAE), and root mean square error (RMSE). The formulas of these evaluation metrics as shown below:

$$MAPE(\%) = \frac{\sum_{i=1}^{n} \frac{|y_i - \hat{y}_i|}{y_i}}{n} \tag{5}$$

Table 2. Details of parameters

Notation	Description	Default
ω	Cost sensitive weight	30
α	The degree of influencing KNR	0.1
t	Predicted time point	1 h
δ	Peak threshold	60 km/h
r	Road segment	Zhubei interchange to Hsinchu interchange

$$MAE = \frac{\sum_{i=1}^{n} |y_i - \hat{y_i}|}{n} \quad (6)$$

$$RMSE = \sqrt{\frac{\sum_{i=1}^{n} (y_i - \hat{y_i})^2}{n}} \quad (7)$$

4.3 Experimental Results

In this section, we first show all the parameters of our DE-SLSTM framework and the default values in Table 2. Next, we show all the experiments and divide the experiments into two categories, internal experiments and external experiments, according to the purpose of each experiment.

Internal Experiments. As for the internal experiments, we first vary some internal parameters in the proposed method to find the best parameters. Note that for each sub-experiment, all the control parameters are set to default values. Next, we determine whether adding weather effect can improve the performance.

Varying Cost Sensitive Weight. Figure 4a shows the results of different cost sensitive weights. Since what we concern for is the peak traffic time, we search for the lowest MAPE in Fig. 4a and observe that the lowest MAPE occurs when the cost sensitive weight is 30. Therefore, we set cost sensitive to 30.

Varying α Value. Figure 4b shows the results of varying α. Since we care more about the peak traffic time, we search the lowest MAPE in Fig. 4b and observe that MAPE is the lowest when the α value is 0.1 and 0.2 separately. Moreover, we set α value to 0.1 because the larger the α value, the K-Nearest Neighbor Regression (KNR) would influence the final result more. And it is better to let our framework be less influenced by KNR.

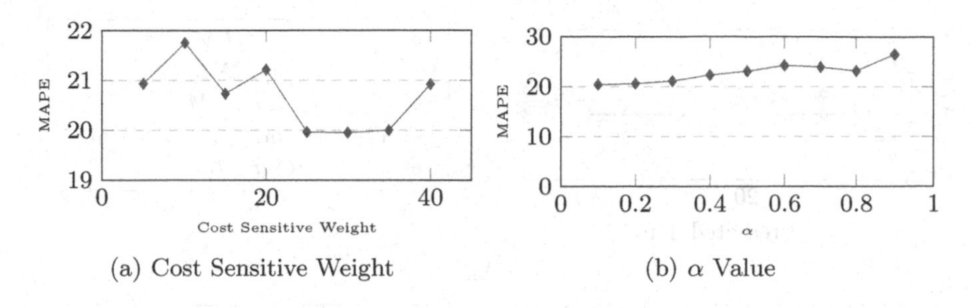

(a) Cost Sensitive Weight

(b) α Value

Fig. 4. Varying parameters (peak traffic time testing data)

Table 3. Number of road segments in each region

Region	Number of road segments
Northern	66
Central	48
Southern	54

Weather Effect. As for the experiment of weather, we choose some road segments and try to clarify that whether our model can perform better with weather effect on these road segments. For the different road segments, National Highway can be divided into three regions, including the northern region, the central region and the southern region. Table 3 shows the number of road segments in each region. We then select fifteen road segments from each region such that there are road segments in equal length in each region; therefore, we totally get forty-five road segments. Table 4 shows that the proposed model performs better when including weather effect for different approaches.

Table 4. Result of weather effect

	With weather effect			Without weather effect		
Metrics	All	Peak	Non-peak	All	Peak	Non-peak
MAPE	5.32%	28.50%	4.87%	5.39%	28.60%	4.97%
MAE	12.36s	128.19s	9.69s	12.60s	129.13s	9.96s
RMSE	29.56s	164.75s	19.43s	29.54s	165.86s	19.33s

External Experiments

Comparing with Different Methods. In the external experiments, we compare our method with other methods on different approaches. The first method is historical average (HA), which predicts the travel time by the average value of the corresponding time periods. The second method is K-Nearest Neighbor

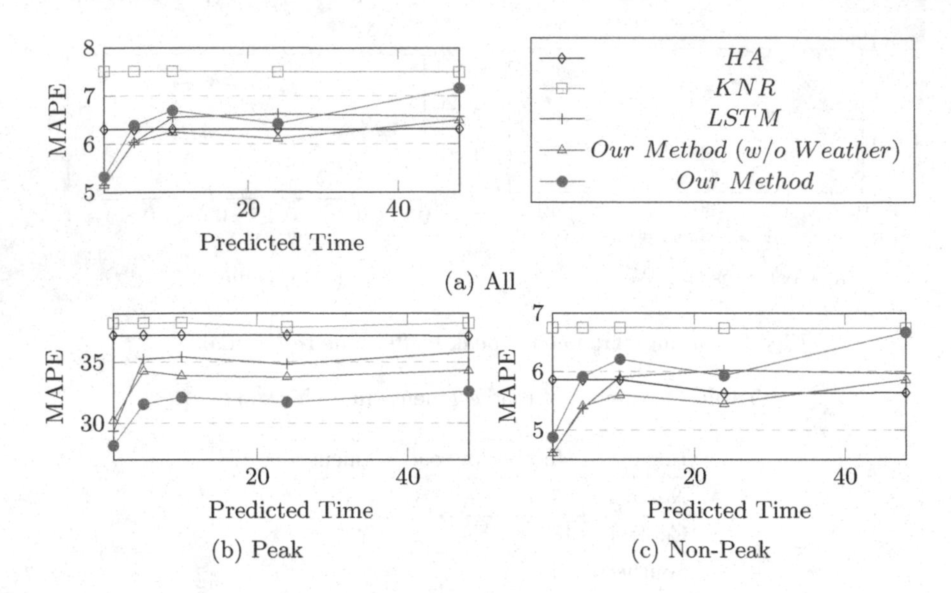

Fig. 5. Three regions combined (MAPE for traffic time tesing data)

Regression (KNR) and the third method is Long Short-Term Memory (LSTM) which uses only a single LSTM layer to do the prediction. And last, the fourth one is our method (w/o KNR & CS), which has the KNR and the cost sensitive ensemble mechanism removed in the proposed framework.

We divide the testing data into two group, peak traffic time and non-peak traffic time testing group. We then compare our method and other methods under three approaches, only peak traffic time, only non-peak traffic time, and all the testing data. In addition, we choose the road segments used in the "Weather Effect" experiment. Hence, we totally have forty-five road segments.

In this experiment, parameters of the proposed method are set to default values except for the predicted time point and the road segment. Figure 5 shows the MAPE of the results of long-term traffic time prediction. We calculate the mean of MAPE of the forty-five road segments we select. Figure 5b shows that our method outperforms other methods on peak traffic time testing data. Nonetheless, Figs. 5a and c show that our method does not outperform other methods with non-peak traffic time testing data. The main reason is the trade-off between the cost sensitive and the performance, as our method focuses on peak traffic time prediction; indeed, it is closer to the daily demand. On the other hand, we can observe that our method (w/o KNR & CS) performs better than LSTM and draw a conclusion that the architecture of stacked LSTM is effective.

Table 5. Selected long roads in each region

Region	Road
Northern	Yuanshan interchange to Hsinchu system interchange
Central	Houli interchange to Beidou interchange
Southern	Hsinying interchange to Kaohsiung terminal

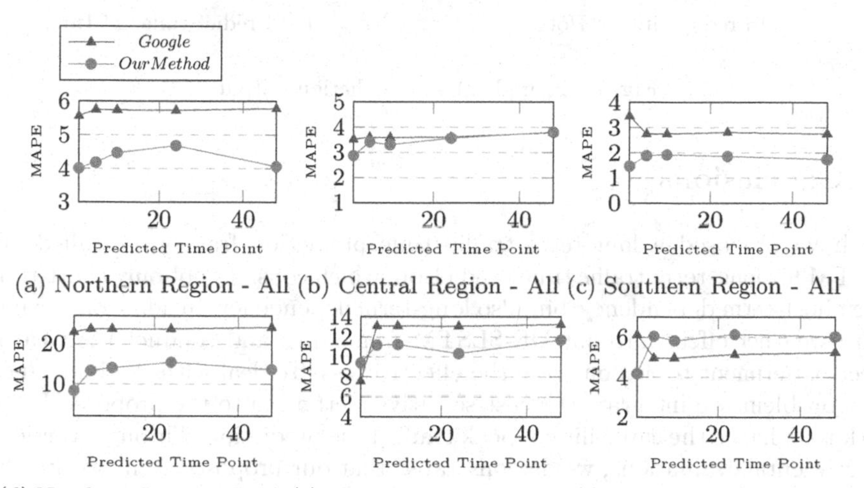

Fig. 6. Compare with Google - northern region

Comparing with Google Map on Long Road. In the final experiment, we compare our method with Google Map on long roads on one-week data which is from 2018/08/01 to 2018/08/08. Table 5 shows the roads we select in each region. As for the Google Map approach, we use Google Map Distance Matrix API and choose the most definite prediction result. As for our method, we combine the road segments that are in the same weather region to form a long road; then, accumulate the prediction results of these long roads as the final prediction.

Figure 6 shows the results of long-term traffic time prediction in different regions comparing with Google Map. It shows that the proposed method outperforms Google Map on all regions except for the southern region on peak traffic time testing data. Since the average peak traffic time ratio in the southern region is lower than other regions, the proposed framework which focuses on the peak traffic time prediction could not get better performance on the metrics. Figure 7 shows the results of the proposed method fit the ground truth better than Google map. In summary, our method is more applicable in long-term traffic time prediction problem.

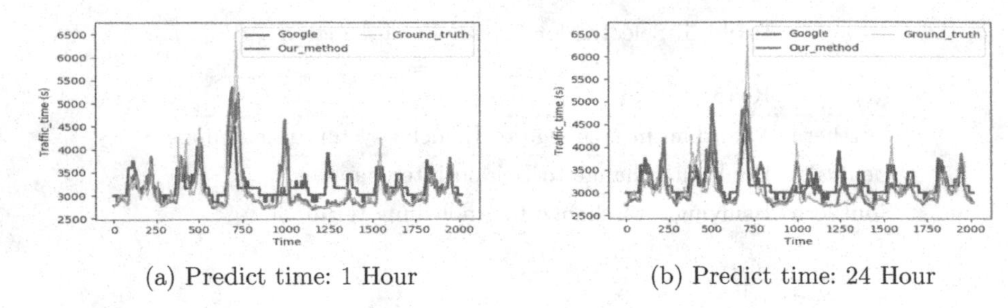

(a) Predict time: 1 Hour (b) Predict time: 24 Hour

Fig. 7. Example of the prediction output

5 Conclusions

We have proposed a long-term traffic time prediction framework called DE-SLSTM for long-term traffic time prediction, which consider not only short-term, long short-term dependency, but also long-term dependency. In addition, we integrate weather effect into our DE-SLSTM framework and conduct the weather effect experiment to demonstrate the effectiveness. To deal with the imbalanced data problem, we integrate the cost sensitive method in to the proposed framework to enhance the capability of peak traffic time prediction. Through a series of experimental evaluations, we demonstrated that our proposed framework delivered excellent performance. To the best of our knowledge, this is the first work that considers deep learning technique on long-term traffic time prediction.

Acknowledgement. This research was partially supported by Ministry of Science and Technology, Taiwan, under grant no. 107-2218-E-009-050.

References

1. Bajwa, S.: An adaptive travel time prediction model based on pattern matching. In: World Congress on ITS 2004 in Nagoya (2004)
2. Billings, D., Yang, J.: Application of the ARIMA models to urban roadway travel time prediction - a case study. In: 2006 IEEE International Conference on Systems, Man and Cybernetics, vol. 3, pp. 2529–2534, October 2006
3. Duan, Y., Lv, Y., Wang, F.-Y.: Travel time prediction with LSTM neural network. In: 2016 IEEE 19th International Conference on Intelligent Transportation Systems (ITSC), pp. 1053–1058, November 2016
4. Hinton, G.E., Osindero, S., Teh, Y.-W.: A fast learning algorithm for deep belief nets. Neural Comput. **18**(7), 1527–1554 (2006)
5. Huang, C., Zhang, J., Zheng, Y., Chawla, N.V.: DeepCrime: attentive hierarchical recurrent networks for crime prediction. In: Proceedings of the 27th ACM International Conference on Information and Knowledge Management, pp. 1423–1432. ACM (2018)
6. Klunder, G., Baas, P., op de Beek, F.: A long-term travel time prediction algorithm using historical data. In: Proceedings of 14th World Congress Intelligent Transportation Systems, pp. 1191–1198 (2007)

7. Li, R., Rose, G., Chen, H., Shen, J.: Effective long-term travel time prediction with fuzzy rules for tollway. Neural Comput. Appl., 1–13 (2017)
8. Myung, J., Kim, D.-K., Kho, S.-Y., Park, C.-H.: Travel time prediction using k nearest neighbor method with combined data from vehicle detector system and automatic toll collection system. Transp. Res. Rec. **2256**(1), 51–59 (2011)
9. Qiao, W., Haghani, A., Hamedi, M.: Short-term travel time prediction considering the effects of weather. Transp. Res. Rec. **2308**(1), 61–72 (2012)
10. Siripanpornchana, C., Panichpapiboon, S., Chaovalit, P.: Travel-time prediction with deep learning. In: 2016 IEEE Region 10 Conference (TENCON), pp. 1859–1862. IEEE (2016)
11. Thai-Nghe, N., Gantner, Z., Schmidt-Thieme, L.: Cost-sensitive learning methods for imbalanced data. In: The 2010 International Joint Conference on Neural Networks (IJCNN), pp. 1–8. IEEE (2010)
12. Ul, S., Bajwa, I., Kuwahara, M.: A travel time prediction method based on pattern matching technique. Publication of, ARRB Transport Research, Limited (2003)
13. van Lint, H.: Reliable travel time prediction for freeways. Netherlands TRAIL Research School (2004)
14. Wang, D., Zhang, J., Cao, W., Li, J., Zheng, Y.: When will you arrive? Estimating travel time based on deep neural networks. In: AAAI (2018)
15. Wu, C.-H., Wei, C.-C., Su, D.-C., Chang, M.-H., Ho, J.-M.: Travel time prediction with support vector regression. In: Proceedings of 2003 IEEE Intelligent Transportation Systems, vol. 2, pp. 1438–1442. IEEE (2003)
16. Yao, H., Tang, X., Wei, H., Zheng, G., Yu, Y., Li, Z.: Modeling spatial-temporal dynamics for traffic prediction. CoRR, abs/1803.01254 (2018)
17. Yi, X., Zhang, J., Wang, Z., Li, T., Zheng, Y.: Deep distributed fusion network for air quality prediction. In: Proceedings of the 24th ACM SIGKDD International Conference on Knowledge Discovery and Data Mining
18. Zhang, H., Wu, H., Sun, W., Zheng, B.: DeepTravel: a neural network based travel time estimation model with auxiliary supervision. arXiv preprint arXiv:1802.02147 (2018)

Arrhythmias Classification by Integrating Stacked Bidirectional LSTM and Two-Dimensional CNN

Fan Liu[1,3](✉), Xingshe Zhou[1], Jinli Cao[2], Zhu Wang[1], Hua Wang[3], and Yanchun Zhang[3]

[1] School of Computer Science, Northwestern Polytechnical University, Xi'an, China
liufant800@mail.nwpu.edu.cn
[2] Department of Computer Science and Information Technology, La Trobe University, Melbourne, Australia
[3] College of Engineering and Science, Victoria University, Melbourne, Australia

Abstract. Classifying different types of arrhythmias based on ECG signal is an important research topic in healthcare. Traditional methods focus on extracting varieties of features from ECG and using them to build a classifier. However, ECG usually presents high inter- and intra-subjects variability both in morphology and timing, hence, it's difficult for predesigned features to accurately depict the fluctuation patterns of each heartbeat. To this end, we propose a novel arrhythmias classification model by integrating stacked bidirectional long short-term memory network (SB-LSTM) and two-dimensional convolutional neural network (TD-CNN). Particularly, SB-LSTM mines the long-term dependencies contained in ECG from both directions to depict the overall variation trend of ECG, while TD-CNN exploits local characteristics of ECG to characterize the short-term fluctuation patterns of ECG. Moreover, we design a discrete wavelet transform (DWT) based ECG decomposition layer and a Sum Rule based intermediate classification result fusion layer, by which ECG can be analyzed from multiple time-frequency resolutions, and the classification results of our model can be more accurate. Experimental results based on MIT-BIH arrhythmia database shows that our model outperforms 3 baseline methods, achieving 99.5% of accuracy, 99.9% of sensitivity and 98.2% specificity, respectively.

Keywords: Arrhythmias classification · Stacked bidirectional LSTM · Convolutional neural network · Wavelet decomposition · Classification result fusion

1 Introduction

Arrhythmias are cardiac conditions caused by the abnormal electrical activities of the heart [1]. During an arrhythmia, the heart can't pump enough blood to the body. Lack of blood flow can damage the brain, heart, and other organs, which usually results in anxiety, dizziness, chest pain, etc. [2]. Without timely treatment, it can lead to serious complications, such as stroke, heart failure, sudden cardiac death (SCD) [3],

Q. Yang et al. (Eds.): PAKDD 2019, LNAI 11440, pp. 136–149, 2019.
https://doi.org/10.1007/978-3-030-16145-3_11

cardiovascular diseases (CVDs) [1], etc. It is reported that about 2200000 people in the US and 4500000 people in EU annually die from arrhythmias, which exceeds the mortality of all cancers combined [4]. The Association for the Advancement of Medical Instrumentation (AAMI) classifies the arrhythmias into 5 categories: non-ectopic (N), ventricular ectopic (V), supraventricular ectopic (S), fusion (F), and unknown (Q) [5]. Each of them shows different symptoms and needs different treatments, therefore, classifying arrhythmias accurately is the prerequisite for effective treatments [6].

Electrocardiogram (ECG), as an easy accessible and non-invasive tool, is the most commonly used physiological signal for arrhythmias diagnosis [22]. By carefully analyzing ECG morphology, different types of heartbeats usually can be distinguished. However, ECG is a kind of non-stationary signal, that is, its morphology changes with respect to time, and these variations present not only between different subjects but also within the same subjects [7], which makes it difficult for physicians to accurately diagnose arrhythmias via visual assessment. In particular, it's reported that licensed general practitioners can only achieve 92% of specificity and 80% of sensitivity when distinguishing atrial fibrillation from a healthy heartbeat [8].

To tackle the drawbacks of visual assessment of ECG, lots of computer-aided methods were proposed [7, 9–16]. They are mainly based on three steps, i.e., feature extraction, feature selection and classification. After removing various kinds of signal noise, several hand-crafted features containing crucial information regarding to the status of the heart are extracted from ECG. Generally, the most commonly used feature extraction techniques include spectral analysis [9, 15], time-frequency analysis [12], hidden Markov model (HMM) [10], higher order statistics (HOS) [11], morphology analysis [7], etc. Then, feature selection techniques such as independent component analysis (ICA) [11, 12, 30], principal component analysis (PCA) [12, 29, 31], and linear discriminant analysis (LDA) [29] are deployed to reduce the dimensionality of the extracted features. Finally, based on these features, classifiers are trained to differentiate different arrhythmia types, for example, support vector machine (SVM) [15], random forest [16], neural networks (NN) [18], ensemble classifiers [19], cluster analysis [20], etc. These methods have notably improved the arrhythmias classification performance, however, it's quite difficult for them to further boost the performance in practice, which is mainly because (1) the noise of ECG signal such as baseline wanders, muscle contraction, power line interference, etc., always distort the ECG waveforms, hence it is difficult to ensure the validity of the values of the extracted features [7]; (2) ECG usually presents high inter- and intra-subjects variability both in morphology and timing, therefore, the predesigned features may not be able to accurately characterize every heartbeat [7].

Compared with the above methods, deep learning is end-to-end method where feature extraction, feature selection and classification are fused together with no need to explicitly extract hand-crafted features [1]. Furthermore, it usually has simpler logical structure but achieves better fitting ability [23]. The long short-term memory (LSTM) network [17] and the convolutional neural network (CNN) [23] are two most promising deep neural network models, and have been successfully applied to many areas such as disease prediction [9–16], face recognition [25], image classification [24] and object recognition [26]. Particularly, LSTM can fully exploit the long dependencies of time

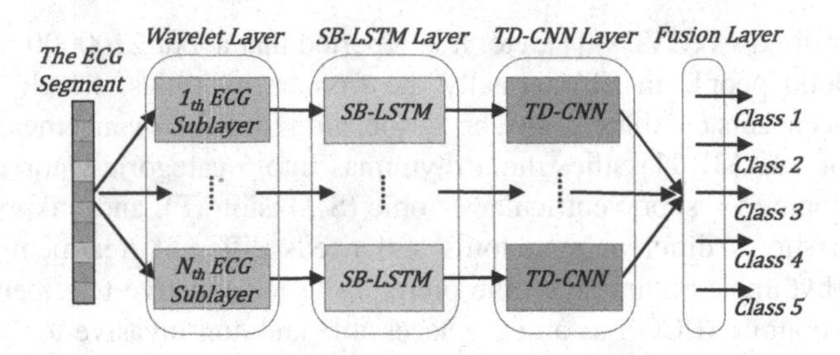

Fig. 1. The proposed arrhythmias classification model framework.

series data, while CNN can effectively mine local characteristics of the data [23]. Although LSTM and CNN have been separately used to classify arrhythmias [1, 17], the performance still need to be improved as they only benefit from just one model.

To achieve better classification performance, we propose a novel arrhythmias classification model by combining LSTM and CNN. It not only can extract more hidden information to accurately model the fluctuation pattern of ECG, but also can obtain more accurate and robust classification results. Our contributions are three-folds:

First, to model the fluctuation pattern of ECG signal more accurately, we propose a novel neural network architecture consisting of a stacked bidirectional LSTM (SB-LSTM) layer and a two-dimensional CNN (TD-CNN) layer, where the outputs of SB-LSTM are fed into TD-CNN as input. In particular, SB-LSTM aims to mine the long-term dependencies of ECG in both directions, by which the overall variation trend of ECG during each heartbeat can be captured. While TD-CNN is especially suitable for extracting short-term characteristics existing in each ECG wave components.

Second, inspired by the benefit of bagging classifiers, we design a DWT-based wavelet layer and a Sum Rule based fusion layer. The former decompose ECG into multiple time-frequency resolutions via DWT, by which more hidden information of ECG can be mined and utilized. Particularly, each ECG resolution is processed by SB-LSTM and TD-CNN sequentially, and gets a classification result separately. The latter resembles traditional voting-based ensemble methods, and is used to assemble the intermediate classification results of each ECG resolution into a final result with the Sum Rule used as fusion strategy. Experimental results show that the utilization of these two layers significantly improves the arrhythmias classification performance.

Third, experimental results based on public MIT-BIH arrhythmia dataset show that our model outperforms three state-of-the-art methods and its accuracy, sensitivity and specificity are 99.5%, 99.9% and 98.2%, respectively. Furthermore, we also compare it with 15 similar network structures, whose results indicate that the structure of our model is the optimal one, and it can give unbiased classification results for each class.

The rest of this paper is organized as follows. Section 2 elaborates the details of our arrhythmias classification model, followed by the experiments and evaluation results in Sect. 3. Finally, we conclude the paper and discuss the future work in Sect. 4.

2 Method

2.1 The Arrhythmias Classification Model Framework

As shown in Fig. 1, the proposed model mainly consists of 4 layers, i.e., Wavelet layer (WL), SB-LSTM layer, TD-CNN layer and Fusion layer (FL). WL contains a wavelet basis and is different from standard deep learning layers. It can decompose ECG into multiple time-frequency resolutions (i.e., wavelet sublayers), which are separately processed by SB-LSTM layer and TD-CNN layer in parallel. SB-LSTM layer is a stacked bidirectional RNN network with LSTM used as cells, which is used to capture the overall variation trend of ECG from both directions, while TD-CNN is used to model the local features of ECG. Due to SB-LSTM and TD-CNN, the long-term and short-term fluctuation pattern of ECG can be fully mined and utilized, besides, each wavelet sublayer obtains a temporary classification result, which is further fused into a final classification result via FL, with the Sum Rule used as fusion strategy.

2.2 The Wavelet Layer (WL)

Inspired by bagging classifiers, in this section we design a special WL to decompose ECG into sublayers via DWT which is a signal processing method often used to extract useful features from non-stationary timing signal [29]. It has two benefits: (1) DWT offers multi-resolution analysis in both time and frequency domains, by which more hidden information of ECG can be mined and utilized; (2) the generated sublayers can be classified separately, and the results are then fused via FL, which takes the advantages of bagging classifiers and hence obtain better classification performance.

Firstly, the ECG signal is decomposed into approximation and detail components at the first level, which is implemented by passing it through a pair of high-pass (HP) and low-pass (LP) filters as follows:

$$Detail(n) = \sum\nolimits_{k=-\infty}^{\infty} x_{(k)}\varphi_h(2n-k) \tag{1}$$

$$Approximation(n) = \sum\nolimits_{k=-\infty}^{\infty} x_{(k)}\varphi_g(2n-k) \tag{2}$$

where $x_{(k)}$ denotes ECG, φ_h and φ_l are HP and LP filters, respectively. The output of HP and LP filters respectively includes the detail coefficients and approximate coefficients of ECG. The popular Daubechies D_6 ('db6') [27] is used as wavelet basis. Afterwards, the above process is iteratively applied to the approximation component at each level until a specified level is reached. Given the frequency of the preprocessed ECG (90 Hz), it is decomposed up to six levels, where the frequency band of the 6th level approximation is 0–1.406 Hz which is mainly the baseline wander and has no useful information. Finally, the ECG is reconstructed into 7 sublayers as follows. The detail coefficients of the 1st, 2nd, 3rd, 4th, 5th, and 6th level are separately reconstructed into a sublayer, with other sub-band coefficients replaced with zeroes. Then, all the detail coefficients in each level are merged together to compose the noise-free ECG sublayer, with the approximate coefficients of the 6th level set to zeroes. The obtained 7 sublayers are then separately processed and classified by SB-LSTM and TD-CNN.

2.3 The Stacked Bidirectional LSTM (SB-LSTM) Layer

Medically, each heartbeat will produce P, Q, R, S, T waves sequentially in ECG [9]. As a continuous signal, apparently, the amplitude of one wave component is related to that of the former one. Particularly, as shown in the Fig. 1 of [1], although same wave components derived from different types of arrhythmias are not that different from each other, the waveform of the whole heartbeat is quite different from each other from an overall perspective. Based on this finding, we use LSTM to model the long-term fluctuation pattern of ECG, in order to improve the classification performance.

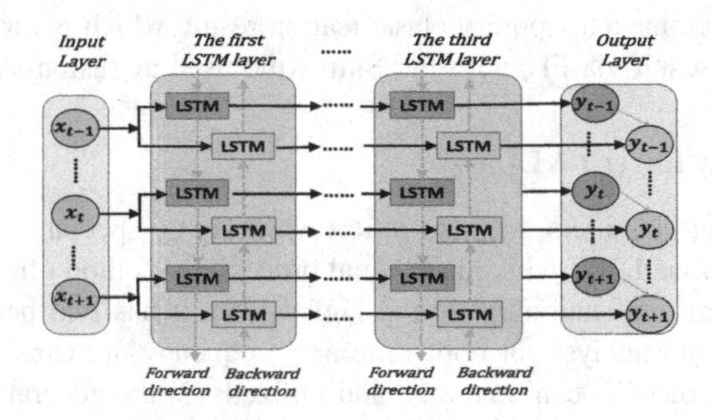

Fig. 2. The structure of the proposed stacked bidirectional LSTM layer.

LSTM is a sub-type of recurrent neural network (RNN) but it solves the vanishing gradient problem of RNN, which is suitable for processing time series signal [7]. In a LSTM cell, three gates, i.e., forget gate, input gate and output gate are used to control the update of the cell's state and the cell's output [17]. Specifically, forget gate is used to discard useless information contained in the cell's past state, input gate determines what information should be added to the cell's current state, and output gate decides what information should be output. Benefiting from these three gates, previous useful information can be held and transferred for a long time. For a given input vector x_t at time t, the cell's output h_t and cell's state c_t can be calculated as follows:

$$f_t = \sigma(W_f \cdot [h_{t-1}, x_t] + b_f) \quad (3)$$

$$i_t = \sigma(W_i \cdot [h_{t-1}, x_t] + b_i) \quad (4)$$

$$o_t = \sigma(W_o \cdot [h_{t-1}, x_t] + b_o) \quad (5)$$

$$c_t = f_t \odot c_{t-1} + i_t \odot tanh(W_c \cdot [h_{t-1}, x_t] + b_c) \quad (6)$$

$$h_t = o_t \odot tanh(c_t) \quad (7)$$

where [] concatenates two matrixes, W and b represent weight and bias of the model respectively, σ is activation function and $\odot$ is element-wise multiplication.

To fully characterize the long-term fluctuation pattern of ECG signal, a stacked bidirectional LSTM network (SB-LSTM) is designed in this study. As shown in Fig. 2, SB-LSTM consists of three LSTM layers where the output of non-last layers is fed into the next layer. Meanwhile, each layer includes 2 LSTM cells with opposite direction, which are used to capture the forward dependency and backward dependency, respectively. Particularly, the number of neurons in LSTM cell is set to 32 after comparing the experimental results where it is set to 16, 32, 64 and 128, respectively. Given a ECG sublayer generated by WL $x = x_1, x_2, \ldots, x_N$, where N is the length of the sublayer, SB-LSTM produce two sets of outputs, i.e., forward results and backward results. Both of them are organized in matrix form (size = N * 32), where the ith row corresponds the outputs at time i, and the j_{th} column represents the outputs of the jth neuron in LSTM cell. Finally, the forward results and backward results are concatenated along the first dimension, and then fed into TD-CNN.

2.4 The Two-Dimensional CNN (TD-CNN) Layer

Besides the overall fluctuation trend of the ECG, the short-term fluctuation patterns contained in local wave component is also useful for distinguishing different types of arrhythmias [1, 23]. Recently, CNN has shown powerful ability to extract local spatial-time features and obtained great success in image recognition field. Therefore, we utilize CNN to extract short-term fluctuation features of ECG in this paper.

Typical CNN structure can be formed as $Input \rightarrow [[Convolutional\ layer] * N \rightarrow Pooling\ layer] * M \rightarrow [Fully\ connected\ layer] * K$, which is shown in Fig. 3. The neurons in convolutional layers and pooling layers are usually organized as matrices (called filters), which sequentially execute convolution or pooling operations by shifting itself in feature map along the vertical and horizontal directions. The convolution operations performed by same filter are like extracting a specific feature from each part of the feature map. Moreover, the extracted features are of shift invariance, that is, the spatial structure relationship contained in the feature map can be reserved, which is helpful for extracting the local characteristics of ECG signal.

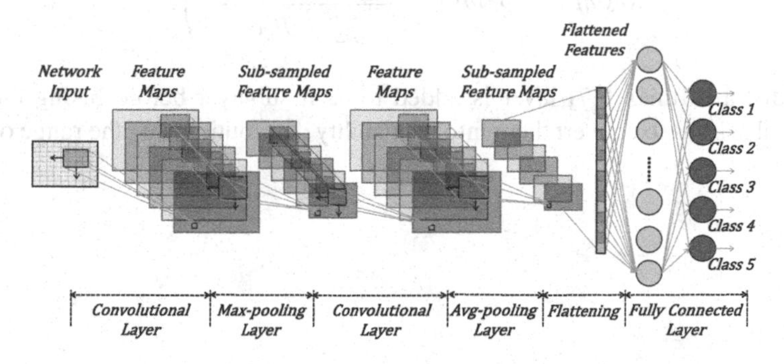

Fig. 3. The structure of the proposed TD-CNN layer.

As shown in Fig. 3, the proposed TD-CNN consists of 2 convolutional layers, 1 max-pooling layer, 1 average-pooling layer and 2 fully-connected layers, which is determined by lots of preliminary experiments conducted on our dataset. In detail, we first perform convolution operations on the input signal by using 32 filters whose size are 4×4 with a stride of 2. Then, the outputs of the first convolutional layer are fed into a max-pooling layer whose receptive field size and stride are set to 4×4 and 4 respectively. Afterwards, a second convolutional layer that is same as the first one but has 64 filters is deployed to obtain higher level representation, followed by an average-pooling layer which has same receptive field size and stride as the former max-pooling layer. The two pooling layers reduce the dimensionality of the feature maps generated by the convolutional layers and hence enable the network to learn higher-level features from wider input windows without enlarging the size of the filters. In particular, the former only retain the most notable feature for each feature map, while the latter averagely reserve the effects of each feature. Finally, the pooled outputs are flattened and concatenated into a one-dimensional vector, and is further fed into two successive fully connected layers, which contains 32 and 5 neurons respectively. Note that the 5 neurons contained in the latter corresponds to 5 arrhythmias types respectively, hence, its outputs can be regarded as the classification results of each sublayer.

2.5 The Fusion Layer (FL)

In our model, the ECG segment corresponding to each heartbeat is decomposed into several sublayers, and each of them is processed by SB-LSTM and then TD-CNN in parallel. That is, each sublayer can obtain an independent prediction result, however, these results may be different from each other. To obtain more robust and accurate prediction result, we design a special fusion layer based on the Sum Rule [7], which resembles traditional voting-based ensemble methods. To be specific, given a sublayer $x_{l(l=1,2,\ldots,\mathrm{N})}$, where N is the number of wavelet sublayers obtained from ECG, $P(\omega_i|x_l)_{(i=1,2,3,4,5)}$ denotes the possibility that x_l belongs to class ω_i, where ω_1, ω_2, ω_3, ω_4 and ω_5 represent five different arrhythmias types respectively. Using the Sum Rule as fusion strategy, the final classification result can be determined as follows:

$$Result = \underset{\omega_i}{argmax}\left(\frac{\sum_{l=1}^{l=N} P(\omega_i|x_l)}{\sum_{i=1}^{i=5}\sum_{l=1}^{l=N} P(\omega_i|x_l)}\right). \tag{8}$$

Note that a softmax [17] layer is added to each sublayer before fusing their prediction results, so as to convert them into probability distribution, i.e., the range of [0, 1].

Table 1. A summary of the obtained heartbeats dataset.

AAMI classes	Subclasses	Num	Total
Non-ectopic (N)	Normal	73214	88680
	Left bundle branch block	8021	
	Right bundle branch block	7215	
	Atrial escape	15	
	Nodal escape	215	
Ventricular ectopic (V)	Aberrated atrial premature	132	2649
	Atrial premature	2434	
	Nodal premature	81	
	Supra-ventricular premature	2	
Supraventricular ectopic (S)	Premature ventricular contraction	6311	6417
	Ventricular escape	106	
Fusion (F)	Fusion of ventricular and normal	776	776
Unknown (Q)	Fusion of paced and normal	956	7983
	Paced	7000	
	Unclassifiable	27	

3 Experiment Evaluation

3.1 Dataset Description and Data Preprocessing

The MIT-BIH arrhythmia dataset was used in this study [21], which consists of 48 half-four excerpts of ECG recordings sampled at 360 Hz. 23 of them were collected from inpatients and contain many normal heartbeats while the others include less common but clinically significant arrhythmias. Based on the ANSI/AAMI EC57:2012 standard [3], each record was annotated by at least two cardiologists independently, and the disagreements were solved by using computer-readable reference annotations.

First, we utilized Pan-Tompkins algorithm to detect R-peaks [1, 7], then the 256 samples centered around the detected R-peak ([−127, 128]) were used to represent each heartbeat [1]. To decrease computational burden, each heartbeat was further subsampled at a frequency of 90 Hz, that is, each heartbeat contains 64 samples. Then, we discarded the first and the last heartbeat of each ECG record since they may lose necessary ECG wave components, and the heartbeats that don't meet the above length requirement. Finally, we extracted a total of 106505 heartbeats, whose details are shown in Table 1. Apparently, it's a quite imbalanced dataset that will inevitably reduce the generalization ability of the model. To this end, we use the Synthetic Minority Over-sampling Technique (SMOTE) [28] to synthesize new heartbeats for the class with fewer instances. Finally, the number of heartbeats of the four minority classes are the same as that of class N (88680), resulting in 443400 beats in total.

3.2 Experimental Setup

The class labels were encoded into one-hot representation, and the loss of the network was computed by employing the categorical cross-entropy function as follows:

$$\mathcal{L} = \frac{1}{\widetilde{N}} \sum\nolimits_{i=1}^{\widetilde{N}} (y_i \log(\widehat{y}_i) + (1 - y_i) log(1 - \widehat{y}_i)), \tag{9}$$

where $\widetilde{N}$ is the number of heartbeat in a batch, while y_i and $\widehat{y}_i$ are the true label and predicted label of the i_{th} heartbeat, respectively. In addition, to avoid shielding negative outputs, the LeakyRelu function [1] was used to activate neurons which enables the utilization of more useful information. Moreover, the weights and biases of our model were initialized to random values, and updated in each iteration by using Adam optimizer [17] with default parameter configuration adopted. Particularly, to avoid over-fitting, a dropout layer with a keep rate of 0.95 was appended after the average-pooling layers. Besides, we decayed the learning rate (initialized as 0.002) exponentially every 1000 iterations by a decay factor of 0.9. Last, we applied L-2 regularization to all layers, with the regularization strength was set to 10^{-4}. When training our model, the balanced dataset was randomly divided into training set, validation set and test set by a ratio of 0.7:0.1:0.2. Intermediate model snapshots were taken every 100 iterations with a mini-batch size of 128, and the snapshot that performed best on validation test was selected as the final model. The dataset was trained 5 times in total.

Accuracy (ACC), sensitivity (SEN) and specificity (SPE) were used to evaluate the proposed model. Specifically, ACC represents the overall performance of our model in correctly classifying heartbeats. SEN measures the ability of our model to not miss abnormal heartbeats, while SPE assesses how good our model is at not misjudging normal heartbeats.

3.3 Evaluation Results

In this section, we first compare the performance of our model with that of some similar network structures, to demonstrate the superiority of our network structure. We then analyze the effect of different network layers on the overall performance. Last, we compare our model with 3 state-of-the-art arrhythmias classification methods.

Comparison with Similar Network Structures. By changing the implementation of each network layer, some similar network structures were created. Specifically, besides SB-LSTM, three other LSTM layers were designed, i.e., unstacked directional LSTM (UD-LSTM), unstacked bidirectional LSTM (UB-LSTM) and stacked directional LSTM (SD-LSTM). Besides TD-CNN, we also designed a one-dimensional CNN layer (OD-CNN). Since the input of OD-CNN should be a one-dimensional vector, only the output of LSTM layer that corresponds to the last timestamp was fed into the OD-CNN. It's notable that only the number of LSTM cell layers, the direction of LSTM and the dimension of the filters contained in CNN were changed when creating these new layers. Especially, we also considered the network structures with or without the wavelet (W) layer and the fusion (F) layer, denoted as with-WF and without-WF, respectively. In total, we got 16 combinations of the wavelet layer, LSTM layer, CNN layer and fusion layer, whose performance is summarized in Table 2.

Table 2. The performance of combinations of different layers.

Network structures			ACC (%)	SEN (%)	SPE (%)
With-WF	UD-LSTM	OD-CNN	91.0	90.9	91.1
		TD-CNN	94.6	94.1	96.6
	UB-LSTM	OD-CNN	93.7	94.0	92.7
		TD-CNN	97.7	98.3	95.6
	SU-LSTM	OD-CNN	95.3	95.4	94.8
		TD-CNN	98.0	98.2	97.3
	SB-LSTM	OD-CNN	96.4	96.6	95.4
		TD-CNN	***99.5***	***99.9***	***98.2***
Without-WF	UD-LSTM	OD-CNN	89.5	88.7	92.7
		TD-CNN	92.0	93.1	87.9
	UB-LSTM	OD-CNN	91.4	92.0	89.0
		TD-CNN	93.8	94.0	93.4
	SU-LSTM	OD-CNN	92.8	93.3	91.0
		TD-CNN	93.3	93.0	94.5
	SB-LSTM	OD-CNN	92.4	91.7	95.4
		TD-CNN	***94.8***	***94.6***	***95.7***

Table 3. The confusion matrix of heartbeats for the balanced dataset.

Balanced dataset		Predicted label					ACC (%)	SEN (%)	SPE (%)
		N	S	V	F	Q			
True label	N	17412	163	3	104	14	99.5	98.2	99.9
	S	56	17667	8	3	2	99.7	99.6	99.7
	V	35	18	17534	141	8	99.7	98.7	99.9
	F	6	1	19	17705	5	99.7	99.8	99.6
	Q	9	9	3	3	17712	99.9	99.9	100

It's obviously that our model obtains the best performance compared with the similar network structures. Concretely, under fixed LSTM layer and CNN layer, the utilization of wavelet layer and fusion layer can significantly improve the classification performance, which may be due to the fact that (1) the wavelet layer can decompose the ECG into multiple time-frequency resolutions, hence, more hidden information can be extracted for classification; (2) the fusion layer can effectively eliminate the bias of each intermediate classification result, so that it can obtain more robust results. In addition, among four kinds of LSTM layer, SBLSTM and UDLSTM achieve the best and worst performance respectively. It is because that UD-LSTM can only extract simple long-term dependencies of ECG from just one direction, while SB-LSTM can mine complex long-term dependencies in both directions, and transform them into higher-level forms, which is conducive to depicting the overall variation trend of ECG more accurately. Moreover, compared with OD-CNN, TD-CNN usually yields better performance, which is probably because that the input of TD-CNN is made up of the outputs of LSTM layer at each moment, therefore, the local temporal relationship of ECG can be exploited to characterize each ECG wave component more accurately.

The confusion matrix of our model is shown in Table 3, which shows that more than 99.2% of the ECG heartbeats are correctly classified. Additionally, all the criteria of each class exceed 99.0%, excepting for the SEN of class N and class V (98.2% and 98.7%), which means that our model is unbiased for each class. Particularly, class N and class Q respectively achieve the worst and the best classification performance among the five classes, which may be explained as follows: as shown in Table 1, class N contains five heartbeat subclasses and three of them (i.e., normal beat, left bundle branch block beat and right bundle branch block beat) have many instances, which makes the characteristics of class N not that typical. However, class Q includes three heartbeat subclasses but two of them (i.e., paced beat and unclassifiable beat) have very few instances, which makes the characteristics of the other subtype dominant, and hence can be identified more easily.

Table 4. The performance of different combinations of network layers.

Combinations of layers	ACC (%)	SEN (%)	SPE (%)
SB-LSTM	91.0	90.7	92.2
TD-CNN	90.0	89.2	92.9
SB-LSTM + TD-CNN	94.8	94.6	95.7
WL + SB-LSTM + TD-CNN + FL	99.5	99.9	98.2

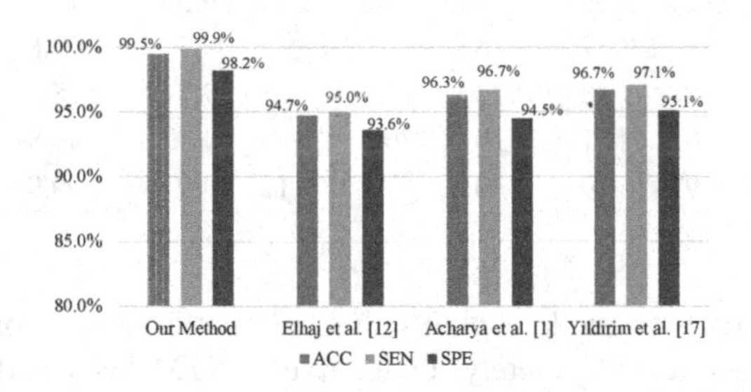

Fig. 4. The performance of the proposed model and baseline methods.

Contribution of Each Network Layer. To investigate the performance growth contributed by each network layer, we set up different combinations of network layers as shown in Table 4, from which three observations are obtained. First, the performance of SB-LSTM layer is similar to that of TD-CNN layer, which means that mining long-term and short-term fluctuation patterns of ECG is equally important for classifying arrhythmias. Second, by integrating SB-LSTM layer and TD-CNN layer together, about 4%–5% performance improvement is obtained, which indicates that more hidden information can be utilized when using this network structure. Third, the utilization of WL and FL further increases the ACC, SEN and SPE to 99.5%, 99.9% and 98.2% respectively, which indicates the effectiveness of the DWT-based ECG decomposition layer and the Sum Rule-based intermediate results fusion layer.

Comparison with State-of-the-Art Model. To evaluate the overall performance of our model, we compare it with three baseline methods proposed in [1, 12] and [17]. In [12], the wavelet sub-band coefficients of ECG and the HOS cumulants were extracted as linear and non-linear features, whose dimensionality were reduced by using PCA and ICA, respectively, based on which a SVM-RBF classifier was built. Acharya et al. [1] developed a 9-layer CNN consisting of 3 one-dimensional convolutional layers, 3 max-pooling layers and 3 fully connected layers, where the ECG segments were directly fed into the network as input. In [17], a model named DBLSTM-WS was designed, which combined wavelet sequences and a deep bidirectional LSTM together, where different wavelet sequences derived from same ECG segment are fused into one feature vector and then fed into the DBLSTM. Note that these three methods were recurred using the balanced dataset created in this study with the best parameter combinations were adopted, to avoid the unfairness caused by the dataset and parameters. As shown in Fig. 4, our model obtains the highest performance, and outperforms the method in [12] by 4.8%, 4.9% and 4.6% in terms of ACC, SEN and SPE, respectively. Furthermore, although both the CNN-based method [1] and the LSTM-based method [17] obtain good results, they are still weaker than our model, which means that extracting only long-term or short-term fluctuation patterns of ECG cannot characterize the ECG signal comprehensively. Moreover, we find that all three deep learning based models are superior to the feature extraction based method, which indicates that deep learning has stronger information extraction and fitting ability.

4 Conclusion and Future Work

In this paper, a novel arrhythmias classification model is proposed based on SB-LSTM and TD-CNN. It's an end-to-end method that doesn't require complex feature extraction and feature selection procedures. Particularly, by combining SB-LSTM and TD-CNN, the long-term and short-term fluctuation patterns of ECG can be characterized more comprehensively, which brings about 4% $\sim$ 5% performance growth. Moreover, the decomposition of ECG and the fusion of intermediate classification results of each ECG sublayer are proved to be useful for mining more hidden information and obtaining more robust classification results. Experimental results shows that our model is the optimal one compared with 15 similar network structures, and yield unbiased classification results for each class. Furthermore, it obtains 99.5% of accuracy, 99.9% of sensitivity and 98.2% of specificity respectively, which outperforms 3 state-of-the-art methods. In the future, we will design a network structure that is simpler than the one proposed in this study but achieves higher classification performance.

Acknowledgements. This work was partially supported by the National Natural Science Foundation of China (No. 61332013, No. 61672161), the National Key Research and Development Program of China (No. 2016YFB1001400), and the China Scholarship Council (No. 201706290110).

References

1. Acharya, U.R., Oh, S.L., Hagiwara, Y., et al.: A deep convolutional neural network model to classify heartbeats. Comput. Biol. Med. **89**, 389–396 (2017)
2. Xiong, Q., Proietti, M., Senoo, K., Lip, G.Y.H.: Asymptomatic versus symptomatic atrial fibrillation: a systematic review of age/gender differences and cardiovascular outcomes. Int. J. Cardiol. **191**, 172–177 (2015)
3. Huikuri, H.V., Castellanos, A., Myerburg, R.J.: Sudden death due to cardiac arrhythmias. New Engl. J. Med. **345**(20), 1473–1482 (2001)
4. Fuster, V., Ryden, L.E., Cannom, D.S., et al.: 2011 ACCF/AHA/HRS focused updates incorporated into the ACC/AHA/ESC 2006 guidelines for the management of patients with atrial fibrillation. J. Am. Coll. Cardiol. **57**(11), e269–e367 (2011)
5. ANSI/AAMI EC57: Testing and Reporting Performance Results of Cardiac Rhythm and ST Segment Measure Algorithms (2012)
6. Martis, R.J., Acharya, U.R., Adeli, H.: Current methods in electrocardiogram characterization. Comput. Biol. Med. **48**, 133–149 (2014)
7. Zhou, F.Y., Jin, L.P., Dong, J.: Premature ventricular contraction detection combining deep neural networks and rules inference. Artif. Intell. Med. **79**, 42–51 (2017)
8. Mant, J., Fitzmaurice, D.A., et al.: Accuracy of diagnosing atrial fibrillation on electrocardiogram by primary care practitioners and interpretative diagnostic software: analysis of data from screening for atrial fibrillation in the elderly (SAFE) trial. BMJ **7616**, 335–380 (2007)
9. Javadi, M., Arani, S.A.A.A., Sajedin, A., Ebrahimpour, R.: Classification of ECG arrhythmia by a modular neural network based on mixture of experts and negatively correlated learning. Biomed. Signal Process. Control **8**(3), 289–296 (2013)
10. Chang, P.C., Lin, J.J., Hsieh, J.C., Weng, J.: Myocardial infarction classification with multi-lead ECG using hidden Markov models and Gaussian mixture models. Appl. Soft Comput. **12**(10), 3165–3175 (2012)
11. Kutlu, Y., Kuntalp, D.: Feature extraction for ECG heartbeats using higher order statistics of WPD coefficients. Comput. Methods Program Biomed. **105**(3), 257–267 (2012)
12. Elhaj, F.A., Salim, N., Harris, A.R., Swee, T.T., Ahmed, T.: Arrhythmia recognition and classification using combined linear and nonlinear features of ECG signals. Comput. Methods Program Biomed. **127**, 52–63 (2016)
13. Jiang, H., Zhou, R., Zhang, L., Wang, H., Zhang Y.: Sentence level topic models for associated topics extraction. World Wide Web. https://doi.org/10.1007/s11280-018-0639-1
14. Peng, M., Zeng, G., Sun, Z., Huang, J., Wang, H., Tian, G.: Personalized app recommendation based on app permissions. World Wide Web **21**(1), 89–104 (2018)
15. Khalaf, A.F., Owis, M.I., Yassine, I.A.: A novel technique for cardiac arrhythmia classification using spectral correlation and support vector machines. Expert Syst. Appl. **42** (21), 8361–8368 (2015)
16. Liu, F., Zhou, X., Wang, Z., Wang, T., Ni, H., Yang, J.: Identifying obstructive sleep apnea by exploiting fine-grained BCG features based on event phase segmentation. In: IEEE BIBE, pp. 293–300 (2016)
17. Yildirim, Ö.: A novel wavelet sequence based on deep bidirectional LSTM network model for ECG signal classification. Comput. Biol. Med. **96**, 189–202 (2018)
18. Liu, F., Zhou, X., Wang, Z., Ni, H., Wang, T.: OSA-weigher: an automated computational framework for identifying obstructive sleep apnea based on event phase segmentation. J. Ambient Intell. Hum. Comput. (2018). https://doi.org/10.1007/s12652-018-0787-2

19. Liu, F., Zhou, X., Wang, Z., Wang, T., Zhang, Y.: Identification of hypertension by mining class association rules from multi-dimensional features. In: ICPR 2018, pp. 3114–3119 (2018)
20. Yeh, Y.C., Chiou, C.W., Lin, H.J.: Analyzing ECG for cardiac arrhythmia using cluster analysis. Expert Syst. Appl. **39**(1), 1000–1010 (2012)
21. Goldberger, A.L., Amaral, L.A.N., Glass, L., et al.: PhysioBank, PhysioToolkit, and PhysioNet: components of a new research resource for complex physiologic signals. Circulation **101**(23), e215–e220 (2000)
22. Liu, F., Zhou, X., Wang, Z., et al.: A light-weight data preprocessing and integrative scheduling framework for health monitoring. In: IEEE-EMBS BHI, pp. 192–195 (2016)
23. Andreotti, F., Carr, O., Pimentel, M.A.F., Mahdi, A., Vos, M.D.: Comparing feature-based classifiers and convolutional neural networks to detect arrhythmia from short segments of ECG. Comput. Cardiol. **44**, 1 (2017)
24. Krizhevsky, A., Sutskever, I., Hinton, G.E.: ImageNet classification with deep convolutional neural networks. In: NIPS, pp. 1097–1105 (2012)
25. Coşkun, M., Uçar, A., Yıldırım, Ö., et al.: Face recognition based on convolutional neural network. In: IEEE MEES, pp. 376–379 (2017)
26. Ren, S., He, K., Girshick, R., Zhang, X., Sun, J.: Object detection networks on convolutional feature maps. IEEE Trans. Pattern Anal. Mach. Intell. **39**(7), 1476–1481 (2017)
27. Singh, B.N., Tiwari, A.K.: Optimal selection of wavelet basis function applied to ECG signal denoising. Digit. Signal Process. **16**(3), 275–287 (2006)
28. Chawla, N.V., Bowyer, K.W., Hall, L.O., Kegelmeyer, W.P.: SMOTE: synthetic minority over-sampling technique. J. Artif. Intell. Res. **16**, 321–357 (2002)
29. Martis, R.J., Acharya, U.R., Min, L.C.: ECG beat classification using PCA, LDA, ICA and discrete wavelet transform. Biomed. Signal Process. Control **8**(5), 437–448 (2013)
30. Wang, Z., Zhou, X., Zhao, W., Liu, F., Ni, H., Yu, Z.: Assessing the severity of sleep apnea syndrome based on ballistocardiogram. PLoS ONE **12**(4), e0175351 (2017)
31. Xie, J., Wang, Z., Yu, Z., Guo, B.: Enabling efficient stroke prediction by exploring sleep related features. In: IEEE UIC, pp. 452–461 (2018)

An Efficient and Resource-Aware Hashtag Recommendation Using Deep Neural Networks

David Kao[1], Kuan-Ting Lai[2(✉)], and Ming-Syan Chen[1]

[1] National Taiwan University, Taipei 10617, Taiwan
dkao@arbor.ee.ntu.edu.tw, mschen@ntu.edu.tw
[2] National Taipei University of Technology, Taipei 10608, Taiwan
ktlai@ntut.edu.tw

Abstract. The goal of this research is to design a system that can predict and recommend hashtags to users when new images are uploaded. The proposed hashtag recommendation system is called HAZEL (HAshtag ZEro-shot Learning). Selecting right hashtags can increase exposure and attract more fans on a social media platform. With the help of the state-of-the-art deep learning technologies such as Convolutional Neural Network (CNN), the recognition accuracy has improved significantly. However, hashtag prediction is still an open problem due to the large amount of media contents and hashtag categories. Using single machine learning method will not be sufficient. To address this issue, we combine image classification and semantic embedding models to achieve the expansion of recommended hashtags. In this research, we show that not all hashtags are equally meaningful, and some are not suitable in recommendation. In addition, by periodically updating semantic embedding model, we ensure that the hashtags being recommended follow the latest trends. Since the recommended hashtags have not received any training examples in the first place, it fulfills the concept of Zero-shot learning. We demonstrate that our system HAZEL can successfully recommend hashtags that are the most relevant to each image input by applying our design to a larger scale of image-hashtag pairs on Instagram.

Keywords: Hashtag recommendation · Zero-shot learning · Deep learning

1 Introduction

There are many different kinds of social media platforms such as Facebook, Instagram, Twitter, and Tumblr, and each of them serves a different function and usage. Some promote text communication, and some encourage media content exchanging. Instagram is one of the most popular social media platforms that provides users the ability to share images and videos easily. Instagram has gained over 800 million active users monthly as of 2017. It already has over 20 billion images in the cloud and is still growing. As a result, one of the most important tasks is to organize these images into coherent categories. Fortunately, the system of hashtagging was developed early on, and many users have adopted the habit of hashtagging their media contents. Any word starting with a "#" symbol in the description area is a hashtag. Hashtags organize media

Q. Yang et al. (Eds.): PAKDD 2019, LNAI 11440, pp. 150–162, 2019.
https://doi.org/10.1007/978-3-030-16145-3_12

contents using keywords that facilitate browsing through relevance. It is possible to have multiple hashtags on a single content.

Unlike major information sharing platforms such as Facebook, Instagram supports only the sharing of images with hashtags and brief descriptions. In our experiments, except that image-hashtag pairs are used as our datasets, there is no additional metadata (e.g. user's information) included. Also, to achieve the goal of this research, we have to first answer the following questions: How can we learn and differentiate new hashtags? How do we recommend reasonable hashtags to users based on their new media content input? How do we ensure the suggested hashtags are popular and current? Since new hashtags are created based on trends that directly depends on the time of year and internet buzzwords, the temporal effects apply to the use of hashtags.

While it is possible to train every hashtag with their associated images, it is highly inefficient due to the constantly evolving nature of social platforms like Instagram. Especially training billions of hashtags with associated images probably requires a supercomputer to execute the task; it is not feasible to train a gigantic machine learning model in a standard laboratory with limited resources. This research aims to address all these limitations and recommend a system that can function using a model trained from defined samples and limited hardware resource.

In this research, we explore a way to recommend hashtags to users using deep neural networks. We downloaded more than 140,000 raw images from the top 100 hashtag categories from Instagram for training and testing. We also collected more than 400,000 corpora (3 million) hashtags. We then utilized three CNN imaging recognition models (ImageNet, Places, and PASCAL VOC) to extract features for the later use of SVM model building. After that, a trained Word2Vec (Skip-gram) model is used to expand the vocabulary options of hashtags. The expansion of hashtag terms fits the concept of zero-shot learning due to not receiving any training examples in the first place. In addition, 60,000 additional images are downloaded based on the recommended hashtags for further verification. We wanted to ensure that the recommended hashtags were relevant to the image contents.

More specifically, to solve the challenging problems mentioned before, our solutions are outlined as follows: (i) Utilize more than one CNN models to abstract more features from the datasets and to gain uniqueness for each class. (ii) Create a semantic embedding space to locate highly probably and its neighboring hashtags. (iii) Update the semantic embedding space periodically in order to recommend hashtags that meet the social networking trends. With these compositions, we are able to recommend relevant hashtags to users for their newly input media contents.

To summarize, our main contribution is a novel idea to recommend multiple hashtags that follow the latest trends to users. The proposed system HAZEL seeks to bridge the gap between users and the hashtagging scenario that changes and evolves constantly. HAZEL saves the hassle of the hashtagging process, and hashtags that are recommended will be popular and up to date. With a snap of a finger, users are ready to upload their media contents with bountiful hashtags to choose from. More importantly, the back-end architecture of HAZEL is an approach to resource-aware hashtag recommendation system. The experimental results are encouraging and with the help of HAZEL, hashtagging will become a user-friendly practice.

2 Preliminaries

2.1 Related Work

Although accessing a large amount of data is easier and cheaper nowadays, it is still difficult to train a multi-million-categories object recognition system. Many researchers are experimenting new ways and leveraging existing technologies in their object recognition projects. Models like zero-shot learning [1–3] and semantic embedding methods [24] can predict the categories that were unseen at the training stage.

Facebook leveraged CNN image classification and embedding models on their hashtag prediction project for Facebook [4]. They fused users' metadata (age, gender, etc.) in their predicting process. Google also conducted a research on visual recognition system that incorporated textual information. Their motivation is to increase the classifying outcomes without training a large scaled imaging model. They showed that their model DeViSE (Deep Visual-Semantic Embedding model) can make predictions about tens of thousands of image labels that are not observed during model training [5, 6].

2.2 Convolutional Neural Network

Neural Network has been used widely in recent years, especially in the fields of artificial intelligent and machine learning. Computer vision becomes one of the popular researching topics, and convolutional neural network is particularly beneficial in recognizing objects and segmentation of images [12–14]. The concept of CNN is to learn informational parameters along the layers from the input. The grid-like topology and matrix multiplication using convolutional dot product is how CNN find its parameters. Another important feature of CNN is called pooling. Pooling allows the reduction of the actual spatial size. It prevents the effect of overfitting and preserves important information only. Common pooling methods include max pooling, average pooling, and L2 norm pooling. Usually pooling layers can be found between two convolutional layers [8, 9]. Generally, millions of parameters are calculated in training a CNN. Two of the popular CNN networks, Residual Network [10, 11] and ZF Net [7], are used in our experiment. CNN trained models ImageNet [15–17], Places [18–21], and PASCAL VOC [22] are used in imaging feature extraction in our research as well.

2.3 Semantic Embedding Model

Semantic or word embedding modeling is a technique to represent words in vector forms. Often, it involves mathematical embedding transformation from one space to the other. For example, textual transformation is a process of mapping from a space with one dimension per word to a lower continuous vector space. Practices like neural networks, dimensional reduction, and probabilistic modeling can be used as ways of mapping [23]. Researchers can utilize these methods in their Natural Language Processing (NLP) work. The most successful method is Word2Vec [24].

There are two types of algorithms in Word2Vec. The first type is called the Continuous Bag-of-Words Model, which it trains each word with its surrounding neighbors in a sentence to construct a model. It can be used to predict the targeted vocabulary when seeing the similar structure of sentences as in the training stage. The second algorithm is called the Skip-gram Model, which the trained model can be used in finding neighboring words to a specific input. In this research, we utilized the Skip-gram model to train our hashtags and it gave us a promising result.

3 Model Architecture

Our proposed model HAZEL utilized a visual-semantic embedding framework similar to [3, 4, 6]. The framework is listed below. Given a visual conceptual embedding function Φ_I(x) and label embedding function Φ_L(y), the hashtag prediction model is of the form:

$$\mathrm{f(x,y)} = \Phi_I(x)'\Phi_L(y) \tag{1}$$

Instead of learning two embedding functions jointly as [6], we choose to use semantic concepts as our embedding space. There are two benefits of using predefined concepts. First, we only need to learn label embedding function, so the learning time is faster than joint learning; second, the embedding function is human interpretable, so we can analyze the content of each hashtag. For the label embedding $\Phi_L(y)$, a simple linear function $\boldsymbol{W}$ is used. The prediction model can be rewritten as:

$$\mathrm{f(x,y)} = \mathbf{W}'\Phi_I(x) \tag{2}$$

To solve the equation, we use the large-margin framework and use the hinge loss:

$$min. \quad \frac{1}{n}\sum_{i=1}^{n}\zeta_i + \lambda\|w\|^2$$

$$s.b.t \quad 1 - y_i(w'\Phi_I(x) - b) \leq \zeta_i \tag{3}$$

where $\zeta_i \geq 0$ for all i. Once we got the predicted hashtag, in the next step we can expand the hashtags using the Word2Vec model.

$$h_{w2v}(f(x,y)) \tag{4}$$

Basically h_{w2v} can be any search algorithm. In our experiment, the k-nearest neighbor is applied to search syntax hashtags. And the cosine distance (cosine similarity) is calculated to rank the relevant hashtags as shown below:

$$s(j,k) = \frac{j \cdot k}{\|j\|\|k\|}$$

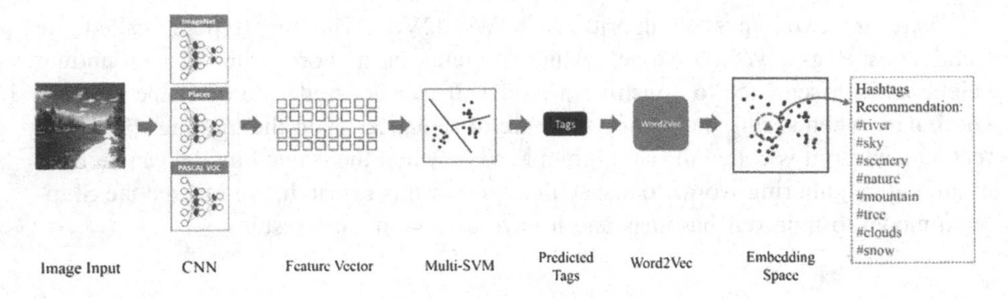

Fig. 1. Overall architecture of the hashtag recommendation system HAZEL.

$$s(j, k_1) > s(j, k_2) > \cdots > s(j, k_{n-1}) > s(j, k_n) \quad (5)$$

A predicted hashtag word vector $j$will have neighboring hashtags $k_1, k_2, \cdots k_n$. By ranking the cosine distance $s(j, k)$, we get the top $n$the most relevant hashtags to the hashtag $j$word vector.

The overall architecture of HAZEL is shown in Fig. 1. To recommend multiple hashtags to users automatically from a single image input, we utilize neural networks to achieve the goal. The recognition and feature extraction of images is done by the convolutional neural networks. We implement three CNN pre-trained models with different networks in extracting features from images. ImageNet1000 model can recognize 1,000 different objects from the images. Places365 model can recognize 365 different scenes or environments of the images, and PASCAL model can recognize 20 different objects and one of them is 'person' or 'human'. Three different pre-trained CNN models are used (ImageNet, Places, and PASCAL) to increase differentiation between hashtags. The total available number of features can be recognized coming from these models are 1,000 + 365 + 20 = 1,385. One thing to take notice of is that each CNN recognition process is done individually. We use ResNet50 trained model for the ImageNet and Places, and ZF Net trained model for the PASCAL. Keras is used for the ImageNet image processing. PyTorch is used for the Places image processing. And Caffe is used for the PASCAL image processing.

Next, we construct an SVM machine learning model that is learned from the extracted features. The Semantic embedding modeling for the vocabularies (hashtags) expansion is done by the Word2Vec Skip-gram model. Our approach shows that with minimal amount of trained image-hashtag pairs is still feasible to suggest great amount of comparable hashtags to users.

4 Data

Two types of datasets will be needed and collected separately in our experiment. That include the image dataset and hashtag dataset. Image features will be used in training an SVM model, and hashtags will be used in training a semantic embedding model. Data will be downloaded in a chronological order from Instagram.

4.1 Image Dataset

To begin our collection of images, we first investigate what hashtags are popular according to their image counts on Instagram. We start off by downloading more than 140,000 raw images from Instagram within top 100 ranked hashtags. The most used hashtag is #love with over 11 billion image counts. Followed by #instagood and #photooftheday. Specifically, Table 1 shows the top 100 hashtags being used on Instagram. We limited the image counts to around 1,400 per hashtag category in the dataset. The image dataset is downloaded through Python-syntax operation with assigned hashtags on Linux environment. We downloaded 140,000 + or 70 + gigabytes of images. We understand that there may be multiple hashtags assigned to an image by users. So we only pick the ones with only one hashtag labeled to prevent overlapping content in our dataset.

Hashtags that express feelings or abstract ideas are difficult for us to perceive its associating images. Since hashtagging can be a lenient gateway on expressing feelings, ideas, and even long paragraphs for users, it is certainly not so definite like labels of specific objects. However, we should not rely on our perception of the hashtag too quickly. Maybe there would be certain characteristics underlying any of the hashtags. Therefore, the role of feature extraction during model construction becomes crucial in revealing what each hashtag is holding.

Table 1. Top 100 Hashtags being used on Instagram and its ROC.

#instago (0.63)	#girl (0.65)	#cool (0.58)	#sunset (0.83)
#bestfriend (0.71)	#work (0.62)	#pink (0.71)	#fitness (0.75)
#winter (0.69)	#bored (0.69)	#smile (0.63)	#igaddict (0.55)
#green (0.75)	#throwback (0.65)	#i (0.58)	#shoes (0.84)
#tbt (0.56)	#photooftheday (0.54)	#food (0.81)	#gym (0.78)
#red (0.67)	#2012 (0.71)	#hair (0.75)	#breakfast (0.89)
#like4like (0.56)	#likeforlike (0.57)	#flowers (0.81)	#travel (0.72)
#party (0.77)	#white (0.63)	#instagramhub (0.54)	#nature (0.78)
#throwbackthursday (0.57)	#webstagram (0.53)	#awesome (0.53)	#dog (0.88)
#tattoo (0.80)	#sky (0.82)	#instalove (0.55)	#school (0.70)
#instacollage (0.60)	#happy (0.57)	#friends (0.67)	#vscocam (0.59)
#swag (0.62)	#landscape (0.83)	#architecture (0.88)	#nice (0.54)
#fashion (0.70)	#fit (0.72)	#photo (0.58)	#funny (0.70)
#family (0.65)	#boy (0.67)	#love (0.51)	#my (0.59)
#birthday (0.72)	#selfie (0.72)	#baby (0.79)	#motivation (0.71)
#eyes (0.73)	#iphoneonly (0.58)	#textgram (0.76)	#hot (0.64)
#black (0.68)	#style (0.68)	#memories (0.62)	#sun (0.73)
#fun (0.58)	#animals (0.84)	#beautiful (0.57)	#instagood (0.50)
#coffee (0.84)	#best (0.57)	#clouds (0.85)	#blessed (0.62)
#lol (0.68)	#nofilter (0.63)	#harrystyles (0.82)	#christmas (0.79)
#yummy (0.90)	#dress (0.84)	#onedirection (0.72)	#followme (0.56)
#cat (0.90)	#summer (0.62)	#weekend (0.58)	#sweet (0.70)
#beach (0.80)	#picstitch (0.57)	#night (0.70)	#cute (0.62)
#health (0.72)	#foodporn (0.88)	#music (0.73)	#followback (0.54)
#vintage (0.74)	#blue (0.66)	#tagsforlikes (0.53)	#art (0.72)

4.2 Hashtag Dataset

After downloading images, hashtag preparation will be the next step in the research. Hashtag crawling is what we called in the process. This time, we are not interested in images that only have one hashtag associate to it. Instead, we hope to download relating hashtags to any given hashtag terms. When the target hashtag is assigned, it tries to download the relevant hashtags of the images on the same web page.

For example, when we are browsing on the webpage www.instagram.com/explore/tags/dog/, it is targeting the #dog hashtag and it brings out all the related images having the tag #dog. At the same time, hashtags that relate to #dog like #puppies, #lovedog, #pet, and more will be shown on the same page. And this is how we can discover its relevant hashtags. In the next iteration, the program can start off with #puppies as a seeding word and start crawling more relevant hashtags by it. We set the iterating process with the top ranked 100 hashtags as seeding words to collect hashtags "corpus" from Instagram. We wrote a Python-syntax code to crawl hashtags corpus on Instagram automatically. The crawling process were left on for over two weeks, and the amount of collection is more than 400,000 corpora. That in total is around 3 million hashtags.

5 Experiments and Results

We recorded the softmax values at the last layer of the CNN networks. Thus, to process feature extraction of 140,000+ images, we obtained a matrix with 140,878 rows 1385 columns when each row represents an image.

At first, we trained an SVM model directly on these features of all images with their hashtags as ground truths. The results were not ideal because the average accuracy is only around 10% and precisions were low. Soon, we realized that image contents within each hashtag do not always have similar features. Hashtags like #tbt, #2012, and #lol have mixtures of all features that would be hard to classify reasonably. That scenario is expected as mentioned in Sect. 4.1.

We also concluded that certain hashtag categories are not ideal in SVM models training because it would depreciate the overall predicted accuracies. In addition, hashtags that are obscure or loosely defined for the media contents are not suitable to use as recommended hashtags. Thus, we limited the hashtag categories to ones that were more definite by measuring their area under the curve (AUC) of receiver operating characteristic (ROC) curve.

5.1 Sampling of Image-Hashtag Pairs

To show that the AUC of the ROC can be an indicator of the consistency of the image-hashtag pairs, we manually selected some hashtag categories for training an SVM model. These hashtags include #birthday, #coffee, #cat, #beach, #landscape, #selfie, #flowers, #architecture, #shoes, and #dog. The average classifying accuracy is around 60% when data split is 7:3 (training : testing). The overall accuracy is not too high due to similar features are shown in these categories. For example, a person can take a picture with their dog, and it will be difficult to define whether it should be tagged as #dog or #selfie. Or perhaps the dog is running in a garden, and it will be difficult to define whether it should be tagged as #dog or #flowers. Therefore, measuring the accuracies of predictions might not be applicable as there are several ways of tagging the same images. However, the averaged AUC for these selected categories are around 88%. Meaning that images in each hashtag category are sharing similar features consistently, and this is what we hope to see. The ROC results can be seen in Fig. 2.

In contrast, we did an experiment on the obscure hashtags with SVM. These hashtags include #lol, #photooftheday, #2012, #cool, #love, #funny, #hot, #instagood, #blessed, and #cute. The overall accuracy is only 20% when data split is 7:3 (training: testing). And the averaged AUC is around 60% as shown in Fig. 3.

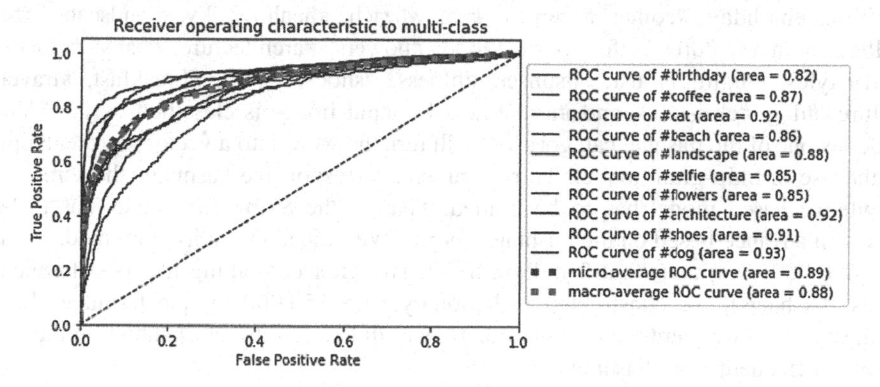

Fig. 2. ROC curve for 10 classes. The averaged AUC is about 88%.

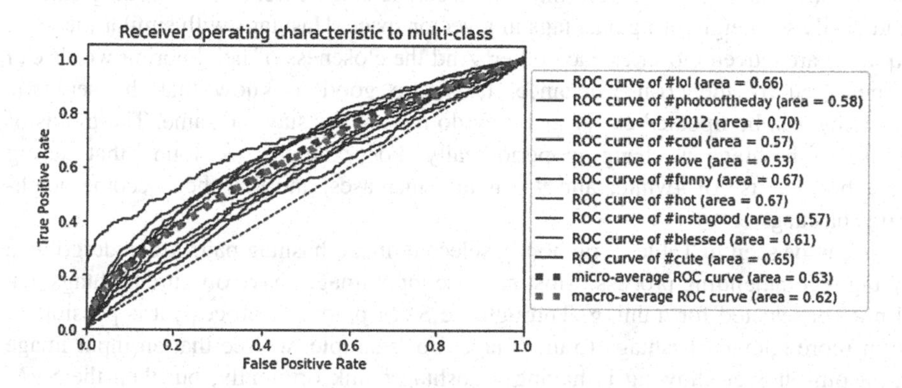

Fig. 3. ROC curve for 10 obscure classes. The averaged AUC is about 60%.

As we can see, the obscure hashtags give a lower performance on ROC as expected. However, these experimental results gave us an idea of determining which hashtags would be good to use for recommended tags. Each hashtag's media content on Instagram could be sampled and tested for their ROCs. If the ones with good ROC performance are found, it could be used as recommended tags. The ROC's AUC could be set as a desired threshold value to eliminate unwanted hashtags. We performed the ROC on the top 100 hashtags on Instagram to justify which hashtags could be used as recommended hashtags. Hashtag with the AUC of 80% or above is highlighted, and the result is shown in Table 1.

5.2 Predicting and Recommending Hashtags

Based on Table 1, we trained our SVM model with selected hashtags that were 70% or above of their ROC's. These hashtag categories include: #bestfriend, #party, #tattoo, #fashion, #birthday, #coffee, #yummy, #cat, #beach, #health, #sky, #landscape, #fit, #selfie, #animals, #dress, #foodporn, #food, #flowers, #architecture, #baby, #clouds, #harrystyles, #night, #music, #sunset, #fitness, #shoes, #gym, #breakfast, #travel, #nature, #dog, #christmas, and #art. When the input image is classified by the SVM model as one of the hashtag categories, it will turn thc word into a vector representation for the use of Skip-gram model. To recommend corresponsive hashtags, the semantic knowledge based model has to be trained. That is, the Skip-gram model has to be trained in advance based on the hashtags corpus. We trained the Skip-gram model with over 400,000+ corpus (about 3 million hashtags). After computing the most frequent words (hashtags), we constructed a dictionary with 150,000 unique hashtags. Surprisingly, many frequent words in the dictionary did match the ones shown in the top 100 most frequent used hashtags.

The recommended hashtags would be suggested in the k-NN algorithm manner. The Skip-gram model is trained by Tensorflow toolsets. The window size of the Skip-gram model is set to 32. Diagram like t-SNE can be constructed. In t-SNE diagram, we could easily see neighboring hashtags in a vector space. Hashtags with similar meaning or quality are indeed closer to each other. And the closeness of neighboring words can be measured by the cosine distance. It is also good to know that the semantic knowledge can be updated easily and they do not always stay the same. The trends of the use of hashtags is changing periodically. For example, we found that during December, the use of #winter and #christmas increases, and that they become neighboring hashtags.

As can be seen in Table 2, randomly selected image-hashtag pairs are undergoing a hashtag recommending process. Most of those input images have obscure hashtags that did not get selected for training. Through the SVM predicted process, it is possible to assign more relevant hashtags to the image. For example, we see that an input image with a pink flower showing is having a hashtag #pink originally, but then the SVM model is able to suggest #flowers and #nature. Or the #sweet one is an image of bread and a knife, and the SVM model is able to predict #breakfast and #yummy accurately.

In addition, we found that the probability with the value 40% or higher is considered a reasonable prediction. Thus, the predicted hashtags with 40% or higher is used as the targeting hashtags for its relevant hashtags. The more predicted hashtags that it had collected, the more hashtags it could recommend to users in the end. Users now have more recommended hashtags options to select from for their image sharing.

5.3 Verification and Inspection

To ensure that users were getting relevant recommended hashtags for their input contents, we inspected images behind these hashtags. We downloaded 300 images for each of the 10 recommended hashtags. The top 20 AUC's hashtag categories in Table 1 were selected for the hashtags recommendation expansion. Therefore, $300 \times 10 \times 20 = 60{,}000$ additional images were downloaded for verification. We used the original trained SVM model to test these newly downloaded images. For example, the relevant recommended hashtags for #architecture are #architecturephotography, #architecture, #architecturelovers, #archilove, #building, #archilovers, #architecturehunter, #arch_daily, #architectural, and #architectlife. Images behind these recommended hashtags do indeed share common features with predicted hashtags most of the time. Experimental results are shown in Fig. 4.

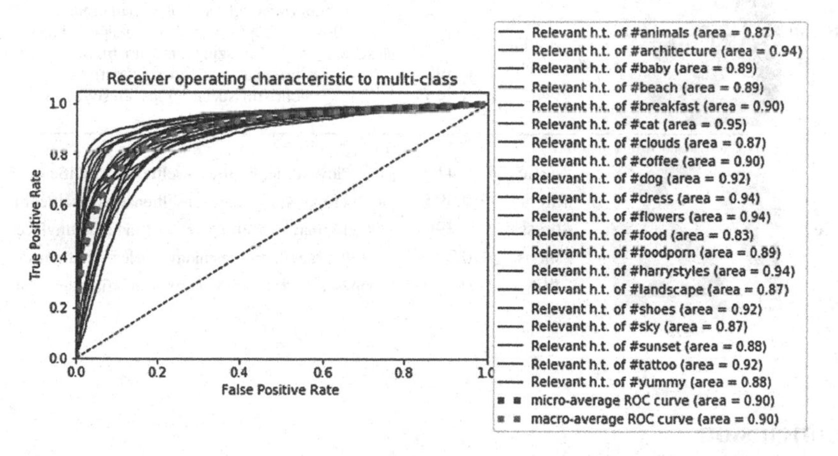

Fig. 4. ROCs of recommended hashtags. The averaged AUC is about 90%.

Table 2. Examples of hashtags recommendation.

Original	Image Input	Predicted Hashtags		Hashtags Recommendation
#pink		#flowers #nature #clouds #health #art	0.634 0.407 0.374 0.363 0.361	#flower, #flowerpop, #flowerpicture, #flowery, #flowering, #natureflower, #prettyflower, #flowersmagic, #purpleflower, #fishingdog, #nature, #naturebeautiful, #naturephotography, #natuurliefhebber, #naturelovers, #natur_perfection
#blue		#travel #architecture #clouds #bestfriend #music	0.471 0.462 0.457 0.435 0.397	#traveling, #internationaltravel, #traveller, #travelblogger, #travelphotography, #architecturephotography, #architechture, #architecturelovers, #archilove, #building, #cloudssky, #cloud, #cloudappreciationsociety, #sky☁, #cloudporn, #bestfriends, #buddy, #bestie
#tbt		#gym #fit #fitness #health #harrystyles	0.447 0.447 0.427 0.390 0.365	#cloud, #cloudappreciationsociety, #sky #gymrats, #gymfit, #gymworkout, #gymlove, #gymlife, #gymtime, #gymislife, #backworkout, #crosstraining, #fitness, #machdichwahr, #fitinspiration, #bcaa, #fitmen, #fit, #crosstraining, #whey, #bcaa, #bodyweight, #regime
#bestfriend		#dog #animals #christmas #bestfriend #baby	0.560 0.436 0.410 0.392 0.382	#dogs, #lifeofdogs, #puppy, #mixdogs, #pupylove, #dogporn, #dogbreeds, #animal, #minizoo, #animallovers, #happyanimal, #animalpost, #catvsdog, #excelent_dogs, #amazingpet, #christmastree, #xmas, #christmastime, #merrychristmas, #christmasdecorations, #festive
#selfie		#selfie #night #dress #shoes #fit	0.416 0.395 0.390 0.382 0.380	#loveselfie, #selfies, #selfiestick, #selfielover, #selfietime, #instaselfies, #selfienation, #selfiegram, #selfiemania, #selfielove, #selfination, #myface, #selfiequeen, #selfiesunday, #selca, #selfiemode, #myselfie, #selfiemood, #mirrorselfie, #selfiegirl

6 Conclusion

The system HAZEL that can predict and recommend hashtags to users when they upload new images on Instagram is achieved in this research. While combining image classification and semantic embedding models, we gain the expansion of recommended hashtags. We also show that not all hashtags are equally meaningful. By evaluating the ROC of the hashtag categories, we can justify the ones that should be kept or eliminated in the hashtag recommendation pool. In addition, by crawling and updating hashtags corpus for training the semantic embedding model periodically, we can ensure that recommended hashtags are popular and in the latest trends. Although the improvement of imaging feature extraction in CNN models can always be made in constructing a better predicting model, our state-of-the-art system HAZEL is highly scalable and could be applied to other social media platforms in recommending relevant hashtags to users automatically.

References

1. Wang, X., et al.: Zero-shot image classification based on deep feature extraction. IEEE Trans. Cogn. Dev. Syst. **10**, 432–444 (2016)
2. Changpinyo, S., et al.: Synthesized classifiers for zero-shot learning. In: Proceedings of the IEEE Conference on Computer Vision and Pattern Recognition (2016)
3. Xian, Y., Schiele, B., Akata, Z.: Zero-shot learning-the good, the bad and the ugly. arXiv preprint arXiv:1703.04394 (2017)
4. Denton, E., et al.: User conditional hashtag prediction for images. In: Proceedings of the 21th ACM SIGKDD International Conference on Knowledge Discovery and Data Mining. ACM (2015)
5. Weston, J., Bengio, S., Usunier, N.: WSABIE: scaling up to large vocabulary image annotation. In: IJCAI, vol. 11 (2011)
6. Frome, A., et al.: DeViSE: a deep visual-semantic embedding model. In: Advances in Neural Information Processing Systems (2013)
7. Zeiler, M.D., Fergus, R.: Visualizing and understanding convolutional networks. In: Fleet, D., Pajdla, T., Schiele, B., Tuytelaars, T. (eds.) ECCV 2014. LNCS, vol. 8689, pp. 818–833. Springer, Cham (2014). https://doi.org/10.1007/978-3-319-10590-1_53
8. Stutz, D.: Understanding convolutional neural networks. In: Seminar Report, Fakultät für Mathematik, Informatik und Naturwissenschaften Lehr-und Forschungsgebiet Informatik VIII Computer Vision (2014)
9. He, K., Sun, J.: Convolutional neural networks at constrained time cost. In: Proceedings of the IEEE Conference on Computer Vision and Pattern Recognition (2015)
10. He, K., et al.: Deep residual learning for image recognition. In: Proceedings of the IEEE Conference on Computer Vision and Pattern Recognition (2016)
11. Targ, S., Almeida, D., Lyman, K.: Resnet in resnet: generalizing residual architectures. arXiv preprint arXiv:1603.08029 (2016)
12. Simonyan, K., Zisserman, A.: Very deep convolutional networks for large-scale image recognition. arXiv preprint arXiv:1409.1556 (2014)
13. Szegedy, C., et al.: Going deeper with convolutions. In: Proceedings of the IEEE Conference on Computer Vision and Pattern Recognition (2015)
14. He, K., et al.: Delving deep into rectifiers: surpassing human-level performance on ImageNet classification. In: Proceedings of the IEEE International Conference on Computer Vision (2015)
15. Deng, J., et al.: ImageNet: a large-scale hierarchical image database. In: IEEE Conference on Computer Vision and Pattern Recognition, CVPR 2009. IEEE (2009)
16. Russakovsky, O., et al.: ImageNet large scale visual recognition challenge. Int. J. Comput. Vis. **115**(3), 211–252 (2015)
17. Krizhevsky, A., Sutskever, I., Hinton, G.E.: ImageNet classification with deep convolutional neural networks. In: Advances in Neural Information Processing Systems (2012)
18. Zhou, B., et al.: Places: a 10 million image database for scene recognition. IEEE Trans. Pattern Anal. Mach. Intell. **40**, 1452–1464 (2017)

19. Zhou, B., et al.: Places: an image database for deep scene understanding. arXiv preprint arXiv:1610.02055 (2016)
20. Zhou, B., et al.: Learning deep features for scene recognition using places database. In: Advances in Neural Information Processing Systems (2014)
21. Xiao, J., et al.: SUN database: large-scale scene recognition from abbey to zoo. In: 2010 IEEE Conference on Computer Vision and Pattern Recognition (CVPR). IEEE (2010)
22. Everingham, M., et al.: The PASCAL visual object classes challenge: a retrospective. Int. J. Comput. Vis. **111**(1), 98–136 (2015)
23. Girshick, R., et al.: Rich feature hierarchies for accurate object detection and semantic segmentation. In: Proceedings of the IEEE Conference on Computer Vision and Pattern Recognition (2014)
24. Mikolov, T., et al.: Efficient estimation of word representations in vector space. In: arXiv preprint arXiv:1301.3781 (2013)

Dynamic Student Classiffication on Memory Networks for Knowledge Tracing

Sein Minn[1(✉)], Michel C. Desmarais[1], Feida Zhu[2], Jing Xiao[3], and Jianzong Wang[3]

[1] Polytechnique Montreal, Montreal, Canada
{sein.minn,michel.desmarais}@polymtl.ca
[2] Singapore Management University, Singapore, Singapore
fdzhu@smu.edu.sg
[3] Ping An Technology (Shenzhen) Co., Ltd., Shenzhen, China
{xiaojing661,wangjianzong347}@pingan.com.cn

Abstract. Knowledge Tracing (KT) is the assessment of student's knowledge state and predicting whether that student may or may not answer the next problem correctly based on a number of previous practices and outcomes in their learning process. KT leverages machine learning and data mining techniques to provide better assessment, supportive learning feedback and adaptive instructions. In this paper, we propose a novel model called Dynamic Student Classification on Memory Networks (DSCMN) for knowledge tracing that enhances existing KT approaches by capturing temporal learning ability at each time interval in student's long-term learning process. Experimental results confirm that the proposed model is significantly better at predicting student performance than well known state-of-the-art KT modelling techniques.

Keywords: Massive open online courses · Knowledge tracing · Key-value memory networks · Student clustering · LSTMs

1 Introduction

Guiding human for solving problems efficiently and effectively is a recurring topic in educational research. Knowledge tracing (KT) gained credibility in this research community to provide appropriate and adaptive guidance in the learning process. KT aims to assess skills that are mastered or not, and use this information to tailor learning experience, whether in MOOCs, in a tutoring system or in web, results to name a few example for applications. For example, when a problem such as "$1 + 2 \times 3.5 =?$" is given to a student, she has to master the skills of *addition* and *multiplication* for solving that problem. The probability of getting a correct answer mainly depends on the mastery level of these two

This work was supported by NSERC Canada, Discovery grant program and Pinnacle lab for analytics at Singapore Management University.

Q. Yang et al. (Eds.): PAKDD 2019, LNAI 11440, pp. 163–174, 2019.
https://doi.org/10.1007/978-3-030-16145-3_13

skills behind that problem. Mastering a skill can be achieved by doing practices on that skill. The goal of knowledge tracing is to track the knowledge state of students based on observed outcomes on their previous practices [5]. This task is also known as student modelling. Research on KT can be traced back to the late 1970s and a wide array of Artificial Intelligence and Knowledge Representation techniques have been explored [3,14]. In environments where the student learns as she interacts with the system, which is specifically the case for learning environments such as MOOCs, modeling student skill mastery involves a temporal dimension. For instance, a sequence problems involving the same skills set may be failed at first, but succeeded later on because the student's skill mastery has increased. Yet, other factors can influence the success outcome, such as the two problem's difficulty level, forgetting, guessing and slipping, and an array of other factors that induce noise if they are not accounted for [11,12].

The dynamic nature of KT in learning environments leads to approaches that have the capacity to model temporal or sequential data. In this paper we propose a novel model for knowledge tracing, Dynamic Student Classified Memory Networks (DSCMN). The model can capture temporal learning ability in student's long-term memory and assess mastery of knowledge state simultaneously. Temporal learning ability refers to the rate of learning of specific skills. It can be tied to phenomena like wheel spinning, where a student fails to learn a skill even after numerous attempts [17]. It relies on an RNN architecture to improve performance prediction. The hypothesis we make is that learning ability can change in time and tracing this factor can help predict future performance.

The rest of this paper is organized as follow. Section 2 reviews the related work on the student modelling techniques for predicting student's performance from data. Section 3 presents the proposed DSCMN model. Section 4 mentioned experimental datasets used. Experimental results are described in Sect. 5 and finally Sect. 6 concludes this work and discusses future avenues of research.

2 Knowledge Tracing

Successful learning environments such as the Cognitive tutors series and the ASSISTments platform rely on some form of KT [6]. In these systems, each problem is labeled with underlying skills required to correctly answer that problem. KT can be seen as the task of supervised sequential learning problem where the model is given student past interactions with the system that includes: skills $S = \{s_1, s_2, \ldots, s_t\}$ along with response outcomes $R = \{r_1, r_2, \ldots, r_t\}$. KT predict the probability of getting a correct answer to the next problem, which mainly depends on mastery of corresponding skill s associated with problems $P = \{p_1, p_2, \ldots, p_t\}$. So we can define the probability of getting correct answer as $p(r_t = 1|s_t, X)$ where $X = \{x_1, x_2, \ldots, x_{t-1}\}$ and $x_{t-1} = (s_{t-1}, r_{t-1})$ is a tuple containing response outcomes r to skill s at time $t-1$. Then, we review here four of the best known state-of-the-art KT modelling methods for estimating student's performance.

2.1 Bayesian Knowledge Tracing (BKT)

BKT is arguably the first model to relax the assumption on static knowledge states. Earlier approaches such as IRT would assume the student does not learn between answers, which is a reasonable assumption for testing, but not for learning environments. BKT was introduced for knowledge tracing within a learning environment [5]. In its original form, it also assumes a single skill is tested per item, but this assumption is relaxed in later work. The data are partitioned by skill and learning a model on each dataset leads to a specific model for each skill s. The standard BKT model is comprised of 4 parameters which are typically learned from the data while building a model for each skill. The model's inferred probability mainly depends on those parameters which are used to predict how a student masters a skill given that student's chronological sequence of incorrect and correct attempts to questions of that skill thus far [1]. To estimate the probability that a student knows the skill given his performance history, BKT needs to have four probabilities: $P(L_0)$, initial probability of mastery of skill L_0; $P(T)$, transition probability from a state of non mastery to mastery; and $P(S)$, slipping, the probability of a wrong answer in spite of mastery, and $P(G)$, guessing, the probability of a correct answer in spite of non mastery.

$$P(L_n|Correct) = \frac{P(L_{n-1})(1-P(S))}{P(L_{n-1})(1-P(S)) + (1-P(L_{n-1}))P(G)} \tag{1}$$

$$P(L_n|Incorrect) = \frac{P(L_{n-1})P(S)}{P(L_{n-1})P(S) + (1-P(L_{n-1}))(1-P(G))} \tag{2}$$

$$P(L_n) = P(L_{n-1}|Outcome) + (1 - P(L_{n-1}|Outcome))P(T) \tag{3}$$

2.2 Deep Knowledge Tracing (DKT)

Similar to BKT, Deep Knowledge Tracing (DKT) [13] works on the skill sequence of attempts but the author leveraged the advantages of neural networks and break the restriction of skill separation and binary state assumption. It takes the previous history of attempts by students and transforms each attempt into one-hot encoded feature vector. Then, those features are fed into a neural network as input and pass information through the hidden layers of the network and onto the output layer. The output layer provides the predicted probability that the student would answer that particular problem correctly in the system.

DKT uses Long Short-Term Memory (LSTM) [8] to represent the latent knowledge space of students along with the number of practices dynamically. The increase in student's knowledge through an assignment can be inferred by utilizing the history of student's previous performance. DKT summarizes a student's knowledge state of all skills in one hidden state in hidden layer. A student's skill mastery state at certain time stamp is defined by the following equations:

$$h_t = \tanh(W_{hx}x_{t-1} + W_{hh}h_{t-1} + b_h), \tag{4}$$

$$p(s_t) \in y_t = \sigma(W_{yh}h_t + b_y), \tag{5}$$

In DKT, both tanh and the sigmoid function are applied element wise and parameterized by an input weight matrix W_{hx}, recurrent weight matrix W_{hh}, initial state h_0, and readout weight matrix W_{yh}. Biases for latent and readout units are represented by b_h and b_y.

2.3 Dynamic Key-Value Memory Network (DKVMN)

DKVMN was proposed an enhancement to DKT that utilizes a neural network module called external memory slots to encode the knowledge state of students and use as key and value components to encode the knowledge state of students [19]. Learning or forgetting of a particular skill are stored in those two components and controlled by read and write operations through additional attention mechanisms. Learning or forgetting of a particular skill is stored in those two components and controlled by read and write operations through additional attention mechanisms.

Unlike DKT, DKVMN performs reading and writing operations to perform local state transitions by avoiding global and unstructured state-to-state transformation in hidden layer. Knowledge state of a student is traced by reading and writing to the value memory slots using correlation weight computed from input skills and the key memory slots. It is comprised of three main steps:

Correlation: The correlation weight of input skill s_t is computed by utilizing the softmax activation of the inner product between k_t and key memory slot $M^k(i)$:

$$w_t = Softmax(k_t^T M^k(i)) \tag{6}$$

where k_t is the continuous embedding vector of s_t and $Softmax(z_i) = e^{z_i}/\sum_j e^{z_j}$ id differentiable. Correlation weight w_t is used in both reading and writing process in later.

Reading: The mastery m_t of s_t is retrieved by weighted sum of values in value memory slots by using w_t:

$$m_t = \sum_{i=1}^{N}(w_t(i) M_t^v(i)) \tag{7}$$

Prediction: The probability of answering the problem with underlying skill $p(s_t)$ is calculated by using mastery level m_t:

$$f_t = \tanh(W_1^T [m_t, k_t] + b_1) \tag{8}$$

$$p(s_t) = \sigma(W_2^T f_t + b_2) \tag{9}$$

Where $\tanh(z_i) = (e^{z_i} - e^{-z_i})/(e^{z_i} + e^{-z_i})$ and $\sigma(z_i) = 1/1 + e^{-z_i}$.

Writing: After the student answers the problem, the model will update the value memory according to response (r_t) of student. A joint embedding of $x_t = (s_t, r_t)$ is converted into embedding values v_t and written to the value memory with same correlation weight w_t used in read process. Erasing is performed before adding new information by using:

$$e_t = \sigma(E^T v_t + b_e), \tag{10}$$

$$\tilde{M_t^v}(i) = M_{t-1}^v(i)[1 - w_t(i)e_t], \tag{11}$$

where 1 is a row-vector of all 1-s. If both the weight at the location and the erase element are 1, the elements of a memory location are reset to zero. No changes are performed in the case of either erase signal or the weight is zero. After erasing previous memory, a_t is used to update each memory slots in value memory.

$$a_t = \tanh(D^T v_t + b_a)^T, \tag{12}$$

$$M_t^v(i) = M_{t-1}^v(i) + w_t(i)a_t, \tag{13}$$

where E and D are the transformation matrix with shape of $d_v \times d_v$. This erase-followed-by-add mechanism allows forgetting and strengthening knowledge states of student learning process [19] which is not able in other RNN based models.

2.4 Deep Knowledge Tracing with Dynamic Student Classification (DKT-DSC)

DKT-DSC was introduced to overcome the problem of short-term learning ability of student when applied to the KT task [10]. During the evaluation of student learning ability, DKT-DSC encodes student's past performance by using the following equation:

$$Correct(s_j)_{1:z} = \sum_{z=1}^{Z} \frac{(s_j = 1)}{|N_j|}, \tag{14}$$

$$Incorrect(s_j)_{1:z} = \sum_{z=1}^{Z} \frac{(s_j = 0)}{|N_j|}, \tag{15}$$

$$R(s_j)_{1:z} = Correct(s_j)_{1:z} - Incorrect(s_j)_{1:z}, \tag{16}$$

$$d_{1:z}^i = (R(s_1)_{1:z}, R(s_2)_{1:z}, \ldots, R(s_n)_{1:z}). \tag{17}$$

in which $Correct(s_j)_{1:z}$ represents the ratio of skill s_j being correctly answered and $Incorrect(s_j)_{1:z}$ for the ratio of incorrectly answered. $d_{1:z}^i$ is the vector of skills mastery for student i on n skills and for time interval 1 to z. $|N_j|$ is the total number of attempts that student i has done on each skill s_j. Evaluating temporal learning ability by assigning students into a group with similar ability c_z at each time interval z by using k-means clustering on encoded data $d_{1:z-1}^i$ [2,9,10] and then the model invokes an RNN to trace her knowledge according to her learning ability c_z at each time interval.

$$h_t = \tanh(W_{hx}[x_{t-1}, s_t, v_t] + W_{hh}h_{t-1} + b_h), \tag{18}$$

$$p(s_t^{c_z}) \in y_t = \sigma(W_{yh}h_t + b_y), \tag{19}$$

where v_t contains success and failure levels of skill s_t until time $t-1$ thus far.The probability of $p(s_t^{c_z}) \in y_t$ represents the probability of getting correctness of problem with associated skill s_t for the student with her temporal learning ability c_z in that time interval z while other models ignore the long-term learning ability in student learning process. DKT-DSC applies temporal value of student's learning ability at each time interval to improve the individualization in long-term knowledge tracing process.

3 Dynamic Student Classification on Memory Networks (DSCMN)

Despite a better accuracy to assess the mastery of skills than DKT, each of the above models has deficiencies for dealing with the KT task. In both DKT and DKVMN, temporal student's long-term learning ability is ignored. So the model cannot evaluate which level of learning ability the student achieved for a given time interval in a long term learning process. In DKT and DKT-DSC, LSTM uses single state vector to encode the temporal information of student knowledge state with corresponding learning ability in a single hidden layer.

To model learning ability, we propose a novel model called Dynamic Student Classification on Memory Networks (DSCMN) that builds upon the advantages of DKVMN and DKT-DSC. DSCMN predicts student performance based on both of evaluated temporal student's long-term ability and assessed mastery of skills simultaneously at each time interval.

Evaluating Temporal Student's Learning Ability: Learning is a process that involves practice: students become proficient through practice. Besides, learning is also affected by the individual's ability to learn, or to become proficient with more or less practice [10].

To detect the regularities and changes of temporal learning ability of a student over series of time intervals in long-term learning process, we need to encode student past performance for predicting her learning ability in the current time interval with DKT-DSC's Eq. 17. The encoded vector of student's past performance is updated after each time interval. The K-means algorithm [9] is used to evaluate the temporal long-term learning ability of students in both training and testing at each time interval z by measuring the Euclidean distance between centroids achieved after training the DKT-DSC process [10] and assigning a nearest cluster label c_z as the long-term learning ability of a student at time z. Evaluation is started after the first 20 attempts and updated after each 20 attempts have been made by a student. For first time interval, every student is assigned with initial learning ability 1 as described in Fig. 1.

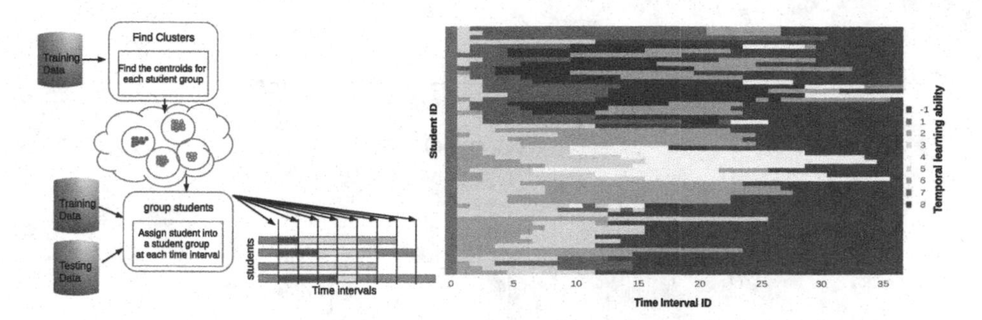

Fig. 1. Evaluation process of student's learning ability (Left) and Evolution of temporal learning ability in long-term learning process of random 56 students in ASSISTments 2009 dataset (Right)

Calculating Problem Difficulty: We measure problem difficulty as one of 10 levels [11,12]. Note that, in this study, the difficulty is associated with problems, not with skills themselves. The difficulty of a problem, $p_j \in D$, is determined as:

$$pd(p_j) = \begin{cases} \delta(p_j, pd), & \text{if } |N_j| \geq 4 \\ pd, & \text{else} \end{cases} \tag{20}$$

where:

$$\delta(p_j, pd) = \frac{\sum_i^{|N_j|} |\{p_{ij} == 0\}|}{|N_j|} \cdot pd \tag{21}$$

and where N_j is the set of students who attempted problem p_j, and p_{ij} is the outcome of the first attempt from student i, to problem p_j. An outcome of 0 is a failure. Constant pd is the problem difficulty (levels) that we wish to retain. It is described in function $\delta(p_j, pd)$ as shown in Eq. (20). Essentially, $\delta(p_j, pd)$ is a function that maps the average success rate of problem p_j onto (10) levels. For problems those do not have responses from at least 4 different students, problems with $|N_j| < 4$ in the dataset, we apply $pd_t = 5$ corresponding to 0.5 difficulty for those problems.

3.1 Assessing Student's Mastery of Skill

To assess the mastery of skill according to temporal learning ability, we use read and write process into two key and value memory slots as like in DKVMN. DSCMN also assess the mastery of skills using the correlation weight computed from the input skill and the key memory. In DSCMN, instead of using embedding values, one-hot encoded inputs are directly fed into memory networks by using Eqs. (6) and (7). Mastery m_t of skill s_t is obtained from reading process before writing x_t to value memory. Then the model writes x_t into value memory by using Eqs. (10) and (12) after the student answered the problem at time t (Fig. 2).

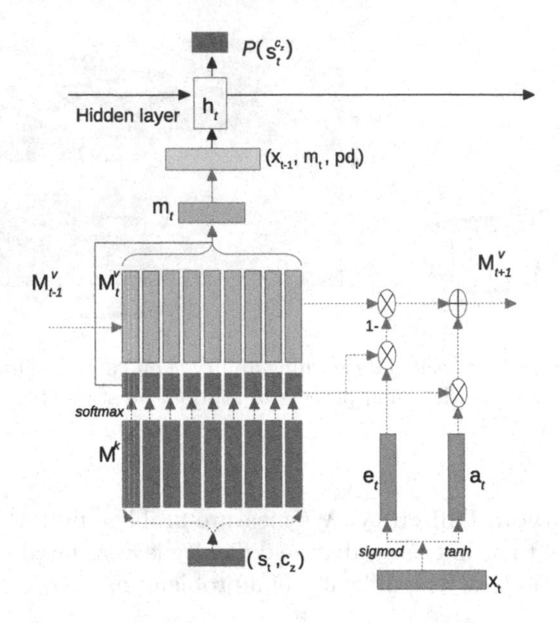

Fig. 2. Architecture of DSCMN

Prediction: The probability of answering the problem with underlying skill $p(s_t)$ of student in temporal learning ability c at time interval z is estimated by feeding previous response and mastery of skill in temporal learning ability of student into additional hidden layer and prediction is performed as follows:

$$h_t = \tanh(W_h[x_{t-1}, m_t, pd_t] + W_{hh}h_{t-1} + b_h), \tag{22}$$

$$p(s_t^{c_z}) \in y_t = \sigma(W_{yh}h_t + b_y), \tag{23}$$

Where c_z is the temporal learning ability of that student at time interval $t \in z$ and $[x_{t-1}, m_t, pd_t]$ encoded x_{t-1} previous response of skill s_{t-1} and mastery of skill s_t with skill id s_t and associated problem difficulty pd_t in temporal learning ability of student i at time interval z. DSCMN possess the ability to assess the mastery of skill based on temporal ong-term learning ability. Prediction is performed by using these factors and stored it in hidden state h_t.

Optimization: To improve the predictive performance of RNN based models, we trained with the cross-entropy loss l between p_t and actual response r_t for all RNN based models as follows:

$$l = \sum_t (r_t \log p_t + (1 - r_t)\log(1 - p_t)), \tag{24}$$

4 Datasets

In order to validate the proposed model, we tested it on four public datasets from two distinct tutoring scenarios in which students interact with a computer-based learning system in the educational settings: (1) ASSISTments[1]: an online tutoring system that was first created in 2004 which engages middle and high-school students with scaffolded hints in their math problem. If students working on ASSISTments answer a problem correctly, they are given a new problem. If they answer it incorrectly, they are provided with a small tutoring session where they must answer a few questions that break the problem down into steps. Datasets are as follows: ASSISTments 2009–2010 (skill builder), ASSISTments 2012–2013, ASSISTments 2014–2015. (2) Cognitive Tutor. Algebra 2005–2006 [4][2]: is a development dataset released in KDD Cup 2010 competition from Carnegie Learning of PSLC DataShop. For all datasets, only first correct attempts to original problems are considered in our experiment. We remove data with missing values for skills and problems with duplicate records. To the best of our knowledge, these are the largest publicly available knowledge tracing datasets (Table 1).

Table 1. Overview of datasets

Dataset	Number of				Description
	Skills	Problems	Students	Records	
Cognitive Tutor	437	15663	574	808,775	KDD Cup 2010 [4]
ASSISTments	123	13002	4,163	278,607	2009–2010 [15]
	198	41918	28,834	2,506,769	2012–2013 [7]
	100	NA	19,840	683,801	2014–2015 [18]

5 Experimental Study

In this experiment, we assume every 20 attempts made by a student is a time interval. The total number of temporal values for student's learning ability used in our experiment is 8 (7 clusters and 1 for initial ability before evaluation in initial time interval for all students) for DKT-DSC and DSCMN. Five fold cross-validations are used to make predictions on all datasets. Each fold involves randomly splitting each dataset into 80% training students and 20% test students of the each datasets. For the input of DKVMN, initial values in both key and value memory are learned in training process. For other models, one hot encoding is applied. Initial values in value memory represents the initial knowledge state as prior difficulty for each skill and is fixed in the testing process.

We implement the all models with Tensorflow and DKT, DKT-DSC and DSCMN share same structure of fully-connected hidden nodes for LSTM hidden layer with the size of 200 for DKT, 200 for DKT-DSC and output size of memory

[1] https://sites.google.com/site/assistmentsdata/.

[2] https://pslcdatashop.web.cmu.edu/KDDCup/downloads.jsp.

networks for DSCMN. For speeding up the training process, mini-batch stochastic gradient descent is used to minimize the loss function. The batch size for our implementation is 32, corresponding 32 to split sequences from each student. We train the model with a learning rate 0.01 and dropout is also applied to avoid over-fitting [16]. We set the number of epochs to 100. All models are trained and tested on the same sets of training and testing students.

For BKT, we use the Expectation Maximization (EM) algorithm and limit the number of iterations to 200. We learn models for each skill and make predictions separately. The results for each skill are averaged.

Table 2. AUC result for all tested datasets. Note that the results of DKT-DSC are slightly different than [10] after fixing bugs in the original code.

Datasets	Model				
	BKT	DKT	DKVMN	DKT-DSC	DSCMN
Cognitive Tutor	64.2 ± 1.0	78.4 ± 0.6	78.0 ± 0.0	79.2 ± 0.5	$\mathbf{86.0 \pm 0.5}$
ASSISTments09	65.1 ± 1.0	72.1 ± 0.5	71.0 ± 0.5	73.5 ± 0.6	$\mathbf{81.2 \pm 0.4}$
ASSISTments12	62.3 ± 0.0	71.3 ± 0.0	70.7 ± 0.1	72.1 ± 0.1	$\mathbf{78.5 \pm 0.1}$
ASSISTments14	61.1 ± 1.0	70.7 ± 0.4	70.0 ± 0.1	$\mathbf{71.6 \pm 0.2}$	71.0 ± 0.01

In Table 2, DSCMN performs significantly better than state-of-the-art models in three datasets. On the Cognitive Tutor dataset, compared with the standard DKT which has an maximum test AUC of 78.4, 79.2 in DKT-DSC and only 78.0 in DKVMN. The DSCMN model can achieve AUC = 86.0, with a notable gain of 10% over the original DKT and DKVMN, and 8% over DKT-DSC. For the ASSISTments09 dataset, DSCMN also achieves about a 10% gain with AUC = 81.2, above DKT-DSC = 78.5, and well above the original DKT, with AUC = 71.3, and DKVMN with AUC = 70.7. On the ASSISTments12 dataset, DSCMN only achieved AUC = 0.71. In the latest ASSISTments14 dataset (which contains more students and less data compared to other three datasets and lacks problem information) DSCNM has AUC slightly lower than DKT-DSC.

Table 3. RMSE result for all tested datasets

Datasets	Model				
	BKT	DKT	DKVMN	DKT-DSC	DSCMN
Cognitive Tutor	0.44 ± 0.00	0.38 ± 0.01	0.38 ± 0.00	0.37 ± 0.03	$\mathbf{0.35 \pm 0.00}$
ASSISTments09	0.47 ± 0.01	0.45 ± 0.00	0.45 ± 0.01	0.43 ± 0.00	$\mathbf{0.40 \pm 0.00}$
ASSISTments12	0.51 ± 0.00	0.43 ± 0.00	0.43 ± 0.00	0.43 ± 0.00	$\mathbf{0.40 \pm 0.00}$
ASSISTments14	0.51 ± 0.00	0.42 ± 0.00	0.42 ± 0.00	0.42 ± 0.00	0.42 ± 0.00

In Table 3, when we compare the models in term of RMSE, BKT is lowest at 0.46 for ASSISTments09, 0.51 for ASSISTments12 and ASSISTments14, and 0.44 for Cognitive Tutor. RMSE results in all dataset is lowest for DSCMN, with 0.40, while all other models are no over 0.43 (except DKT in the Cognitive Tutor dataset and DSCMN in ASSISTments14). According to these results, DSCMN shows better performance than DKT-DSC and significantly better than other models in Cognitive Tutor, ASSISTments09, ASSISTments12 but a little lower than DKT-DSC in ASSISTments14.

6 Conclusion and Future Work

In this paper, we propose a new model, DSCMN, which can predict the student performance by gathering information from skills, problems and student: mastery level of skills of student on various problems at each time step, along with student learning ability at each time interval.

Experiments with four datasets show that the proposed model performs better in predictive performance than state-of-the-art KT models. Dynamic evaluation of student's temporal learning ability at each time interval plays a critical role and helps DSCMN capture more variance in the data, leading to more accurate predictions.

In our future work, we plan to adapt this model to problems associated with multiple skills and apply it in the recommendation of related problems.

References

1. d Baker, R.S.J., Corbett, A.T., Aleven, V.: More accurate student modeling through contextual estimation of slip and guess probabilities in Bayesian knowledge tracing. In: Woolf, B.P., Aïmeur, E., Nkambou, R., Lajoie, S. (eds.) ITS 2008. LNCS, vol. 5091, pp. 406–415. Springer, Heidelberg (2008). https://doi.org/10.1007/978-3-540-69132-7_44
2. Ball, G., Hall Dj, I.: A novel method of data analysis and pattern classification. Isodata, a novel method of data analysis and pattern classification. Technical report 5ri, project 5533 (1965)
3. Brown, J.S., Burton, R.R.: Diagnostic models for procedural bugs in basic mathematical skills. Cogn. Sci. **2**(2), 155–192 (1978)
4. Corbett, A.: Cognitive computer tutors: solving the two-sigma problem. User Model. **2001**, 137–147 (2001)
5. Corbett, A.T., Anderson, J.R.: Knowledge tracing: modeling the acquisition of procedural knowledge. User Model. User-Adapt. Interact. **4**(4), 253–278 (1994)
6. Desmarais, M.C., Baker, R.S.: A review of recent advances in learner and skill modeling in intelligent learning environments. User Model. User-Adapt. Interact. **22**(1–2), 9–38 (2012)
7. Feng, M., Heffernan, N., Koedinger, K.: Addressing the assessment challenge with an online system that tutors as it assesses. User Model. User-Adapt. Interact. **19**(3), 243–266 (2009)
8. Hochreiter, S., Schmidhuber, J.: Long short-term memory. Neural Comput. **9**(8), 1735–1780 (1997)

9. MacQueen, J., et al.: Some methods for classification and analysis of multivariate observations. In: Proceedings of the Fifth Berkeley Symposium on Mathematical Statistics and Probability, Oakland, CA, USA, vol. 1, pp. 281–297 (1967)
10. Minn, S., Yu, Y., Desmarais, M.C., Zhu, F., Vie, J.J.: Deep knowledge tracing and dynamic student classification for knowledge tracing. In: IEEE International Conference on Data Mining (2018)
11. Minn, S., Zhu, F., Desmarais, M.C.: Improving knowledge tracing model by integrating problem difficulty. In: IEEE International Conference on Data Mining, Ph.D. Forum (2018)
12. Pardos, Z.A., Heffernan, N.T.: KT-IDEM: introducing item difficulty to the knowledge tracing model. In: Konstan, J.A., Conejo, R., Marzo, J.L., Oliver, N. (eds.) UMAP 2011. LNCS, vol. 6787, pp. 243–254. Springer, Heidelberg (2011). https://doi.org/10.1007/978-3-642-22362-4_21
13. Piech, C., et al.: Deep knowledge tracing. In: Advances in Neural Information Processing Systems, pp. 505–513 (2015)
14. Polson, M.C., Richardson, J.J.: Foundations of Intelligent Tutoring Systems. Psychology Press, London (2013)
15. Razzaq, L., et al.: The assistment project: blending assessment and assisting. In: Proceedings of the 12th Annual Conference on Artificial Intelligence in Education, pp. 555–562 (2005)
16. Srivastava, N., Hinton, G., Krizhevsky, A., Sutskever, I., Salakhutdinov, R.: Dropout: a simple way to prevent neural networks from overfitting. J. Mach. Learn. Res. **15**(1), 1929–1958 (2014)
17. Wan, H., Beck, J.B.: Considering the influence of prerequisite performance on wheel spinning. In: International Educational Data Mining Society (2015)
18. Xiong, X., Zhao, S., Van Inwegen, E., Beck, J.: Going deeper with deep knowledge tracing. In: EDM, pp. 545–550 (2016)
19. Zhang, J., Shi, X., King, I., Yeung, D.Y.: Dynamic key-value memory networks for knowledge tracing. In: Proceedings of the 26th International Conference on World Wide Web, pp. 765–774. International World Wide Web Conferences Steering Committee (2017)

Targeted Knowledge Transfer for Learning Traffic Signal Plans

Nan Xu[1], Guanjie Zheng[2], Kai Xu[3], Yanmin Zhu[1](✉), and Zhenhui Li[2]

[1] Shanghai Jiao Tong University, Shanghai, China
{xunannancy,yzhu}@sjtu.edu.cn

[2] Pennsylvania State University, University Park, USA
{gjz5038,jessieli}@psu.edu

[3] Shanghai Tianrang Intelligent Technology Co., Ltd, Shanghai, China
kai.xu@tianrang-inc.com

Abstract. Traffic signal control in cities today is not well optimized according to the feedback received from the real world. And such an inefficiency in traffic signal control results in people's waste of time in commuting, road rage in the traffic jam, and high cost for city operation. Recently, deep reinforcement learning (DRL) approaches shed lights to better optimize traffic signal plans according to the feedback received from the environment. Most of these methods are evaluated in a simulated environment, but can not be applied to intersections in the real world directly, as the training of DRL relies on a great amount of samples and takes a long time to converge. In this paper, we propose a batch learning framework where the targeted transfer reinforcement learning (*TTRL-B*) is introduced to speed up learning. Specifically, a separate unsupervised method is designed to measure the similarities of traffic conditions to select the suitable source intersection for transfer. The proposed framework allows batch learning and this is the first work to consider the impact of slow learning in RL on real-world applications. Experiments on real traffic data demonstrate that our model accelerates learning with good performance.

Keywords: Deep reinforcement learning · Transfer learning · Traffic signal control

1 Introduction

Traffic congestion is one of the most severe issues in cities today. Part of the reason is that the current traffic signal system is not efficient. Current traffic signal control systems such as SCATS [9] and SCOOT [7] adjust traffic signals locally according the loop sensor data at the intersection and they do not optimize globally based on the feedback received from the real world. Recent

N. Xu—Work done during an internship at Tianrang.

Q. Yang et al. (Eds.): PAKDD 2019, LNAI 11440, pp. 175–187, 2019.
https://doi.org/10.1007/978-3-030-16145-3_14

attempts using Deep Reinforcement Learning (DRL) have shown more effective results [4,13,24,25]. Compared with traditional transportation approaches, DRL approaches can learn and adjust traffic signal policy based on the feedback received from the environment.

However, if we directly apply DRL to traffic signal control problem, we face two key challenges: (1) the training of a DRL model usually requires millions of samples [12], but we usually have very limited data on a new real-world intersection; (2) the principle of RL is trial-and-error and such error may cause severe implications in the real world. Therefore, we ask a critical question: how can we transfer the knowledge learnt from other intersections to this new intersection so we can try to reduce the error and speed up the learning process?

Transfer learning [14,19] and meta-learning [21,22] have been widely used to transfer knowledge from similar tasks to speed up the learning of target tasks. Recently, researchers apply this idea in DRL to play games [15,18,20]. In these problems, agents learning from different games separately will act as teachers to distill knowledge in various ways, e.g., policy regression [15,18] and high-level feature representation regression [15]. A student model may take over these knowledge and adapt itself while interacting with the new environment. However, such a useful approach has never been investigated in traffic signal control scenario.

In this paper, we propose a transfer learning model for traffic signal control on a series of intersections. Our model is an organic combination of three steps: (1) source task selection; (2) model and sample transfer; (3) a batch learning framework.

We first select proper source tasks for target using the similarity of embeddings of traffic volume variation. This can effectively avoid the negative transfer [15]. Previous methods either use domain knowledge [3] or rely on a joint learning model with a task classifier and a RL agent [20]. In our problem setting, using traffic data to measure the similarity is more accurate than using domain knowledge. We also find training a joint model requires much more data samples and it is significantly slow. *Second, we transfer the model and samples to the target intersection*: besides employing the model which mimics the teachers' actions as the pretrained model for the new intersection, we further refer to the teachers' samples to regulate the parameter update when applied to new intersections. *Third, we adopt a batch learning framework to further improve the knowledge distillation.* In each round, well-tuned transfer models are saved in a teacher pool. In the next round, these transferred models will also play the role of teachers. This will keep on distillating the knowledge to its most concise representation.

Our contributions can be summarized as follows:

- This is the first work to consider the effective transfer of RL algorithms trained on simulated traffic to the real-world traffic. This is essential to reduce the mistakes to be made in the real world.
- We propose an elegant transfer learning framework with unsupervised teacher selection and batch learning.

- We conduct comprehensive experiments on the real-world traffic datasets from Hangzhou, China. We show that our proposed method outperforms the baselines and each component of the proposed method makes its own contribution.

2 Related Work

2.1 Approaches for Traffic Signal Control

Traditional Transportation Approaches. The current road traffic is mainly managed by systems with two kinds of control: fixed-time [11,23] or vehicle-actuated signals [2,23]. Fixed-time control gives a fixed cycle and green ratio split, while vehicle-actuated determines the time to change signals according to a specific rule (e.g., whether the number of vehicles on the red direction is larger than a threshold). Some other transportation practice [17] also suggests to use the historical traffic volume to compute the cycle and green ratio split, in order to minimize the total travel time under certain traffic volume assumptions. However, those methods all depend heavily on either manually crafted rules or unrealistic assumptions. The policy that achieves good performance on one intersection cannot be applied to another efficiently, either.

Reinforcement Learning Approaches. RL approaches have been proved to achieve better performance in traffic signal control in recent studies. Early studies [1,25] used tabular methods to compute the reward for discrete state-action pairs. Unfortunately, continuous traffic attributes or high-dimensional features were never fully exploited. Recent deep reinforcement learning methods [4,8,13,24] further utilize the continuous traffic features to solve the problem. However, all these methods treat intersections as individuals, in which model parameters are learned from scratch. As a result, experience accumulated on previous intersections can not be utilized to speed up the learning on new intersections. This will result in slow learning and economic loss in real practice.

2.2 Methods for Knowledge Transfer

Transfer learning [14,19] and meta learning [21,22] are methodologies that people proposed to share the knowledge among tasks to boost the performance or speed up the learning. With transfer algorithms as key components in both methods, meta learning concentrates more on a continual stream of tasks while transfer learning may reasonably focus on a single pair of related tasks. Recently, they have been proved to benefit RL learning practices in many game tasks, e.g., Atari [15,18], Minecraft [20], etc. However, little efforts have been made to transfer the learning of traffic signal control problems to mitigate the real traffic congestion problem. Compared to the other transfer learning problems, learning to control traffic signals is cost-sensitive so that the transfer source and target need to be more carefully selected and the transferred knowledge needs to be better represented to avoid negative transfer [3,15,20]. Therefore, we need to develop a new transfer learning framework in this paper.

3 Problem Definition

In a single intersection with four-way traffic, there is a signal to direct the traffic. There are two kinds of traffic light settings and we call them phases, i.e., *Green-Horizon* (green light on the horizontal direction and red light on the vertical direction), *Red-Horizon* (red light on the horizontal direction and green light on the vertical direction).

Projecting the situation to the RL definitions, the traffic condition on this intersection, such as the position and speed of each vehicle, is treated as the environment. An agent is trained to decide whether to change the signal to the next phase (action is 1) or keep the current phase (action is 0). In each time slot, the agent takes an action, and receives a reward from the environment. Then, the agent updates the model after a certain period.

Problem 1. The goals of this paper are:

- Design a RL algorithm to control the traffic signal to minimize the total travel time of vehicles.
- Transfer the knowledge accumulated in learned intersections to the target intersections to speed up agent learning.

4 Method

Our model is a transfer learning solution to speed up learning in target tasks with experience accumulated in source tasks. In this section, we will first introduce a non-transfer RL method *IntelliLight* for signal control. Then we show the transfer properties of our model in three aspects: (1) source task selection; (2) model and sample transfer; (3) the batch learning transfer framework.

4.1 Non-transfer Reinforcement Learning Solution

Our signal control model *TTRL-B* follows the agent design and the network structure of model *IntelliLight* [24]. This non-transfer model is a DQN [12] solution and has two additional techniques, i.e., Memory Palace and Phase Gate, to enhance model performance. The agent takes the action with the maximum long-term reward and updates at the i-th iteration according to the following loss function:

$$L(\theta_i) = \mathbb{E}_{(s_t,a_t,r_t,s_{t+1})\sim U(\mathcal{D})}\left[\left(r + \gamma \max_{a_{t+1}} Q(s_{t+1}, a_{t+1}; \theta_i^-) - Q(s_t, a_t; \theta_i)\right)^2\right], \quad (1)$$

in which γ is the discount factor, θ_i, θ_i^- are the parameters of the Q-network at i-th iteration for action prediction and for target computation, respectively, $\mathcal{D}$ is the pool of stored samples.

4.2 *TTRL-B*: Targeted Transfer Reinforcement Learning in a Batch Learning Framework

To control the signal for a target traffic flow, the proposed model first looks for the most similar flows by analyzing their distance from the target in the embedding space. Then the model for the target task is built with weights initialized via model guidance and keeps on updating with sample guidance. To control signals on a set of intersections, *TTRL-B* will create batches of target tasks to form a batch learning framework.

Source Task Selection Based on Traffic Embedding. One policy, that successfully eases a congested intersection, plays a instruction role for controlling another high-traffic intersection. Given historical traffic condition of an intersection without deterministic labels, we treat targeted source selection as an unsupervised task where traffic similarities are measured by their distance from each other in an embedding space. Traffic flows are time series data with the number of passing vehicles over a certain time interval in each direction periodically recorded. To represent flow data of an arbitrary length by a fixed-dimensional vector, we build a long short-term memory (LSTM) [6] autoencoder to produce a dense representation that captures the road volume variation along time. In particular, the autoencoder consists of one encoder that first maps the sequence input to a fixed-dimensional vector, followed by one decoder that inversely reconstructs the original sequence. The reconstruction loss between the original and the generated sequence is minimized and we finally extract the state vector of the encoder at the final time step as the traffic representation. Note that the flow information on the intersection to be controlled is unknown, we replace it with the historical traffic data on this intersection from the same time period in the identical workday (or weekday) to represent the upcoming traffic condition.

As the euclidean distance among vectors is widely adopted for their similarity calculation [10,26], we calculate such distance between the target flow representation with that of the candidate source flows, each of which is controlled by an agent with a rich accumulation of samples and experience. k among the source candidates, which are closest to the target in the embedding space, are selected and their respective agents will transfer knowledge to the target agent.

Transfer Reinforcement Learning

Model Guidance. Given a set of source flows $F_1, \ldots, F_k$, the first step is to train a single network that can control signals of the source flows under the supervision of a set of DQN agents $A_1, \ldots, A_k$, which were once responsible for the source tasks. Agent A_i has a sample pool $\mathcal{D}_i^S = \{(s_t, Q^S(s_t, a)\}$, where the sample from the t-th time step consists of the current state s_t, and a vector $Q(s_t, a)$ of unnormalized Q-values with one value per action. The target network is trained with a mean-squared-error loss (MSE) that would match Q-values between the source and target network:

$$L_{MSE}(\theta) = \sum_{i=1}^{i=k} \sum_{(s_t, Q^S(s_t,a)) \in \mathcal{D}_i^S} \left\| Q^S(s_t, a) - Q^T(s_t, a) \right\|_2^2, \tag{2}$$

where $Q^S(s_t, a)$ is sampled from $\{\mathcal{D}_i^S | 1 \leq i \leq k\}$ to represent the Q-value predicted by the source network, $Q^T(s_t, a)$ is the Q-value predicted by the target network parameterized by θ.

For knowledge transfer from the source tasks to the target task, it is possible to replace MSE with other frequently adopted loss functions, e.g., negative log likelihood loss (NLL) [18], cross-entropy loss [15], Kullback-Leibler divergence (KL) [5,18], etc.

As the traffic signal control tasks have the identical state and action space, we directly use the weights of the previously trained target network as an instantiation for a new DQN model that will be trained on the target task. We call such supervised training of the target network as model guidance. Since the source and target flows are very close to each other in traffic embedding, model guidance from source agents will be effective in signal control on the target intersection.

Sample Guidance. Previous Experiments show that knowledge transfer from source tasks via model initialization does not always have significant positive effects on the target task [15]. Meanwhile, it has been pointed out when DQN algorithm was first proposed, that deep reinforcement learning tends to be unstable or even diverge for several causes: one is the correlations in the sequence of observations, another is the fact that small updates to Q may significantly change the policy and the data distribution [12]. Applying model guidance alone is likely to cause the same instability problem to DRL in the very beginning of the training, where samples accumulated from the new task are consecutive, limited and biased. One of the approaches for DQN to removing correlations in the observation sequence is to randomize over the data through experience replay. However, sample accumulation for replay memory needs plenty of time followed with great cost in signal control domain. Hence realizing experience replay based on samples from the target task does not benefit the model learning at the very beginning.

We introduce another transfer method called sample guidance, where the replay memory is filled with sufficient samples collected from the source agents' learning process prior to training on target tasks. Through sample guidance, the participation of source networks on the target network is not limited to the parameter initialization, but extended to every subsequent update. Based on the basic DQN update listed in Eq. 1, we define the parameter update for *TTRL-B* at the i-th iteration with sample guidance as follows.

$$\begin{aligned} L(\theta_i) = \mathbb{E}&_{(s_t, a_t, r_t, s_{t+1}) \sim U(\mathcal{D}^T, \mathcal{D}^S)} \\ &\left[\left(r + \gamma \max_{a_{t+1}} Q^T(s_{t+1}, a_{t+1}; \theta_i^-) - Q^T(s_t, a_t; \theta_i) \right)^2 \right], \end{aligned} \tag{3}$$

where samples are drawn uniformly at random from both the source and target sample pools, i.e., $\mathcal{D}^S = \{\mathcal{D}_i^S | 1 \leq i \leq k\}$ and $\mathcal{D}^T$, respectively.

A Batch Learning Framework. For a city with all the signals on roads controlled by traditional transportation systems, there is no experience in signal control by RL agents for transfer learning. To resolve such a cold-start problem, we accumulate experience in mediating synthetic flows for fast adaption of RL models to real-world traffic flows.

We believe that knowledge transfer from synthetic to real-world data is better than non-transfer but not the optimal. Synthetic data can hardly mimic every transportation characteristics, while two real flows can have a lot in common, e.g., similar volume trend in daytime, north-east arterial roads, etc. Knowledge transfer between the most similar real-world flows should always be advocated and realized.

Instead of transferring experience of signal control in synthetic flows to all of the real-world intersections, the target intersections are batch selected so that the current batch of roads has an unprecedented amount of source flow candidates than those in the previous batches. In particular, we utilize the Gaussian Mixture Model (GMM) [16] to group all the target flows in C clusters according to their traffic data embedding. Every time we pick the centroid traffic flow in each cluster as one of the target task in the current batch. After determining the source tasks for each target task, *TTRL-B* extracts samples from the source sample pools to learn a DQN model in a supervised way. Learning as well as evaluating on the target traffic flows is conducted with the initial model guidance and the sample guidance in each network update. After the end of each batch, the number of source flow candidates as well as their accumulated experience expands for the next batch of target flows.

5 Experiments

We conduct experiments on a simulation platform SUMO (Simulation of Urban MObility)[1]. All the compared algorithms are employed to control the traffic signal on isolated four-way intersections.

5.1 Datasets

Synthetic Data. Vehicles arrive at the approach at uniform rate in the four directions. We utilize 13 different arrival rates which range from 25 to 550 vehicles/hour/lane.

[1] http://sumo.dlr.de/index.html.

Table 1. Performance evaluated by 2 transfer measures: 1st hour and overall average travel time (in seconds).

Model	Off-peak Hours		Peak Hours	
	1st hour	Overall	1st hour	Overall
IntelliLight	70.52	52.92	49.59	75.35
TTRL-B	**35.68**	**32.14**	**34.31**	**70.72**

Real-World Data. We collect the traffic volume data from loop sensors during 04/01/2018-04/30/2018, in Hangzhou, China. There are 48 intersections in total, 22 of which have most sensor undamaged. As the number of vehicles passing one intersection varies dramatically throughout day, evaluation from each passenger's standpoint over a 24-h time span is not fair for models with good performance on low-density traffic. Therefore, we extract two 5-h segments from the whole-day traffic, i.e., Off-peak Hours and Peak Hours, and treat them as two separate datasets for a comprehensive model evaluation. Specifically, Off-peak Hours contains continuous traffic flows during which the maximum hourly volume is smaller than 350, while Peak Hours covers those above 350. The average hourly traffic per lane for Off-peak Hours and Peak Hours hours after division is 110.5 and 393.4 respectively.

5.2 Compared Methods

We compare the following models to illustrate the benefits of the proposed batch learning framework for targeted transfer. All hyperparameters of the baselines are carefully tuned.

- *IntelliLight* [24]: a recent solution for signaling on the basis of DQN, but with a phase-gated structure to enhance performance.
- *TTRL-B*: our **b**atchwise **t**argeted **t**ransfer **r**einforcement **l**earning based on *IntelliLight*, it maintains an expanding source pool where experience of controlling both synthetic and real-world traffic is accumulated in each batch.

5.3 Evaluation Metric

Average Travel Time (Duration). Travel time for a vehicle is defined as the time that one car spends from entering the approaching lane until leaving the intersection. We use the average travel time to evaluate different methods.

Transfer Evaluation. To measure the effects of transfer, we follow the two metrics suggested in [19]: jumpstart and transfer reward. Under this scenario, these measurements correspond to 1st hour performance and overall performance.

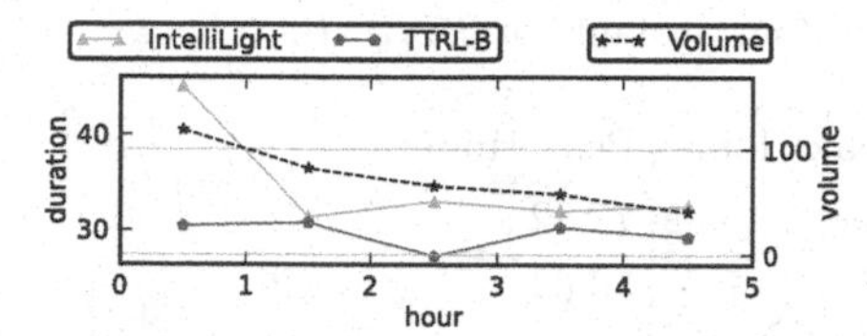

(a) Hourly volume of the intersection and duration of the passing vehicles.

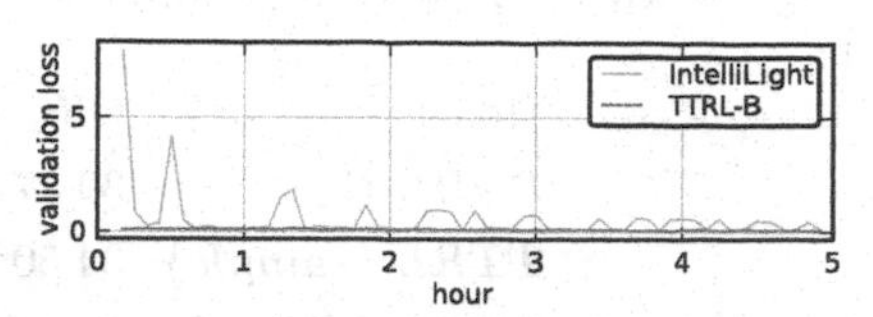

(b) Validation loss during training non-transfer and transfer learners.

Fig. 1. Case study of non-transfer and transfer models in Off-peak Hours on Moganshan Road and Wenyi Road in Hangzhou on April 2nd, 2018. We use this sampled intersection throughout this paper for case study.

5.4 Overall Performance

The results on real-world data are shown in Table 1. As expected, *IntelliLight* with randomly initialized parameters results in long travel time in the 1st hour and the performance gradually improves after 5-h training. In contrast, *TTRL-B* shows quick adaptivity, with a lower travel time obtained in the 1st hour and in the whole testing process as well. To better demonstrate the fast convergence and adaptability of the proposed model, we show the travel time of vehicles and the validation loss along time for the non-transfer and transfer RL models in Fig. 1a. Compared to the non-transfer model *IntelliLight*, *TTRL-B* always mediates the traffic better from the very start to the end with extremely low loss.

5.5 Variants of Our Model

To test effectiveness of the components in our model, we conduct experiments with the following variational models of *TTRL-B*:

- *RTRL-B*: a non-targeted transfer learner, which selects the source tasks randomly regardless of their similarity with the target.
- *TTRL-{sample}*: a targeted transfer learner in which sample guidance is removed deliberately.
- *TTRL-{model}*: a targeted transfer learner that lacks model guidance in knowledge transfer.
- *TTRL*: a targeted transfer learner without the batch learning framework, so that each target task only has a fixed number of source candidates whose experience is limited in synthetic traffic.

As shown in Table 2, none of the four variants can achieve comparable performance as *TTRL-B*. *RTRL-B* shows inferior performance as the experience from random source flows is not necessarily beneficial to the target. Model guidance alone (*TTRL-{sample}*) works fine only in Off-peak Hours (compared to *IntelliLight* in Table 1), while *TTRL-{sample}* gets trapped in serious negative

Table 2. Overall performance of four variants of *TTRL-B*.

Model	Off-peak Hours	Peak Hours
RTRL-B	39.47	77.82
TTRL-{sample}	34.50	76.64
TTRL-{model}	50.04	82.63
TTRL	33.27	73.42
TTRL-B	**32.14**	**70.72**

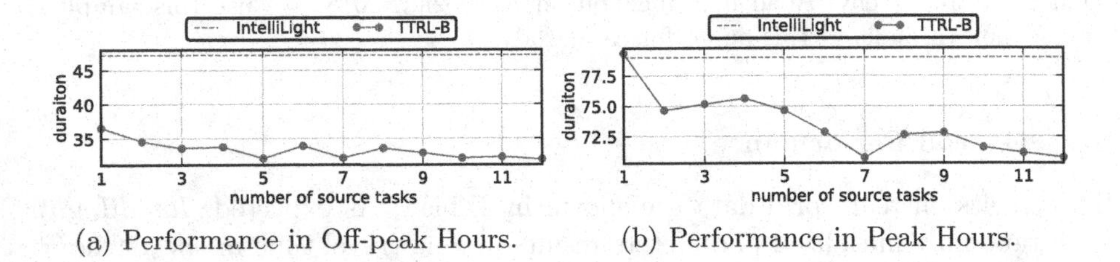

(a) Performance in Off-peak Hours. (b) Performance in Peak Hours.

Fig. 2. Parameter sensitivity of model *TTRL-B* in the number of source tasks.

transfer in both off-peak and peak hours. It has been proved that sample guidance and model guidance should be combined. Without the batch learning structure, *TTRL* also shows inferior results than *TTRL-B*, due to the never-expanded, experience-limited source pool.

5.6 Parameter Sensitivity

As shown in Fig. 2, our method achieves the best performance when experience from 7 source tasks are used to train the model. But generally, it is not sensitive to the number of source tasks as the travel time on intersections controlled by *TTRL-B* is always far below that of the non-transfer model *IntelliLight*.

5.7 Case Study of the Batch Learning Framework

To show the efficiency of knowledge transfer in the batchwise way in detail, we compare *TTRL-B* with the plain targeted transfer learner *TTRL*, which only has experience guidance from synthetic flows. In Fig. 3, we show the comparison of *TTRL-B* and*TTRL* in three aspects: traffic embedding, volume trends and periodical performance.

To visualize the relationships between flows, we map the high-dimensional traffic embedding in a 2-dimensional space. Figure 3a shows that both of two targeted transfer learners select source tasks that deal with traffic flows in a relative small euclidean distance to the target's flow. *TTRL-B* differs from *TTRL* as the former retains the most similar synthetic sources selected by *TTRL* and adds some close real-world ones. Based on the volume trend of source flows

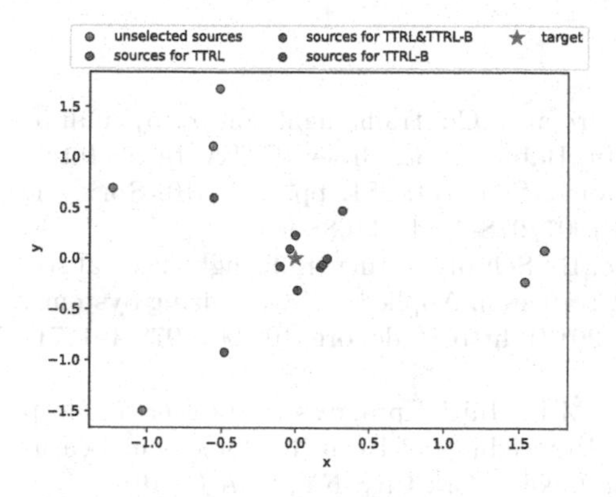

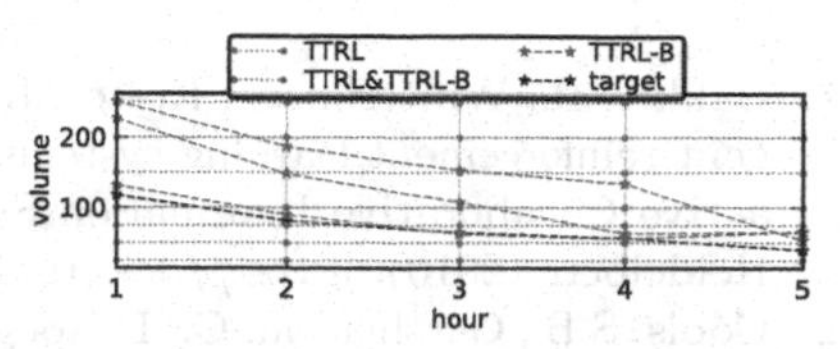

(b) Volume trends of selected flows.

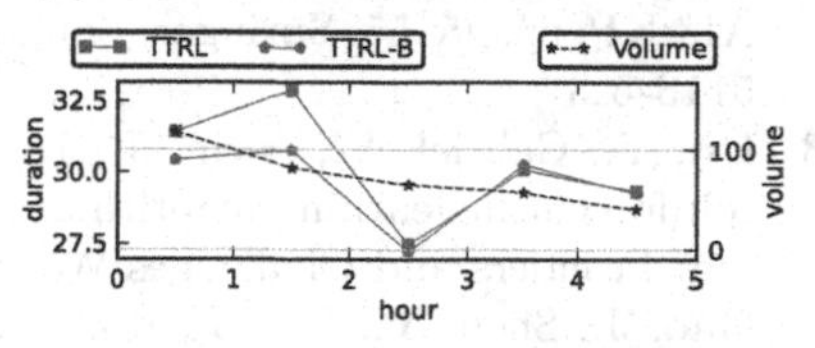

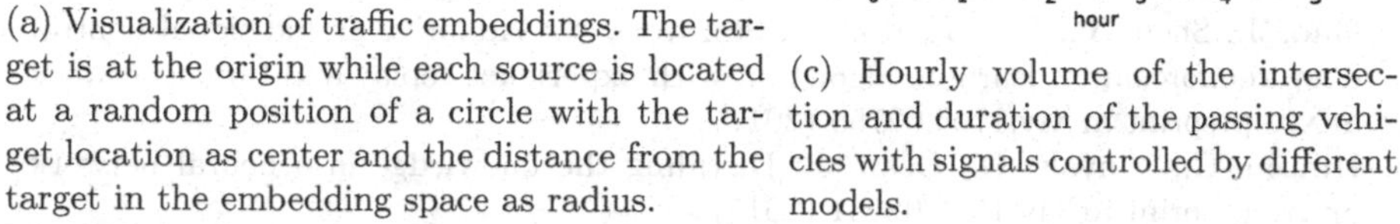

(a) Visualization of traffic embeddings. The target is at the origin while each source is located at a random position of a circle with the target location as center and the distance from the target in the embedding space as radius.

(c) Hourly volume of the intersection and duration of the passing vehicles with signals controlled by different models.

Fig. 3. Case study to analyze benefits of the batch learning framework on one real-world intersection.

along time in Fig. 3b, real-world sources selected by *TTRL-B* seem more reasonable than synthetic ones, as they have many transportation characteristics in common, which can be captured by embeddings, e.g., the tendency of traffic load, the number of vehicles in the same time interval, etc. In Fig. 3c, selecting source tasks in the batchwise way further proves effective when controlling signals according to their guidance: *TTRL-B* shows an obvious advantage over *TTRL* in the jumpstart and overall performance in our sampled intersection.

6 Conclusion

In this paper, we solve the problem of using RL to do the traffic signal control on new intersections. Compared with traditional methods, we propose a batchwise targeted transfer framework, which can significantly speed up the convergence and achieve lower vehicles' travel time with much fewer training samples from the new intersection. This will avoid the high cost of traffic jam when directly applying RL algorithms in real world intersections. Our extensive experiments have shown that our method outperforms the baselines and each component makes contribution to the performance boost. We are going to extend our work to more real scenarios by considering multi-phase (e.g., turning vehicles) and multi-intersection traffic signal control for the future work.

References

1. Bakker, B., Whiteson, S., Kester, L., Groen, F.C.: Traffic light control by multiagent reinforcement learning systems. In: Babuška, R., Groen, F.C.A. (eds.) Interactive Collaborative Information Systems. SCI, vol. 281, pp. 475–510. Springer, Heidelberg (2010). https://doi.org/10.1007/978-3-642-11688-9_18
2. Cools, S.B., Gershenson, C., D'Hooghe, B.: Self-organizing traffic lights: a realistic simulation. In: Prokopenko, M. (ed.) Advances in Applied Self-Organizing Systems. AI&KP, pp. 45–55. Springer, London (2013). https://doi.org/10.1007/978-1-4471-5113-5_3
3. Du, Y., Gabriel, V., Irwin, J., Taylor, M.E.: Initial progress in transfer for deep reinforcement learning algorithms. In: Proceedings of Deep Reinforcement Learning: Frontiers and Challenges Workshop, New York City, NY, USA (2016)
4. Gao, J., Shen, Y., Liu, J., Ito, M., Shiratori, N.: Adaptive traffic signal control: deep reinforcement learning algorithm with experience replay and target network. arXiv preprint arXiv:1705.02755 (2017)
5. Hinton, G., Vinyals, O., Dean, J.: Distilling the knowledge in a neural network. arXiv preprint arXiv:1503.02531 (2015)
6. Hochreiter, S., Schmidhuber, J.: Long short-term memory. Neural Comput. **9**(8), 1735–1780 (1997)
7. Hunt, P., Robertson, D., Bretherton, R., Winton, R.: Scoot - a traffic responsive method of coordinating signals. Technical report (1981)
8. Liu, M., Deng, J., Xu, M., Zhang, X., Wang, W.: Cooperative deep reinforcement learning for traffic signal control (2017)
9. Lowrie, P.: SCATS, Sydney co-ordinated adaptive traffic system: a traffic responsive method of controlling urban traffic (1990)
10. Lu, W., Hou, J., Yan, Y., Zhang, M., Du, X., Moscibroda, T.: MSQL: efficient similarity search in metric spaces using SQL. The VLDB J.-Int. J. Very Large Data Bases **26**(6), 829–854 (2017)
11. Miller, A.J.: Settings for fixed-cycle traffic signals. J. Oper. Res. Soc. **14**(4), 373–386 (1963)
12. Mnih, V., et al.: Human-level control through deep reinforcement learning. Nature **518**(7540), 529 (2015)
13. Mousavi, S.S., Schukat, M., Howley, E.: Traffic light control using deep policy-gradient and value-function-based reinforcement learning. Intell. Transp. Syst. (ITS) **11**(7), 417–423 (2017)
14. Pan, S.J., Yang, Q., et al.: A survey on transfer learning. IEEE Trans. Knowl. Data Eng. **22**(10), 1345–1359 (2010)
15. Parisotto, E., Ba, J.L., Salakhutdinov, R.: Actor-mimic: deep multitask and transfer reinforcement learning. arXiv preprint arXiv:1511.06342 (2015)
16. Reynolds, D.: Gaussian mixture models. In: Li, S.Z., Jain, A. (eds.) Encyclopedia of Biometrics, pp. 827–832. Springer, Boston (2015). https://doi.org/10.1007/978-0-387-73003-5_196
17. Roess, R.P., Prassas, E.S., McShane, W.R.: Traffic Engineering. Pearson/Prentice Hall, Upper Saddle River (2004)
18. Rusu, A.A., et al.: Policy distillation. arXiv preprint arXiv:1511.06295 (2015)
19. Taylor, M.E., Stone, P.: Transfer learning for reinforcement learning domains: a survey. J. Mach. Learn. Res. **10**(July), 1633–1685 (2009)
20. Tessler, C., Givony, S., Zahavy, T., Mankowitz, D.J., Mannor, S.: A deep hierarchical approach to lifelong learning in minecraft. In: AAAI, vol. 3, p. 6 (2017)

21. Thrun, S., Pratt, L.: Learning to Learn. Springer, New York (2012). https://doi.org/10.1007/978-1-4615-5529-2
22. Wang, J.X., et al.: Learning to reinforcement learn. arXiv preprint arXiv:1611.05763 (2016)
23. Webster, F.V.: Traffic signal settings. Technical report (1958)
24. Wei, H., Zheng, G., Yao, H., Li, Z.: IntelliLight: a reinforcement learning approach for intelligent traffic light control. In: ACM SIGKDD International Conference on Knowledge Discovery & Data Mining (KDD), pp. 2496–2505 (2018)
25. Wiering, M.: Multi-agent reinforcement learning for traffic light control. In: Machine Learning: Proceedings of the Seventeenth International Conference (ICML 2000), pp. 1151–1158 (2000)
26. Zhang, Z., Huang, K., Tan, T.: Comparison of similarity measures for trajectory clustering in outdoor surveillance scenes. In: 18th International Conference on Pattern Recognition, ICPR 2006, vol. 3, pp. 1135–1138. IEEE (2006)

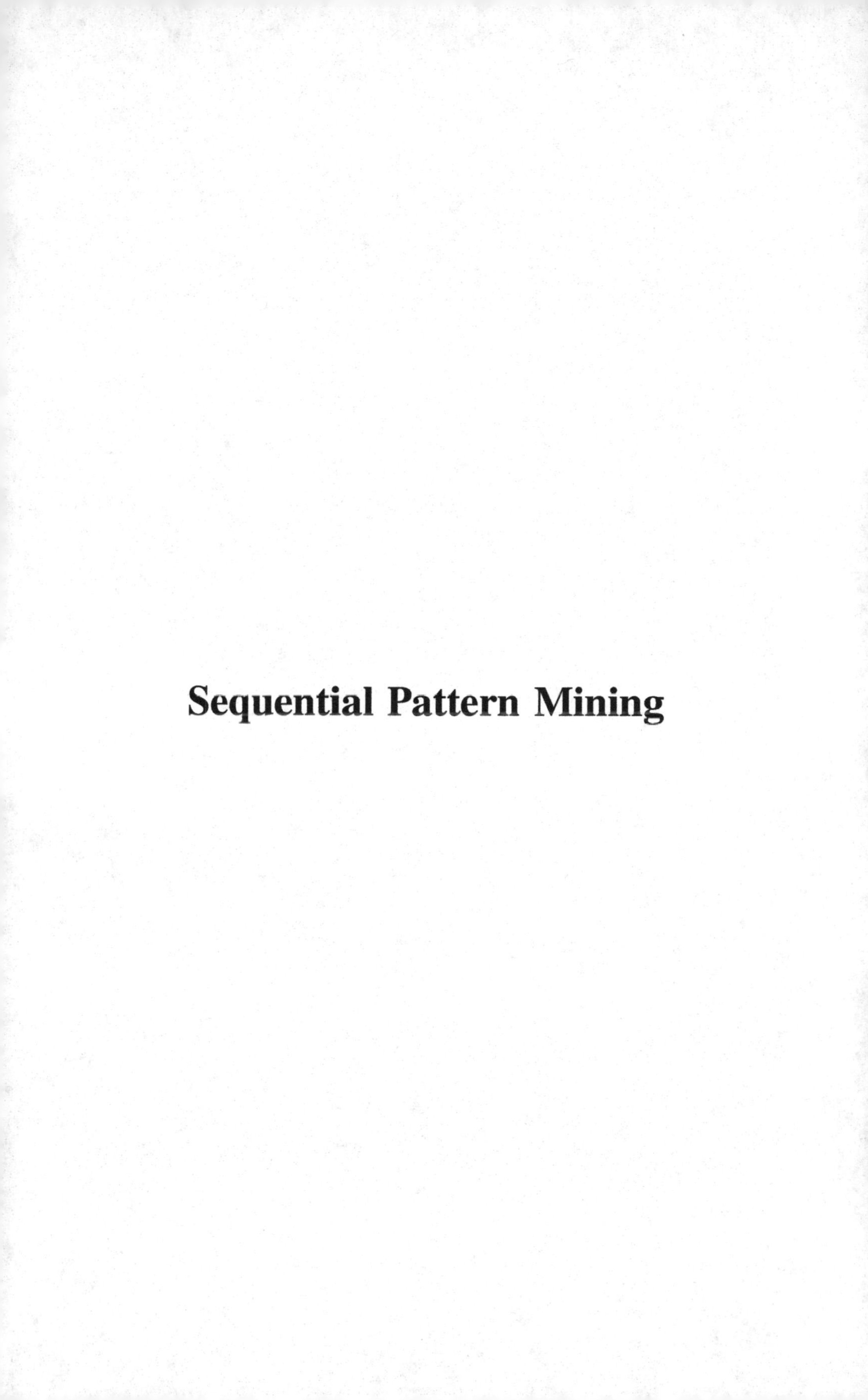

Sequential Pattern Mining

Efficiently Finding High Utility-Frequent Itemsets Using Cutoff and Suffix Utility

R. Uday Kiran[1,2(✉)], T. Yashwanth Reddy[3], Philippe Fournier-Viger[4], Masashi Toyoda[2], P. Krishna Reddy[3], and Masaru Kitsuregawa[2,5]

[1] National Institute of Information and Communications Technology, Tokyo, Japan
[2] The University of Tokyo, Tokyo, Japan
{uday_rage,toyoda,kitsure}@tkl.iis.u-tokyo.ac.jp
[3] International Institute of Information Technology-Hyderabad, Hyderabad, India
yashwanth.t@research.iiit.ac.in, pkreddy@iiit.ac.in
[4] Harbin Institute of Technology (Shenzhen), Shenzhen, China
philfv8@yahoo.com
[5] National Institute of Informatics, Tokyo, Japan

Abstract. High utility itemset mining is an important model with many real-world applications. But the popular adoption and successful industrial application of this model has been hindered by the following two limitations: (*i*) computational expensiveness of the model and (*ii*) infrequent itemsets may be output as high utility itemsets. This paper makes an effort to address these two limitations. A generic high utility-frequent itemset model is introduced to find all itemsets in the data that satisfy user-specified *minimum support* and *minimum utility* constraints. Two new pruning measures, named *cutoff utility* and *suffix utility*, are introduced to reduce the computational cost of finding the desired itemsets. A single phase fast algorithm, called High Utility Frequent Itemset Miner (HU-FIMi), is introduced to discover the itemsets efficiently. Experimental results demonstrate that the proposed algorithm is efficient.

Keywords: Data mining · Itemset mining · Utility itemset

1 Introduction

High Utility Itemset Model (HUIM) is an important knowledge discovery technique in data mining. It aims to discover all interesting itemsets whose *utility* in a transactional database is no less than a user-specified *minimum utility* ($minUtil$) constraint. The utility of an *itemset* is the summation of its utilities in all the transactions. The classic application of HUIM is market-basket analysis. It consists of analyzing which sets of items purchased by customers generate a sufficient revenue for a retailer. An example of utility itemset generated in the Yahoo! JAPAN retail data[1] is:

$$\{Nintendo3Ds\_game,\ Playstation4\_game\}\quad [utility = 604,231\ ¥].$$

[1] More details of this dataset are presented in latter parts of this paper.

Q. Yang et al. (Eds.): PAKDD 2019, LNAI 11440, pp. 191–203, 2019.
https://doi.org/10.1007/978-3-030-16145-3_15

The above utility itemset says that the *income* (or *utility*) generated from the simultaneous purchases of 'nintendo3Ds_game' and 'playstation4_game' is 604,231 ¥. This information may be useful to the retailer, because it disproves the general assumption of expecting little revenue from customers purchasing games for competing products simultaneously. HUIM has many other applications, such as website click stream analysis, cross-marketing and bio-medical applications [2].

The popular adoption and successful industrial application of HUIM has been hindered by the following two obstacles:

1. **High computational cost of the model:** The itemsets generated by *utility* measure do not satisfy the convertible anti-monotone, convertible monotone, or convertible succinct properties [5]. To reduce the search space in the itemset lattice, most algorithms [2] initially find secondary items by pruning all items that have a *local utility*[2] less than $minUtil$. The *local utility* is a convertible anti-monotonic measure, which represents an upper bound on the *utility* of itemsets. Next, primary items are generated by ordering the secondary items in ascending order of their *local utility.* Finally, all high utility itemsets are generated by recursively mining the projected databases of each primary item. Since the construction of projected databases requires a database scan, the computational cost of an high utility itemset mining algorithm depends primarily on the number of primary items. We have observed that finding primary items based on the *local utility* order is inefficient (or computationally expensive), because such an order often generate a large number of primary items, thereby increasing the number of database scans. More importantly, finding high utility itemsets using the *local utility* order makes the model impracticable in many real-world sparse databases.
2. **Infrequent itemsets may be generated as the utility itemsets:** Since HUIM determines the interestingness of an itemset without taking into account its *support* within the data, uninteresting itemsets with very low *support* may be generated as the utility itemsets. In our empirical study, we have observed that a significant portion of high utility itemsets generated by HUIM appeared seldomly in the data. It is because the *utility* measure is sensitive to items with high external utility values. Additionally, directly pushing the minimum support constraint into existing HUIM algorithms [2] is not an effective solution to this problem. It is because such algorithms cannot exploit the relationship between the *support* and *utility* measures to reduce the search space effectively.

This paper makes an effort to address these two problems. We propose a generic High Utility-Frequent Itemset Model (HU-FIM) to find all high utility-frequent itemsets in the data that satisfy a user-specified *minimum support* ($minSup$) and $minUtil$ values. Using $minSup$ facilitates pruning uninteresting itemsets having low *support* in the data. The itemsets generated by the proposed model

[2] Since the *local utility* measure generalizes the TWU measure by taking into account itemsets, we use the former measure throughout this paper for brevity.

do not satisfy the convertible anti-monotonic property, convertible monotonic property, or convertible succinct property. Two new pruning measures, "*cutoff utility*" (*CU*) and "*suffix utility*" (*SU*), have been introduced to reduce the search space and the computational cost of HU-FIM. The *CU* measure tries to reduce the search space by exploiting the relationship between the $minSup$ and external utility of an item. It states that if the *utility* of an item is less than the product of its *external utility* and $minSup$, then neither the item nor its supersets will be high utility-frequent itemsets. Given **a list of items in** *utility* **descending order**, the *SU* of an item represents the sum of utilities of remaining items. This measure states that if the sum of *utility* and *SU* of an item in the *utility* ordered list is less than $minUtil$, then the mining algorithm can be terminated as no more high utility-frequent itemsets will be generated from the data. Thus, *SU* can be used to reduce the search space of the model effectively. A single phase algorithm, called High Utility-Frequent Itemset Miner (HU-FIMi), is proposed to find high utility-frequent itemsets efficiently. HU-FIMi is based on EFIM [9]. Experimental results demonstrate that HU-FIMi can discover the desired itemsets efficiently in sparse databases.

The rest of the paper is organized as follows. Related work is presented in Sect. 2. Section 3 introduces HU-FIM. Section 4 briefly describes EFIM. Section 5 introduces the proposed algorithm. Experimental results are reported in Sect. 6. Section 7 concludes the paper with future research directions.

2 Related Work

FIM is an important model in data mining. The main limitation of this model is that it ignores the crucial information regarding the importance of items and their occurrence $frequency$ in every transaction. Yao et al. [7] introduced HUIM by taking into account the importance of items and their occurrence $frequency$ in every transaction. To circumvent the fact that the utility is not anti-monotonic and to find all high utility itemsets, several HUIM algorithms (e.g. Two-Phase [4] and UP-Growth+ [6]) have employed *local utility* to reduce the search space.

Recently, single phase algorithms (e.g. d2HUP [3] and EFIM [9]) to mine high utility itemsets were developed to avoid the problem of candidate generation. These algorithms use upper bounds that are tighter than the *local utility* to prune the search space and can immediately obtain the exact utility of any itemset to decide if it should be output. Zhang et al. [8] conducted an empirical study on various HUIM algorithms and concluded that EFIM has consistently shown better performance over other algorithms. In this paper, we push the proposed pruning measures into EFIM to efficiently find the desired itemsets.

3 Proposed Model: High Utility-Frequent Itemset

Let $I = \{i_1, i_2, \cdots, i_m\}$, $m \geq 1$, be a set of items. Each item $i_j \in I$ is associated with a positive number $p(i_j)$ known as **external utility**. The external utility of an item represents its relative importance to the user. The **utility database**,

UD, is a set of all items in I and their respective external utility values. That is, $UD = \{(i_1, p(i_1)),\ (i_2, p(i_2), \cdots, (i_m, p(i_m))\}$. A **transactional database** is a set of transactions $D = \{T_1, T_2, \cdots, T_n\}$ such that for each $T_c \in D$, $T_c \subseteq I$ and T_c has a unique identifier $c \in \mathbb{Z}^+$ called its transaction-identifier (or *tid*). Every item $i_j \in T_c$ has a positive number $q(i_j, T_c)$, called its **internal utility**. The internal utility of an item generally represents its *frequency* in a transaction.

Table 1. Market-basket database

tid	Items	*tid*	Items
1	$(a,2),(b,3),(f,2)$	5	$(a,1),(b,2),(c,1)$ $(d,4),(g,2)$
2	$(a,2),(c,1),(d,3),(e,2)$	6	$(c,3),(d,2),(f,3),$ $(e,1)$
3	$(a,3),(b,1),(h,2)$	7	$(b,3),(d,4)$
4	$(c,2),(d,3),(e,1)$		

Table 2. Price of items

Item	*price*	Item	*price*
a	200	e	200
b	300	f	500
c	200	g	200
d	400	h	300

Example 1. Consider the market-basket (or transactional) database shown in Table 1. The set of all items in the database, i.e. $I = \{a, b, c, d, e, f, g, h\}$. The *prices* (or external utilities) for all items in Table 1 are shown in Table 2. Let the unit for these *prices* be Japanese Yen. The first transaction in Table 1 indicates that a customer has purchased 2 quantities of item a, 3 quantities of item b, and 2 quantities of item f. These quantities represent the internal utilities of the items appearing in the first transaction.

Definition 1 ***(Utility of an item in a transaction).*** *The utility of an item i_j in a transaction T_c, is denoted as $u(i_j, T_c)$, represents the product of its internal and external utility values. That is, $u(i_j, T_c) = p(i_j) \times q(i_j, T_c)$.*

Example 2. Continuing with the previous example, the *utility* (or *income*) of an item a in the first transaction, i.e., $u(a, T_1) = p(a) \times q(a, T_1) = 2 \times 200 = 400$ ¥.

Definition 2 ***(Utility of an itemset in a transaction).*** *Let $X \subseteq I$ be an itemset. An itemset is a k-itemset if it contains k items. The utility of an itemset X in a transaction T_c, denoted as $u(X, T_c) = \Sigma_{i_j \in X} u(i_j, T_c)$ if $X \subseteq T_c$.*

Example 3. The set of items 'a' and 'b', i.e., $\{a, b\}$ (or 'ab' in short) is an itemset. This is a 2-itemset. The utility (or *income*) of ab in T_1, $u(ab, T_1) = u(a, T_1) + u(b, T_1) = 400 + 900 = 1300$ ¥.

Definition 3 ***(Utility of an itemset in a database).*** *The utility of an itemset X in the database D, denoted as $u(X) = \Sigma_{T_c \in g(X)} u(X, T_c)$, where $g(X)$ is the set of transactions containing X.*

Example 4. In Table 1, ab has appeared in T_1, T_3 and T_5. Therefore, $g(x) = \{T_1, T_3, T_5\}$. The *utility* (or *income*) of ab in each of these transactions is $u(ab, T_1) = 1300$ ¥, $u(ab, T_3) = 900$ ¥ and $u(ab, T_5) = 800$ ¥. Therefore, the utility (or *income*) of ab in the database is $u(ab) = 1300 + 900 + 800 = 3000$ ¥.

Definition 4 ***(Frequent itemset).*** *The support of an itemset X in D, denoted as $s(X) = |g(X)|$, where $|g(X)|$ represents the total number of transactions containing X in D. An itemset X is said to be* ***frequent*** *if $s(X) \geq minSup$, where $minSup$ represents the user-specified minimum support.*

Example 5. The *support* of ab in Table 1, i.e., $s(ab) = |g(ab)| = |\{T_1, T_3, T_5\}| = 3$. If $minSup = 3$, then ab is a frequent itemset because $s(ab) \geq minSup$.

Definition 5 ***(High utility-frequent itemset).*** *A frequent itemset X is a high utility-frequent itemset if its $u(X) \geq minUtil$, where $minUtil$ represents the user-specified minimum utility value. A high utility-frequent itemset X is expressed as X $[support = s(X),\ utility = u(X)]$.*

Example 6. If the user-specified $minUtil = 2600$, then the frequent itemset ab is a high utility-frequent itemset because $u(ab) \geq minUtil$. This itemset is expressed as ab $[support = 3,\ utility = 3{,}000$ ¥$]$.

Definition 6 ***(Problem definition).*** *Given a transactional database D and a utility database UD, the problem of HU-FIM is to find all itemsets in D that have support $\geq minSup$ and utility $\geq minUtil$. The support of an itemset can be expressed in percentage of $|D|$. It is interesting to note that $HUIM$ is a special case of the problem HU-FIM when $minSup = 0$.*

4 EFIM and Its Limitations

Fournier-Viger et al. [9] introduced EFIM to find high utility itemsets. Since the proposed algorithm extends EFIM to find high utility-frequent itemsets, we briefly describe the steps of the latter algorithm. First, EFIM scans the database and calculates the *local utility* (see Definition 9) for all items. Next, secondary items are generated by pruning all items that have a *local utility* less than $minUtil$. These **secondary items are later sorted in *local utility* ascending order**. Let $\succ$ denote this sorted list of secondary items. Next, the *sub-tree utility* (see Definition 10) for all secondary items is determined by performing another scan on the database. Next, items with *sub-tree utility* no less than $minUtil$ are considered as primary items. Finally, for each primary item, a projected database consisting of primary and secondary items is constructed and mined recursively to find all high utility itemsets.

Definition 7 ***(Items that can extend an itemset).*** *Let α be an itemset. Let $E(\alpha)$ denote the set of all items that can be used to extend α according to the depth-first search, that is $E(\alpha) = \{z,\ z \in I \wedge\ z \succ x,\ \forall\ x \in \alpha\}$.*

Example 7. Consider the database from Table 1. Let $\succ$ be the alphabetical order and $\alpha = \{a\}$. Then $E(\alpha) = \{b, c, d, e, f, g, h\}$.

Definition 8 ***(Remaining utility).*** *The remaining utility of X in a transaction T_c is defined as $re(X, T_c) = \Sigma_{i \in T_c \wedge i \succ x \forall x \in X} u(i, T_c)$.*

Example 8. Remaining utility of ac in T_5 is $re(ac, T_5) = 4 \times 400 + 2 \times 200 = 2000$.

Definition 9 ***(local utility).*** *For an itemset α and item $z \in E(\alpha)$, the LocalUtilty of z with respect to α is $lu(\alpha, z) = \Sigma_{T \in g(\alpha, \{z\})}[u(\alpha, T) + re(\alpha, T)]$.*

Definition 10 ***(Sub-tree utility).*** *For an itemset α and item z that can extend α according to the depth-first search ($z \in E(\alpha)$, the Sub-tree utility of z with respect to α is $stu(\alpha, z) = \Sigma_{T \in g(\alpha \cup \{z\})}[u(\alpha, T) + u(z, T) + \Sigma_{i \in T \wedge i \in E(\alpha \cup \{z\})} u(i, T)]$.*

Example 9. Continuing with the previous example, let $\alpha = a$ and c be an item that can extend α in depth-first. The *sub-tree utility* of c with respect to α is $stu(\alpha, c) = (((2 \times 200) + (1 \times 200) + (3 \times 400 + 2 \times 200)) + ((1 \times 200) + (1 \times 200) + (4 \times 400 + 2 \times 200))) = 4600$.

The limitations of EFIM are as follows (i) EFIM finds high utility itemsets without taking into account their *support*. As a result, itemsets with low frequency can be identified as high utility itemsets. (ii) Since creating a projected database for a primary item requires scanning the data, the computational cost of EFIM mostly depends on the number of primary items. We have observed that EFIM is computationally expensive (or impractical on very large databases). It is because the *local utility* order of items generates too many primary items, thus increasing the number of database scans.

5 Proposed Algorithm

HU-FIMi is a single phase algorithm that extends EFIM [9] to find high utility-frequent itemsets in a transactional database. To achieve better performance over EFIM, HU-FIMi exploits different ordering of items and introduces two new pruning measures to reduce the computational cost of finding the desired itemsets. The algorithm HU-FIMi is presented in Algorithms 1 and 2. The algorithm has the following steps: (i) finding secondary items, i.e., items whose supersets can be high utility-frequent itemsets, (ii) finding candidate items from secondary items and (iii) finding primary items from candidate items, and (iv) finding all high utility-frequent itemsets by recursively mining all primary items. We briefly explain each of these steps along with the pruning measures.

5.1 Finding Secondary Items

Since the *utility* measure is neither an monotonic nor anti-monotonic function, high utility-frequent itemsets generated by the proposed model do not satisfy the

Algorithm 1. HU-FIMi

1: **input** : D: a transaction database, $minUtil$: a user-specified threshold, $minSup$: a user-specified threshold
2: **output** : the set of high utility-frequent itemsets
3: Let α denote an itemset that needs to be extended. Initially, set $\alpha = \phi$;
4: Scan the database D to determine the TWU, *support* and *utility* for every item $i_j \in I$. Bin-arrays [9] can be used to efficiently calculate the TWU, *support* and *utility* of items.
5: $Secondary(\alpha) = \{i | i \in I \wedge lu(\alpha, i) \geq minUtil \wedge s(i) \geq minSup \wedge u(i) \geq cu(i)\}$;
6: Let $\succ$ be the total order of *utility* descending values on $Secondary(\alpha)$;
7: Scan D to remove each item $i \notin Secondary(\alpha)$ from the transactions, sort items in each transaction according to $\succ$, and delete empty transactions;
8: Sort transactions in D according to $\succ_T$;
9: Calculate *suffix utility* for each item $i \in Secondary(\alpha)$.
10: Let $pi(\beta)$ denote the set of all items in $Secondary(\alpha)$ that have $u(i) + su(i) \geq minUtil$;
11: Calculate sub-tree utility for all items in $pi(\beta)$ by scanning the database D once using utility-bin array;
12: $Primary(\alpha) = \{z \in pi(\beta) | stu(\alpha, z) \geq minUtil\}$;
13: $RecursiveSearch(\alpha, D, Primary(\alpha), Secondary(\alpha), minUtil, minSup)$;

anti-monotonic property. To reduce the search space, we initially find secondary items consisting of all items whose supersets may be high utility-frequent itemsets. The secondary items (see Definition 12) are identified by calculating each items' *local utility* (see Definition 9), *cutoff utility* (see Definition 11) and *support*. Since *cutoff utility* is a new pruning measure, we define this measure.

Definition 11. *The* cutoff utility *of an item i_j, denoted as $cu(i_j)$, is the product of its external utilty and $minSup$. That is, $cu(i_j) = p(i_j) \times minSup$.*

Example 10. Consider the item 'g' in Table 1. The external utility of 'g,' is $p(g) =$ 200. If the user-specified $minSup = 3$, then 'g' should appear at least in three transactions with internal utility of 1. Therefore, the cutoff utility that item 'g' must have to be a high utility-frequent itemset is 600 ($= p(g) \times minSup$).

The pruning of items using *cutoff utility* is given in Property 1.

Property 1. For an item i_j, if $u(i_j) < cu(i_j)$, then neither i_j nor its supersets can be high utility-frequent itemsets.

Example 11. The *utility* of g in Table 1 is $u(g) = 400$. Since $u(g) < cu(g)$, 'g' and its supersets cannot be utility-frequent itemsets.

Definition 12 ***(Secondary item).*** *An item $i_j \in I$ is a secondary item if $lu(i_j) \geq minUtil$, $u(i_j) \geq cu(i_j)$ and $s(i_j) \geq minSup$.*

Example 12. All secondary items generated from Table 1 are a, b, c, d and e. The items f, g and h are not secondary items because $s(f) < minSup$, $u(g) < cu(g)$ and $lu(h) < minUtil$, respectively.

Algorithm 2. RecursiveSearch

1: **input** : α: an itemset, $\alpha - D$; the α projected database, $Primary(\alpha)$: the primary items of α, $Secondary(\alpha)$: the secondary items of α, the minutil threshold, the minSup threshold
2: **output**: the set of high utility-frequent itemsets that are extensions of α
3: **for** each $item\ i \in Primary(\alpha)$ **do**
4: $\beta = \alpha \cup \{i\}$;
5: Scan α-D to calculate $u(\beta)$, $s(\beta)$ and create β-D; //uses transaction merging from EFIM
6: **if** $u(\beta) \geq minutil$ && $s(\beta) \geq minSup$ **then**
7: output β
8: **end if**
9: **if** $s(\beta) \geq minSup$ **then**
10: Calculate $stu(\beta, z)$ and $lu(\beta, z)$ for all item z $\in Secondary(\alpha)$ by scanning β-D once, using two utility-bin array;
11: $Primary(\beta) = \{\text{z} \in Secondary(\alpha) | stu(\beta, z) \geq minutil\}$;
12: $Secondary(\beta) = \{\text{z} \in Secondary(\alpha) | lu(\beta, z) \geq minutil\}$;
13: Search(β, β-D, $Primary(\beta)$, $Secondary(\alpha)$, $minutil$, $minSup$);
14: **end if**
15: **end for**

5.2 Finding Candidate Items

The secondary items generated in the above step constitute both high utility-frequent items and uninteresting items. To reduce the computational cost of finding the desired itemsets, we need to identify those secondary items whose depth-first search in the itemset lattice (or projected databases) will result in finding all high utility-frequent itemsets. To find such items, we introduce a new pruning measure, called *suffix utility* (see Definition 13), by exploiting items' *utility* descending order. The *suffix utility* facilitates defining a **novel terminating condition**, which does not exist in any of the previous utility itemset mining algorithms. That is, if the sum of *utility* and *suffix utility* of an item is less than *minUtil*, then the mining process can be terminated as no further desired itemsets will be generated (see Lemma 1).

The *suffix utility* of an item i_j is $su(i_j) = u(i_{j+1}) + su(i_{j+2})$. If i_{j+2} represents the last time in the sorted list, then $su(i_{j+2}) = 0$. Thus, the time complexity of the *suffix utility* measure is $O(1)$. Definition 14 defines **candidate items** generated using both *utility* and *suffix utility* measures.

Property 2 **(Additive property).** The *utility* of an itemset X will always be less than or equal to sum of *utility* of its items. That is, $u(X) \leq \sum_{i_j \in X} u(i_j)$.

Definition 13 ***(Suffix utility).*** *Let* $S = \{i_1, i_2, \cdots, i_k\} \subseteq I$ *be an ordered list of secondary items such that* $u(i_1) \geq u(i_2) \geq \cdots \geq u(i_k)$. *The* suffix utility *of an item* $i_j \in S$*, denoted as* $su(i_j)$*, is the sum of utilities of remaining items in the list. That is,* $su(i_j) = \sum_{p=j+1}^{|S|} u(i_p)$. *For the last item in* S*,* $su(i_k) = 0$.

Example 13. The secondary items in *utility* descending order are d, b, a, c and e. The *suffix utility* of d is $su(d) = u(b) + u(a) + u(c) + u(e) = 6500$ ¥.

Property 3. If $u(i_j)+su(i_j) < minUtil$, $i_j \in S$, then neither i_j nor the supersets generated from its projected database will be high utility-frequent itemsets.

Property 4. The suffix utility is a monotonically decreasing function. That is, $su(i_j) \geq su(i_k)$, where $j < k$ and $i_j, i_k \in S$.

Lemma 1. *Let $S = \{i_1, i_2, \cdots, i_k\} \subseteq I$ be an ordered list of secondary items in utility descending order, i.e., $u(i_1) \geq u(i_2) \geq \cdots \geq u(i_k)$. For an item $i_j \in S$, if $u(i_j) + su(i_j) < minUtil$, then no more highly utility-frequent itemsets will be generated from the projected databases of the remaining items in S.*

Proof. According to Property 3, if $u(i_j)+su(i_j) < minUtil$, then no more highly utility-frequent itemsets will be generated from the projected database of i_j. Now let us consider another item i_k, $1 \leq j < k \leq |S|$. Since S is in *utility* descending order, $u(i_k) \leq u(i_j)$ and $su(i_k) \leq su(i_j)$ (see Property 4). Thus, $u(i_k)+su(i_k) \leq u(i_j) + su(i_j) < minutil$. Thus, neither i_k nor the itemsets generated from its projected database will be high utility-frequent itemsets. Hence proved.

Definition 14 ***(candidate item).*** *A secondary item $i_j \in S$ is a candidate item if $u(i_j) + su(i_j) \geq minUtil$.*

5.3 Finding Primary Items

In most cases, the candidate items generated in the previous step form the primary items. However, in a few cases, especially when mining itemsets at low *minUtil* values, suffix utility is inadequate. It is because most secondary items will be generated as candidate items. To handle such cases, we generate primary items by calculating the *sub-tree utility* for candidate items. The calculation of *sub-tree utility* is a computationally expensive step because it needs a database scan. We recommend eliminating this step when finding high utility-frequent itemsets at high *minUtil* values. In this paper, we are providing this step for completeness. (**Please note that the primary items generated by EFIM and HI-FIMi algorithms can be different as both algorithms employ different ordering of secondary items.**)

Definition 15 ***(Primary item).*** *Let C denote the set of candidate items. A candidate item $i_j \in C$ is a primary item if $stu(i_j) \geq minUtil$.*

5.4 Recursive Mining of Primary Items

For each primary item, construct its projected database, and recursively mine the projected database until the projected database is empty.

6 Experimental Results

In this section, we first show that HU-FIMi performs better than EFIM. Later we evaluate only the performance of HU-FIMi as there exists no algorithm to find high utility-frequent itemsets. Both algorithms were written in C++ and executed on a machine with 1.5 GHz processor and 4 GB RAM. The experiments have been conducted using synthetic (**T10I4D100K** and **Retail**) and real-world (**Yahoo**) databases. The **T10I4D100K** database was generated using the SPMF toolkit [1]. This database contains 870 items and 100,000 transactions. The minimum and maximum transaction lengths of this database are 1 and 29, respectively. The **Retail** database was also provided by SPMF toolkit. The internal and external utilities of the items are synthetically generated by SPMF toolkit. This database contains 16,470 items and 88,162 transactions. The minimum and maximum transaction lengths are 1 and 76, respectively. The **Yahoo** database represents a portion of retail data generated by Yahoo! JAPAN. It contains 7,290 items and 93,113 transactions. The minimum and maximum transaction lengths are 1 and 24, respectively.

To evaluate the EFIM and HU-FIMi algorithms, we find high utility-frequent itemsets by setting $minSup = 0$. Due to page limitation, we present the experimental results only for Yahoo database. Fig. 1(a) show the number of primary items generated by EFIM and HU-FIMi algorithms. It can be observed that HU-FIMi has generated less primary items for any given $minUtil$ value (less is preferable as each primary item requires a database scan). Figure 1(b) show the number of nodes explored in the itemset lattice to find the desired itemsets. It can be observed that the proposed algorithm has explored relatively few nodes compared to EFIM. Figure 1(c) shows the total runtime of EFIM and HU-FIMi algorithms to find all high utility itemsets. It can be observed that HU-FIMi's runtime is shorter than EFIM. The performance improvement of HU-FIMi over EFIM is mainly due to the $suffix$ $utility$, which facilitates finding all high utility itemsets with relatively few primary items. Figure 1(d) shows the memory consumed by EFIM and HU-FIMi on the Yahoo database. It is observed that HU-FIMi has consumed slightly more memory than EFIM. It is because HU-FIMi has to additionally store the *support* of each itemset.

Figures 2(a)–(c) show the number of primary items generated in various databases at different $minSup$ and $minUtil$ values. It can be observed that increase in $minSup$ and/or $minUtil$ results in decrease of primary items, because many items have failed to satisfy the increased $minUtil$ or $minSup$ values.

Figures 3(a)–(c) show the number of high utility-frequent itemsets generated by HU-FIMi in various databases at different $minUtil$ and $minSup$ values. It can be observed that an increase in $minSup$ and/or $minUtil$ results in a decrease of high utility-frequent itemsets. It is because many itemsets fail to satisfy the increased $minUtil$ or $minSup$ constraints. Another important observation in these figures (especially in Yahoo database) is that when $minSup$ is slightly increased from 0 to 0.01 (%), their is a significant drop in the number of high utility-frequent itemsets. It is because items with high external utility values were generating too many high utility itemsets when combined with other items for $minSup = 0$. Table 3 presents some of the interesting high utility-frequent itemsets generated in Yahoo database.

Figures 4(a)–(c) show the runtime requirements of HU-FIMi in various databases for different $minUtil$ and $minSup$ values. It can be observed that an increase in $minUtil$ and/or $minSup$ results in a decrease in runtime. It is because of the number of primary items is reduced, which significantly influences the runtime requirements of HU-FIMi. (Memory requirements of HU-FIMi also showed similar affect as the runtime. Due to page limitation, we are unable to present these results.)

Table 3. A few interesting itemsets found in the Yahoo database

Itemset	Utility (¥)
{face_care:essences, face_care:skin_lotions}	389,476
{ladies:long_sleeve, ladies:knit_sweater:other}	105,994
{ladies:skirt_pants:other, ladies:tops:other}	150,970

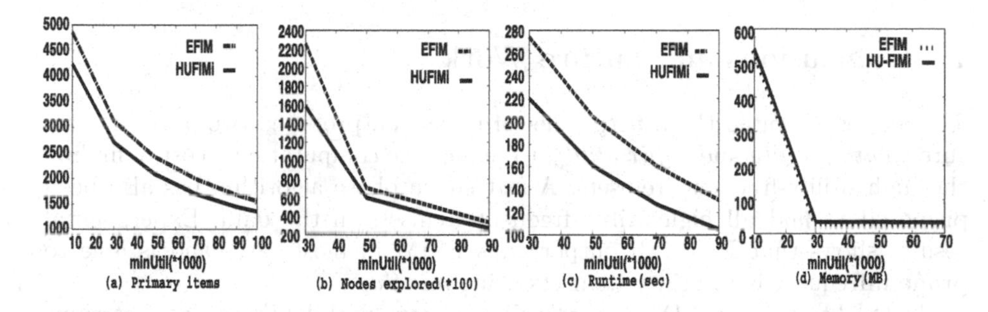

Fig. 1. Performance evaluation of EFIM and HUFIM algorithms on Yahoo database

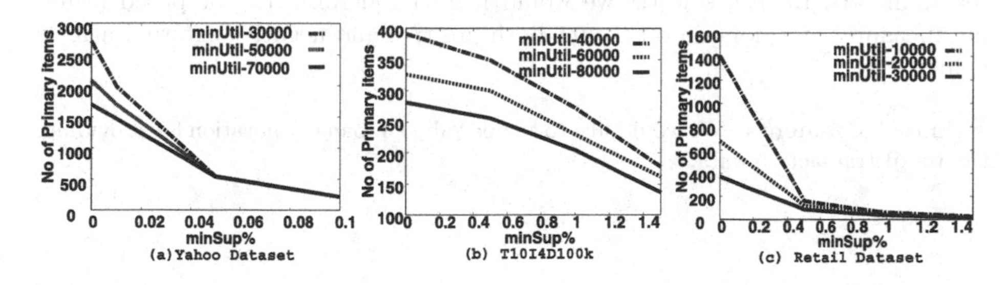

Fig. 2. Primary items generated at different $minSup$ and $minUtil$ values

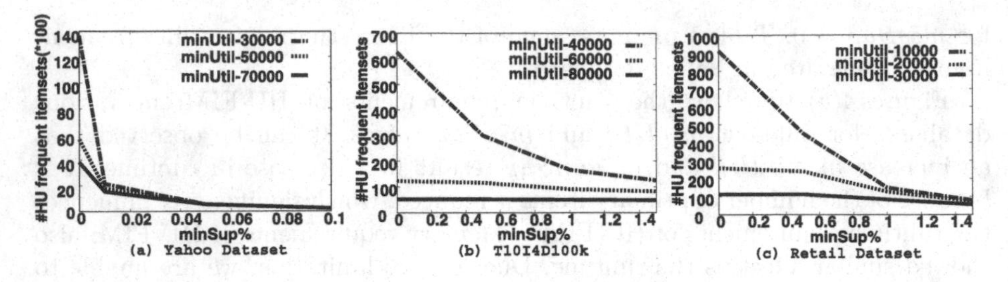

Fig. 3. Itemsets generated at different *minSup* and *minUtil* values

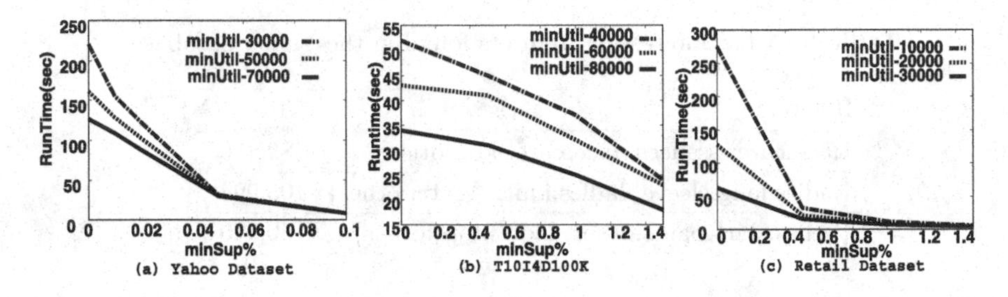

Fig. 4. Runtime of HU-FIMi at different *minSup* and *minUtil* values

7 Conclusions and Future Work

This paper exploited the *utility* order of items and proposed two pruning measures, *cutoff utility* and *suffix utility*, to reduce the computational cost of finding the high utility-frequent itemsets. A fast single phase algorithm has also been proposed to find all high utility-frequent itemsets in the data. Experimental results shows that HU-FIMi outperforms EFIM in most cases and is able to prune infrequent high utility itemsets using *minSup*.

In the literature, HUIM was studied in incremental databases, data streams and uncertain databases. It is interesting to investigate how to extend the proposed pruning measures to discover the desired itemsets in such databases. The proposed pruning measures consider items' external utility values as positive real numbers. In future work, we would like to generalize the proposed pruning measures by taking into account both positive and negative external utility values.

Acknowledgements. We would like to thank Yahoo Japan Corporation for providing the retail transaction data.

References

1. Fournier-Viger, P., Gomariz, A., Gueniche, T., Soltani, A., Wu, C.W., Tseng, V.S.: SPMF: a Java open-source pattern mining library. J. Mach. Learn. Res. **15**(1), 3389–3393 (2014)
2. Gan, W., Lin, J.C.W., Fournier-Viger, P., Chao, H.C., Hong, T.P., Fujita, H.: A survey of incremental high-utility itemset mining. Wiley Interdiscip. Rev.: Data Min. Knowl. Discov. **8**(2), e1242 (2018)
3. Liu, J., Wang, K., Fung, B.C.: Direct discovery of high utility itemsets without candidate generation. In: ICDM, pp. 984–989. IEEE (2012)
4. Liu, Y., Liao, W., Choudhary, A.: A two-phase algorithm for fast discovery of high utility itemsets. In: Ho, T.B., Cheung, D., Liu, H. (eds.) PAKDD 2005. LNCS (LNAI), vol. 3518, pp. 689–695. Springer, Heidelberg (2005). https://doi.org/10.1007/11430919_79
5. Pei, J., Han, J., Wang, W.: Constraint-based sequential pattern mining: the pattern-growth methods. J. Intell. Inf. Syst. **28**(2), 133–160 (2007)
6. Tseng, V.S., Shie, B.E., Wu, C.W., Yu, P.S.: Efficient algorithms for mining high utility itemsets from transactional databases. IEEE Trans. Knowl. Data Eng. **25**(8), 1772–1786 (2013)
7. Yao, H., Hamilton, H.J., Butz, C.J.: A foundational approach to mining itemset utilities from databases. In: SIAM, pp. 482–486 (2004)
8. Zhang, C., Almpanidis, G., Wang, W., Liu, C.: An empirical evaluation of high utility itemset mining algorithms. Expert Syst. with Appl. **101**, 91–115 (2018)
9. Zida, S., Fournier-Viger, P., Lin, J.C.W., Wu, C.W., Tseng, V.S.: EFIM: a fast and memory efficient algorithm for high-utility itemset mining. Knowl. Inf. Syst. **51**(2), 595–625 (2017)

How Much Can A Retailer Sell? Sales Forecasting on Tmall

Chaochao Chen(✉), Ziqi Liu, Jun Zhou, Xiaolong Li, Yuan Qi, Yujing Jiao, and Xingyu Zhong

Ant Financial Services Group, Hangzhou 310099, China
{chaochao.ccc,ziqiliu,jun.zhoujun,xl.li,yuan.qi,yujing.jyj, xingyu.zxy}@antfin.com

Abstract. Time-series forecasting is an important task in both academic and industry, which can be applied to solve many real forecasting problems like stock, water-supply, and sales predictions. In this paper, we study the case of retailers' sales forecasting on Tmall—the world's leading online B2C platform. By analyzing the data, we have two main observations, i.e., *sales seasonality* after we group different groups of retails and a *Tweedie distribution* after we transform the sales (target to forecast). Based on our observations, we design two mechanisms for sales forecasting, i.e., seasonality extraction and distribution transformation. First, we adopt Fourier decomposition to automatically extract the seasonalities for different categories of retailers, which can further be used as additional features for any established regression algorithms. Second, we propose to optimize the Tweedie loss of sales after logarithmic transformations. We apply these two mechanisms to classic regression models, i.e., neural network and Gradient Boosting Decision Tree, and the experimental results on Tmall dataset show that both mechanisms can significantly improve the forecasting results.

Keywords: Sales forecasting · Tweedie distribution · Distribution transform · Seasonality extraction

1 Introduction

Time-series forecasting is an important task in both academic [4] and industry [16], which can be applied to solve many real forecasting problems including stock, water-supply, and sales predictions. In this paper, we study the forecasts of retailers' future sales at Tmall.com[1], one of the world's leading online business-to-customer (B2C) platform operated by Alibaba Group. The problem is essentially important because accurate estimation of future sales for each retailer can help evaluate and assess the potential values of small businesses, and help discover potentials for further investment.

[1] https://en.wikipedia.org/wiki/Tmall.

C. Chen and Z. Liu—Equal contribution.

Q. Yang et al. (Eds.): PAKDD 2019, LNAI 11440, pp. 204–216, 2019.
https://doi.org/10.1007/978-3-030-16145-3_16

However, accurately estimating each retailer's future sales on Tmall could be way challenging in several reasons. First of all, naturally different goods or products are sold with strong seasonal properties. For example, most of fans are sold in summer, while most of heaters are sold in winter, i.e. we can observe strong seasonal properties on different groups of retailers. Secondly, the distribution of sales among all retailers demonstrates an over-tail power law distribution , i.e., the retailers' sales spread a lot. Naively ignore such issues could make the performance much worse.

In this paper, we analyze and summarize the characteristics of the sales data at Tmall.com, and propose two mechanisms to improve forecasting performance. On one hand, we propose to extract seasonalities from groups of retailers. Specifically, we characterize the seasonal evolutions of sales by first clustering retailers into groups and then study the seasonal series as decompositions from a series of Fourier basis functions. The results of our approach can be utilized as features, and simply added to the feature space of any established machine learning toolkits, e.g., linear regression, neural network, and tree-based model. On the other hand, we place distribution transformations on the original retailers' sales. Specifically, we observe that the distribution over retailers' sales follows a Tweedie distribution after we transform retailers' sales by logarithmic. Thus, we propose to optimize Tweedie loss for regression on the logarithmic transformed sales data instead of other losses on the original ones. Empirically, we show that the proposed two mechanisms can significantly improve the performance of predicting retailers' future sales by applying them into both neural networks and tree based ensemble models.

We summarize our main contributions as follows:

- By analyzing the Tmall data, we obtain two observations, i.e., *sales seasonality* after we group different categories of retailers and a *Tweedie distribution* after we transform the original sales.
- Based on our observations, we design two general mechanisms, i.e., seasonality extraction and distribution transform, for sales forecasting. Both mechanisms can be applied to most existing regression models.
- We apply the proposed two mechanisms into two popular existing regression models, i.e., neural network and Gradient Boosting Decision Tree (GBDT), and the experimental results on Tmall dataset demonstrate that both mechanisms can significantly improve the forecasting results.

2 Data Analysis and Problem Definition

In this section, we first describe the sales data and features on Tmall. We then analyze the seasonality and distribution of sales data. Finally, we give the sales forecasting problem a formal definition.

2.1 Sales Data and Feature Description

Tmall.com is nowadays one of the largest business-to-customer (B2C) E-commerce platform. It has more than 180,000 retailers. Among of those retailers,

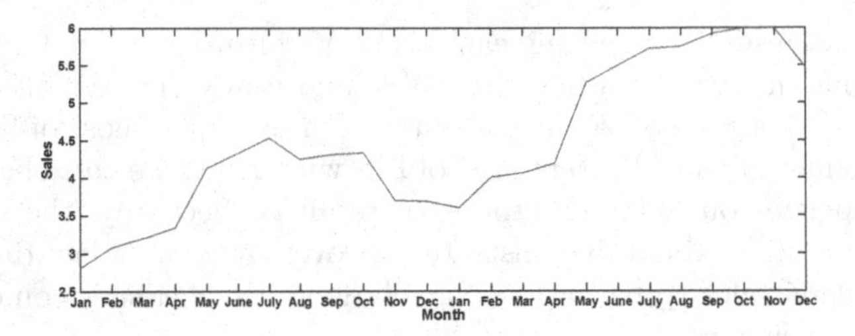

Fig. 1. GMV on Tmall. The horizontal axis denotes the months between Jan. 2015 to Dec. 2016, and the vertical axis denotes the GMV on Tmall where we omit the scale.

there could be giant retailers like Apple.com, Prada, and together with small businesses. The platform is selling hundreds of thousands products in diverse categories, e.g., 'furniture', 'snack', and 'entertainment'.

Besides category information, the other features of retailers on Tmall can be mainly divided into three types: (1) The basic features that are able to reflect the marketing and selling capability of each retailer. For example, the amount of advertisement investment, the number of buyers, the rating/review given by the buyers, and so on. (2) The high-level features that are generated from historical sales and basic feature data. Suppose a retailer i generates a series of sales data, e.g., $y_{i,t-2}, y_{i,t-1}, y_{i,t}, y_{i,t+1}$. We are currently at time t and want to forecast $y_{i,t+1}$. Then $y_{i,t}$ can be taken as a feature which indicates the sale amount of previous period, $y_{i,t} - y_{i,t-1}$ is a feature that indicates the increasing speed of the sales, and $(y_{i,t} - y_{i,t-1}) - (y_{i,t-1} - y_{i,t-2})$ is a feature that denotes the accelerated speed of the sales. Similarly, we can generate other high-level features, e.g., the number of buyers, using the basic features available. (3) The seasonality features that are generated by using other machine learning techniques, which aim to capture the seasonal property of different retailers. We will present how to generate these features in Sect. 3.1.

2.2 Seasonality Analysis

The retailers' sales tend to have different seasonality due to the seasonal items they sell. Take vegetables for example, tomatoes and cucumbers are usually sold more in summer, while celery cabbage is likely sold more in winter. Although the GMV demonstrate seasonal properties, as is shown in Fig. 1, the analysis of the seasonality related to the gross merchandise volume is relative meaningless for the prediction on each retailer. In contrast, the seasonality analysis on each single retailer makes the analysis cannot generalize well in the future. Instead, we further investigate the seasonalities in different groups of retailers.

By analyzing the sales data on Tmall, we observe different seasonal patterns on different categories of retailers. Figure 2 shows the sales of four different categories of retailers, i.e., 'Women's Wearing', 'Men's Wearing', 'Snack', and 'Meat', where we use two year's sales data from January 2015 to December 2016. We can observe that, 'Women's Wearing' and 'Men's Wearing' show quite similar

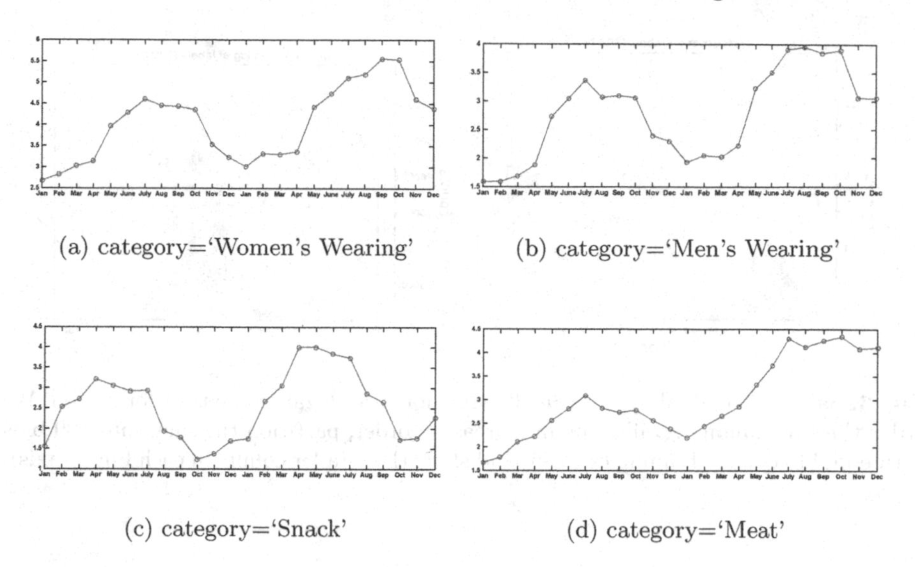

(a) category='Women's Wearing' (b) category='Men's Wearing'

(c) category='Snack' (d) category='Meat'

Fig. 2. Sales seasonality of different categories on Tmall. The horizontal axis denotes the months between Jan. 2015 to Dec. 2016, and the vertical axis denotes the total sale among of retailers in each category where we omit the scale.

seasonal patterns, i.e., they both reach peak in summer (July or August) and decline to nadir in winter (January). On the contrary, 'Snack' and 'Meat' show different seasonal patterns. In summary, the seasonalities under different categories could differ quite a lot. Thus if we can somehow partition the retailers into appropriate groups, the shared seasonality among retailers in one group could be statistically useful for characterizing each retailer in the group. Given a group of retailers, how can we characterize the seasonality for the group remains to be solved. We will discuss our approaches in Sect. 3.1.

2.3 Sale Amount Analysis

The sales of each retailer over time-series could be much challenging. To illustrate this, we show the histograms over sales in Fig. 3 (left). It shows that the sales could be much diverse across over all the retailers. In practice, this is very hard to formalize as a trivial regression problem because the errors on those sales from giant retailers could dominate the loss, e.g. least squared loss.

Instead, after we do a logarithmic transformation on the sales of each retailer, we found that the histogram appears to be a clear Tweedie distribution, i.e. Fig. 3 (right), which will be further described in details in Sect. 3.2. As a result, such transformation on the dependent variables makes our forecasting much easier. Note that, there are always some retailers' sale around zero. This is because some shops on Tmall will close or forced to be closed by Tmall due to some reason from time to time, and correspondingly, some shop will be newly opened or reopen. Consequently, some retailers' sales are around zero.

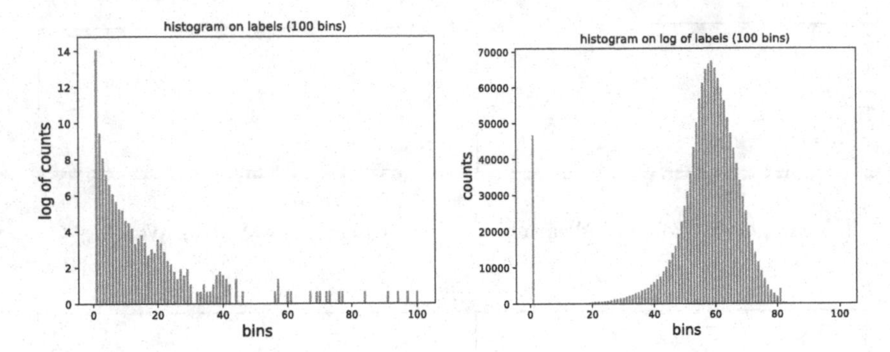

Fig. 3. Sales on Tmall obey Tweedie distribution after logarithmic transformation. We order the sales among retailers in an increasing order, partition the sales into 100 bins with equal frequency binning (x-axis), and show the retailer counts in each bin (y-axis).

2.4 Problem Definition

Assuming any retailer in Tmall.com as i, at month t, We formalize the sales forecasts problem as a regression problem, i.e. given the features of each retailer $x_{i,<t} \in \mathbb{R}^d$, where d is the feature dimensionality and $< t$ denotes the months before t, and the known sales of each retailer $y_{i,<t+1}$, we want to learn a function: $f : x_{i,t} \mapsto y_{i,t+1}$, where $x_{i,t}$ denotes the features of retailers i at month t and $y_{i,t+1}$ denotes its sales at month $t+1$. That is, give the features of any retailer at month t, we want to predict their corresponding sales at month $t+1$.

3 Model Design and Implementation

In this section, we will present our designed two mechanisms, i.e., seasonality extraction and distribution transform, for sales forecasting.

3.1 Seasonality over Groups of Retailers

As we reported in Sect. 2.2, the seasonalities under different categories could differ quite a lot, therefore, the remaining problem is that how should we partition the retailers into appropriate groups. Instead of manually partition retailers, we adopt clustering methods for time-series data [14] to do so. Specifically, we group the retailers by using the basic and high-level features described in Sect. 2.1, so that retailers that have similar features are grouped together.

After we partition retailers into groups, we adopt discrete Fourier transform to automatically extract the seasonality for retailers in different groups. Formally, assuming a group of sellers with expected amount of sales annotated as $\tilde{y}(t)$ at time t, thus results into a series of expected sales as observations, i.e.

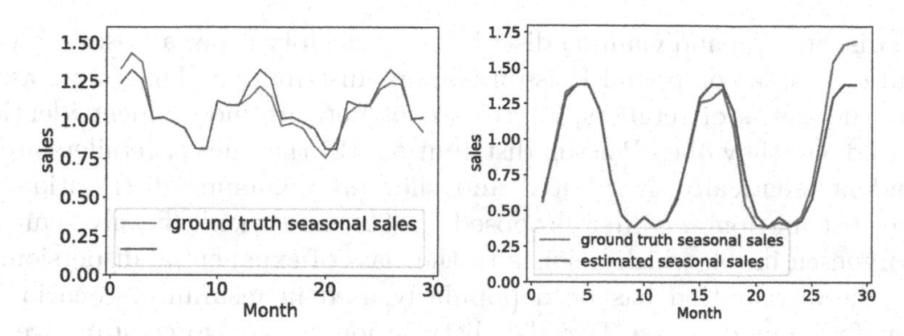

Fig. 4. Seasonality extraction results for two groups of retailers. The left one is mainly the group of retailers who sell purses, and the right one is mainly the group of retailers who sell accessories. In both figures, we use the first 15 months data to learn the parameters in Eq. (1), and further use them to predict the seasonality of all the months' data (note that we also omit the scale of the sale amount).

$\{\tilde{y}(0), \ldots, \tilde{y}(t), \ldots, \tilde{y}(T)\}$. Each periodic function $\tilde{y}(\cdot)$ can be expanded by the Fourier series, which is a linear combination of infinite sines and cosines,

$$\tilde{y}(t) = a_0 + \sum_{n=1}^{\infty} a_n \cos(nt) \sum_{n=1}^{\infty} b_n \sin(nt), \quad (1)$$

where $\{a_i, b_i | i \in \{0, \ldots, n, \ldots, \infty\}\}$ are parameters to be optimized. As a result, the function $\tilde{y}(\cdot)$ can be represented by a Fourier basis.

We now show the results of extracted seasonalities on different groups of retailers. We randomly select two groups of retailers and show their seasonalities and estimates for sales in Fig. 4, where we find the two groups of retailers mainly sell purses and accessaries, respectively. In Fig. 4, we use the first 15 months' data to learn the parameters in Eq. (1), and further use them to predict the seasonality of all the months' data. It is obvious that our extracted seasonality is very close to the real one in both groups. Similarly, we can extract seasonality for other features, e.g., the number of buyers and the among of advertisement investment, by using the same method.

In practice, we use two types of features extracted from such seasonal patterns: (1) the seasonal values of the target we want the extract, e.g., sales and the number of buyers, in a window of 12 months centered around the month t. (2) the variation, i.e. the difference among those seasonal values. Hopefully, such seasonality or trend measures for each group of sellers can be fed into any classifiers, so as to characterize the seasonal patterns for each seller. We will empirically study the effectiveness of these seasonality features in experiments.

3.2 Tweedie Loss for Regression

As we described in Sect. 2.3, based on our observation, the sales on Tmall will clearly obey Tweedie distribution after a logarithmic transformation. From Fig. 3 (right), we see that the sales after logarithmic transformation is a combination of

Poisson distribution and Gamma distribution, which is a special case of Tweedie distribution, i.e., a compound Poisson-Gamma distribution. That is, we assume that (1) the status of retailers, i.e., closed or not, are independent identically distributed and they obey Poisson distribution; (2) the sales of retailers are also independent identically distributed and they obey Gamma distribution. The Tweedie distribution was first proposed in [22], and then officially named by Bent Jorgensen in [9], which belongs to the class of exponential dispersion.

Tweedie distribution has been popularly used in insurance scenarios [23]. We now formally describe Tweedie distribution in sale forecasting scenario. Suppose Let N be a Poisson random variable denoted by $Pois(\lambda)$, and let $Z_i, i = 0, 1, 2, \ldots, N$ be independent identically distributed gamma random variables denoted by $Gamma(\alpha, \gamma)$ with mean $\alpha\gamma$ and variance $\alpha\gamma^2$. We also assume that N is s independent of Z_i. Define a random variable Z by

$$Z = \begin{cases} 0, & \text{if } N = 0, \\ Z_1 + Z_2 + \ldots + Z_N, & \text{if } N > 0. \end{cases} \tag{2}$$

We can see from Eq. (2) that Z is the Poisson sum of independent Gamma random variables, which is also called compound Poisson-Gamma distribution. In sales forecasting scenarios, Z can be viewed as the total number of retailers, N as the opened retailers, and Z_i as the sale amount of retailers i. Note that the distribution of Z has a probability mass at zero, i.e., $Pr(Z = 0) = exp(-\lambda)$. The existing research has proven that, if we reparametrize $(\lambda, \alpha, \gamma)$ by

$$\lambda = \frac{1}{\phi}\frac{\mu^{2-p}}{2-p}, \alpha = \frac{2-p}{1-p}, \qquad \gamma = \phi(\rho - 1)\mu^{\rho-1},$$

Eq.(2) then becomes the form of a Tweedie model $Tw(\mu, \phi, \rho)$ with $1 < \rho < 2$ and $\mu > 0$. Here, the boundary cases $\rho \rightarrow 1$ and $\rho \rightarrow 2$ correspond to the Poisson and the gamma distributions, respectively. The compound Poisson-Gamma distribution with $1 < \rho < 2$ can be seen as a bridge between the Poisson and the Gamma distributions.

The log-likelihood of this Tweedie model for the sale y of a retailer is

$$L(y|\mu, \phi, \rho) = \frac{1}{\phi}\left(y\frac{1}{\phi}\frac{\mu^{1-p}}{1-p} - \frac{1}{\phi}\frac{\mu^{2-p}}{2-p}\right) + log(a(y, \phi, \rho)), \tag{3}$$

where the normalizing function $a(\cdot)$ can be written as

$$a(y, \phi, \rho) = \begin{cases} \frac{1}{y}\sum_t W_t(y, \phi, \rho), & \text{for } y > 0, \\ 1, & \text{for } y = 0, \end{cases}$$

and $\sum_t W_t$ is an example of Wright's generalized Bessel function [22].

After that, given the parameter ρ for Tweedie model, the other parameters can be efficiently solved by using maximum log-likelihood approach [23]. The Tweedie model can be naturally combined with most existing regression models, e.g., NN and GBDT. That is, we can train a Tweedie loss NN model or GBDT model instead of the models with other losses, e.g., square loss [23]. Obviously, the results of Tweedie loss regression are much better than those of other loss regression, e.g., square loss, as will be shown in experiments. This is because Tweedie loss fits the real sales distribution after logarithmic transformation of sales, as is shown in Fig. 3 (right).

4 Empirical Study

In this section, we first describe the dataset and the experimental settings. Then we report the experimental result by comparing with various state-of-the-art sales forecasting techniques. We finally analyze the effect of Tweedie distribution parameter (ρ) on model performance.

4.1 Dataset

Features. As we described in Sect. 2, the features of retailers mainly contain three types, i.e., the basic features, high-level features that are generated from historical sales and basic feature data, and the seasonality features that are generated by using other machine learning techniques. This includes 189 features in total, where there are 79 basic features, 102 high-level features, and 8 seasonality features as we discussed in Sect. 3.1.

Samples. We choose the samples (retailers) during Jan. 2015 and Dec. 2016 on Tmall. Note that we only focus on forecasting the relative small retailers whose monthly sale amount is under a certain range (300,000). Because, in practice, the sales of big retailers are very stable, and it is meaningless to forecast their sales. After that, we have 783,340 samples. We use the samples in 2015 as training data, the samples from Jan. 2016 to June 2016 as validation, and the samples from July 2016 to Dec. 2016 as test data.

4.2 Experimental Settings

Evaluation Metric. Most existing research use error-based metric, e.g., Mean Average Error (MAE) and Root Mean Square Error (RSME), to evaluate the performance in time-series forecasting [1,6]. However, these metrics are way sensitive to those retailers whose sales are large. As we can see in Fig. 1, the sales on Tmall spread a lot. In practice, the forecasting precision of the retailers with small sales counts the same as the ones with big sales. Therefore, we propose to use Relative Precision (RP) for sales forecasting on Tmall, which is defined as

$$RP@p = \frac{\sum_{i=1}^{N} \mathbb{1}\left(\frac{|y_i - \hat{y}_i|}{y_i} < p\right)}{N}, \tag{4}$$

where N is the total number of retailers, y_i as the real sale and $\hat{y}_i$ as the forecasted sale, $p \in [0, 1]$, and $\mathbb{1}(\cdot)$ is the indicator function that equals to 1 if the expression in it is true and 0 otherwise.

Table 1. Comparison result on test data.

Model	NN	NN-S	NN-T	**NN-ST**	GBDT	GBDT-S	GBDT-T	GBDT-ST
RP@0.1	0.1693	0.1723	0.3236	**0.3338**	0.1719	0.1859	0.3159	**0.3263**
RP@0.2	0.1933	0.1987	0.3484	**0.3534**	0.2095	0.2242	0.3394	**0.3520**
RP@0.3	0.2603	0.2657	0.3950	**0.3956**	0.2681	0.2816	0.3821	**0.3966**

As we can see from Eq. (4), RP is actually the percentage of the retailers whose forecasting error is in a certain range p. Intuitively, the smaller p is, the smaller RP will be. Because one has higher demanding for the forecasting performance when p is smaller.

Comparison Methods. Our proposed mechanisms, i.e., seasonality extraction and distribution transform, has the ability to generalize to most existing regression algorithms. To prove this, we apply the mechanisms into two popular regression models, i.e., Neural Network (NN) and Gradient Boosting Decision Tree (GBDT). We summarize all the methods, including ours, as follow:

- **NN** has been used to do time-series forecasting and proven effective where we use square loss [1,20].
- **NN-S** uses extra our proposed seasonal feature in Sect. 3.1 for NN, and its comparison with NN will prove the effectiveness of seasonality extraction.
- **NN-T** uses our proposed Tweedie-loss in Sect. 3.2 for NN, and its comparison with NN will prove the effectiveness of our proposed Tweedie-loss regression after sale distribution transform.
- **NN-ST** extra uses our proposed seasonal feature in Sect. 3.1 for NN-T, which is the application of our proposed two mechanisms in NN.
- **GBDT** is developed for additive expansions based on any fitting criterion, which belongs to a general gradient-descent 'boosting' paradigm and suits for regression tasks with many types of loss functions, e.g., least-square loss, Huber loss, and Tweedie loss [7]. Specifically, we use the GBDT algorithms implemented on Kunpeng [26]—a distributed learning system that is popularly used in Alibaba and Ant Financial, where we also use square loss.
- **GBDT-S** uses extra seasonal feature for GBDT, similar as NN-S.
- **GBDT-T** uses Tweedie-loss for GBDT, similar as NN-T.
- **GBDT-ST** uses extra seasonal feature for GBDT-T, similar as NN-ST.

Parameter Setting. For NN, we use a three-layer network, with Rectified Linear Unit (ReLU) as active functions, and optimized with Adam [12] (learning rate as 0.1). For GBDT, we set tree number as 120, learning rate as 0.3, and regularizer of ℓ_2 norm as 0.5. We will study the effect of parameter ρ of Tweedie regression in Sect. 4.4.

4.3 Comparison Results

We summarize the comparison results in Table 1, and have the following comments.

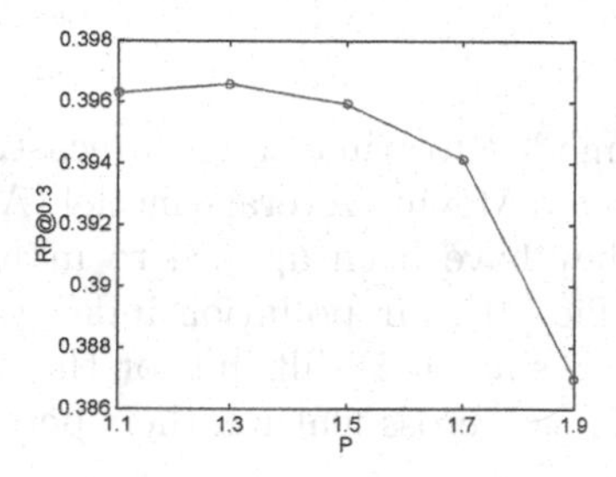

Fig. 5. Effect of Tweedie distribution parameter (ρ) on GBDT-ST on validate data.

(1) The forecasting performance of NN and GBDT are close, and the performance of GBDT is slightly higher than NN. This is because GBDT can naturally consider the complicate relationship, e.g., cross feature, between features. **(2)** Our proposed seasonality extraction mechanism can clearly improve the forecasting performance of both NN and GBDT. For example, GBDT-S improves the forecasting performance of GBDT by 8.14% in terms of RP@0.1, and GBDT-ST further improves the forecasting performance of GBDT-T by 3.29%. **(3)** Our proposed distribution transform mechanism can significantly improve the forecasting performance of both NN and GBDT. For example, NN-T improves the forecasting performance of NN by 91.14% in terms of RP@0.1, and NN-ST improves the forecasting performance of NN-S by 93.73% in terms of RP@0.1 **(4)** In summary, our proposed two mechanisms consistently improve the forecasting performances of both NN and GBDT models. Specifically, NN-ST improves the forecasting performance of NN by 97.14%, 82.82%, 51.98% in terms of RP@0.1, RP@0.2, and RP@0.3 respectively. And, GBDT-ST improves the forecasting performance of GBDT by 89.82%, 68.10%, 47.93% in terms of RP@0.1, RP@0.2, and RP@0.3 respectively. The results not only demonstrate the effectiveness of our proposed mechanisms, but also indicate the generalizability of them.

4.4 Effect of Tweedie Distribution Parameter (ρ)

As described in Sect. 3.2, the Tweedie distribution parameter (ρ) bridges the Poisson and the Gamma distributions, and the boundary cases $\rho \to 1$ and $\rho \to 2$ correspond to the Poisson and the Gamma distributions, respectively. The effect of Tweedie distribution parameter (ρ) on GBDT-ST is shown in Fig. 5, where we use the validate data. From it, we find that GBDT-ST achieves the best performance when $\rho = 1.3$. This indicates that the real sales data on Tmall fit the Tweedie distribution when $\rho = 1.3$.

5 Related Works

In this section, we will review literatures on time-series forecasting, which are mainly in two types, i.e., linear model and non-liner model.

5.1 Linear Model

The most popular linear models for time-series forecasting are linear regression and Autoregressive Integrated Moving Average model (ARIMA) [8]. Due to their efficiency and stability, they have been applied to many forecasting problems, e.g., wind speed [10], traffic [21], air pollution index [13], electricity price [3], and Inflation [19]. However, since it is difficult for them to consider complicate relations between features, e.g., cross feature, their performance are limited.

5.2 Non-linear Model

Non-linear models are also adopted for time-series forecasting. The most popular ones are Support Vector Machine (SVM), neural network, and tree-based ensemble models. For example, SVM are applied to financial forecasting [11] and wind speed forecasting [15]. Neural network are also used in financial marketing forecasting [2] and electric load forecasting [18]. Recently, Gradient Boosting Decision Tree (GBDT) are also adopted to forecast traffic flow [24].

Moreover, model ensemble is also popular for time-series forecasting. For example, ARIMA and SVM were combined to forecast stock price [17]. Hybrid ARIMA and NN models were also used for time-series forecasting [5,25].

In this paper, we do not focus on the choices of regression models. Instead, based on our observation, we focus on extracting seasonality information and transforming label for better forecasting performance. Our proposed seasonality extraction and label distribution transform can be applied into most forecasting models, including NN and GBDT.

6 Conclusions

In this paper, we studied the case of retailers' sales forecasting on Tmall—the world's leading online B2C platform. We first observed sales seasonality after we group different categories of retailers and Tweedie distribution after we transform the sales. We then designed two mechanisms, i.e., seasonality extraction and distribution transform, for sales forecasting. For seasonality extraction, we first adopted clustering method to group the retailers so that each group of retailers have similar features, and then applied Fourier transform to automatically extract the seasonality for retailers in different groups. For distribution transform mechanism, we used Tweedie loss for regression instead of other losses that do not fit the real sale distribution. Our proposed two mechanisms can be used as add-ons to classic regression models, and the experimental results showed that both mechanisms can significantly improve the forecasting results.

References

1. Ahmed, N.K., Atiya, A.F., Gayar, N.E., El-Shishiny, H.: An empirical comparison of machine learning models for time series forecasting. Econometr. Rev. **29**(5–6), 594–621 (2010)
2. Azoff, E.M.: Neural Network Time Series Forecasting of Financial Markets. Wiley, Hoboken (1994)
3. Bianco, V., Manca, O., Nardini, S.: Electricity consumption forecasting in italy using linear regression models. Energy **34**(9), 1413–1421 (2009)
4. Box, G.E., Jenkins, G.M., Reinsel, G.C., Ljung, G.M.: Time Series Analysis: Forecasting and Control. Wiley, Hoboken (2015)
5. Cadenas, E., Rivera, W.: Wind speed forecasting in three different regions of Mexico, using a hybrid ARIMA-ANN model. Renew. Energy **35**(12), 2732–2738 (2010)
6. Carbonneau, R., Laframboise, K., Vahidov, R.: Application of machine learning techniques for supply chain demand forecasting. Eur. J. Oper. Res. **184**(3), 1140–1154 (2008)
7. Friedman, J.H.: Greedy function approximation: a gradient boosting machine. Ann. Stat. 1189–1232 (2001)
8. Hannan, E.J.: Multiple Time Series, vol. 38. Wiley, Hoboken (2009)
9. Jorgensen, B.: Exponential dispersion models. J. Roy. Stat. Soc. Ser. B (Methodol.) **49**, 127–162 (1987)
10. Kavasseri, R.G., Seetharaman, K.: Day-ahead wind speed forecasting using f-ARIMA models. Renew. Energy **34**(5), 1388–1393 (2009)
11. Kim, K.J.: Financial time series forecasting using support vector machines. Neurocomputing **55**(1–2), 307–319 (2003)
12. Kingma, D.P., Ba, J.: Adam: a method for stochastic optimization. arXiv preprint arXiv:1412.6980 (2014)
13. Lee, M.H., Rahman, N.H.A., Latif, M.T., Nor, M.E., Kamisan, N.A.B., et al.: Seasonal ARIMA for forecasting air pollution index: a case study. Am. J. Appl. Sci. **9**(4), 570–578 (2012)
14. Liao, T.W.: Clustering of time series data—a survey. Pattern Recogn. **38**(11), 1857–1874 (2005)
15. Liu, D., Niu, D., Wang, H., Fan, L.: Short-term wind speed forecasting using wavelet transform and support vector machines optimized by genetic algorithm. Renew. Energy **62**, 592–597 (2014)
16. Makridakis, S., Hibon, M.: The M3-competition: results, conclusions and implications. Int. J. Forecast. **16**(4), 451–476 (2000)
17. Pai, P.F., Lin, C.S.: A hybrid ARIMA and support vector machines model in stock price forecasting. Omega **33**(6), 497–505 (2005)
18. Park, D.C., El-Sharkawi, M., Marks, R., Atlas, L., Damborg, M.: Electric load forecasting using an artificial neural network. IEEE Trans. Pow. Syst. **6**(2), 442–449 (1991)
19. Pufnik, A., Kunovac, D., et al.: Short-term forecasting of inflation in Croatia with seasonal ARIMA processes. Technical report (2006)
20. Qi, M., Zhang, G.P.: Trend time-series modeling and forecasting with neural networks. IEEE Trans. Neural Netw. **19**(5), 808–816 (2008)
21. Sun, H., Liu, H., Xiao, H., He, R., Ran, B.: Use of local linear regression model for short-term traffic forecasting. Transp. Res. Rec.: J. Transp. Res. Board **1836**, 143–150 (2003)

22. Tweedie, M.: An index which distinguishes between some important exponential families. In: Statistics: Applications and New Directions: Proceedings of Indian Statistical Institute Golden Jubilee International Conference, pp. 579–604 (1984)
23. Yang, Y., Qian, W., Zou, H.: Insurance premium prediction via gradient tree-boosted Tweedie compound Poisson models. J. Bus. Econ. Stat. **36**, 1–15 (2018)
24. Yinga, X., Jungangb, C.: Traffic flow forecasting method based on gradient boosting decision tree (2017)
25. Zhang, G.P.: Time series forecasting using a hybrid arima and neural network model. Neurocomputing **50**, 159–175 (2003)
26. Zhou, J., et al.: KunPeng: parameter server based distributed learning systems and its applications in Alibaba and ant financial. In: SIGKDD, pp. 1693–1702 (2017)

Hierarchical LSTM: Modeling Temporal Dynamics and Taxonomy in Location-Based Mobile Check-Ins

Chun-Hao Liu[1(✉)], Da-Cheng Juan[2], Xuan-An Tseng[1], Wei Wei[2], Yu-Ting Chen[3], Jia-Yu Pan[2], and Shih-Chieh Chang[1,4]

[1] National Tsing Hua University, Hsinchu, Taiwan
newgod1992@gapp.nthu.edu.tw, killerjack003@gmail.com, scchang@cs.nthu.edu.tw

[2] Carnegie Mellon University, Pittsburgh, USA
x@dacheng.info, {wewei,jypan}@cs.cmu.edu

[3] University of California, Los Angeles, Los Angeles, USA
ytchen@cs.ucla.edu

[4] Electronic and Optoelectronic System Research Laboratories, ITRI, Hsinchu, Taiwan

Abstract. "Is there any pattern in location-based, mobile check-in activities?" "If yes, is it possible to accurately predict the intention of a user's next check-in, given his/her check-in history?" To answer these questions, we study and analyze probably the largest mobile check-in datasets, containing 20 millions check-in activities from 0.4 million users. We provide two observations: "*work-n-relax*" and "*diurnal-n-nocturnal*" showing that the intentions of users' check-ins are strongly associated with time. Furthermore, the category of each check-in venue, which reveals users' intentions, has structure and forms taxonomy. In this paper, we propose **Hierarchical LSTM** that takes both (a) check-in time and (b) taxonomy structure of venues from check-in sequences into consideration, providing accurate predictions on the category of a user's next check-in location. Hierarchical LSTM also projects each category into an embedding space, providing a new representation with stronger semantic meanings. Experimental results are poised to demonstrate the effectiveness of the proposed Hierarchical LSTM: (a) Hierarchical LSTM improves *Accuracy@5* by 4.22% on average, and (b) Hierarchical LSTM learns a better taxonomy embedding for clustering categories, which improves *Silhouette Coefficient* by 1.5X.

Keywords: Long Short-Term Memory · Location-Based Social Network · Point of Interest · Behavior model

D.-C. Juan, W. Wei, Y.-T. Chen and J.-Y. Pan—Recently working at Google, Mountain View, CA, USA.

Q. Yang et al. (Eds.): PAKDD 2019, LNAI 11440, pp. 217–228, 2019.
https://doi.org/10.1007/978-3-030-16145-3_17

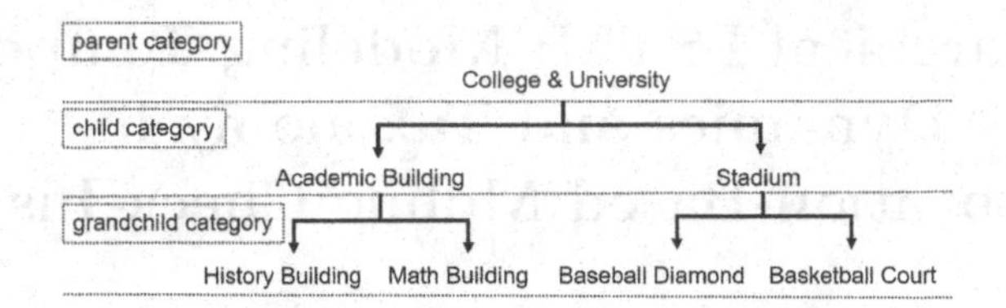

Fig. 1. Example taxonomy of hierarchical category in Foursquare and Jiepang datasets. One parent category in a taxonomy contains several child categories. Similarly, one child category contains several grandchild categories.

1 Introduction

"Is there any pattern in location-based, mobile check-in activities?" "If yes, is it possible to accurately predict the intention of a user's next check-in, given his/her check-in history?" These questions serve as the motivations for this work.

Location-Based Social Networks (LBSN) are rising due to the ubiquity of GPS-equipped smart phones. In a LBSN, users bridge the gap between the physical world and the online social networks by checking in their footprints on the visited venues, referred as Point Of Interests (POIs). The category of a POI is often associated with certain activities related to a user's intention [4,6,9,12]. Furthermore, a user's next intention and activity can be modeled and even predicted by analyzing these temporal check-in sequences on POIs. For example, if a person checks in at the office during the daytime, after work he/she may check in at a bar or a restaurant, and eventually checks in at home. Understanding and modeling these temporal dynamics of users' intentions or behaviors enable many useful applications, such as recommendation systems which are widely deployed in many products [14].

To better understand users' intentions and preferences, we study the public check-in logs from Foursquare and Jiepang [1] containing 20 million check-in activities from 0.4 million users in total. The categories of venues from these datasets are hierarchical and form a taxonomy shown in Fig. 1. For example, parent category "*College & University*" includes child categories "*Academic Building*" and "*Stadium*" while child category "*Academic Building*" also includes grandchild categories "*History Building*" and "*Math Building*".

One challenge here is how to model both the taxonomy and long sequences of LBSN data at the same time. For modeling sequences, one popular and effective approach is Long Short-Term Memory (LSTM) [3]—famous for its superior ability to preserve sequence information over time. The order of check-in sequences is the key to modeling some subtle intention from a user; for example, a person takes metro to company in the morning, eats lunch at a restaurant, has a teatime in a coffee shop, and chats with friends at a bar after work; therefore, the expansion of these logs is exactly fitting to LSTM's characteristic of sequence modeling. We aim at predicting the next category of POI which a user is interested in by expanding user check-in logs as input sequences of LSTM to model users' activity preferences. Furthermore, LSTM networks have been successfully applied to predict the semantic relatedness [2] and capture syntactic structure over the sentence [10]. Similarly, we use LSTM to capture the relationship between parent category and child category in taxonomy structure and project each category into an embedding space, providing a new representation with stronger semantic meanings.

This paper brings the following contributions:

- We analyze two large-scale Location-Based Social Networks datasets from Foursquare and Jiepang, and provide two observations: "*work-n-relax*" and "*diurnal-n-nocturnal*" showing that the intentions of users' check-ins are strongly associated with time.
- We propose a novel and effective model: **Hierarchical LSTM** that accurately predicts the next POI category. Hierarchical LSTM also captures the hierarchical structure of POI categories.
- Experimental results show that, on average, Hierarchical LSTM outperforms state-of-the-art approaches by 4.22% on *Accuracy@5* metric.
- Furthermore, Hierarchical LSTM learns a more effective taxonomy embedding as a vector representation for clustering categories. Experimental results show that Hierarchical LSTM improves *Silhouette Coefficient* by 1.5X, compared with the embedding learned by vanilla LSTM.

2 Problem Definition

2.1 Datasets: Foursquare and Jiepang

We analyze the public check-in posts from the Foursquare and Jiepang websites. The specifications of both datasets are as follows: Foursquare dataset contains over 11 million check-in activities at 560 thousand venues collected from 56 thousand users in the United States from February 2010 to January 2011; Jiepang dataset contains over 8 million check-in activities at 87 thousand venues collected from 382 thousand users in China from December 2010 to March 2013.

The venues in the datasets are marked with hierarchical categories, as illustrated in Fig. 1. There are 312 child categories within 12 parent categories in the Foursquare dataset and 51 child categories within 7 parent categories in the Jiepang dataset. In this paper, we only use two levels of the taxonomy structure (parent and child categories) in the Hierarchical LSTM, because some information of the grandchild categories are incomplete.

2.2 Observations

User activities have been found to have *"Temporal Correlation"*. For example, D. Yang et al. [11] observed that people usually go to a coffee shop or a burger joint between 13:00 to 14:00 on a weekday, stay at a bar between 21:00 to 22:00 on Friday and go to the gym or outdoor places between 16:00 and 17:00 on the weekend. We analyzed the correlation of the check-in time and category, and also found two observations of Temporal Correlation.

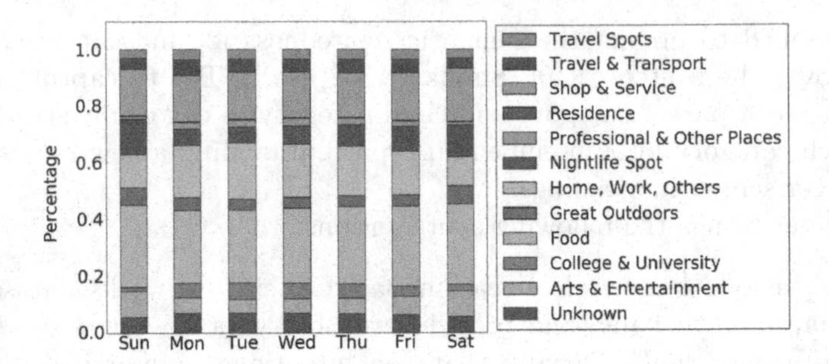

Fig. 2. Percentage of each parent category in a week after we normalize the actual check-in count in Foursquare dataset. The percentage of category associated with "work" in weekdays is larger than the ones on weekends. On the contrary, the percentage of category associated with "relax" on weekends is larger than the ones in weekdays. For example, on weekdays, the percentage of the categories "*College & University*" is much larger than they are on weekends. And On weekends, the percentage of the categories "*Great Outdoors*" is much larger than they are on weekdays

Observation 1. *"Work-n-relax" pattern: weekday check-ins are associated more with "work", and check-ins on weekend are relatively more related to "entertainment & relax" (see Fig. 2).*

Observation 2. *"Diurnal-n-nocturnal" pattern: check-in venues during the day can be very different from the ones during the night (see Fig. 3).*

2.3 Problem Formulation

The problem formulation can be described as: "*Given a user's check-in sequence, predict the child category of the POI of his/her next check-in.*"

Specifically, given a sequence of τ check-ins, we want to predict the child category of POIs in the next check-in, *i.e.*, the $(\tau+1)^{th}$ check-in as label y. Mathematically, this prediction problem can be expressed as:

$$y = f(\{x_{P,j}, x_{C,j}, x_{T,j}\}_{j=1\ldots\tau}), \tag{1}$$

where $x_{P,j}$, $x_{C,j}$, and $x_{T,j}$ are denoted as the parent category, the child category, and the check-in time respectively.

The goal here is to find a function f that takes $x_{P,j}$, $x_{C,j}$ and $x_{T,j}$ ($j = 1$ to τ) as inputs to predict y. Note that we predict the child category instead of the parent category. As Sect. 2.1 mentioned, there are much more child categories than parent categories. Predicting a child category provides the "fine-grained" intention or preference of a user, which is very important when designing a recommendation system. For example, predicting the next POI category as "*Chinese Food Restaurant*" reveals more intention or preference of a user, compared to its parent category "*Food.*"

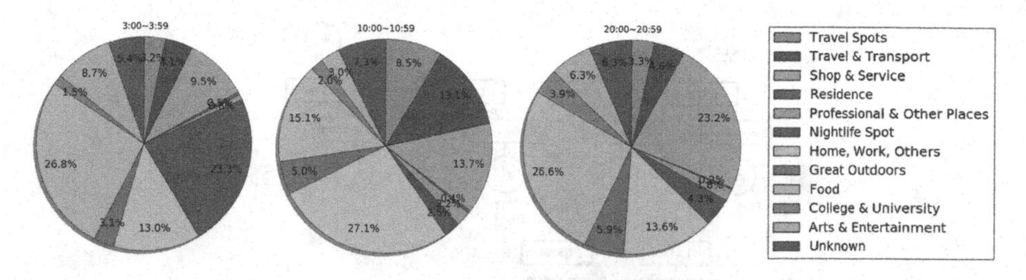

Fig. 3. Pie charts of the parent categories at time intervals 3:00 to 3:59, 10:00 to 10:59 and 20:00 to 20:59, which have a strong contrasting pattern to each other in Foursquare dataset. The percentage of the checked-in categories by "diurnal" users in daytime is larger than the ones at night, and vice versa. For example, during 3:00 to 3:59, most people enjoy their nightlife with the category "*Nightlife Spot*" accounts for over 23%, which is much larger than during the daytime. At 10 a.m., commuters take transportations to work place or school so the check-in counts of category "*Travel & Transport*" and "*Home, Work, Others*" increase.

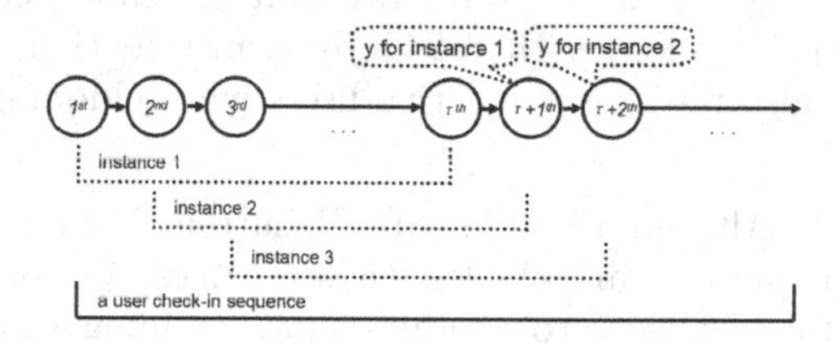

Fig. 4. One user check-in sequence is partitioned into several instances. Each instance is with length of τ: instance 1 contains the 1st to the τth check-in, instance 2 is the 2nd to $(\tau+1)^{th}$ check-in, and so on. Given an instance, the goal is to predict the child category of POI in the next check-in—for example, given instance 1, the goal is to predict the child category of POI in the $(\tau+1)^{th}$ check-in.

2.4 Data Preprocessing

We model the prediction problem from Eq. (1) as a multi-class classification and construct the training and testing datasets accordingly.

For each user's check-in sequence, we partition it into several instances of length τ, *e.g.*, instance 1 contains the 1st to the τth check-ins, instance 2 is the 2nd to the $(\tau+1)^{th}$ check-ins, and more generally, instance i is i^{th} to $(i+\tau-1)^{th}$ check-ins. Figure 4 illustrates this partitioning procedure.

Each instance from the partitioning procedure also has a label y, which is an index value corresponding to the child category of $(\tau+1)^{th}$ check-in. For example, in the Foursquare data set, we use the numbers between 0 to 311 to represent the 312 child categories. Similarly, the values of the predictive features $x_{P,j}$ and $x_{C,j}$ are index values representing the parent category and child category, respectively. The feature $x_{T,j}$ is the check-in time in Eq. (1) which represents the weekday in a week and the hour in a day.

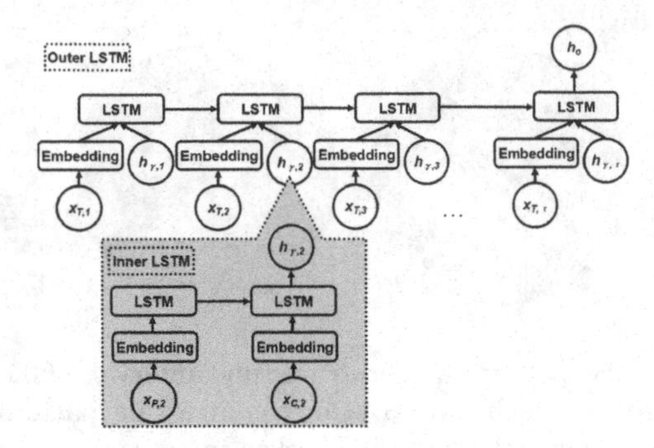

Fig. 5. The structure of Hierarchical LSTM. Two LSTM layers are used in this model: *Inner LSTM* and *Outer LSTM*. Inner LSTM outputs taxonomy embedding $h_{\gamma,j}$ as the input of Outer LSTM.

When preprocessing the data, we require that the time period between two consecutive check-ins is no longer than 24 h to ensure the tight relation between each check-in. This also filters out inactive users who seldom check in during a week.

Instance Length τ**.** Although LSTM (and Recurrent Neural Networks in general) are suited to handle variable length sequences, we partition an input sequence into several instances to produce more training samples (*e.g.* if the length of an input sequence is l, it can be partitioned into $l - \tau + 1$ instances). We note that, since Hierarchical LSTM is based on LSTM, it is also suitable for arbitrary length of input sequences.

Now, the question is: How to choose the sequence length τ to get accurate prediction? When τ is small, we get shorter instances from both active and inactive users, and we can extract more training samples. When τ is large, we only get longer instances from active users, but we extract fewer training samples. Therefore, the selection of τ is the first important question we need to face. We provide the best experimental results in Sect. 4.3 for the value of τ we selected. Overall, an instance has 3 kinds of features ($x_{P,j}$, $x_{C,j}$ and $x_{T,j}$) $\times$ τ (sequence length per instance) $= 3\tau$ predictive features.

3 Methodology

3.1 Hierarchical LSTM

We illustrate the details of Hierarchical LSTM with Fig. 5. First, we extract input features ($x_{P,j}$, $x_{C,j}$, $x_{T,j}$) where $j = 1$ to τ and the label y from each instance, as described in Sect. 2.4.

Then, we feed $x_{P,j}$, $x_{C,j}$ and $x_{T,j}$ into different *embedding layers* (Eq. (2)), and turn indexes of category and check-in time into dense vectors of fixed size, to produce the embedding vectors $e_{P,j}$, $e_{C,j}$, and $e_{T,j}$.

$$e_{P,j} = Emb(x_{P,j}); \quad e_{C,j} = Emb(x_{C,j}); \quad e_{T,j} = Emb(x_{T,j}) \tag{2}$$

Next, the embedding vectors $e_{P,j}$ and $e_{C,j}$ are sequentially fed into the *Inner LSTM* to output a taxonomy embedding $h_{\gamma,j}$ which represents the hierarchical relationship from parent category to child category (Eq. (3)). The Inner LSTM is proposed to capture the semantic relation between the parent and the child categories. We do not include the check-in time in the Inner LSTM because it doesn't have semantic dependency with the categories.

$$h_{\gamma,j} = LSTM(e_{P,j} \rightarrow e_{C,j}) \tag{3}$$

Then, the taxonomy embedding $h_{\gamma,j}$ and the embedding vector $e_{T,j}$ are concatenated and fed into *Outer LSTM* to capture the sequence information and get the last output as the internal vector h_o (Eq. (4)). Outer LSTM feeds the internal vector h_o to a *softmax layer* to make the final decision of the next POI form the N child categories.

$$h_o = LSTM(h_{\gamma,j}, e_{T,j}), \text{ where } j = 1 \text{ to } \tau \tag{4}$$

The output of the softmax layer can be used to represent a probability distribution over N different possible outcomes. Eq. (5) is the predicted probability for the ith child category given an internal vector h_o in Eq. (4) and the weight matrix W_i where $i = 1$ to N.

$$P(y = i|h_o) = \frac{exp(W_i h_o)}{\sum_{l=1}^{N} exp(W_l h_o)} \tag{5}$$

We pick the output of the softmax layer with the highest probability for Accuracy@1 and top k probabilities for Accuracy@k. Note that, at each time step, Inner LSTM takes one sequence of $\{e_{P,j}, e_{C,j}\}$ as input and calculates a taxonomy embedding $h_{\gamma,j}$, which is then passed to the Outer LSTM. Overall, the Outer LSTM takes τ sequences of $\{h_{\gamma,j}, e_{T,j}\}$ as input and eventually calculates the internal vector h_o. Algorithm 1 shows the pseudo code of making a prediction using the Hierarchical LSTM.

4 Experimental Result

4.1 Experimental Setup

In this work, when training and testing models, *Cross Entropy* is used to measure the difference between the truth class y (represented as an one-hot vector of length N) and the distribution of predicted classes $\hat{y}$.

To best train the proposed Hierarchical LSTM and other state-of-the-art approaches, we search for the best hyperparameters by using 10-fold cross validation. Then we evaluate the performance of each model on the test set. After the data preprocessing (Sect. 2.4), there are 297938 venues, 26692 users in the Foursquare data set and 15645 venues, 1141 users in the Jiepang data set. We split the data for training and testing, and divide the training samples for 10-fold cross validation. At the end, we train on 422184 samples, validate on 52773

Algorithm 1. Prediction using Hierarchical LSTM.

Input: $\{x_{P,j}, x_{C,j}, x_{T,j}\}$, where $j = 1...\tau$
Output: The Next Check-In Child Category Of A User
1 **for** $j \leftarrow 1$ *to* τ **do**
2 $\quad e_{P,j} \leftarrow Emb(x_{P,j})$;
3 $\quad e_{C,j} \leftarrow Emb(x_{C,j})$;
4 $\quad e_{T,j} \leftarrow Emb(x_{T,j})$;
5 $\quad$ **for** *each level l in taxonomy* **do**
6 $\quad\quad InnerLSTM(e_{l,j})$;
7 $\quad\quad$ /*parent category is the 1th level, so $e_{1,j} = e_{P,j}$*/
8 $\quad\quad$ /*child category is the 2nd level, so $e_{2,j} = e_{C,j}$*/
9 $\quad h_{\gamma,j} \leftarrow InnerLSTM$;
10 $\quad OuterLSTM(Concatenate(h_{\gamma,j}, e_{T,j}))$;
11 $h_o \leftarrow OuterLSTM$;
12 $\hat{y} \leftarrow Softmax(h_o)$;
13 **return** $\hat{y}$;

samples and test on 52773 samples from the Foursquare data set. Similarly, we train on 38536 samples, validate on 4817 samples and test on 4817 samples from the Jiepang data set.

For the evaluation metrics, we report *Accuracy@1* and *Accuracy@5* to provide a comprehensive study on the performance evaluation of different models.

4.2 Models Compared and Previous Work

We compare the performance of Hierarchical LSTM with some state-of-the-art approaches, which are variants of the basic LSTM model as follows:

- **SCP-RNN** [13] is a framework for click prediction based on Recurrent Neural Networks. We adopt SCP-RNN to our application by replacing the user behavior sequences with the child category $x_{C,j}$ to achieve the goal of predicting users' next POIs. We re-implement SCP-RNN by a single layer LSTM with embedding vectors the $e_{C,j}$ of $x_{C,j}$ (Eq. (2)) as input.
- **ST-RNN** [5] considers not only temporal but also spatial dependency in user's behavior sequences into prediction. We adopt this work to our application by replacing these temporal and spatial sequences with $x_{T,j}$ and child category $x_{C,j}$ to achieve the goal of predicting users' next POIs. We re-implement ST-RNN by a single layer LSTM and concatenating the embedding vectors $e_{C,j}$ and $e_{T,j}$ of $x_{C,j}$ and $x_{T,j}$ (Eq. (2)) as input.
- **ST-RNN+P** is an extension of ST-RNN, which we proposed as an enhanced baseline. This model improves on the original ST-RNN model that, beside using the spatial sequences $x_{C,j}$ as input, we additionally add parent category $x_{P,j}$ as an input feature to improve the accuracy. Figure 6 illustrates the proposed ST-RNN+P. We implement ST-RNN+P by a single-layer LSTM

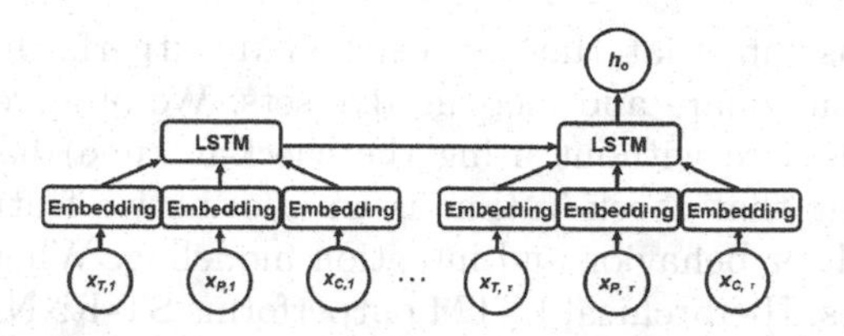

Fig. 6. ST-RNN+P is a single layer LSTM with the concatenation of the embedding vectors $e_{C,j}$, $e_{P,j}$ and $e_{T,j}$ (*i.e.* output of embedding layer when feeding $x_{C,j}$, $x_{P,j}$ and $x_{T,j}$) as input. The only difference compared with Hierarchical LSTM is that ST-RNN+P doesn't use *Inner LSTM* to capture the relation between parent category $x_{P,j}$ and child category $x_{C,j}$.

Table 1. Experimental results on both Foursquare and Jiepang datasets with the selected τ values. Notice that the proposed Hierarchical LSTM consistently outperforms the previous state-of-the-art approaches over the τ values.

Foursquare	$\tau = 35$	
Model	Accuracy@1	Accuracy@5
SCP-RNN [13]	20.62% (baseline)	46.33% (baseline)
ST-RNN [5]	21.02% (+1.94%)	46.33% (+0.00%)
ST-RNN+P	20.97% (+1.70%)	46.93% (+1.30%)
Hierarchical LSTM	**21.87% (+6.06%)**	**48.06% (+3.73%)**
Jiepang	$\tau = 45$	
Model	Accuracy@1	Accuracy@5
SCP-RNN [7]	36.09% (baseline)	65.25% (baseline)
ST-RNN [5]	36.00% (−0.25%)	65.38% (+0.02%)
ST-RNN+P	36.28% (+0.53%)	66.29% (+1.59%)
Hierarchical LSTM	**37.99% (+5.27%)**	**68.20% (+4.52%)**

that takes the concatenation of the embedding vectors $e_{C,j}$, $e_{P,j}$ and $e_{T,j}$ of $x_{C,j}$, $x_{P,j}$ and $x_{T,j}$ (Eq. (2)) as input. The only difference compared with Hierarchical LSTM is that ST-RNN+P doesn't use *Inner LSTM* to capture the relation between parent category $x_{P,j}$ and child category $x_{C,j}$.

We optimize the baseline models above to the best of our knowledge. These models also are trained via 10-fold cross validation and grid-search to find the best hyperparameters.

4.3 Result Summary

Table 1 shows the best experimental results from different state-of-the-art models and Hierarchical LSTM with the selected τ values. For a comprehensive comparison, we calculate *Accuracy@1*, *Accuracy@5* and the improvement from baseline.

The results demonstrate that Hierarchical LSTM outperforms state-of-the-art models on both Foursquare and Jiepang datasets. We observe that SCP-RNN (the only approach here without using the check-in time) has the worst performance, indicating that check-in time is an informative feature which should always be included for behavior and intention modeling. When using the same set of input features, Hierarchical LSTM outperforms ST-RNN+P in every evaluation metric.

4.4 Taxonomy Embedding Analysis

To get some insights about the better performance of Hierarchical LSTM, we analyze the difference between the taxonomy embedding $h_{\gamma,j}$ of Hierarchical LSTM and the embedding vectors $e_{C,j}$ of ST-RNN+P, which both represent the child categories. Figure 7 shows the t-SNE [8] projection of the $h_{\gamma,j}$ and $e_{C,j}$ vectors onto X-axis and Y-axis. The dots with the same color in the t-SNE graph indicate the different child categories from the same parent category in Foursquare dataset. Figure 7(a) is messy and Fig. 7(b) is organized according to the colors (i.e., the parent categories). Inner LSTM performs well on matching the hierarchical relationship of the taxonomy path from the parent category to the child category.

Cluster Metrics. In addition, we provide an evaluation metrics: *Silhouette Coefficient* [8]. The vectors from Hierarchical LSTM and ST-RNN+P are normalized before the calculation. *Silhouette Coefficient* is calculated by considering the mean of the intra-cluster distance α and the mean of the nearest-cluster distance β for each sample; mathematically: $\frac{\beta-\alpha}{max(\alpha,\beta)}$.

The value of *Silhouette Coefficient* is in $[-1; 1]$. A value closer to 1 indicates better clustering. *Silhouette Coefficient* of Hierarchical LSTM is 0.034 while ST-RNN+P is only -0.065, which means Inner LSTM is better in clustering child categories.

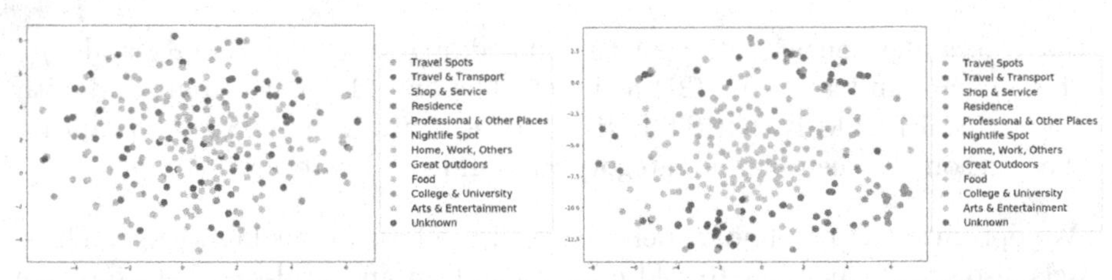

(a) T-SNE for ST-RNN+P with the *Silhouette Coefficient* of -0.065 (project to original embedding space).

(b) T-SNE for Hierarchical LSTM with the *Silhouette Coefficient* of 0.034 (project to original embedding space).

Fig. 7. T-SNE graphs for embedding vectors $e_{C,j}$ of ST-RNN+P and taxonomy embedding $h_{\gamma,j}$ of Hierarchical LSTM from Foursquare dataset. The dots with the same color in the graphs indicate the child categories from the same parent category. (Color figure online)

Meaningful Embedding. In Fig. 8(a), we labeled the projected taxonomy embedding $h_{\gamma,j}$ (the same plot as shown in Fig. 7(b)) with their corresponding child category names. We observed that the embedding vectors obtained from the proposed Hierarchical LSTM model successfully captures the semantics of the categories. In particular, child categories of POIs with similar semantics will have similar embedding vectors, even if these child categories belong to different parent categories as specified in the given data.

For example, in the Foursquare dataset, the child category *"Library"* has an embedding vector similar to that of the child category *"College Library"* (Fig. 8(b)), even though one belongs to the parent category *"Home, Work, Others"* and the other belongs to the parent category *"College & University"*. The proposed Nested LSTM model is able to handle the imperfect/redundant human categorization in the real-world dataset and generate meaningful embedding vectors to the categories.

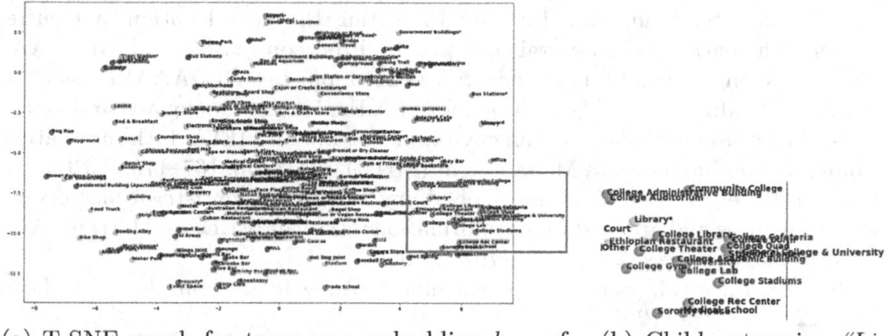

(a) T-SNE graph for taxonomy embedding $h_{\gamma,j}$ of Hierarchical LSTM from Foursquare dataset and labeled the name of child category.

(b) Child categories *"Library"* and *"College Library"*.

Fig. 8. Hierarchical LSTM generates meaningful embedding vectors of POIs. The colors of the dots represent the parent categories (please refer to the legends in Fig. 7(b)). (Color figure online)

5 Conclusion

In this paper, we first analyze two large LBSN datasets and provide two observations: "*work-n-relax*" and "*diurnal-n-nocturnal.*" Then we propose Hierarchical LSTM to predict the category of the next POI where a user will check in. Thanks to Hierarchical LSTM, we now can answer the two motivational questions: "Is there any pattern in location-based, mobile check-in activities?" "If yes, is it possible to accurately predict a user's next check-in intention, given his/her check-in history?" Experimental results show that, Hierarchical LSTM achieves *Accuracy@5* 48.96% and 68.20% on the Foursquare and Jiepang datasets, respectively. The taxonomy embedding learned by Hierarchical LSTM achieves *Silhouette Coefficient* of 0.034 which outperforms other state-of-the-art approaches.

The Inner LSTM captures the hierarchical relationship of categories on taxonomy (parent category—child category), and has the capability of better clustering child category.

References

1. Jiepang Website (2018). https://jiepang.com/
2. Bjerva, J., Bos, J., Van der Goot, R., Nissim, M.: The meaning factory: formal semantics for recognizing textual entailment and determining semantic similarity. In: Proceedings of the 8th International Workshop on Semantic Evaluation (SemEval 2014), pp. 642–646 (2014)
3. Hochreiter, S., Schmidhuber, J.: Long short-term memory. Neural Comput. **9**(8), 1735–1780 (1997)
4. Lian, D., Xie, X.: Collaborative activity recognition via check-in history. In: Proceedings of the 3rd ACM SIGSPATIAL International Workshop on Location-Based Social Networks, pp. 45–48. ACM (2011)
5. Liu, Q., Wu, S., Wang, L., Tan, T.: Predicting the next location: a recurrent model with spatial and temporal contexts. In: Proceedings of the Thirtieth AAAI Conference on Artificial Intelligence, AAAI 2016, pp. 194–200. AAAI Press (2016)
6. Noulas, A., Mascolo, C., Frias-Martinez, E.: Exploiting foursquare and cellular data to infer user activity in urban environments. In: 2013 IEEE 14th International Conference on Mobile Data Management (MDM), vol. 1, pp. 167–176. IEEE (2013)
7. Palangi, H., et al.: Deep sentence embedding using long short-term memory networks: analysis and application to information retrieval. IEEE/ACM Trans. Audio Speech Lang. Process. **24**(4), 694–707 (2016)
8. Pedregosa, F., et al.: Scikit-learn: machine learning in Python. J. Mach. Learn. Res. **12**, 2825–2830 (2011)
9. Pianese, F., An, X., Kawsar, F., Ishizuka, H.: Discovering and predicting user routines by differential analysis of social network traces. In: 2013 IEEE 14th International Symposium and Workshops on a World of Wireless, Mobile and Multimedia Networks (WoWMoM), pp. 1–9. IEEE (2013)
10. Socher, R., Lin, C.C., Manning, C., Ng, A.Y.: Parsing natural scenes and natural language with recursive neural networks. In: Proceedings of the 28th International Conference on Machine Learning (ICML-2011), pp. 129–136 (2011)
11. Yang, D., Zhang, D., Zheng, V.W., Yu, Z.: Modeling user activity preference by leveraging user spatial temporal characteristics in LBSNs. IEEE Trans. Syst. Man Cybern.: Syst. **45**(1), 129–142 (2015)
12. Ye, J., Zhu, Z., Cheng, H.: What's your next move: user activity prediction in location-based social networks. In: Proceedings of the 2013 SIAM International Conference on Data Mining, pp. 171–179. SIAM (2013)
13. Zhang, Y., et al.: Sequential click prediction for sponsored search with recurrent neural networks. In: AAAI 2014, pp. 1369–1375 (2014)
14. Zhao, S., King, I., Lyu, M.R.: A survey of point-of-interest recommendation in location-based social networks. CoRR abs/1607.00647 (2016). http://arxiv.org/abs/1607.00647

Recovering DTW Distance Between Noise Superposed NHPP

Yongzhe Chang[1,2(✉)], Zhidong Li[1,3], Bang Zhang[1], Ling Luo[1,3], Arcot Sowmya[2], Yang Wang[1,3], and Fang Chen[1,3]

[1] Data 61 CSIRO, Sydney, Australia
{yongzhe.chang,zhidong.li,bang.zhang,ling.luo,yang.wang, fang.chen}@data61.csiro.au
[2] University of New South Wales, Sydney, Australia
{yongzhe.chang,arcot.sowmya}@unsw.edu.au
[3] University of Technology Sydney, Sydney, Australia
{zhidong.li,ling.luo,yang.wang,fang.chen}@uts.edu.au

Abstract. Unmarked event data is increasingly popular in temporal modeling, containing only the timestamp of each event occurrence without specifying the class or description of the events. A sequence of event is usually modeled as the realization from a latent intensity series. When the intensity varies, the events follow the Non-Homogeneous Poisson Process (NHPP). To analyze a sequence of such kind of events, an important task is to measure the similarity between two sequences based on their intensities. To avoid the difficulties of estimating the latent intensities, we measure the similarity using timestamps by Dynamic Time Warping (DTW), which can also resolve the issue that observations between two sequences are not aligned in time. Furthermore, real event data always has superposed noise, e.g. when comparing the purchase behaviour of two customers, we can be mislead if one customer visits market more often because of some occasional shopping events. We shall recover the DTW distance between two noise-superposed NHPP sequences to evaluate the similarity between them. We proposed two strategies, which are removing noise events on all possibilities before calculating the DTW distance, and integrating the noise removal into the DTW calculation in dynamic programming. We compare empirical performance of all the methods and quantitatively show that the proposed methods can recover the DTW distance effectively and efficiently.

Keywords: Dynamic Time Warping · Non-Homogeneous Poisson Process · Noise · Dynamic programming · Bayesian prior

1 Introduction

The temporal event sequence is an increasingly popular data type that records event occurrences in time domain. In studies of stochastic processes, a sequence

Q. Yang et al. (Eds.): PAKDD 2019, LNAI 11440, pp. 229–241, 2019.
https://doi.org/10.1007/978-3-030-16145-3_18

is a realization of a Poisson Process (PP) based on an intensity function. As one of the most important tasks when analyzing PP, measuring the similarity between two sequences of events has been widely applied in many applications including financial portfolios [6,12], hospital treatments [16], and asset failure prediction [9]. However, traditional distance measurements based on time to time alignment (i.e. data at time t in one sequence is compared with data at the same time in the other sequence) is infeasible because of the randomness of event intervals. Smoothing is a solution but it leads to information loss in finer resolution. To obtain the distance in finer resolution, Dynamic Time Warping (DTW) algorithm can be used to measure the distance between two unaligned sequences of events.

DTW algorithm is designed to find the minimum distance between sequences without the restriction of time to time alignment. It is well known that the choice of representation of the time series is important when using DTW [8,14]. The ideal representation for temporal event sequences is to use the intensity. However, the intensity for Non-Homogeneous Poisson Process (NHPP) is latent if parametric model is not assumed [15]. Therefore, it may suffer from the model selection problem. There are three alternatives to represent sequences realized from NHPP: first we use binary label on the time slot, recording an event occurred or not, then applying DTW to this representation will align all '0' to '0' and '1' to '1' but the output distance may be largely underestimated. Second, we can use the event count in the given time window. However, the variance of the count increases with the length of the sequence. To avoid these issues, in this paper, we use the third representation, which is the timestamp of each event.

The observed sequences can be superposed by stochastic noise, which means that except for the original NHPP (referred as the clean sequence), there also exists a PP that stochastically generates events as noise. This common phenomenon has been observed in many applications. For example, when comparing two patients' records of hospital visits to check whether they have a similar disease, it can be misleading if one patient visits more often because of another concurrent disease. The current DTW algorithm does not consider the existence of noise, instead, all events in the sequence are used to measure the distance. As a result, the DTW distance between noise superposed NHPP cannot represent the actual DTW distance. In our solution, we try to recover the DTW distance between clean sequences given the superposition of a noise sequence. It is worth noting that the proposed method can be used to recover the distance for any NHPP with arbitrary unknown intensity function. The noise is from reasonably known homogeneous Poisson process (HPP) with commonly used noise setting, that is, the events are generated according to an intensity of HPP. However, we cannot know the exact values of the intensity, so we assume that the intensity is generated from a random distribution as a Bayesian prior. Such noise can be seen from many real applications, for example, acoustic engineering, telecommunications, and statistical forecasting [2,4].

The key challenge of recovering DTW distance from noise-superposed NHPP is to identify which event is noise. The key idea is to consider the probability of an event being noise or not and utilize it to obtain a practical and probabilistically sound DTW distance measure. To solve this issue, two strategies can be considered. One is to remove noise events from the whole sequence with certain probability and then calculate the DTW distance. For this strategy, we discuss three methods, including (1) traversing over all possible subsets of noise and calculating the probability; (2) only traversing the average case based on noise parameters; (3)using a sampling method to approximate the traverse and probability. However, the computational costs of these methods are high. The other strategy, which we have adopted in this paper, is to embed a noise removal step into the DTW calculation in an efficient way using dynamic programming.

In summary, the novelties in this paper are three folds: (1) We can recover the DTW distance between noise superposed NHPP without estimating the arbitrary intensity function for the NHPP sequences. (2) We combine a probability based method with DTW distance computation. (3) We propose a more efficient method to recover the DTW distance based on dynamic programming.

2 Related Work

Distance between sequences of events has wide applications with clustering as an example. [9] and [5] have proposed to use Bayesian nonparametric prior and Hawkes process to cluster event sequences by the likelihood of timestamps, where more similar sequences have larger likelihood to have the same parameter so that they are in the same cluster. The covariance between sequences of four types of events (stock price up and down, volume up and down) has been measured [1]. There are two main drawbacks in these methods. First, they have to assume the intensity function, which is not required in our solutions. Second, when applying to the real data, noise is the major concern that may largely influence the final clustering result. Our methods can recover the distance even with noise.

There are other methods to obtain the distance between multiple dimensional data, a more general data type that includes time series, but cannot be directly applied to solving in our problem. For example, Kernel methods [11] and covariance based methods such as connectedness [6] that can measure the relation between sequences of extreme events in stock market. These methods are based on a time (or dimension) aligned comparison while distance comparison of DTW is based on the optimized alignment.

DTW is a popular distance measure defined for time series [13] by measuring the similarity between two temporal sequences that vary in frequency and length. In relation to our method, stochastic DTW [10] has been proposed to use probability to select the path in DTW, where, to our knowledge, using DTW on intensity-unknown NHPP that is superposed with noise has not been discussed before.

3 Methods

In the beginning we shall show how to apply DTW to sequences of times stamps. Then we will consider two strategies to recover DTW from noise superposed sequences. The first strategy is by calculating the DTW after removing noise points and the second is to integrate the probability to remove noise into the recursion of DTW.

3.1 DTW Distance on Sequences of Timestamps

Given two sequences of timestamps $\mathbf{x} = (x_i; i = 1 \ldots n)$ and $\mathbf{y} = (y_j; j = 1 \ldots m)$, our aim is to calculate the distance between $\mathbf{x}$ and $\mathbf{y}$.

Then given a cost matrix (e.g. $|x_i - y_j|$) $\mathbf{\Delta}_{n \times m} := \delta(x_i, y_j)$ that stores the distance values between each pair of points in two sequences, and an alignment matrix $\mathbf{A} \in \mathcal{A}_{n,m}$, where $\mathbf{A} = a_{i,j} \in \{0, 1\}$ that shows all the possible alignments in matching two sequences, the final distance can be defined as the inner product of $D_{\mathbf{A}} = \langle \mathbf{A}, \mathbf{\Delta} \rangle$. Among all the distance values, the DTW distance is the minimum $D_{\mathbf{A}}$ given all paths. In defining a path, we have: $\sum_{\beta,\gamma \in \{0,1\}} a_{i-\beta,j-\gamma} \geq 2$ if $a_{i,j} = 1$, and $a_{1,1} = a_{n,m} = 1$.

Since the recursion algorithm based on dynamic programming can solve the optimization problem, we write $dtw(x_i, y_j)$ as the DTW distance between two sequences with lengths i and j. In Fig. 1(b), the recursion algorithm can be written as:

$$\mathrm{dtw}(x_i, y_j) = \delta(x_i, y_j) + \min(\mathrm{dtw}(x_i, y_{j-1}), \mathrm{dtw}(x_{i-1}, y_{j-1}), \mathrm{dtw}(x_{i-1}, y_j)). \quad (1)$$

In our problem, the noise is also a sequence of points with timestamps that are mixed into the clean sequences $\mathbf{x}$ and $\mathbf{y}$, following the NHPP with latent intensity μ_x and μ_y. When dealing with one sequence, we write it as μ for simplicity. There is no mark on noise so we do not know which point is the noise. Then the objective is to recover the possible DTW distance between two clean sequences, given two sequences $\mathbf{x}$ and $\mathbf{y}$ with noise.

3.2 Remove Noise Before DTW Calculation

Recovering DTW distance means to get the true DTW distance by eliminating the influence of noise points. To achieve this, before performing DTW, we need to remove the noise events. We use a vector $\mathbf{z_x} = (z_i \in \{0, 1\})$ to indicate whether the event i is noise or not. Here $\mathbf{z_x}$ has the same length as $\mathbf{x}$. We omit $\mathbf{x}$ for simplicity since we only discuss one sequence. We then denote $s = \sum_i z_i$. Since we do not know which point is the real noise, we assume that any point can be noise with a probability, which leads to multiple realizations of $\mathbf{z}$. Given each realization of $\mathbf{z}$, a clean sequence $\mathbf{x}^* \in \mathbb{R}^{1 \times (n-s)}$ can be obtained by $\mathbf{x}^* = (x_i | z_i = 1)$, then a DTW distance can be retrieved based on $\mathbf{x}^*$ and $\mathbf{y}^*$. Here the multiple realizations of $\mathbf{z}$ lead to possible variants of the recovered DTW distance. To obtain the varying DTW distance and the corresponding probability, we introduce four methods, including **traversing method**, **averaging method**, **sampling method** and the **dynamic DTW method**.

To recover DTW distance, an intuitive method is to remove noise first. In this case the noise points are selected from n points so all the possible selections of $\mathbf{z}$ are limited to 2^n. In the traversing method, we traverse all selections and calculate the DTW distance for each realization of $\mathbf{z}$. Each selection has a probability and it is used as the weight to combine the obtained DTW distances. The probability of selecting a $\mathbf{z}$ is:

$$P(\mathbf{z}; \mu T) = \int P(\mathbf{z}|s)\text{Poisson}(s|\mu, T)ds = \int \frac{1}{g}\text{Poisson}(s|\mu, T)ds, \quad (2)$$

where $g = \binom{n}{s}$, and T is the length of a sequence. Since the noise intensity μ is latent, we select a distribution for μ as a Bayesian prior (usually a gamma distribution is chosen), which derives the selection probability:

$$P(\mathbf{z}; \theta, \lambda) = \int P(\mathbf{z}; \mu, T)\text{Gamma}(\mu; \theta, \lambda)d\mu, \quad (3)$$

where θ and λ are the hyperparameters, θ is the shape parameter and λ is the scale parameter. Moreover, it can be seen that the only part related to $\mathbf{z}$ is $P(\mathbf{z}|s)$ (the likelihood term $P(s|\mu, T) = \mu_x^s e^{-\mu_x T}$ does not include event location), so the probability of all selections given s are the same. From the equation above, it is easy to find out that $P(\mathbf{z}|\alpha, \beta)$ are all the same for different realisations, so the weights can be ignored.

Although this method can help us obtain the true distribution of DTW distance, it cannot be applied when the sequence is long (usually infeasible when $n > 20$) due to the expensive calculation cost.

Then, a simplified approximation of the traverse method is to use the average case of s as the approximation, then consider all the selections of $\mathbf{z}$ by fixing s using the average as $E(s) = \theta\lambda$. The prior distribution is assumed to be a gamma distribution: $\text{Gamma}(\theta, \lambda)$. Then all the DTW distances based on $E(s)$ are traversed to obtain the distribution of DTW distance. In this method, only $\binom{n}{E(s)T}$ DTW distances need to be calculated and averaged. We denote this method as **Averaging**.

The two methods mentioned above are either infeasible or over simplified, so we will then introduce a balanced method, in which we sample each point of $\mathbf{z}$ to approximate the distribution of $\mathbf{z}$. This method is based on sequential sampling. Given that z_i is sampled as the last noise point, we can sample whether the point x_{i+q} is noise (that is, $z_{i+q} = 1$) or not, using the cumulative probability. The compound probability of exponential distribution and gamma distribution is exactly the Lomax distribution [7]. Therefore, the cumulative probability is: $z_{i+q} \sim \text{Ber}\left(\text{Lomax}_{\text{cdf}}\left(x_{i+q} - x_i; \theta, \lambda\right)\right)$, and we denote this method as **Sampling**.

3.3 Integrating Noise Removal Probability to DTW

The previous methods link noise removal and DTW algorithms by directly inputting a clean sequence into DTW. We propose a method to integrate noise removal and optimization with recursion. In the original DTW recursion, to calculate the value of $dtw(x_i, y_j)$, three DTW distances from previous calculations are used, as it is shown in Fig. 1(b).

Moreover, we consider the probability that all the consecutive $c = 0 \ldots C$ points in the precedent are noise. That is, given c, with certain probability, all the points $x_{i-c} \ldots x_{i-1}$ are noise but x_{i-c-1} is not noise. We use 5 points ($i = 5$) as an example in Fig. 1(a). We can see that all combinations of selected noise points are included by setting c from 0 to 3 (without loss of generality, the first point is assumed to be not a noise).

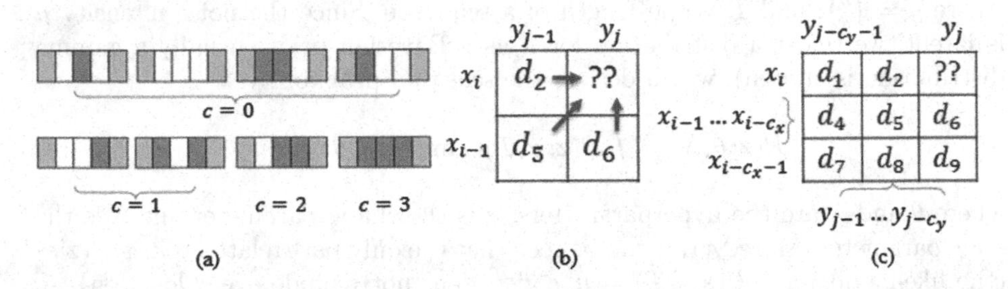

Fig. 1. (a) All possible situations of noise points given 5 points, where the first and last points are not noise. All green points are noise, and yellow points are not noise. (b) The original DTW path determination when calculating DTW for the grid. (c) Grids for previous DTWs ($d_1 - d_9$), the middle row and column can contain c_x, c_y grids. (Color figure online)

We explain how to calculate $\text{dtw}(x_i, y_j)$ in the example in Fig. 1(b) by fixing c_x and c_y for sequences $\mathbf{x}$ and $\mathbf{y}$ respectively. In the example, we use $x_{i-1,\ldots,i-c_x}$ to represent that all the consecutive c_x points could be noise on sequence $\mathbf{x}$. That is, whether $\sum_{k=i-1,\ldots,i-c_x} z_k = 0$ or c_x the corresponding probability is $P(c_x) = P(\sum_{k=i-1,\ldots,i-c_x} z_k = c_x)$. Similarly, consecutive c_y points could be noise on sequence $\mathbf{y}$ with probability $P(c_y)$. Then we consider the DTW with the given c_x and c_y as:

$$\text{dtw}_{|c_x,c_y}(x_i, y_j) = \delta(x_i, y_j) + D_m, \tag{4}$$

where $D_m = \min(d_1, d_7, d_9)$ in the example of Fig. 1(c). It is noticeable that the equations still hold for the special case of $c_x = 0$, where $D_m = \min(d_1, d_4, d_6)$. In this case, $x_{i-c_x-1} = x_{i-1}$ so that we have $d_4 = d_7$ and $d_6 = d_9$. Similarly, when $c_y = 0$, we have $D_m = \min(d_2 = d_1, d_8 = d_7, d_9)$. If both $c_x = 0$ and $c_y = 0$, then $D_m = \min(d_2 = d_1, d_5 = d_7, d_6 = d_9)$ and it becomes the original DTW recursion without noise. To generalize the results, we can write D_m as:

$$\begin{aligned} D_m = \min(&\text{dtw}(x_i, y_{j-c_y-1}), \\ &\text{dtw}(x_{i-c_x-1}, y_{j-c_y-1}), \text{dtw}(x_{i-c_x-1}, y_j)). \end{aligned} \tag{5}$$

In this case D_m can be represented by a function of c_x, c_y and it can be written into $D_m(c_x, c_y)$.

Then we consider multiple values for c_x and c_y instead of fixed values, so (4) can be written as:

$$\begin{aligned}\text{dtw}(x_i, y_j) &= E_{c_x,c_y}(\text{dtw}_{c_x,c_y}(x_i, y_j)) \\ &= \delta(x_i, y_j) + \sum_{c_x=0}^{C_x} P(c_x) \sum_{c_y=0}^{C_y} (P(c_y) D_m(c_x, c_y))\end{aligned} \tag{6}$$

Probability of $P(c_x)$: Let us see how to obtain the probabilities in equation (6), using $P(c_x)$ as an example. For consecutive c_x noise points, the probability can be calculated by:

$$P(c_x) = (1 - P_{i-c_x-1}) \prod_{k=i-c_x}^{i-1} P_k, \tag{7}$$

where P_k denotes the probability that point x_k is noise. Since the clean sequence is non-homogeneous, the probability is changing through time.

P_k can be written into $P(z_k = 1|N(\tau), \lambda, \theta)$, where $N(\tau)$ is the number of points observed in given time τ before the point k (between the time $[x_k - \tau, x_k)$). Simplify $N(\tau)$ as N, the points are indexed by m where $m = 1...N$. First, the probability that $m \leq N$ noise events is $P(m|u, N) = \frac{\text{Poisson}(m;u)}{B}$ where $B = \sum_{i=0}^{N} \text{Poisson}(i; u)$. Here u is the total intensity in τ that can be written as $u = \mu\tau$. Given m points, there are $\binom{N}{m}$ ways of different combinations of noise points. To fix one point as noise, there are $\binom{N-1}{m-1}$ combinations. Then we have the likelihood probability that one point can be noise as:

$$P(z_k = 1|\mu\tau, N) = \sum_{m=1}^{N} P(m|u, N) \frac{\binom{N-1}{m-1}}{\binom{N}{m}}. \tag{8}$$

Given the gamma prior, we have:

$$P_k = P(z_k = 1|N, \lambda, \theta) = \sum_{m=1}^{N} \int P(z_k = 1|\mu\tau, N) \text{Gamma}(\mu|\lambda, \theta) d\mu. \tag{9}$$

It is difficult to get the analytic form for the integral in (9) so we used Monte Carlo (MC) integral to estimate, by randomly sampling μ. The MC method is slow and not suitable to be used in our dynamic programming otherwise we need to conduct MC for the DTW calculation on each pair of the points. However, given a fixed τ, we can take (9) as a function of N ($P_k = F(N)$), so that we can learn the pre-estimate of $F(N)$ and use it to look up the results in the dynamic optimization. The function of N, which is a gamma distribution given λ and θ is shown in Fig. 2.

Selecting C: We will determine C for $\mathbf{x}$ by setting $C_x = C_x^*$. Generally, the value C_x^* is up to $i - 2$ for x_i, when all points (except x_1) before i are noise points. However, $i - 2$ can be large when the sequence is long. It is unnecessary

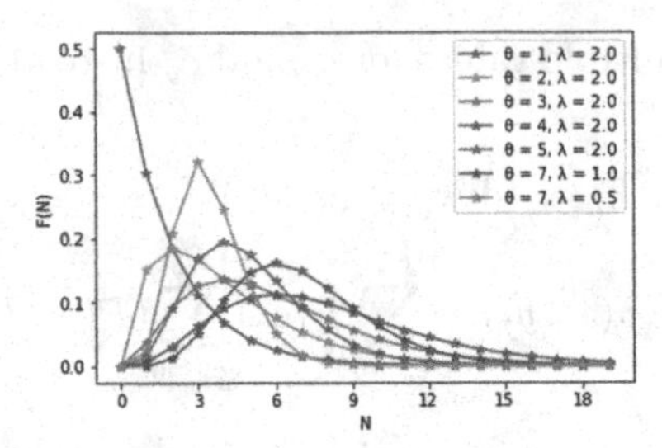

Fig. 2. The value of P_k with different gamma hyperparameters θ and λ.

to use $i-2$ so we need to determine C_x^*. As we have shown in equation (6), when we increase C_x by 1, the newly added component is:

$$\begin{aligned} F(C_x+1) &= dtw_{|C_x+1}(x_i, y_j) - dtw_{|C_x}(x_i, y_j) \\ &= P(C_x+1)\sum_{c_y}(P(c_y)D_m(C_x+1, c_y)). \end{aligned} \tag{10}$$

Here if we only consider the weight for the newly added component we have:

$$\frac{F(C_x+1)}{F(C_x)} \approx \frac{P(C_x+1)}{P(C_x)} = \frac{(1-P_{i-C_x-2})P_{i-C_x-1}}{1-P_{i-C_x-1}}. \tag{11}$$

The ratio is P_k if we consider the average case that all the probabilities are the same, by setting $\tau = x_n - x_1$. The total weights of using C_x^* is $1 - P_k(C_x^*+1)$. Therefore, given an acceptable rate α to approximate $C_x = i-2$, C_x^* can be selected based on:

$$C_x^* = \min(i-2, \lceil log_{P_k}(1-\alpha) - 1\rceil). \tag{12}$$

Normally P_k is less than 0.5. In this case, the computational cost is still low as C_x^* is generally less than 6 when $\alpha = 99\%$.

4 Experiments

In this section, we present the results of the proposed methods on both synthetic data and real data, and compare the results on different settings to demonstrate the general applicability of the proposed method.

4.1 Synthetic Data

In synthetic data experiments, we empirically test whether the proposed methods can recover DTW distance from noise superposed NHPP sequences.

Data Preparation: First we generate a function, by creating a random number from 0 to 10 in each 10 time units. Then using two such time series as the intensity of two sequences. These two sequences are regarded as clean sequences. Using DTW, we can calculate the distance (D_P) between them as the baseline. Then two HPP sequences are generated with intensity $\mu = \mu_x = \mu_y$. They are superposed to the clean sequences by some offsets to make sure that the first event is from the clean sequence. After superposition, two sequences ($\mathbf{x}$ and $\mathbf{y}$) are obtained. We can calculate DTW distance (D_N) between $\mathbf{x}$ and $\mathbf{y}$ with length T. Then all our proposed methods are applied to $\mathbf{x}$ and $\mathbf{y}$. The corresponding distances are denoted as **Averaging**: D_A, **Sampling**: D_S, **Dynamic**: D_D.

Experiment Setup: The recovered DTW distance by all methods are shown. The purpose is to show that DTW distances from our methods can be shorter than the baseline D_P, comparing with D_N. Experiments are carried out to test how the performance changes when μ is altered. For $\mu = 0.1, 0.2, 0.3, 0.4$, we set the hyperparameter (θ, λ) as (0.5, 0.2), (0.5, 0.4), (1, 0.3), (2, 1). When $\mu = 0.5$, the number of noise roughly equals to the number of points in the clean sequences. We also tested the performance by altering the length of sequence.

Quantitative Results: The results are shown in Table 1. The different settings of μ disclosed that when the noise ratio is high, our methods are more useful, comparing with directly applying the DTW on noise superposed NHPP. In the test, by altering the length we found that D_D is more stable than other methods when the sequence is short.

Table 1. Recovered DTW distance for different T and μ. $\mu = 0.2$ when altering T, and $T = 100$ when altering μ.

	T					μ			
	20	30	40	60	80	0.1	0.2	0.3	0.4
D_P	9	25	28	52	121	191	193	187	176
D_N	10	19.2	20.5	48.3	129	192	220	251	310
D_A	13	20.2	18.8	58.3	113	207	210	201	186
D_S	15	23.3	21.5	55.6	125	214	212	206	194
D_D	9	23.2	27.7	59.2	114	209	215	192	196

4.2 Classification on Real Data

Data Preparation: In order to test whether the recovered DTW distance is useful in a real application, we apply it on the classification task. The dataset was obtained based on [3], which was designed to test most DTW algorithms. Because of the page limitation, we shall show the results on part of the whole dataset. The names of all datasets are shown in the first column of Table 2. However, the sequences in the dataset are continuous time series rather than PP. We modify the sequences by using them as the intensities (I) to generate events. All values in the sequences are uniformly normalized to the range 0.2 to 0.8 using the maximum and minimum. Then the normalized sequence is used as the intensity of a NHPP to generate the events, after which the sequences of events are superposed with noise with a random intensity. The average number of events and noise events are shown in Table 2 as N_P and N_N, and the length of each sequence is T.

Experiment Setup: The classification task is performed by supervised 1-NN, using the provided training set and test set. Based on the recovered DTW distance, we find the sequence in the training set that is closest to a test sequence. Then the label is determined by the sequence found. We evaluate 7 DTW distances, and presented their accuracies in Table 2. Specifically, the results include 5 distances used in Sect. 4.1 and 2 additional distances by using DTW on I (D_I) and on empirical intensity (D_E). For D_I, we assume the actual intensity I is given and compute DTW on I. As to D_E, this is designed to demonstrate that our methods are better than using a guessed intensity. We use the window-wise estimation, in which the window size is 3 time units, to obtain the empirical intensity. Then the empirical intensity is used to obtain DTW (D_E). It is worth nothing that both D_P and D_I are obtained on sequences from the original dataset without superposed noise, so their accuracies can be considered as ideal cases for reference.

Quantitative Results: Unsurprisingly, the true intensity is the best representation for DTW distance, where the accuracies of D_I are the highest. However, using the intensity is sensitive when it is latent, as an example, the accuracy of using the empirical intensity D_E is significantly lower than others. It should be noticed that our methods should be compared to D_P since we may lose information when generating events from the obtained intensity. As we expected, D_A, D_S, and D_D can achieve higher accuracies. On average, D_S has the best performance which is about 1% higher than D_D. However, D_S requires a large number of samplings to gain the performance, which is much slower than D_D.

4.3 Case Study for Customer Behaviour Segmentation

Customers' behaviour is important in business data research, so we used our methods to segment different types of customers (households) based on their shopping behaviours (transactions in super markets). The data comes from Dunnhumby[1] which is the world's first customer data science platform. We will mainly use four fields in our experiment: unique household id, transactions from different households, time of transactions and store number. We use 802 households with complete demographic data and their transactions in 1 year in our experiment and use 1 day as a time unit. We treat the transactions for each household as a NHPP. In these transactions, however, we can image some of them may not reflect the life style as people can go to supermarket just because of promotion. This may cause a great deviation in similarity learning on customer behaviour pattern. To get a more accurate result, we treat this kind of event as noise (but still keep them as latent) by using our methods for similarity calculation and behaviour segmentation.

[1] https://www.dunnhumby.com/.

Table 2. Dataset description and classification accuracy using different DTW distances.

	Description			Accuracy						
	N_P	N_N	T	D_P	D_I	D_A	D_S	D_D	D_N	D_E
Beef	81	26	471	.85	.93	**.80**	.76	.70	.63	.13
Plane	69	30	144	.81	.96	.73	**.74**	.68	.56	.10
Coffee	286	54	286	.91	1	.68	**.78**	.68	.36	.07
50Words	126	35	271	.82	.93	.81	**.83**	**.83**	.49	.09
Adiac	152	66	177	.79	.96	**.83**	.81	.82	.59	.05
Arrowhead	95	25	252	.89	.95	.72	.80	**.81**	.58	.08
BeetleFly	306	131	513	.85	.96	.79	**.85**	.82	.6	.11
Cartrain	158	57	578	.85	.91	.72	.80	**.81**	.26	.11
CBF	129	41	129	.84	.91	.52	.71	**.77**	.35	.11
Computers	433	169	721	.92	.90	.59	.75	**.79**	.38	.15
CricketX	201	42	301	.90	.93	.69	**.73**	.72	.45	.19
DiatomSizeReduction	117	22	346	.90	.94	.66	**.79**	.76	.59	.11
DistalPhalanxTW	40	11	81	.88	.89	.67	.70	**.75**	.26	.05
Earthquakes	168	63	513	.81	.89	.79	**.81**	.79	.49	.09
ECG200	87	15	97	.82	.87	.68	.79	**.80**	.33	.01
FaceAll	108	29	131	.80	.93	.71	**.76**	.72	.54	.15
Fish	343	141	464	.88	.93	.77	**.80**	.77	.43	.12
FordA	298	55	501	.92	.92	.73	.79	**.80**	.44	.19
Average accuracy				.857	.929	.717	**.776**	.762	.463	.111

For the segmentation, we consider that households who have similar shopping habits are more likely to have similar demographic information. We use annual income, household composition and the number of the family as the features to cluster different households into 12 clusters and set the clustering results as the ground truth.

In our study, we compare the clustering results using the three methods we proposed (D_A, D_S, D_D) to the clustering results using the original transactions without considering the noise (D_P). Before this, we used grid search and 10-fold cross-validation to obtain the optimal hyperparameters $(\theta = 3, \lambda = 2.0)$, though which we found that when the value of hyperparameters approaches to the optimal one, the clustering accuracy does not change much. The result of each method is obtained and organised into a confusion matrix, then is aggregated as $Accuracy = (TruePositives + TrueNegatives)/TotalCases$, in Fig. 3. The accuracy rate from confusion matrix shows that our model has a much higher clustering accuracy comparing to customers' background. The results show that the transaction records contain noise and it may degrade the clustering result.

Our methods can largely improve the results by considering the noise in NHPP with the assumption for learning intensities.

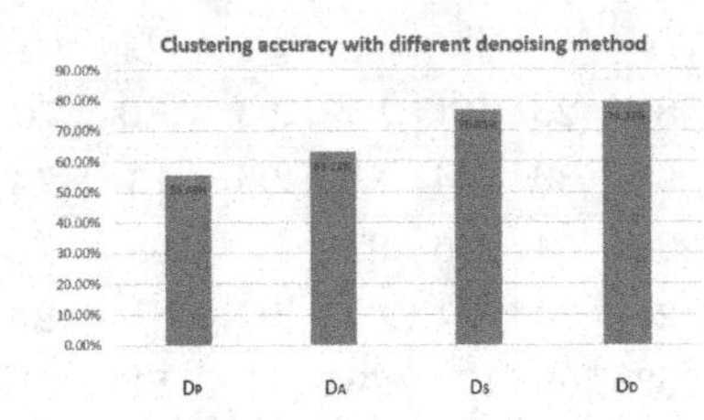

Fig. 3. Clustering accuracy for customer behaviour segmentation calculated by confusion matrix.

5 Conclusion

In this paper, we discussed the solutions to measure the DTW distance between noise superposed NHPP, including removing noise before DTW and integrating the probability of noise removal into DTW. The traversing method is infeasible for long sequences, so we implement the approximated methods. The averaging method can largely reduce computational cost when the distribution is narrow, but its performance is not superior. The Sampling method is restricted by the iteration of sampling itself, more iterations can improve the performance but it needs more resource. Therefore, we recommend to apply the dynamic method, although there are more approximation steps, it is much more efficient and stable in performance.

References

1. Bacry, E., Muzy, J.F.: First-and second-order statistics characterization of Hawkes processes and non-parametric estimation. IEEE Trans. Inf. Theory **62**(4), 2184–2202 (2016)
2. Bokhari, S., Geltner, D., van de Minne, A.: A Bayesian structural time series approach to constructing rent indexes: an application to Indian office markets (2017)
3. Chen, Y., et al.: The UCR time series classification archive. www.cs.ucr.edu/~eamonn/time_series_data (2015)
4. Del Giudice, V., De Paola, P., Forte, F., Manganelli, B.: Real estate appraisals with bayesian approach and Markov chain hybrid Monte Carlo method: an application to a central urban area of Naples. Sustainability **9**(11), 2138 (2017)
5. Du, N., Farajtabar, M., Ahmed, A., Smola, A.J., Song, L.: Dirichlet-Hawkes processes with applications to clustering continuous-time document streams. In: Proceedings of the 21th ACM SIGKDD International Conference on Knowledge Discovery and Data Mining, pp. 219–228. ACM (2015)

6. Ganeshapillai, G., Guttag, J., Lo, A.: Learning connections in financial time series. In: International Conference on Machine Learning, pp. 109–117 (2013)
7. Ghitany, M., Al-Awadhi, F., Alkhalfan, L.: Marshall-Olkin extended Lomax distribution and its application to censored data. Commun. Stat.-Theory Methods **36**(10), 1855–1866 (2007)
8. Guan, X., Huang, C., Liu, G., Meng, X., Liu, Q.: Mapping rice cropping systems in Vietnam using an NDVI-based time-series similarity measurement based on DTW distance. Remote Sens. **8**(1), 19 (2016)
9. Lin, P., Zhang, B., Guo, T., Wang, Y., Chen, F.: Interaction point processes via infinite branching model. In: AAAI, pp. 1853–1859 (2016)
10. Nakagawa, S., Nakanishi, H.: Speaker-independent English consonant and Japanese word recognition by a stochastic dynamic time warping method. IETE J. Res. **34**(1), 87–95 (1988)
11. Nasrabadi, N.M.: Pattern recognition and machine learning. J. Electron. Imaging **16**(4), 049901 (2007)
12. Rakthanmanon, T., et al.: Searching and mining trillions of time series subsequences under dynamic time warping. In: Proceedings of the 18th ACM SIGKDD International Conference on Knowledge Discovery and Data Mining, pp. 262–270. ACM (2012)
13. Sakoe, H., Chiba, S.: Dynamic programming algorithm optimization for spoken word recognition. IEEE Trans. Acoust. Speech Sig. Process. **26**(1), 43–49 (1978)
14. Wang, X., Mueen, A., Ding, H., Trajcevski, G., Scheuermann, P., Keogh, E.: Experimental comparison of representation methods and distance measures for time series data. Data Min. Knowl. Discov. **26**(2), 275–309 (2013)
15. Wu, J.: Reliability analysis for small wind turbines using Bayesian hierarchical modelling (2017)
16. Xu, H., Luo, D., Zha, H.: Learning Hawkes processes from short doubly-censored event sequences. arXiv preprint arXiv:1702.07013 (2017)

ATNet: Answering Cloze-Style Questions via Intra-attention and Inter-attention

Chengzhen Fu(✉), Yuntao Li(✉), and Yan Zhang(✉)

Department of Machine Intelligence, Peking University, Beijing, China
{fuchengzhen,liyuntao,zhy.cis}@pku.edu.cn

Abstract. This paper proposes a novel framework, named ATNet, for answering cloze-style questions over documents. Our model, in the encoder phase, projects all contextual embeddings into multiple latent semantic spaces, with representations of each space attending to a specific aspect of semantics. Long-term dependencies among the whole document are captured via the intra-attention module. A gate is produced to control the degree to which the retrieved dependency information is fused and the previous token embedding is exposed. Then, in the interaction phase, the context is aligned with the query across different semantic spaces to achieve the information aggregation. Specifically, we compute inter-attention based on a sophisticated feature set. Experiments and ablation studies demonstrate the effectiveness of ATNet.

Keywords: Question answering · Intra-attention · Inter-attention

1 Introduction

Benefiting from the rapid development of deep learning techniques, researchers have achieved promising results on cloze-style question answering tasks. Although significant progress has been made in cloze-style question answering, much remains to be done to solve several critical problems.

In the *encoder phase*, previous models, such as Deep LSTM Reader [7], apply the recurrent neural networks to gain some dependency between adjacent words. However, long-term dependencies are still very hard to preserve even using the advanced memory cell structures like long short-term memory network (LSTM) [9] or gated recurrent units (GRU) [4]. In the *interaction phase*, attention mechanisms, borrowed from the machine translation literature, are introduced to guide the extraction of information relevant to the query. A major downside for existing attention scoring functions is that, importance scores are mostly computed based on the individual representations for query and document, with no interactive terms being considered.

In this paper, we propose an end-to-end **at**tention based **net**work structure, named **ATNet**, which deals with the above challenges in the following aspects.

First, in the *encoder phase*, self-attention mechanism, also called intra-attention, is performed on top of LSTM or GRU. It provides complementary

Q. Yang et al. (Eds.): PAKDD 2019, LNAI 11440, pp. 242–252, 2019.
https://doi.org/10.1007/978-3-030-16145-3_19

information to the distance-aware dependencies, thus relives some long-term memorization burden from sequential structures. Besides, we equip it with a novel *gating* operation. Unlike existing models [3,23], we do not simply use the weighted summation as outputs, but instead evolve token representations via a fusion gate. The gate is applied to control the degree to which the previous token embedding is exposed and the retrieved dependency information is fused.

Second, in the *interaction phase*, ATNet offers the following improvements to the previously popular inter-attention paradigms. On the one hand, given that the QA task consists of heterogeneous queries and various document topics, we project encoded embeddings for the query and the document into multiple latent spaces respectively. We hypothesize that the representation for each space will attend to a specific aspect of the semantics. On the other hand, we propose a fine-grained feature set to compute inter-attention. It is performed on each of these projected versions of query and document representations in parallel.

To summarize, our main contributions are three folds:

(1) the inter-attention part is equipped with an multi-space attention scoring function, which is defined by a sophisticated feature set.
(2) the intra-attention part uses a gate to dynamically choose how much of the dependency information needs to be reserved to evolve the token representations;
(3) Both extensive experiments and ablation studies show the effectiveness of model.

2 Related Work

2.1 LSTM with Attention

These models aim at computing a **joint document-query representation**, which is used to rank the candidate answers. This includes the **DeepLSTM Reader** [7] which processes the concatenated (*document, query*) pair by employing a Deep LSTM cell with skip connections to obtain the joint representation; the **Attentive Reader** [2,7] which computes the query-aware document vector as the weighted sum of the token embeddings based on aligning scores calculated by the attention scoring function; and the **Impatient Reader** [7] which allows the model to recurrently accumulate information from the document as it sees each query token, ultimately outputting a final joint document query representation for the answer prediction.

2.2 Pointer-Style Attention Sum

Unlike previous works that using a joint representation to estimate the answer, **AS reader** [10] directly pick the answer from the document, which is motivated by the Pointer Network [24]. An attention over the document is obtained by computing dot products between the query embeddings and contextual embeddings, and successively normalizing the weight matrix using the softmax function.

Then, an aggregation scheme named pointer-sum attention is further applied to sum the word's attention across all the occurrences. Inspired by AS reader, the **Attention-over-Attention (AoA) Reader** [5] exploit mutual information between the document and query based on query-to-document attention and document-to-query attention.

2.3 Self-attention

Besides, some models propose to use **self-attention aligning** on top of the above mentioned sequential structures. It allows modeling dependencies without regard to distance, which has been successfully applied in a variety of tasks including reading comprehension, abstractive summarization, learning task-independent sentence representations, machine translation and language understanding [3,13,15,18,23].

Our model tightly integrates previous ideas related to self-attention. Moreover, we equip it with a novel *gating* operation, which controls the dependency information that flows into or out of original representations, acting as a flexible information filter.

2.4 Multi-hop Architecture

All the above mentioned models use a single-hop architecture. The effectiveness of multi-hop reasoning and attentions have also been explored so far in the literature. Many extensions of **Memory Networks** [1,12,21] shows a multi-hop architecture with an explicit memory and a recurrent attention mechanism for reading the memory can achieve good performance on QA tasks. **Neural Semantic Encoders (NSE)** [14] extends MemNets by introducing a write operation which can evolve the memory over time during the course of reading. The **GA Reader** [6] allows the query to directly interact with each dimension of the token embeddings at the semantic-level, and is applied layer-wise as information filters during the multi-hop process.

3 ATNet

3.1 Contextual Encoding Representations

We obtain $x_1^c, \ldots, x_m^c \in \mathbb{R}^d$ for the context document and $x_1^q, \ldots, x_n^q \in \mathbb{R}^d$ for the query via an embedding matrix $\boldsymbol{E} \in \mathbb{R}^{d \times |V|}$. Then, a bidirectional GRU is applied to encode the context,

$$\begin{aligned} \vec{h}_i &= \overrightarrow{\text{GRU}}(\vec{h}_{i-1}, x_i^c), i = 1, \ldots, m \\ \overleftarrow{h}_i &= \overleftarrow{\text{GRU}}(\overleftarrow{h}_{i-1}, x_i^c), i = m, \ldots, 1 \end{aligned} \tag{1}$$

Finally, we obtain two contextual encoded representations: $\boldsymbol{C} = \left\{c_i = [\vec{h}_i; \overleftarrow{h}_i]\right\}_{i=1}^m \in \mathbb{R}^{2h \times m}$ for the context and $q = q_n \in \mathbb{R}^{2h}$ for the query.

3.2 Intra-attention Aligner

We further apply intra-attention to align C with itself, which relates words from different positions without regard to their distance.

Figure 1 provides an overview of intra-attention. A self-coattention matrix $\boldsymbol{B} \in \mathbb{R}^{m \times m}$ for the document is defined as,

$$B_{i,j} = \begin{cases} c_i^{\mathrm{T}} c_j, & i \neq j \\ -\infty, & \text{otherwise} \end{cases} \tag{2}$$

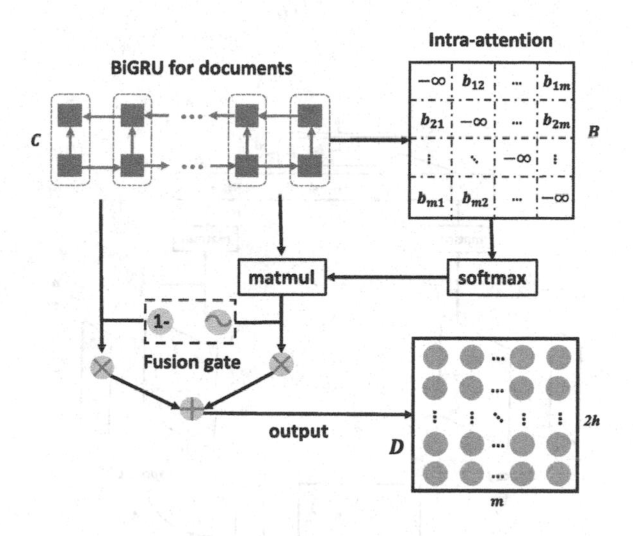

Fig. 1. Intra-attention.

where $B_{i,j}$ indicates the relevance between the i-th word and the j-th word. Note that, we disable the attention of each token to itself in case of the word being aligned with itself. We then compute an attended vector $\bar{c}_i \in \mathbb{R}^{2h}$ for i-th word as follows,

$$\begin{aligned} b_i &= \operatorname{softmax}(\boldsymbol{B}_{i,:}) \\ \bar{c}_i &= \boldsymbol{C} \cdot b_i^{\mathrm{T}}, \quad \forall i \in [1, \ldots, m] \end{aligned} \tag{3}$$

Gated Connection (GC). To efficiently fuse attended information $\bar{c}_i$ into original word c_i, we propose a simple gating operation to complete information integration. The fusion gate is computed as

$$g_i = \operatorname{sigmoid}(\boldsymbol{W}_g [\bar{c}_i; c_i] + b_g) \tag{4}$$

We use g_i and $1-g_i$ as the gated weights to assemble $\bar{c}_i$ and c_i. The integrated information is computed by a weighted sum as:

$$d_i = g_i \odot \bar{c}_i + (1 - g_i) \odot c_i \tag{5}$$

Hence, $\boldsymbol{D} = \{d_i\}_{i=1}^{m} \in \mathbb{R}^{2h \times m}$ represents obtained self-aware document representations after intra-attention.

3.3 Inter-attention Aligner

Figure 2 provides an overview of inter-attention. The projected embeddings in each space are represented as:

$$\begin{aligned} \boldsymbol{D}^l &= F_d^l(\boldsymbol{D}) \\ q^l &= F_q^l(q) \end{aligned} \tag{6}$$

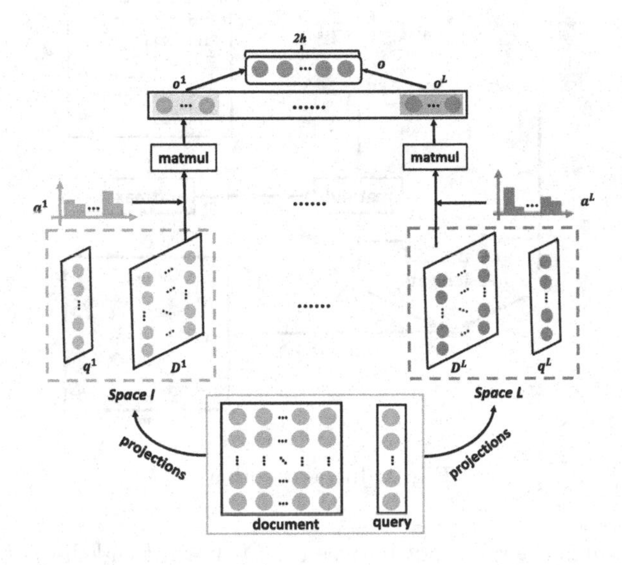

Fig. 2. Inter-attention.

where F_d^l and F_q^l are two projection functions for the document and query in the l-th space respectively, which are set to be a single-layer perceptron with *relu* as the activation function.

Note that, $\boldsymbol{D}^l = \{d_i^l\}_{i=1}^{m} \in \mathbb{R}^{p^l \times m}$ and $q^l \in \mathbb{R}^{p^l}$, where p^l denotes the size of vectors in the l-th space. **To control the model complexity, we constrain that the dimension for each space is the same and the total dimensions of all spaces equivalent to that of original embeddings**, i.e., $p^1 = \ldots = p^L = 2h/L$, where L denotes the number of spaces.

extAdditive Attention. Formally, the attention score vector in the l-th space is defined as a^l. Traditional attention scoring functions take a feature set as input and produces a scalar score as weight. Previous attention mechanisms have one characteristic that, the feature set only includes the individual representations for query and document, but no interactive terms are incorporated. For instance, *additive attention* is associated with

$$a_i^l = (w^l)^T \tanh(\boldsymbol{W}_d^l \, d_i^l + \boldsymbol{W}_q^l \, q^l) \tag{7}$$

where w^l is a weight vector.

In this paper, we explore some novel strategies to compute a_i^l from the intuition that if the i-th word in the document has a low contribution to the overall semantic, it will gather less dependency information in the self-attention phase. The dependency information can be denoted as the difference between d_i^l and c_i approximately. Therefore, we compute a feature as follows:

$$f_{\mathrm{att_i}}^l = |d_i^l - \boldsymbol{W}_c^l c_i| \odot q^l \tag{8}$$

On the one hand, if the difference is small or even close to zero, the importance of corresponding word should be small. On the other hand, if the difference (a vector) is not similar to the representation of q^l, the corresponding word is probably of less importance in measuring semantic similarity between queries and documents. From these two points, we think $f_{\mathrm{att_i}}^l$ is a good feature to measure word importance.

We first define a large feature set that captures a variety of similarities between documents and queries,

$$s(d_i^l, q^l) = [d_i^l \odot q^l, |d_i^l - q^l|, f_{\mathrm{att_i}}^l, d_i^l, q^l] \tag{9}$$

Then scoring function is defined in form of *additive attention*. Thus, we named it *extAdditive attention*.

$$a_i^l = (w^l)^T \tanh(\boldsymbol{W}_s^l \, s(d_i^l, q^l) + b^l) \tag{10}$$

We then obtain a normalized weight vector $\tilde{a}^l \in \mathbb{R}^m$.

$$\tilde{a}^l = \mathrm{softmax}(a^l) \tag{11}$$

Let o^l denotes the corresponding summarized context vector in the l-th space. The computation is defined in Eq.(12),

$$o^l = \boldsymbol{D} \cdot \tilde{a}^l \tag{12}$$

We refine the final query-aware context representation as

$$o = \boldsymbol{W_o}(\mathrm{concat}^{(1)}(o^1, \ldots, o^L)) \tag{13}$$

where $\boldsymbol{W_o}$ is a feed-forward network with one hidden layer, ensuring that o keeps the same shape (*2h-dimensional*) as the input. L indicates the number of subspaces.

3.4 Answer Prediction Module

The system adds a softmax function on top of the final query-aware context vector and adopts a negative log-likelihood objective for training.

$$p = \text{softmax}(\boldsymbol{W}_a o), \tag{14}$$

The entity with maximum probability which appear in the passage is the answer.

4 Experiments

4.1 Experimental Setups

Word Embedding Layer. All tokens are initialized with the 300-dimensional pre-trained GloVe word embeddings [16]. and updated during training. Tokens that are not covered by GloVe are replaced with a randomly initialized UNK embedding.

Contextual Encoding Layer. We use hidden size h = 256 for CNN, CBT-NE and 384 for Daily Mail, CBT-CN. The number of latent semantic subspaces L is 4.

Training. We adopt Adam for optimization [11], with an initial learning rate of 0.001 and mini-batches of 32. We set GRU-dropout probability to 0.1 [20] and the gradient clipping threshold to 5.

Datasets. CNN and Daily Mail datasets[1] are constructed with web-crawled CNN and Daily Mail news data [7]. The next two datasets are formed from two different subsets of the Children's Book Test[2] (CBT) [8].

4.2 Overall Results

Since we have reviewed all the baseline models in Sect. 2, we do not further discuss them in detail in this section. Notably, for baseline models, we report results presented in previously published works. Compared with prior works as shown in Table 1, ATNet brings nearly 0.8% absolute improvements over the best previous single model GA Reader on the CNN and Daily Mail testsets. ATNet also stays on par with the second-best baseline (AoA Reader with the assistance of the reranking strategy) when evaluated on the CBT-NE datasets. Moreover, on the CBT-CN test sets, it leads to an improvement of 1.4% over the most competitive model AoA Reader, which demonstrates its effectiveness.

[1] http://cs.nyu.edu/~kcho/DMQA/.
[2] http://www.thespermwhale.com/jaseweston/babi/CBTest.tgz.

Table 1. Performance comparison on four benchmark datasets.

Model	Acc (%)							
	CNN		Daily Mail		CBT-NE		CBT-CN	
	Valid	Test	Valid	Test	Valid	Test	Valid	Test
Deep LSTM Reader [7]	55.0	57.0	63.3	62.2	-	-	-	-
Attentive Reader [7]	61.6	63.0	70.5	69.0	-	-	-	-
Impatient Reader [7]	61.8	63.8	69.0	68.0	-	-	-	-
MemNets [8]	63.4	66.8	-	-	70.4	66.6	64.2	63.0
AS Reader [10]	68.6	69.5	75.0	73.9	73.8	68.6	68.8	63.4
Stanford AR [2]	72.4	72.4	-	-	-	-	-	-
Iterative Attention [19]	72.6	73.3	-	-	75.2	68.6	72.1	69.2
EpiReader [22]	73.4	74.0	-	-	75.3	69.7	71.5	67.4
BiDAF [17]	76.3	76.9	80.3	79.6	-	-	-	-
AoA Reader [5]	73.1	74.4	-	-	77.8	72.0	72.2	69.4
AoA Reader + Reranking [5]	-	-	-	-	**79.6**	74.4	75.7	73.1
NSE [14]	-	-	-	-	78.2	73.2	74.3	71.9
GA Reader (+ feature, fix $L(w)$) [6]	76.7	77.4	80.0	79.3	78.5	**74.9**	74.4	70.7
GA Reader (update $L(w)$) [6]	77.9	77.9	81.5	80.9	76.7	70.1	69.8	67.3
ATNet	**78.4**	**78.7**	**82.3**	**81.7**	77.7	74.2	**75.9**	**74.5**

5 Ablation Study

5.1 Effectiveness of Self-attention Module

Table 2 shows the accuracy on CNN and CBT-CN by removing self-attention aligning. Applying self-attention to ATNet improves the results by a remarkable margin of nearly 1.4%. In contrast to RNN which models the document in sequential manner, self-attention mechanism captures inner interactions regardless of the distance. Thus, long-range dependency information is fully incorporated into representations. It can be observed that, removing the *gate* operation leads to a reduction of about 0.6% and 0.8% on the CNN and CBT-NE testsets. The fusion gate function is responsible for storing and filtering information dynamically.

Table 2. Results with and w/o self-attention.

Attention	Acc on Testsets (%)	
	CNN	CBT-CN
W/o self-attention	77.4	73.1
ATNet (w/o gate)	78.1	73.7
ATNet	78.7	74.5

Table 3. Results with three attentions.

Attention	Acc on Testsets (%)	
	CNN	CBT-CN
Additive	77.9	73.7
Multiplicative	77.5	73.5
extAdditive	78.7	74.5

5.2 Effectiveness of extAttention

Next we look at the question of how to compute the alignment scores in the interactive phase. We compare three variants of operations, including *additive, multiplicative and extAdditive.* Results in Table 3 empirically demonstrate that the extAdditive attention does significantly better than the other two. It justifies our motivation to allow the query to interact with each token embedding based on a fine-grained feature set.

6 Case Study

We **use the bilinear scoring function** to visualize *self-attention weight distribution* over the context tokens with a heat map. For simplicity, we use the average values across multiple spaces to represent the self-coattention matrix.

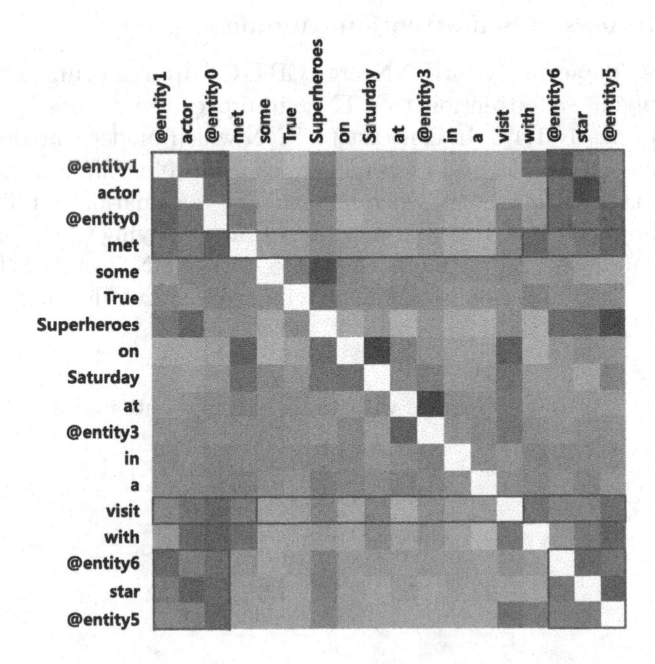

Fig. 3. A visualized example of self-attention.

We have several observations[3] in Fig. 3. First, several semantically similar phrases have been successfully aligned with each other, e.g., *Captain America actor Chris Evans* (@entity1 actor @entity0) and *Guardians of the Galaxy star Chris Pratt* (@entity6 star @entity5), both of which are highly related phases. Second, we observe that the self-attention attends to a distant dependency of the verb *met*, completing the phrase *@entity1 met...with...@entity5*. Attentions shown for the word *visit* also demonstrate that the self-attention aligning enables the model to capture the dependency between *visit* (indicating the event) with *@entity1, @entity5* (indicating two participants'names), which provides crucial clues to answer the query.

7 Conclusion

In this paper, we propose Attention-Net, which completes the cloze-style question answering via intra-attention and inter-attention. We demonstrate that both gated intra-attention and multi-space based inter-attention are integral parts of ATNet by ablation studies. We also show empirically that the proposed *extadditive attention* is superior to both additive and multiplicative attention. In the future, we will extend ATNet to the multi-hop architecture.

Acknowledgments. The authors would like to thank the anonymous reviewers for their valuable comments and helpful suggestions. This work is supported by NSFC under Grant No. 61532001, and MOE-ChinaMobile program under Grant No. MCM20170503.

References

1. Bordes, A., Usunier, N., Chopra, S., Weston, J.: Large-scale simple question answering with memory networks. arXiv preprint arXiv:1506.02075 (2015)
2. Chen, D., Bolton, J., Manning, C.D.: A thorough examination of the CNN/daily mail reading comprehension task. In: Proceedings of the 54th Annual Meeting of the Association for Computational Linguistics (Volume 1: Long Papers), vol. 1, pp. 2358–2367 (2016)
3. Cheng, J., Dong, L., Lapata, M.: Long short-term memory-networks for machine reading. In: Proceedings of the 2016 Conference on Empirical Methods in Natural Language Processing, pp. 551–561 (2016)
4. Chung, J., Gulcehre, C., Cho, K., Bengio, Y.: Empirical evaluation of gated recurrent neural networks on sequence modeling. arXiv preprint arXiv:1412.3555 (2014)
5. Cui, Y., Chen, Z., Wei, S., Wang, S., Liu, T., Hu, G.: Attention-over-attention neural networks for reading comprehension. In: Proceedings of the 55th Annual Meeting of the Association for Computational Linguistics (Volume 1: Long Papers), vol. 1, pp. 593–602 (2017)
6. Dhingra, B., Liu, H., Yang, Z., Cohen, W., Salakhutdinov, R.: Gated-attention readers for text comprehension. In: Proceedings of the 55th Annual Meeting of the Association for Computational Linguistics (Volume 1: Long Papers), vol. 1, pp. 1832–1846 (2017)

[3] Meanings for entities: **entity1**: Captain America; **entity0**: Chris Evans; **entity3**: Seattle Children's Hospital; **entity6**: Guardians of the Galaxy; **entity 5**: Chris Pratt.

7. Hermann, K.M., et al.: Teaching machines to read and comprehend. In: Cortes, C., Lawrence, N.D., Lee, D.D., Sugiyama, M., Garnett, R. (eds.) Advances in Neural Information Processing Systems, vol. 28, pp. 1693–1701. Curran Associates Inc. (2015). http://papers.nips.cc/paper/5945-teaching-machines-to-read-and-comprehend.pdf
8. Hill, F., Bordes, A., Chopra, S., Weston, J.: The goldilocks principle: reading children's books with explicit memory representations. arXiv preprint arXiv:1511.02301 (2015)
9. Hochreiter, S., Schmidhuber, J.: Long short-term memory. Neural Comput. **9**(8), 1735–1780 (1997)
10. Kadlec, R., Schmid, M., Bajgar, O., Kleindienst, J.: Text understanding with the attention sum reader network. In: Proceedings of the 54th Annual Meeting of the Association for Computational Linguistics (Volume 1: Long Papers), vol. 1, pp. 908–918 (2016)
11. Kingma, D.P., Ba, J.: Adam: a method for stochastic optimization. arXiv preprint arXiv:1412.6980 (2014)
12. Kumar, A., et al.: Ask me anything: dynamic memory networks for natural language processing. In: International Conference on Machine Learning, pp. 1378–1387 (2016)
13. Lin, Z., et al.: A structured self-attentive sentence embedding. arXiv preprint arXiv:1703.03130 (2017)
14. Munkhdalai, T., Yu, H.: Neural semantic encoders. In: Proceedings of the Conference Association for Computational Linguistics Meeting, vol. 1, p. 397. NIH Public Access (2017)
15. Paulus, R., Xiong, C., Socher, R.: A deep reinforced model for abstractive summarization. arXiv preprint arXiv:1705.04304 (2017)
16. Pennington, J., Socher, R., Manning, C.: Glove: global vectors for word representation. In: Proceedings of the 2014 Conference on Empirical Methods in Natural Language Processing (EMNLP), pp. 1532–1543 (2014)
17. Seo, M., Kembhavi, A., Farhadi, A., Hajishirzi, H.: Bidirectional attention flow for machine comprehension. arXiv preprint arXiv:1611.01603 (2016)
18. Shen, Y., Huang, P.S., Gao, J., Chen, W.: Reasonet: learning to stop reading in machine comprehension. In: Proceedings of the 23rd ACM SIGKDD International Conference on Knowledge Discovery and Data Mining, pp. 1047–1055. ACM (2017)
19. Sordoni, A., Bachman, P., Trischler, A., Bengio, Y.: Iterative alternating neural attention for machine reading. arXiv preprint arXiv:1606.02245 (2016)
20. Srivastava, N., Hinton, G., Krizhevsky, A., Sutskever, I., Salakhutdinov, R.: Dropout: a simple way to prevent neural networks from overfitting. J. Mach. Learn. Res. **15**(1), 1929–1958 (2014)
21. Sukhbaatar, S., Weston, J., Fergus, R., et al.: End-to-end memory networks. In: Advances in Neural Information Processing Systems, pp. 2440–2448 (2015)
22. Trischler, A., Ye, Z., Yuan, X., Bachman, P., Sordoni, A., Suleman, K.: Natural language comprehension with the epireader. In: Proceedings of the 2016 Conference on Empirical Methods in Natural Language Processing, pp. 128–137 (2016)
23. Vaswani, A., et al.: Attention is all you need. In: Advances in Neural Information Processing Systems, pp. 6000–6010 (2017)
24. Vinyals, O., Fortunato, M., Jaitly, N.: Pointer networks. In: Advances in Neural Information Processing Systems, pp. 2692–2700 (2015)

Parallel Mining of Top-*k* High Utility Itemsets in Spark In-Memory Computing Architecture

Chun-Han Lin[1], Cheng-Wei Wu[2], JianTao Huang[2], and Vincent S. Tseng[3(✉)]

[1] Department of Computer Science and Information Engineering, National Cheng Kung University, Tainan, Taiwan, ROC
[2] Department of Computer Science and Information Engineering, National Ilan University, Yilan, Taiwan, ROC
[3] Department of Computer Science, National Chiao Tung University, Hsinchu, Taiwan, ROC
vtseng@cs.nctu.edu.tw

Abstract. *Top-k high utility itemset* (abbr. *Top-k HUI*) *mining* aims at efficiently mining *k* itemsets having the highest utility without setting the minimum utility thresholds. Although some studies have been conducted on top-*k* HUI mining recently, they mainly focus on centralized databases and are not scalable for big data environments. To address the above issues, this paper proposes a novel framework for *parallel mining of top-k high utility itemsets in big data*. Besides, a new algorithm called *PKU* (*Parallel Top-K High Utility Itemset Mining*) is proposed for parallel mining of top-*k* HUIs on Spark in-memory platform. It adopts MapReduce architecture to divide the whole mining task into several independent subtasks, and takes good use of Spark in-memory computing technology for efficiently processing data in parallel. Moreover, several novel strategies are also proposed for pruning the redundant candidates such that the execution time and memory usage in the mining process are reduced greatly. The proposed PKU algorithm inherits several advantages of Spark, including low communication cost, fault tolerance, and high scalability. Experimental results on both real and synthetic datasets show that PKU has good scalability and performance on large datasets with outperforming several benchmarking algorithms.

Keywords: Top-*k* high utility itemset · In-memory computing · Big data · MapReduce · Spark platform

1 Introduction

High utility itemset (abbr. *HUI*) *mining* is an important technology in various fields. However, a critical problem of HUI mining is that it is not easy for users to set appropriate minimum utility thresholds. In HUI mining, users need to set a minimum utility threshold before performing the mining algorithms. Nevertheless, users cannot precisely predict or control the number of extracted itemsets by the threshold. For example, if the threshold is set too high, it is likely that only few HUIs or no HUI may be found, and it is hard for users to utilize the mining results due to insufficient information. On the contrary, if the threshold is set too low, an explosive number of

Q. Yang et al. (Eds.): PAKDD 2019, LNAI 11440, pp. 253–265, 2019.
https://doi.org/10.1007/978-3-030-16145-3_20

HUIs will be found, and it may lead the mining algorithms to suffer from very high computational costs and very large search space.

To address the issues mentioned above, the concept of *top-k high utility itemset mining* [11] was introduced, where *k* is a user-specified parameter that used to control the number of patterns to be discovered. In the scenario of top-*k* HUI mining, the users just need to set a parameter *k* and execute the top-*k* HUI mining algorithm once. Due to flexibility of top-*k* HUI mining in many real-life applications, various algorithms [6, 9, 11] are proposed. Although these algorithms are pioneers for top-*k* HUI mining, they are developed for running on a single machine and may not scalable enough for handling big data.

To efficiently process big data, *Apache Software Foundation* [12] proposes a parallel computing framework called *Hadoop* [13], which is composed of storage section and processing section. The storage section of Hadoop is called *HDFS* (*Hadoop Distributed File System*). It is a scalable distributed file system that replicates the data across multiple nodes for reaching high reliability. The processing section of Hadoop refers to the MapReduce framework. The algorithms developed based on MapReduce framework can take the nice property of key-value pairs to distribute the mining tasks among multiple computers. However, Hadoop is a disk-based architecture and it may spend a lot of time on I/O, which degrades the overall performance for processing.

In view of this, another parallel computing platform called *Spark* [15] was proposed. A novel data structure called *RDD* (*Resilient Distributed Dataset*) was applied to manage the in-memory data distributed among the nodes in cluster. If a part of RDD is lost during the mining process, the lost part can be reconstructed by using the previous RDD, and thereby the fault tolerance can be ensured. On the contrary, Hadoop relies on the replication of data in disk. During the progress of MapReduce iterations, Hadoop will write the intermediate data into disk, and read them back afterwards. However, for Spark, it can directly keep these data in memory, leading to large reduction of I/O costs. The several advantages of Spark motivate us to design a new framework for parallel mining of top-*k* HUIs on Spark platform.

To the best of our knowledge, the concept of top-*k* HUI mining has not yet been incorporated with Spark in-memory computing architecture. In view of this, this paper proposes a novel framework for *parallel mining of top-k high utility itemsets in Spark in-memory computing architecture*. This work is the first work that incorporates the top-*k* HUI mining with Spark in-memory computing architecture. We propose an efficient algorithm named *PKU* (*Parallel Top-K High Utility Itemset Mining*) for efficiently discovering the complete top-*k* HUIs using Spark in-memory computing architecture. During the mining process of the algorithm, several dynamic threshold raising strategies are proposed for effectively raising the border minimum utility thresholds. Besides, we address the issue of large search space problem by proposing two novel strategies, named *DLUP* (*Discarding Local Unpromising Items in Parallel*) and *MCTP* (*Merge Conditional Transactions in Parallel*) respectively. They can effectively reduce the size of conditional databases, leading to less required computing resources in each MapReduce pass. Moreover, the algorithm is implemented with Spark RDD structures for largely reducing the I/O costs on disks. Finally, extensive experiments on both real and synthetic datasets are conducted to evaluate the performance of PKU. Experiments show that PKU has good scalability on large datasets and outperforms several benchmarking algorithms [4, 5, 11].

2 Preliminary

Given a finite set of distinct *items* $I^* = \{I_1, I_2..., I_M\}$. Each item $I_i \in I^*(1 \leq i \leq M)$ has a positive number *EU*(I_i), called its *external utility* (e.g., unit profit). A database $D = \{T_1, T_2, ..., T_N\}$ is a set of *transactions*, where each *transaction* $T_j(1 \leq j \leq N)$ is a subset of I^* and has a unique identifier, call its *TID*. An itemset *X* of length *L* is a set of *L* distinct items, which is also called *L-itemset*. Besides, each item *I* in the transaction $T_j(1 \leq j \leq N)$ has a positive integer called its *internal utility* (e.g., quantity), which is denoted as $IU(I_i, T_j)$.

Definition 1 (Contain). An itemset *X* is said to *be contained* in a transaction *T* (or *T* contains *X*), iff *X* is a subset of *T*, denoted by $X \subseteq T$ The set of all the transactions containing *X* is denoted as $T_s(X)$.

Definition 2 (The utility of an item in a transaction). The *utility of an item* $I \in I^*$ in a transaction $T \in D$ is defined as $u(I, T) = EU(I) \times IU(I, T)$.

Definition 3 (The utility of an itemset in a transaction). The *utility of an itemset X* in a transaction *T* is defined as $u(X, T) = \sum_{I \in X} EU(I) \times IU(I, T)$.

Definition 4 (The utility of an itemset in a database). The *utility of an itemset X in a database D* is defined as $u(X) = \sum_{T \in T_{S(X)}} u(X, T)$.

Definition 5 (Transaction utility and total utility). The *transaction utility of a transaction T* is defined as $TU(T) = \sum_{I \in T} u(I, T)$. The total utility of a database *D* is defined as $\eta(D) = \sum_{T \in D} TU(T)$.

Definition 6 (The relative utility of an itemset in a database). The *relative utility* of an itemset *X* in a database *D* is defined as $ru(X) = u(X)/\eta(D)$.

Definition 7 (The complete set of high utility itemsets). Given a user-specified *minimum utility threshold* δ. The complete set of HUIs in *D* is denoted as $f_H(D, \delta)$.

Definition 8 (Top-*k* high utility itemset). An itemset *X* is called *top-k high utility itemset* (abbr. *top-k HUI*) in a database *D* iff there are less than *k* itemsets having a utility greater than *u*(*X*) in $f_H(D, 0)$.

Definition 9 (Optimal minimum utility threshold). Let *H* be the complete set of top-*k* HUIs in *D*. A minimum utility threshold δ^* is called *optimal minimum utility threshold* iff there does not exist another threshold δ such that $\delta \geq \delta^*$ and $|f_H(D, \delta)| \geq k$. If $|H| \geq k$, then $\delta^* = min\{u(X)|X \in H\}$ [11].

3 Related Work

3.1 High Utility Itemset Mining

In general, HUI mining algorithms can be categorized into two types: *two-phase* and *one-phase*. The main idea of the *two-phase* algorithms is that they attempt to generate candidates of HUIs in phase I, and then perform an additional database scan in phase II to calculate exact utility of each candidate generated in phase I. After that, all the HUIs can be identified from the generated candidates in phase I. During the mining process of HUI mining algorithms, *transaction-weighted downward closure* (abbreviated as *TWDC*) property [3] is usually used to effectively prune the search space. Many HUI mining algorithms are two-phase algorithms, including *Two-Phase* [3], *IHUP* [1] and *UP-Growth* [8]. *HUI-Miner* [4] is a typical *one-phase* algorithm. A special data structure called *utility-list* is developed to store the utility information of items. All the HUIs and their utilities can be directly obtained through the utility-list structure. The performance evaluation in [4] also shows that the efficiency of HUI-Miner is generally better than UP-Growth. However, the above mentioned algorithms are designed for HUI mining, instead of top-*k* HUI mining.

3.2 Top-*K* High Utility Itemset Mining

TKU [11] is a two-phase algorithm for mining top-*k* HUIs. In the scenario of top-*k* HUI mining, users do not need to set minimum utility thresholds. Instead, only a parameter *k* is used to control the number of patterns to be discovered. The TKU algorithm generally consists of two phases. In phase I, TKU incorporates four novel strategies *PE*, *NU*, *MD* and *MC* to raise the minimum utility thresholds. In phase II, all the top-*k* HUIs are identified from the candidates generated in phase I by the *SE* strategy. Afterwards, the REPT [6] algorithm is proposed, which incorporates several new strategies to reduce the computational costs for mining top-*k* HUIs. Although REPT achieves better performance than TKU on some types of datasets, it is still a two-phase algorithm and may generate too many candidates on large-scale datasets. Recently, a novel algorithm named *TKO* [9] is proposed to mine top-*k* HUIs in only one phase. It is the first algorithm that incorporates the concept of utility-list [4] for top-*k* HUI mining. The experiments in [9] show that the performance of TKO is generally better than REPT and TKU.

3.3 Parallel Mining of High Utility Patterns

Although many studies have focused on mining high utility patterns in centralized databases, they may not be salable enough for big data environments, where multiple machines are required. *DTWU-Mining* [10] and *FUM-D* [7] algorithms are proposed for parallel mining HUIs from distributed databases. Although they are parallel mining algorithms, they do not support fault tolerance and fault recovery mechanisms in the mining process. If one of the cluster nodes suddenly crashes, they may fail to complete the whole mining task and produce incorrect mining results.

To overcome the issues mentioned above, *PHUI-Growth* [5], a MapReduce-based algorithm, is first proposed to parallel mine HUIs from distributed databases. It is implemented based on the Hadoop platform. PHUI-Growth generally consists of three phases: (1) *mining phase*, (2) *database transformation phase* and (3) *mining phase*. In mining phase, TWUs [3] of items in the database are calculated by a MapReduce pass. In database transformation phase, TWDC property is applied to remove unpromising items [3] from the transactions. Remaining promising items in transactions are sorted in ascending order of TWU. In mining phase, several MapReduce passes are performed to discover HUIs from the transformed transactions. Although PHUI-Growth is the first algorithm for mining HUIs in parallel based on Hadoop, it is not developed for top-*k* HUI mining. Besides, when several MapReduce passes are performed, it may involve expensive disk I/O costs, and consequently degrading the overall mining performance. As surveyed above, no study has been proposed for mining top-*k* HUIs in parallel based on Spark platform [15].

4 The Proposed Method

This section introduces the proposed *PKU* (*Parallel Top-K High Utility Itemset Mining*) algorithm for parallel discovering top-*k* HUIs in Spark in-memory computing architecture [15]. The input of PKU includes a database *D* with external and internal utilities, a user-specified desired number of itemsets *k*. Notice that, in this framework, there is no need for users to provide the minimum utility threshold. In HUI mining, algorithms can use the minimum utility threshold to largely prune the search space. However, in top-*k* HUI mining, the minimum utility threshold is not given in advance. Therefore, PKU adopts a dynamically adjusted internal variable called *border minimum utility threshold*, denoted as θ, for search space pruning. The value of θ is initialized to 0, and it will be gradually raised during the mining process. To parallel discover top-*k* HUIs, there are three main stages designed in PKU: (1) *Pre-Evaluation in Parallel* (abbr. *PEP*), (2) *Reorganize Transactions in Parallel* (abbr. *RTP*), and (3) *Mining Patterns in Parallel* (abbr. *MPP*). After the mining process, PKU outputs the complete set of top-*k* HUIs in *D*.

4.1 Pre-evaluation in Parallel

The *PEP* stage includes three steps. In the first step, it performs the *FindOneItems* function to find all the items and their utilities through a MapReduce pass. The *FindOneItems* function works as follows. In Map phase, each transaction *T* in *D* is processed by a Mapper. For each transaction *T* received by the Mapper, the algorithm visits each item in *T*. For each visited item *I* in *T*, the algorithm outputs a key-value pair $\langle I, u(I,T) \rangle$. In Reduce phase, all the key-value pairs having the same key are fed into the same Reducer. Let $\langle K, SetValue = [v_1, v_2, v_3, \ldots] \rangle$ be the data received by a Reducer, where the key *K* is an item and *SetValue* stores a set of values having the same key as *K*. Then, the algorithm calculates the utility of the item *K* by summing up all the values in *SetValue* into a variable *ItemUtility*, and outputs a key-value pair $\langle K, ItemUtility \rangle$. The results of each Reducer are outputted and collected into the set *Set1*. Figure 1(a) shows an example for the process of *FindOneItems*. In Map phase, as for the transaction T_1, the algorithm outputs three key-value pairs ⟨A, 4⟩, ⟨B, 15⟩, and ⟨C, 4⟩ (Table 2).

Table 1. An example database.

Tid	Transaction
T1	(A, 2) (B, 5) (C, 2)
T2	(A, 2) (B, 2) (C, 2) (E, 1) (G, 1)
T3	(A, 1) (D, 3) (E, 3) (F, 3)
T4	(B, 1) (D, 3) (F, 2)

Table 2. A profit table for the database of Table 1.

Item	A	B	C	D	E	F	G
Unit Profit	2	3	2	3	4	1	2

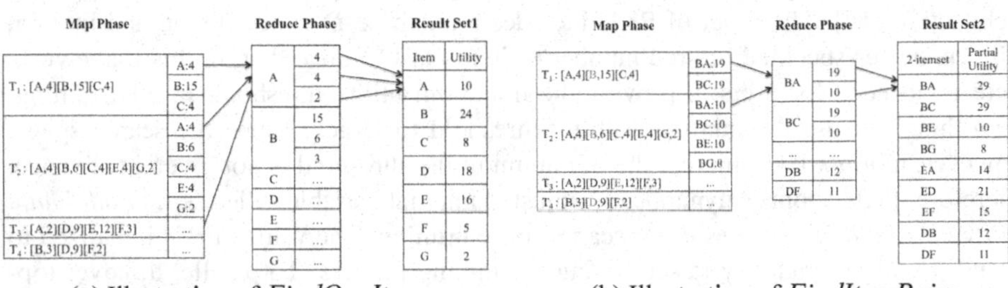

(a) Illustration of *FindOneItems*. (b) Illustration of *FindItemPairs*.

Fig. 1. An example of the PEP strategy.

In Reduce phase, as for the key A, the summation of all the values in *SetValue* of key A is (4 + 4 + 2) = 10, and then the algorithm outputs a key-value pair ⟨A, 10⟩ in *Set1*.

In the second step of *PEP*, the *FindItemPairs* function is called. It performs a MapReduce pass to find some 2-itemsets and their partial utilities. In Map phase, each transaction *T* in *D* is processed by a Mapper. For each transaction *T* received by the Mapper, the algorithm finds the item *I* having the highest utility in *T*. Then, the algorithm visits each of other items in *T*. For each visited item *J* in *T* ($J \neq I$), the algorithm outputs a key-value pair $\langle K, V \rangle$, where $K = I \cup J$ and $V = u(I, T) + u(J, T)$. In Reduce phase, all the key-value pairs having the same key are fed into the same Reducer. Let $\langle K, SetValue = [v_1, v_2, v_3, \ldots] \rangle$ be the data received by the Reducer, where the key *K* is an 2-itemset and *SetValue* stores a set of values having the same key as *K*. Then, the algorithm calculates the utility of the item *K* by summing up all the values in *SetValue* into a variable *PairUtility*, and outputs a key-value pair $\langle K, PairUtility \rangle$. The results of each Reducer are outputted and collected into the set *Set2*. Figure 1(b) shows an example for the process of *FindItemPairs*. In Map phase, as for the transaction T_1, the algorithm outputs two key-value pairs ⟨{BA}, 4⟩ and ⟨{BC}, 15⟩. In Reduce phase, as for the key {BA}, the summation of all the values in *SetValue* of the key {BA} is (19 + 10) = 29, and then the algorithm outputs ⟨{BA}, 29⟩ and puts it into *Set2*.

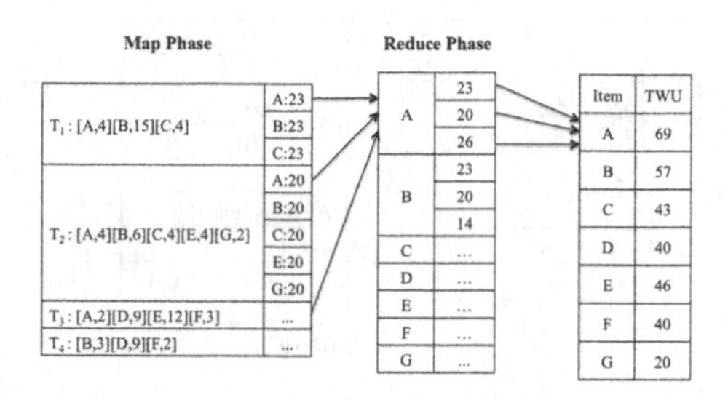

Fig. 2. Illustration of *FindItemTWU.*

TID	Reorganized Transaction
T_1'	[C,4][B,15][A,4]
T_2'	[C,4][E,4][B,6][A,4]
T_3'	[D,9][F,3][E,12][A,2]
T_4'	[D,9][F,2][B,3]

TID	Conditional Transaction
C_1	$\langle\{\emptyset\}$, 0:([C, 4], [B, 15], [A, 4])$\rangle$
C_2	$\langle\{\emptyset\}$, 0:([C, 4], [E, 4], [B, 6], [A, 4])$\rangle$
C_3	$\langle\{\emptyset\}$, 0:([D, 9], [F, 3], [E, 12], [A, 2])$\rangle$
C_4	$\langle\{\emptyset\}$, 0:([D, 9], [F, 2], [B, 3])$\rangle$

Fig. 3. Example of conditional transactions.

Let the set *Set3* be *Set1* ∪ *Set2*, in the third step of *PEP*, the algorithm sorts itemsets in *Set3* in descending order of their utilities by using *SortByKey* function supported by Spark. Then, if the total number of itemsets in *Set3* is higher than *k*, the threshold θ is raised to the utility of the *k*-th itemset in *Set3*.

Figure 1 shows an example for the process of the *PEP* strategy when $k = 3$. As shown in Fig. 1, there are 7 and 9 itemsets in *Set1* and *Set2*, respectively. After sorting itemsets in *Set3* = *Set1* ∪ *Set2*, since |*Set3*| is larger than *k*. the threshold θ is raised to the utility of the third itemset in *Set3*, which is 24.

4.2 Reorganize Transactions in Parallel

The *RTP* stage includes two steps. In the first step, it performs the *FindItemTWU* function to find all the items with their TWUs through a MapReduce pass. The *FindItemTWU* function works as follows. In Map phase, each transaction *T* in *D* is handled by a Mapper. For each transaction *T* received by the Mapper, the algorithm visits each item in *T*. For each visited item *I* in *T*, the algorithm outputs a key-value pair $\langle I, TU(T)\rangle$, where *TU*(*T*) is the transaction utility of *T*. In Reduce phase, all the key-value pairs having the same key as *I* are fed into the same Reducer. Let $\langle K, SetValue = [v_1, v_2, v_3, \ldots]\rangle$ be the data received by the Reducer, where *K* is an item and *SetValue* is s set of values having the same key as *K*. Then, the algorithm calculates the TWU of the item *K* by summing up all the values in *SetValue* into a variable *ItemTWU*, and outputs a key-value pair $\langle K, ItemTWU\rangle$. The results of Reducers are collected into *ItemTWUSet*, and the *FindItemTWU* function outputs *ItemTWUSet*.

In the second step of *RTP*, the algorithm applies the *transaction-weighted downward closure property* [9, 11] to remove items having a TWU less than θ from the transactions. These items are called *unpromising items* since they are unpromising to be a part of any top-*k* HUIs.

Property 1 (*Transaction-weighted Downward Closure Property*) [9, 11]. For any itemset *X*, if *TWU*(*X*) is smaller than the current border minimum utility threshold θ, *X* and all its supersets are not top-*k* HUIs.

After removing unpromising items for each transaction, the remaining items in each transaction are sorted in TWU ascending order. Figure 2 gives an example for *FindItemTWU*. Since the current threshold θ is 24, and *TWU*(*G*) = 20. By Property 1, G is

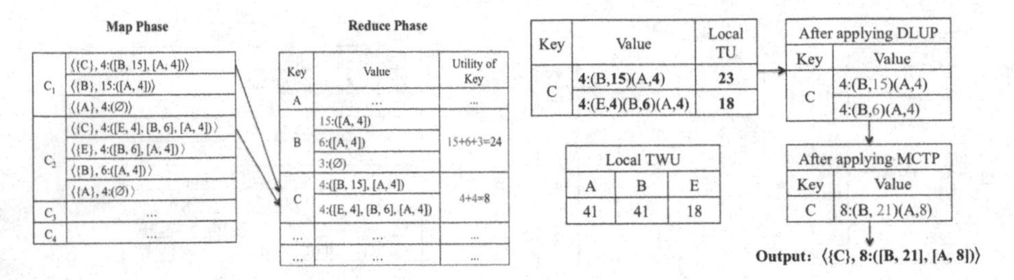

Fig. 4. Generation of PKHUIs in *MPP*.

Fig. 5. Examples for DLUP and MCTP.

an unpromising item and can be removed from the transactions. The transactions after the above reorganization process are called *recognized transactions*. For each reorganized transaction *T*', each item *I* in *T*' is attached with its utility in the original transaction *T*. The example is shown in Fig. 3.

Then, the algorithm transforms each reorganized transaction into a special form called *conditional transaction*. A conditional transaction *C* is of the form $\langle P, Q : R\rangle$, where *P* is an itemset and called *prefix*, *Q* is a value called *prefix utility*, and *R* is called *element list* and stores a set of elements. Each element in *R* consists of an item and a utility value. Then, each reorganized transaction *T*' is transformed to a conditional transaction *C* by the following way. First, the prefix and prefix utility of *C* are set to $\varnothing$ and 0, respectively. Then, the element list is set to *T*'. Figure 3 shows an example for the transformation process.

4.3 Mining Patterns in Parallel

The MPP stage mainly consists of three steps. In the first step, it adopts a *pattern-growth* approach to discover top-*k* HUIs using MapReduce iterations. In the *L*-th MapReduce iteration, it will generate potential top-*k* HUIs (abbr. *PKHUIs*) of length L and new conditional transactions. Each iteration works as follows. In Map phase, each conditional transaction is handled by a Mapper. For each conditional transaction $\alpha = \langle X, V : ([I_1, u_1], [I_2, u_2], \ldots, ([I_n, u_n]\rangle$ received by the Mapper, the algorithm visits each element in its utility-list structure. For the *i*-th visited element [I_i, u_i], the algorithm outputs a new conditional transaction $\beta = \langle X \cup I_i, V + u_i : ([I_{i+1}, u_{i+1}], [I_{i+2}, u_{i+2}], \ldots, [I_{i+n}, u_{i+n}])\rangle$. For example, considering the first conditional transaction $C_1 = \langle\varnothing, 0 : ([\text{C}, 4], [\text{B}, 15], [\text{A}, 4])\rangle$ in Fig. 3. In Map phase, the algorithm will generate three new conditional transactions for it, which are $\langle\{\text{C} \cup \varnothing\}, (0+4) : ([\text{B}, 15], [\text{A}, 4])\rangle, \langle\{\text{B} \cup \varnothing\}(0+15) : ([\text{A}, 4])\rangle$, and $\langle\{\text{A} \cup \varnothing\}, (0+4) : (\varnothing)\rangle$.

The prefix of each generated conditional transaction is treated as a key, and its prefix utility and element list form the value. In Reduce phase, all the key-value pairs having the same key are collected into the same Reducer. For all the key-value pairs having the same key *X*, their prefix utilities are summed up. The result is the utility of *X* in the database *D*. Figure 4 shows an example for the above process. After the calculation, if the utility of *X* is higher than the current threshold θ, *X* is put into a list called *PKL* (PKHUI List).

In the second step of MPP, the algorithms sorts all the itemsets in PKL in descending order of their utilities. If the number of itemsets in *PKL* is higher than k and the utility of the k-th itemset in PKL is higher than θ, then θ is raised to the utility of the k-th itemset in *PKL*. After raising the threshold, all the itemsets having a utility less than θ are removed from PKL.

Then, we propose two strategies for further enhancing the performance of PKU. The two proposed strategies are applied during the process of Reduce phase. The first strategy is called *DLUP* (*Discarding Local Unpromising Items in Parallel*), which works as follows. The algorithm calculates the *local transaction utility* for each conditional transaction. The local transaction utility of a conditional transaction $C = \langle X, V : ([I_1, u_1], [I_2, u_2], \ldots, ([I_n, u_n]\rangle$ is defined as $LTU(C) = V + \sum_{i=1}^{n} u_i$. Let I be an item, we use the notation $I \in C$ to represent that I appears in the element list of C. Besides, let *CT*(*X*) be the set of all the conditional transactions having the same prefix X, then the *local TWU of the item I* is defined as $\sum_{i \in C \wedge C \in CT(X)} TU(C)$. If the TWU of an item I in *CT*(*X*) is less than the current threshold θ, it can be removed from the conditional transactions. The second proposed strategy is called *MCPT* (*Merge Conditional Transactions in Parallel*). For any two conditional transactions having the same prefix, if all the items in their element lists are the same, then the two utility lists can be merged into one. Figure 5 shows the examples for DLUP and MCPT. After applying the DLUP and MCPT strategies, the algorithm outputs the newly generated conditional transactions as the input of the next iteration of MapReduce process. The algorithm completes until no PKHUIs are generated. When the algorithm completes, all the top-k HUIs are discovered and maintained in *PKL*.

5 Experimental Results

In this section, we evaluate the performance of the proposed algorithm *PKU*. Because PKU is the first algorithm designed for parallel mining of top-k HUIs, we compare it with *TKU* [11], *HUI-Miner* [4], and *PHUI-Growth* [5]. The characteristics of these algorithms are shown in Table 3. In order to compare these algorithms, we consider the optimal parameters for HUI mining and discovering the same amount of patterns as PKU.

All of the compared algorithms are implemented in Java. The experiments of PKU and PHUI-Growth are conducted on a master computer with a 3.40 GHz Intel Core Processor and 32 GB memory, running Ubuntu 14.04, and ten slave computers with a 2.7 GHz Intel Celeron Processor and 8 GB memory, running Ubuntu 14.04. The version of Hadoop is 2.6.0, and the version of Spark is 1.5.0. The experiments of TKU and HUI-Miner are conducted on a computer with a 3.40 GHz Intel Core Processor and 32 GB memory, running Ubuntu 14.04. Four real-life datasets and one synthetic dataset are used to evaluation the performance of the algorithms.

Table 3. The characteristics of the algorithms.

Mining Task	Algorithm	Parallel	Hadoop	Spark
HUI mining	HUI-Miner	×	×	×
	PHUI-Growth	√	√	×
Top-*k* HUI mining	TKU	×	×	×
	PKU	√	×	√

Table 4. The characteristics of all the datasets.

Dataset	|*D*|	*N*	*T*	Type
Mushrooms	8,124	118	23.0	Dense
Chainstore	1,112,949	46,086	7.2	Large
Chainstore5x	5,565,015	46,086	7.2	Large
Mushrooms20x	162,480	118	23.0	Large
T12I10N10 K|D|1,000 K	1,000,000	10,000	10.0	Synthetic

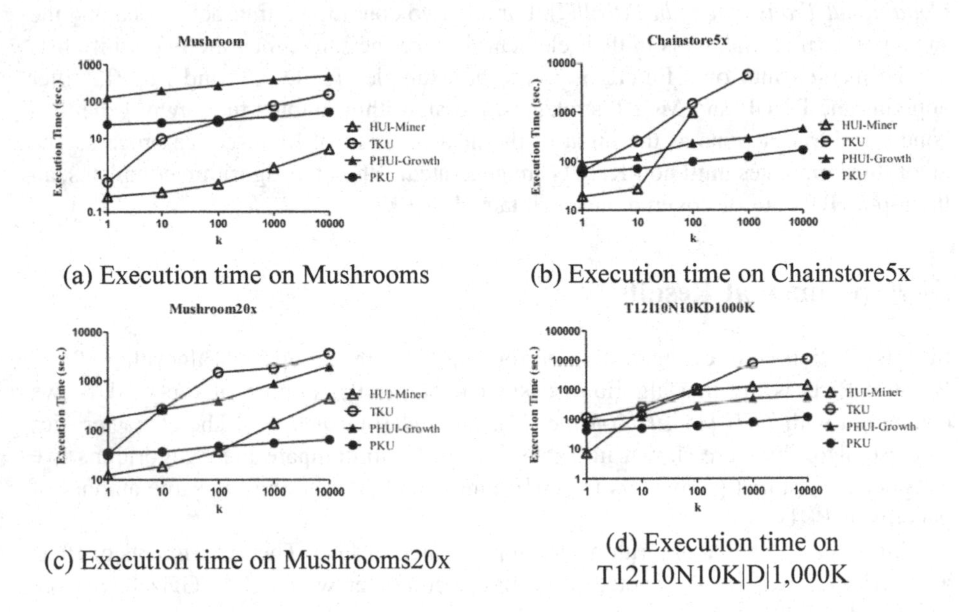

(a) Execution time on Mushrooms

(b) Execution time on Chainstore5x

(c) Execution time on Mushrooms20x

(d) Execution time on T12I10N10K|D|1,000K

Fig. 6. Execution time on different datasets of the algorithm.

The four real-life datasets used in the experiments are *Mushrooms* [2], Chainstore [2] and a larger dataset *Chainstore5x* where all the transactions in Chainstore are duplicated five times to form. Similarly, we also duplicated all the transaction in Mushrooms twenty times to form a large and dense dataset named *Mushrooms20x*. The synthetic dataset *T12I10N10 K|D|1,000 K* was generated from IBM data generator [14] where *T* is the average length of transactions, *I* is the average of maximal potential frequent itemsets, *N* is the number of distinct items and |*D*| is the number of transactions. The characteristics of all the datasets are shown in Table 4.

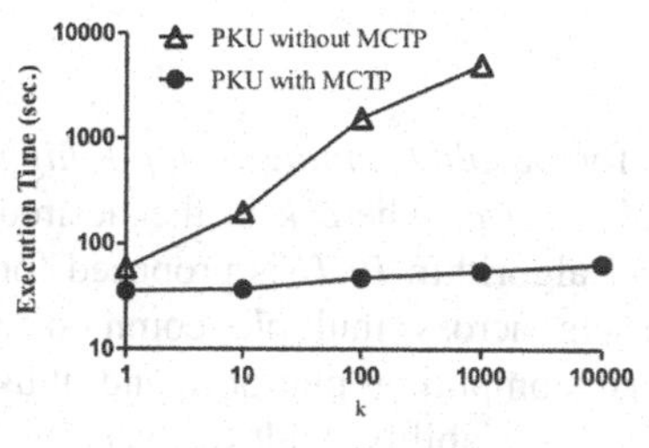

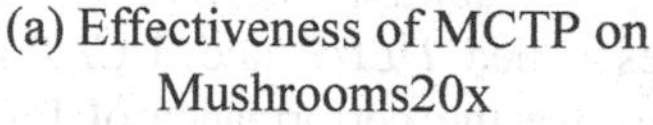
(a) Effectiveness of MCTP on Mushrooms20x

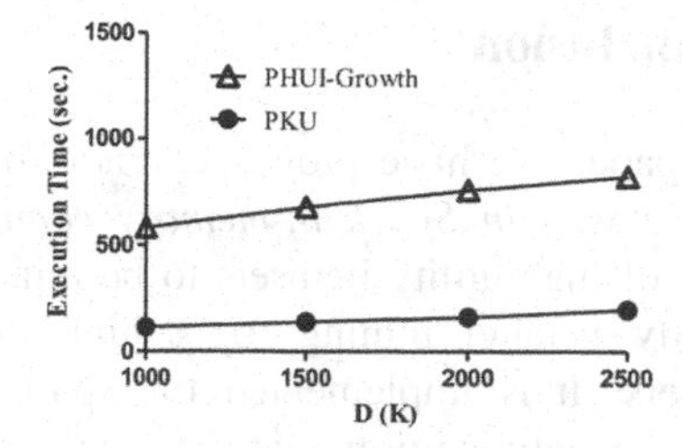

(b) Execution time under varied number of transactions

Fig. 7. Execution time for varied pruning strategies and under varied $|D|$.

Figure 6(a) shows the experiments on the Mushrooms dataset. When the parameter k is less than 100, we observe that TKU is faster than PKU because of the additional communication costs among nodes. However, when the parameter k increases, the overheads of the algorithms start to increase, and the advantages of parallel framework reduce the influence of the increasing overheads. In general, the performance of PKU is worse than that of HUI-Miner and TKU on small datasets, especially when parameter k is small. However, PKU always outperforms the optimal case of PHUI-Growth.

On the other hand, PKU possesses the ability to handle dense datasets and achieve a good performance. On the Chainstore5x dataset as shown in Fig. 6(b), when the value of k is small, we can observe that HUI-Miner can reach a better performance. However, when the parameter k increases, the execution times of HUI-Miner is much longer than other algorithms. Besides, when k = 10,000, both TKU and HUI-Miner are unable to complete. On the contrary, parallel algorithms (i.e., PHUI-Growth and PKU) can achieve better performance. Moreover, PKU is more efficient than PHUI-Growth. Figure 6(c) shows that the execution time of PKU slightly increases when the parameter k increases. When the parameter k is no less than 1,000, PKU has the best performance. Moreover, when k = 10,000, PKU is over 50 times faster than TKU. Finally, we utilize T12I10N10 K|D|1,000 K, a synthetic dataset with 1 million transactions, to evaluation the performance of PKU as shown in Fig. 6(d). In Fig. 6(d), it can be seen that HUI-Miner shows the best performance when the parameter k is equal to 1. However, when the parameter k is larger than 10, PKU possesses the best performance. In general, the experimental results show that PKU outperforms other algorithms on large datasets.

Figure 7(a) shows the execution times of *PKU* with *MCTP* and *PKU* without *MCTP* on Mushrooms20x dataset. It can be seen that *PKU* with *MCTP* outperforms *PKU* without *MCTP* substantially. When the parameter k is set to 1,000, the execution time of PKU with *MCTP* (i.e. 55.089 s) is about 90 times fewer than *PKU* without *MCTP* (i.e. 5016.044 s) to complete the mining process. Figure 7(b) shows the experiment of PKU which $|D|$ (i.e., number of transactions) varies from 1,000,000 to 2,500,000. From the figure, we can observe that the execution time is approximately proportional to the $|D|$. Even when the number of transactions reaches 2,500,000, PKU still has a good performance. Moreover, the execution time of PKU is from 4 to 5 faster than that of PHUI-Growth.

6 Conclusion

In this paper, we have proposed a new framework for *parallel mining of top-k high utility itemsets in Spark in-memory computing architecture*, where *k* is the desired number of high utility itemsets to be mined. A novel algorithm *PKU* is proposed for efficiently parallel mining top-*k* high utility itemsets across multiple commodity computers. It is implemented on Spark in-memory computing platform and thus inherits several nice properties of Spark, including high scalability, fault recovery and low communication overheads. Two novel strategies called *DLUP* and *MCPT* are proposed to reduce the search space and greatly improve the performance of PKU. Empirical evaluations on both real-life and synthetic datasets show that the performance of PKU outperforms several benchmarking algorithms [4, 5, 11] in utility mining fields. For example, on the Chainstore5x dataset, when *k* is set to 10,000, both TKU [11] and HUI-Miner [4] are unable to complete, while the proposed PKU still has good performance. Moreover, PKU possesses better performance than PHUI-Growth [5] with optimal minimum utility thresholds. To the best of our knowledge, this is the first work that explores the new research topic on parallel mining of top-*k* high utility itemsets on Spark.

Acknowledgement. This work is supported in part by Ministry of Science and Technology, Taiwan, ROC under grant no. 104-2221-E-009-128-MY3, 107-2218-E-009-050 and 107-2218-E-197-002.

References

1. Ahmed, C.F., Tanbeer, S.K., Jeong, B., Lee, Y.: Efficient tree structures for high utility pattern mining in incremental databases. IEEE Trans. Knowl. Data Eng. **21**, 1708–1721 (2009)
2. Fournier-Viger, P., Gomariz, A., Gueniche, T., Soltani, A., Wu, C., Tseng, V.S.: SPMF: a java open-source pattern mining library. J. Mach. Learn. Res. **15**, 3389–3393 (2014)
3. Liu, Y., Liao, W., Choudhary, A.: A fast high utility itemsets mining algorithm. In: Proceedings of the 1st International Workshop on Utility-Based Data Mining, pp. 90–99 (2005)
4. Liu, M., Qu, J.: Mining high utility itemsets without candidate generation. In: Proceedings of the 21st ACM International Conference on Information and Knowledge Management, pp. 55–64 (2012)
5. Lin, Y., Wu, C., Tseng, V.S.: Mining high utility itemsets in big data. In: Proceedings of the Pacific-Asia Conference on Knowledge Discovery and Data Mining, pp. 649–661 (2015)
6. Ryang, H., Yun, U.: Top-k high utility pattern mining with effective threshold raising strategies. Knowl.-Based Syst. **76**, 109–126 (2015)
7. Subramanian, K., Kandhasamy, P., Subramanian, S.: A novel approach to extract high utility itemsets from distributed databases. Comput. Inform. **31**, 1597–1615 (2012)
8. Tseng, V.S., Shie, B., Wu, C., Yu, P.S.: Efficient algorithms for mining high utility itemsets from transactional databases. IEEE Trans. Knowl. Data Eng. **25**, 1772–1786 (2013)
9. Tseng, V.S., Wu, C., Fournier-Viger, P., Yu, P.S.: Efficient algorithms for mining top-k high utility itemsets. IEEE Trans. Knowl. Data Eng. **28**, 54–67 (2016)

10. Vo, B., Nguyen, H., Ho, T.B., Le, B.: Parallel method for mining high utility itemsets from vertically partitioned distributed databases. In: Proceedings of the 13th International Conference on Knowledge-Based and Intelligent Information and Engineering Systems, pp. 251–260 (2009)
11. Wu, C., Shie, B., Tseng, V.S., Yu, P.S.: Mining top-k high utility itemsets. In: Proceedings of the 18th ACM SIGKDD International Conference on Knowledge Discovery and Data Mining, pp. 78–86 (2012)
12. Apache Software Foundation. http://www.apache.org/
13. Hadoop. http://hadoop.apache.org/
14. IBM Quest Data Mining Project, Quest Synthetic Data Generation Code. (https://sourceforge.net/projects/ibmquestdatagen/)
15. Spark. http://spark.apache.org/

Weakly Supervised Learning

Robust Semi-supervised Multi-label Learning by Triple Low-Rank Regularization

Lijuan Sun, Songhe Feng(✉), Gengyu Lyu, and Congyan Lang

School of Computer and Information Technology, Beijing Jiaotong University, Beijing, China
{17112082,shfeng,18112030,cylang}@bjtu.edu.cn

Abstract. Multi-Label Learning (MLL) deals with the problem when one instance is associated with multiple labels simultaneously. Previous methods have shown promising performance by effectively exploiting the semantic correlations among different labels. However, most of the existing methods may not be robust to the situation when the training instances are labeled with noisy or incomplete labels, which are common in reality. In this paper, we propose Robust Semi-Supervised Multi-Label Learning by Triple Low-Rank Regularization approach to address this problem. Specifically, a linear self-representative model is firstly introduced to recover the possibly noisy label matrix by exploiting the label correlations. Then, our method develops a low-rank pairwise similarity matrix to capture the global relationships among labeled and unlabeled samples by taking advantage of Low-Rank Representation (LRR). In addition, by utilizing the pairwise similarity matrix defined above, we construct the graph Laplacian regularization to acquire geometric structural information from both labeled and unlabeled samples. Moreover, the proposed method concatenate the prediction models for different labels into a matrix, and introduces the matrix trace norm to capture the correlations and control the model complexity. Experimental studies across a wide range of benchmark datasets show that our method achieves highly competitive performance against other state-of-the-art approaches.

Keywords: Multi-label learning · Triple low-rank regularization · Semi-supervised learning · Graph Laplacian regularization

1 Introduction

Traditional supervised learning often assumes that each instance is associated with a single label. In reality, one object usually has multiple labels simultaneously. For example, in document topic analysis, a document may belong to multiple topics. Conventional supervised learning is out of its capability to cope with this problem, and Multi-Label Learning (MLL) [18] that deals with instances

Q. Yang et al. (Eds.): PAKDD 2019, LNAI 11440, pp. 269–280, 2019.
https://doi.org/10.1007/978-3-030-16145-3_21

associated with a set of labels could address this problem. Current studies on MLL always assume that each training sample is associated with a complete label assignment. However, many training samples may only be labeled with a partial set of labels, or the labels can be randomly corrupted in practice [19]. In addition, how to capture the inherent correlations correctly between different labels by making use of the finite labeled data is still under investigation.

To effectively deal with this problem, we present a novel algorithm called Robust Semi-Supervised Multi-Label Learning by Triple Low-Rank Regularization. Specifically, given the fact that the training labels which always contain noisy labels or missing labels [12], our method provides a linear self-recovery model on label matrix by introducing a low rank constrained label coefficient matrix. Then, we develop a low-rank pairwise similarity matrix to capture the global relationships among samples by taking advantage of Low-Rank Representation(LRR). Furthermore, by utilizing the pairwise similarity matrix that we got earlier, we construct the graph Laplacian regularization that is a smooth operator to maintain a local geometric structure on both labeled and unlabeled samples. Meanwhile, our method concatenates prediction models for different labels into a matrix, and introduces the low-rank constrained matrix to capture the correlations among different labels. What is noteworthy is that the work of [5] takes the low-rank property of feature mapping matrix, graph Laplacian and missing labels into account as well. However, our method adopts a more sophisticated approach to recover the noisy label matrix during the training stage as compared with [5]. Extensive experimental results on real-world data sets validate the effectiveness of our model against other competitive algorithms.

2 Related Work

In this section, we briefly review the related work on exploiting label correlations. Existing approaches can be roughly grouped into three major categories based on the order of correlations being considered. First-order approaches tackle MLL problem by decomposing it into a series of independent binary classification problems such as SVM [1] or Naive Bayes [17]. Despite their conceptual simplicity, these approaches could be less effective due to their regardless of label coincidence. Second-order approaches tackle MLL problem by exploiting pairwise relationships between the labels. Representatives include ranking based approaches [7] which consider correlations between pairs of labels and work by transforming the task into a ranking problem to order the relevant labels in front of irrelevant labels. Nevertheless, label correlations might be much more complex than second-order in reality. High-order approaches try to discover high-order relationships among the labels to address multi-label learning problem. One straightforward approach is to model interactions among all class labels by considering influences of all other labels on each label. Representative approaches include assuming linear combination [3], nonlinear mapping [11], or shared subspace [8] over the whole label space. There is no doubt that high-order approaches have stronger correlation-modeling capabilities than the first-order and second-order counterparts. However, these approaches would be more computationally expensive and less scalable.

3 Problem Formulation

In this section, we first describe the concept that will be used throughout the paper. Then, we introduce the proposed method in detail. In the rest of this section, we discuss how to solve the optimization problem by an iterative optimization algorithm.

3.1 Notations

Let $\mathbf{X}_\ell = [\mathbf{x}_1, \mathbf{x}_2, \ldots, \mathbf{x}_n] \in \mathbb{R}^{d\times n}$ denote the labeled feature matrix, where d is the dimension of the feature vector and n is the number of the labeled samples. And we define the matrix of unlabeled samples as $\mathbf{X}_u = [\mathbf{x}_{n+1}, \mathbf{x}_{n+2}, \ldots, \mathbf{x}_{n+m}] \in \mathbb{R}^{d\times m}$, where m is the number of the unlabeled samples. We assume $m \gg n$ without loss of generality. Subsequently, we use $\mathbf{X} = [\mathbf{X}_\ell, \mathbf{X}_u] \in \mathbb{R}^{d\times(n+m)}$ to represent the whole training set. Accordingly, we use $\mathbf{Y}_\ell = [\mathbf{y}_1, \mathbf{y}_2, \ldots, \mathbf{y}_n] \in \{0,1\}^{k\times n}$ to represent the label assignments for the corresponding labeled samples.

3.2 The Regularization Framework

Our goal is to use the samples in $\mathbf{X}$ to train a new MLL approach and to predict the labels of these unlabeled samples. For the i-th label, the goal is to learn a linear function f_i where $\mathbf{w}_i$ is the model parameter for the i-th label. Here we restrict the prediction function f_i to linear function for simplicity, i.e., $f_i(\mathbf{X}) = \mathbf{w}_i^\top \mathbf{X}$. Define $\mathbf{W} = [\mathbf{w}_1, \mathbf{w}_2, \ldots, \mathbf{w}_k] \in \mathbb{R}^{d\times k}$ denote the model parameters for all labels. We first employ the loss function $\mathcal{L}(\mathbf{X}_\ell, \mathbf{Y}_\ell; \mathbf{W})$ given only the labeled training data as:

$$\mathcal{L}(\mathbf{X}_\ell, \mathbf{Y}_\ell; \mathbf{W}) = \frac{1}{2}\|\mathbf{Y}_\ell - \mathbf{W}^\top \mathbf{X}_\ell\|_F^2$$

Intuitively, the target is to seek a matrix $\mathbf{W}$ for minimizing the loss function $\mathcal{L}(\mathbf{X}_\ell, \mathbf{Y}_\ell; \mathbf{W})$. Nevertheless, we have to think over the model complexity and overfitting problem. In many real-world problems, labels are relevant among different class labels. We assume that the prediction functions in $\mathbf{W}$ are linearly dependent to effectively capture the correlation. Formally, we formulate minimizing the loss function problem as low-rank pursuit on the matrix $\mathbf{W}$ and the optimization problem is defined as:

$$\min_{\mathbf{W}} \frac{1}{2}\|\mathbf{Y}_\ell - \mathbf{W}^\top \mathbf{X}_\ell\|_F^2 + \lambda\|\mathbf{W}\|_* \quad (1)$$

Here, $\|.\|_*$ denotes the nuclear norm of a matrix, i.e., the sum of the singular values of the matrix.

Unfortunately, errors often caused by noisy training label matrix $\mathbf{Y}_\ell$ in typical community-contributed images applications. Therefore, these noisy labels must be refined for achieving satisfactory multi-label learning performance if we use them as training labels. We reconstruct $\mathbf{Y}_\ell$ by employing a linear self-recovery model. $\boldsymbol{\Omega} \in \mathbb{R}^{m\times m}$ is a coefficient matrix, and the similarity between the i-th label and j-th label was indicated by the element which locates in the i-th row and j-th column. In order to narrow down the solution for the matrix $\boldsymbol{\Omega}$, we ponder the following criteria. Firstly, the new matrix $\mathbf{Y}_\ell$ that refined by the coefficient matrix $\boldsymbol{\Omega}$ should be similar to the observed noisy matrix $\mathbf{Y}_\ell$. We add this constraint by penalizing the difference between $\mathbf{Y}_\ell$ and $\boldsymbol{\Omega}\mathbf{Y}_\ell$ with a Frobenius Norm, and we prefer the solution $\boldsymbol{\Omega}$ with small $\|\boldsymbol{\Omega}\mathbf{Y}_\ell - \mathbf{Y}_\ell\|_F^2$. Secondly, $\boldsymbol{\Omega}\mathbf{Y}_\ell$ should have the ability to keep the intrinsic structure among feature vectors. Formally, we formulate the proposed approach as follows:

$$\begin{aligned} &\min_{\boldsymbol{\Omega}} \frac{\beta}{2}\|\boldsymbol{\Omega}\mathbf{Y}_\ell - \mathbf{Y}_\ell\|_F^2 + \gamma\|\boldsymbol{\Omega}\|_* \\ &s.t.\, \boldsymbol{\Omega} \geq 0, diag\left(\boldsymbol{\Omega}\right) = 0 \end{aligned} \tag{2}$$

Here, we assume the coefficient matrix Ω be low rank which again enforces the similarity between different labels implicitly. Subsequently, we got the following optimization problem:

$$\begin{aligned} &\min_{\mathbf{W},\boldsymbol{\Omega}} \frac{1}{2}\|\boldsymbol{\Omega}\mathbf{Y}_\ell - \mathbf{W}^\top\mathbf{X}_\ell\|_F^2 + \lambda \parallel \mathbf{W}\|_* \\ &+ \frac{\beta}{2}\|\boldsymbol{\Omega}\mathbf{Y}_\ell - \mathbf{Y}_\ell\|_F^2 + \gamma\|\boldsymbol{\Omega}\|_* \\ &s.t.\, \boldsymbol{\Omega} \geq 0, diag\left(\boldsymbol{\Omega}\right) = 0 \end{aligned} \tag{3}$$

However, existing methods always ignore the situation that there are a large number of unlabeled samples in the training sets. Consequently, we utilize graph Laplacian based manifold regularization [15] into the learning process be able to acquire geometric structural information from unlabeled samples. We construct the graph Laplacian regularization by taking advantage of Low-Rank Representation(LRR). In order to capture the global relationships among samples, we introduce the pairwise similarity matrix $\mathbf{S}$ and the similarity between the i-th label and j-th label was indicated by the element which locates in the i-th row and j-th column. Unlike [13] simply uses a k-nearest neighbor graph to model the local geometry structure in the feature space, we take advantage of LRR to construct a pairwise similarity matrix $\mathbf{S}$ in the first stage. Given that $\mathbf{X} \in \mathbb{R}^{dx(n+m)}$ represents the feature space of both labeled and unlabeled samples, where each column corresponds to an sample, each sample can be viewed as a linear combination of basis from a dictionary $\mathbf{A} = [\mathbf{a}_1, \mathbf{a}_2, \ldots, \mathbf{a}_M] \in \mathbb{R}^{d\times M}$. LRR encodes each sample by a linear combination of the basis in $\mathbf{A}$ as follow:

$$\mathbf{X} = \mathbf{AZ}$$

where $\mathbf{Z} \in \mathbb{R}^{(n+m)\times(n+m)}$ is the coefficient matrix with each $\mathbf{Z}(.,i) \in \mathbb{R}^{(n+m)}$ being the representation coefficient vector for sample $\mathbf{x}_i$ with respect to $n+m$ samples. Here, we set $\mathbf{A} = \mathbf{X}$ in this paper. LRR enforces $\mathbf{Z}$ to be low rank and solves the following optimization problem:

$$\min_{\mathbf{Z}} \frac{1}{2}\|\mathbf{XZ} - \mathbf{X}\|_F^2 + \eta\|\mathbf{Z}\|_*$$

Since LRR jointly finds the low-rank coefficient matrix $\mathbf{Z}$ for all samples in $\mathbf{X}$, we adapt $\mathbf{Z}$ to define a feature-based graph whose weighted adjacent matrix is $\mathbf{S}$, where $\mathbf{S} = (|\mathbf{Z}| + |\mathbf{Z}^\top|)/2$.

In the second stage, we utilize the pairwise similarity matrix $\mathbf{S}$ to acquire the graph Laplacian regularization. Intuitively, if two samples $\mathbf{x}_i$ and $\mathbf{x}_j$ are close in the feature space, then (i.e. $\mathbf{W}^\top\mathbf{x}_i$ and $\mathbf{W}^\top\mathbf{x}_j$) are also close to each other in the label space. Therefore, we minimize the correlation of two samples by using the constraint as follows:

$$\begin{aligned}
&\frac{1}{2}\sum_{i=1}^{n+m}\sum_{j=1}^{n+m}\mathbf{S}_{ij}\|\frac{\mathbf{W}^\top\mathbf{x}_i}{\sqrt{\mathbf{E}^{ii}}} - \frac{\mathbf{W}^\top\mathbf{x}_j}{\sqrt{\mathbf{E}^{jj}}}\|_2^2 \\
&= \mathrm{Tr}\left[\mathbf{W}^\top\mathbf{X}\left(\mathbf{E}^{-\frac{1}{2}}(\mathbf{E}-\mathbf{S})\mathbf{E}^{-\frac{1}{2}}\right)\mathbf{X}^\top\mathbf{W}\right] \\
&= \mathrm{Tr}\left[\mathbf{W}^\top\mathbf{XLX}^\top\mathbf{W}\right]
\end{aligned} \tag{4}$$

where $\mathbf{L} = \mathbf{E}^{-\frac{1}{2}}(\mathbf{E}-\mathbf{S})\mathbf{E}^{-\frac{1}{2}}$ is the graph Laplacian matrix and $\mathbf{E}$ is a diagonal matrix with $\mathbf{E}^{ii} = \sum_{j=1}^{n+m}\mathbf{S}_{ij}$.

Combining the above criteria, the final formulation of the proposed method is defined as:

$$\begin{aligned}
&\min_{W,\mathbf{\Omega},\mathbf{Z}} \frac{1}{2}\|\mathbf{\Omega Y}_\ell - \mathbf{W}^\top\mathbf{X}_\ell\|_F^2 + \lambda\|\mathbf{W}\|_* \\
&+ \frac{\beta}{2}\|\mathbf{\Omega Y}_\ell - \mathbf{Y}_\ell\|_F^2 + \gamma\|\mathbf{\Omega}\|_* + \alpha\mathrm{Tr}\left[\mathbf{W}^\top\mathbf{XLX}^\top\mathbf{W}\right] \\
&+ \frac{1}{2}\|\mathbf{XZ} - \mathbf{X}\|_F^2 + \eta\|\mathbf{Z}\|_* \\
&s.t.\, \mathbf{\Omega} \geq 0,\ diag(\mathbf{\Omega}) = 0,\ \mathbf{L} = \mathbf{E}^{-\frac{1}{2}}(\mathbf{E}-\mathbf{S})\mathbf{E}^{-\frac{1}{2}}, \\
&\mathbf{E}_{ii} = \sum_{j=1}^{n+m}\mathbf{S}_{ij},\ \mathbf{S} = (|\mathbf{Z}| + |\mathbf{Z}^\top|)/2
\end{aligned} \tag{5}$$

where α, β, γ, η and λ are trade-off parameters, $\mathbf{\Omega}$ is the low rank coefficient matrix for label matrix $\mathbf{Y}_\ell \in \{0,1\}^{k\times n}$, $\mathbf{Z}$ is the low rank constrained matrix for the sample matrix $\mathbf{X} \in \mathbb{R}^{d\times(n+m)}$ and $\mathbf{W}$ is the low rank feature mapping matrix. The framework will be more robust to noises and outliers. At the same time, it can preserve the local and global structural consistency. Moreover, this is our superiority by using triple trace norm regularization for robust semi-supervised multi-label learning.

3.3 Optimization

The optimization problem in (5) is convex and involves the trace norm which is non-smooth that cannot have a closed form solution. Firstly, we used an Augmented Lagrange Multiplier (ALM) to get $\mathbf{Z}$, and then we adopt an Accelerated Proximal Gradient algorithm(APG) [9] by iteratively updating $\mathbf{W}$ and $\boldsymbol{\Omega}$. In the following, we introduce the proposed update rules in detail.

Update $\boldsymbol{\Omega}$ When Fixed W. Next, with fixed $\mathbf{W} = \mathbf{W}_t$ at the t-th iteration, the subproblem that minimizes (5) over $\boldsymbol{\Omega}$ can be written as:

$$\min_{\boldsymbol{\Omega}} \frac{1}{2}\|\boldsymbol{\Omega}\mathbf{Y}_\ell - \mathbf{W}^\top \mathbf{X}_\ell\|_F^2 + \frac{\beta}{2}\|\boldsymbol{\Omega}\mathbf{Y}_\ell - \mathbf{Y}_\ell\|_F^2 + \gamma\|\boldsymbol{\Omega}\|_* \tag{6}$$

We can further simplify the above optimization problem as the following problem equivalently:

$$G(\boldsymbol{\Omega}) = g(\boldsymbol{\Omega}) + \gamma\|\boldsymbol{\Omega}\|_* \tag{7}$$

where

$$g(\boldsymbol{\Omega}) = \frac{1}{2}\|\boldsymbol{\Omega}\mathbf{Y}_\ell - \mathbf{W}^\top \mathbf{X}_\ell\|_F^2 + \frac{\beta}{2}\|\boldsymbol{\Omega}\mathbf{Y}_\ell - \mathbf{Y}_\ell\|_F^2 \tag{8}$$

Accordingly, we adopt the APG algorithm to solve it as follows:

$$\Omega_t = \arg\min_{\boldsymbol{\Omega}} \frac{1}{2\theta_t}\|\boldsymbol{\Omega} - \boldsymbol{\Omega}_t\|_F^2 + \gamma\|\boldsymbol{\Omega}\|_* \tag{9}$$

where

$$\boldsymbol{\Omega}_t^{'} = \boldsymbol{\Omega}_{t-1} - \theta_t \nabla g(\boldsymbol{\Omega}_{t-1}) \tag{10}$$

with θ_t is the step size. Firstly, we compute the singular value decomposition (SVD) of $\theta_t^{'}$, and then utilize soft-thresholding in the singular value of $\theta_t^{'}$. Thenceforth we get the optimal solution as follows: $\boldsymbol{\Omega}_t = U\sum_{\eta\theta_t} V^\top$, where $\boldsymbol{\Omega}_t^{'} = U\sum V^\top$ is the SVD of $\boldsymbol{\Omega}_t^{'}$. $\sum_{\eta\theta_t}$ is a diagonal matrix and its diagonal elements can be obtained according to $\left(\sum_{\eta\theta_t}\right)_{ii} = \max\{0, \sum_{ii} - \eta\theta_t\}$. Notice that the gradient of $\theta_t^{'}$ can be computed as following equation shows:

$$\nabla g(\boldsymbol{\Omega}) = (1+\beta)\boldsymbol{\Omega}\mathbf{Y}_\ell\mathbf{Y}_\ell^\top - \mathbf{W}_t^\top\mathbf{X}_\ell\mathbf{Y}_\ell^\top - \beta\mathbf{Y}_\ell\mathbf{Y}_\ell^\top \tag{11}$$

Update W When Fixed $\boldsymbol{\Omega}$. When we fix $\boldsymbol{\Omega} = \boldsymbol{\Omega}_t$, discard the entries which are not relevant to $\mathbf{W}$, and then $\mathbf{W}$ will be decided by solving the following problem:

$$\min_{\mathbf{W}} \frac{1}{2}\|\boldsymbol{\Omega}_t\mathbf{Y}_\ell - \mathbf{W}^\top \mathbf{Y}_\ell\|_F^2 + \lambda\|\mathbf{W}\|_* + \alpha \mathrm{Tr}\left[\mathbf{W}^\top \mathbf{X}\mathbf{L}\mathbf{X}^\top \mathbf{W}\right] \tag{12}$$

We first convert (12) to the following equivalent problem:

$$\min_{\mathbf{W}} F(\mathbf{W}) = f(\mathbf{W}) + \lambda\|\mathbf{W}\|_* \tag{13}$$

where

$$f(\mathbf{W}) = \frac{1}{2}\|\boldsymbol{\Omega}_t\mathbf{Y}_\ell - \mathbf{W}^\top\mathbf{X}_\ell\|_F^2 + \alpha\mathrm{Tr}\left[\mathbf{W}^\top\mathbf{X}\mathbf{L}\mathbf{X}^\top\mathbf{W}\right] \tag{14}$$

which is also a convex and smooth function. Equation(9) should be converted as following:

$$\mathbf{W}_t = \arg\min_{\mathbf{W}} \frac{1}{2\xi_t}\|\mathbf{W} - \mathbf{W}_t^{'}\|_F^2 + \lambda\|\mathbf{W}\|_* \tag{15}$$

where

$$\mathbf{W}_t^{'} = \mathbf{W}_{t-1} - \xi_t\nabla f(\mathbf{W}_{t-1}) \tag{16}$$

and ξ_t is step size. We get the definition of the diagonal matrix $\sum_{\lambda\xi_t}$ on the basis of $(\sum_{\lambda\xi_t})_{ii} = \max\{0, \sum_{ii} - \lambda\xi_t\}$, when we compute the singular value decomposition of $\mathbf{W}$.

$$\nabla f(\mathbf{W}) = -\mathbf{X}_\ell\mathbf{Y}_\ell^\top\boldsymbol{\Omega}_t^\top + \mathbf{X}_\ell\mathbf{X}_\ell^\top\mathbf{W} + 2\alpha\mathbf{X}\mathbf{L}\mathbf{X}^\top\mathbf{W} \tag{17}$$

The entire optimization procedure will be terminated when both $\mathbf{W}$ and $\boldsymbol{\Omega}$ obtain the optimal solution. Despite the algorithm does not guarantee a global optimum, we found it perform well in our experiments.

4 Experiments

In this part, we conduct extensive experiments to validate the effectiveness of the proposed algorithm. To step further, we compare our method against state of-the-art multi-label learning approaches on four datasets. We also perform our method by using training samples with missing labels to ensure that the performance of the proposed approach is statistically significant. In addition, we conduct experiments to study the sensitivity of the proposed algorithm to parameter α, β, γ and λ. Meanwhile, we also execute the experiment to study the influence of limited number of training samples on IAPRTC-12 dataset. Due to page limit, we cannot report all results.

Table 1. Statistics summary for the experimental datasets. The bottom two rows are given in the format mean/maximum.

	ESPGame	IAPRTC-12	PASCAL VOC2007	NUS-WIDE
No.of images	20,770	19,627	9,963	31,570
Vocabulary size	268	291	399	430
Tags per image	4.69/15	5.72/23	4.2/35	12.53/114
Image per tag	363/5,059	386/5,534	53/2,095	920/8,397

4.1 Datasets

To examine the performance of the proposed approach, we perform extensive experiments on four data sets, including ESPGame [10], IAPRTC-12 [6], PASCAL VOC2007 [5] and NUS-WIDE [4]. The first three datasets are directly acquired from [4] and the last dataset is gathered by Lab for Media Search in the National University of Singapore [4]. Table 1 summarizes the statistics of the datasets in detail. ESPGame and IAPRTC-12 image datasets are two bag-of-words models with 1000 visual words, which represent the visual content of images by using based on densely sampled SIFT descriptors. We extract three types of image features from PASCAL VOC2007 dataset. We use six types of low-level visual features to represent each image for the NUS-WIDE dataset.

4.2 Evaluation Criteria

In the experiments, we exploit the Average Precision ($AP@K$) and Average Recall ($AR@K$) scores to evaluate the performance of multi-label learning method in this paper. The evaluation metrics are defined as follows:

$$AP@K = \frac{1}{n_t} \sum_{i=1}^{n_t} \frac{N_c(i)}{K} \tag{18}$$

$$AR@K = \frac{1}{n_t} \sum_{i=1}^{n_t} \frac{N_c(i)}{N_g(i)} \tag{19}$$

where K is the number of truncated tags, n_t is the number of test instances, $N_c(i)$ is the number of correctly annotated tags for the i-th test instance, $N_g(i)$ is the number of tags assigned to the i-th instance.

4.3 Comparison with the State-of-the-art Algorithms

Five competing algorithms are involved as a comparison, namely Binary Reference Model based on RBF Kernel (BR-R)[1], Multi-Label k-Nearest Neighbor (ML-KNN) [17], Matrix Completion using Side Information (MAXIDE) [14], Multi-Label Learning with Label Specific Features (LIFT) [16], and Efficient Multi-Label Ranking Method (MLR) [2].

On each dataset in Table 1, we randomly sample 50% of the dataset as training data $\mathbf{X}$, of which 20% instances are selected as labeled data $\mathbf{X}_\ell$ and 80%instances are unlabeled data $\mathbf{X}_\mu$, respectively. The remaining 50% of the dataset are used as testing data $\mathbf{T}$. Each experiment is repeated for 5 times on a different splitting of training and testing data, and we report the average result. All the above methods are implemented in Matlab and run on Windows with 16G memory and 3.6 GHz CPU.

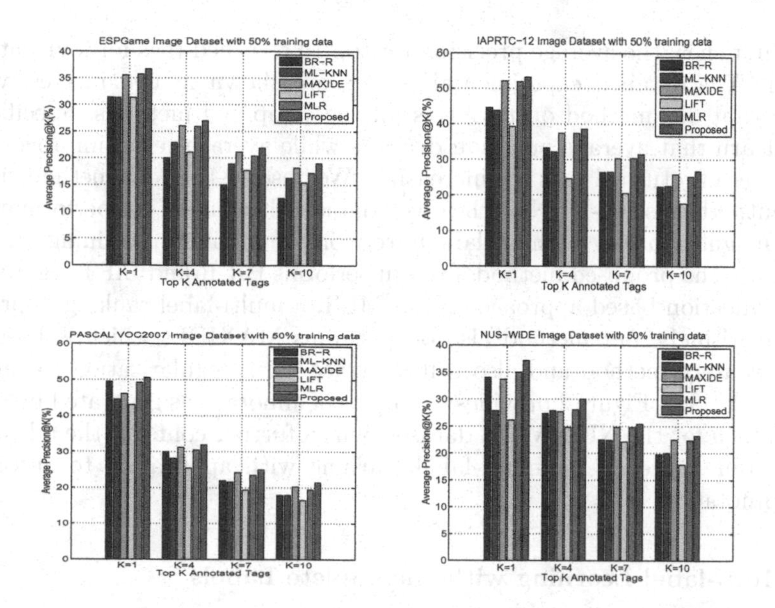

Fig. 1. Experimental result of the Average precision for proposed method on ESPGame, IAPRTC-12, PASCAL VOC2007 and NUS-WIDE datasets.

Ground Truth	grass flower path house tree lake mountain snow sky beautiful wood	dog snow white ground horn spot hat red christmas pug clothes collar	person clothes tree sky ground bike grass street buildings play house	cars house buildings tree road sign bridge green grass light red clay traffic streetlight	fire light fireworks boat lake buildings tree sky city night yellow board
BR-R	grass flower path house tree lake sky *clouds becah color*	*cat* snow white ground horn *grass toy person water* red	person clothes tree sky *beach sea* grass street buildings *clouds*	cars house *rain beach* road buildings *night* tree grass *man*	fire light fireworks *car water* buildings *black street* board *grass*
MLKNN	grass flower house tree lake beautiful *color* sky *nature clouds*	dog snow white *toy black grass dress clouds water* pug	person clothes tree ky *beach yellow* grass street *sea clouds*	cars *blue* buildings *clouds water* grass *brown window* traffic *box*	fire light fireworks *water black road grass colorful black* night
MAXIDE	grass *red* house sky *clouds color beach nature clouds blue*	dog snow *grass black water flower brown garden* pug *toy*	person tree sky *lake yellow* grass street *garden clouds city*	cars *water box* buildings *brown window clouds street boat*	fire *car water* buildings *black street colorful sign wood* city
LIFT	tree flower *garden* sky *clouds color nature blue water*	dog *dress black* snow *water flower brown tree garden brown*	person grass *yellow* sky *clouds city sea beach green beautiful*	*night clouds sky brown* cars *water street* traffic *window box*	fire *car water black street bridge sign wood* city night
MLR	grass flower house lake *green clouds sea color beach* sky	dog snow white *toy black grass* clothes *water* pug red	person clothes *green beautiful clouds* sky *beach sea* grass street	cars *night clouds* road *brown* buildings *box blue* green light	fire light fireworks boat *house* lake *clolrful grass black* city
Proposed	grass flower house lake tree *clouds* sky *color beach sea*	dog snow white spot red *black grass water toy flower*	person clothes tree sky *beach sea* grass street buildings *beautiful*	cars house buildings tree road sign *night* green light traffic	fire light fireworks boat *house* lake *clolrful* tree *black* city

Fig. 2. Examples of test images from the NUS-WIDE dataset with top 10 annotations generated by different methods. The correct tags are highlighted by blue bold font whereas the incorrect ones are highlighted by italic font. (Color figure online)

We first show the average precision for top 10 returned tags for four datasets in Fig. 1. The detailed experimental results are shown in two images, which indicate that our method outperforms all the compared methods. Specifically, we can learn that average precision declines while average recall improves with the number of returned tags are increasing. We observe that our method significantly outperforms ML-KNN on the given datasets because of the performance of nearest-neighbor based methods largely rely on the number of training samples. In addition, the proposed method also outperforms BR-R and LIFT algorithms, two classification based approaches, and MLR, a multi-label ranking approach. More surprisingly, our method also outperforms MAXIDE, which is based on matrix recovery method and also utilizes trace norm regularization to capture label correlations. Figure 2 provides examples of annotations generated by different approach for the NUS-WIDE dataset, which further confirms the advantage of using our approach for multi-label learning with application to automatic image annotation.

4.4 Multi-label Learning with Incomplete Labels

We conduct experiments on PASCAL VOC2007 dataset to verify the effectiveness of our method for incomplete labels. So we randomly select only 20%, 40%, and 60% of the assigned labels for training examples. Figure 3 respectively show the average precision results of different multi-label learning algorithms on PASCAL VOC2007.

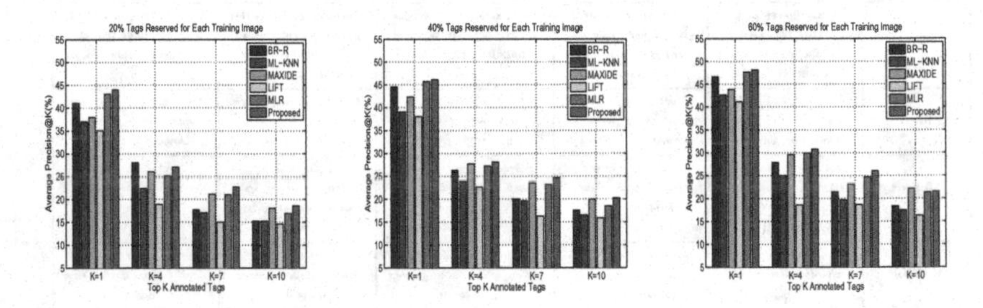

Fig. 3. Performance of the proposed method on the PASCAL VOC2007 dataset with incomplete image tags, where the number of observed tags is varied from 20%, 40% to 60%.

Apparently, our method is more resilient to the missing labels in all methods on both the two datasets: on the PASCAL VOC2007 dataset, it only experiences a 2.03% drop in average precision when the number of observed labels decreases from 60% to 20%, while the other four baseline methods suffer from 4% to 8% loss for AP@4. Clearly, we can learn that the missing labels could greatly affect the annotation performance due to all methods drops as the number of observed annotations decreases. Although Fig. 3 reports and reflects that the proposed method is more effective in multi-label learning with incomplete labels.

4.5 Parameters Sensitivity Analysis

We study the influences of the four parameters α, β, γ and λ for the proposed method on the ESPGame dataset. Before the experiment, we first initialize λ as 0.1. For the rest parameters, we fix one parameter and then vary the other two parameters from 0.001 to 10, respectively. The experimental results are shown in Fig. 4 which are measured by average precision. It can be seen that the performance of our algorithm varies when the parameters (α, β, γ) change. Therefore we should safely set them in a wide range of practice. From this figure, we can notice that better performances are gained when $\alpha = 0.01$, $\beta = 1$, $\lambda = 0.1$ and $\gamma = 0.01$.

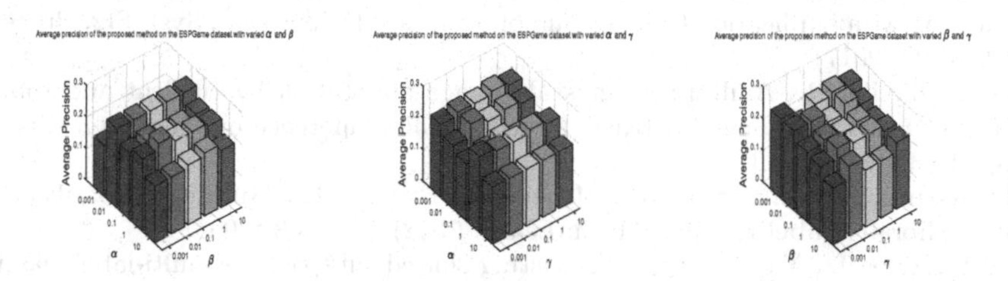

Fig. 4. Average precision of the proposed method on ESPGame dataset with varied parameters.

5 Conclusion

In this paper, we present a novel approach to address the drawbacks of existing MLL methods in a principled manner. The proposed algorithm concatenates prediction models for different tags into a matrix, and introduces the low rank constrained matrix to capture correlations between different labels and control the model complexity. Our approach also utilizes graph Laplacian regularization to maintain a local geometric structure on both labeled and unlabeled samples. Meanwhile, our method introduces a low rank constrained sample coefficient matrix to capture the relationships between samples. Moreover, our algorithm exploits a low rank constrained label coefficient matrix again to build a linear self-recovery model for labels are incomplete or noisy. Finally, we conduct extensive experiments to validate the effectiveness of our algorithm. For future work, we would like to extend the framework by explicitly taking the limited number of training samples into consideration.

Acknowledgements. This work was supported in part by the National Natural Science Foundation of China (Nos. 61872032), in part by the Fundamental Research Funds for the Central universities (2018YJS038, 2017JBZ108).

References

1. Boutell, M., Luo, J., Shen, X., Brown, C.: Learning multi-label scene classification. Pattern Recogn. **37**(9), 1757–1771 (2004)
2. Bucak, S., Jin, R., Jain, A.: Multi-label learning with incomplete class assignments. In: IEEE Conference on Computer Vision and Pattern Recognition, pp. 2801–2808 (2011)
3. Cheng, W., Hullermeier, E.: Combining instance-based learning and logistic regression for multilabel classification. Mach. Learn. **76**(2), 211–225 (2009)
4. Chua, T., Tang, J., Hong, R., Li, H., Luo, Z., Zheng, Y.: NUS-WIDE: a real-world web image database from National University of Singapore. In: ACM International Conference on Image and Video Retrieval, p. 48 (2009)
5. Feng, S., Lang, C.: Graph regularized low-rank feature mapping for multi-label learning with application to image annotation. Multidimension. Syst. Sig. Process. **11**, 1–22 (2017)
6. Feng, Z., Jin, R., Anil, J.: Large-scale image annotation by efficient and robust Kernel metric learning. In: IEEE International Conference on Computer Vision, pp. 1609–1616 (2013)
7. Furnkranz, J., Hullermeier, E., Mencia, E., Brinker, K.: Multilabel classification via calibrated label ranking. Mach. Learn. **73**(2), 133–153 (2008)
8. Ji, S., Tang, L., Yu, S., Ye, J.: Extracting shared subspace for multi-label classification. In: ACM SIGKDD International Conference on Knowledge Discovery and Data Mining, pp. 381–389 (2008)
9. Ji, S., Ye, J.: An accelerated gradient method for trace norm minimization. In: Proceedings of Annual International Conference on Machine Learning, pp. 457–464 (2009)
10. Matthieu, G., Thomas, M., Jakob, V., Cordelia, S.: Tagprop: discriminative metric learning in nearest neighbor models for image auto-annotation. In: IEEE International Conference on Computer Vision, pp. 309–316 (2009)
11. Montanes, E., Senge, R., Barranquero, J., Quevedo, J., Coz, J., Hullermeier, E.: Dependent binary relevance models for multi-label classification. Pattern Recogn. **47**(3), 1494–1508 (2014)
12. Sheng, L., Yun, F.: Robust multi-label semi-supervised classification. In: IEEE International Conference on Big Data, pp. 27–36 (2017)
13. Wang, X., Feng, S., Lang, C.: Semi-supervised dual low-rank feature mapping for multi-label image annotation. Multimed. Tools Appl. **8**, 1–20 (2018)
14. Xu, M., Jin, R., Zhou, Z.: Speedup matrix completion with side information: application to multi-label learning. In: Advances in Neural Information Processing Systems, pp. 2301–2309 (2013)
15. Yin, M., Gao, J., Lin, Z.: Laplacian regularized low-rank representation and its applications. IEEE Trans. Pattern Anal. Mach. Intell. **38**(3), 504–517 (2016)
16. Zhang, M., Wu, L.: Lift: multi-label learning with label-specific features. IEEE Trans. Pattern Anal. Mach. Intell. **37**(1), 107–120 (2015)
17. Zhang, M., Zhou, Z.: ML-KNN: a lazy learning approach to multi-label learning. Pattern Recogn. **40**(7), 2038–2048 (2007)
18. Zhang, M., Zhou, Z.: A review on multi-label learning algorithms. IEEE Trans. Knowl. Data Eng. **26**(8), 1819–1837 (2014)
19. Zhang, Y., Zhang, Z., Qin, J., Zhang, L., Li, B., Li, F.: Semi-supervised local multi-manifold isomap by linear embedding for feature extraction. Pattern Recogn. **76**, 662–678 (2018)

Multi-class Semi-supervised Logistic I-RELIEF Feature Selection Based on Nearest Neighbor

Baige Tang[1] and Li Zhang[1,2](✉)

[1] School of Computer Science and Technology, Joint International Research Laboratory of Machine Learning and Neuromorphic Computing, Soochow University, Suzhou 215006, Jiangsu, China
bgtang@stu.suda.edu.cn

[2] Provincial Key Laboratory for Computer Information Processing Technology, Soochow University, Suzhou 215006, Jiangsu, China
zhangliml@suda.edu.cn

Abstract. The multi-class semi-supervised logistic I-RELEIF (MSLIR) algorithm has been proposed and showed its feature selection ability using both labeled and unlabeled samples. Unfortunately, MSLIR is poor when predicting labels for unlabeled samples. To solve this issue, this paper presents a novel multi-class semi-supervised logistic I-RELEIF based on nearest neighbor (MSLIR-NN) for multi-class feature selection tasks. To generate better margin vectors for unlabeled samples, MSLIR-NN uses the nearest neighbor scheme to first predict the labels of unlabeled samples and then calculates their margin vectors according to these estimated labels. Experimental results demonstrate that MSLIR-NN can improve the prediction accuracy of unlabeled data.

Keywords: Logistic I-RELIEF · Feature selection · Multi-class classification · Semi-supervised · Nearest neighbor

1 Introduction

In many fields such as data mining and machine learning, we usually need to deal with high-dimensional data which may contain a large number of irrelevant and redundant features. These features would lead to the sparsity of data distribution in the feature space and be a hindrance to data analysis tasks. The rapid growth of data dimension not only increases the computational cost and memory consumption, but also affects the classification performance of classifiers. In order to improve learning performance, a variety of data dimensionality reduction methods have been produced, among which feature selection is one of the most effective techniques for processing high-dimensional data [1, 2].

The main goal of feature selection is to select an optimal feature subset, which contains most useful information in original features and has the greatest correlation with classification tasks. Based on the optimal feature subset, the training

Q. Yang et al. (Eds.): PAKDD 2019, LNAI 11440, pp. 281–292, 2019.
https://doi.org/10.1007/978-3-030-16145-3_22

time of classifiers can be effectively shorten, and the learning performance could be enhanced further [3–5]. In recent years, the technology of feature selection is of diversity. A lot of feature methods have been proposed [9–11,15,17,18]. Here, we consider RELIEF-based methods.

Kira and Rendell proposed the famous Relief algorithm in 1992, which uses the Euclidean distance as a metric to select features with great weights [6]. In 1994, Kononenko presented the extended algorithm RELIEF-F to solve the problem of multi-class classification [7]. On the basis of RELIEF, Sun et al. proposed an iterative RELEIF (I-RELIEF) algorithm to alleviate the deficiencies of RELEIF by exploring the framework of the expectation-maximization algorithm [8]. In order to better estimate feature weights, Sun et al. also proposed a logistic I-RELIEF (LIR) algorithm, which optimizes I-RELIEF in the form of logistic regression [16]. All of the above RELIEF-based algorithms are supervised feature selection ones, which can only use data with class labels. However, these methods cannot have a good performance when there exist few labeled data and a large number of unlabeled data. To remedy it, a semi-supervised logistic I-RELEIF (SLIR) method was presented [16]. In SLIR, both labeled and unlabeled data are used to calculate margin vectors of samples. However, SLIR was designed only for binary classification tasks. Tang et al. developed a multi-class semi-supervised logistic I-RELIEF (MSLIR) algorithm [19]. MSLIR designs a novel scheme to find margin vectors of unlabeled samples by calculating all possible candidate margin vectors and picking an optimal one under the condition of current feature weights. However, although MSLIR can implement multi-class feature selection in semi-supervised learning and get better classification performance than LIR, the supervised learning method, the prediction performance of unlabeled samples is unsatisfactory.

In order to solve the above issue, we propose a multi-class semi-supervised logistic I-RELEIF based on nearest neighbor (MSLIR-NN) for multi-class feature selection. In MSLIR-NN, the nearest neighbor scheme is adopted to assign pseudo labels to unlabeled samples according to labeled data in each iteration. In this case, unlabeled samples with pseudo labels could be treated as labeled ones. Thus, the margin vectors of labeled and unlabeled samples could be calculated easily. MSLIR-NN has a smaller computational complexity than MSLIR when calculating margin vectors of unlabeled samples. In experiments, support vector machine (SVM) and nearest neighbor (NN) classifiers are used to ensure the fairness of the classification results, respectively. Experimental results show that MSLIR-NN greatly improves the prediction performance of unlabeled data and enhances the performance of classifiers.

The rest part of this paper is organized as follows. The proposed method is described in detail in Sect. 2. The connections of MSLIR-NN to other related work are also discussed. Section 3 gives and analyzes experimental results. Section 4 concludes this paper.

2 Proposed Method: MSLIR-NN

In this section, we design a novel multi-class semi-supervised feature selection method, MSLIR-NN which adopts the nearest neighbor scheme to assign pseudo labels to unlabeled samples in each iteration. In doing so, unlabeled samples with pseudo labels could be treated as labeled ones. Thus, the margin vectors of labeled and unlabeled samples could be calculated easily. In the following, we describe MSLIR-NN in detail and discuss its connections to MSLIR and SLIR.

2.1 Margin Vectors

Assume that there is a labeled sample set $D_l = \{(\mathbf{x}_i^l, y_i^l)\}_{i=1}^L$ and an unlabeled sample set $D_u = \{\mathbf{x}_i^u\}_{i=1}^U$, where $\mathbf{x}_i^l \in \mathbb{R}^I$, $y_i^l \in \{1, 2, \ldots, c\})$, $\mathbf{x}_i^u \in \mathbb{R}^I$, I is the number of original features, c is the class number, L and U represent the number of labeled and unlabeled samples, respectively. Generally, $L \ll U$.

It is well known that one of main differences of RELIEF-based methods is the way of calculating margin vectors of samples. Without loss of generality, let $\mathbf{z}_i^l$ and $\mathbf{z}_i^u$ be margin vectors of the labeled sample $\mathbf{x}_i^l$ and the unlabeled sample $\mathbf{x}_i^u$, respectively. It is easy to generate the margin vectors of labeled samples in semi-supervised RELIEF-based methods. Similar to MSRIL [19], the margin vector $\mathbf{z}_i^l$ of the labeled sample $\mathbf{x}_i^l$ can be expressed as follows:

$$\begin{aligned} \mathbf{z}_i^l = &\sum_{\mathbf{x}_k^l \in M_i} P(\mathbf{x}_k^l = NM(\mathbf{x}_i^l)|\mathbf{w})|\mathbf{x}_i^l - \mathbf{x}_k^l| \\ &- \sum_{\mathbf{x}_k^l \in H_i} P(\mathbf{x}_k^l = NH(\mathbf{x}_i^l)|\mathbf{w})|\mathbf{x}_i^l - \mathbf{x}_k^l| \end{aligned} \tag{1}$$

where $\mathbf{w}$ is the feature weight vector, the set $M_i = \{\mathbf{x}_k^l | (\mathbf{x}_k^l, y_k^l) \in D_l, y_i^l \neq y_k^l, k = 1, \cdots, L, y_k^l \in \{1, \cdots, c\}\}$ contains all labeled samples that have different labels from $\mathbf{x}_i^l$, the set $H_i = \{\mathbf{x}_k^l | (\mathbf{x}_k^l, y_k^l) \in D_l, y_i^l = y_k^l, k = 1, \cdots, L, y_k^l \in \{1, \cdots, c\}\}$ contains all labeled samples that have the same label as $\mathbf{x}_i^l$, $P(\mathbf{x}_k^l = NM(\mathbf{x}_i^l)|\mathbf{w})$ and $P(\mathbf{x}_k^l = NH(\mathbf{x}_i^l)|\mathbf{w})$ are the probabilities that the sample $\mathbf{x}_k^l$ is the nearest miss and the nearest hit of $\mathbf{x}_i^l$, respectively, $NM(\mathbf{x}_i^l)$ represents the nearest miss (the nearest neighbor of sample $\mathbf{x}_i^l$ from a different class) of $\mathbf{x}_i^l$, and $NH(\mathbf{x}_i^l)$ the nearest hit (the nearest neighbor of sample $\mathbf{x}_i^l$ from the same class) of $\mathbf{x}_i^l$.

For semi-supervised RELIEF-based methods, it is the key and difficulty that how to define the margin vectors of unlabeled samples. Before calculating them, we first predict the pseudo labels of unlabeled data. According to the information contained in the labeled set D_l, we use the nearest neighbor scheme to predict the pseudo labels of unlabeled data in the set D_u. Note that $\mathbf{w}$ changes as iterations. For any unlabeled sample, its neighborhood is metabolic under different weight conditions. In other words, $\mathbf{w}$ has an effect on the procedure of searching nearest neighbors. Thus, we search nearest neighbors of unlabeled samples in the weighted feature space instead of the original input space. Then, we extend

D_u to the set $\hat{D}_u = \{(\mathbf{x}_i^u, \hat{y}_i^u)\}_{i=1}^U$ with pseudo labels $\hat{y}_i^u$ for $\mathbf{x}_i^u$. Similar to (1), we define the margin vector $\mathbf{z}_i^u$ of the unlabeled sample $\mathbf{x}_i^u$:

$$\mathbf{z}_i^u = \sum_{\mathbf{x}_k^l \in M_i'} P(\mathbf{x}_k^l = NM(\mathbf{x}_i^u)|\mathbf{w})|\mathbf{x}_i^u - \mathbf{x}_k^l| - \sum_{\mathbf{x}_k^l \in H_i'} P(\mathbf{x}_k^l = NH(\mathbf{x}_i^u)|\mathbf{w})|\mathbf{x}_i^u - \mathbf{x}_k^l| \quad (2)$$

where the set $M_i' = \{\mathbf{x}_k^l | (\mathbf{x}_k^l, y_k^l) \in D_l, \hat{y}_i^u \neq y_k^l, k = 1, \cdots, L, y_k^l \in \{1, \cdots, c\}\}$ contains all labeled samples that have different labels from $\mathbf{x}_i^u$, and the set $H_i' = \{\mathbf{x}_k^l | (\mathbf{x}_k^l, y_k^l) \in D_l, \hat{y}_i^u = y_k^l, k = 1, \cdots, L, y_k^l \in \{1, \cdots, c\}\}$ contains all labeled samples that have the same label as $\mathbf{x}_i^u$.

2.2 Optimization Problem

After obtaining margin vectors of all samples, the optimization of MSLIR-NN can be described as:

$$\min_{\mathbf{w}} \|\mathbf{w}\|_1 + \alpha \sum_{i=1}^{L} \log(1 + \exp(-\mathbf{w}^T \mathbf{z}_i^l)) + \beta \sum_{i=1}^{U} \log(1 + \exp(-\mathbf{w}^T \mathbf{z}_i^u)) \quad (3)$$
$$s.t. \ \ \mathbf{w} \geq 0$$

where $\|\cdot\|_1$ is the 1-norm, the regularization parameters $\alpha \geq 0$ and $\beta \geq 0$ control the importance of labeled and unlabeled samples, respectively.

To eliminate the constraint of $\mathbf{w} \geq 0$, let $\mathbf{w} = [v_1^2, \cdots, v_I^2]^T$ and $\mathbf{v} = [v_1, \cdots, v_I]$. Substituting $\mathbf{v}$ into (3), we can make it to an unconstraint optimization problem and have

$$\min_{\mathbf{v}} J = \|\mathbf{v}\|_2^2 + \alpha \sum_{i=1}^{L} \log(1 + \exp(-\sum_{d=1}^{I} v_d^2 z_{id}^l)) + \beta \sum_{i=1}^{U} \log(1 + \exp(-\sum_{d=1}^{I} v_d^2 z_{id}^u)) \quad (4)$$

where $\mathbf{z}_i^l = [z_{i1}^l, \cdots, z_{iI}^l]$ and $\mathbf{z}_i^u = [z_{i1}^u, \cdots, z_{iI}^u]$. (4) can be solved by using the gradient descent method. The derivation of J to $\mathbf{v}$ can be written as follows:

$$\frac{\partial J}{\partial v_k} = 2v_k - \alpha \sum_{i=1}^{L} \frac{\exp(-\sum_{d=1}^{I} v_d^2 z_{id}^l)(2v_k z_{ik}^l)}{1 + \exp(-\sum_{d=1}^{I} v_d^2 z_{id}^l)} - \beta \sum_{i=1}^{U} \frac{\exp(-\sum_{d=1}^{I} v_d^2 z_{id}^u)(2v_k z_{ik}^u)}{1 + \exp(-\sum_{d=1}^{I} v_d^2 z_{id}^u)} \quad (5)$$

Let

$$Q = \alpha \sum_{i=1}^{L} \frac{\exp(-\sum_{d=1}^{I} v_d^2 z_{id}^l)(v_k z_{ik}^l)}{1 + \exp(-\sum_{d=1}^{I} v_d^2 z_{id}^l)} + \beta \sum_{i=1}^{U} \frac{\exp(-\sum_{d=1}^{I} v_d^2 z_{id}^u)(v_k z_{ik}^u)}{1 + \exp(-\sum_{d=1}^{I} v_d^2 z_{id}^u)}$$

Then we can update v_k by

$$v_k \leftarrow v_k - \eta(v_k - Q) \tag{6}$$

where $\eta > 0$ is the learning rate.

2.3 Algorithm and Complexity Analysis

The algorithm description of MSLIR-NN is shown in Algorithm 1. Given a labeled dataset D_l and an unlabeled dataset D_u, the weight vector $\mathbf{w}$ is updated iteratively. Note that the weight vector in the t-th iteration is denoted as $\mathbf{w}_{(t-1)}$. First, under the current feature weights, MSLIR-NN computes the weighted labeled samples $\mathbf{x}^{l*}$ and unlabeled $\mathbf{x}^{u*}$, that is: $\mathbf{x}^{l*}$=$\mathbf{x}^l \circ \mathbf{w}_{(t)}$ and $\mathbf{x}^{u*} = \mathbf{x}^u \circ \mathbf{w}_{(t-1)}$, where $\circ$ denotes the element-by-element multiplication. The pseudo labels of unlabeled samples are determined in the weighted sample space using the NN scheme. Then, the margin vectors of $\mathbf{x}^l$ and $\mathbf{x}^u$ are calculated by (1) and (2), respectively. Finally, $\mathbf{w}$ is obtained by solving (3). MSLIR-NN alternatively modifies the weight vector until convergence.

The computational complexity of MSLIR-NN mainly includes three parts: the calculation of margin vectors for labeled samples, the calculation of margin vectors for unlabeled samples, and the solution to the optimization problem (3). The computational complexity of calculating margin vectors for labeled samples is identical to that of MSLIR and SLIR, which is $O(dL^2)$ without considering the calculation of probability terms, where d is the dimension of samples, and L is the number of labeled samples. For calculating of margin vectors for unlabeled samples, the computational complexity in MSLIR-NN is about $O(dUL)$, where U is the number of unlabeled samples. For the last part, MSLIR-NN has the same computational complexity as MSLIR and SLIR.

2.4 Connections to Related Work

Four RELEIF-based methods, LIR, SLIR, MSLIR and MSLIR-NN use the logistic regression formulation to optimize the feature weight vector $\mathbf{w}$. We discuss the connections of MSLIR-NN to LIR, SLIR, MSLIR in the following.

MSLIR-NN is a semi-supervised learning method as well as SLIR and MSLIR, LIR is designed for supervised learning. Base on LIR, SLIR introduces a term about unlabelled samples into the objective function. MSLIR changes the objective function of SLIR, which makes a balance calculation between labeled and unlabeled samples. Although MSLIR-NN has the same optimization function as MSLIR, MSLIR-NN has a different way for computing margin vectors of unlabelled samples.

SLIR, MSLIR and MSLIR-NN all adopt the way of calculating margin vectors of labeled samples in LIR. It is intuitive for SLIR to get margin vectors of unlabeled samples since SLIR deals with only binary classification tasks. For a given unlabeled sample, MSLIR first calculates all possible candidate margin vectors and takes an optimal one in the weighted feature space as its margin

Algorithm 1. MSLIR-NN

Input: Labeled dataset $D_l = \{(\mathbf{x}_i^l, y_i^l)\}_{i=1}^L \subset R^I \times \{1, 2, \ldots, c\}$; unlabeled dataset $D_u = \{\mathbf{x}_i^u\}_{i=1}^U \subset R^I$, regularization parameters α and β, the iteration number T, and the stop criterion θ.

Output: Feature weight $\mathbf{w}$.

1 **begin**
2   **Initialization**: Set $\mathbf{w}_{(0)} = [1, 1, \ldots, 1]^T$, $t = 1$, and $\rho = 1 + \theta$;
3   **while** $t \leq T$ && $\rho > \theta$ **do**
4   Compute the weighted samples $\mathbf{x}_i^{l*}$ and $\mathbf{x}_i^{u*}$: $\mathbf{x}_i^{l*} = \mathbf{x}_i^l \circ \mathbf{w}_{(t-1)}$ and $\mathbf{x}_i^{u*} = \mathbf{x}_i^u \circ \mathbf{w}_{(t-1)}$, where $\circ$ denotes the element-by-element multiplication;
5   Predict pseudo labels of unlabeled samples $\mathbf{x}_i^{u*}$ by using weighted data, $i = 1, \cdots, U$;
6   Compute $\mathbf{z}_i^l$ by (1) and $\mathbf{z}_i^u$ by (2);
7   Solve the optimization problem (3) using the gradient descent method to find $\mathbf{v}$;
8   Compute $\mathbf{w}_{(t)} = [v_1^2, \ldots, v_I^2]^T$;
9   Let $\rho = \|\mathbf{w}_{(t)} - \mathbf{w}_{(t-1)}\|$;
10     $t = t + 1$;
11   **end**
12   $\mathbf{w} = \mathbf{w}_{(t)}$;
13   **Return** $\mathbf{w}$.
14 **end**

vector. The computational complexity of MSLIR is $O(cdLU)$ when calculating margin vectors of unlabel samples. MSLIR-NN first assigns a pseudo label for the unlabeled sample and then directly calculate its margin vector as label samples. Compared to MSLIR, MSLIR-NN has a lower complexity, or $O(dLU)$, which is independent of the class number c and identical to SLIR.

3 Experiments

We conduct extensive experiments to demonstrate the efficiency and effectiveness of MSLIR-NN. Ten UCI datasets [20] including Pendigits, Satimage, Waveform, Wine, Vehicle, Iris, Breast, Heart, Wdbc and Pima are adopted, where the first six datasets are multi-class, and the rest four ones are binary. All datasets are randomly divided into training and test subsets, and the training subsets contain labeled and unlabeled samples. A brief description of datasets is listed in Table 1, where "#Training" and "#Test" represent the number of training and test samples, respectively, "#Labeled" and "#Unlabeled" are the number of labeled and unlabeled samples in a training set, "#Feature" represents the dimension of samples, and "#Class" indicates the number of categories in datasets. For each dataset, we add 100 additional noise features which are independently Gaussian distributed. We normalize all features with the original data.

Table 1. Description of ten UCI datasets.

Data sets	#Training		#Test	#Feature	#Class
	#Labeled	#Unlabeled			
Pendigits	20	7474	3498	16(+100)	10
Satimage	30	4405	2000	36(+100)	6
Waveform	30	470	4500	21(+100)	3
Wine	10	40	128	13(+100)	3
Vehicle	20	180	646	18(+100)	4
Iris	10	40	100	4(+100)	3
Breast	20	179	500	9(+100)	2
Heart	40	130	133	13(+100)	2
Wdbc	20	149	400	30(+100)	2
Pima	30	158	580	8(+100)	2

In our experiments, the compared methods include RELIEF-F, LIR, SLIR, MSLIR and MSLIR-NN. Both RELIEF-F and LIR use only labeled data, while SLIR, MSLIR and MSLIR-NN use both labeled and unlabeled data. The classification performance is tested on the same test subsets. In LIR, the regularization parameter λ and learning rate η are 10 and 0.03, respectively. In MSLIR-NN, MSLIR and SLIR, the parameters α and β are 10 and 0.1, respectively, and the learning rate is the same as that of LIR.

To eliminate the effect of statistical error, each algorithm runs 10 times for each dataset, and takes the average result as the final one. In order to ensure the reliability of experimental results, the nearest neighbor (NN) and support vector machine (SVM) classifiers are used in experiments. Here the Gaussian kernel and regularization parameters in SVM are selected by the grid search method, where both vary from 2^{-10} to 2^{10}.

3.1 Experiments on Multi-class Datasets

For multi-classification tasks, we compare the proposed algorithm with other three methods MSLIR, LIR and RELIEF-F. Experiments are implemented on the Pendigits, Satimage, Waveform, Wine, Vehicle and Iris datasets. The performance of these feature selection algorithms are evaluated by the classification accuracy with selected features, and the final experimental results obtained by SVM are shown in Fig. 1. From Fig. 1, we can observe that the classification accuracy of MSLIR-NN is the best among compared methods, which indicates that features selected by MSLIR-NN have a greater correlation with the label information. In Figs. 1(a) and (c), it can be observed that the classification accuracy of the four algorithms increases as increasing the number of features, which demonstrates that the chosen features are all useful features for classification and the noisy features are excluded. In Figs. 1(b), (d), (e) and (f), the classification accuracy tends to be steady or decreasing, which indicates that feature selection is useful. In other words, not all original features are related to classification tasks. For example, in the Wine dataset, when the number of selected features is eight, MSLIR-NN achieves the highest classification accuracy of 98%, and the last four selected features are likely to be irrelevant features.

The performance curves obtained by NN are shown in Fig. 2. From these figures, we can have similar conclusions as those from Fig. 1. We give the best average accuracies and the corresponding standard deviations of four methods in Table 2, where the best results are bolded. Compared with the other three methods, MSLIR-NN has a higher classification accuracy and smaller standard deviation, reflecting that our method has better stability than the previous MSLIR.

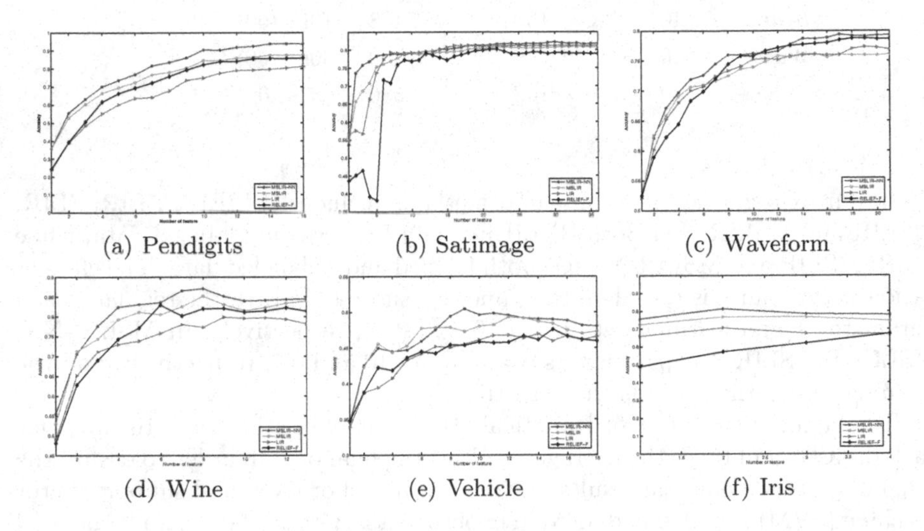

(a) Pendigits (b) Satimage (c) Waveform

(d) Wine (e) Vehicle (f) Iris

Fig. 1. Classification accuracy of SVM using four feature selection methods on multi-class datasets: (a) Pendigits, (b) Satimage, (c) Waveform, (d) Wine, (e) Vehicle and (f) Iris.

Table 2. Classification accuracy and standard deviations (%) of NN using four feature selection methods on multi-class datasets

Data sets	MSLIR-NN	MSLIR	LIR	RELIEF-F
Pendigits	**97.74** ± 0.10	92.22 ± 17.45	90.05 ± 6.13	89.49 ± 4.92
Satimage	**88.55** ± 0.00	79.18 ± 4.23	83.07 ± 1.82	84.21 ± 1.25
Waveform	**70.85** ± 2.90	68.38 ± 2.81	67.17 ± 4.65	69.74 ± 3.26
Wine	**87.11** ± 3.85	85.23 ± 5.37	79.92 ± 8.92	82.97 ± 9.21
Vehicle	**51.35** ± 6.93	48.98 ± 7.41	45.74 ± 8.93	47.77 ± 6.99
Iris	**92.30** ± 5.23	86.10 ± 19.91	79.70 ± 25.79	81.20 ± 23.88

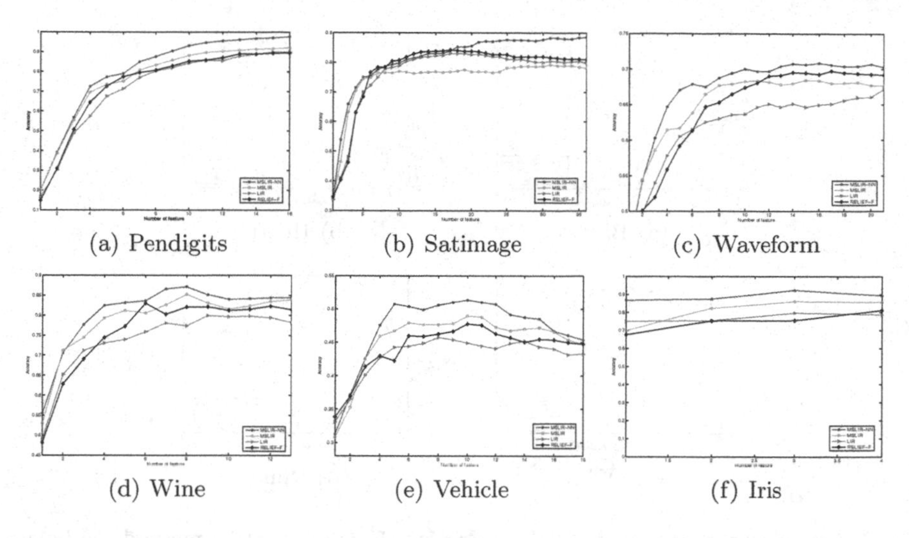

(a) Pendigits (b) Satimage (c) Waveform

(d) Wine (e) Vehicle (f) Iris

Fig. 2. Classification accuracy of NN using four feature selection methods on multi-class datasets: (a) Pendigits, (b) Satimage, (c) Waveform, (d) Wine, (e) Vehicle and (f) Iris.

3.2 Experiments on Binary Datasets

Similar to MSLIR, MSLIR-NN can also be applied to binary classification tasks. Since SLIR is only applicable to binary classification tasks, we compare MSLIR-NN and MSLIR with it on Breast, Heart, Wdbc and Pima datasets.

The experimental results obtained by SVM are given in Fig. 3. We can see that MSLIR-NN is much better than both MSLIR and SLIR, especially in Figs. 3(a), (c) and (d).

The performance curves obtained by NN are shown in Fig. 4. MSLIR-NN still has an advantage over other two methods. We list the best average classification accuracies and corresponding standard deviations in Table 3. Obviously, MSLIR-NN has the best performance among three methods on four binary datasets, followed by MSLIR. Compared to MSLIR, MSLIR-NN is improved 2.03% accuracy on Heart and 5.73% the accuracy on Pima, respectively.

Table 3. Classification accuracy and standard deviations (%) of NN using three feature selection methods on binary datasets

Data sets	MSLIR-NN	MSLIR	SLIR
Breast	**94.74** $\pm$ 1.12	93.86 $\pm$ 1.88	93.00 $\pm$ 3.48
Heart	**74.51** $\pm$ 4.48	72.48 $\pm$ 4.12	69.02 $\pm$ 6.70
Wdbc	**87.43** $\pm$ 3.65	87.18 $\pm$ 2.58	84.30 $\pm$ 7.34
Pima	**67.20** $\pm$ 3.37	61.47 $\pm$ 3.72	60.81 $\pm$ 5.24

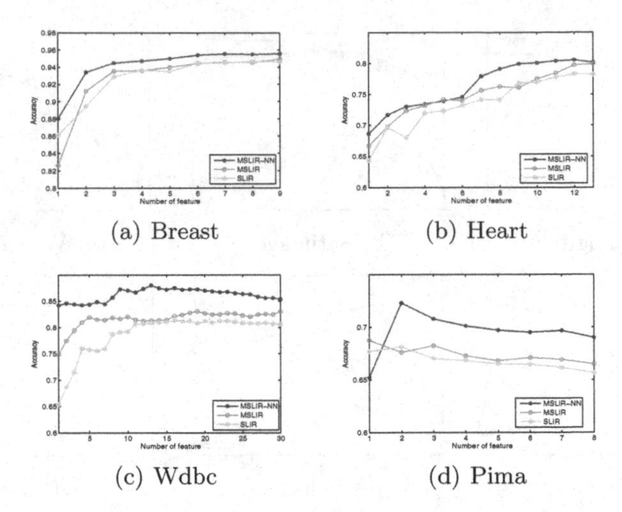

(a) Breast (b) Heart

(c) Wdbc (d) Pima

Fig. 3. Classification accuracy of SVM using four feature selection methods on binary datasets: (a) Breast, (b) Heart, (c) Wdbc, (d) Pima.

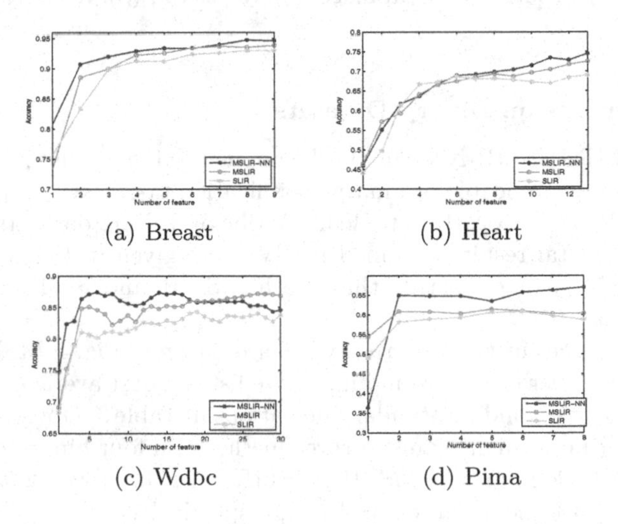

(a) Breast (b) Heart

(c) Wdbc (d) Pima

Fig. 4. Classification accuracy of NN using four feature selection methods on binary datasets: (a) Breast, (b) Heart, (c) Wdbc, (d) Pima.

3.3 Comparison of MSLIR-NN and MSLIR

MSLIR-NN and MSLIR have the same objective function, and different ways for constructing margin vectors of unlabeled samples. Here, we compare them in two aspects, the prediction ability on unlabeled samples in training subsets and the running time of feature selection.

Table 4 lists the accuracy on unlabeled samples and running time of feature selection for two methods. We can see that MSLIR-NN is significantly better than MSLIR on the prediction performance, which indicates that our new proposed

Table 4. Comparison of MSLIR-NN and MSLIR algorithms

Data sets	Accuracy (%)		Running time (sec.)	
	MSLIR-NN	MSLIR	MSLIR-NN	MSLIR
Pendigits	**30.33**	2.98	**55.26**	64.00
Satimage	**48.57**	2.77	**9.86**	31.70
Waveform	**63.15**	6.00	**63.01**	59.93
Wine	**62.00**	5.00	**2.87**	2.48
Vehicle	**35.78**	18.00	**18.93**	10.78
Iris	**84.25**	6.00	**0.84**	5.99
Breast	**90.61**	64.75	**1.75**	14.68
Heart	**67.31**	46.08	**49.59**	53.12
Wdbc	**83.22**	63.75	**1.48**	19.42
Pima	**57.53**	35.70	**0.78**	44.78

algorithm can use a few labeled data to predict the label of unlabeled data. Thus we can calculate the margin vectors of unlabeled samples more accurately, which can improve the stability of algorithm. The computational complexity of algorithms can be reflected by the running time of feature selection. Obviously, MSLIR-NN is much faster than MSLIR on all ten datasets, which supports our analysis about computational complexity in Sect. 2.3.

4 Conclusions

In this paper, we propose MSLIR-NN based on MSLIR for multi-class semi-supervised feature selection by introducing the nearest neighbor scheme. MSLIR-NN has a less complexity than MSLIR, and can improve the accuracy of label prediction for unlabeled data which contributes to the calculation way of margin vectors of unlabeled samples. Extensive experiments are performed on binary and multi-class classification tasks. Two classical classifiers NN and SVM are used to implement classification after feature selection has finished. On multi-class datasets, MSLIR-NN is superior to supervised methods LIR and RELEIF-F, and the semi-supervised method MSLIR. On the binary datasets, MSLIR-NN performs the best among three semi-supervised methods. In experiments of comparison with MSLIR, MSLIR-NN unfolds its ability in predicting labels of unlabeled samples and speedability. In general, MSLIR-NN can extract useful features and achieve better performance.

Acknowledgements. This work was supported in part by the National Natural Science Foundation of China under Grant No. 61373093, by the Soochow Scholar Project of Soochow University, and by the Six Talent Peak Project of Jiangsu Province of China.

References

1. Guyon, I., Elisseeff, A.: An introduction to variable and feature selection. J. Mach. Learn. Res. **3**(6), 1157–1182 (2003)
2. Zhao, Z., Wang, L., Liu, H., Ye, J.: On similarity preserving feature selection. IEEE Trans. Knowl. Data Eng. **25**(3), 619–632 (2013)
3. Benabdeslem, K., Hindawi, M.: Efficient semi-supervised feature selection: constraint, relevance, and redundancy. IEEE Trans. Knowl. Data Eng. **26**(5), 1131–114326 (2014)
4. Zhang, D., Chen, S., Zhou, Z.H.: Constraint score: a new filter method for feature selection with pairwise constraints. Pattern Recogn. **41**(5), 1440–1451 (2008)
5. Sheikhpour, R., Sarram, M.A., Gharaghani, S., et al.: A survey on semi-supervised feature selection methods. Pattern Recogn. **64**(C), 141–158 (2016)
6. Kira, K., Rendell, L.A.: The feature selection problem: traditional methods and a new algorithm. In: Tenth National Conference on Artificial Intelligence, pp. 129–134 (1992)
7. Kononenko, I.: Estimating attributes: analysis and extensions of RELIEF. In: European Conference on Machine Learning on Machine Learning, pp. 171–182 (1994)
8. Sun, Y.: Iterative RELIEF for feature weighting: algorithms, theories, and applications. IEEE Trans. Pattern Anal. Mach. Intell. **29**, 1035–1051 (2007)
9. Cheng, Z.D., Zhang, Y.J., Fan, X., Zhu, B.: Study on discriminant matrices of commonly used fisher discriminant functions. Acta Autom. Sinica **36**(10), 1361–1370 (2010)
10. Chen, L.F., Liao, H.Y.M., Ko, M.T., Lin, J.C., Yu, G.J.: A new LDA based face recognition system which can solve the small sample size problem. Pattern Recogn. **33**(10), 1713–1726 (2000)
11. Peng, H., Long, F., Ding, C.: Feature selection based on mutual information: criteria of max-dependency, max-relevance, and min-redundancy. IEEE Comput. Soc. **27**(8), 1226 (2005)
12. Mitra, P., Murthy, C.A., Pal, S.K.: Unsupervised feature selection using feature similarity. IEEE Trans. Pattern Anal. Mach. Intell. **24**(3), 301–312 (2002)
13. He, X., Cai, D., Niyogi, P.: Laplacian score for feature selection. In: International Conference on Neural Information Processing Systems, vol. 18, pp. 507–514 (2005)
14. Bishop, C.M.: Pattern Recognition and Machine Learning. Springer, New York (2006)
15. Zeng, Z., Wang, X.D., Zhang, J., Wu, Q.: Semi-supervised feature selection based on local discriminative information. Neurocomputing **173**(P1), 102–109 (2016)
16. Cheng, Y., Cai, Y., Sun, Y., Li, J.: Semi-supervised feature selection under the Logistic I-RELIEF framework. In: International Conference on Pattern Recognition, pp. 1–4 (2008)
17. Zhao, Z., Liu, H.: Semi-supervised feature selection via spectral analysis. In: SIAM International Conference on Data Mining, SIAM 2007, pp. 641–646. SIAM, Minneapolis (2007)
18. Xu, J., Tang, B., He, H., Man, H.: Semi-supervised feature selection based on relevance and redundancy criteria. IEEE Trans. Neural Netw. Learn. Syst. **28**(9), 1974–1984 (2016)
19. Tang, B., Zhang, L.: Semi-supervised feature selection based on logistic I-RELIEF for multi-classification. In: Geng, X., Kang, B.-H. (eds.) PRICAI 2018. LNCS (LNAI), vol. 11012, pp. 719–731. Springer, Cham (2018). https://doi.org/10.1007/978-3-319-97304-3_55
20. UCI Machine Learning Repository. http://archive.ics.uci.edu/ml/datasets.html

Effort-Aware Tri-Training for Semi-supervised Just-in-Time Defect Prediction

Wenzhou Zhang[1], Weiwei Li[2], and Xiuyi Jia[1,3](✉)

[1] School of Computer Science and Engineering, Nanjing University of Science and Technology, Nanjing 210094, China
jiaxy@njust.edu.cn

[2] College of Astronautics, Nanjing University of Aeronautics and Astronautics, Nanjing 210016, China

[3] State Key Laboratory for Novel Software Technology, Nanjing University, Nanjing 210023, China

Abstract. In recent years, just-in-time (JIT) defect prediction has gained considerable interest as it enables developers to identify risky changes at check-in time. Previous studies tried to conduct research from both supervised and unsupervised perspectives. Since the label of change is hard to acquire, it would be more desirable for applications if a prediction model doesn't highly rely on the label information. However, the performance of the unsupervised models proposed by previous work isn't good in terms of *precision* and $F1$ due to the lack of supervised information. To overcome this weakness, we try to study the JIT defect prediction from the semi-supervised perspective, which only requires a few labeled data for training. In this paper, we propose an Effort-Aware Tri-Training (EATT) semi-supervised model for JIT defect prediction based on sample selection. We compare EATT with the state-of-the-art supervised and unsupervised models with respect to different labeled rates. The experimental results on six open-source projects demonstrate that EATT performs better than existing supervised and unsupervised models for effort-aware JIT defect prediction.

Keywords: Defect prediction · Just-in-time · Tri-training · Effort-aware

1 Introduction

In order to produce software of high quality, developers have to spend a lot of effort testing and debugging the software. Software defect prediction [15] can infer code segments that may contain defects, which can help developers effectively save testing time and reduce the cost of software development. Most defect prediction studies focused on predicting the defect at coarse granularity level, such as file, package, or module [7,14,20]. In recent years, there is an increasing

Q. Yang et al. (Eds.): PAKDD 2019, LNAI 11440, pp. 293–304, 2019.
https://doi.org/10.1007/978-3-030-16145-3_23

interest in defect prediction at change level. Such prediction was first proposed by Mockus and Weiss [19], and referred as just-in-time (JIT) defect prediction by Kamei et al. [12]. JIT defect prediction tries to identify defect-prone ("risky") software changes, whose data is collected by combining information extracted from the version archive (such as Git) with bug reports. Compared with traditional software defect prediction, such prediction has many advantages. As stated in Ref. [12], they can be summarized as the following three points: (1) Predictions are made at a fine granularity. (2) Predictions can be associated with the developers. (3) Predictions are made earlier. Furthermore, different changes require different amount of effort to review. Kamei et al. [12] proposed effort-aware JIT defect prediction using lines of code (LOC) to indicate the amount of effort required to review changes. This kind of work is more practical as it could help find more defect-inducing changes with the same inspection cost.

Previous JIT defect prediction studies mainly carried out based on supervised or unsupervised models. Many supervised strategies have been proposed to predict the defect-prone changes, such as linear regression [12], logistic regression [8], ensemble learning [21] and other approaches [5,6]. These supervised-based studies have achieved significant high performance for JIT defect prediction. However, the acquisition of labeled data (i.e., whether a change will induce defect or not) is usually quiet expensive and time-consuming. For many new projects, developers have spent a lot of time and effort testing unlabeled changes when the supervised models are capable of predicting new change and the trained model doesn't make the best use of its value. In previous work, many studies [21,23] have leveraged unsupervised model for effort-aware JIT defect prediction and found that many unsupervised models could outperform supervised model in terms of ACC[1] and P_{opt}[2]. But there are many drawbacks in these works. As stated in Ref. [8], many highly ranked changes are false alarms and they don't outperform supervised model when the $F1-score$ is considered. Different from unsupervised models, semi-supervised methods attempt to exploit the intrinsic data distribution information disclosed by the unlabeled data and the information is usually considered to be helpful for learning [4,27]. Many studies have been carried out for the semi-supervised defect prediction at coarser granularity [2,10,13,17,18,24], which shows that the performance of defect prediction can be improved by using unlabeled data. To verify whether unlabeled data is still helpful for JIT defect prediction, we leverage tri-training [25] method to exploit unlabeled data and propose an effort-aware tri-training (EATT) method for semi-supervised JIT defect prediction. Specifically, (1) EATT uses an effort-aware indicator to produce the final hypothesis. (2) Instead of subsampling the unlabeled data randomly, EATT takes the predicted confidence as well as the review budget into account. (3) EATT employs a greedy strategy to rank changes

[1] ACC denotes the recall of defect-inducing changes when using 20% of the entire effort to inspect the top ranked changes.

[2] P_{opt} is the normalized version of the effort-aware performance indicator based on the concept of the "code-churn-based" Alberg diagram. More details could be found in Sect. 4.5.

according to their tendency to be defect-prone of unit effort. To the best of our knowledge, this is the first time to leverage semi-supervised method for effort-aware JIT defect prediction.

The main contributions of this paper are as follows:

(1) We investigate the predictive effectiveness of sample-based semi-supervised models in effort-aware JIT defect prediction and propose an effort-aware tri-training method for semi-supervised JIT defect prediction.
(2) We perform an in-depth evaluation on existing supervised and unsupervised models with cross-validation under different labeled rates.
(3) We compare EATT with several supervised, unsupervised and semi-supervised methods under different evaluation indicators. The experimental results show that EATT can significantly improve the performance of effort-aware JIT defect prediction.

The rest of this paper is organized as follows. Section 2 introduces the related work on defect prediction. Section 3 describes the over-all framework of our method and presents the detail of our approach. Section 4 provides the experimental setup and the research questions in our study. Section 5 gives the answers of these questions and reports the experimental results in detail. Section 6 concludes this paper.

2 Related Work

In this section, we will introduce the related work on JIT defect prediction and some semi-supervised learning applications in traditional defect prediction.

2.1 Just-in-Time Defect Prediction

JIT defect prediction is proposed by Kamei et al. [12], they performed a large-scale empirical study on six open source projects and five commercial projects. Their results showed that they can predict defect-inducing changes with 68% accuracy and 64% recall. Kamei et al. [11] explored cross-project models in the context of JIT prediction, but the JIT models rarely performed well in a cross-project context. Yang et al. [23] found that simple unsupervised models could achieve higher recall than supervised models. Fu et al. [6] improved the related unsupervised models by pruning weaker predictors away. Huang et al. [8] found the unsupervised model required developers to inspect numerous changes and then proposed an improved supervised model. Furthermore, Yang et al. [21, 22] leveraged deep learning techniques and ensemble learning to predict defect-prone changes. Liu et al. [16] found that code churn is a neglected metric and built unsupervised model based on it. Chen et al. [5] proposed a multi-objective optimization based supervised method MULTI.

2.2 Semi-supervised Learning in Traditional Defect Prediction

Semi-supervised model is a machine learning technique trained by utilizing a few labeled and abundant unlabeled data. Semi-supervised approach for software defect prediction has attracted considerable researchers' interest and many approaches have been proposed recently. There are many kinds of methods such as constraint-based semi-supervised clustering [2], sample-based approaches [10,13,17], labeled propagation approach [24], and preprocessing strategies [10,18]. These studies show that semi-supervised methods can improve the performance of the model in traditional defect prediction. To this end, we investigate whether semi-supervised models are still effective when it comes to effort-aware JIT defect prediction. From previous studies [3,25], the unlabeled data could be helpful if they are exploited properly. Many strategies [26] have been proposed to exploit the disagreements among the learners during the semi-supervised learning process. All these studies show that the sample-based semi-supervised methods are capable for effort-aware JIT defect prediction. Since the semi-supervised methods only need a few labeled data, it will be more desirable for application if they still have good performances.

3 Effort-Aware Tri-Training

JIT defect prediction tries to predict whether a committed change is defect-prone or not. Based on the previous study [12], the JIT datasets are relatively imbalanced and most change measures are highly skewed. We adopt the same method as Ref. [12] to deal with these problems.

The tri-training method was proposed by Zhou and Li [25], which attempts to exploit unlabeled data using three classifiers. Although good performance could be obtained by using traditional semi-supervised methods, they are not suitable for effort-aware JIT defect prediction as such methods are designed for classification tasks and do not generalize to effort-aware ranking tasks. For JIT defect prediction, the effort is measured with the size of change (i.e., the total number of modified lines) [12]. In our work, we follow the previous work [5,12,23] only use 12 metrics (excluding LA and LD) to build semi-supervised model, and the LA (lines of code added) and LD (lines of code deleted) will be combined to make up the effort value in predicting the risk value of changes. We propose an effort-aware tri-training (EATT) method for JIT defect prediction and the overall framework is presented in Fig. 1.

3.1 Problem Formulation

Let L denote the labeled software changes (i.e., whether these changes will induce defect or not are already known) and U denote the unlabeled software changes. Let X denote the new changes whose labels we try to predict and the function $Effort(x)$ denote the amount of effort required by the change x ($x \in X$), i.e., the number of modified lines. Assume there are three classifier h_1, h_2 and h_3, EATT tries to exploit the unlabeled data by employing these classifiers and taking the inspect effort into account.

3.2 Model Evaluation

In order to discover more defect-inducing changes under given inspection budget, the recall of defect-inducing changes with a fixed effort is used to produce the final hypothesis in our work. Although the classification accuracy may decrease, EATT could find more defect-inducing changes by ranking the changes according to their defect density.

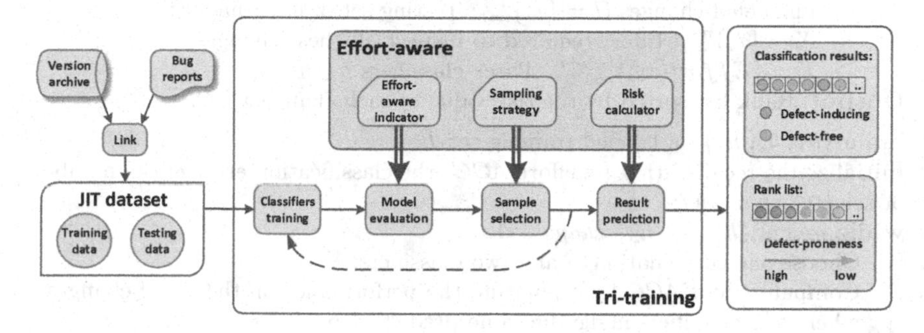

Fig. 1. Illustration of EATT for JIT defect prediction. (1) The JIT dataset is extracted from the combination of version archive and bug reports. (2) Then our semi-supervised model is trained by utilizing the EATT method. (3) Finally, the model outputs classification or ranking results.

3.3 Sample Selection

Let e^t and $|L^t|$ denote the classification error rate and the number of changes introduced into the labeled data in the t-th round. $L' \subseteq U$ represents the set of changes on which two classifiers have same prediction results. From the study of [1,25], as long as the number of samples introduced per round satisfies $|L^t| < e^{t-1}|L^{t-1}|/e^t$, the negative impact of introducing noise can be compensated by the newly sufficient labeled examples. In the actual learning process, $|L'|$ is usually larger than the $e^{t-1}|L^{t-1}|/e^t$. We prefer to select the changes that have higher confidence and extremism effort rather than randomly as previous studies [16,23] show that "smaller" changes tend to be more proportionally defect-prone.

3.4 Result Prediction

For a new change x, the label (i.e., defect-inducing or defect-free) is predicted by majority voting. When it comes to the effort-aware JIT defect prediction, the risk value $R(x)$ is predicted as follows:

$$R(x) = \frac{\sum_i p_i(x)}{Effort(x)}. \tag{1}$$

Where $p_i(x)$ denotes the probability to be defect-inducing predicted by h_i. By adding the probabilities to be defect-inducing predicted by three classifiers, the confidence of all three classifiers can be reflected on the rank list.

The pseudo-code of the EATT is shown in Algorithm 1.

Algorithm 1. Effort-Aware Tri-Training for JIT Defect Prediction

Input: Training set of labeled changes $L = \{x_i, y_i\}_{i=1}^{l}$; Training set of unlabeled changes $U = \{x_i\}_{i=1}^{u}$; Testing set of new changes $X = \{x_i\}_{i=1}^{n}$; Effort required to inspect all these changes $E = \{Effort(x_i)\}_{i=1}^{l+u+n}$; Three classifiers h_1, h_2, h_3;

Output: Rank list sorted by the risk value of each change $x \in X$.

```
1  Train {h1, h2, h3} on labeled training set L;
2  Initialize the recall with a fix effort ACC; the classification error e; the number
   of selected changes l;
3  while any of {h1, h2, h3} changes do
4      Choose one combination of any two classifiers.
5      Compute e and ACC by evaluating the performance on the set of changes
       where two classifiers make the same prediction on L;
6      if ACC is greater than that in the last round then
7          Get the set of changes L' where the corresponding two classifiers make
           the same prediction on U;
8          Compute the number of changes to be selected;
9          Subsample the L' by selecting high confident changes;
10     end
11     Update ACC, e and l;
12     Retrain the remaining classifier on the new labeled data L ⋃ L';
13 end
14 Compute the risk value R(x) of each change x ∈ X by Eq.(1).
```

4 Experiment Setup

In this section, we first describe the statistics of datasets used in our experiments. Then, we introduce the baseline models in our study. Next, we present the evaluation methods used to analyze datasets. After that, we describe the performance indicators used to evaluate the performance of defect prediction models. Finally, the research questions are presented.

4.1 Datasets

In this work, we conduct semi-supervised learning experiments on six large open source projects. These projects include Bugzilla (BUG), Columba (COL), Eclipse JDT (JDT), Eclipse Platform (PLA), Mozilla (MOZ) and PostgreSQL (POS), which are shared by Kamei et al. [12]. Table 1 summarizes the statistics of the studied projects, including the total number of changes, the period of the project and the percent of defect-inducing changes.

Table 1. Statistics of the studied projects.

Project	Period	#Changes	%Defect
BUG	1998/08/26–2006/12/16	4620	36%
COL	2002/11/25–2006/07/27	4455	30%
JDT	2001/05/02–2007/12/31	35386	14%
PLA	2001/05/02–2007/12/31	64250	14%
MOZ	2000/01/01–2006/12/17	98275	5%
POS	1996/07/09–2010/05/15	20431	25%

4.2 Baseline Models

In this study, we compare EATT against the state-of-the-art supervised models and unsupervised models. The supervised model including EALR [12], TLEL [21], CBS [8], OneWay (OW) [6], MULTI [5], and the unsupervised model including LT [23] and CCUM [16]. EALR is a linear regression based effort-aware JIT defect prediction model. TLEL applies tow-layer ensemble strategy to build JIT defect prediction model. CBS is a supervised model by sorting changes after classifying. OneWay is a supervised model based on the simple unsupervised learners. MULTI is a supervised JIT defect prediction model based on multi-objective optimization. The LT model is an unsupervised model by ranking changes according to the measure *LT*. CCUM is built by ranking changes according to code churn (i.e., LA+LD). In addition, we also compare EATT against other sample-based semi-supervised methods in traditional defect prediction including CoForest (CF) [13] and FTF [17]. CoForest labels changes based on the random forest method and produce the final hypothesis. FTF is a semi-supervised learner based on self-training.

4.3 Classifiers Selection

In our experiment, three different classifiers are selected as our base models. They are logistic regression, support vector machine and random forest classifier. The reasons for selecting these classifiers are three-fold. First, they are popular and mature classification techniques and easy to be implemented. Second, there is a large disagreement between these classifiers while their performance are comparable, which is considered beneficial for learning [26]. Third, these classifiers are also widely used in previous JIT defect prediction researches [5,8,12,21,22]. Our implementation[3] is based on *python*. All the parameters of the classifiers use the default values provided by *sklearn* package. In order to make a fair comparison, the parameters of classifiers adopted by other baseline models are the same.

[3] The data and code used in this paper are available at https://github.com/NJUST-IDAM/EATT.

4.4 Evaluation Strategy

In our experiment, we use the 10 times 10-fold cross-validation, which is also used in previous works [5,16,23]. As we try to apply semi-supervised method for JIT defect prediction, the training data is a little different from previous work by removing the label of training data randomly. In particular, for 10 times 10-fold cross-validation, the training data is created by selecting a small rate of samples as labeled data while the other as unlabeled data. After re-sampling the training data, we select samples under labeled rates at 0.1 and 0.2, which are also adopted in traditional semi-supervised defect prediction [2,24]. Note that the number of labeled positive samples (i.e., defect-inducing changes) is the same as that of labeled negative samples (i.e., defect-free changes). After that, the training data consists of a few labeled data and abundant unlabeled data.

4.5 Performance Indicators

Five performance indicators are used to evaluate the performance of effort-aware JIT defect prediction models. They are *precision*, *recall*, $F1$, ACC and P_{opt}.

The *precision*, *recall* and $F1$ are commonly-used indicators to evaluate classification performance and also used by many previous studies [12,16,21,22]. The ACC and P_{opt} are used to evaluate ranking performance by taking the effort into account [5,6,12,16,23]. ACC denotes the recall of defect-inducing changes when using 20% of the entire effort to inspect the top ranked changes. P_{opt} is defined as $1 - \Delta_{opt}$, where Δ_{opt} is the area between the effort-based cumulative lift chart of the optimal model and the prediction model.

4.6 Research Questions

To evaluate the effectiveness of our semi-supervised method EATT, we investigate the following three research questions.

- RQ1: Could the performance of supervised model be improved with the help of unlabeled data?
- RQ2: How about the performance of EATT when compared to the unsupervised method?
- RQ3: Does EATT perform better than other sample-based semi-supervised approaches?

5 Experimental Results and Analysis

In this section, we answer our research questions and report the experimental results.

RQ1: Could the performance of supervised model be improved with the help of unlabeled data?

To answer this question, we compare EATT with five state-of-the-art supervised methods under different evaluation strategies as described in Sect. 4.4.

Tables 2, 3 show the comparison results of different methods. The number in gray cell indicates that the corresponding model has an obvious advantage over EATT method with respect to corresponding evaluation indicator according to the Cliff's δ [9], where $|\delta| \geq 0.147$. All the supervised and semi-supervised models are trained based on the same labeled data and the semi-supervised models also utilize the unlabeled data.

Table 2. The performance of the compared semi-supervised models and some supervised models

Indicator	Project	0.1 labeled rate						0.2 labeled rate					
		supervised			semi			supervised			semi		
		CBS	**TLEL**	**OW**	**CF**	**FTF**	**EATT**	**CBS**	**TLEL**	**OW**	**CF**	**FTF**	**EATT**
Precision	BUG	.5143	.5162	.3556	.5045	**.5761**	.5330	.5226	.5319	.3522	.5075	**.6037**	.5424
	COL	.4804	.4582	.2663	.4326	**.5221**	.4893	.4828	.4684	.2657	.4434	**.5254**	.4947
	JDT	.2569	.2436	.1155	.2123	**.2871**	.2707	.2531	.2501	.1153	.2168	**.2980**	.2726
	MOZ	.1299	.1275	.0319	.1080	**.1711**	.1436	.1331	.1321	.0330	.1073	**.1751**	.1483
	PLA	.2563	.2644	.1172	.2343	**.3186**	.2702	.2578	.2714	.1157	.2409	**.3296**	.2716
	POS	.4831	.4502	.1889	.4141	**.5409**	.4953	.4892	.4575	.1830	.4304	**.5534**	.5026
	AVG	.3535	.3433	.1792	.3176	**.4027**	.3670	.3564	.3519	.1775	.3244	**.4142**	.3720
Recall	BUG	**.6465**	.6219	.4231	.4602	.4320	.6232	**.6566**	.6447	.4197	.4987	.4667	.6436
	COL	.6252	**.6498**	.6398	.4591	.4823	.5905	.6193	**.6610**	.6523	.5077	.5032	.6016
	JDT	.6442	**.6579**	.5372	.4840	.4735	.6201	.6435	**.6665**	.5407	.5027	.4849	.6279
	MOZ	.6211	**.6880**	.2711	.5292	.5470	.6263	.6204	**.6984**	.3043	.5535	.5636	.6331
	PLA	**.7012**	.6952	.5010	.5104	.5098	.6951	.6959	.6988	.5049	.5262	.5136	**.7018**
	POS	.6262	**.6574**	.4970	.5140	.5220	.6150	.6223	**.6679**	.5245	.5085	.5347	.6193
	AVG	.6441	**.6617**	.4782	.4928	.4944	.6284	.6430	**.6729**	.4911	.5162	.5111	.6379
*F*1	BUG	.5717	.5625	.3744	.4775	.4905	**.5718**	.5811	.5816	.3725	.5015	.5242	**.5870**
	COL	**.5414**	.5362	.3743	.4429	.4973	.5319	.5410	**.5470**	.3756	.4714	.5111	.5410
	JDT	.3669	.3553	.1892	.2945	.3563	**.3764**	.3631	.3635	.1878	.3026	.3686	**.3798**
	MOZ	.2147	.2150	.0567	.1784	**.2600**	.2335	.2190	.2222	.0592	.1791	**.2669**	.2402
	PLA	.3753	.3830	.1892	.3207	**.3913**	.3890	.3761	.3908	.1877	.3300	**.4011**	.3915
	POS	.5445	.5339	.2704	.4542	.5297	**.5477**	.5470	.5427	.2702	.4573	.5430	**.5543**
	AVG	.4358	.4310	.2424	.3614	.4208	**.4417**	.4379	.4413	.2422	.3736	.4358	**.4490**
ACC	BUG	.5436	.5006	.4231	.5794	.5727	**.7632**	.5484	.5164	.4197	.6157	.6330	**.7594**
	COL	.5215	.5217	.6398	.6192	.5867	**.7986**	.5217	.5386	.6523	.6649	.6594	**.8058**
	JDT	.5450	.5351	.5372	.6084	.5860	**.7449**	.5469	.5522	.5407	.6466	.6447	**.7540**
	MOZ	.4441	.4986	.2711	.5333	.5242	**.6239**	.4454	.5120	.3043	.5602	.5639	**.6265**
	PLA	.6144	.5961	.5010	.6449	.6463	**.7840**	.6121	.6059	.5049	.6783	.6929	**.7856**
	POS	.4975	.4976	.4970	.5786	.5621	**.7175**	.4953	.5079	.5245	.6097	.6104	**.7239**
	AVG	.5277	.5250	.4782	.5940	.5796	**.7387**	.5283	.5388	.4911	.6292	.6340	**.7425**
P_{opt}	BUG	.7120	.5953	.7444	.7023	.7077	**.9242**	.7302	.6183	.7419	.7569	.7819	**.9247**
	COL	.6219	.5984	.8361	.7182	.6863	**.9336**	.6194	.6121	.8495	.7706	.7626	**.9349**
	JDT	.6547	.6156	.7831	.7239	.6945	**.8847**	.6525	.6271	.7837	.7608	.7486	**.8881**
	MOZ	.6216	.6165	.6396	.6897	.6751	**.8212**	.6221	.6283	.6485	.7269	.7211	**.8228**
	PLA	.7043	.6588	.7691	.7394	.7277	**.8993**	.7022	.6664	.7715	.7758	.7793	**.9005**
	POS	.6310	.6018	.7870	.7201	.6934	**.9035**	.6250	.6122	.8017	.7588	.7531	**.9057**
	AVG	.6576	.6144	.7599	.7156	.6974	**.8944**	.6586	.6274	.7661	.7583	.7578	**.8961**

From the results showed in Tables 2, 3, CBS and TLEL have a good performance over EATT with respect to *recall*, but when considering the harmonic mean of *precision* and *recall* (i.e., $F1$), they have no obvious advantage anymore. When it comes to effort-aware JIT defect prediction, EATT outperforms all the supervised models on six datasets in terms of ACC and P_{opt}.

Table 3. The performance of the compared unsupervised models and other supervised models

Indicator	Project	0.1 labeled rate					0.2 labeled rate				
		Supervised		Unsupervised		Semi	Supervised		Unsupervised		Semi
		EALR	MULTI	LT	CCUM	EATT	EALR	MULTI	LT	CCUM	EATT
ACC	BUG	.3870	.6960	.4788	.7516	**.7632**	.3905	.6960	.4744	.7465	**.7594**
	COL	.3897	.6960	.6036	.7982	**.7986**	.4031	.6960	.6043	.8048	**.8058**
	JDT	.1953	.6260	.5583	.7302	**.7449**	.2167	.6260	.5724	.7404	**.7540**
	MOZ	.1534	.5110	.3587	.5883	**.6239**	.1570	.5110	.3649	.5914	**.6265**
	PLA	.2998	.6840	.5150	.7606	**.7840**	.2951	.6840	.5150	.7601	**.7856**
	POS	.2698	.6130	.5231	.6985	**.7175**	.2749	.6130	.5304	.7044	**.7239**
	AVG	.2825	.6380	.5063	.7212	**.7387**	.2896	.6380	.5102	.7246	**.7425**
P_{opt}	BUG	.6986	.8830	.7516	.9154	**.9242**	.7139	.8830	.7494	.9140	**.9247**
	COL	.6003	.8800	.8275	**.9339**	.9336	.6035	.8800	.8283	.9346	**.9349**
	JDT	.4918	.8290	.7905	.8781	**.8847**	.4959	.8290	.7943	.8807	**.8881**
	MOZ	.4662	.7570	.6559	.8044	**.8212**	.4680	.7570	.6584	.8051	**.8228**
	PLA	.5557	.8530	.7735	.8897	**.8993**	.5538	.8530	.7741	.8900	**.9005**
	POS	.5126	.8430	.8003	.8950	**.9035**	.5204	.8430	.8039	.8964	**.9057**
	AVG	.5542	.8410	.7665	.8861	**.8944**	.5593	.8410	.7681	.8868	**.8961**

Therefore, we can conclude that with the help of unlabeled data, EATT is comparable with the supervised models for predicting defect-inducing changes and significantly outperforms than almost all the supervised models for effort-aware JIT defect prediction.

RQ2: How about the performance of EATT when compared to the unsupervised method?

To answer this question, we compare EATT with two unsupervised models, including LT and CCUM. Since the two unsupervised models are not specifically trained for predicting binary results, we haven't compared them on indicators *precision*, *recall* and *F*1.

Table 3 shows the performance of LT and CCUM against EATT. As can be seen, EATT can further improve the performance by using a few labeled data. Furthermore, it is worth noting that although EATT only slightly increases the performance under *ACC* and P_{opt}, it is still comparable to supervised models when the truly effort isn't available, while CCUM heavily relies on the predefined effort (i.e., LOC) as foundation for sorting, which may not be the same in the real world.

Overall, the above analysis shows that EATT are more capable for practical application when predicting either defect-inducing changes or risk value of changes.

RQ3: Does EATT perform better than other sample-based semi-supervised approaches?

To answer this question, we compare EATT with CoForest and FTF, which are sample-based semi-supervised models for traditional defect prediction. We use the similar strategy to predict the risk value of defect-inducing changes, that is, dividing the predicted probability to be defect-inducing by the effort required to inspect this change.

From the results shown in Table 2, FTF is significantly better than EATT in terms of *precision* according to cliff's δ. But EATT has a better average score over six projects in terms of $F1$ as FTF performs not so good with respect to *recall*.

When it comes to effort-aware JIT defect prediction, CoForest and FTF all have great improvement as the labeled rate increased, but EATT can achieve high performance even though the labeled data is insufficient. In addition, EATT always significantly outperforms CoForest and FTF under ACC and P_{opt} by applying different evaluation strategies according to cliff's δ.

Overall, the above observations show that EATT still has better performance when compared with other sample-based semi-supervised models.

6 Conclusion

In this paper, we investigate the predictive effectiveness of semi-supervised learning for effort-aware JIT defect prediction. Based on tri-training, we propose an effort-aware tri-training (EATT) method for JIT defect prediction. Several comparison experiments are conducted to demonstrate the effectiveness of our proposed method. The experimental results on six projects show that EATT has a great advantage over supervised and unsupervised models. On the one hand, EATT can significantly improve the ability of detecting defect-inducing changes while ensuring high prediction accuracy. It outperforms almost all supervised and unsupervised models in terms of ACC and P_{opt}. On the other hand, EATT still has the advantage as unsupervised models do, requiring only a few labeled data for training. Therefore, semi-supervised learning is more desirable for practical applications since it can effectively combine the advantages of supervised learning and unsupervised learning.

Acknowledgment. This paper is supported by the National Natural Science Foundations of China (Grant Nos. 61773208, 71671086), the Natural Science Foundation of Jiangsu Province (Grant No. BK20170809) and the China Postdoctoral Science Foundation (Grant No. 2018YFB1003902).

References

1. Angluin, D., Laird, P.D.: Learning from noisy examples. Mach. Learn. **2**(4), 343–370 (1987)
2. Arshad, A., Riaz, S., Jiao, L., Murthy, A.: Semi-supervised deep fuzzy c-mean clustering for software fault prediction. IEEE Access **6**, 25675–25685 (2018)
3. Blum, A., Mitchell, T.M.: Combining labeled and unlabeled data with co-training. In: Proceedings of COLT, pp. 92–100 (1998)
4. Chapelle, O., Scholkopf, B., Zien, A.: Semi-supervised learning. IEEE Trans. Neural Netw. **20**(3), 542–542 (2006)
5. Chen, X., Zhao, Y., Wang, Q., Yuan, Z.: MULTI: multi-objective effort-aware just-in-time software defect prediction. Inf. Softw. Tech. **93**, 1–13 (2018)

6. Fu, W., Menzies, T.: Revisiting unsupervised learning for defect prediction. In: ESEC/FSE, pp. 72–83 (2017)
7. Hata, H., Mizuno, O., Kikuno, T.: Bug prediction based on fine-grained module histories. In: ICSE, pp. 200–210 (2012)
8. Huang, Q., Xia, X., Lo, D.: Supervised vs unsupervised models: a holistic look at effort-aware just-in-time defect prediction. In: ICSME, pp. 159–170 (2017)
9. Romano, J., Kromrey, J.D., Coraggio, J., Skowronek, J., Devine, L.: Exploring methods for evaluating group differences on the NSSE and other surveys: are the t-test and Cohen's d indices the most appropriate choices. In: Annual Meeting of the Southern Association for Institutional Research (2006)
10. Jiang, Y., Li, M., Zhou, Z.: Software defect detection with rocus. J. Comput. Sci. Technol. **26**(2), 328–342 (2011)
11. Kamei, Y., Fukushima, T., McIntosh, S., Yamashita, K., Ubayashi, N., Hassan, A.E.: Studying just-in-time defect prediction using cross-project models. Empir. Softw. Eng. **21**(5), 2072–2106 (2016)
12. Kamei, Y., et al.: A large-scale empirical study of just-in-time quality assurance. IEEE Trans. Softw. Eng. **39**(6), 757–773 (2013)
13. Li, M., Zhang, H., Wu, R., Zhou, Z.: Sample-based software defect prediction with active and semi-supervised learning. Autom. Softw. Eng. **19**(2), 201–230 (2012)
14. Li, W., Huang, Z., Li, Q.: Three-way decisions based software defect prediction. Knowl.-Based Syst. **91**, 263–274 (2016)
15. Li, Z., Jing, X., Zhu, X.: Progress on approaches to software defect prediction. IET Softw. **12**(3), 161–175 (2018)
16. Liu, J., Zhou, Y., Yang, Y., Lu, H., Xu, B.: Code churn: a neglected metric in effort-aware just-in-time defect prediction. In: ESEM, pp. 11–19 (2017)
17. Lu, H., Cukic, B., Culp, M.V.: An iterative semi-supervised approach to software fault prediction. In: PROMISE, pp. 15:1–15:10 (2011)
18. Lu, H., Cukic, B., Culp, M.V.: Software defect prediction using semi-supervised learning with dimension reduction. In: ASE, pp. 314–317 (2012)
19. Mockus, A., Weiss, D.M.: Predicting risk of software changes. Bell Labs Tech. J. **5**(2), 169–180 (2000)
20. Song, Q., Jia, Z., Shepperd, M.J., Ying, S., Liu, J.: A general software defect-proneness prediction framework. IEEE Trans. Softw. Eng. **37**(3), 356–370 (2011)
21. Yang, X., Lo, D., Xia, X., Sun, J.: TLEL: a two-layer ensemble learning approach for just-in-time defect prediction. Inf. Softw. Tech. **87**, 206–220 (2017)
22. Yang, X., Lo, D., Xia, X., Zhang, Y., Sun, J.: Deep learning for just-in-time defect prediction. In: QRS, pp. 17–26 (2015)
23. Yang, Y., et al.: Effort-aware just-in-time defect prediction: simple unsupervised models could be better than supervised models. In: FSE, pp. 157–168 (2016)
24. Zhang, Z., Jing, X., Wang, T.: Label propagation based semi-supervised learning for software defect prediction. Autom. Softw. Eng. **24**(1), 47–69 (2017)
25. Zhou, Z., Li, M.: Tri-training: exploiting unlabeled data using three classifiers. IEEE Trans. Knowl. Data Eng. **17**(11), 1529–1541 (2005)
26. Zhou, Z., Li, M.: Semi-supervised learning by disagreement. Knowl. Inf. Syst. **24**(3), 415–439 (2010)
27. Zhu, X.: Semi-supervised learning. In: Encyclopedia of Machine Learning and Data Mining, pp. 1142–1147 (2017)

One Shot Learning with Margin

Xianchao Zhang[1,2], Jinlong Nie[1,2], Linlin Zong[1,2], Hong Yu[1,2], and Wenxin Liang[3(✉)]

[1] School of Software, Dalian University of Technology, Dalian 116620, China
[2] Key Laboratory for Ubiquitous Network and Service Software of Liaoning Province, Dalian 116024, China
[3] School of Software Engineering, Chongqing University of Posts and Telecommunications, Chongqing 400065, China
wxliang@cqupt.edu.cn

Abstract. One shot learning is a task of learning from a few examples, which poses a great challenge for current machine learning algorithms. One of the most effective approaches for one shot learning is metric learning. But metric-based approaches suffer from data shortage problem in one shot scenario. To alleviate this problem, we propose one shot learning with margin. The margin is beneficial to learn a more discriminative metric space. We integrate the margin into two representative one shot learning models, prototypical networks and matching networks, to enhance their generalization ability. Experimental results on benchmark datasets show that margin effectively boosts the performance of one shot learning models.

Keywords: One shot learning · Metric learning · Meta learning

1 Introduction

One shot learning [14,31] is a task that algorithms are able to learn new classes with only a few examples, and it is an ability that human beings naturally have. For instance, human can recognize "elephant" by providing only one example. Whereas current machine learning algorithms require at least hundreds of examples to learn a new concept. One shot learning is proposed to investigate algorithms with such kind of ability. In recent years, a large number of studies [3,14,19,20,22,23,26,28,30,31] have been proposed on the one shot learning problem. Among these studies, metric-based approaches, such as matching networks [31] and prototypical networks [28], show outperforming results.

The goal of metric-based approaches is to learn a mapping function. The function maps examples into a low dimensional space. In the space, examples

Supported by National Science Foundation of China (No. 61632019; No. 61876028; No. 61806034) and Foundation of Department of Education of Liaoning Province (No. L2015001).

Q. Yang et al. (Eds.): PAKDD 2019, LNAI 11440, pp. 305–317, 2019.
https://doi.org/10.1007/978-3-030-16145-3_24

in the same class cluster around each other, while examples belong to different classes distribute far away. The quality of learned mapping determines the effectiveness of metric learning algorithms.

However, due to the shortage of examples in one shot learning, learning a high-quality mapping is rather hard. A few examples cannot describe a class accurately. Based on that, we suggest improving the low dimensional representation of examples benefits one shot learning. A series of literature on metric learning [6,16,17,27,32] imply that margin-based loss often learns metric spaces with more distinct cluster structure. Moreover, the margin can enhance the generalization ability of the classifier from the training set to the testing set [2]. Inspired by these ideas, we propose one shot learning with margin. By introducing the margin, we expect that one shot learning models learn a more discriminative metric space. Thus the generalization error is reduced.

In this paper, we propose a margin-based loss function for one shot learning called multi-way contrastive loss. The loss explores the relationship between an example and multiple examples. The loss minimizes the inner-class distance and maximizes the intra-class distance. We further integrate the proposed multi-way contrastive loss into two representative one shot learning models, prototypical networks and matching networks. Experiments validate that the proposed margin-based loss effectively boosts the performance of prototypical networks and matching networks.

2 One Shot Learning with Margin

2.1 One Shot Learning

One shot learning is a task in which a classifier must accommodate new classes not appeared in training, given only a few examples of each of these classes. Simply re-train the model on new data would severely overfit. The basic idea of current solutions is to train a meta-classifier which can adapt to new classes with a few examples. One shot learning consists of two stages: training and evaluating. Training dataset and evaluating dataset share no common labels, to ensure the one shot constraint is strictly satisfied.

Training. A widely adopted training scheme for one shot learning is episode training strategy [31]. The episode training strategy consists of many episodes. For each episode, a support set is sampled from training dataset. The support set contains N labeled examples $S = \{(x_1, y_1), \ldots, (x_N, y_N)\}$. $x_i \in \mathbb{R}^D$ denotes an example of D-dimension, and $y_i \in \{1, \ldots, K\}$ is the respective label. Given a query example $\hat{x}$, one shot learning models predict the right label of $\hat{x}$. If the support set contains k examples from each of n classes, i.e., $N = k \times n$, the task is called n-way k-shot task. Typically, k is a small number such as 1 or 5.

Specifically, metric-based approaches attack n-way k-shot task by learning an embedding. The learning process can be summarized to two steps. Firstly, all examples are mapped to a new space. Neural networks are a good choice for mapping function because of its strong expression power. Secondly, a non-parametric method is conducted on the results of the mapping function to classify query examples. In the above two steps, learning proceeds by updating the parameters in the mapping function. Therefore, the quality of the learned mapping determines the effectiveness of one shot learning algorithm.

Evaluating. Evaluating adopts the same form as episode training, except that the data is sampled from evaluating dataset. Specifically, a support set and a query set are both sampled from evaluating dataset. One shot learning models need to infer the labels of the examples in the query set based on the examples in the support set.

Analysis. Due to the lack of examples in one shot scenario, a few examples lead to poor class estimation. For instance, in prototypical networks, a single example may produce a poor estimation of class prototype. To alleviate this problem, we propose to learn a discriminative metric space. In other words, we enforce the representation of examples to be discriminative. Examples in the same class are close, while the distance of examples in the different class is large.

2.2 One Shot Learning with Margin

A series of literature on metric learning [6,16,17,27,32] imply that margin-based loss often learns a more discriminative metric space. Moreover, the margin in loss function can enhance the generalization ability of the classifier [2]. Inspired by these ideas, we propose one shot learning with margin. However, existing margin-based loss cannot be directly applied to one shot learning. Therefore, we design a novel loss for one shot learning scenario, which is multi-way contrastive loss.

The proposed multi-way contrastive loss is derived from contrastive loss [6]. Multi-way contrastive loss recruits multiple examples as reference for each update. In this case, an input example is being compared against multiple examples from multiple classes. And the input example should be distinguishable from examples with different labels, while being close to examples with the same label.

In short, the proposed loss explores the relationship between an example (query example) and a set of examples (examples in the support set). To be specific, let $\hat{x}$ denote the query example, $S = \{x_1, \ldots, x_N\}$ denotes the examples in the support set. $\mathbf{1}$ is an indicator of N dimension, the i-th element is defined as follows,

$$\mathbf{1}_i = \begin{cases} 1 \text{ if } \hat{x} \text{ and } x_i \text{ belong to the same class,} \\ 0 \text{ otherwise.} \end{cases} \tag{1}$$

Multi-way contrastive loss consists of two terms, the one acts to pull examples in the same class together, and the other which acts to push differently labeled examples further apart. These two terms have competing effects, since the first

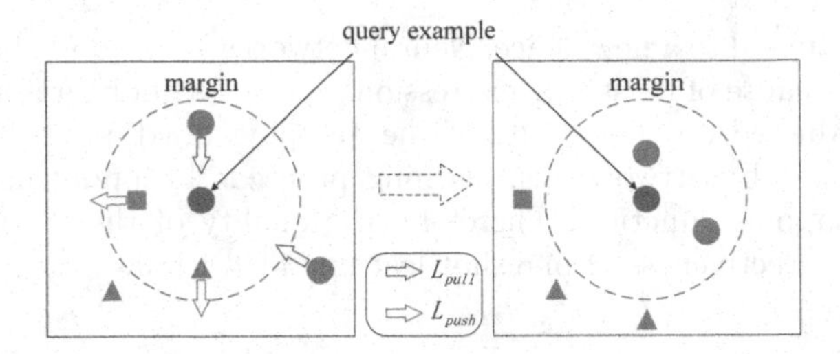

Fig. 1. Illustration of an update of multi-way contrastive loss. Solid objects of different shape indicate examples of different class. Red object in center denotes the query example, blue objects are examples in the support set. Arrows indicate the pull and push force generated by optimizing the cost function. (Color figure online)

is reduced by shrinking the distances between examples while the second is generally reduced by magnifying them. The pulling force tries to minimize the distance between the query example $\hat{x}$ and support examples S in the same class,

$$L_{pull} = \sum_{i=1}^{N} \mathbf{1}_i d_\theta^2(\hat{x}, x_i), \tag{2}$$

where $d_\theta(\cdot,\cdot)$ refers to the distance between two examples, and θ denotes the parameters of distance metric.

The pushing force constrains the distance between the query example $\hat{x}$ and support examples S in different classes to be larger than a margin,

$$L_{push} = \sum_{i=1}^{N} (1 - \mathbf{1}_i) \left[m - d_\theta(\hat{x}, x_i)\right]_+^2, \tag{3}$$

where m is a margin parameter imposing the distance between examples from different classes to be larger than m, and $[x]_+$ denotes the operation of $max(x, 0)$. $d_\theta(\cdot,\cdot) \geq m$ leads to $L_{push} = 0$, otherwise $L_{push} > 0$ and loss will backpropagate to enlarge their distance, through updating θ.

The multi-way contrastive loss function is a weighted combination of L_{pull} and L_{push},

$$L_{MCL} = \frac{1}{N}\left(\alpha \sum_{i=1}^{N} \mathbf{1}_i d_\theta^2(\hat{x}, x_i) + (1-\alpha) \sum_{i=1}^{N} (1 - \mathbf{1}_i) \left[m - d_\theta(\hat{x}, x_i)\right]_+^2\right), \tag{4}$$

where α is a weighting parameter to balance pulling force and pushing force, and it is tuned via cross validation. The gradient of the first summation term in our loss generates a pulling force on support examples with the same label as the query example, while the second summation term imposes a margin between examples labeled differently with the query example. Figure 1 illustrates the update of multi-way contrastive loss. The proposed multi-way contrastive loss can be easily integrated into one shot learning models. Two cases are provided in Sect. 3.

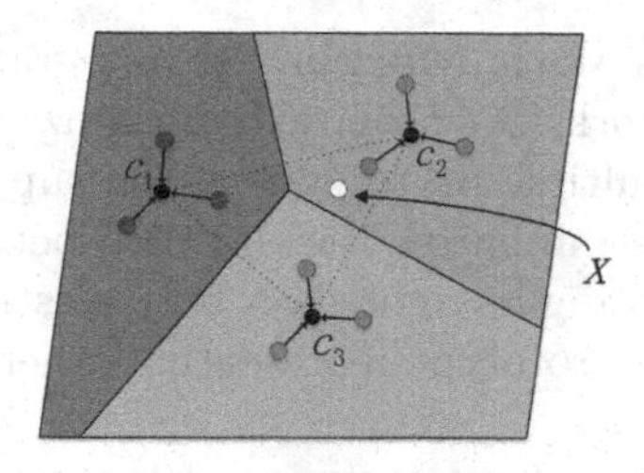

Fig. 2. Structure of prototypical networks. Prototypes c_k are the mean of embedded examples in the support set for each class, and a query example $\hat{x}$ is classified via a softmax over distances to class prototypes: $p_\theta(\hat{y} = k|\hat{x}) \propto \exp(-d(f_\theta(\hat{x}), c_k))$.

3 Case Study

In this section, we briefly introduce two representative one shot learning models, prototypical networks and matching networks. And we present solutions to integrate the proposed margin-based loss into them.

3.1 Prototypical Networks

Model. Prototypical networks [28] are newly proposed frameworks for one shot learning. They are composed by an embedding module and a classification module. The embedding module is a simple convolution network. The classification module represents each class in the support set by a prototype. And the query example is labeled as the nearest prototype. Figure 2 shows the structure of prototypical networks.

Let $f_\theta : \mathbb{R}^D \rightarrow \mathbb{R}^M$ denotes the embedding network with parameters θ. f_θ maps all examples into an M-dimensional space, and θ is the parameters to be optimized. Each prototype c_k is the mean vector of the embedded examples in the support set belonging to its class,

$$c_k = \frac{1}{|S_k|} \sum_{(x_i, y_i) \in S_k} f_\theta(x_i), \tag{5}$$

where S_k denotes the set of support examples labeled with class k, $|S_k|$ is the size of S_k. Given a distance function $d : \mathbb{R}^M \times \mathbb{R}^M \rightarrow [0, +\infty)$, prototypical networks produce a distribution over classes for a query example $\hat{x}$ based on a softmax over distances to the prototypes in the embedding space,

$$p_\theta(\hat{y} = k|\hat{x}) = \frac{\exp\left(-d(f_\theta(\hat{x}), c_k)\right)}{\sum_{k'} \exp\left(-d(f_\theta(\hat{x}), c_{k'})\right)}. \tag{6}$$

Learning proceeds by minimizing the cross entropy loss $L_{PN} = -\log p_\theta(\hat{y} = k|\hat{x})$ of the true class k via stochastic gradient descent. When predicting, a query example $\hat{x}$ is labeled as the class of the nearest prototype, that is,

$$\hat{y} = \arg\min_k d(f_\theta(\hat{x}), c_k). \tag{7}$$

Prototypical Networks with Margin. To integrate multi-way contrastive loss into prototypical networks, we propose to optimize the distance between the query example and the multiple prototypes in the support set. In other words, multi-way contrastive loss minimizes the distance between the query example and the prototype representing its right class, and constrain the distance between the query example and the prototype representing other classes to be larger than a margin at the same time.

Let $\{c_1, \ldots, c_K\}$ denotes the prototypes of the support set. $\hat{x}$ is the query example. $\mathbf{1}$ is an indicator of K dimension, the i-th element is defined as follows,

$$\mathbf{1}_i = \begin{cases} 1 \text{ if } \hat{x} \text{ and } c_i \text{ belong to the same class,} \\ 0 \text{ otherwise.} \end{cases} \tag{8}$$

By investigating the relationship between the query example and the prototypes, we formulate prototypical networks with margin as follows,

$$L_{PN\text{-}M} = \frac{1}{K}\left(\alpha \sum_{i=1}^{K} \mathbf{1}_i d^2(f_\theta(\hat{x}), c_i) + (1-\alpha) \sum_{i=1}^{K} (1-\mathbf{1}_i)\left[m - d(f_\theta(\hat{x}), c_i)\right]_+^2\right). \tag{9}$$

The loss in (9) is used for training. When predicting, a query example $\hat{x}$ is labeled according to (7).

3.2 Matching Networks

Model. Matching networks [31] use attention mechanism [1] to infer the label of the query example. The attention mechanism takes two examples as input, and return a scaler as attention output. The output scaler denotes the similarity of input examples. Then, matching networks classify query examples by similarity weighting method. For a query example $\hat{x}$, the attention is a softmax over the distance, i.e.,

$$a(\hat{x}, x_i) = \frac{\exp\left(-d(f_\theta(\hat{x}), f_\theta(x_i))\right)}{\sum_{x_i \in S} \exp\left(-d(f_\theta(\hat{x}), f_\theta(x_i))\right)}, \tag{10}$$

where S is the support set, f_θ is an embedding function.

Then the label of the query example $\hat{x}$ computes as follows,

$$\hat{y} = \arg\max_k \sum_{(x_i, y_i) \in S_k} a(\hat{x}, x_i). \tag{11}$$

Matching Networks with Margin. To integrate multi-way contrastive loss into matching networks, we revise the training loss of matching networks as follows,

$$\begin{aligned} L_{MN\text{-}M} = \frac{1}{N}\Bigg(&\alpha \sum_{i=1}^{N} \mathbf{1}_i d^2(f_\theta(\hat{x}), f_\theta(x_i)) \\ &+ (1-\alpha) \sum_{i=1}^{N} (1 - \mathbf{1}_i)\left[m - d(f_\theta(\hat{x}), f_\theta(x_i))\right]_+^2\Bigg). \end{aligned} \tag{12}$$

where $\mathbf{1}_i$ is 1 when $\hat{x}$ and x_i belong to the same class, otherwise 0. The loss in (12) learns to impose a margin between the query example and support examples in different class. In evaluating, a query example is labeled according to (11).

4 Experiments

4.1 Settings

In one shot learning, we split training dataset and evaluating dataset according to example labels. Let C denotes the set of classes in dataset D. Training set labels C_{train} and evaluating set labels C_{eval} are two subsets of C, and $C_{train} \cup C_{eval} = C$, $C_{train} \cap C_{eval} = \emptyset$. Then training dataset D_{train} contains examples which have labels $c \in C_{train}$, and evaluating dataset D_{eval} contains examples which have labels $c \in C_{eval}$. Algorithms learns to transfer knowledge from training dataset to evaluating dataset [21,34,35]. If validation is required, the whole dataset should be split into three non-intersect subsets, for training, validation, and evaluation respectively.

To evaluate our algorithm, we perform one shot learning experiments on two datasets, Omniglot [13] and *mini*ImageNet [31]. Both are specially proposed for one shot scenario. We compared a number of alternative models, including siamese networks [11], neural statistician [3], MAML [4], meta-learner LSTM [22], matching networks (MN) [31], matching networks with margin (MN-M), prototypical networks (PN) [28], and prototypical networks with margin (PN-M).

4.2 Results on Omniglot

Omniglot [13] comprises of 1623 characters collected from 50 alphabets. Each character has 20 images drawn by different people. We follow the procedure of Vinyals et al. [31] by resizing the grayscale images to 28 × 28. Our embedding architecture mirrors that used by Vinyals et al. [31] and is composed of four convolutional blocks. Each block comprises of a 64-filter 3×3 convolution, a batch normalization layer [8] with decay rate 0.99, a ReLU nonlinearity and a 2×2 max-pooling layer. When applied to the 28 × 28 Omniglot images this architecture results in a 64-dimensional output space. Our models are trained with Adam [9]. The initial learning rate is set as 0.001, and we cut it into half every 2000 episodes. No regularization is used other than batch normalization. To evaluate models on the 5-way task and the 20-way task, we train it with the 20-way 1-shot task and the 60-way 5-shot task respectively. We split Omniglot dataset into three parts randomly, 1000 classes for training, 200 classes for validation, and the remaining for evaluation.

Results on Omniglot are shown in Table 1. Under all conditions including 5-way and 20-way as well as 1-shot and 5-shot, prototypical networks with margin (PN-M) outperforms all the others. By introducing margin, PN-M outperforms original prototypical networks (PN), and matching networks with margin (MN-M) outperforms original matching networks (MN). These results confirm that the

Table 1. One shot classification accuracies on Omniglot. †Results reported by Snell et al. [28]. ‡Results reported by Finn et al. [4].

Model	5-way accuracy		20-way accuracy	
	1-shot	5-shot	1-shot	5-shot
SIAMESE NETWORKS‡ [11]	97.3%	98.4%	88.2%	97.0%
NEURAL STATISTICIAN† [3]	98.1%	99.5%	93.2%	98.1%
MAML‡ [4]	98.7%	99.9%	95.8%	98.9%
MN† [31]	98.1%	98.9%	93.8%	98.5%
MN-M	**99.1 ± 0.2%**	**99.3 ± 0.2%**	**96.2 ± 0.2%**	**98.9 ± 0.1%**
PN† [28]	98.8%	99.7%	96.0%	98.9%
PN-M	**99.5 ± 0.1%**	**99.9 ± 0.1%**	**97.5 ± 0.3%**	**99.3 ± 0.2%**

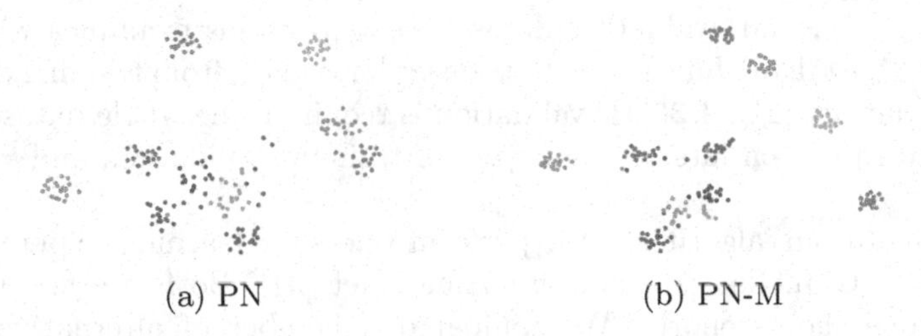

(a) PN (b) PN-M

Fig. 3. The t-SNE visualization of the embedding learned by the respective model. A subset of Tengwar script (an alphabet in evaluation set of Omniglot dataset) is shown, different color indicates different class. Best viewed in color. (Color figure online)

margin improved the generalization of one shot learning models. In particular, PN-M achieves near-perfect performance on 5-way 1-shot task, 5-way 5-shot task, and 20-way 5-shot task.

In addition, we compared the embedding learned by PN and PN-M. We choose a subset of characters in evaluation set of Omniglot, and embed them to 64-dimensional vectors with the trained embedding network in the respective model. Then we visualize the vectors using t-SNE [18]. The results are reported in Fig. 3. It is evident that the margin helps to learn a more distinct cluster structure. The embedding learned by PN-M shows compact intra-class variations and separable inter-class differences.

4.3 Results on *mini*ImageNet

The *mini*ImageNet dataset, originally proposed by Vinyals et al. [31], is derived from the larger ILSVRC-12 dataset [24]. 60,000 color images of size 84×84 are contained in *mini*ImageNet, including 100 classes and 600 images for each class.

We use the same four convolutional blocks in our Omniglot experiments as our embedding network, but it results in a 1600-dimensional output space due to the increased size of the input images. No learning rate decay is adopted here. We rely on the class split used by Ravi and Larochelle [22]. These splits use 64 classes for training, 16 for validation, and 20 for evaluation. To evaluate model on 5-way 1-shot task and 5-way 5-shot task, we train it with 20-way 5-shot task.

Table 2. One shot classification accuracies on *mini*ImageNet. The ± shows 95% confidence intervals over tasks. †Results reported by Snell et al. [28]. ‡Results reported by Finn et al. [4].

Model	5-way accuracy	
	1-shot	5-shot
META-LEARNER LSTM† [22]	$43.44 \pm 0.77\%$	$60.60 \pm 0.71\%$
MAML‡ [4]	$48.70 \pm 1.84\%$	$63.11 \pm 0.92\%$
MN† [31]	$43.56 \pm 0.84\%$	$55.31 \pm 0.73\%$
MN-M	$\mathbf{47.52 \pm 0.92\%}$	$\mathbf{63.72 \pm 1.03\%}$
PN† [28]	$49.42 \pm 0.78\%$	$68.20 \pm 0.66\%$
PN-M	$\mathbf{51.62 \pm 0.76\%}$	$\mathbf{70.24 \pm 0.81\%}$

Table 3. One shot classification accuracies of PN-M on Omniglot with varying α and margin m.

	$\alpha = 0.01$	$\alpha = 0.1$	$\alpha = 0.2$	$\alpha = 0.5$
$m = 0.1$	95.28%	96.56%	96.66%	95.69%
$m = 0.2$	96.28%	96.91%	96.38%	95.31%
$m = 0.5$	96.88%	97.31%	96.31%	95.47%
$m = 1.0$	86.31%	93.25%	94.09%	93.06%

Results on *mini*ImageNet are given in Table 2. In general, the performance of prototypical networks with margin (PN-M) surpasses all the others. Specifically, PN-M outperforms prototypical networks (PN) by 2% in terms of accuracy. Similarly, matching networks with margin (MN-M) gains more than 4% improvement over matching networks (MN). Overall, the results demonstrate that models with margin have superior generalization on one shot learning tasks.

4.4 Parameter Study

The proposed margin-based loss contains two hyperparameters, the margin parameter m and the weighting parameter α. To study the influence of the hyperparameters, we perform the grid search strategy on these 2 hyperparameters. For the weighting parameter $\alpha \in [0, 1]$, we evaluate $\alpha \in \{0.01, 0.1, 0.2, 0.5\}$. For the margin parameter m, since we normalized the learned embedding by l_2 normalization, we check $m \in \{0.1, 0.2, 0.5, 1.0\}$. We analyze both hyperparameters on Omniglot and *mini*ImageNet. The learning rate is fixed at 0.001. For Omniglot, the model is trained and evaluated with 20-way 1-shot task. For *mini*ImageNet, the model is trained with 20-way 5-shot task and evaluated with 5-way 1-shot task. Results on Omniglot and *mini*ImageNet are shown in Tables 3 and 4 respectively.

Table 4. One shot classification accuracies of PN-M on *mini*ImageNet with varying α and margin m.

	$\alpha = 0.01$	$\alpha = 0.1$	$\alpha = 0.2$	$\alpha = 0.5$
$m = 0.1$	31.58%	42.66%	42.54%	47.20%
$m = 0.2$	38.40%	49.34%	50.02%	50.80%
$m = 0.5$	38.24%	50.30%	51.52%	51.10%
$m = 1.0$	38.10%	49.20%	50.56%	51.46%

From Tables 3 and 4, we can see that our model achieves stably good performance with a wide range of both hyperparameters. The margin $m = 0.5$ works well on Omniglot and *mini*ImageNet. As for weighting parameter α, the best α is the one balanced pulling force and pushing force. According to experiments, we recommend to search around $\alpha = \frac{2l}{m}$, where l denotes the number of examples in the support set sharing the same label as the query example, and m indicates the number of examples in the support set labeled differently as the query one.

5 Related Work

5.1 One Shot Learning

One shot learning deals with the problem of learning from a few examples, and our work shares some similarity with those studies. Prototypical networks [28] and matching networks [31] are two representative one shot learning models. We evaluate the effectiveness of one shot learning with margin in these two networks. Ren et al. [23] proposed semi-supervised prototypical networks for few-shot learning, which makes use of unlabeled examples to improve estimation of class prototypes in prototypical networks. Triantafillou and Zemel [30] defined a training objective aimed to optimize overall relative orderings of the query points simultaneously. For a query example, support examples in the same class must rank ahead of support examples in other classes.

There are many other meta-learning based methods related to our work. Ravi and Larochelle [22] attacked one shot learning through optimization. An LSTM-based [7] meta-learner model is proposed to learn the optimization algorithm which is used to train another learner classifier. The neural statistician [3] uses an extension of variational autoencoder [10] to learn representations of datasets. The model produces a representation for each class. And the right label of the query example is the class whose representation has minimal KL-divergence with the representation of the query example. Finn et al. [4] introduced a meta-learning method trains for a representation that can be quickly adapted to a new task, via a few gradient steps.

5.2 Metric Learning

Metric learning [12,15,33] is to learn a distance metric that preserves the distance relationship among the data. N-pair loss [29] use multiple negative examples to achieve better convergence. But only one positive example is contained in the query set of N-pair loss. Triplet loss [27] is calculated on triplets, which consists of three examples with two different labels. However, our multi-way contrastive loss is calculated on an arbitrary number of examples. Large margin nearest neighbor (LMNN) [32] classification considers both pulling force and pushing force as the final loss. The push term of LMNN uses triplet loss of all triplets in the dataset. Neighborhood Components Analysis (NCA) [5] learns a Mahalanobis distance to maximize K-nearest-neighbour's leave-one-out accuracy. Salakhutdinov and Hinton [25] extended NCA by replacing learnable Mahalanobis distance with a neural network, to endow the algorithm with stronger learning ability.

6 Conclusion

In this paper, we analyze the bottleneck of metric-based one shot learning. To improve the generalization of these models, we introduce margin in one shot learning. And we proposed a novel margin-based loss called multi-way contrastive loss. The proposed loss effectively learns to push away inter-class examples, while pulling intra-class examples together. Meanwhile, we have described the methods to integrate multi-way contrastive loss into prototypical networks and matching networks. The one shot learning models with margin can learn an embedding with more appropriate cluster structure than the original models do. Extensive experiment results on several benchmark datasets show that the margin boosts the generalization of one shot learning models on data scarcity tasks.

References

1. Bahdanau, D., Cho, K., Bengio, Y.: Neural machine translation by jointly learning to align and translate. arXiv preprint arXiv:1409.0473 (2014)
2. Cortes, C., Vapnik, V.: Support-vector networks. Mach. Learn. **20**(3), 273–297 (1995)
3. Edwards, H., Storkey, A.: Towards a neural statistician. In: International Conference on Learning Representations (ICLR) (2017)
4. Finn, C., Abbeel, P., Levine, S.: Model-agnostic meta-learning for fast adaptation of deep networks. In: International Conference on Machine Learning (ICML), pp. 1126–1135 (2017)
5. Goldberger, J., Hinton, G.E., Roweis, S.T., Salakhutdinov, R.R.: Neighbourhood components analysis. In: Advances in Neural Information Processing Systems (NIPS), pp. 513–520 (2005)
6. Hadsell, R., Chopra, S., LeCun, Y.: Dimensionality reduction by learning an invariant mapping. In: Proceedings of the IEEE Conference on Computer Vision and Pattern Recognition (CVPR), vol. 2, pp. 1735–1742. IEEE (2006)
7. Hochreiter, S., Schmidhuber, J.: Long short-term memory. Neural Comput. **9**(8), 1735–1780 (1997)

8. Ioffe, S., Szegedy, C.: Batch normalization: accelerating deep network training by reducing internal covariate shift. arXiv preprint arXiv:1502.03167 (2015)
9. Kingma, D.P., Ba, J.: Adam: a method for stochastic optimization. arXiv preprint arXiv:1412.6980 (2014)
10. Kingma, D.P., Welling, M.: Auto-encoding variational bayes. arXiv preprint arXiv:1312.6114 (2013)
11. Koch, G., Zemel, R., Salakhutdinov, R.: Siamese neural networks for one-shot image recognition. In: ICML Deep Learning Workshop, vol. 2 (2015)
12. Kulis, B., et al.: Metric learning: a survey. Found. Trends® Mach. Learn. **5**(4), 287–364 (2013)
13. Lake, B., Salakhutdinov, R., Gross, J., Tenenbaum, J.: One shot learning of simple visual concepts. In: Proceedings of the Annual Meeting of the Cognitive Science Society, vol. 33 (2011)
14. Lake, B.M., Salakhutdinov, R., Tenenbaum, J.B.: Human-level concept learning through probabilistic program induction. Science **350**(6266), 1332–1338 (2015)
15. Liu, H., Zhang, X., Zhang, X., Cui, Y.: Self-adapted mixture distance measure for clustering uncertain data. Knowl.-Based Syst. **126**, 33–47 (2017)
16. Liu, W., Wen, Y., Yu, Z., Li, M., Raj, B., Song, L.: SphereFace: deep hypersphere embedding for face recognition. In: Proceedings of the IEEE Conference on Computer Vision and Pattern Recognition (CVPR), vol. 1 (2017)
17. Liu, W., Wen, Y., Yu, Z., Yang, M.: Large-margin softmax loss for convolutional neural networks. In: International Conference on Machine Learning (ICML), pp. 507–516 (2016)
18. van der Maaten, L., Hinton, G.: Visualizing data using t-SNE. J. Mach. Learn. Res. **9**(Nov), 2579–2605 (2008)
19. Mishra, N., Rohaninejad, M., Chen, X., Abbeel, P.: A simple neural attentive meta-learner. In: International Conference on Learning Representations (ICLR) (2018)
20. Nichol, A., Achiam, J., Schulman, J.: On first-order meta-learning algorithms. arXiv preprint arXiv:1803.02999 (2018)
21. Pan, S.J., Yang, Q.: A survey on transfer learning. IEEE Trans. Knowl. Data Eng. **22**(10), 1345–1359 (2010)
22. Ravi, S., Larochelle, H.: Optimization as a model for few-shot learning. In: International Conference on Learning Representations (ICLR) (2017)
23. Ren, M., et al.: Meta-learning for semi-supervised few-shot classification. In: International Conference on Learning Representations (ICLR) (2018)
24. Russakovsky, O., et al.: ImageNet large scale visual recognition challenge. Int. J. Comput. Vis. **115**(3), 211–252 (2015)
25. Salakhutdinov, R., Hinton, G.: Learning a nonlinear embedding by preserving class neighbourhood structure. In: Artificial Intelligence and Statistics, pp. 412–419 (2007)
26. Santoro, A., Bartunov, S., Botvinick, M., Wierstra, D., Lillicrap, T.: Meta-learning with memory-augmented neural networks. In: International Conference on Machine Learning (ICML), pp. 1842–1850 (2016)
27. Schroff, F., Kalenichenko, D., Philbin, J.: FaceNet: a unified embedding for face recognition and clustering. In: Proceedings of the IEEE Conference on Computer Vision and Pattern Recognition (CVPR), pp. 815–823 (2015)
28. Snell, J., Swersky, K., Zemel, R.: Prototypical networks for few-shot learning. In: Advances in Neural Information Processing Systems (NIPS), pp. 4080–4090 (2017)
29. Sohn, K.: Improved deep metric learning with multi-class N-pair loss objective. In: Advances in Neural Information Processing Systems (NIPS), pp. 1857–1865 (2016)

30. Triantafillou, E., Zemel, R., Urtasun, R.: Few-shot learning through an information retrieval lens. In: Advances in Neural Information Processing Systems (NIPS), pp. 2252–2262 (2017)
31. Vinyals, O., Blundell, C., Lillicrap, T., Wierstra, D., et al.: Matching networks for one shot learning. In: Advances in Neural Information Processing Systems (NIPS), pp. 3630–3638 (2016)
32. Weinberger, K.Q., Saul, L.K.: Distance metric learning for large margin nearest neighbor classification. J. Mach. Learn. Res. **10**(Feb), 207–244 (2009)
33. Yang, L., Jin, R.: Distance metric learning: a comprehensive survey. Mich. State Univ. **2**(2), 4 (2006)
34. Zhang, X., Zhang, X., Liu, H.: Self-adapted multi-task clustering. In: IJCAI, pp. 2357–2363 (2016)
35. Zhang, X., Zhang, X., Liu, H., Liu, X.: Multi-task clustering through instances transfer. Neurocomputing **251**, 145–155 (2017)

DeepReview: Automatic Code Review Using Deep Multi-instance Learning

Heng-Yi Li[1], Shu-Ting Shi[1], Ferdian Thung[2], Xuan Huo[1], Bowen Xu[2], Ming Li[1(✉)], and David Lo[2]

[1] National Key Laboratory for Novel Software Technology, Nanjing University, Nanjing 210023, China
{lihy,shist,huox,lim}@lamda.nju.edu.cn

[2] School of Information Systems, Singapore Management University, Singapore, Singapore
{ferdiant.2013,bowenxu.2017}@phdis.smu.edu.sg, davidlo@smu.edu.sg

Abstract. Code review, an inspection of code changes in order to identify and fix defects before integration, is essential in Software Quality Assurance (SQA). Code review is a time-consuming task since the reviewers need to understand, analysis and provide comments manually. To alleviate the burden of reviewers, automatic code review is needed. However, this task has not been well studied before. To bridge this research gap, in this paper, we formalize automatic code review as a multi-instance learning task that each change consisting of multiple hunks is regarded as a bag, and each hunk is described as an instance. We propose a novel deep learning model named DeepReview based on Convolutional Neural Network (CNN), which is an end-to-end model that learns feature representation to predict whether one change is approved or rejected. Experimental results on open source projects show that DeepReview is effective in automatic code review tasks. In terms of F1 score and AUC, DeepReview outperforms the performance of traditional single-instance based model TFIDF-SVM and the state-of-the-art deep feature based model Deeper.

Keywords: Software mining · Machine learning · Multi-instance learning · Automatic code review

1 Introduction

Software Quality Assurance (SQA) is essential in software development. Software code review [16] is an important inspection of code changes written by an independent third-party developer in order to identify and fix defects before integration. Effective code review can largely improve the software quality.

However, code review is a very time-consuming task that the reviewer needs to spend much time to understand, analyze and provide comments for the code review request. Additionally, with the rapid growth of software, the number of

Q. Yang et al. (Eds.): PAKDD 2019, LNAI 11440, pp. 318–330, 2019.
https://doi.org/10.1007/978-3-030-16145-3_25

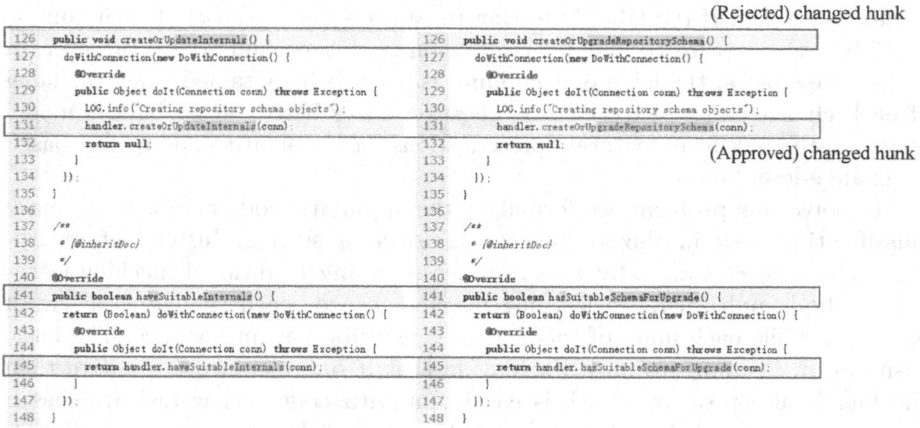

(Rejected) changed hunk

```
126 public void createOrUpdateInternals() {
127   doWithConnection(new DoWithConnection() {
128     @Override
129     public Object doIt(Connection conn) throws Exception {
130       LOG.info("Creating repository schema objects");
131       handler.createOrUpdateInternals(conn);
132       return null;
133     }
134   });
135 }
136
137 /**
138  * {@inheritDoc}
139  */
140 @Override
141 public boolean haveSuitableInternals() {
142   return (Boolean) doWithConnection(new DoWithConnection() {
143     @Override
144     public Object doIt(Connection conn) throws Exception {
145       return handler.haveSuitableInternals(conn);
146     }
147   });
148 }
```

```
126 public void createOrUpgradeRepositorySchema() {
127   doWithConnection(new DoWithConnection() {
128     @Override
129     public Object doIt(Connection conn) throws Exception {
130       LOG.info("Creating repository schema objects");
131       handler.createOrUpgradeRepositorySchema(conn);
132       return null;
133     }
134   });
135 }
136
137 /**
138  * {@inheritDoc}
139  */
140 @Override
141 public boolean hasSuitableSchemaForUpgrade() {
142   return (Boolean) doWithConnection(new DoWithConnection() {
143     @Override
144     public Object doIt(Connection conn) throws Exception {
145       return handler.hasSuitableSchemaForUpgrade(conn);
146     }
147   });
148 }
```

(Approved) changed hunk

Fig. 1. An example of rejected change `JdbcRepository.java` of review request 26657 from Apache. This change contains four hunks and only one hunk is rejected.

review requests are growing, which leads to a heavier burden on code reviewers. Therefore, automatic code review is important to alleviate the burden of reviewers.

Recently, some studies have been proposed to improve the effectiveness of code review [1,16]. Thongtanunam et al. [16] revealed that 4%–30% of reviews have code-reviewer assignment problems. They proposed a code reviewer recommendation approach named REVFINDER to solve it by leveraging the file location information. Ebert et al. [1] proposed to identify the factors that confuse reviewers and understand how confusion impacts the efficiency of code reviewers. However, the task of automatic code review has not been well studied previously.

Considering the above issues, an automated approach is needed, which is able to help reviewers to review the code submitted by developers. Usually, a review request submitted by developers contains some changes of source code in the form of `diff` files and textual descriptions indicating the intention of the change. Notice that each change may contain multiple change hunks and each hunk corresponds to a set of continuous lines of code. For example, Fig. 1 shows the change in the file `JdbcRepository.java` of review request 26657 from Apache project. This change contains four hunks. One of the most common ways to analyze this change is to combine all hunks together and generate a unified feature to represent the change. However, this method may lead to two problems. First, the hunks appearing in each change may be discontinuous and unrelated to one another. Directly combining the hunks together may generate misleading feature representations, leading to a poor prediction performance. Second, when the change is rejected, not every hunk in the change is rejected. Some hunks have no issues and can be approved by reviewers. So the approved hunks and the rejected hunks should not be processed together for feature extraction. Therefore, separately generating features from each individual hunk in automatic code review is needed. If the label (referring to *approved* or *rejected*) of each hunk is available, we can directly build classification models on hunk data. However, in code review tasks, the label of each hunk is hard to be obtained while the label of each change can be extracted. A question arises here, can we build a model to generate hunk-level feature representations for automatic code review based on change-level labels?

To solve this problem, we formulate the automatic code review as a binary classification task in the *multi-instance learning* setting. Instead of regarding each change as an individual instance following traditional machine learning method, multi-instance learning method regards each change as a bag of instances while each hunk of the change is described as an instance. The basic assumption in multi-instance learning is that if one instance is positive then the bag is also positive, which is consistent with code review task whereas if one hunk is rejected then the change is also rejected. In our paper, we propose a deep learning model named DeepReview based on Convolutional Neural Network (CNN) via multi-instance learning, which is able to automatically learn semantic features from each hunk and predict if one change is approved or rejected. Additionally, in order to obtain the features that capture the difference of code changes, DeepReview firstly recovers the code before change (old source code) and after change (new source code) according to the `diff` markers. These snippers are then fed in to the deep model to generate feature presentation, based on which the label of each change is predicted. We conduct experiments on large datasets collected from open source Apache projects for evaluation. The results in terms of widely-used metrics AUC and F1 score indicate that DeepReview is effective in automatic code review and outperforms previous state-of-the-art feature representation methods previously used for related software engineering tasks.

The contributions of our work are several folds:

- We are the first to study automatic code review task as multi-instance learning task. One change always contains multiple hunks, where each hunk is described as an instance and the change can be represented by a set of instances. Experiment results on five large datasets show that the proposed multi-instance model is effective in automatic code review tasks.
- We propose a novel deep learning model named DeepReview based on Convolutional Neural Network (CNN), which learns semantic feature representation from source code change and change descriptions, to predict if one change is approved or rejected.

2 The DeepReview Approach

In this section, we introduce the details of applying DeepReview for automatic code review. The goal of this task is to predict if one code change of review request submitted by developers is approved or rejected. The general process of automatic code review based on machine learning model is illustrated in Fig. 2.

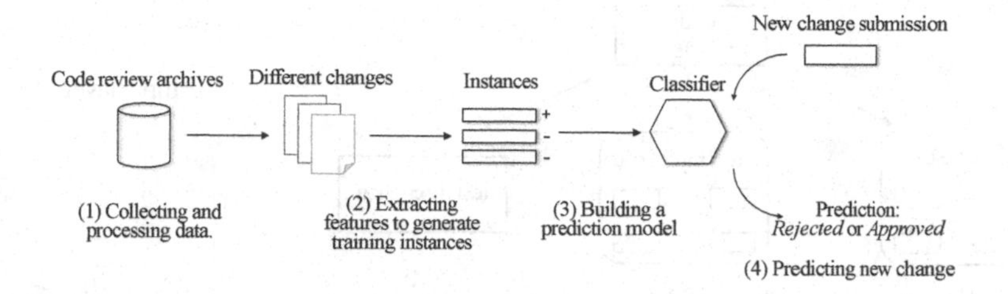

Fig. 2. The general automatic code review process based on machine learning model.

The automatic code review prediction process mainly contains several parts:

- Collecting data from code review systems and processing data.
- Generating feature representations of the input data.
- Training a classifier based on the generated features and labels.
- Predicting if a new change is approved or rejected.

In the following subsections, we first introduce the general framework of DeepReview in Subsect. 2.1, and the data processing is reported in Subsect. 2.2. The core parts of DeepReview is elaborated in Subsects. 2.3 and 2.4.

2.1 The Framework of DeepReview

We introduce some notations of our framework. Let $\mathcal{C}^o = \{c_1^o, c_2^o, \ldots, c_N^o\}$ and $\mathcal{C}^n = \{c_1^n, c_2^n, \ldots, c_N^n\}$ denotes the collection of old code and new code. Let $\mathcal{D} = \{d_1, d_2, \ldots, d_N\}$ denotes the collection of change descriptions, where N is the number of changes. In this paper, we formalize the code review as a learning task, which attempts to learn a prediction function $f : \mathcal{X} \mapsto \mathcal{Y}$. $x_i \in \mathcal{X} = (c_i^o, c_i^n, d_i)$ denotes each change, where c_i^o and c_i^n denotes the i-th old code (before changed) and new code (after changed) respectively. Here $c_i^o = \{h_{i1}^o, h_{i2}^o, \ldots, h_{im}^o\}$ and $c_i^o = \{h_{i1}^n, h_{i2}^n, \ldots, h_{im}^n\}$ contains multiple hunks and m is the number of hunks. d_i denotes the text description of i-th change. $y_i \in \mathcal{Y} = \{1, 0\}$ indicates whether the change is approved or rejected.

We instantiate the code review prediction model by constructing a multi-instance learning based deep neural network named DeepReview. The general framework of DeepReview is illustrated in Fig. 3. The DeepReview model contains three parts: input layers, instance feature generation layers and multi-instance based prediction layers.

In the DeepReview model, each hunk of source code change is regarded as an instance. In the input layers, the source code and text description of each instance is encoded as feature vectors and then are fed into the neural network for processing. The details of data processing in the input layers will be discussed in Subsect. 2.2. Then the encoded data of each instance is fed into instance feature generation layers. In these layers, DeepReview utilizes different convolutional neural networks (CNN) to extract features from the source code input and the textual description input. The convolutional neural networks for programming

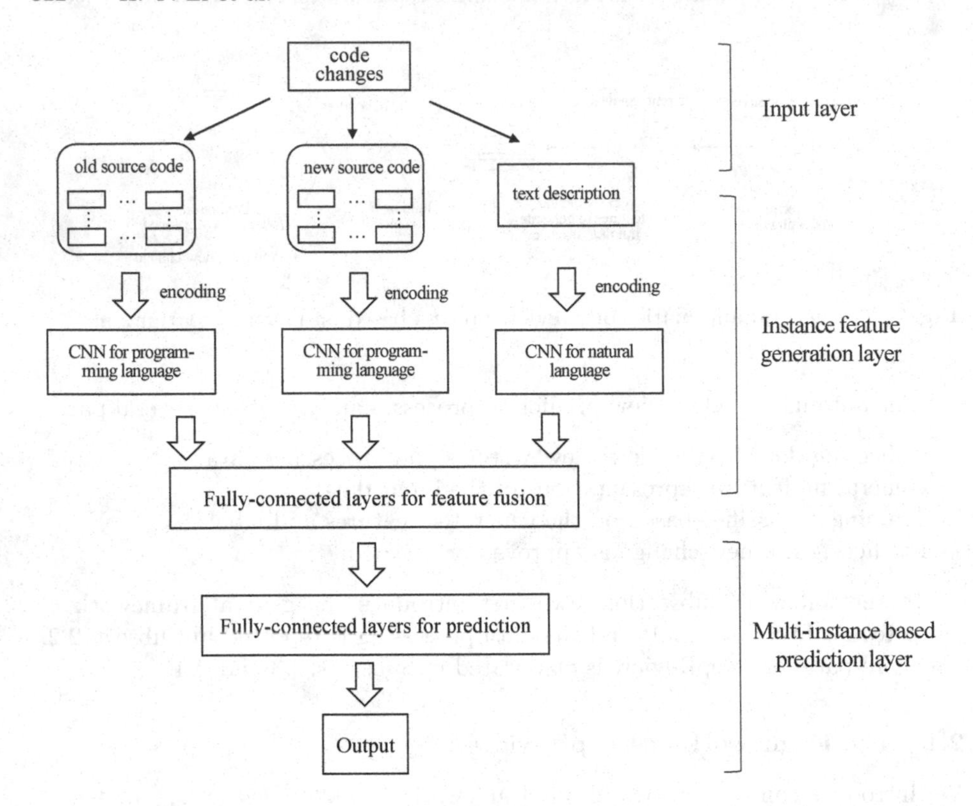

Fig. 3. The general framework of DeepReview for automatic code review prediction. The DeepReview model contains three parts: input layer, instance feature generation layer, multi-instance based prediction layer.

language processing (called PCNN) is carefully designed respecting to the characteristics of source code, which is similar to the network structure in [4]. The convolutional neural networks for textual description processing (called NCNN) is a standard way in [6]. Then the generated middle-level features of old code, new code and textual descriptions of each instance are fused to learn a unified feature representation via fully-connected networks mapping. Finally, after generating unified feature representations, the DeepReview model make a prediction for each change via the multi-instance learning way in the multi-instance based prediction layers.

2.2 Data Processing

The datasets used for automatic code review is the changed source code submitted by developers, which always appears in form of `diffs` and contains both source code and `diff` markers (e.g., + stands for adding a line, - stands for deleting a line). The main features in code changes are the difference between the code before changed and after changed. So in data preprocessing shown in

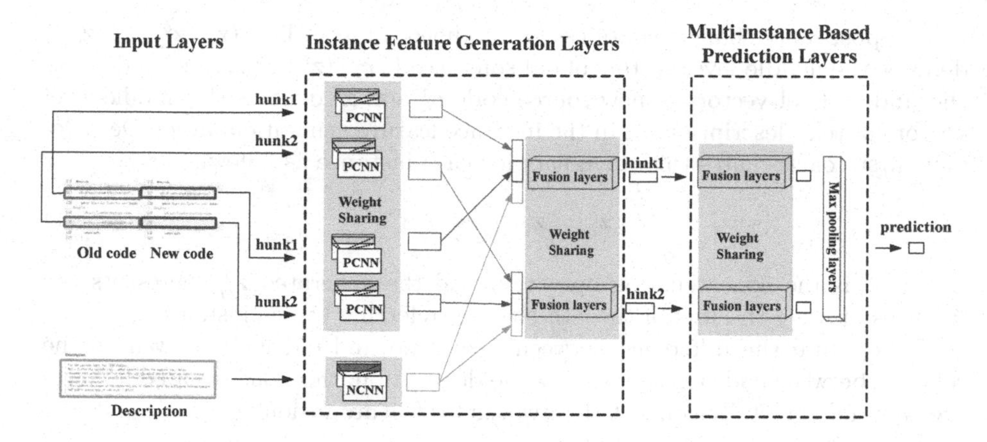

Fig. 4. Automatic code review by DeepReview. When a change is processed for prediction, three parts of the change (old code, new code and text descriptions) are firstly encoded as feature vectors to feed into deep model. Then three parts of convolutional neural networks are followed to extracte semantic features for source code and text description separately. After that a fully-connected network is used to get fusion feature for hunks. Finally, another fully-connected network and a max-pooling layer is connected to generate a prediction indicating approved or rejected of the change.

the left part of Fig. 4, we extract both old code (before changed) and new code (after changed) from `diffs` as input. We also use the change descriptions since they contain the goal of this change and are helpful to improve the prediction performance.

After splitting `diff` files into old code, new code and text description, a pre-trained word2vec [10] technique is used to encode every token as vector representations (e.g., a 300 dimension vector), which has been shown effective in processing textual data and widely used in text processing tasks [6,10]. In a similar way, we split descriptions as words and encode them as vector representations too. All these vector representations are sent into the deep neural network to learn the semantic features.

2.3 Instance Feature Generation Layer

DeepReview takes old source code (before change) and new source code (after change) along with the text descriptions as inputs. Noticing that the source code and text descriptions are with different structures. Therefore we use PCNN network for code and NCNN network for text to extract feature, respectively.

As aforementioned, each change will contain multiple hunks and different hunks are individual instance, therefore the instance features should be extracted separately by the same neural network. In other words, the weight of PCNN is shared for all code hunks. In this way, we can get unbiased feature representations for each hunk with both old code and new code.

Suppose one change contains m modified hunks. Let $(\mathbf{z}^o_{i1}, \mathbf{z}^o_{i2}, \ldots, \mathbf{z}^o_{im})$ denotes the middle-level vectors of old source code c^o_i, $(\mathbf{z}^n_{i1}, \mathbf{z}^n_{i2}, \ldots, \mathbf{z}^n_{im})$ denotes the middle-level vectors of new source code c^n_i and $\mathbf{z}^t_i$ denotes the middle-level vectors of text description d_i. In the instance feature generation layers, DeepReview first concatenates this three part for each instance as following:

$$\mathbf{z}^h_{ij} = \mathbf{z}^o_{ij} \odot \mathbf{z}^n_{ij} \odot \mathbf{z}^t_i \tag{1}$$

where $\odot$ is the concatenating operation and the generated $\mathbf{z}^h_{ij}$ represents the features of the j-th hunk of the i-th change (referring to one instance).

To capture the difference between new code and old code as well as the relation between code change and change description, this concatenated features are then fed into fully-connected networks for feature fusion.

2.4 Multi-instance Based Prediction Layer

In the prediction layers, we first make a prediction for each hunk (also called instance) using fully-connected networks following a sigmoid layer based on the generated hunk representations. Similarly, all the fully-connected networks are shared weights to each hunk so that the generated prediction does not have bias. The output prediction of each hunk $\mathbf{p_i} = (p_{i1}, p_{i2}, \ldots, p_{im})$ is generated.

In the multi-instance setting, if any instance is positive (rejected), the bag is also positive (rejected). So the maximum value of predictions for hunks is used for predicting the label of each change. Then, a max-pooling layer is employed to get the final prediction for the change, that is $\hat{p_i} = \max\{\mathbf{p}\}$.

Specifically, the parameters of the convolutional neural networks layers can be denoted as $\Theta = \{\theta_1, \theta_2, \ldots, \theta_l\}$ and the parameters of the fully-connected networks layers can be denoted as $W = \{\mathbf{w}_1, \mathbf{w}_2, \ldots, \mathbf{w}_3\}$. Therefore, the loss function implied in DeepReview is:

$$\mathcal{L}(\Theta, W) = -\sum_{i=1}^{N}(c_a y_i \log \hat{p_i} + c_r(1 - y_i)\log(1 - \hat{p_i})) + \lambda\Omega(f) \tag{2}$$

where $\mathcal{L}$ is a cross-entropy loss, $\Omega(f)$ is the regularization term which imposes regularization (e.g., L_2 regularization) on the weights of model, and λ is the trade-off parameter balancing these two terms. c_a denotes the cost of incorrectly predicting a rejected change as approved and c_r denote the cost of incorrectly predicting a approved change as rejected. This objective function can be effectively optimized by SGD (Stochastic gradient descent) algorithm.

3 Experiments

To evaluate the effectiveness of DeepReview, we conduct experiments on thousands of code reviews from open source software projects and compare with several state-of-the-art code review methods.

3.1 Experiment Settings

The datasets used in our experiment are from Apache[1] Code Review Board, which are also analyzed by prior studies on code reviews [13, 14]. We downloaded all reviews on October 2017 and selected only code reviews in which the reviewers highlighted the line numbers that they have issues with, totally 1,011 code reviews. We further extracted five repositories with the largest number of involved files in the collected code reviews – the different datasets and their statistics are shown in Table 1. For each repository, we have more than 1,000 involved files and at least 3,500 hunks.

Table 1. Statistics of our data sets.

Datasets	#changes	#hunks	#rejected
cloudstack-git	1,682	6,171	128
aurora	1,161	6,762	168
drill-git	1,015	3,575	43
accumulo	1,011	5,798	152
hbase-git	1,009	6,702	140

As indicated by Table 1, the number of rejected hunks is only a small part of all hunks and the datasets are very imbalanced. Therefore, we use F1 to evaluate the performance; F1 has been widely used in imbalanced learning settings. Additionally, we record the AUC, which is a non-parametric method to evaluate model performance and is unaffected by class imbalance. The evaluation metrics used in our experiments were adopted to evaluate many approaches that automate various software engineering tasks [4, 5, 9, 12, 17].

We compare the proposed model DeepReview with following baseline methods and some of its variants:

- TFIDF-LR [2], which uses Term Frequency-Inverse Document Frequency (TFIDF) feature to represent source code changes and Logistic Regression (LR) for classification.
- TFIDF-SVM, which uses TFIDF features to represent source code changes and Support Vector Machine (SVM) for classification.
- Deeper [20], one of the state-of-the-art deep learning models on software engineering, which extracts deep features from changes with DBN models and then apply Logistic Regression (LR) for classification.
- Deeper-SVM, a slight variant of Deeper, which uses DBN model for feature extraction and then apply Support Vector Machine for classification.
- DeepReview-SingleInstance, one variant of DeepReview, which does not consider the multi-instance setting and concatenate the all hunks together as one instance for input.

[1] https://reviews.apache.org/r/.

- DeepReview-diff, one variant of DeepReview, which does not separate the code change and taking diff marks and diff code as input.

The settings of DeepReview and its variants are introduced here: in the convolution layers, we use activation function $\sigma(x) = max(x, 0)$. Also, we set the size of convolution windows as 2 and 3 with 100 feature maps each.

3.2 Experiment Results

For each dataset, 10-fold cross validation is repeated 5 times and we report the average value of all compared methods in order to reduce the evaluation bias. We also apply the statistic test to evaluate the significance of DeepReview. Pairwise t-test at 95% confidence level is conducted.

We firstly compare our proposed model DeepReview with several traditional non-multi instance models. One of the most common methods is to employ Vector Space Model (VSM) to represent the changes. In addition, we compare DeepReview with latest deep learning based models Deeper [20] on software engineering, which applies Deep Believe Network for semantic feature extraction. The results are shown in Tablcs 2 and 3. The highest results of each repository is highlighted in bold. The compared methods that are significantly inferior than our approach will be marked with "∘" and significantly better than our approach be marked with "•".

Table 2. The performance comparison in terms of F1 on all methods.

Datasets	TFIDF-LR	TFIDF-SVM	Deeper	Deeper-SVM	DeepReview
accumulo	0.219∘	0.231∘	0.208∘	0.199∘	**0.444**
aurora	0.202∘	0.214∘	0.352∘	0.298∘	**0.436**
cloudstack-git	0.252∘	0.276∘	0.392∘	0.257∘	**0.497**
drill-git	0.213∘	0.235∘	0.277∘	0.226∘	**0.414**
hbase-git	0.235∘	0.257∘	0.182∘	0.142∘	**0.463**
Avg.	0.224∘	0.243∘	0.282∘	0.224∘	**0.451**

Table 3. The performance comparison in terms of AUC on all methods.

Datasets	TFIDF-LR	TFIDF-SVM	Deeper	Deeper-SVM	DeepReview
accumulo	0.635∘	0.678∘	0.697∘	0.705∘	**0.746**
aurora	0.577∘	0.629∘	0.687∘	0.566∘	**0.758**
cloudstack-git	0.755∘	0.827∘	0.825∘	0.637∘	**0.870**
drill-git	0.676∘	0.725∘	0.639∘	0.571∘	**0.761**
hbase-git	0.685∘	0.751	0.597∘	0.547∘	**0.758**
Avg.	0.666∘	0.722∘	0.689∘	0.605∘	**0.779**

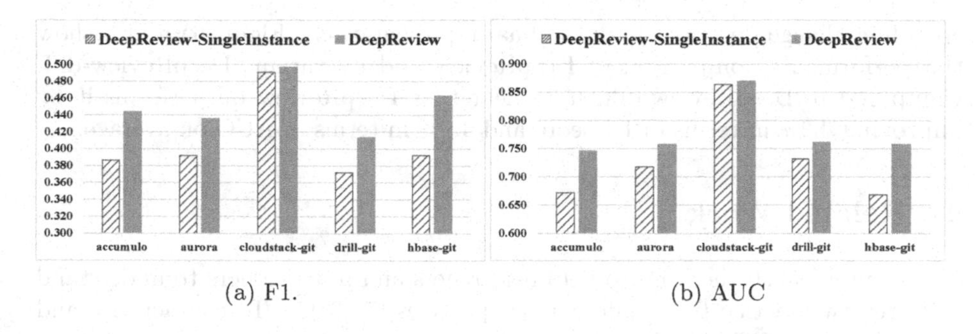

(a) F1. (b) AUC

Fig. 5. F1 and AUC of the compared methods on five datasets.

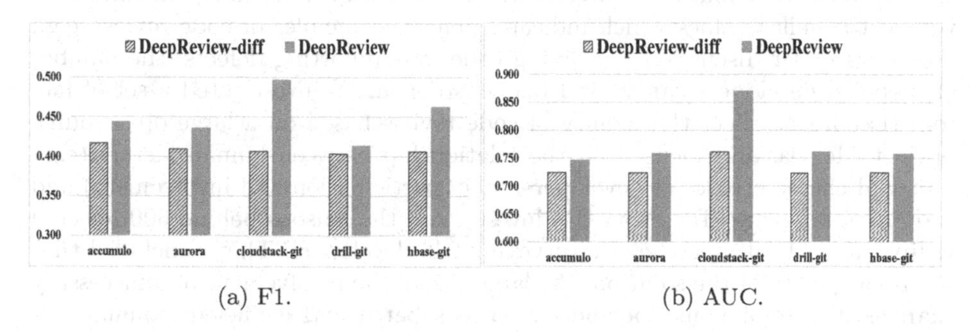

(a) F1. (b) AUC.

Fig. 6. F1 and AUC of the compared methods on five datasets.

As indicated in Tables 2 and 3, DeepReview achieves the best performance on all datasets in terms of F1 score. On average, DeepReview can lead to AUC value 0.779, which is significant better than the value achieves by TFIDF-LR (0.666), TFIDF-SVM (0.722). When compared with Deeper and its variant Deeper-SVM, it can be easily find that DeepReview achieves the best F1 score and AUC value. On average, the superiority of DeepReview to other deep feature based methods is statistically significant. In conclusion, the proposed DeepReview is effective in automatic code review prediction, which indicates that DeepReview can learn better features than traditional hand-crafted features or previous deep learning based features.

To evaluate the effectiveness of applying multi-instance learning strategy for code review, we compare our model to traditional single-instance learning model, named DeepReview-SingleInstance. Figure 5a and b show the performance comparison of DeepReview and a variant DeepReview-SingleInstance. It can be observed that DeepReview achieves higher AUC value and F1 score than DeepReview-SingleInstance on all datasets, indicating that multi-instance learning approach is effective in code review task.

To evaluate the effectiveness of applying both source code before and after changes to model the difference features of change, we compare another variant of DeepReview, named DeepReview-diff. We use the same network structure to extract the features of code in `diffs` and fuse it with the features of corresponding change description as the final representations. Figure 6a and b show the performance comparison of DeepReview and its variant DeepReview-diff. Compared to DeepReview-diff, it is clear that DeepReview outperforms it by improving 4.2% in terms of F1 score and 4.7% in terms of AUC on average.

4 Related Work

Many empirical studies aim to help researchers and practitioners to understand code review practice from different perspectives [7,13,15]. To characterize and understand the differences between a diverse set of software projects, Rigby et al. [13] found that many characteristics of code review have independently converged to similar values which indicates general principles of code review, e.g., reviewers prefer discussion and fixing code over reporting defects, the number of involved developers can vary. Kononenko et al. [7] investigated a set of factors that might affect the quality of code review based on a large open-source project Mozilla, and focused on the relationship between human factors (e.g., personal characteristics of developers, team participation and involvement) and code review quality. Tao et al. [15] investigated the reasons behind 300 rejected Eclipse and Mozilla patches by surveying 246 developers. They concluded that the poor quality of the solution, the large size of the involvement of unnecessary changes, the ambiguous documentation of a patch and inefficient communication. Moreover, Thongtanunam et al. [16] revealed that 4%–30% of reviews have code-reviewer assignment problem. Thus, they proposed a code-reviewer recommendation approach REVFINDER to solve the problem by leveraging the file location information. The intuition is that files that are located in similar file paths would be managed and reviewed by experienced code-reviewers. Zanjani et al. [21] also studied on code reviewer recommendation problem and they proposed an approach cHRev by leveraging the specific information in previously completed reviews (i.e., quantification of review comments and their recency).

Recently, deep learning has been applied in software engineering. For example, Yang et al. applied Deep Belief Network (DBN) to learn higher-level features from a set of basic features extracted from commits (e.g., lines of code added, lines of code deleted, etc.) to predict buggy commits [20]. Xu et al. applied word embedding and convolutional neural network (CNN) to predict semantic links between knowledge units in Stack Overflow (i.e., questions and answers) to help developers better navigate and search the popular knowledge base [19]. Lee et al. applied word embedding and CNN to identify developers that should be assigned to fix a bug report [8]. Mou et al. [11] applied tree based CNN on abstract syntax tree to detect code snippets of certain patterns. Huo et al. [3,4] applied learned unified semantic feature based on bug reports in natural language and source code in programming language for bug localization tasks. Wei et al. [18] proposed deep feature learning framework AST-based LSTM network for functional clone detection, which exploits the lexical and syntactical information.

5 Conclusion

In this paper, we are the first to formulate code review as a multi-instance learning task. We propose a novel deep learning model named DeepReview for automatic code review, which takes raw data of a changed code containing multiple hunks along with the textual descriptions as inputs and predicts if one change is approved or rejected. Experimental results on five open source datasets show that DeepReview is effective and outperforms the state-of-the-art models previously proposed for other automated software engineering tasks.

Acknowledgment. This research was supported by National Key Research and Development Program (2017YFB1001903) and NSFC (61751306).

References

1. Ebert, F., Castor, F., Novielli, N., Serebrenik, A.: Confusion detection in code reviews. In: ICSME, pp. 549–553 (2017)
2. Gay, G., Haiduc, S., Marcus, A., Menzies, T.: On the use of relevance feedback in IR-based concept location. In: ICSM, pp. 351–360 (2009)
3. Huo, X., Li, M.: Enhancing the unified features to locate buggy files by exploiting the sequential nature of source code. In: IJCAI, pp. 1909–1915 (2017)
4. Huo, X., Li, M., Zhou, Z.H.: Learning unified features from natural and programming languages for locating buggy source code. In: IJCAI, pp. 1606–1612 (2016)
5. Jiang, T., Tan, L., Kim, S.: Personalized defect prediction. In: ASE, pp. 279–289 (2013)
6. Kim, Y.: Convolutional neural networks for sentence classification. In: EMNLP, pp. 1746–1751 (2014)
7. Kononenko, O., Baysal, O., Guerrouj, L., Cao, Y., Godfrey, M.W.: Investigating code review quality: do people and participation matter? In: ICSME, pp. 111–120 (2015)
8. Lee, S., Heo, M., Lee, C., Kim, M., Jeong, G.: Applying deep learning based automatic bug triager to industrial projects. In: ESEC/FSE, pp. 926–931 (2017)
9. Menzies, T., Greenwald, J., Frank, A.: Data mining static code attributes to learn defect predictors. IEEE TSE **33**(1), 2–13 (2007)
10. Mikolov, T., Sutskever, I., Chen, K., Corrado, G.S., Dean, J.: Distributed representations of words and phrases and their compositionality. In: NIPS, pp. 3111–3119 (2013)
11. Mou, L., Li, G., Zhang, L., Wang, T., Jin, Z.: Convolutional neural networks over tree structures for programming language processing. In: AAAI, pp. 1287–1293 (2016)
12. Nam, J., Pan, S.J., Kim, S.: Transfer defect learning. In: ICSE, pp. 382–391 (2013)
13. Rigby, P.C., Bird, C.: Convergent contemporary software peer review practices. In: FSE, pp. 202–212 (2013)
14. Rigby, P.C., German, D.M., Storey, M.A.: Open source software peer review practices: a case study of the apache server. In: ICSE, pp. 541–550 (2008)
15. Tao, Y., Han, D., Kim, S.: Writing acceptable patches: an empirical study of open source project patches. In: ICSME, pp. 271–280 (2014)

16. Thongtanunam, P., Tantithamthavorn, C., Kula, R.G., Yoshida, N., Iida, H., Matsumoto, K.I.: Who should review my code? A file location-based code-reviewer recommendation approach for modern code review. In: SANER, pp. 141–150 (2015)
17. Wang, S., Liu, T., Tan, L.: Automatically learning semantic features for defect prediction. In: ICSE, pp. 297–308 (2016)
18. Wei, H.H., Li, M.: Supervised deep features for software functional clone detection by exploiting lexical and syntactical information in source code. In: IJCAI, pp. 3034–3040 (2017)
19. Xu, B., Ye, D., Xing, Z., Xia, X., Chen, G., Li, S.: Predicting semantically linkable knowledge in developer online forums via convolutional neural network. In: ASE, pp. 51–62 (2016)
20. Yang, X., Lo, D., Xia, X., Zhang, Y., Sun, J.: Deep learning for just-in-time defect prediction. In: QRS, pp. 17–26 (2015)
21. Zanjani, M.B., Kagdi, H., Bird, C.: Automatically recommending peer reviewers in modern code review. IEEE TSE **42**(6), 530–543 (2016)

Multi-label Active Learning with Error Correcting Output Codes

Ningzhao Sun, Jincheng Shan, and Chenping Hou(✉)

National University of Defense Technology, Changsha, China
nz.sun@hotmail.com, njusjc@sina.cn, hcpnudt@hotmail.com

Abstract. Due to the demand of practical problems, multi-label learning has become an important research where each instance belongs to multiple classes. Compared with single-label problem, the labeling cost for multi-label one is rather expensive because of the diversity and non-uniqueness of the labels. Therefore, the active learning which reduces the cost by selecting the most valuable data to query the labels attracts a lot of interests. Although several multi-label active learning (MLAL) methods were proposed, they often identify the label merely through a classifier via one-versus-all (OVA) strategy for each class, which makes the classification model very fragile, thus having a serious impact on the later selection criteria. In this paper, we utilize a new multi-label Error Correcting Output Codes (ECOC) method which determines the label of an instance on each class by combining multiple classifiers. This makes our classification model has a good ability of error-correcting and thus ensures the effectiveness of evaluation information in the selection process. Then we combine two effective selection strategies, the margin prediction uncertainty and label cardinality inconsistency, to complement each other and select the most informative instance. Based on this combination, we propose a novel MLAL framework, termed Multi-label Active Learning with Error Correcting Output Codes (MAOC). Experiments on multiple benchmark multi-label datasets demonstrate the efficacy of the combination in proposed approach.

Keywords: Active learning · Multi-label classification · Error Correcting Output Codes

1 Introduction

In a learning system, we usually need enough examples to train a high performance strong model. Nevertheless, in many real word scenarios, there are a small amount of labeled data but a large amount of unlabeled data. While labeling is usually expensive especially for multi-label data that a sample may belong to multiple labels at the same time. For example, an image can be labeled as "boat" and "water" simultaneously if it contains both objects. In the text category, press documents about the presidential election can cover both politics and

Q. Yang et al. (Eds.): PAKDD 2019, LNAI 11440, pp. 331–342, 2019.
https://doi.org/10.1007/978-3-030-16145-3_26

economics. This makes it harder to label the multi-label data than traditional single-label data. Thus, using a small amount of labeled data to train an accurate model becomes a significant challenge. Active learning is aimed at selective instance marking and reducing the marking of well-trained predictive models, so it is particularly important for multi-label classification.

In multi-label active learning, the main concern is how to select effective data. Two common strategies are often used to measure the unified informativeness of unlabeled instances, the max-margin prediction uncertainty strategy and the label cardinality inconsistency strategy, which exploit uncertainty and diversity in the instance space, respectively [1]. However, both the strategies have their pros and cons, so we need to combine these two strategies in order to play to their strengths at the same time.

Meanwhile, in traditional active learning, multi-label SVM classification was widely utilized. However, the robustness of traditional SVM algorithm is not very good, as it identifies the label of each class by just a classifier. To this problem, the ECOC method utilized in this paper transforms the multi-label problem into a collection of multi-label random selections, is a good solution.

In this paper, we combine the active learning and an effective ECOC framework. We use ECOC to obtain soft labels for each data, which can either classify the data directly through a threshold label or rank all the labels for each data. Based on the soft labels after classification, we get the uncertainty and diversity of each data, respectively. Eventually we can get the selection information of each data and select the most valuable instance to query. In fact, it is the high performance of the new ECOC method that makes the active selection strategy we use more reliable. Because the sample selected through a more robust classifier will have more accurate information. After each query, we update the multi-label model step by step on the basis of new labeled data.

The rest of this paper is organized as follows. Section 2 briefly introduces some related works. Section 3 presents the new ECOC classification model. Section 4 presents the multi-label active selection strategy. Section 5 shows the experimental set-up and discusses the results, followed by the conclusion in Sect. 6.

2 Related Work

Active learning is a machine learning framework making full use of enough unlabeled data, which aims to reduce the labeling effort and cost required for training high-quality predictive models. Useful supervised information is queried iteratively from the oracle. The key of it is to develop an effective query strategy. Over the past decades, a number of active selection criteria have been developed. In [2], The sample was selected by maximizing the likelihood of labeled and selected samples while the uncertainty of unlabeled samples was minimized. For the multi-class problem, the method proposed in [3] adopted entropy to evaluate the uncertainty of unlabeled samples, and applied diversity criteria to make information selection diversity as much as possible. However, these methods cannot be applied directly to multi-label learning tasks.

In most cases, multi-label active learning algorithms decompose the task into a set of binary classification problems. For example, in [4], the uncertainty of each label is measured, and then combined to form the uncertainty measurement of each sample. In addition, a SVM classifier is trained for each label, and the sample that results in the maximum reduction of expected loss is selected. Similarly, in [5], the expected loss reduction of SVM based on independent training was used as the selection criteria by introducing additional regression models to predict the number of labels to be assigned to each sample. In [1], the information of a sample was proposed by combining the inconsistency of the label cardinality and the separation boundary with the trade-off parameter.

Most active learning algorithms are designed to query all the label assignments of the selected instances, Huang et al. proposed a two-dimensional approach in [6] that queries instance-label pairs based on a label ranking model; in other words, it selects one label c and an instance $\mathbf{x}$, and queries the oracle if $\mathbf{x}$ should be assigned to label c. While most of the existing multi-label active learning methods use multi-label SVM classification, which leads to the similarity between sub-classifiers and thus difficult to achieve good results. The active learning model used in this paper is extended from [1] which combines the label cardinality inconsistency and the separation margin and uses the SVM classification. However, we have made some improvements in classifier and parameter selection.

3 Revisiting ECOC via Multi-label SVM Classification

In the traditional multi-label active learning, most people use the multi-label SVM classification to obtain the model. It is a conceptually simple and computationally efficient solution. However, the similarity between classifiers has a serious impact on the SVM model, especially a large label set exists. In this paper, we employ a new method, Revisiting ECOC for Multi-Label Classification (RECOC), which can be regarded as an extension of the traditional multi-label SVM classification. To be specific, we employ the OVA encoding on a number of small-sized random selection.

Let $\{(\mathbf{x}_1, \mathbf{y}_1), (\mathbf{x}_2, \mathbf{y}_2), \cdots, (\mathbf{x}_{n_l}, \mathbf{y}_{n_l}), \mathbf{x}_{n_l+1}, \cdots, \mathbf{x}_n\}$ be the training data set, where $\mathcal{L} = \{(\mathbf{x}_i, \mathbf{y}_i)\}_{i=1}^{n_l}$ is a small set of labeled instances and $\mathcal{U} = \{\mathbf{x}_i\}_{i=n_u}^{n}$ is a large pool of unlabeled instances. Each instance $\mathbf{x}_i$ is a q-dimensional feature vector. Suppose there are m possible labels in all, $\mathbf{y}_i = [y_{i1}, y_{i2}, \cdots, y_{im}]^\top \in \{-1, +1\}^m$ is the class label of $\mathbf{x}_i$. The instance $\mathbf{x}_i$ is assigned into j-th label if $y_{ij} = +1$, otherwise, the instance does not belong to the j-th label.

Firstly, let's review the multi-label SVM classification which transforms a multi-label classification problem into a set of independent binary classification problems via the OVA strategy. For the j-th class of a m labels SVM classification, the binary SVM training is a standard quadratic optimization problem:

$$\begin{aligned} \min_{\mathbf{w}_j, b_j, \{\xi_{ij}\}} \quad & \frac{1}{2} \parallel \mathbf{w}_j \parallel + C \sum_{i=1}^{n} \xi_{ij} \\ \text{subject to} \quad & \mathbf{y}_{ij}(\mathbf{w}_j^\top \mathbf{x}_i) \geq 1 - \xi_{ij},\ \xi_{ij} \geq 0,\ \forall i \end{aligned} \tag{1}$$

In the RECOC method, we turn the original problem into a series of multi-label random selections with $k (k \leq m)$ classes. Specifically, we will randomly select k classes from the all m classes. We denote all possible random selections as R_k, and we can get a total of $|R_k| = C_m^k$ random selections. Then we choose d random selections from R_k denoted as $\Phi = \{\phi_i\}_{i=1}^d$. For each random selection ϕ_i, we will train a multi-label SVM classification via OVA scheme thus we can get $k \times d$ binary SVM classifiers. In ϕ_i, we do not select all instances when training the SVM classification because of not all instances falling into any of the selected classes. These samples are not much useful in training the classification in ϕ_i. We only select the instances that belong to at least one of the k classes.

Next, let's think about how do we get the predictive labels for unlabeled instances. The vector we get above is not the prediction label because there are repeating classes. In a nutshell, we obtain the soft label of the sample in each class by using the method of proportion of positive label in all the prediction results that contain this class. Due to the unavoidable error, different classifiers may have different results for the prediction of the same sample on the same label. But intuitively, the more classifiers predict this sample as a positive class, the more likely it is to be a positive one.

In particular, for an unlabeled instance $\mathbf{x}$, we suppose there are p_i classifiers that predict it as positive sample on i-th class and q_i classifiers which need to treat i-th class as positive for training. Then we can get the soft label of the sample in i-th class: $\widetilde{y}_i = p_i/q_i$. Let's denote the set of the random selections that contain the i-th class as Φ_i, the result can be expressed as:

$$\widetilde{y}_i = \frac{\sum_{h_i \in \Phi_i} I[h_i(\mathbf{x}) = 1]}{|\Phi_i|} \tag{2}$$

And $\widetilde{\mathbf{y}} = (\widetilde{y}_1, \widetilde{y}_2, \cdots, \widetilde{y}_m)$ is the soft label vector of the instance $\mathbf{x}$. In the last, we can select a threshold value $\widetilde{y}_0$, when the soft label value is greater than or equal to the threshold value, it is marked as positive, otherwise marked as negative. In this paper, we fix the threshold value $\widetilde{y}_0 = 0.5$. Let $f_i(\mathbf{x}) = \widetilde{y}_i - \widetilde{y}_0$ and $f(\mathbf{x}) = (f_1(\mathbf{x}), f_2(\mathbf{x}), \cdots, f_m(\mathbf{x}))$. Finally, the prediction label for instance $\mathbf{x}$ is the form of $\hat{\mathbf{y}} = sign(f(\mathbf{x}))$.

If we set the parameters $k = m$ and $d = 1$, we will find that the RECOC classifier degrades to a multi-label SVM classifier via OVA strategy, so we say the former is a promotion of the latter. However, compared with the traditional multi-label SVM classifier, we only utilize a few classes when each binary SVM classifier is trained in the RECOC method which allows for differentiation between different classifiers so there will be some error-correcting function.

4 The Algorithm

In this section, we review two pool-based selection strategies: max-margin uncertainty sampling and label cardinality inconsistency [1], which consider the perspective of label forecasting and label statistics respectively. We aim to label the most informative instance from the unlabeled pool $\mathcal{U}$ iteratively, then move it to the labeled set $\mathcal{L}$ and retrain the classification model on the incremental $\mathcal{L}$. Same as Sect. 3, we set the label vector $\mathbf{y}_i \in \{1, -1\}^m$.

4.1 Max-Margin Uncertainty Sampling

Perhaps the uncertainty sampling is the simplest and most commonly used active query framework. The central idea of this framework is that the active learner queries the instance about which it is least certain how to label [7]. In single label problem, the instance closest to the classification boundary is regarded as the most uncertain instance for binary SVM classification. For multi-label SVM classification, this strategy can be applied directly to each binary classification, and then taking the minimum [8] or average [9] over all classes.

Specifically, for an unlabeled instance $\mathbf{x}_i$, we get the soft labels $f(\mathbf{x}_i) = (f_1(\mathbf{x}_i), f_2(\mathbf{x}_i), \cdots, f_m(\mathbf{x}_i))$ with the RECOC classification. Denote the set of predicted positive labels as $\hat{\mathbf{y}}_i^+$ and the set of predicted negative labels as $\hat{\mathbf{y}}_i^-$. We define the separation margin over instance $\mathbf{x}_i$ as:

$$sep\_margin(\mathrm{x}_i) = \min_{s \in \hat{\mathbf{y}}_i^+} f_s(\mathbf{x}_i) - \max_{t \in \hat{\mathbf{y}}_i^-} f_t(\mathbf{x}_i) = \min_{s \in \hat{\mathbf{y}}_i^+} |f_s(\mathbf{x}_i)| + \min_{t \in \hat{\mathbf{y}}_i^-} |f_t(\mathbf{x}_i)| \quad (3)$$

Then we define the margin prediction uncertainty as the inverse separation margin [1]:

$$u(\mathbf{x}) = \frac{1}{sep\_margin(\mathbf{x})} \quad (4)$$

Intuitively, a good SVM classification model should keep the positive and negative samples as far from the interface as possible. Thus the most uncertain instance is going to get the minimum value of separation margin, i.e. the maximum margin. Similarly, we apply this idea to the RECOC classification prediction uncertainty.

4.2 Label Cardinality Inconsistency

As we defined above, the separation margin is an effective uncertainty sampling. However, it is defined based on the predicted labels only, not existing ones. Under normal circumstances, the unlabeled and labeled samples are all drawn from the same underlying distribution so that their label cardinality are not too different. While, in practice, the results of our training will inevitably be biased, which may lead to a large difference of label cardinality between the unlabeled and labeled samples, that is, too many or too few positive tags are obtained. Figure 1 demonstrates such a toy example with five classes. For predicted results, the separation margin of the example instance $\mathbf{x}$ is large so that the uncertainty of $\mathbf{x}$ is not large, but it is opposite actually. This suggests that there is a large error in the mere use of margin prediction uncertainty.

Consider that the unlabeled and labeled samples come from the same distribution we mentioned above, we introduce the label cardinality inconsistency to measure the prediction uncertainty over an unlabeled instance from the label space perspective. For an unlabeled instance $\mathbf{x}$, it is defined as the Euclidean distance between the predicted positive label and the label cardinality of labeled instances $\mathcal{D}$ [1]:

$$c(\mathbf{x}) = \left\| \sum_{j=1}^{m} I[\hat{\mathbf{y}}_j > 0] - \frac{1}{n_l} \sum_{k=1}^{n_l} \sum_{j=1}^{m} I[\mathbf{y}_{kj} > 0] \right\|_2 \tag{5}$$

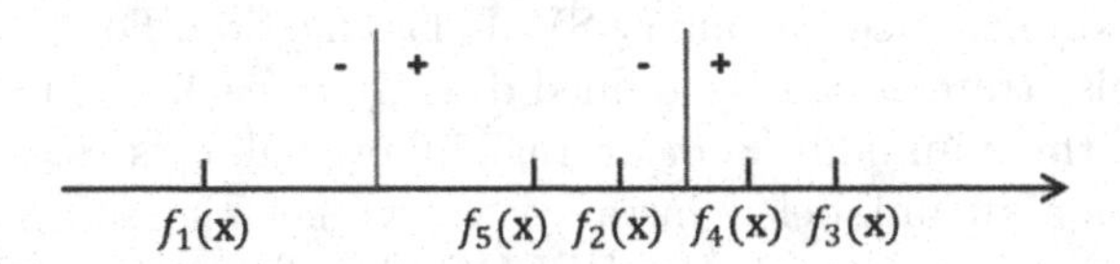

Fig. 1. The ordered prediction values across five classes over instance $\mathbf{x}$. The predicted separation line between positive and negative labels is marked as red line and the true separation line between positive and negative labels is marked as blue line. (Color figure online)

Obviously,the larger value of the label cardinality inconsistency, the more likely the sample is to be incorrectly marked.

4.3 Active Selection

In many cases, the two active learning strategies we reviewed above are complementary. So we use a weighted form to combine the advantages of these two strategies:

$$q(\mathbf{x}, \beta) = u(\mathbf{x})^{\beta} \cdot c(\mathbf{x})^{1-\beta} \tag{6}$$

where β is a trade-off parameter which balances the relative importance of the two measures.

Finally the instance selection on $\mathcal{U}$ can be conducted by

$$\mathbf{x}_* = \arg\max_{\mathbf{x} \in \mathcal{U}} \; q(\mathbf{x}, \beta) \tag{7}$$

In [1], the author set a parameter set B, and adaptively select the best parameter in each iteration of the active learning. However, it requires training the classification for each $\beta \in B$, which is very time consuming for the RECOC classification. In addition, we find in the experiment that for a specific dataset, there is little difference between using a fixed optimal parameter and adaptive selection the best parameter. We thus chose the previous method on each dataset. The overall active learning procedure is described in Algorithm 1.

Algorithm 1. The MAOC algorithm

Input:
labeled set $\mathcal{L}$, unlabeled set $\mathcal{U}$, the parameter k and d in the RECOC classifier and the trade-off parameter β.
Repeat:
Train RECOC classifier F^0 on $\mathcal{L}$.
for each $\mathbf{x}_i \in \mathcal{U}$ **do**
Compute $u(\mathbf{x}_i)$ and $c(\mathbf{x}_i)$.
end for
Select instance $\mathbf{x}_* = \arg\max_{\mathbf{x} \in \mathcal{U}} q(\mathbf{x}, \beta)$.
Query the label vector $\mathbf{y}_*$ for $\mathbf{x}_*$.
Update $\mathcal{L}$ and $\mathcal{U}$: $\mathcal{L} \rightarrow \mathcal{L} \cup (\mathbf{x}_*, \mathbf{y}_*)$, $\mathcal{U} \rightarrow \mathcal{U} \backslash \mathbf{x}_*$.
Retrain F with all instances in $\mathcal{L}$.
until enough instances are queried.

5 Experiments

5.1 Experiments Setup

We evaluate empirical performance of the proposed algorithm in the following six benchmark datasets: corel5k [10], image [11], medical [12], scene [13], tmc2007 [14] and yeast [15]. These datasets cover a number of areas and most of them are available at MULAN project[1]. Detailed characteristics of these datasets are summarized in Table 1, including associated domains, number of instances, number of labels, feature space dimensionality and label cardinality (LC), where LC is the average number of labels per instance.

Table 1. Statistics on datasets

Dataset	Domin	Instance	Label	Feature	LC
Corel5k	Image	5000	374	499	3.52
Image	Image	2000	5	294	1.24
Medical	Text	978	45	1499	1.25
Scene	Image	2407	6	294	1.07
Tmc2007	Text	28596	22	500	2.16
Yeast	Biology	2417	14	103	4.23

For each experiment, we first randomly partitioned the data into two parts with equal size, one part is taken as test set and the other part as the unlabeled pool for active selection. Because of an initial model required before active learning, we randomly sample 10% instances from the unlabeled pool as initial labeled data on the datasets corel5k, image, scene and yeast while 20% on the

[1] http://mulan.sourceforge.net/datasets.html.

dataset medical (Because the label/sample of the data set is relatively large). In experiments, the parameters k and d affect the performance and speed of the algorithm, we fix $k = 3$ and d equals the minimum of 3 m and C_m^k. We find that this parameter pair is the minimum to ensure the performance of the algorithm. We repeated the random data partition for 20 times, and average results over the 20 repeats are reported.

In the experiments, we will compare the proposed approach with four multi-label active learning approaches:

- **Random:** the baseline which randomly selects query instances.
- **Adaptive:** the method proposed in [1], which considers both the max-margin prediction uncertainty and the label cardinality inconsistency.
- **AUDI:** the method proposed in [6], which considers uncertainty and diversity when selecting instance-label pairs.
- **QUIER:** the method proposed in [16], which selects instance-label pairs based on informativeness and representativeness.
- **MAOC:** the method proposed in this paper.

We use one-versus-all linear SVM (implemented with LIBLINEAR [17]) as the classification model for evaluating the compared approaches Random, Adaptive, AUDI and QUIER. For the MAOC, we use the LIBLINEAR software to learn base binary classifiers. And we set C = 10 as default for all the approaches. At each iteration of active learning, we select one instance or one instance-label pair in each the active learning methods based on their own strategy, and then add it into the labeled data. After one instance (for AUDI and QUIER, m instance-label pairs) queried, a new classification model on the labeled data will be trained and evaluated the performance on the test data.

We evaluate the performances of the trained classifiers on the multi-label data set with the macro-F1 and micro-F1 score, which are commonly used in multi-label learning. [1,6,16], Both of them locate in [18].

- Macro-F1

$$\text{Micro-F1} = \frac{1}{n}\sum_{i=1}^{n}\frac{\sum_{k=1}^{m} I[\mathbf{y}_{ik}] = 1 \cdot I[\hat{\mathbf{y}}_{ik}] = 1}{\sum_{k=1}^{m}(I[\mathbf{y}_{ik}] = 1 + I[\hat{\mathbf{y}}_{ik}] = 1)}. \tag{8}$$

- Micro-F1

$$\text{Micro-F1} = \frac{\sum_{i=1}^{n}\sum_{k=1}^{m} I[\mathbf{y}_{ik}] = 1 \cdot I[\hat{\mathbf{y}}_{ik}] = 1}{\sum_{i=1}^{n}\sum_{k=1}^{m}(I[\mathbf{y}_{ik}] = 1 + I[\hat{\mathbf{y}}_{ik}] = 1)}. \tag{9}$$

Where $\mathbf{y}_{ik}$ and $\hat{\mathbf{y}}_{ik}$ are the true and predicted label of the ith instance on the kth class respectively, and $I[\cdot]$ is the indicator function.

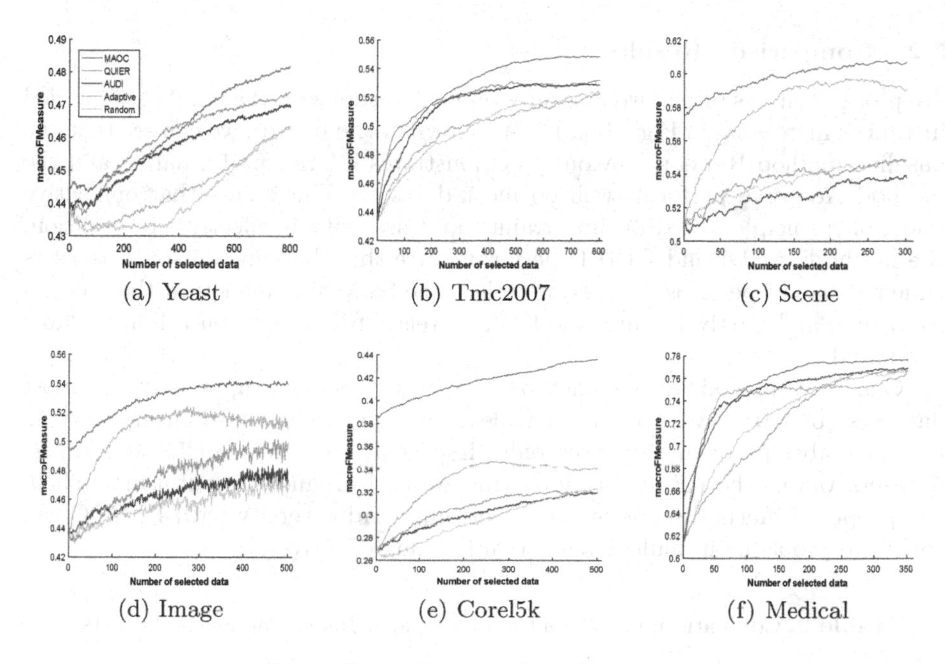

(a) Yeast (b) Tmc2007 (c) Scene

(d) Image (e) Corel5k (f) Medical

Fig. 2. Comparison result on Macro-F1.

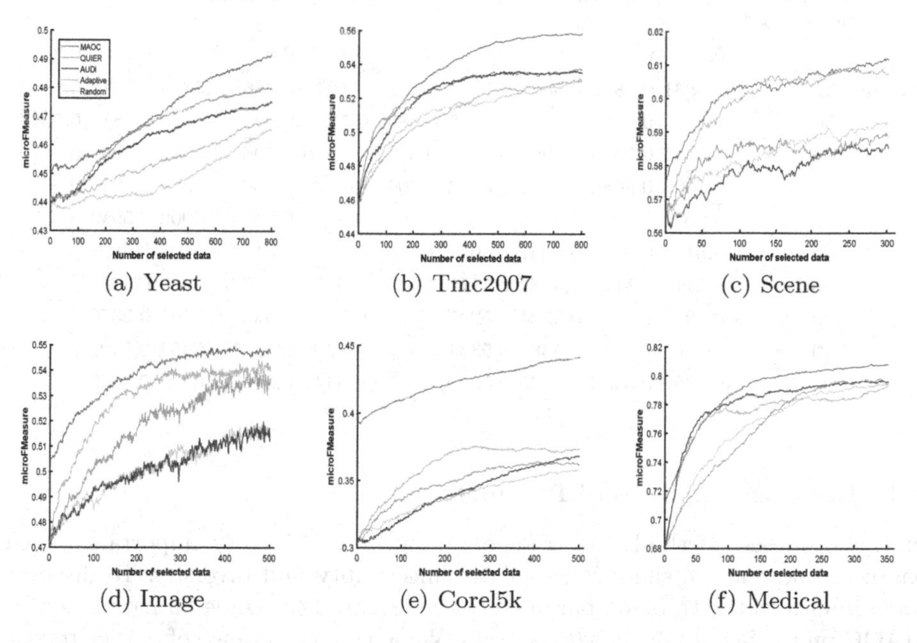

(a) Yeast (b) Tmc2007 (c) Scene

(d) Image (e) Corel5k (f) Medical

Fig. 3. Comparison result on Micro-F1.

The Macro-F1 and Micro-F1 utilize predictions of all instances on the whole label set simultaneously. Obviously, for both the metrics, the larger the value, the better model we obtain.

5.2 Comparison Results

We plot the curves of the two measures on all datasets with the number of queried instances increasing in Figs. 2 and 3. As shown in the figures, we can see that the baseline method Random obviously demonstrates inferior performance, and the method Adaptive performs well on most datasets. This shows that our active selection principle based on uncertainty and diversity is effective. In addition, the methods AUDI and QUIER perform worse than baseline on some datasets which is partly perhaps because we select the SVM parameters in the method Adaptive, and partly because the label correlation is lost when a limited label is queried.

Generally speaking, our method demonstrates the superiority on most datasets compared with other active learning approaches, no matter querying instance-label pairs or instances only. Especially, on the Corel5k and Image datasets, our method always achieves the best performance. To put it in nutshell, the proposed method merging the uncertainty and diversity with RECOC can solve the problems in multi-label active learning effectively.

Table 2. Comparison of different parameters on Image and Scene datasets.

Data	Selected instances	macroF measure					microF measure				
		0.1	0.3	0.5	0.7	0.9	0.1	0.3	0.5	0.7	0.9
Scene	25	0.5831	0.5832	0.5838	0.5817	0.5810	0.5874	0.5862	0.5826	0.5810	0.5809
	50	0.5890	0.5887	0.5889	0.5845	0.5856	0.5934	0.5924	0.5868	0.5834	0.5843
	100	0.5984	0.5960	0.5963	0.5935	0.5901	0.6040	0.5963	0.5919	0.5890	0.5870
	200	0.6008	0.6035	0.6042	0.6003	0.6004	0.6077	0.6004	0.5988	0.5949	0.5951
	300	0.6060	0.6063	0.6073	0.6022	0.6015	0.6115	0.6019	0.6005	0.5969	0.5959
Image	50	0.5106	0.504	0.5067	0.5077	/	0.5277	0.5172	0.5194	0.5203	/
	100	0.5117	0.5124	0.5172	0.5175	/	0.5254	0.5260	0.5280	0.5284	/
	200	0.5255	0.5262	0.5324	0.5297	/	0.5374	0.5383	0.5401	0.5376	/
	300	0.5262	0.5323	0.5381	0.5331	/	0.5390	0.5425	0.5454	0.5406	/
	500	0.5329	0.5433	0.5395	0.5409	/	0.5441	0.5507	0.5470	0.5469	/

5.3 Influence Analysis of Parameter

In the proposed method, the trade-off parameter β is very important which balances the relative significance of the uncertainty and diversity. To discover the influence of the trade-off parameter, we evaluated the trade-off parameter for MAOC on the Scene and Image datasets. We selected the range of parameters as $\{0.1, 0.3, 0.5, 0.7, 0.9\}$. The other settings are same to the previous experiments. The phenomenon to notice is that the difference is so slight when β equals 0.7 and 0.9 on the image dataset that we ignore the situation where $\beta = 0.9$.

Table 2 shows the average results in 20 runs of macro-F1 and micro-F1 values, with different percent of unlabeled data used as queries. We can observe that different parameters are selected as the best option for different datasets. For example, the smaller the parameter β, the better results that the proposed method obtains on the Scene dataset, which indicates the uncertainty plays a crucial role. While it is not on another dataset.

5.4 Compared with Traditional ECOC Method

To further elucidate the motivation of our proposed method, we replaced the RECOC classification with traditional multi-label ECOC classification, which is usually adopted in the state-of-the-art methods. The results obtained by RECOC and traditional ECOC are shown in some different percent of selected data in Fig. 4. Tr-ECOC in the figure denotes the traditional ECOC method.

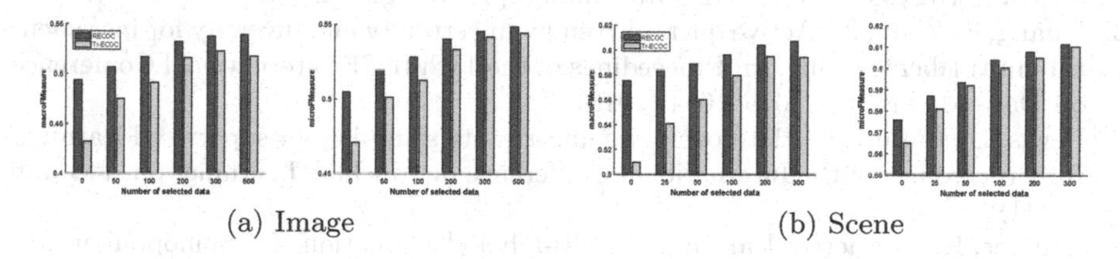

(a) Image (b) Scene

Fig. 4. Comparison with traditional ECOC on two datasets.

We can observe that the RECOC classification performs better than traditional ECOC classification based on the same active selection. Because RECOC classification is more capable of dealing with class imbalance problems. In a word, the method based on RECOC may get a better classification model so that our active learning approach can perform well.

6 Conclusion

A successful active learning approach needs both the effective classification model and the good selection criterion. In this paper, we proposed a novel multi-label active learning method, combining a new ECOC classification with a traditional active learning strategy which utilizes the max-margin prediction uncertainty and the label cardinality diversity. The experiments on multiple multi-label classification datasets from different application areas shows that the proposed multi-label active learning method outperforms state-of- the-art methods in most cases. In our future work, we plan to more efficiently select the trade-off between uncertainty and diversity. In addition, we plan to combine the RECOC classification with other active selection criterion.

Acknowledgments. This work was supported by the National Natural Science Foundation of China (No. 61473302, 61503396). Chenping Hou is the corresponding author of this paper.

References

1. Li, X., Guo, Y.: Active learning with multi-label SVM classification. In: Proceedings of the 23rd International Joint Conference on Artificial Intelligence, pp. 1479–1485 (2013)
2. Guo, Y., Schuurmans, D.: Discriminative batch mode active learning. In: Proceedings of Advances in Neural Information Processing Systems, pp. 593–600 (2008)
3. Yang, Y., Ma, Z., Nie, F., Chang, X., Hauptmann, A.G.: Multi-class active learning by uncertainty sampling with diversity maximization. Int. J. Comput. Vis. **113**(2), 113–127 (2014)
4. Li, X., Wang, L., Sung, E.: Multi-label SVM active learning for image classification. In: International Conference on Image Processing, pp. 2207–2210 (2004)
5. Yang, B., Sun, J., Wang, T., Chen, Z.: Effective multi-label active learning for text classification. In: Proceedings of the 15th ACM SIGKDD International Conference on Knowledge Discovery and Data Mining, pp. 917–926 (2009)
6. Huang, S., Zhou, Z.: Active query driven by uncertainty and diversity for incremental multi-label learning. In: Proceedings of the 13th IEEE International Conference on Data Mining, pp. 1079–1084 (2013)
7. Lewis, D., Catlett, J.: Heterogeneous uncertainty sampling for supervised learning. In: Proceedings of the International Conference on Machine Learning, pp. 148–156 (1994)
8. Brinker, K.: On active learning in multi-label classification. In: Spiliopoulou, M., Kruse, R., Borgelt, C., Nürnberger, A., Gaul, W. (eds.) From Data and Information Analysis to Knowledge Engineering. STUDIES CLASS, pp. 206–213. Springer, Heidelberg (2006). https://doi.org/10.1007/3-540-31314-1_24
9. Singh, M., Curran, E., Cunningham, P.: Active learning for multi-label image annotation. In: Proceedings of the 19th Irish Conference on Artificial Intelligence and Cognitive Science, pp. 173–182 (2009)
10. Duygulu, P., Barnard, K., de Freitas, J.F.G., Forsyth, D.A.: Object recognition as machine translation: learning a lexicon for a fixed image vocabulary. In: Heyden, A., Sparr, G., Nielsen, M., Johansen, P. (eds.) ECCV 2002. LNCS, vol. 2353, pp. 97–112. Springer, Heidelberg (2002). https://doi.org/10.1007/3-540-47979-1_7
11. Zhang, M., Zhou, Z.: ML-KNN: a lazy learning approach to multi-label learning. In: Pattern Recognition, pp. 2038–2048 (2007)
12. Pestian, J.P., et al.: A shared task involving multi-label classification of clinical free text. In: Proceedings of the Workshop on BioNLP 2007: Biological, Translational, and Clinical Language Processing, pp. 97–104 (2007)
13. Boutell, M., Luo, J., Shen, X., Brown, C.: Learning multi-label scene classification. Pattern Recogn. **37**, 1757–1771 (2004)
14. Srivastava, A., Zane-Ulman, B.: Discovering recurring anomalies in text reports regarding complex space systems. In: IEEE Aerospace Conference (2005)
15. Elisseeff, A., Weston, J.: A kernel method for multi-labelled classification. In: Advances in Neural Information Processing Systems, pp. 681–687 (2001)
16. Huang, S., Jin, R., Zhou, Z.: Active learning by querying informative and representative examples. IEEE Trans. Pattern Anal. Mach. Intell. **36**(10), 1936–1949 (2014)
17. Fan, R., Chang, K., Hsieh, C., Wang, X., Lin, C.: LIBLINEAR: a library for large linear classification. J. Mach. Learn. Res. **9**, 1871–1874 (2008)
18. Zhang, M., Zhou, Z.: A review on multi-label learning algorithms. IEEE Trans. Knowl. Data Eng. **26**(8), 1819–1837 (2014)

Dynamically Weighted Multi-View Semi-Supervised Learning for CAPTCHA

Congqing He, Li Peng(✉), Yuquan Le, and Jiawei He

College of Computer Science and Electronic Engineering, Hunan University, Changsha, China
{hecongqing,rj_lpeng,leyuquan,hejiawei}@hnu.edu.cn

Abstract. With the development of Optical Character Recognition and artificial intelligence technologies, the security of Behavioral Completely Automated Public Turing test to tell Computers and Humans Apart (CAPTCHA) has become an increasingly difficult task. In order to prevent malicious attacks and maintain network security, most existing works on CAPTCHA are to construct a fine binary classifier model but are not yet capable of detecting new attack means during confrontation. This motivates us to propose a Dynamically Weighted Multi-View Semi-Supervised Learning, dubbed as DWMVSSL method, to relieve this problem. More specifically, our proposed method extracts hidden patterns from multiple perspectives and updates the view weighting dynamically which can constantly detect new attack means. In addition, due to existing some redundant feature in views, we design a Filter Artificial Bee Colony method, named as FABC for feature selection which can efficiently reduce the impact of high dimensional features. The experimental results show that, compared the existing representative baseline methods, our DWMVSSL method can effectively detecting new attacks on confrontation.

Keywords: CAPTCHA · Semi-supervised learning · Multi-view · Feature selection

1 Introduction

CAPTCHA is a widely used security defense mechanism which is utilized by service providers to determine whether the entity interacting with their system is human or robot [1]. Generally, it can be divided into two types: Random CAPTCHA and Behavioral CAPTCHA. With the development of Optical Character Recognition (OCR) and the emergence of artificial intelligence technologies, the security of Random CAPTCHA based on character (number, letter) recognition is declining and easy to get hacked. However, Behavioral CAPTCHA makes lots of analysis on human behaviors and then recognizes human or machine by algorithms. Therefore, Behavioral CAPTCHA is less vulnerable to attack compared with Random CAPTCHA. Behavioral CAPTCHA has recently received loads of attention because of its maneuverability and simplicity.

Q. Yang et al. (Eds.): PAKDD 2019, LNAI 11440, pp. 343–354, 2019.
https://doi.org/10.1007/978-3-030-16145-3_27

With the development of machine learning and deep learning, many methods have been successfully applied for CAPTCHA and have achieved good performance. In this paper, we mainly divide them into three categories, supervised learning, unsupervised learning and semi-supervised learning. Many supervised learning methods, such as Decision Tree [2] and Boosting [3] have been wildly used to recognize human or robot. However, these methods need mountains of label dataset and take lots of time to mark unlabeled dataset. What'smore, simple labeled dataset is considered as obsolete, and the accuracy of supervised learning drops dramatically for new type of attacks. Unsupervised learning is another technique used to improve the detection rate of the CAPTCHA. However, unsupervised learning depends on manual assignment of cluster numbers, which leads to lower accuracy in prediction. Many new attacks have been developed during the confrontation, and a small number of labeled dataset with a large number of unlabeled dataset could make semi-supervised learning [4] to be the better choices for improving the accuracy of CAPTCHA. Existing semi-supervised learning such as Co-Training [10], Tri-Training [11], Co-Forest [12] are representative methods. Nevertheless, these methods cannot be solved well in detecting new attacks during confrontation due to they not consider collecting information from multiple perspectives, which have been proved beneficial by prior works [13]. However, multiple perspectives bring noise information in each iteration of semi-supervised learning, which lead to reduce the effect of views.

Inspired by the above observations, in this paper we propose a Dynamically Weighted Multi-View Semi-Supervised Learning method. In each iteration, the method can absorb the information from multiple perspectives as much as possible by enduing different weights to each perspective, which can relieve the perspective absorbing error information during iterations. In addition, our proposed method extracts hidden patterns from multiple perspectives and updates the view weights dynamically which can constantly detect new attack means.

To summarize, the main contributions of this paper are:

- We propose a Dynamically Weighted Multi-View Semi-Supervised Learning (DWMVSSL) method that not only handles CAPTCHA problems better, but also detects new attacks on confrontation.
- Due to the fact that each perspective has some redundant information and traditional feature selection methods take high computational cost or cannot obtain optimal subset in Behavioral CAPTCHA. Inspired by this, our proposed Filter Artificial Bee Colony (FABC) method for feature selection, which can shorten the time to find the best subsets.
- Comparing with the representative baselines on a real-world dataset, the experimental results show that our model have state-of-the-art performance and strong generalization ability in CAPTCHA.

2 Related Work

With the development of CAPTCHA, many methods have been successfully applied in the field of CAPTCHA and have achieved good performance. In this paper, we mainly divide them into three categories, supervised learning, unsupervised learning and semi-supervised learning. Many supervised learning methods, such as Decision Tree [2] and Boosting [3] have been wildly used to recognize human or robot. However, these methods need mountains of label dataset and take lots of time to mark unlabeled dataset manually. What'smore, simple labeled dataset is considered as obsolete, and the accuracy of supervised learning technologies drop dramatically for new types of attacks. Unsupervised learning is another technique, which is used to improve the detection rate of the CAPTCHA. However, unsupervised learning depends on manual assignment of cluster numbers, which leads to lower accuracy in prediction. Many new attacks have been developed during the confrontation, and a small number of labeled dataset with a large number of unlabeled dataset could make semi-supervised learning techniques [4] to be the better choices for improving the accuracy of CAPTCHA. Existing semi-supervised learning such as Co-Training [10], Tri-Training [11], Co-Forest [12] are representative methods. Co-Training, which assumes that the data set has two sufficient and redundant views. However, this method is hard to satisfy the sufficient and redundant views. In order to relax the constraints of Co-Training, Zhou et al. [11] propose a Tri-Training algorithm that doesn't require sufficient and redundant views, and doesn't require the use of different types of classifiers. Hereafter, Li et al. [12] extend the Tri-Training and propose a Co-Forest algorithm that could play a better role in ensemble learning. Our proposed method belongs to semi-supervised learning.

Compared to these semi-supervised learning methods, our proposed method shares several common features with theirs: (1) We are all semi-supervised learning approaches and (2) we are all iterating unlabeled dataset. Nevertheless, our work is different from theirs in several features at least: (1) Most of these works are not considered from multiple perspectives and (2) our proposed method achieves better performance in CAPTCHA task. In the existing works, the most relevant works to us is Sindhwani et al. [14]. Compared to these multi-view semi-supervised learning methods, our proposed method shares several common features with theirs: Our work also use multi-view semi-supervised learning framework. Nevertheless, our work is different from theirs in several features at least: (1) We weight the perspective by its representational ability during the iterations and (2) our proposed method detects new attacks in CAPTCHA task.

3 Method

In this section, we first describe our proposed Filter Artificial Bee Colony method to reduce the impact of high dimensional features, and then propose a Dynamically Weighted Multi-View Semi-Supervised Learning method to detect new attack means for CAPTCHA.

3.1 Filter Artificial Bee Colony Method

Due to the fact that our proposed DWMVSSL method has some redundant features in each perspective, we explore a feature selection method for CAPTCHA.

Artificial Bee Colony (ABC) algorithm [7] is an optimization algorithm based on the intelligent behaviour of honey bee swarm which has capability for exploring optimal subset, and has been wildly used for feature selection. The ABC algorithm includes four phases: initialization phase, employed bees phase, onlooker bees phase and scout bees phase. In addition, the employed bees phase describes the employed bees behaviour for finding a better food source within the neighbourhood of the food source in their minds which is critical for explore the best solution.

Standard ABC [7] algorithm has strong capability for exploration but poor at exploitation, especially both exploration and exploitation are necessary for a population-based optimization algorithm. Simultaneously, the Fast Correlation-Based Filter (FCBF) [5,8] algorithm is a filter method for feature selection which is hard to obtain optimal subset. Based on previous works, we combine FCBF with ABC algorithm, and then propose a novel Filter Artificial Bee Colony (FABC) algorithm for feature selection. The proposed method modifies the formula Eq. (2) of employed bees phase and proposes a novel formula which is adapted for feature selection. Specifically, our proposed FABC method employs FCBF to generate the approximate optimal solution S_{best} at first. In order to improve the exploitation, we then use Eq. (1) to generate a subset S'_{list}, and this subset chooses relevant features from the candidate solutions x_i which can guide the search of candidate direction. We modify the solution search equation described by Eq. (3). The formulas are shown as follows:

$$SU(X,Y) = 2\left[\frac{IG(X|Y)}{H(X)+H(Y)}\right] \tag{1}$$

$$v_i = x_{i,j} + \phi_{i,j} * (x_{k,j} - x_{i,j}) \tag{2}$$

$$v_i = \alpha * S'_{list} + (1-\alpha) * S_{best} + \phi_{i,j} * (x_{k,j} - x_{i,j}). \tag{3}$$

where X and Y represent features variables and labels variables respectively. $H(X)$ is the entropy of a variable X, $IG(X|Y)$ is called information gain, and $SU(X,Y)$ is symmetrical uncertainty. α is uniformly distributed vectors within the range $[0,1]$. $\phi_{i,j}$ is a random number within the range $[-1,1]$. x_k is the neighbor of x_i, then the formula produce the new candidate solution v_i.

3.2 Dynamically Weighted Multi-View Semi-Supervised Learning

The architectural overview of the proposed DWMVSSL method is introduced in this section. As shown in Fig. 1, our proposed method first extracts features from multiple views, this will be introduced in Sect. 4, and then employs FABC algorithm for feature selection. Afterwards, our proposed DWMVSSL method trains the model from multiple views and predicts the unlabeled dataset. Furthermore, we cluster the probabilities of unlabeled dataset and add the highest confident dataset to training dataset for each iteration.

For a given unlabeled dataset $X = \{x_t | x_t \in \Re^D, i = 1, \cdots, t, \cdots, T\}$ and number of views V. $Y_{(T \times V)}$ is the predicted probability of X by classifier from

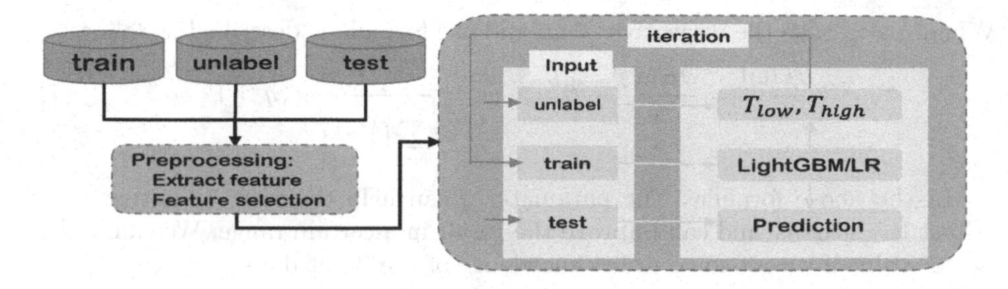

Fig. 1. The architectural overview of our proposed method.

V views, where $Y^v = \{y_1^v, \cdots, y_t^v, \cdots, y_T^v | y_t^v \in (0,1)\}$ denotes the prediction for unlabeled dataset from the v-th view. In order to choose unlabeled dataset with highest confidence, we categorize unlabeled dataset into three groups, Y_{low} is the samples which clustering center is closest to 0, and Y_{high} is the samples which clustering center is closest to 1, otherwise, Y_{mid}.

In order to choose unlabeled dataset with highest confidence, we design an objective function as follows:

$$F(\mu, c, \omega) = \sum_{k=1}^{K} \sum_{t=1}^{T} \mu_{kt}^{p} \sum_{v=1}^{V} \omega_v^q ||c_k^v - y_t^v||^2. \tag{4}$$

Where k, t, v satisfy $\forall t, \sum_{k=1}^{K} \mu_{kt} = 1; \forall k, 0 < \sum_{t=1}^{T} \mu_{kt} < T; \forall v, \sum_{v=1}^{V} \omega_v = 1, \omega_v \geq 0$ respectively. K, T, V represent number of cluster centers K, number of dataset T, number of views V respectively. μ_{kt} is a membership matrix which represents y_t belongs to membership of class k probability. ω_v denote the weight of dataset from the v-th view. $\|c_k^v - y_t^v\|^2$ is the Euclidean distance.

The goal of the Eq. (4) is to find the optimal ω, μ, c, so the objective function should be minimized as show follows:

$$\min_{\{\omega\}_{v=1}^{V}, \{\mu_k\}_{t=1}^{T}, \{c_v\}_{k=1}^{K}} F(\mu, c, \omega) \tag{5}$$

We design an EM-like iteration, which contains two stages, in the first stage, we update the view weight ω by fixing the membership matrix μ and the clustering center c. In the second stage, we update the membership matrix μ and the clustering center c by fixing the view weight ω.

Updating the View Weighting ω: In this stage, to search for the optimal the view weight, we design a Lagrangian formula based the constraint of the extremum $\sum_{v=1}^{V} \omega_v = 1$ to update the view weight vector ω as follows.

$$L(\omega) = F(\omega) + \beta(\sum_{v=1}^{V} \omega_v - 1) \tag{6}$$

When $q > 1$, take the derivative to ω, and the formula is described as follows.

$$\frac{\partial L(\omega)}{\partial \omega_v} = \frac{\partial F(\omega)}{\partial \omega_v} + \beta \Longrightarrow w_v = \frac{1}{\sum_{\hat{v}=1}^{V} (\frac{G_v}{G_{\hat{v}}})^{\frac{1}{q-1}}}, q > 1 \tag{7}$$

In the above formulas, the parameter q can help the view weighting ω to adjust the sparsity and can improve the result in a certain range. We can find a better value if we get some priori knowledge of the input data.

Updating Membership Matrix μ and Clustering Center c: In this stage, to search for the optimal membership matrix μ and clustering center c. Similar to the updating of the view weighting, we design a Lagrangian formula based the constraint of the extremum $\sum_{k=1}^{K} \mu_{kt} = 1$ to update the membership vector μ and clustering center c as follows.

$$L(\mu, \lambda, c) = F(\mu, c) + \sum_{t=1}^{T} \lambda_t (\sum_{k=1}^{K} \mu_{kt} - 1) \tag{8}$$

When $p > 1$, take the derivative to μ, and we can get the formula as follows.

$$\frac{\partial L(\mu, \lambda, c)}{\partial \mu_{kt}} = \frac{\partial F(\mu, \omega, c)}{\partial \mu_{kt}} + \lambda_t \Longrightarrow \mu_{kt} = \frac{1}{\sum_{\hat{k}=1}^{K} (\frac{W_{kt}}{W_{\hat{k}t}})^{\frac{1}{p-1}}} \tag{9}$$

Where the parameter p represents the fuzzy coefficient of FCM [15] and can affect the accuracy of classification.

Then we update the clustering center c according to in the above formulas. take the derivative to c, and then setting $\frac{\partial L(\mu, c)}{\partial c_k} = 0$, we can get the formula as follows.

$$\frac{\partial L(\mu, c)}{\partial c_k} = \sum_{t=1}^{T} \mu_{kt}^{p} \sum_{v=1}^{V} \omega_v^q [-2A(y_t^v - c_k^v)] = 0 \tag{10}$$

$$c_k^v = \frac{\sum_{t=1}^{T} \mu_{kt}^{p} \sum_{v=1}^{V} \omega_v^q y_t^v}{\sum_{t=1}^{T} \mu_{kt}^{p} \sum_{v=1}^{V} \omega_v^q} \tag{11}$$

3.3 The Complete Algorithm

Our proposed method is described in Algorithm 1. The algorithm first extracts features from each view, and then select optimal subset features by FABC. Secondly, the probability matrix $Y_{(T \times V)}$ is obtained by using base learner to predict dataset probability from multiple views, and initialize the weights for each view. Afterwards, our proposed DWMVSSL method clusters the probability matrix $Y_{(T \times V)}$, and categorizes the unlabeled dataset into three groups, i.e., when $\|c_k\| < \epsilon$, the dataset T_{low} belong to label 0, when $\|c_k - 1\| < \epsilon$, the dataset T_{high} belong to label 1, otherwise, the dataset T_{mid}. We add highest confidence dataset T_{low} and T_{high} to training dataset.

Algorithm 1. Framework of DWMVSSL method

Input:
Tr: Labeled dataset($x_i, y_i|1 \le i \le N$); Ts: Unlabeled dataset($x_t|1 \le t \le T$);
V: Number of views; Predefined thresholds: ε, γ;
Output:
The F1-score of test dataset.
1: Extract features from multi-view, and employ FABC for feature selection;
2: **repeat**
3: Obtain probability matrix $Y_{(T \times V)}$ by classifier Tr;
4: Initialize weighted multi-view $\omega_v = \frac{1}{V}$;
5: Generate cluster $F(Y_{(T \times V)})$;
6: **repeat**
7: Update the view weight ω_v;
8: Update the subjection vector μ and center of cluster c ;
9: **until** $L_k - L_{k-1} \le \gamma$.
10: $T_{low} = Ts_{\|c_k\|<\varepsilon}, T_{high} = Ts_{\|c_k - 1\|<\varepsilon}$,otherwise,$T_{mid}$;
11: $Tr_{new} = Tr + T_{low} + T_{high}$;
12: **until** Convergence or ($T_{low} or T_{high} = \oslash$).

4 Experiments

In this section, we empirically evaluate the efficiency of our proposed method. In Sect. 4.1, we describe data specification which includes dataset introduction and feature extraction. In Sect. 4.2, we describe several baselines for feature selection and several representative semi-supervised learning baselines. Then compare our proposed FABC method with baselines of feature selection in Sect. 4.3. In Sect. 4.4, we compare our proposed DWMVSSL method with representative baseline methods.

4.1 Data Specification

Since there are no publicly available datasets in previous works for Behavioral CAPTCHA task, we collect dataset published by a Behavioral CAPTCHA product[1]. We construct two datasets with different attacks, denoted as part-one, part-two. The part-one suffers simple attack means and the part-two confronts with more complex attack means. The training dataset is imbalanced which includes 400 robot's samples and 2600 human's samples. The positive and negative ratio of unlabeled dataset and test dataset is similar as the training dataset. In addition, we design F1-score to measure the performance. The statistics of our datasets are reported in Table 1.

Specifically, given a dataset consists of N instances, where move trajectory $m_n = (x_n, y_n, t_n)$ represents the coordinates and move time of the mouse during the movement, and target trajectory $T_n = (x_{(n,last)}, y_{(n,last)})$ means that the

[1] https://drive.google.com/open?id=1snepgqYUMBoTXWIPwPmumiieLqJKYPM_.

Table 1. The statistics of different datasets

Datasets	Train dataset	Unlabeled dataset	Test dataset
part-one	3,000	100,000	5,000
part-two	3,000	100,000	5,000

end point of the target, label is 0 and 1 which represent robot trajectory and human trajectory respectively.

In this paper, we extract features from $x - t$, $y - t$, $xy - t$ views such as speeds, accelerations, angles, speed deviation, time interval, angles deviation, distance, location. Taking an example of $xy - t$ view, the x, y Euclidean step sizes are represented as $\Delta x = \|x_{n+1} - x_n\|$, $\Delta y = \|y_{n+1} - y_n\|$ respectively, and the time interval is computed by $\Delta t = t_{n+1} - t_n$. What'smore, a speed Δv is calculated by $\frac{\Delta x}{\Delta t}$, an acceleration Δa is calculated by $\frac{\Delta v}{\Delta t}$, an angle θ_t is the angle between $\frac{\Delta y}{\Delta x}$, a speed deviation is calculated by $\Delta v_{n+1}/v_n$, the Euclidean distance is computed by $\sqrt{\Delta y^2 + \Delta x^2}$.

4.2 Baselines

We first compare the proposed FABC method with several classic baselines for feature selection: Artificial Bee Colony (ABC) algorithm is an optimization algorithm which can be efficiently employed to solve engineering problems with high dimensionality. Binary particle swarm optimisation (BPSO) [6,9] is another optimization algorithm which chooses a small number of features and achieves high classification accuracy. Fast Correlation-Based Filter (FCBF) is an algorithm which decouples relevance analysis and redundancy analysis.

Furthermore, we compare the proposed DWMVSSL method with several representative semi-supervised learning baselines: Self-Training is a simple semi-supervised algorithm which augmenst the original training set with a set of automatic predictions. Tri-Training generates three classifiers from the original labeled example set, and these classifiers are then refined using unlabeled examples. Co-Forest is a semi-supervised learning method that could play a better role in ensemble learning.

4.3 Result and Discussions on Feature Selection

In order to explore the performance of our proposed FABC method, we compare it with several baselines for feature selection. The experimental parameters are set as follows: For BPSO, the number of particles is set to 20, dimensionality of particle is equal to number of features in each view, and the maximum iterations is set to 30. For ABC, the population size is set to 20, and the maximum iteration number is set to 100. For FCBF, we set the selection of threshold $\lambda = 0$. For the proposed FABC method, the parameter α is set to 0.4.

As shown in Table 2, we compare these methods (in seconds) from running times and selecting number of features. We can observe that FABC is simultaneously faster than ABC and BPSO, these can attribute as follows: FABC employs FCBF to generate the approximate optimal solution, and uses Eq. (1) to generate a subset which can improve the time of decide the optimal direction. What'smore, compared with ABC and BPSO, FABC can obtain similar number of features. ABC, BPSO and the proposed FABC method are all learning algorithms and can search best feature subsets towards the right direction, which is different from FCBF. FCBF not only runs less time, but also has fewer features. The reason is that FCBF is a filter method of feature selection.

Table 2. Running time (in second) and number of features for each view

Test data (set 1)	Running time (seconds)				Number of feature selection				
	FABC	ABC	BPSO	FCBF	Full set	FABC	ABC	BPSO	FCBF
$View_1$	11.43	17.61	15.27	0.18	106	56	58	53	28
$View_2$	9.68	14.80	13.22	0.15	81	39	40	40	16
$View_3$	7.23	10.85	10.05	0.07	56	28	30	33	8

Table 3 shows the F1-score of all the methods from three views. In order to evaluate the performance of the full set, BPSO, ABC, FCBF and FABC, we employ Light Gradient Boosting Machine (LightGBM) as our base learner. Compared with the full set, all feature selection methods can obtain better performance except FCBF, and the F1-score of FCBF is not even better than the full set. These can attribute as follows: BPSO, ABC and FABC are all learning methods and can search best feature subsets. These methods remove redundant features and improve the performance. However, FCBF selects a small number of features and removes useful features which is critical for detecting attacks. Furthermore, FABC can achieve similar performance as ABC, and outperforms BPSO with a significant margin on three views from different test dataset. the reason is that the FABC and the ABC are flexible and robust optimization algorithms, can be used efficiently in the confrontation process.

4.4 Result and Discussions on DWMVSSL

In order to evaluate the performance of the proposed DWMVSSL method on Behavioral CAPTCHA task, we choose two kinds of unlabeled datasets to training our model, and then predict test dataset respectively. We compare our proposed DWMVSSL method with representative semi-supervised learning (self-training, Tri-training, Co-forest). These baselines are trained from multiple views and choose the best view as our result. For a fair comparison, these baselines are set the total iterations to 30 in our experiments, which is same as the proposed DWMVSSL method's setting. What'smore, all methods are employed LightGBM and Logistic regression (LR) as base learner.

Table 3. F1-score of feature selection algorithm for each view

Test dataset	View	Full set	FABC	ABC	BPSO	FCBF
part-one	$view_1$	82.91	**88.14**	88.04	87.73	87.48
	$view_2$	82.63	86.32	**86.35**	86.35	75.97
	$view_3$	77.04	**83.28**	81.28	81.03	73.25
part-two	$view_1$	73.29	**86.23**	85.16	85.08	70.32
	$view_2$	75.34	84.30	**84.44**	83.92	70.93
	$view_3$	70.58	77.30	**78.50**	75.24	70.07

In the part-one dataset's scenario, as shown in Fig. 2, all the methods are employed LR as base learner. We can observe that all the methods have an improvement of the performance when adding more and more iterations. However, our proposed method still outperforms other baselines at the beginning of iteration, the reason is that our proposed method can obtain highest confident unlabeled dataset more quickly. Finally, our proposed method and other baselines can obtain similar performance, the main reason is that all the methods can learn the correct unlabeled dataset during the iteration when suffer simple attack means. Correspondingly, as shown in Fig. 3, all the methods are employed LightGBM as base learner. We can find that our proposed method outperforms other baselines with a significant margin when adding more and more iterations, and has a higher peak value than other baselines. These can attribute as follows: LightGBM is an ensemble learning model which can cooperate with our proposed method obtaining an improve performance.

Figure 4 shown the part-two dataset's scenario and all the methods are employed LR as base learner. The F1-score of the our proposed method increases more quickly and then keeps stable at about 91.8%. What'smore, the Tri-Training can obtain better performance than our proposed method when the iterations is increased to 20, the reason can attribute as follows: our proposed method has a higher peak value when the iteration is set to 9, and unlabeled dataset is not sufficient. In addition, Fig. 5 shown the part-two dataset's scenario and all the methods are employed LightGBM as base learner. The F1-score of our proposed method outperforms other baselines with a significant margin when adding more and more iterations, and has a higher peak value than other baselines. Because of LightGBM is an ensemble learning model which can cooperate with our proposed method obtaining an improve performance. In general, our proposed method outperforms than other baselines when confronted with more complex attack means which illustrates that our proposed method can obtain competitive performance and detect new attacks means on confrontation.

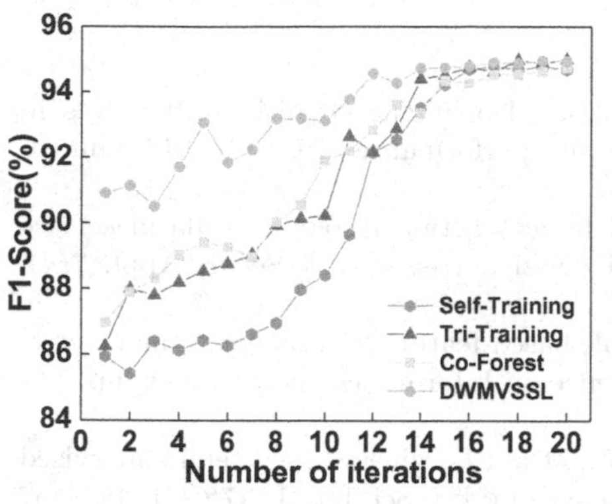

Fig. 2. Performance on the part-one test dataset (with LR as base learner).

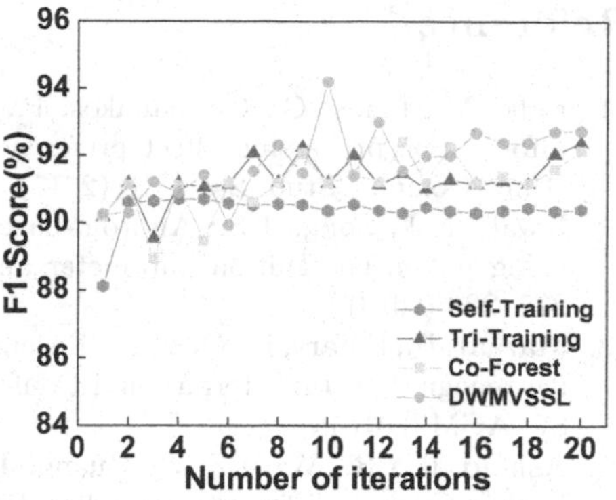

Fig. 3. Performance on the part-one test dataset (with LightGBM as base learner).

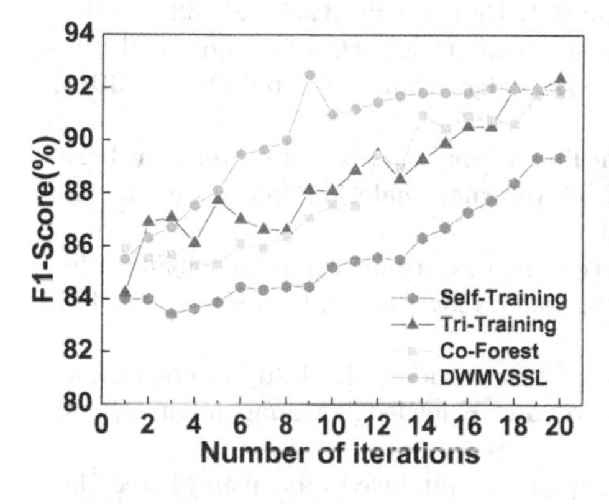

Fig. 4. Performance on the part-two test dataset (with LR as base learner).

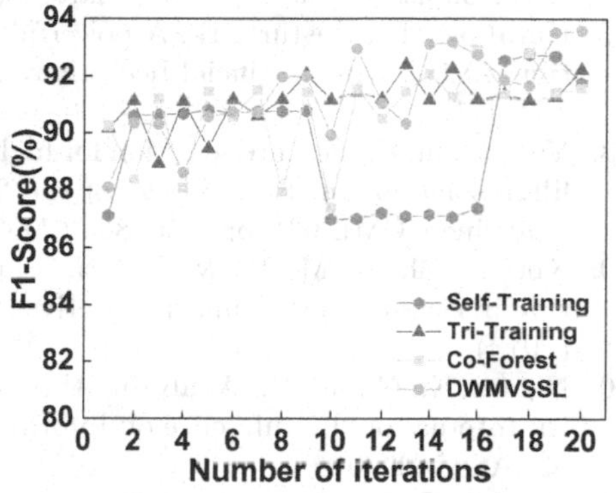

Fig. 5. Performance on the part-two test dataset (with LightGBM as base learner).

5 Conclusion

In this paper, we explore the method of detecting new attack means during confrontation for CAPTCHA. To alleviate the problem, we introduce Dynamically Weighted Multi-View Semi-Supervised Learning method. In particular, our proposed method extracts hidden patterns from multiple perspectives and dynamically updates the view weighting which can constantly detect new attack means. Experiments on the real-world datasets show that our method can achieve competitive performance.

In the future, we only utilize several uncomplicated attacks, while there exist more complex attacks. We plan to research on the possibility of applying on more complex dataset by our proposed DWMVSSL method.

References

1. Belk, M., Fidas, C., Germanakos, P., et al.: Do human cognitive differences in information processing affect preference and performance of CAPTCHA? Int. J. Hum.-Comput. Stud. **84**, 1–18 (2015)
2. Kwak, N.J., Song, T.S.: Android-based human action recognition alarm service using action recognition parameter and decision tree. Int. J. Secur. Appl. **7**(4), 277–286 (2013)
3. Mazaar, H., Emary, E., Onsi, H.: Ensemble based-feature selection on human activity recognition. In: International Conference on Informatics and Systems, pp. 81–87. ACM (2016)
4. Ashfaq, R.A.R., Wang, X.Z., Huang, J.Z., et al.: Fuzziness based semi-supervised learning approach for intrusion detection system. Inf. Sci. Int. J. **378**(C), 484–497 (2017)
5. Yu, L., Liu, H.: Eficient feature selection via analysis of relevance and redundancy. J. Mach. Learn. Res. **5**(12), 1205–1224 (2004)
6. Chuang, L.Y., Chang, H.W., Tu, C.J., et al.: Improved binary PSO for feature selection using gene expression data. Comput. Biol. Chem. **32**(1), 29–38 (2008)
7. Karaboga, D., Basturk, B.: A powerful and efficient algorithm for numerical function optimization: artificial bee colony (ABC) algorithm. J. Global Optim. **39**(3), 459–471 (2007)
8. Yu, L., Liu, H.: Feature selection for high-dimensional data: a fast correlation-based filter solution. In: Proceedings of the 20th International Conference on Machine Learning (ICML-03), pp. 856–863 (2003)
9. Xue, B., Zhang, M., Browne, W.N.: Particle swarm optimization for feature selection in classification: a multi-objective approach. IEEE Trans. Cybern. **43**(6), 1656 (2013)
10. Nigam, K., Ghani, R.: Analyzing the effectiveness and applicability of co-training. In: International Conference on Information and Knowledge Management, pp. 86–93. ACM (2000)
11. Zhou, Z.H., Li, M., et al.: Tri-training: exploiting unlabeled data using three classifiers. IEEE Trans. Knowl. Data Eng. **17**(11), 1529–1541 (2005)
12. Li, M., Zhou, Z.H.: Improve Computer-Aided Diagnosis With Machine Learning Techniques Using Undiagnosed Samples. IEEE Press (2007)
13. Zhu, S., Sun, X., Jin, D.: Multi-view semi-supervised learning for image classification. Neurocomputing **208**, 136–142 (2016)
14. Sindhwani, V., Niyogi, P., Belkin, M.: A co-regularization approach to semi-supervised learning with multiple views. In: Proceedings of ICML Workshop on Learning with Multiple Views, pp. 74–79. Citeseer (2005)
15. Bezdek, J.C., Ehrlich, R., Full, W.: FCM: the fuzzy c-means clustering algorithm. Comput. Geosci. **10**(2–3), 191–203 (1984)

Recommender System

A Novel Top-N Recommendation Approach Based on Conditional Variational Auto-Encoder

Bo Pang[1], Min Yang[2], and Chongjun Wang[1(✉)]

[1] State Key Laboratory for Novel Software Technology, Nanjing University, Nanjing, China
bpang@smail.nju.edu.cn, chjwang@nju.edu.cn
[2] Software Institute, Jilin University, Jilin, China
yangmin5516@mails.jlu.edu.cn

Abstract. Personalized recommendation has continuously received attention due to its great commercial value in business. Recently variational auto-encoder is employed in top-N recommendation for its effectiveness in deep collaborative filtering. The key challenge of model-based collaborative filtering is to develop effective latent factors representations with user-item interaction records. In this paper, we present a new class of conditional variational auto-encoders (CVAEs) that utilizes the fact of similar users tending to associate with each other on purchasing preference. This type of conditional variational auto-encoder concentrates on learning with label verification signals to ensure an exclusive latent mean factor for users with the same labels. Moreover, to handle complex multi-label combinations, we extend the model with a split-merge framework by learning labels of different conditional attributes separately and then merge the results from multiple prediction pools. Extensive experiments are conducted on two real-life datasets to simulate both user-based and item-based recommendation scenarios. Experimental results are favorable when comparing with the state-of-art methods.

Keywords: Recommender systems · Collaborative filtering · Variational auto-encoder

1 Introduction

In the era of information explosion, recommendation systems are of significantly importance in our daily life. Different from search engines, these systems provide users with information and content automatically according to their respective characteristics [18]. To be more specific, given the observed user-item interaction records, these systems try to figure out the user behavior patterns behind and make a personalized item recommendation list to each user. As the items grow in size rapidly these years, top-N recommendation task [2,14,15,19] plays a more important role than traditional rating prediction in helping users interact with larger amounts of interested items directly and efficiently.

Q. Yang et al. (Eds.): PAKDD 2019, LNAI 11440, pp. 357–368, 2019.
https://doi.org/10.1007/978-3-030-16145-3_28

Collaborative filtering is then a widely investigated approach both in scientific researches and industrial applications. Its intuition is simple but effective, which based on a fact that if users show similarity tastes on a set of items, then they are more likely to give similar ratings on another set of items. The model-based methods [3,5,20], which exploits the low-dimensional subspace representations of the users and items, have largely dominated the collaborative filtering researches due to their simplicity and effectiveness when compared with the neighbourhood-based methods [6,16]. Among several deep learning models, variational auto-encoders have been shown to be superior for the top-N recommendation [10–12]. Typically, variational auto-encoder based methods receive a user's preference on all items as input and reconstruction them in the output ending at one time. It is clear that a preference probability representation on all items for each specific user is acquired efficiently and thus well-suited for modeling this problem. Besides, VAE generates new data by modeling the underlying probability distribution of input data so that the diversity of recommendations can be controlled by sampling multiple results from that distribution.

Despite their state-of-the-art performances, the further improvements of recommendation quality have been largely limited by the utilizing method of side information. A powerful representation learning ability results in meaningful latent factors and helps to improve the recommendation accuracy in model-based collaborative filtering. However, lots of literatures have simply resorted to stacking network layers or more complicated structures to blend the side information into model. Therefore, we exploit the possibilities of leveraging these auxiliary labels as a learning signal in this work. Through label verification, the model is guided to be proactive about mining the latent relationships both between user-item and user-label. Another challenge is the complex multi-label combinations in real-world scenarios. Actually, personalized recommendation is a sparsity data mining task due to the user-item interaction records is relatively small comparing to the whole pool [7]. With the aggregation of conditional labels, the dimension of this feature grows high and thus introduces another sparsity data problem. It results in a less or even never appeared combinations in both training and testing stage, so that roughly aggregating labels interferes model learning and damage recommendation performance.

To address the first problem, we propose a novel class of CVAE [8] for output representation learning and label verification. In other words, the distribution of high dimensional output space is modeled as a generative part conditioned on the input observation, while label verification forces our model to distinguish and cluster similar users in the latent mean subspace. In addition, a split-merge framework is proposed to alleviate the second problem, where we utilize the generating property of CVAE and the idea of bagging. We firstly separate condition labels by attributes and train multiple models. Each sub model understands the task from different perspectives but all of them can generate predictions with roughly equivalent accuracy. Then we randomly pick results from the predictions pool and merge for further improvement.

The contributions of this paper are summarized as follows:

- We propose a novel solution in learning a CVAE by introducing label verification, where the projection of the latent mean factor shows our model successfully learn close representations for users with same labels.
- We demonstrate the effectiveness of our split-merge framework in handling complex condition combinations by further performance improvements.
- We simulate user-based and item-based recommendation by two real-life datasets and evaluate on them.

2 Preliminary

2.1 Variational Auto-encoder

Variational auto-encoder [9] consists of two parts: a recognition model and a generative model. The recognition model, known as an encoder, encodes input data X to latent representations z. The generative model then decodes the latent representations z to generate meaningful outputs. The optimization objective is the sum of the reconstruction loss of input data and the KL-divergence between the variational posterior and the prior:

$$\log P(X) - D_{KL}[Q(z|X) \| P(z|X)] = E[\log P(X|z)] - D_{KL}[Q(z|X) \| P(z)] \quad (1)$$

In other words, the inputs X are described by $\log P(X)$ under some error $D_{KL}[Q(z|X) \| P(z|X)]$. Considering the intractability of finding the exact distribution, a more practical method is to estimate the lower bound instead. Then the model could be established by maximizing likelihood estimation on the mappings from latent variable to data $\log P(X|z)$ and minimizing the differences between the predefined simple distribution $Q(z|X)$ and the true latent distribution $\log P(Z)$:

$$\log P(X) \geq E[\log P(X|z)] - D_{KL}[Q(z|X) \| P(z)] = \mathcal{L}_{recon} + \mathcal{L}_{\mu,\sigma^2} \quad (2)$$

2.2 Problem Description

Supposed there are users $U = \{1, \ldots, M\}$ and items $I = \{1, \ldots, N\}$, let matrix $R^{M \times N}$ denotes the user-item interaction records matrix. While referring to implicit feedback, matrix R is filled with binarized value instead of ratings, where $R_{mn} = 1$ denoting user m has click or view history on item n while $R_{mn} = 0$ denoting not.

Taking user-based recommendation as an example, each user $m \in U$ can be represented by a vector on its interactions over all items $x_m = (R_{m1}, \ldots, R_{mn})$. Full users set U is then divided into three subsets, known as training set U_{train}, validation set U_{val} and test set U_{test}. The variational auto-encoder based collaborative filtering receives entire U_{train} as training data and tunes model hyperparameters through U_{val}. To test and evaluate, each $x_{te} \in U_{test}$ is taken part of click histories deliberately as held-out set. The model use remaining click histories to learn a necessary user-level representation and reconstruct each user's full click history as output. Then the model is evaluated by examining how well it ranks those held-out records.

3 Proposed Method

This section presents the details of how to perform label verification in representation learning and handle complex condition combinations.

3.1 CVAE Model

Whereas variational auto-encoder essentially models latent variables and data directly, conditional variational auto-encoder models latent factors conditioned to some given attributes. Input X is then encoded to a distribution $Q(z|X,c)$ owning zero mean and unit variance, which prevents noise from being zero and ensure the model is capable of generation simultaneously. Also, reconstruction outputs X' can be generated by sampling from that Gaussian distribution. This procedure is commonly known as amortized inference [1] in the variational auto-encoder. In terms of the existing of condition c, the original evidence lower bound can be rewritten as follows:

$$\log P(X|c) = E[\log P(X|z,c)] - D_{KL}[Q(z|X,c)||P(z)] \tag{3}$$

As we mentioned above, we expect entities with the same condition label c can own their unique latent mean factor. Taking user-based recommendation as an example, if age information is acquired as an argument condition, then our model try to learn exclusive mean factor for children and adults' groups, where latent variance factor remains unit. In other words, recommendation results are then sampled and reconstructed from two different distributions that conditioned on age information. To achieve this goal, side information C should be encoded in our network as a vector the same with users' representations. Figure 1 is an illustration of our model.

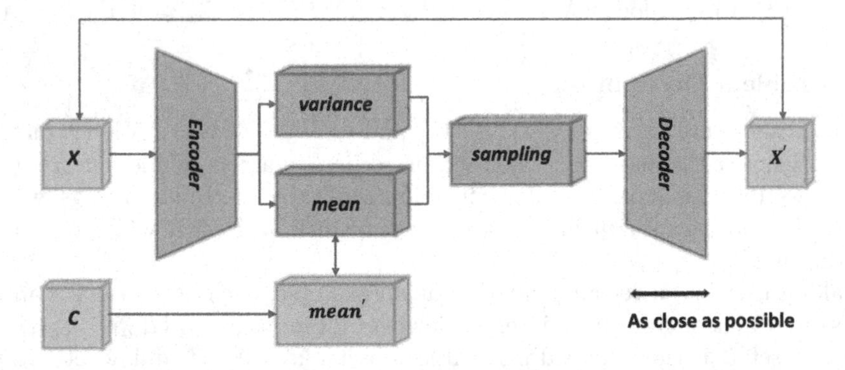

Fig. 1. An overview of the proposed CVAE model. A unidirectional arrow denotes input-output relationships between modules and a bidirectional arrow denotes the optimization objective.

We introduce label verification process by applying three changes to the optimization objective. Firstly, KL divergence loss is integrated with encoded side information to force the model to distinguish entities with different condition labels:

$$\mathcal{L}_{\mu,\sigma^2} = \frac{1}{2}\sum_{i=1}^{k}\left[\left(\mu_i - \mu_{X_c}\right)^2 + \sigma_i^2 - \log\sigma_i^2 - 1\right] \tag{4}$$

where k is the dimension of latent sampled factor, μ_i and σ_i are the latent mean and unit variance of the approximate posterior. Meanwhile μ_{X_c} is the new latent mean factor for entities with a specific condition label c, and this value can be learned automatically.

Secondly, we employ the softmax as the activation function of the output layer and categorical cross-entropy as the reconstruction loss of input data:

$$\mathcal{L}_{recon} = \sum_{i=1}^{n} y^{(i)} \log h_\theta(x_i) + (1 - y_i)\log(1 - h_\theta(x_i)) \tag{5}$$

where n is number of the entire items, $h_\theta(x_i) = e^{x_i}/\sum_{i=1}^{n} e^{x_i}$ is the softmax function. The outputs of softmax are actually the probabilities of each possible classification labels, where each output probability is greater than or equal to 0 and the sum is up to 1. It is quite suitable in modeling the top-N recommendation task since each probability can be recognized as the degree of preference on all items for each specific user. Through applying a categorical cross-entropy, the model can allocate a larger probability value to items that a user may pay more attention to and meanwhile reduce the probability value on other uninterested items. Another benefit is the probabilistic outputs are more stable to fit our proposed split-merge framework in combing prediction results.

Thirdly, β-VAE [4] proves that using a single hyper parameter β to balance latent channel capacity and independence constraints with reconstruction accuracy works well in image generation tasks. Similarly, we employ a hyper parameter β in our optimization objective to balance reconstruction loss of the inputs X and the KL-divergence loss between the variational posterior and the prior:

$$\log P(X) = \mathcal{L}_{recon} + \beta \cdot \mathcal{L}_{\mu,\sigma^2} \tag{6}$$

where we empirically set an initial value at 0.01 for β and then anneal it to 1.

- **Learning CVAE:** We adopt stochastic gradient descent to train the network. While the loss on validation set decreases, we repeat feeding batches of data and update parameters by computing their gradients. Besides, we apply the reparametrization trick to isolate the sampling process, thus make the gradient can be back-propagated.

3.2 The Split-Merge Framework for Multiple Conditions

Since our CVAE is designed for the simplest case, this method soon encounter a bottleneck when users or items are labeled with multiple times in practical.

Supposed we have n conditions $c = c_1, c_2, \ldots, c_n$, where c_i can be expressed by a one-hot vector with m dimension. This side information feature vector is then expanded to a complex combination of all labels. While there exists $m * n$ conditions combinations in total, actually only a small part of the combinations will appear. In the training stage, this cause our data inputs more unbalanced. As for the testing stage, model generalization ability is then reduced by some less or even never appeared combinations. Therefore, a split-merge framework is proposed to address this problem. The core idea of this method is learning conditions separately based on their attributes. Besides, we also utilize the generating property of CVAEs and the idea of bagging to increase the diversity of prediction results and recommendation performance. This framework is then described in Algorithm 1.

Algorithm 1. The Split-Merge Framework for Multiple Conditions (User-based)

Input: The training user set U_{train}, the validation user set U_{val}, the test user set U_{test}, the condition number n, the conditions $C = \{c1, \ldots, c_n\}$, user-item interaction matrix $R^{M \times N}$

Output: Recommend K items for each user in U_{test}

1: Split $R^{M \times N}$ by rows with three user subset U_{train}, U_{val} and U_{test}
2: Randomly select part of item-click history of each user in U_{test} to hidden as hold-out testing
3: $Y = [\,]$
4: **for all** attribute c_i in Conditions C **do**
5: Transform c_i to an one-hot feature vector A_i
6: Train CVAE on U_{train} with A_i
7: Evaluate on U_{val} and get a validation score w_i
8: Make predictions on U_{test} and append to Y
9: **end for**
10: Combine probability predictions from different CVAE with $\sum_{i=1}^{n} w_i Y_i / n$
11: **return** K items with top-K highest probability

As shown in Line 1–2, input user-item interaction matrix is firstly divided into training, validating and testing sets. After data preparation, condition labels are split by attributes and then used for training multiple models. In Line 3–9, each model generates a set of prediction lists respectively by repeated sampling, which stands for users' tastes on items varies according to their current state. Noted that in line 7 we evaluate these models and record their performances as weights for further merging. In Line 10, prediction results are selected from each sub model and combined in a linear way. At last, a final top-K recommendation list on the whole items is obtained after ranking the aggregated preference probability of each item.

4 Experiments

- **Datasets:** We select two widely-used datasets to evaluate our method. As shown in Table 1, plenty of user and item information can be regarded as condition labels. We transform the explicit data to implicit form as [12], where each record is marked as 0 or 1 to indicate whether the user has rated the item. We use the first dataset to simulate user-based recommendation and divide users into an 80/10/10 split. Different from the item-based recommendation, we provide users with items instead of recommending potential buyers for each item. As for the second dataset, we split training and testing samples based on items.

Table 1. Statistics of our datasets.

Datasets	MovieLens-1M	MovieLens-2k
Rating#	1,000,209	855,598
User#	6040	2113
Item#	3706	10197
Density	4.7%	3.9%
User Information	Age, Gender, Occupation	Tag
Item Information	Title, Genre	Genre, Country

- **Metrics:** In Top-k recommendation, hit ratio is a commonly used indicator to measure the recall rate. Formally, $w(i)$ denotes the item ranking at i, and I_u denotes the set of held-out items that user u has clicked on. Its calculation formula for each user u is as follows:

$$Recall@(u,K) = \frac{\sum_{i=1}^{K} \|w(i) \in I_u\|}{\min(K, \|I_u\|)} \tag{7}$$

Discounted cumulative gain (DCG) for each user u can also be calculated, where $\mathbb{I}[\cdot]$ is the scoring function. To make DCG comparable across users, we normalize the gain to a number between 0.0 and 1.0. The normalization is accomplished by dividing each user's DCG with the so-called Ideal DCG, which is the best DCG results among all the users.

$$DCG@(u,K) = \sum_{i=1}^{K} \frac{\mathbb{I}[w(i) \in I_u]}{log_2(i+1)} \tag{8}$$

- **Implementation Details:** In encoder, we set the dimension of the intermediate layer to 64 and both latent mean and variance layer to 32. In decoder, the dimension of sampling layer is 32 while the intermediate layer is 64. We apply relu activation for all the intermediate layers and softmax activation for the output layer. As for the optimizer, we use Adam with a reducing learning rate begin at 0.01 and with a dropping factor at 0.1. The batch size is set to 151 in ML-1m dataset and 202 in ML-2k dataset.

4.1 The Projection of Latent Feature

In this section, we aim to demonstrate that the condition information is indeed encoded in the latent mean factor after our training procedure. To this end, we perform t-SNE [13] visualizations on the mean factor obtained with and without label verification. To be more specific, we project the latent mean factor of each user to a point onto a two-dimensional plane and label them with corresponding colors so that we can visualize the result easily. We take a CVAE with Age labels as an example. There are seven colors in figures that correspond to the seven values of this type of condition labels. Figure 2(a) shows the distribution of points remain out of order after training. We can hardly find the trend that users with same condition label are grouped together through augmenting variational auto-encoders on network structures. In contrast, from Fig. 2(b) we can clearly observe that points of similar users get closer and have a tendency to be clustered. Such rearrangements are exposed among most of users while some still remain disperse. It is normal because there always exists someone who behave different with others in the group but show similar behavior patterns to people from other groups. From this point of view, we can draw the conclusion that our label verification is effective.

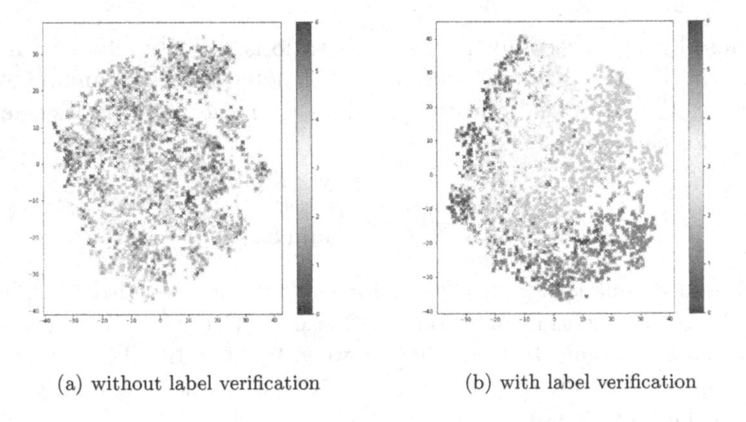

(a) without label verification (b) with label verification

Fig. 2. t-SNE visualizations of latent mean vector on Dataset ML-1M. (Color figure online)

4.2 The Impact of Side Information

After the t-SNE visualization test, it is curious to see whether stacking much more condition labels in one model is beneficial to the recommendation performance. For this purpose, we firstly test our CVAE without any labels and gradually blend different types of side information in model to investigate their impacts. The results are summarized in Tables 2 and 3. From this test, we learn that the performance improvement through mining information from the condition labels varies, which implies labels of different attributes differ in helping model learn to distinguish users. Another observation is loading multiple labels one time degrades the model performance. We then compare the results under the usage of our split-merge framework, where all the recommendation indicators have increased significantly. We attribute this improvement mainly to our split-merge framework, which emphasizes the importance of the recommendation diversity. Each sub model is forced to concentrate on learning only a small aspect of the full features. This leads to multiple models with competitive performance but actually understanding the task from different point of views. When we combine the prediction results, it performs better without doubts.

Table 2. Test performance in terms of multiple condition on dataset ML-1M.

Model	Recall@10	Recall@25	Recall@50	NDCG@50
VAE	0.3178	0.3409	0.4211	0.2269
CVAE with Sex	0.3151	0.3470	0.4277	0.2719
CVAE with Age	0.3238	0.3531	0.4332	0.2798
CVAE with Occupation	0.3172	0.3437	0.4222	0.2803
CVAE with Sex, Age, Occupation	0.3132	0.3425	0.4273	0.2839
Split-Merge CVAE	**0.3470**	**0.3750**	**0.4500**	**0.2843**

Table 3. Test performance in terms of multiple condition on dataset ML-2k.

Model	Recall@10	Recall@25	Recall@50	NDCG@50
VAE	0.2699	0.3561	0.4603	0.1425
CVAE with Genre	0.2801	0.3588	0.4674	0.1570
CVAE with Country	0.2763	0.3608	0.4600	0.1431
CVAE with Genre, Country	0.2758	0.3584	0.4656	0.1435
Split-Merge CVAE	**0.3025**	**0.3773**	**0.4808**	**0.1597**

4.3 Performance Comparison

To make a further comparison, we select four state-of-art collaborative filtering methods that targets top-N preference recommendation:

- **CDAE** [19]: A denoising auto-encoder which is augmented by feeding an user-item preference vector to the input.
- **MULT-VAE** [12]: A variational auto-encoder with multinomial likelihood and use Bayesian inference for parameter estimation.
- **CVAE-CF/JVAE-CF** [10]: It augments the variational auto-encoder by applying variant architectures to tackle collaborative filtering. We implement CVAE-CF and JVAE-CF by adding additional latent variable to extract high-level features associated with auxiliary information as the paper describes. All baseline parameters are build and set according to the respective papers, and the other parts of the experiment settings are kept the same.

Table 4. Performance comparison on dataset ML-1M.

Model	Recall@10	Recall@25	Recall@50	NDCG@10	NDCG@25	NDCG@50
CDAE	0.3180	0.3361	0.4105	0.3176	0.2669	0.2183
MULT-VAE	0.3178	0.3409	0.4211	0.3181	0.2701	0.2269
CVAE-CF	0.3282	0.3483	0.4253	0.3317	0.2891	0.2665
JVAE-CF	0.3194	0.3423	0.4256	0.3396	**0.2963**	**0.2794**
Split-Merge CVAE	**0.3470**	**0.3750**	**0.4500**	**0.3478**	0.2905	0.2543

Table 5. Performance comparison on dataset ML-2k.

Model	Recall@10	Recall@25	Recall@50	NDCG@10	NDCG@25	NDCG@50
CDAE	0.2694	0.3536	0.4512	0.2031	0.1713	0.1346
MULT-VAE	0.2699	0.3561	0.4603	0.2038	0.1736	0.1425
CVAE-CF	0.2876	0.3679	0.4672	0.2110	0.1898	0.1554
JVAE-CF	0.2881	0.3623	0.4617	0.2109	0.1768	0.1546
Split-Merge CVAE	**0.3025**	**0.3773**	**0.4808**	**0.2211**	**0.1922**	**0.1597**

Tables 4 and 5 summarize the results of our proposed methods and three state-of-art methods on two datasets. These two datasets simulate two recommended scenarios of user-based and item-based respectively. In both cases, our method outperforms the related methods on recall and normalized discounted cumulative gain metrics across different truncate value K. Combined with previous analysis, this numeric evaluation result aligns our expectation. Interestingly, we also observe that normalized discounted cumulative gain is declining as the K increases and the advantage of the proposed method is very limited or even worse when arriving 50, which implies the different sizes of held-out set may disturb the measurement. For some inactive users or items, the discounted cumulative gain is relatively at a low level not because of the poor prediction but the lack of interaction records.

5 Related Works

Traditional works for recommender systems focus on explicit feedback and rating predictions. Along with the development of personalized recommendation technology, researchers gradually recognize the importance of implicit feedback data and top-N recommendation for their applicability in practice [2,15,19]. Increasingly works [3,20] apply deep neural networks to model the latent non-linearity of the data, where brings a marked switch from hand-crafted features to directly learn from raw inputs. Auto-encoder is then a type of neural network for unsupervised learning in efficient data representation. [17] stems from the successes of auto-encoder in vision and speech tasks and apply it to collaborative filtering in recommendation. Later [19] extends the work by denoising auto-encoder and measure their performances based on the top-N results. [21] expands the auto-encoder framework for movie recommendation by blending side information in and show significant improvements over traditional methods.

Variational auto-encoder can be regarded as a variation of auto-encoder, which makes strong assumptions on the distribution of latent variables. [10–12] are collaborative filtering methods based on variational auto-encoder and also the most relevant studies to ours. [11] considers the influence of both explicit ratings and content in multimedia scenarios. Inspired by this work, we notice that variational auto-encoder can be extended in recommendation and side information may help improve recommendation quality. Then [10] presents multiple variational approaches for collaborative filtering to deal with auxiliary information. However, it mainly concentrates on how to encompass variational auto-encoder through augmenting structures, while failing to consider auxiliary information could be utilized in a more elegant way. [12] concentrates on likelihood functions chosen and the regularization hyper-parameter tuning study. Although it is the state-of-art work on variational auto-encoder for top-N recommendation with implicit feedback, it still leaves investigation on conditional variational auto-encoder models as future works.

6 Conclusion

This paper proposes an expanded variational auto-encoder recommendation framework based on multiple condition labels. Side information is firstly blended into the conditional variational auto-encoder and then a split-merge framework is adopted to further improve the recommendation performance. Instead of simply stacking network layers, we leverage those condition labels as a part of optimization objective to help model distinguish and cluster users and items in latent subspace. Experimental results on two public datasets indicate the effectiveness. In future, we are to investigate joint training to reduce time consuming.

Acknowledgements. This paper is supported by the National Key Research and Development Program of China (Grant No. 2016YF- B1001102), the National Natural Science Foundation of China (Grant Nos. 61502227, 61876080), the Collaborative Innovation Center of Novel Software Technology and Industrialization at Nanjing University.

References

1. Blei, D.M., Kucukelbir, A., McAuliffe, J.D.: Variational inference: a review for statisticians. J. Am. Stat. Assoc. **112**(518), 859–877 (2017)
2. He, X., Chua, T.S., He, Z., Liu, Z., Song, J., Jiang, Y.G.: NAIS: neural attentive item similarity model for recommendation. IEEE Trans. Knowl. Data Eng. **22**(1), 1 (2018)
3. He, X., Liao, L., Zhang, H., Nie, L., Hu, X., Chua, T.S.: Neural collaborative filtering. In: WWW 2017, pp. 173–182 (2017)

4. Higgins, I., et al.: β-VAE: learning basic visual concepts with a constrained variational framework. In: ICLR 2017 (2017)
5. Hu, Y., Koren, Y., Volinsky, C.: Collaborative filtering for implicit feedback datasets. In: ICDM 2008, pp. 263–272 (2008)
6. Karypis, G., Deshpande, M.: Item-based top-n recommendation algorithms. ACM Trans. Inf. Syst. **22**(1), 143–177 (2004)
7. Karypis, G., Riedl, J., Konstan, J.A., Sarwar, B.M.: Analysis of recommendation algorithms for e-commerce. In: ECRA 2000, pp. 158–167 (2000)
8. Kingma, D.P., Mohamed, S., Rezende, D.J., Welling, M.: Semi-supervised learning with deep generative models. In: NIPS 2014, pp. 3581–3589 (2014)
9. Kingma, D.P., Welling, M.: Auto-encoding variational bayes. In: ICLR 2014 (2014)
10. Lee, W., Song, K., Moon, I.C.: Augmented variational autoencoders for collaborative filtering with auxiliary information. In: CIKM 2017, pp. 1139–1148 (2017)
11. Li, X., She, J.: Collaborative variational autoencoder for recommender systems. In: SIGKDD 2017, pp. 305–314 (2017)
12. Liang, D., Krishnan, R.G., Hoffman, M.D., Jebara, T.: Variational autoencoders for collaborative filtering. In: WWW 2018, pp. 689–698 (2018)
13. van der Maaten, L., Hinton, G.: Visualizing data using t-SNE. J. Mach. Learn. Res. **9**(2605), 2579–2605 (2008)
14. Ning, X., Karypis, G.: SLIM: sparse linear methods for top-n recommender systems. In: ICDM 2011, pp. 497–506 (2011)
15. de Rijke, M., Zhao, X., Chen, Y.: Top-N recommendation with high-dimensional side information via locality preserving projection. In: SIGIR 2017, pp. 985–988 (2017)
16. Sarwar, B.M., Karypis, G., Konstan, J.A., Riedl, J.: Item-based collaborative filtering recommendation algorithms. In: WWW 2001, pp. 285–295 (2001)
17. Sedhain, S., Menon, A.K., Sanner, S., Xie, L.: AutoRec: autoencoders meet collaborative filtering. In: WWW 2015, pp. 111–112 (2015)
18. Wu, D., Lu, J., Zhang, G., Mao, M., Wang, W.: Recommender system application developments: a survey. Decis. Support Syst. **74**(C), 12–32 (2015)
19. Wu, Y., DuBois, C., Zheng, A.X., Ester, M.: Collaborative denoising auto-encoders for top-n recommender systems. In: WSDM 2016, pp. 153–162 (2016)
20. Xue, H.J., Dai, X., Zhang, J., Huang, S., Chen, J.: Deep matrix factorization models for recommender systems. In: IJCAI 2017, pp. 3203–3209 (2017)
21. Yi, B., Shen, X., Zhang, Z., Shu, J., Liu, H.: Expanded autoencoder recommendation framework and its application in movie recommendation. In: SKIMA 2016, pp. 298–303 (2016)

Jaccard Coefficient-Based Bi-clustering and Fusion Recommender System for Solving Data Sparsity

Jiangfei Cheng[1] and Li Zhang[1,2(✉)]

[1] Department of Computer Science and Technology, Soochow University, Suzhou, China
20165227039@stu.suda.edu.cn
[2] Provincial Key Laboratory for Computer Information Processing Technology, Soochow University, Suzhou, China
zhanglim1@suda.edu.cn

Abstract. Recommender systems have been very common and useful nowadays, which recommend suitable items to users by predicting ratings for items. The most used collaborative filtering recommender system suffers from the sparsity issue due to insufficient data. To cope with this issue, we propose a Jaccard Coefficient-based Bi-clustering and Fusion (JC-BiFu) method for Recommender system. JC-BiFu uses density peak clustering for both users and items, and then makes estimations for missing values in the user-item rating matrix when finding the similar users. Finally, we utilize both users and items to generate the final predictions. Experimental analysis shows that our approach can improve the performance of user recommendations at the extreme levels of sparsity in user-item rating matrix.

Keywords: Collaborative filtering · Recommender system · Jaccard coefficient · Cluster · Data sparsity

1 Introduction

The rapid and fast expansion of Internet has significantly changed people's traditional perspective on shopping, reading and entertainment by providing huge amounts of information and bringing huge convenience. However, as information in internet has dramatically increased, it is difficulty to filter the information that people really desired. Recommender systems (RSs), which can play an expert role in assisting us to find the information comforting to our interests. In particular, given the user purchased or rating profiles, recommender systems provide a ranked list of items that we may be interested in.

Generally speaking, there are two kinds of approaches for RSs: content-based and collaborative filtering-based (CF) approaches [1,2]. Recently, CF-based methods have become popular and widely used in lots of domains, such

Q. Yang et al. (Eds.): PAKDD 2019, LNAI 11440, pp. 369–380, 2019.
https://doi.org/10.1007/978-3-030-16145-3_29

as movies [3], music [4], and news [5]. The CF-based methods in RSs make a recommendation list according to the similar users or items with the target user or item. A widely accepted taxonomy divides CF-based methods into model-based and memory-based method [6]. Model-based methods use the information in RS to construct a model that generates the recommendations. However, the memory-based methods calculate the similarity between users or items and then select several similar users or items known as the neighbors to generate the recommendations [7]. To generate more accuracy predictions, many new similarity measures for CF have been proposed [9–11]. The traditional Pearson correlation coefficient-based method [8] does not consider the size of common users. To solve this problem, a weighted Pearson correlation coefficient method has been proposed [9], which captures the user confidence. The confidence of two users would be greater and greater with the increase number of common rated items. Jamali and Ester introduced a similarity measure based on the sigmoid function and proposed the TrustWalker model [10]. This model can weaken the similarity of small common items among users. The adjusted cosine similarity measure [11] was proposed to make up the shortage of traditional cosine similarity, however, it dose not consider the preference of user ratings.

As we know, memory-based method can give a considerable recommend accuracy. Since the scale of data in internet is becoming larger and larger, the data sparsity and cold start issues in CF are getting more and more severe. The data sparsity issue is a situation that the size of the user-item matrix is extremely high and there are lots of missing values in the user-item matrix. In other words, users only rated few items and even had no ratings on all items. The cold start issue is a particular case of data sparsity. In that situation, it can be very hard to make reliable recommendations, which seriously reduces the performance of RSs. To solve the data sparsity issue, some techniques have been successfully applied to RSs, such as dimensionality reduction methods [12–14]. RS can also use clustering techniques to improve the prediction quality and reduce sparsity issue. This paper focuses on the clustering-based methods which are typical to form clusters of items, users or both [15–18]. In [15], Ji et al. proposed a method combined co-clustering and Radial Basis Function network to predict the missing value in the rating matrix. In [16], Zhu et al. analyze the scalable CF using clustering technology. In [17], George et al. proposed a method that involved simultaneous clustering of users and items. In [18], Zhang et al. proposed a bi-clustering and fusion method called BiFu. The missing values in the rating matrix can be smoothed by the user or item clusters to eliminate the data sparsity. But BiFu only considers the user or item information to predict the missing values, and ignores the rating diversity.

To remedy this, we propose a Jaccard Coefficient-based Bi-clustering and fusion (JC-BiFu) method for CF to solve the data sparsity issue. JC-BiFu introduces the item popularity using Jaccard coefficients. First, we cluster both users and items using the density peak method according to the user-item rating matrix. Then we select the most similar cluster for both the target user and item, so we can make estimations for the unrated entries in the rating matrix to solve the sparsity issue. Finally, we can make prediction for the target user according to the similar neighbors.

The remaining of this paper is organized as follows. Section 2 introduces our proposed method JC-BiFu in detail. Section 3 reports experiment settings and results. Section 4 concludes our work with future directions.

2 Jaccard Coefficient-Based Bi-clustering and Fusion

In this section, we introduce our motivation by giving an example and then illustrate our proposed method JC-BiFu in detail.

2.1 Motivation

In memory-based approaches, the implicit (click, view times and purchased) and explicit (ratings) feedback are the basis to make recommendations by estimating ratings for items that have not been rated by users [19]. The feedback information shows the preference of users for items, which is stored in the user-item rating matrix. Let n be the number of users and m be the number of items. An $n \times m$ user-item rating matrix represents all the users' preference for items, where the (i, j)-th entry of this matrix R stands for the i-th user's rating for the j-th item. If the user has not rated the item yet, the rating value would be 0.

Various rating-based methods have been used to compute the similarity $sim(u, u')$ between two users u and u' in RSs. The two most popular methods [20] are cosine and Pearson correlation coefficient (PCC), which are defined as follows:

$$sim_c(u, u') = \frac{\sum_i r_{u,i} r_{u',i}}{\sqrt{\sum_i r_{u,i}^2} \sqrt{\sum_i r_{u',i}^2}} \tag{1}$$

$$sim_p(u, u') = \frac{\sum_i (r_{u,i} - \bar{r}_u)(r_{u',i} - \bar{r}_{u'})}{\sqrt{\sum_i (r_{u,i} - \bar{r}_u)^2} \sqrt{\sum_i (r_{u,i} - \bar{r}_{u'})^2}} \tag{2}$$

where $sim_p(u, u')$ and $sim_c(u, u')$ represent the Pearson similarity and cosine similarity between users u and u', respectively, $r_{u,i}$ denotes the rating of user u on item i, $\bar{r}_u$ is the average rating of user u, $i \in I_u \cap I_{u'}$ that is the common items both rated by users u and u' and I_u is the set of items rated by user u. From (1) and (2) we can see that when calculating the similarity between two users, we need find the common items both rated by the two users. However, we can hardly find the common items since the user-item matrix is very sparse. Without adequate and sufficient data, the similarity cannot reflect the correlation between users, which would be illustrated by an example in Table 1 where are 3 users and the corresponding rating on 7 items. If we use PCC (2) to measure the similarity between users, we can have $sim_p(u_1, u_2) = 1$ and $sim_p(u_1, u_3) = 0.5$. Obviously, the similarity between user u_1 and u_2 is greater than that between u_1 and u_3. There is only one common rating item that users u_1 and u_2 have, while we take a set $\{i_1, i_3, i_6, i_7\}$ to calculate $sim(u_1, u_3)$. The latter could better reflect the

Table 1. The user-item rating matrix.

	i_1	i_2	i_3	i_4	i_5	i_6	i_7
u_1	2	*	3	3	1	1	4
u_2	*	2	*	5	*	*	*
u_3	5	*	3	*	*	3	3

* represents the user did not rate the item.

similarity between users. We cannot get good recommendations when we cannot find good neighbors. In a word, data sparsity has a terrible influence in CF recommender systems.

2.2 JC-BiFu

To solve the data sparsity issue, we propose a Jaccard Coefficient-based Bi-clustering and Fusion method. We introduce it in detail.

Clustering. Cluster analysis aims at classifying data points into categories on the basis of their similarity. BiFu [18] takes k-means to cluster users and items. It is well known that k-means is an unstable method and cannot automatically find the correct number of cluster centroids. In [21], a fast clustering method based on density peak (we call it CDP) have been proposed, which is able to automatically find the correct number of cluster centroids. In this way, it can avoid the parameter issue to improve the clustering accuracy. If we use CDP to cluster users or items to find similar neighbors for the target user or item, we can find better neighbors and get better recommendations.

CDP has a basis assumption that cluster centers have a relatively large local density and are surrounded by neighbors with a lower local density. For each user u, we need to compute two quantities: its local density ρ_u, and its distance δ_u from user u' with a higher density. Both these quantities depend only on the distances $d_{uu'}$ between users. The local density ρ_u of user u is defined as:

$$\rho_u = \sum_{u'} f(d_{uu'} - d_c) \tag{3}$$

where d_c is a cutoff distance determined empirically, and $f(x) = \begin{cases} 0, & \text{if } x < 0 \\ 1, & \text{otherwise} \end{cases}$.
The distance δ_u is measured by computing the minimum distance between the user u and any other users u' with higher density:

$$\delta_u = \min_{u:\rho_{u'} > \rho_u} (d_{uu'}) \tag{4}$$

For the user u with the highest density, we conventionally take $\delta_u = \max_{u'} (d_{uu'})$.

For the user set $U = \{u_1, u_2, \cdots, u_n\}$, and item set $I = \{i_1, i_2, \cdots, i_m\}$, JC-BiFu uses CDP with cosine similarity and partitions both U and I to get k_u user clusters $U_1, U_2, \cdots, U_{k_u}$ and k_i item clusters $I_1, I_2, \cdots, I_{k_i}$, respectively, if the value of cosine is less than 0, let it be the absolute value.

Estimating Missing Values. In the user-item rating matrix, lots of values are missing owing to data sparsity in the cold-start settings [18]. We can hardly find co-rating items for two users when calculating the similarity between them because of the data sparsity issue. To alleviate the influence of data sparsity, the missing values in the user-item rating matrix need to be estimated. Some estimation methods have been proposed in [18,19,22].

In [18], BiFu predicts the missing values based on users information without considering the rating diversity. To consider the user and item information simultaneously, we adopt the estimated method in [19] and apply the Jaccard coefficeient [23] to measure the similarity between two items. For items i and j, the Jaccard coefficient can be defined as:

$$J(i,j) = \frac{|U_i \cap U_j|}{|U_i \cup U_j|} \tag{5}$$

where U_i is the set of users that have rated item i, and $|\cdot|$ is the cardinal number of a set $\cdot$. The numerator of (5) is the number of users that have co-rated both item i and j while the denominator is the number of users that have rated item i or j. If $U_i \cap U_j = \emptyset$, $J(i,j) = 0$; $J(i,j) = 1$ when $U_i = U_j$. Obviously, $0 \leqslant J(i,j) \leqslant 1$. The larger $J(i,j)$ is, the more similar the two items is. If the user u has not rated the item i, the estimation of missing value $\widetilde{r}_{u,i}$ can be computed by:

$$\widetilde{r}_{u,i} = \bar{r}_u + \frac{\sum_{i,j \in I_k} J(i,j)(r_{u,j} - \bar{r}_u)}{\lambda + \sum_{i,j \in I_k} J(i,j)} \tag{6}$$

where I_k is the item cluster that contains item i and j and $\lambda > 0$ is a parameter to smooth the prediction and can avoid the denominator to be 0. Intuitively, ratings with larger J will contribute more to the prediction. So far, we can obtain a new user-item rating matrix $\widetilde{R}$ with each entry $\widetilde{r}_{u,i} > 0$.

Evaluation. In this phase, we mainly aim to calculate the final rating of the target user u_t on the target item i_t. First of all, we need to look for the most similar user and item clusters for the target user u_t and item i_t, and then we need to find several similar users and items for u_t and i_t in the selected clusters, respectively. Finally, we calculate the final prediction. In the following, we describe the evaluation procedure in detail.

The similarity between the target user u_t and a user cluster is defined as:

$$sim(u_t, U_k) = \frac{\sum_{j \in M_p} (\widetilde{r}_{u_t,j} - \bar{\widetilde{r}}_{u_t}) (\sum_{u' \in U_k} \frac{\widetilde{r}_{u',j} - \bar{\widetilde{r}}_{u'}}{|U_k|})}{\sqrt{\sum_{j \in M_p} (\widetilde{r}_{u_t,j} - \bar{\widetilde{r}}_{u_t})^2} \sqrt{\sum_{j \in M_p} (\sum_{u' \in U_k} \frac{\widetilde{r}_{u',j} - \bar{\widetilde{r}}_{u'}}{|U_k|})^2}} \tag{7}$$

where M_p is the set of items which have been co-rated by user u_t and users in the cluster U_k, $\widetilde{r}_{u_t,j}$ is the rating the target user u_t on item j, and $\bar{\widetilde{r}}_{u_t}$ is the average rating of user j after estimation. After all the similarity between the target user and user clusters are calculated, we need to select the most similar user cluster U_{k^*} from all the clusters:

$$U_{k^*} = \arg\max_{k=1,\ldots,k_u} sim(u_t, U_k). \tag{8}$$

where k_u is the number of the user cluster. Once the most similar user cluster U_{k^*} is confirmed, we need to extract a certain number of the target user's most similar users S_{u_t} from the most similar user cluster U_{k^*}.

And the similarity between item and item cluster is defined as:

$$sim(i_t, I_k) = \frac{\sum_{v\in N_p}(\widetilde{r}_{v,i_t} - \bar{\widetilde{r}}_{i_t})(\sum_{i'\in I_k}\frac{\widetilde{r}_{v,i'} - \bar{\widetilde{r}}_{i'}}{|I_k|})}{\sqrt{\sum_{v\in N_p}(\widetilde{r}_{v,i_t} - \bar{\widetilde{r}}_{i_t})^2}\sqrt{\sum_{v\in U_p}(\sum_{i'\in I_k}\frac{\widetilde{r}_{v,i'} - \bar{\widetilde{r}}_{i'}}{|I_k|})^2}} \tag{9}$$

where N_p is the user set that users co-rated the item i and items in cluster I_k and $\bar{\widetilde{r}}_{i_t}$ is the average rating of the target item i_t after estimation. After all the similarity between the target item and item clusters are calculated, we need to select the most similar item cluster I_{k^*} from all the clusters:

$$I_{k^*} = \arg\max_{k=1,\ldots,k_i} sim(i_t, I_k). \tag{10}$$

where k_i is the number of item cluster. After confirming the most similar item cluster I_{k^*}, we need to extract a certain number of the target item's most similar items S_{i_t} from the most similar item cluster I_{k^*}.

When predicting the rating that target user u_t on the target item i_t, JC-BiFu utilizes user-based approach by using the similar users and items simultaneously to make the predictions more precise. Each rating is weighted by the corresponding similarity $sim(u, v)$.

$$\hat{r}_{u_t,i_t} = \bar{\widetilde{r}}_{u_t} + \gamma \cdot \frac{\sum_{u\in S_{u_t}}\sum_{i\in S_{i_t}} sim(u_t, u)(\widetilde{r}_{u,i} - \bar{\widetilde{r}}_u)}{\sum_u sim(u_t, u)} \tag{11}$$

where γ is the weight during the computation for user-based approach. The similarity between the target user and similar user $sim(u_t, u)$ is calculated by Pearson correlation coefficient and cosine respectively. Algorithm 1 shows the procedure of JC-BiFu.

3 Experiments

In this section, the evaluation metrics and the datasets used are described to verify the accuracy and efficiency of JC-BiFu. Then, the experimental results will be given. All experiments are conducted in MATLAB R2014b on a PC with an Intel Core i7 processor with 16 GB RAM.

Algorithm 1. JC-BiFu.

Input: User-item rating matrix R
Output: Predictions $\hat{r}_{u_t,i_t}$ of the target user on the target item
1: Clustering for both users and items
2: Calculating the distance matrices for users and items, respectively
3: Using CDP based on the distance matrices
4: Estimating the missing values
5: Calculating the Jaccard coefficient matrix J of items
6: Making estimations based on the Jaccard coefficient matrix J
7: Evaluation
8: Calculating the similarity between the target user and each user cluster according to (7)
9: Calculating the similarity between the target item and each item cluster according to (9)
10: Selecting the most similar user cluster U_{k^*} and item cluster I_{k^*} according to (8) and (10), respectively
11: Making final predictions according to (11)

3.1 Datasets

In our experiments, we evaluate our algorithm on three datasets MovieLens-100k, MovieLens-1M and FilmTrust [26]. MovieLens-100k and MovieLens-1M consist of 100,000 ratings of 943 users on 1,682 movies and 1,000,209 ratings of 6,040 users on 3,952 movies, respectively, and each user has rated at least 20 movies. Every rating of the two datasets is a positive integer on a 5-star scale. FilmTrust consists of 1,058 users and 2,071 movies. In this dataset every rating is a positive float value on the range of 0.5 to 4. In experiments, we only utilize the user-item rating information and ignore the redundant information (e.g., users' personality, items' features). Table 2 shows the statistical characteristics of these datasets, where the sparsity level of these datasets is calculated by the method in [24]:

$$Sparsity\ level = 1 - \frac{\#Rated\ entries}{\#Total\ entries} \tag{12}$$

Table 2. Statistical of all dataset

Datasets	FilmTrust	MovieLens-100K	MovieLens-1M
Number of users	1,508	943	6,040
Number of items	2,071	1,682	3,952
Number of ratings	35,497	100,000	1,000,209
Total Mean rating	3.0027	3.59299	3.5816
Sparsity level	1.14%	6.3%	4.47%

3.2 Evaluation Metrices

We utilize mean absolute error (MAE) and root mean square error (RMSE) to measure the recommendation accuracy, which is frequently used for measuring the differences between predicted ratings and users' real ratings. MAE and RMSE are defined as:

$$MAE = \frac{1}{N_p}\sum_{u,i}|\hat{r}_{u,i} - r_{u,i}| \tag{13}$$

$$RMSE = \sqrt{\frac{1}{N_p}\sum_{u,i}(\hat{r}_{u,i} - r_{u,i})^2} \tag{14}$$

where N_p denotes the number of predictions. The smaller the MAE and RMSE are, the better the recommendation accuracy.

3.3 Parameters Analysis

In the experiment, we use 70% of the whole dataset for training and the remaining for testing. JC-BiFu makes use of parameters λ to make estimation of the missing values in the user-item matrix R and γ to generate the final predictions. Thus, we design experiments on MovieLens-100K to check how much the parameters affect the performance of JC-BiFu when using PCC and cosine as the similarity measure which is defined as (1) and (2), respectively. The parameter of λ varies in the set $\{10, 20, \cdots, 100\}$ and γ varies in the set $\{0.1, 0.2, \cdots, 1\}$.

Figure 1 shows the results of JC-BiFu with PCC. When the value of λ increases, JC-BiFu can get a better accuracy and becomes stable as shown in Fig. 1(a). While with the increasement of γ, JC-BiFu becomes worse as shown in Fig. 1(b).

Figure 2 gives the results of JC-BiFu with the cosine similarity. From Fig. 2(a), we can have the same conclusion as Fig. 1(a). In Fig. 2(b), JC-BiFu has a better performance and becomes stale with the increasement of γ which is different from Fig. 1(b).

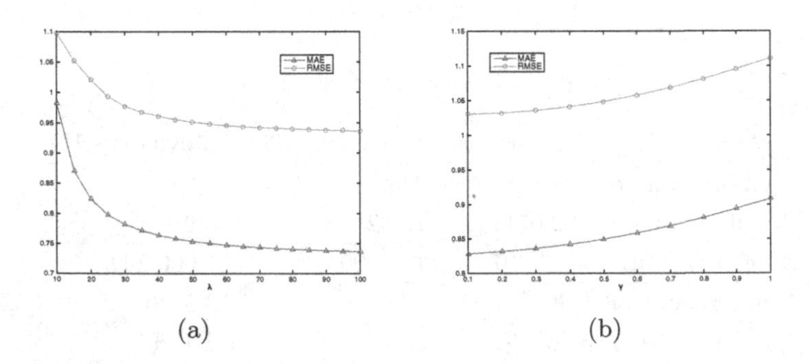

(a) (b)

Fig. 1. MAE and RMSE for sensitivity of parameters λ and γ in MovieLens-100k with PCC.

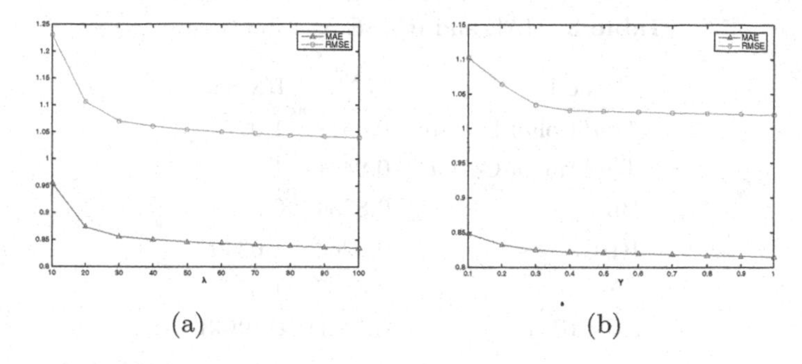

(a) (b)

Fig. 2. MAE and RMSE for sensitivity of parameters λ and γ in MovieLens-100k with cosine.

3.4 Performance

In the phase, we compare JC-BiFu with other related methods on three commonly used RS datasets listed in Table 2.

According to the results mentioned above, JC-BiFu can have a better performance when using PCC with $\lambda = 100$ and $\gamma = 0.1$. The other compared methods are described as follows:

- Traditional Pearson CF: User-based CF with Pearson correlation coefficient for user similarity.
- Traditional Cosine CF: User-based CF with cosine for user similarity.
- JAC [25]: A method with a Jaccard-based similarity to improve performance in RS.
- RBRA [19]: A rating-based algorithm to solve the sparsity issue in RS which uses triple to represent user or item.
- BiFu [18]: A bi-clustering method with k-means by Pearson correlation for user similarity.

Table 3 shows MAE and RMSE of all six methods on the FilmTrust dataset. We can see that JC-BiFu can achieve the best performance among the six methods. For MAE, JC-BiFu is the best (0.7931), followed by RBRA (0.8264) and BiFu method (0.8354). Compared to the second best BiFu, RMSE of JC-BiFu increases by 5.5%.

Tables 4 and 5 show the performance of all six methods on the MovieLens-100K and MovieLens-1M datasets, respectively. Similar to Table 3, JC-BiFu has the best performance compared to the other methods. On MovieLens-100K, the MAE and RMSE of JC-BiFu increase by 12.2% and 8.0%, respectively. On MovieLens-1M, the MAE and RMSE of JC-BiFu increase by 6.0% and 6.5%.

From the results all above, we can see that our JC-BiFu can have the best performance with the lowest MAE and RMSE.

Table 3. MAE and RMSE on FilmTrust.

Methods	MAE	RMSE
Traditional-Pearson	0.8911	1.0271
Traditional-Cosine	0.8491	1.0729
BiFu	0.8354	0.9885
RBRA	0.8264	1.0340
JAC	0.8807	1.0223
JC-BiFu	**0.7931**	**0.9363**

Table 4. MAE and RMSE on Movielens-100K.

Methods	MAE	RMSE
Traditional-Pearson	0.8823	1.0909
Traditional-Cosine	0.8724	1.0860
BiFu	0.8359	1.0196
RBRA	0.8479	1.0387
JAC	0.8538	1.0734
JC-BiFu	**0.7452**	**0.9431**

Table 5. MAE and RMSE on Movielens-1M.

Methods	MAE	RMSE
Traditional-Pearson	0.8917	1.0955
Traditional-Cosine	0.9090	1.1318
BiFu	0.8835	1.0242
RBRA	0.8754	1.0704
JAC	0.8946	1.1683
JC-BiFu	**0.8262**	**0.9613**

4 Conclusion

In this paper, we propose an efficient rating-based recommender algorithm, JC-BiFu which uses the density peak clustering method to cluster the user-item rating matrix, and estimates the missing values for sparsity data to cope with the sparsity problem in cold-start settings. We conduct our experiments on three common used recommender datasets MovieLen-100K, MovieLens-1M and FilmTrust. Experimentally, the recommendation quality in terms of RMSE and MAE shows that JC-BiFu generates more accurate recommendations with less prediction error compare with other methods.

However, JC-BiFu can be further improved. If there exists a brand new user that have not rate any items, or a brand new item that has not been rated by any users, JC-BiFu cannot make good recommendations. We plan to improve our method to handle this situation.

Acknowledgement. This work was supported in part by the National Natural Science Foundation of China under Grant No. 61373093, by the Soochow Scholar Project, by the Six Talent Peak Project of Jiangsu Province of China, and by the Collaborative Innovation Center of Novel Software Technology and Industrialization.

References

1. Burke, R.: Hybrid recommender systems: survey and experiments. User Model. User-Adapt. Interact. **12**(4), 331–370 (2002)
2. Tuzhilin, A.: Towards the next generation of recommender systems. In: ICEBI-10 (2010)
3. Odić, A., Tkalčič, M., Tasič, J.F., Košir, A.: Predicting and detecting the relevant contextual information in a movie-recommender system. Interact. Comput. **25**(1), 74–90 (2013)
4. Benzi, K., Kalofolias, V., Bresson, X., Vandergheynst, P.: Song recommendation with non-negative matrix factorization and graph total variation. In: IEEE International Conference on Acoustics, Speech and Signal Processing, pp. 2439–2443 (2016)
5. Beam, M.: Automating the news: how personalized news recommender system design choices impact news reception. Commun. Res. **41**(8), 1019–1041 (2013)
6. Bobadilla, J., Ortega, F., Hernando, A.: Recommender systems survey. Knowl.-Based Syst. **46**(1), 109–132 (2013)
7. Thorat, P.B., Goudar, R.M., Barve, S.: Survey on collaborative filtering, content-based filtering and hybrid recommendation system. Int. J. Comput. Appl. **110**(4), 31–36 (2015)
8. Resnick, P., Iacovou, N., Suchak, M., Bergstrom, P., Riedl, J.: GroupLens: an open architecture for collaborative filtering of netnews. In: ACM Conference on Computer Supported Cooperative Work, pp. 175–186 (1994)
9. Herlocker, J.L., Konstan, J.A., Borchers, A., Riedl, J.: An algorithmic framework for performing collaborative filtering. ACM SIGIR Forum **51**(2), 227–234 (1999)
10. Jamali, M., Ester, M.: TrustWalker: a random walk model for combining trust-based and item-based recommendation. In: ACM SIGKDD International Conference on Knowledge Discovery and Data Mining, pp. 397–406 (2009)
11. Ahn, H.J.: A new similarity measure for collaborative filtering to alleviate the new user cold-starting problem. Inf. Sci. **178**(1), 37–51 (2008)
12. Sarwar, B.: Application of dimensionality reduction in recommender systems - a case study (2000)
13. Zhang, Z.: Sparsity, robustness, and diversification of recommender systems. Dissertations and Thesis - Gradworks (2014)
14. Ranjbar, M., Moradi, P., Azami, M., Jalili, M.: An imputation-based matrix factorization method for improving accuracy of collaborative filtering systems. Eng. Appl. Artif. Intell. **46**(PA), 58–66 (2015)
15. Ji, Y., Hong, W., Qi, J.: Missing value prediction using co-clustering and RBF for collaborative filtering. In: International Conference on Cloud Computing and Big Data, pp. 350–353 (2016)

16. Zhu, R.L., Gong, S.J.: Analyzing of collaborative filtering using clustering technology. In: ISECS International Colloquium on Computing, communication, control, and Management Proceedings, pp. 57–59 (2009)
17. George, T., Merugu, S.: A scalable collaborative filtering framework based on co-clustering. In: IEEE International Conference on Data Mining, p. 4 (2005)
18. Zhang, D., Hsu, C.H., Chen, M., Chen, Q., Xiong, N., Lloret, J.: Cold-start recommendation using bi-clustering and fusion for large-scale social recommender systems. IEEE Trans. Emerg. Top. Comput. **2**(2), 239–250 (2014)
19. Xie, F., Xu, M., Chen, Z.: RBRA: a simple and efficient rating-based recommender algorithm to cope with sparsity in recommender systems. In: International Conference on Advanced Information NETWORKING and Applications Workshops, pp. 306–311 (2012)
20. Ricci, F., Rokach, L., Shapira, B., Kantor, P.B.: Recommender Systems Handbook. Springer, US (2011). https://doi.org/10.1007/978-0-387-85820-3
21. Rodriguez, A., Laio, A.: Clustering by fast search and find of density peaks. Science **344**(6191), 1492 (2014)
22. Lemire, D., Maclachlan, A.: Slope one predictors for online rating-based collaborative filtering. In: Computer Science, pp. 21–23 (2007)
23. Anand, D., Bharadwaj, K.K.: Utilizing various sparsity measures for enhancing accuracy of collaborative recommender systems based on local and global similarities. Expert Syst. Appl. **38**(5), 5101–5109 (2011)
24. Sarwar, B., Karypis, G., Konstan, J., Riedl, J.: Analysis of recommendation algorithms for e-commerce. In: ACM Conference on Electronic Commerce, pp. 158–167 (2000)
25. Ayub, M., Ghazanfar, M.A., Maqsood, M., Saleem, A.: A Jaccard base similarity measure to improve performance of CF based recommender systems. In: International Conference on Information NETWORKING, pp. 1–6 (2018)
26. Guo, G., Zhang, J., Yorke-Smith, N.: A novel Bayesian similarity measure for recommender systems. In: Proceedings of the 23rd International Joint Conference on Artificial Intelligence (IJCAI), pp. 2619–2625 (2013)

A Novel KNN Approach for Session-Based Recommendation

Huifeng Guo[2], Ruiming Tang[2], Yunming Ye[1,3](✉), Feng Liu[1,3], and Yuzhou Zhang[2]

[1] Harbin Institute of Technology, Shenzhen, China
yeyunming@hit.edu.cn, liufeng@stmail.hitsz.edu.cn
[2] Noah's Ark Lab, Huawei, China
{huifeng.guo,tangruiming,zhangyuzhou3}@huawei.com
[3] Shenzhen Key Laboratory of Internet Information Collaboration, Shenzhen, China

Abstract. The KNN approach, which is widely used in recommender systems because of its efficiency, robustness and interpretability, is proposed for session-based recommendation recently and outperforms recurrent neural network algorithms. It captures the most recent co-occurrence information of items by considering the interaction time. However, it neglects the co-occurrence information of items in the historical behavior which is interacted earlier than others and cannot discriminate the impact of vertices with different popularity. Due to these observations, this paper presents a novel KNN approach to address these issues for session-based recommendation. Specifically, a diffusion-based similarity method is proposed for incorporating the popularity of items, and the candidate selection method is proposed to capture more co-occurrence information of items in the same session efficiently. Comprehensive experiments are conducted to demonstrate the effectiveness of our KNN app roach over the state-of-the-art KNN approach for session-based recommendation on three benchmark datasets.

Keywords: Diffusion model · Session-based recommendation · Nearest neighbor

1 Introduction

With the development of Internet, there have been many web applications. However, the recommender systems of many applications, particularly those of small retailers, do not track the visit information of all users over a period of time. Moreover, the cookies are also unavailable due to the technology reliability and privacy concerns [5,9]. Even if the visit information of users can be tracked, the number of sessions for a specific user in a small application site is limited and the behavior of users mostly shows session-based traits. Therefore, session-based

Work done while Huifeng Guo at Harbin Institute of Technology.

Q. Yang et al. (Eds.): PAKDD 2019, LNAI 11440, pp. 381–393, 2019.
https://doi.org/10.1007/978-3-030-16145-3_30

Table 1. The session data example.

Session id	i	i	i	i	j	j	k	k	k	l	l
Item id	$\alpha 1$	$\alpha 2$	$\alpha 3$	$\alpha 6$	$\alpha 3$	$\alpha 4$	$\alpha 2$	$\alpha 3$	$\alpha 4$	$\alpha 3$	$\alpha 5$
Time	0	1	2	3	4	5	6	7	8	9	10

recommendation, where the task is predicting the next action of a user given the action sequence in the current session, is critical for recommender systems.

Session-based recommendation is a special case of sequential learning, such as basket prediction [11] and music playlist generation [1]. Traditional recommendation methods [3,4,8] are not well suited for sequential learning. Therefore, several approaches [7,10] are proposed to capture both sequential and personalized information. However, these methods all require **long-term** user history behavior and usually lead to poor performance when these information is unavailable. In order to address this limitation, a recurrent neural network, named as GRU4Rec [5], is used for session-based recommendation. Nevertheless, it suffers from several limitations: (1) Limited ability of capturing co-occurrence information of items: (2) Tremendous parameters that to be trained.

Compared with other methods, the KNN approach is widely used in recommender systems because of its efficiency, robustness and interpretability [1,9]. Recently, a contextual KNN approach (CKNN) is proposed for session-based recommendation and **outperforms** GRU4Rec [5] on benchmark datasets according to the reported result [6]. Although CKNN works well in session-based recommendation, it still has two limitations: (1) Lacking the ability to distinguish the influence of items with different popularity. (2) CKNN cannot guarantee the ratio of relevant sessions of different items clicked in current session, especially the last action, in the relevant session set of current session.

In order to address these limitations in CKNN approach for session-based recommendation, we propose a novel CKNN approach in this paper. Specifically, a diffusion-based similarity method is proposed for incorporating graph structure information and two candidate selection strategies are designed to guarantee the ratio of relevant sessions related to last click of current session in the relevant session set of current session and the ratio of relevant sessions related different items clicked in current session.

2 Our Approach

In order to illustrate the procedure of session-based recommendation, Table 1 presents an example data. This example includes 11 session-item interactions, which is consisting of Session id, Item id and time, respectively.

The session-item interactions presented in Table 1 are able to be represented as a **session-item bipartite network** $\mathcal{G} = \{\mathcal{S}, \mathcal{I}, \mathcal{E}\}$, which consists of a session set $\mathcal{S}$, an item set $\mathcal{I}$ and an interaction set $\mathcal{E}$. The number of sessions and items are denoted as m and n, the cardinality of $\mathcal{E}$ is l. The adjacency matrix of $\mathcal{G}$ can

Table 2. The map from session id to the pair of item id and interaction time.

Session id	(Item id, interaction time)
i	(α1, 00), (α2, 01), (α3, 02), (α6, 03)
j	(α3, 04), (α4, 05)
k	(α2, 06), (α3, 07), (α4, 08)
l	(α3, 09), (α5, 10)

Table 3. The map from item id to the pair of session id and interaction time.

Item id	(Session id, interaction time)
α1	(i, 00)
α2	(i, 01), (k, 06)
α3	(i, 02), (j, 04), (k, 07), (l, 09)
α4	(j, 05), (k, 08)
α5	(l, 10)
α6	(i, 03)

be denoted as $A \in R^{m \times n}$, where $a_{xi} = 1$ if $(x, i) \in \mathcal{E}$ and 0 otherwise. Moreover, we define the degree of session x and item i as $\mathbf{d}_x$ and $\mathbf{d}_i$, respectively.

In the rest of this section, the procedure of the contextual KNN (CKNN for short) approach for session-based recommendation is presented.

2.1 Contextual KNN Approach

Before staring, we need construct two dictionaries, namely Map_{S2I} and Map_{I2S}. Specifically, Map_{S2I} is a map from session to the pairs of item and interaction time, Map_{I2S} is another map from item to the pairs of session and interaction time. The Map_{S2I} and Map_{I2S} of the example data that introduced in Table 1 are presented in Tables 2 and 3, respectively. The pipeline of CKNN approach is described as follows (which is also presented in the bottom of Fig. 1):

- **At step 0**, a session-based recommendation is triggered when a user clicks some item. **At step 1**, we need to find **RL(x)**, which is the relevant session set related to items in the current session x. For example in Fig. 1, the relevant session set related to the current session $x = \{\alpha_4, \alpha_1\}$ is $\{j, k, i\}$ (session j, k are related to item α_4 and session i is related to item α_1).
- **At step 2**, we select the most recent k_{recent} sessions from the relevant session set as the recent session set **RC(x)**, because focusing on the most recent events has shown to be effective in the domains of e-commerce and news recommendations [2]. For example in Fig. 1, $k_{recent} = 2$ recent sessions are chosen and the recent session set of current session x is $\{j, k\}$.

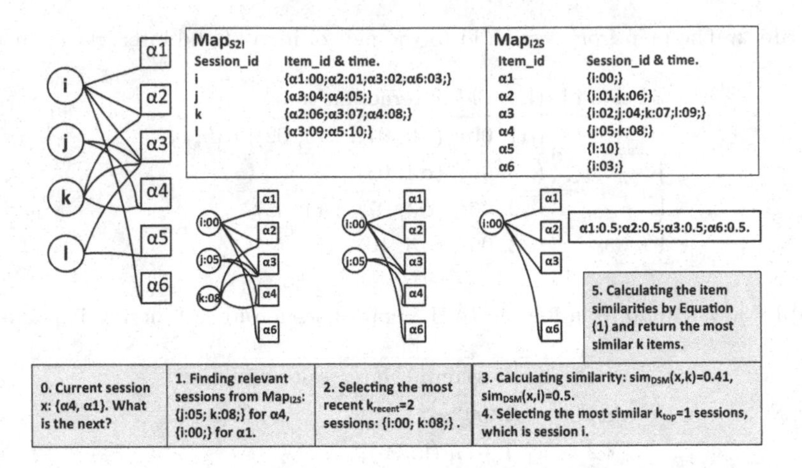

Fig. 1. Examples for bipartite network construction and our session-based KNN algorithm. Note that, $Maps_{S2I}$ (or Map_{I2S}) is a map, where the key is session id (or item id) and the value is the set of item id (or session id)-timestamp pairs. Specifically, the smaller the timestamp, the earlier the interaction time is (03 is more recent than 00).

- **At step 3**, the similarity score $sim(x, j)$ between the current session x and each session j in RC(x) (selected by step 2) is calculated. After that, the most similar k_{top} sessions are selected as the nearest neighbor session set, denoted as $NN(x)$, for current session x.
- **Finally at step 4**, based on the nearest neighbor session set $NN(x)$, and the similarities with the current session x, the score of a recommendable item α for the current session x is

$$score_{KNN}(\alpha, x) = \sum_{j \in NN(x)} sim(x, j) \times 1_j(\alpha), \tag{1}$$

where $1_j(\alpha) = 1$ represents session j containing item α and 0 otherwise. Then the most similar k items $result_x$ are returned.

2.2 Candidate Selection

Focusing on the most recent events has shown to be effective in the domains of e-commerce and news recommendation [2]. Therefore, [6] proposes an algorithm, denoted as **Original** for distinction, to select the most k_{recent} recent relevant sessions from the related session set $RL(x)$ of current session x. However, this algorithm still suffers following limitations: **(1) Last click**: The sessions containing the last items may not be included because the interaction time of such sessions may be earlier than the others. So it is not guaranteed to include the relevant sessions of last click. **(2) Other clicks**: Same as **Last click**, the relevant sessions for other clicks will be excluded when the interaction time are earlier.

Focusing on the Last Click. In order to focus on the last click item, we propose a strategy to guarantee the ratio of last item's recent relevant sessions in the recent session set RC. The basic idea of our strategy is to find the recent session set from two sources: **(1) Recent relevant sessions for last click**: Assuming i is the last item, we select the most recent $[k_{recent} \times p_i]$[1] sessions from RL_i as RC_i, where $\mathbf{p}_i$ is the ratio of RC_i in $RC(x)$ according to different strategies. **(2) Recent relevant sessions for other clicks**: Selecting the most recent $[k_{recent} \times (1 - p_i)]$ sessions of $x \setminus i$ (other items in current session) as $RC(x \setminus i)$ from $RL(x \setminus i)$, which has already been found previously. **Finally**: Merging the recent session set of last item and of the other items in current session x as the recent session set of x: $RC(x) = RC_i \cup RC(x \setminus i)$.

Assuming the influence of items with different popularity is same, we propose **E**qual **P**robability **C**andidate **S**election (EPCS for short). So $\mathbf{p}_i = [\frac{k_{recent}}{|x|}]$, where $|x|$ is the number of elements in current session x.

To make it easier to understand, we take an example to illustrate the difference between the **Original** algorithm proposed by [6] and the **EPCS** algorithm in this paper. Recall the items of current session x are $\{\alpha_4, \alpha_1\}$, where α_1 is the last interacted item. Therefore, the $RL(x)$ is $\{j : 05, k : 08, i : 00\}$ (where $RL_{\alpha_4} = \{j, k\}$ and $RL_{\alpha_1} = \{i\}$). If we set $k_{recent} = 2$, session i will be excluded because the interaction time of $\{i : 00\}$ is in the third place. Therefore, there is no session relevant to last item α_1 in the recent session set. On the contrary, EPCS **guarantees the ratio** of RC_i[2] in $RC(x)$.

Focusing on All Clicks. In fact, not only the last click is important, but also the co-occurrence of other clicks is helpful for recommendation. To guarantee the ratio of every item's recent relevant sessions in $RC(x)$, we can select the recent session set for each item that contained in current session, then aggregate these sets together as the recent session set of current session x. However, selecting the recent session set for each item is very expensive in terms of time cost, because querying and sorting are required every time. Recall the candidate selection algorithm introduced in Sect. 2.2, the recent session set of other clicks $(x \setminus i)$ is selected at previous step. So we can select $[k_{recent} \times (1 - p_i)]$ elements from $(x \setminus i)$ randomly. This approach has **two advantages**: (1) Random selection is efficient; (2) Random selection guarantees the ratio of every item's relevant sessions in $RC(x)$. Based on the **random** selection strategy and **EPCS**, we propose another strategy and name it as **EPCSR**.

2.3 Diffusion-Based Similarity

In addition to candidate selection, similarity calculation is also important for the performance of CKNN approach. In CKNN algorithm, some interactions are discarded when constructing the recent session set of current session. So

[1] [ ] indicates rounding.

[2] RC_i is the set of recent sessions related to item i.

the degree of items in recent session set is smaller than their real degree in the original network $\mathcal{G}$. As a result, using items' original degree will over-affect the similarity metric. To address this limitation, we propose a **D**iffusion-based **S**imilarity **M**ethod (noted as DSM for short). In DSM, we adopt an exponential function of items' degree to control the impact of items' popularity for similarity calculation. Specifically, the importance of item i is denoted as d_i^β, where β is a hyper-parameter and the importance of item degree is increasing when β becomes larger. The DSM similarity between current session x and target session j is denoted as:

$$Sim_{DSM}(x, j, \lambda, \beta) = \frac{1}{d_x^\lambda \times d_j^{1-\lambda}} \sum_{i=0}^{n} \frac{a_{xi} \cdot a_{ji}}{d_i{}^\beta}. \tag{2}$$

It's easy to find that DSM is a similarity framework which is a generalization of several existing similarity methods. For instance, DSM is Mass Diffusion [12] when $\lambda = 1$, $\beta = 1$, and is cosine when $\lambda = 0.5$, $\beta = 0$, etc.

3 Experiment

In this section, we conduct experiments to answer the following research questions: **RQ1:** Does the candidate selection strategies improve the performance of session-based recommendation? **RQ2:** How do the hyper-parameters λ and β in DSM influence the session-based recommendation? **RQ3:** How does our approach perform, compared to other KNN approaches for session-based recommendation?

3.1 Experiment Setting

Datasets. The effectiveness of our proposed approach is evaluated on three datasets, including RSC, RSCW and Atom. The recommendation task is to predict the kth item knowing the previous $k-1$ items in a session with length n, where $k \in [2, n]$.

Particularly, Atom is a dataset containing music playlists from artofthemix.org. The playlists in Atom dataset have no timestamp information therefore the authors of [6] assigns each playlist with a timestamp uniformly at random, under the assumption that the whole dataset is of 31 consecutive days. The training set of Atom is the first 30 days' data while the last day's data is the test set.

RSC and RSCW are 2 variants of the ACM RecSys 2015 Challenge dataset (RSC15) as used in [6]. RSC15 contains the sessions of items in 182 consecutive days. Note that, all of the session-item interactions in RSC and RSCW are associated with timestamps. In RSC, the first 181 days' data is identified as training set, while the last day's data is left as test set. And the RSCW dataset is constructed by selecting five subsets of 91 consecutive days' data from the original RSC15. In each of such subsets, the first 90 days' data is the training set,

Table 4. Dataset characteristics.

	RSC	RSCW	Atom
Sessions	8M	4M	82K
Avg. length	3.97	3.92	11.48
Items	37K	34K	54K
With timestamp	Yes	Yes	No

while the last day's data is the test set. Hence, RSCW dataset contains five sub-datasets, in each of which the recommendation problem is set as in RSC dataset. The performance on RSCW is averaged on the five sub-datasets, to minimize the risk that the obtained results are sensitive to the splitting strategy.

The statistic information of the three datasets are shown in Table 4. The statistics of RSCW is averaged on its five sub-datasets.

Metrics. In our experiments, we adopt three evaluation metrics: **Hit Rate (HR)**, **Mean Reciprocal Rank (MRR)** and **Coverage**. In the following, the test dataset is denoted as $Test$, the cardinality of the test dataset is denoted as $|Test|$, $R_i@L$ is the recommendation list at length $L = 20$ for current session i, I is the item set and its cardinality is denoted as $|I|$.

$$HR@L = \frac{1}{|Test|} \sum_{i \in Test} hit_i, \tag{3}$$

where $hit_i = 1$ when the ground truth item of current session i is recommended in $R_i@L$.

$$MRR@L = \frac{1}{|Test|}\left(\sum_{j \in Test} \frac{1}{rank_j} \right), \tag{4}$$

where $rank_j$ indicates the rank of item which is interacted in sample j. MRR is a widely used metric in information retrieval, where the rank of the ground truth item in $R_i@L$ is valued.

$$Coverage@L = \frac{1}{|I|} \left| \bigcup_{j \in Test} R_j@L \right|. \tag{5}$$

The coverage describes the percentage of recommended items in Top-L places of all the samples' recommendation lists over all the candidate items.

Baselines. We conduct experiments to compare the following approaches:

- **IKNN** [5] proceeds in an item-centric manner. In IKNN, the most similar items of current item are selected through an item-item similarity matrix, which has been established based on session-item records.

- **CKNN-cosine-Original** [6] proceeds in an session-centric manner. Specifically, it adopts cosine as the similarity metric and **Original** as the candidate selection method. It is the state-of-the-art KNN approach for session-based recommendation.
- **CKNN-{MD, HC, MDHC, DSM}-Original:** To compare the performance of different diffusion-based similarity methods, we conduct the experiments of CKNN approaches equipped with MD, HC, MDHC and DSM similarity metrics, and the **Original** candidate selection method.
- **CKNN-{cosine, DSM}-{EPCS, EPCSR}:** To compare different candidate selection strategies, we conduct the experiments of CKNN approaches equipped with EPCS and EPCSR under both cosine and DSM.

For the approaches of CKNN, we set the number of the most recent sessions $k_{recent} = 1000$ and the number of the most similar sessions $k_{top} = 500$ according to the parameters which achieve the best performance in [6].

3.2 The Performance of Candidate Selection Strategy

In this section, we present two experiments to evaluate the effectiveness and efficiency of the different candidate selection strategies respectively. Noted that, both λ and β in DSM are set as 0.5.

The Effectiveness of Candidate Selection Strategy. Figure 2 shows the performance of CKNN-DSM(0.5, 0.5) and CKNN-cosine under different candidate selection strategies in terms of HR@20, MRR@20 and Coverage@20 on RSC dataset. Because EPCS guarantees the ratio of relevant sessions containing the last item in the recent relevant session set, the items co-occurred with the last item are captured. As a result, the accuracy and diversity of both DSM(0.5, 0.5) and CKNN are improved on RSC dataset when we adopt EPCS. In addition, EPCSR is able to focus on all clicks in current session x and guarantee the ratio of every click's recent relevant sessions in $RC(x)$. Therefore, on the basis of EPCS, EPCSR achieves the best results on RSC dataset.

The Efficiency of Candidate Selection Strategy. The EPCS and EPCSR is more efficient than the Original approach [6]. The reasons are as follows: (1) The recent relevant session set of the other items in current session has been found previously, only a little more calculation is needed; (2) Selecting the recent session set of a single item (i.e., the last item in the session) is much easier and requires less computation than finding that of a set of items. As a result, the efficiency of these three strategies is ranked as follows: EPCSR>EPCS>Original.

3.3 The Study of λ and β in DSM

In this section, we conduct experiments to study the impact of hyper-parameter λ and β in DSM on RSC dataset, where both hyper-parameters are ranged in $[0, 1]$.

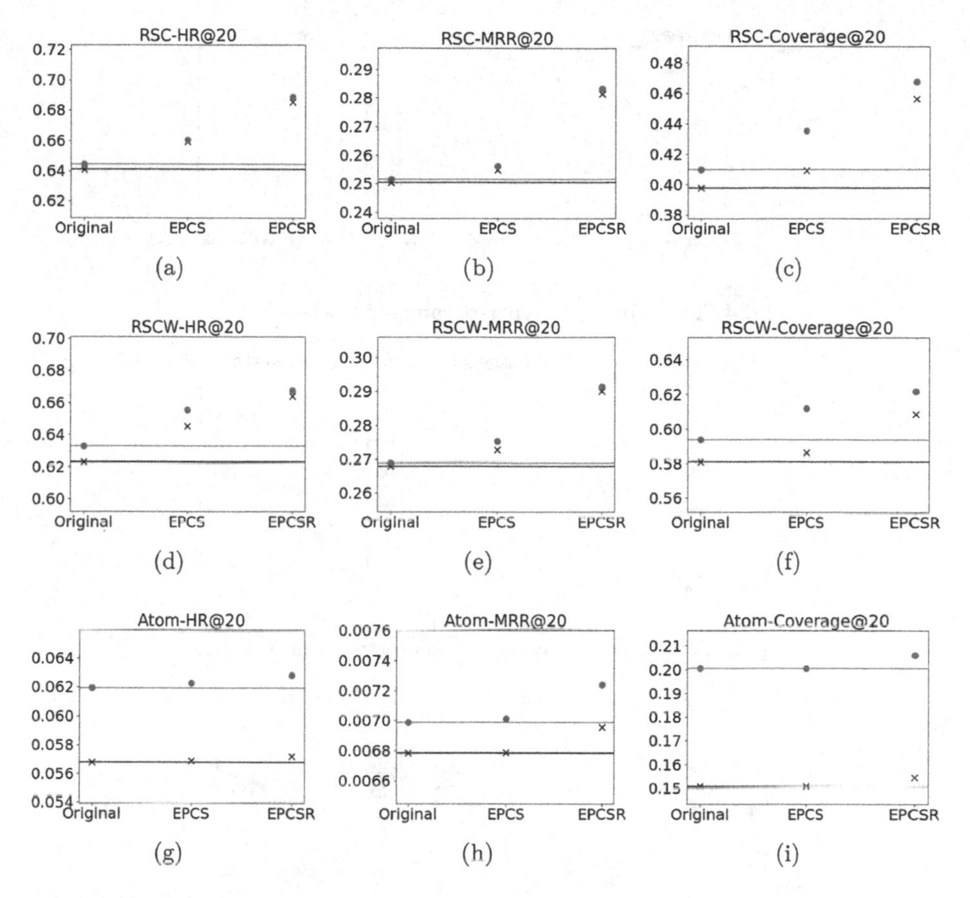

Fig. 2. The performance of different candidate selection strategies. Note: the red solid circle and blue × represent the result of CKNN algorithm when using DSM(0.5, 0.5) and cosine respectively. (Color figure online)

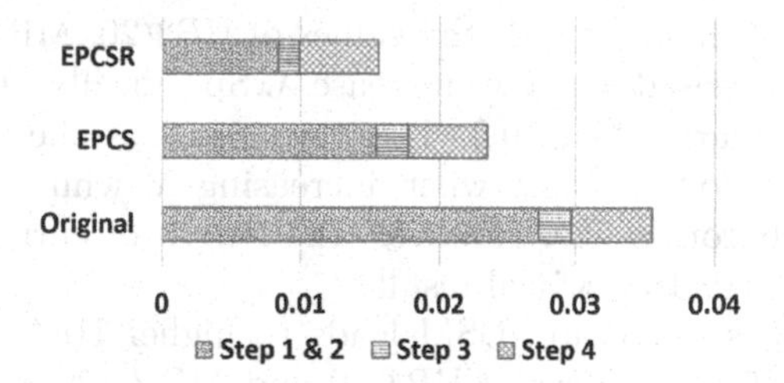

Fig. 3. The running time comparison.

Figures 4, 5 and 6 present the impact of λ, β and both of λ and β respectively. In these figures, from left to right are HR@20, MRR@20 and Coverage@20. The observations are summarized as follows (Fig. 3):

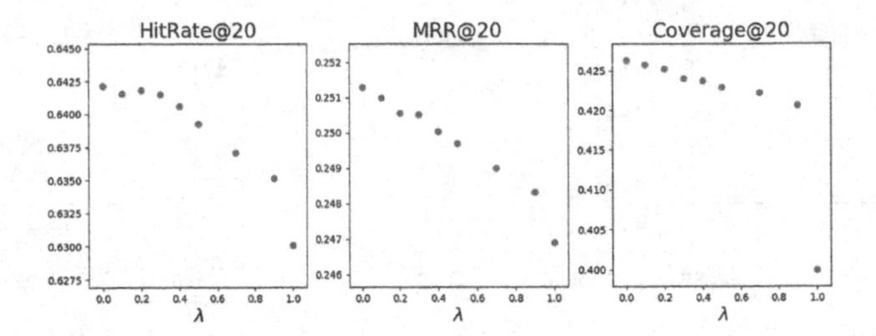

Fig. 4. The impact of hyper-parameter λ when $\beta = 1$.

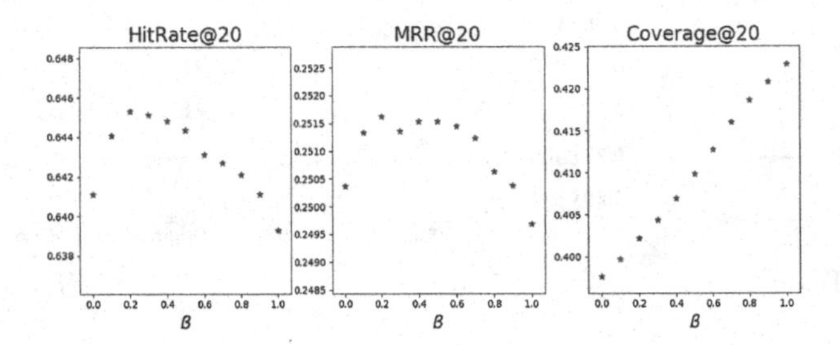

Fig. 5. The impact of hyper-parameter β when $\lambda = 0.5$.

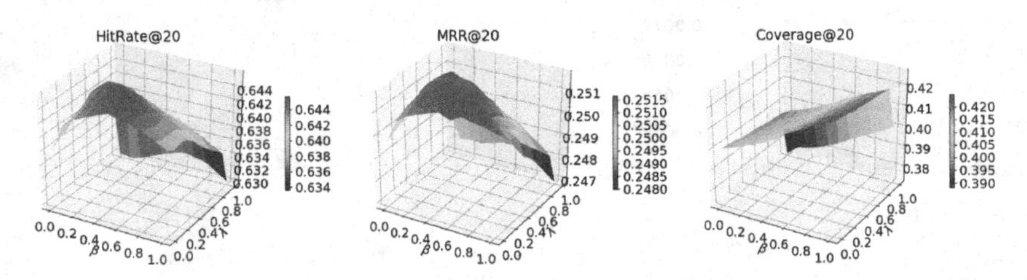

Fig. 6. The impact of hyper-parameters λ and β.

- As shown in Fig. 4, fixing $\beta = 1$, the values of HR@20, MRR@20 and Coverage@20 are all decreased when we increase λ. Specifically, these metrics drop rapidly when λ is larger than 0.5. The reason is that the impact of current session's length becomes larger when increasing λ, while that of compared session's length becomes less. However, the length of current session makes no sense, which leads to a trivial result.
- If the value of β is moderate, DSM leads to higher HR@20 and MRR@20. As shown in Fig. 5, the values of HR@20 and MRR@20 are increasing as β becomes larger from 0, while declining when β is greater than 0.5. In addition, the impact of item's popularity becomes greater as β increases, as a result, the coverage@20 increases.
- Figure 6 presents that DSM achieves the highest value in terms of HR@20 and MRR@20 when β and λ are all around 0.5. Under the same setting, DSM reaches a relative high value in terms of Coverage@20. Although DSM is able to obtain the highest value in terms of Coverage@20 when $\beta = 1$,

Table 5. The performance on all datasets.

Datasets	Algorithms	HR@20	MRR@20	Coverage@20
RSC	IKNN	0.5129	0.2051	**0.6267**
	CKNN-cosine-Original	0.6411	0.2504	0.3976
	CKNN-HC-Original	0.6422	0.2513	**0.4263**
	CKNN-MD-Original	0.6301	0.2469	0.3999
	CKNN-MDHC-Original	0.6393	0.2497	0.4229
	CKNN-DSM(0.5, 0.5)-Original	**0.6444**	**0.2515**	0.4099
	CKNN-cosine-EPCSR	0.6854	0.2815	0.4563
	CKNN-DSM(0.5, 0.5)-EPCSR	**0.6888**	**0.2834**	**0.4678**
RSCW	IKNN	0.4736	0.1975	**0.7590**
	CKNN-cosine-Original	0.6234	0.2679	0.5810
	CKNN-HC-Original	0.6289	0.2674	**0.6151**
	CKNN-MD-Original	0.6238	0.2645	0.5884
	CKNN-MDHC-Original	0.6296	0.2665	0.6077
	CKNN-DSM(0.5, 0.5)-Original	**0.6329**	**0.2688**	0.5939
	CKNN-cosine-EPCSR	0.6641	0.2900	0.6089
	CKNN-DSM(0.5, 0.5)-EPCSR	**0.6678**	**0.2915**	**0.6131**
Atom	IKNN	0.0260	0.0066	**0.6065**
	CKNN-cosine-Original	0.0568	0.0068	0.1509
	CKNN-HC-Original	0.0520	0.0065	**0.2357**
	CKNN-MD-Original	0.0546	0.0069	0.1767
	CKNN-MDHC-Original	0.0534	0.0067	0.2319
	CKNN-DSM(0.5, 0.5)-Original	**0.0620**	**0.0070**	0.2006
	CKNN-cosine-EPCSR	0.0572	0.0070	0.1546
	CKNN-DSM(0.5, 0.5)-EPCSR	**0.0628**	**0.0072**	**0.2061**

other metrics are poor. Therefore, we set $\beta = \mathbf{0.5}$ and $\lambda = \mathbf{0.5}$ in the following experiments to obtain both high accuracy and good diversity.

3.4 Overall Performance

According to the results of different approaches on the three datasets in Table 5, the following conclusions are observed:

- Due to ignoring the contextual information, IKNN achieves the worse performance in terms of HR@20 and MRR@20 on three datasets. Because the **limited accuracy of IKNN**, the best result in terms of Coverage@20 does not matter to the recommendation task.

- Compared with the CKNN algorithms equipped with other similarity metrics and Original candidate selection method, **CKNN-DSM-Original achieves better accuracy** (i.e., HR@20 and MRR@20) when we set $\lambda = 0.5$ and $\beta = 0.5$. It is because DSM(0.5, 0.5) incorporates reasonable graph information when calculating the session similarities.
- Compared with **Original** strategy, the proposed **EPCSR** improves the performance on all three metrics. Specifically, CKNN-DSM(0.5, 0.5)-EPCSR outperforms CKNN-cosine-Original (which is the **state-of-the-art KNN** approach [6]) by 7.4%, 7.1% and 10.6% in terms of HR@20 (13.2%, 8.8% and 5.9% in terms of MRR@20, 17.7%, 5.5% and 36.6% in terms of Coverage@20) on RSC, RSCW and Atom datasets.

4 Conclusions

In this paper, we introduced a new contextual KNN approach for session-based recommendation, which incorporates the power of diffusion-based similarity method DSM and candidate selection method EPCSR. It gains performance improvement from these advantages: (1) By adopting DSM, the session-item graph structure is utilized in the procedure of similarity calculation. (2) Through guaranteeing the ratio of different clicks' recent relevant sessions in the recent session set of current session, EPCSR is able to capture the items that co-occurred with different historical clicked items in the same session efficiently. We conducted extensive experiments on three benchmark datasets to compare the effectiveness of our approach and the state-of-the-art KNN approaches for session-based recommendation. Our experimental results demonstrate that our approach obtains better performance.

Acknowledgement. This research was supported in part by NSFC under Grant No. U1836107, and National Key R&D Program of China under Grant No. 2018YFB0504905.

References

1. Bonnin, G., Jannach, D.: Automated generation of music playlists: survey and experiments. ACM Comput. Surv. **47**(2), 26:1–26:35 (2014)
2. Covington, P., Jay, A., Sargin, E.: Deep neural networks for Youtube recommendations. In: ACM RecSys, pp. 191–198 (2016)
3. Guo, H., Tang, R., Ye, Y., Li, Z., He, X.: DeepFM: a factorization-machine based neural network for CTR prediction. In: IJCAI, pp. 1725–1731 (2017)
4. Guo, H., Tang, R., Ye, Y., Li, Z., He, X.: A graph-based push service platform. In: Candan, S., Chen, L., Pedersen, T.B., Chang, L., Hua, W. (eds.) DASFAA 2017. LNCS, vol. 10178, pp. 636–648. Springer, Cham (2017). https://doi.org/10.1007/978-3-319-55699-4_40
5. Hidasi, B., Karatzoglou, A., Baltrunas, L., Tikk, D.: Session-based recommendations with recurrent neural networks. CoRR abs/1511.06939 (2015)

6. Jannach, D., Ludewig, M.: When recurrent neural networks meet the neighborhood for session-based recommendation. In: ACM RecSys, pp. 306–310 (2017)
7. Kabbur, S., Ning, X., Karypis, G.: FISM: factored item similarity models for top-n recommender systems. In: SIGKDD, pp. 659–667 (2013)
8. Koren, Y., Bell, R.M., Volinsky, C.: Matrix factorization techniques for recommender systems. IEEE Comput. J. **42**(8), 30–37 (2009)
9. Ludewig, M., Jannach, D.: Evaluation of session-based recommendation algorithms. User Model. User-Adapt. Interact. **28**(4–5), 331–390 (2018)
10. Rendle, S., Freudenthaler, C., Schmidt-Thieme, L.: Factorizing personalized Markov chains for next-basket recommendation. In: WWW, pp. 811–820 (2010)
11. Yap, G., Li, X., Yu, P.S.: Effective next-items recommendation via personalized sequential pattern mining. In: DASFAA, pp. 48–64 (2012)
12. Zhou, T., Ren, J., Medo, M., Zhang, Y.: Bipartite network projection and personal recommendation. Phys. Rev. E **76**(2), 046115 (2007)

A Contextual Bandit Approach to Personalized Online Recommendation via Sparse Interactions

Chenyu Zhang[1], Hao Wang[2(✉)], Shangdong Yang[1], and Yang Gao[1]

[1] State Key Laboratory for Novel Software Technology, Nanjing University, Nanjing 210023, China
zhangcy@smail.nju.edu.cn, shangdong007@gmail.com, gaoy@nju.edu.cn
[2] Inception Institute of Artificial Intelligence, Abu Dhabi, UAE
hao.wang@inceptioniai.org

Abstract. Online recommendation is an important feature in many applications. In practice, the interaction between the users and the recommender system might be sparse, i.e., the users are not always interacting with the recommender system. For example, some users prefer to sweep around the recommendation instead of clicking into the details. Therefore, a response of 0 may not necessarily be a negative response, but a non-response. It comes worse to distinguish these two situations when only one item is recommended to the user each time and few further information is reachable. Most existing recommendation strategies ignore the difference between non-responses and negative responses. In this paper, we propose a novel approach, named SAOR, to make online recommendations via sparse interactions. SAOR uses positive and negative responses to build the user preference model, ignoring all non-responses. Regret analysis of SAOR is provided, experiments on both real and synthetic datasets also show that SAOR outperforms competing methods.

Keywords: Online recommendation · Sparse interaction · Contextual bandit

1 Introduction

Recently, *online recommendation* has become a key feature in many practical applications such as e-commerce services, news services, and streaming services, etc. [16]. When using such applications, users are essentially, and probably unknowingly, *interacting* with recommender systems. Online recommendation problems have some unique characteristics.

1. *Continual interaction*. The user continually interacts with the recommender system, generating an endless *stream* of interaction data.

H. Wang—This work was done when this author was an assistant professor at Nanjing University.

Q. Yang et al. (Eds.): PAKDD 2019, LNAI 11440, pp. 394–406, 2019.
https://doi.org/10.1007/978-3-030-16145-3_31

2. *Implicit interaction.* The user does not directly indicate how much s/he likes or dislikes a recommended item.
3. *Sparse interaction.* It is impractical to assume that the user is always interacting with the recommender system. That means, to the recommender system, a non-response does not necessarily mean a negative response. We argue that such discrimination is crucial in computing recommendations.

In particular, *a non-response is used to update the recommender system as if it is a negative response, since the recommended item is believed to be unattractive to the target user.* However, as explained earlier, sparse interaction is overlooked in the field of recommender systems. Some users used to ignore the recommender system very often, even if the recommended items are in fact of much interest. In this case, it is unfair to blame the recommender system for making a bad recommendation. Any update (penalty) introduced into the learning process could potentially be misleading. In addition, computational resources (time, memory, etc.) used to implement the update are probably wasted.

In this paper, we propose a novel approach, SAOR (sparsity-aware online recommender), taking consideration of continual, implicit and sparse interactions. SAOR makes probabilistic estimations on whether the user is interacting or not, by reasonably assuming that similar items are similarly attractive. Based on such estimations, SAOR uses positive and negative responses to build the user preference model, ignoring all non-responses.

The main contributions made in this work include:

- We propose SAOR, a novel online recommendation algorithm, as a solution to the problem. SAOR is able to distinguish between non-responses and negative responses. By ignoring non-responses, SAOR realizes a more accurate and efficient learning.
- The regret analysis of SAOR is provided. Besides, experiments on both real and synthetic datasets show that SAOR outperforms competing methods.

The rest of this paper is organized as follows. Section 2 gives the problem formulation. Section 3 describes our approach in detail. Then, a regret analysis is presented in Sect. 4. Extensive experimental results are shown in Sect. 5. Section 6 reviews the related work, and finally Sect. 7 concludes the paper.

2 Problem Formulation and Methodology

The entire recommender service consists of two independent modules, a content provider and a recommender system. The content provider refreshes the pool of candidate items at a certain frequency. At time t, for an active user u, the recommender system presents to her one item x_t out of the pool of K items, $C_t = \{x_{1,t}, x_{2,t}, \cdots, x_{K,t}\}$, in the hope that s/he might be interested in that recommended item. The recommender system then observes the user's response towards the recommended item x_t. The user's response y_t is either *click* the item ($y_t = 1$) or not ($y_t = 0$). This 0/1 response may be used by the recommender system to update its internal models. This process then continues for time $t+1$. The goal of the recommender system is to make a good sequential decision $\langle x_1, x_2, \cdots, x_t, \cdots \rangle$. If x_t^* is an optimal choice at time t, then the recommender system aims at minimizing the *regret* up to time T,

$$R_T = \mathbb{E}\left[\sum_{t=1}^{T} y_t(x_t^*)\right] - \mathbb{E}\left[\sum_{t=1}^{T} y_t(x_t)\right].$$

To establish a computational model, we consider the content of an item $x_{j,t}$ as an n-dimensional feature vector $\boldsymbol{x}_{j,t} \in [0,1]^n$. A user u is considered as a pair $u = (\pi, \boldsymbol{\theta})$. The *preference* $\boldsymbol{\theta} \in [0,1]^n$ is an n-dimensional weighting vector. The *attention* π is an arbitrary mapping from $\mathcal{T}$, the time domain, to $\{0,1\}$. $\pi_t = \pi(t)$ indicates whether the user is interacting with the recommender system at time t, which may not necessarily depend on the preference $\boldsymbol{\theta}$ or any item content $\boldsymbol{x}$. When being recommended an item $\boldsymbol{x}$, the user u is assumed to click $\boldsymbol{x}$ with probability $\pi \cdot \boldsymbol{x}^\top \boldsymbol{\theta}$. That is, when the user is inattentive (i.e., $\pi = 0$), s/he will simply ignore $\boldsymbol{x}$; otherwise (i.e., $\pi = 1$), the probability of clicking is linearly determined by the item content $\boldsymbol{x}$ and the user preference $\boldsymbol{\theta}$.

Methodology. The key to a good sequential decision $\langle \boldsymbol{x}_1, \boldsymbol{x}_2, \cdots, \boldsymbol{x}_t, \cdots \rangle$ is clearly a good user model $\hat{u} = (\hat{\pi}, \hat{\boldsymbol{\theta}})$ for choosing item $\boldsymbol{x}_t$ from the candidate set C_t. Given an oracle attention model $\hat{\pi}$ (i.e., suppose that the recommender system knows when the user is interacting and when is not), one may ignore all non-responses and $\hat{\boldsymbol{\theta}}$ can be effectively learned from positive and negative responses using existing techniques (e.g., [3,12]). To obtain a good attention model $\hat{\pi}$, the key insight of our approach is two-fold:

1. If a recommended item $\boldsymbol{x}_t$ is sufficiently attractive but not positively responded, then it is likely that the user is inattentive; and
2. If a recommended item $\boldsymbol{x}_t$ is similar to some item a that has recently been positively responded, then it is likely that $\boldsymbol{x}_t$ is sufficiently attractive.[1]

In this way, we may estimate a good $\hat{\pi}$ by estimating the probability of $\pi = 0$ conditioned on the item contents and the user's recent history. This can then help to distinguish true negative responses from non-responses. After filtering out non-responses, rest of the data can be used to learn a more accurate user preference model $\hat{\boldsymbol{\theta}}$.

3 Our Approach

As explained earlier in Sect. 2, for a user u with attention π and preference $\boldsymbol{\theta}$, the key in our methodology is to obtain a good estimation $\hat{\pi}$ of π and $\hat{\boldsymbol{\theta}}$ of $\boldsymbol{\theta}$.

[1] The reason why we consider a *recent*, instead of the *entire*, history is that a user's interests may change with time in general but remain focused in a short period [10]. This assumption is in fact a basis of many item-based recommendation algorithms (see, for example, [17]).

In the following, Sects. 3.1 and 3.2 present the technical details of estimating $\hat{\pi}$ and learning $\hat{\boldsymbol{\theta}}$, respectively. Finally, Sect. 3.3 presents SAOR, our proposed algorithm.

3.1 Estimating the User Attention

For a user u, we consider the temporal list of items that s/he has ever clicked, i.e., his/her *click history* $\mathcal{H} = \langle \boldsymbol{x}_1, \boldsymbol{x}_2, \cdots, \boldsymbol{x}_t \rangle$, assuming that there are t such items in total. For each $\boldsymbol{x}_j$ in the list $\mathcal{H}$, we compute the minimum Euclidean distance between $\boldsymbol{x}_j$ and the m items directly prior to $\boldsymbol{x}_j$, i.e., we compute

$$d_j = \min_{i=1,2,\cdots,m} \|\boldsymbol{x}_j - \boldsymbol{x}_{j-i}\|_2, \quad (j > m) \tag{1}$$

where m is the *recency parameter*. Although the actual π of a user u may be complicated, it can be estimated as $\Pr\{d \geq d_{t+1} \,|d \sim \mathcal{D}_{t,m}\}$, where $\mathcal{D}_{t,m}$ is the distribution of all *small* values in the time series $S_{t,m} = \langle d_{m+1}, d_{m+2}, \cdots, d_t \rangle$.

Since interest drifts are unforeseeable in nature, they should be considered as outliers and excluded from the distribution model $\mathcal{D}_{t,m}$. To that end, we use the standard technique of Fourier transform.[2] Noise reduction can be implemented by ignoring component waves of high frequencies.

After noise reduction, the multiset $S_{t,m}$ contains only those distance values that are "small" enough and should observe a unimodal distribution over the range $[0, 1]$.[3] We thus fit the data into a beta distribution, which has a probability density function of the form $f(x|\alpha, \beta) = \frac{1}{B(\alpha,\beta)} x^{\alpha-1}(1-x)^{\beta-1}$, where $B(\alpha, \beta) = \int_0^1 v^{\alpha-1}(1-v)^{\beta-1}dv$. If $\hat{\mu}$ and $\hat{\sigma}$ are the sample mean and sample standard deviation of $S_{t,m}$, respectively, then the beta distribution is determined by

$$\begin{cases} \hat{\alpha} = -\hat{\mu}(\hat{\sigma}^2 + \mu^2 - \mu)/\hat{\sigma}^2, \\ \hat{\beta} = (\hat{\mu} - 1)(\hat{\sigma}^2 + \hat{\mu}^2 - \hat{\mu})/\hat{\sigma}^2. \end{cases} \tag{2}$$

Using the estimated beta distribution, we calculate

$$\Pr\{d \geq d_{t+1} \,|d \sim \mathcal{D}_{t,m}\} = \frac{\int_{d_{t+1}}^1 v^{\alpha-1}(1-v)^{\beta-1}dv}{\int_0^1 v^{\alpha-1}(1-v)^{\beta-1}dv}. \tag{3}$$

Algorithm 1 summarizes the discussion of this section. Lines 1–3 are the process of Fourier fitlering. We use standard fast Fourier transform (FFT) and inverse fast Fourier transform (IFFT) for time-efficiency considerations [7]. Lines 4–5 are to compute $\Pr\{\pi_{t+1} = 0 \,|d_{t+1}\}$.

[2] See, for example, https://en.wikipedia.org/wiki/Fourier_transform.

[3] Here and hereafter, we may assume that the maximum possible distance $d_{max} = 1$, without loss of generality. This is because, when $d_{\max} > 1$, a simple rescaling can transform all the data into $[0, 1]$.

Algorithm 1. ESTIMATEATTENTION(S, d)

Input: Distance time series S; distance d
Output: $\hat{\pi}$
1 Do fast Fourier transform $F \leftarrow \text{FFT}(S)$
2 $F \leftarrow F$ with high frequencies filtered
3 Do inverse fast Fourier transform $S \leftarrow \text{IFFT}(F)$
4 Estimate beta distribution $\mathcal{D}$ from S using Eq. 2
5 Compute $p \leftarrow \Pr\{z \geq d \,|\, z \sim \mathcal{D}\}$ using Eq. 3
6 **return** 0 w.p. p or 1 w.p. $1-p$

3.2 Learning the User Preference

Now during the entire history $\langle(\boldsymbol{x}_1, y_1), (\boldsymbol{x}_2, y_2), \cdots, (\boldsymbol{x}_t, y_t)\rangle$, suppose without loss of generality that we have $\hat{\pi}_j = 1, j = 1, 2, \cdots, t$. Recall that, in this case, the user's decision on whether clicking an item $\boldsymbol{x}$ is determined by $\boldsymbol{x}^\top\boldsymbol{\theta}$ (see Sect. 2). At time t, it suffices to solve the equation $\boldsymbol{X}_t^\top\hat{\boldsymbol{\theta}}_t = \boldsymbol{y}_t$ for an estimated preference $\hat{\boldsymbol{\theta}}_t$, where $\boldsymbol{X}_t = (\boldsymbol{x}_1, \boldsymbol{x}_2, \cdots, \boldsymbol{x}_t)$ is an $n \times t$ matrix and $\boldsymbol{y}_t = (y_1, y_2, \cdots, y_t)^\top$. Ridge regression can be applied as

$$\hat{\boldsymbol{\theta}}_t = \boldsymbol{A}_t^{-1}\boldsymbol{X}_t\boldsymbol{y}_t, \tag{4}$$

where $\boldsymbol{A}_t \stackrel{\text{def}}{=} \boldsymbol{X}_t\boldsymbol{X}_t^\top + \boldsymbol{I}_n$ and $\boldsymbol{I}_n$ is the n-dimensional identity matrix.

Lemma 1 (Lemma 1 of [5]**).** *For a total number of T time steps, for any $\boldsymbol{x}_{t+1}$ of the K candidate items, the inequality*

$$\left|\boldsymbol{x}_{t+1}^\top\hat{\boldsymbol{\theta}}_t - \mathbb{E}\left[y_{t+1}(\boldsymbol{x}_{t+1})\right]\right| \leq (1+\alpha)\sqrt{\boldsymbol{x}_{t+1}^\top\boldsymbol{A}_t^{-1}\boldsymbol{x}_{t+1}} \tag{5}$$

holds w.p. at least $1 - \frac{\delta}{T}$, where $\alpha = \sqrt{\frac{1}{2}\ln\frac{2TK}{\delta}}$. □

Lemma 1 means that, with a sufficiently large t (i.e., with sufficient training data), $\hat{\boldsymbol{\theta}}_t$ can be a good model for generating recommendations.

Online Update. If a new offer-response record $(\boldsymbol{x}_{t+1}, y_{t+1})$ is to be included to compute an updated preference model $\hat{\boldsymbol{\theta}}_{t+1}$, the following online update rule can be used:

$$\begin{cases} \boldsymbol{A}_{t+1} \leftarrow \boldsymbol{A}_t + \boldsymbol{x}_{t+1}\boldsymbol{x}_{t+1}^\top, \\ \boldsymbol{b}_{t+1} \leftarrow \boldsymbol{b}_t + y_{t+1}\cdot\boldsymbol{x}_{t+1}, \\ \hat{\boldsymbol{\theta}}_{t+1} \leftarrow \boldsymbol{A}_{t+1}^{-1}\boldsymbol{b}_{t+1}, \end{cases} \tag{6}$$

where $\boldsymbol{b}_t \stackrel{\text{def}}{=} \boldsymbol{X}_t\boldsymbol{y}_t$.

3.3 Putting Everything Together

There is a so-called *exploration-exploitation tradeoff* when choosing $\boldsymbol{x}_{t+1}$. Let $\boldsymbol{x}_{\max} \in C_{t+1}$ be the item that maximizes $\boldsymbol{x}^\top \hat{\boldsymbol{\theta}}_t$. The *upper confidence bound* (UCB) is a way to implement the exploration-exploitation tradeoff [4]. Specifically, the UCB strategy predicts the user's response to item $\boldsymbol{x}_{j,t+1}$ by

$$\hat{y}_{j,t+1} = \boldsymbol{x}_{j,t+1}^\top \hat{\boldsymbol{\theta}}_t + \alpha \sqrt{\boldsymbol{x}_{j,t+1}^\top \mathbf{A}_t^{-1} \boldsymbol{x}_{j,t+1}}. \tag{7}$$

Then the item that maximizes $\hat{y}_{j,t+1}$ is recommended.

Putting all the above discussions together, Algorithm 2 gives an overview of our approach. Lines 1–3 initialize the algorithm. A small *warm-start parameter* k is used to guarantee that any statistical tool is applied with sufficient data. In practice, the recommender system may simply recommend random or most popular items until there are k clicks. The endless loop of Lines 5–19 is the online process. Lines 6–9 generates a recommendation using the UCB strategy. Lines 10–19 update the current model(s). In particular, a response of 0 is simply ignored if EstimateAttention (Algorithm 1) believes that the user is likely noninteractive.

Algorithm 2. SAOR

```
   Input: exploration factor α; recency parameter m; warm-start parameter k
 1 Set t ← k
 2 H_t ← the temporal list of clicked items up to time t
 3 S_{t,m} ← the time series from H_t using recency m
 4 Construct A_t and b_t, and compute θ_t using Eq. 4
 5 while true do
 6     Observe candidates x_{1,t+1}, x_{2,t+1}, ··· , x_{K,t+1}
 7     for each x_{j,t+1} do
 8         Calculate ŷ_{j,t+1} using Eq. 7 with α
 9     Recommend x_{t+1} that has the maximum ŷ_{j,t+1}
10     Compute d_{t+1} using Eq. 1 with x_{t+1} and S_{t,m}
11     Observe the user's response y_{t+1}
12     if y_{t+1} = 0 then
13         π̂ ← EstimateAttention(S_{t,m}, d_{t+1})
14         if π̂ = 0 then continue
15     else
16         H_{t+1} ← Append x_{t+1} to H_t
17         S_{t+1,m} ← Append d_{t+1} to S_{t,m}
18     Compute A_{t+1}, b_{t+1}, and θ̂_{t+1} using Eq. 6
19     t ← t + 1
```

4 Regret Analysis

This section analyzes the regret bound of SAOR. We can see clearly from Algorithm 2, that regrets may due to either a suboptimal recommendation $\boldsymbol{x}_{t+1}$ (Line 9) or a mistakenly estimated $\hat{\pi}$ when $y_{t+1} = 0$ (Line 13). Specifically, we distinguish the following two types of regrets.

1. The *Type-I regret* is the regret from the time steps t when $y_t = 1$, or $\hat{\pi}_t = \pi_t$ when $y_t = 0$;[4] and
2. The *Type-II regret* is the regret from the time steps t when $y_t = 0$ and $\hat{\pi}_t \neq \pi_t$.

We shall first bound the two types of regret in each time step, respectively, and then prove an overall regret bound by combining the results.

Theorem 1 (Bound of single-step type-I regret). *Let t be a time step on which $y_t = 1$ or $\hat{\pi}_t = \pi_t$. Assume that there are in total T time steps and at each time step there are K candidate items, then the type-I regret at time t is bounded by*

$$R_t^I \leq (1+\alpha)\sqrt{\boldsymbol{x}_t^\top \boldsymbol{A}_t^{-1} \boldsymbol{x}_t},$$

w.p. at least $1 - \frac{\delta}{T}$, where $\alpha = \sqrt{\frac{1}{2} \ln \frac{2TK}{\delta}}$.

Proof. When $y_t = 1$ or $\hat{\pi}_t = \pi_t$, the module EstimateAttention (Algorithm 1) correctly recognizes positive and negative responses, and ignores non-responses also. This reduces to the case where every record $(\boldsymbol{x}_t, y_t)$ is either positive or negative, which has already been studied by, for example, SupLinUCB [5]. Let $\boldsymbol{\theta}_t^*$ be the true preference at time t. Then, $R_t^I = \left| \boldsymbol{x}_t^\top \hat{\boldsymbol{\theta}}_t - \boldsymbol{x}_t^\top \boldsymbol{\theta}^* \right|$. Directly applying Lemma 1 completes this proof. □

To obtain a bound for single-step type-II regret, we need the following lemma.

Lemma 2 (Lemma 12 of [1]). *Let $\boldsymbol{A}$, $\boldsymbol{B}$ and $\boldsymbol{C}$ be positive semi-definite matrices such that $\boldsymbol{A} = \boldsymbol{B} + \boldsymbol{C}$. Then,*

$$\sup_{x \neq 0} \frac{\boldsymbol{x}^\top \boldsymbol{A} \boldsymbol{x}}{\boldsymbol{x}^\top \boldsymbol{B} \boldsymbol{x}} \leq \frac{\det(\boldsymbol{A})}{\det(\boldsymbol{B})}.$$

Now we give the bound of type-II regret.

Theorem 2 (Bound of single-step type-II regret). *Let t be a time step on which $y_t = 0$ and $\hat{\pi}_t \neq \pi_t$. Assume that there are in total T time steps and at each time step there are K candidate items, then the type-II regret at time t is bounded by*

$$R_t^{II} \leq 2(1+\alpha)\sqrt{\boldsymbol{x}_t^\top \boldsymbol{A}_t^{-1} \boldsymbol{x}_t},$$

w.p. at least $1 - \frac{\delta}{T}$, where $\alpha = \sqrt{\frac{1}{2} \ln \frac{2TK}{\delta}}$.

---

[4] We do not consider the case where the user mistakenly clicks some item.

Proof. When $y_t = 0$ and $\hat{\pi}_t \neq \pi_t$, there are two cases: (1) SAOR mistakenly ignores $(\boldsymbol{x}_t, y_t)$ while it is in fact a negative response, and (2) SAOR mistakenly updates using $(\boldsymbol{x}_t, y_t)$ while it is in fact a non-response. In this proof we only consider the first case, since the other case can be similarly analyzed.

Consider the matrix $\boldsymbol{A}_t$, and let $\boldsymbol{A}_\tau$ $(\tau < t)$ be the latest version of the matrix before time t. As the effect of one single feature vector to $\boldsymbol{A}_t$ is limited, there exists a $0 < \gamma < 1$ such that $\det(\boldsymbol{A}_t) \leq (1+\gamma)\det(\boldsymbol{A}_\tau)$.

Then,

$$R_t^{II} = \left|\boldsymbol{x}_t^\top \boldsymbol{\theta}_\tau - \boldsymbol{x}_t^\top \boldsymbol{\theta}^*\right| \leq (1+\alpha)\sqrt{\boldsymbol{x}_t^\top \boldsymbol{A}_\tau^{-1} \boldsymbol{x}_t} < 2(1+\alpha)\sqrt{\boldsymbol{x}_t^\top \boldsymbol{A}_t^{-1} \boldsymbol{x}_t}.$$

□

Next, we use Theorems 1 and 2 to derive the overall regret bound of SAOR. We shall use the following technical lemma.

Lemma 3 (Lemma 3 of [5]). *Let* $s_{t,\boldsymbol{x}_t} \stackrel{def}{=} \sqrt{\boldsymbol{x}_t^\top \boldsymbol{A}_T^{-1} \boldsymbol{x}_t}$. *Assume that all the records in a total number of* T *time steps,* $\{(\boldsymbol{x}_1, y_1), (\boldsymbol{x}_2, y_2), \cdots, (\boldsymbol{x}_T, y_T),\}$, *can be divided into* $\ln T$ *sets of independent samples,* $\Psi_1, \Psi_2, \cdots, \Psi_{\ln T}$, *where each* Ψ_j *contains at least 2 records. Then, for any* Ψ_j *we have*

$$\sum_{t \in \Psi_j} s_{t,\boldsymbol{x}_t} \leq \sqrt{n \cdot |\Psi_j| \cdot \ln |\Psi_j|},$$

where n *is dimensionality of item feature vectors.* □

The following theorem gives the overall regret bound of SAOR.

Theorem 3 (Regret bound of SAOR). *For any small constant* $\delta > 0$, *the regret of SAOR up to time* T *is bounded by*

$$R_T = O\left(\sqrt{nT \ln^3 \frac{KT \ln T}{\delta}}\right)$$

with probability at least $1 - \frac{\delta}{T}$.

Proof. As the regret of single-step type-I (*Theorem* 1) and single-step type-II (*Theorem* 2) are of the same asymptotic order, the total regret of SAOR can be calculated as

$$R_T \leq \sum R_I^t + \sum R_t^{II} \leq \sum_{t=1}^{T} 2(1+\alpha)s_{t,\boldsymbol{x}_t} = \sum_{j=1}^{\ln T} \sum_{t \in \Psi_j} 2(1+\alpha)s_{t,\boldsymbol{x}_t}.$$

Applying Lemma 3 with $\alpha = \sqrt{\frac{1}{2} \ln \frac{2TK}{\delta}}$, we get an overall regret bound of $O\left(\sqrt{nT \ln^3 \frac{KT \ln T}{\delta}}\right)$. □

5 Experiments

In this section, the proposed SAOR algorithm is evaluated in several ways. First, our algorithm is compared with some other typical contextual bandit algorithms in effectiveness. Then, the algorithm's sensitivity of parameters is also studied.

Dataset and Evaluation Method. To verify the performance of our algorithm in real applications, we use the real data set *Yahoo! R6A*[5]. The data set is filtered to get 420,0230 records from 200 users. An unbiased offline evaluation method [13] is used to compare bandit algorithms in a reliable way, that is, if an algorithm chooses item $x_{i,t}$ and the randomly chosen item in the data set is x_t, the RoC of each algorithm in T rounds can be evaluated as

$$RoC(T) = \frac{\sum_{i=0}^{T} \mathbb{1}(x_{i,t} = x_t \; and \; r_{t,x_t} = 1)}{\sum_{i=0}^{T} \mathbb{1}(x_{i,t} = x_t)}.$$

Also, we use synthetic data to further verify the performance of our algorithm. In synthetic data, the preference $\boldsymbol{\theta}_u$ is constructed for each user, so that we can know the optimal item x_t^* at round t. Then if an algorithm chooses item $x_{i,t}$ at round t, the accuracy of each algorithm in T rounds can be evaluated as

$$Accuracy(T) = \frac{\sum_{i=0}^{T} \mathbb{1}(x_{i,t} = x_t^*)}{T}.$$

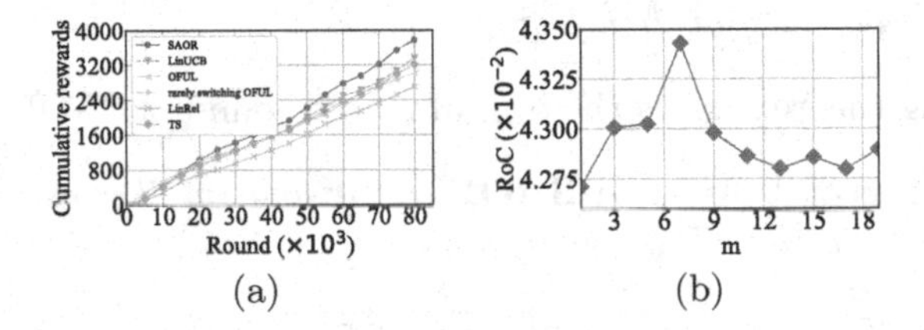

Fig. 1. (a) Cumulative rewards along with the number of rounds; (b) RoC along with parameter m.

Competing Algorithms. We are aware of several recent studies on online CF. However, instead of regret minimization, the term online there is about system-level supports for real-time responses. Their problem settings and focuses are clearly different from ours, thus the results are not directly comparable. Our work follows the line of contextual bandit research; hence we compare our method with the state-of-the-art bandit solutions. The following algorithms are compared with our algorithm:

- LinUCB [12]: The original LinUCB is applied as a baseline in the experiment, every record is used to update the model at each round.

[5] https://webscope.sandbox.yahoo.com/.

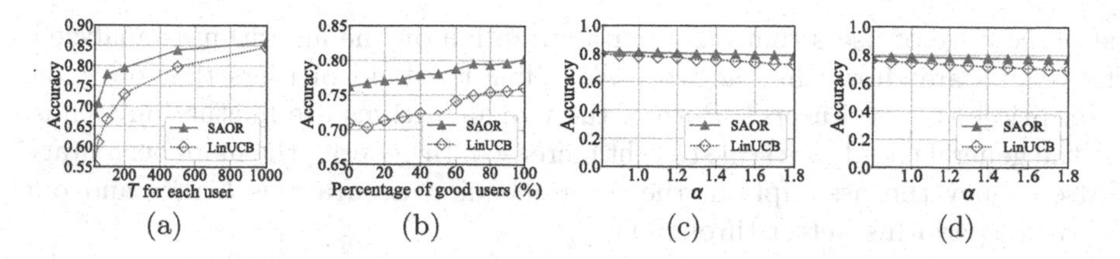

Fig. 2. (a) Accuracy along with the recommending times ($\alpha = 1.7$); (b) Accuracy along with the percentage of "good users" ($\alpha = 1.3$); (c) Analysis of the effect of parameter α ($T = 200$); (d) Analysis of the effect of parameter α ($Percentage = 50\%$).

- LinRel [3]: LinRel is similar to LinUCBin problem setting but different in the form of regularization.
- OFUL and its variant [1]: The OFUL algorithm maintains a confidence set for $\boldsymbol{\theta}$. A rarely switching OFUL algorithm was also proposed to update periodically for saving computation.
- TS [2]: Thompson Sampling for contextual bandits uses Thompson Sampling to sample $\boldsymbol{\theta}$ from Gaussian distribution.

Experiment Analysis. Each algorithm compared in this paper, including our SAOR, requires parameters. To compare all algorithms fairly, we choose the optimal parameter for each algorithm. The growth of cumulative rewards along with the number of rounds is plotted in Fig. 1. It's obvious that SAOR outperforms all the other compared algorithms. It should be noted that we cannot really expect RoCs to be high in a practical online recommender system. Low RoCs are normal, and they create regrets. That's why the lines in Fig. 1 all look more linear than logarithmic. Nonetheless, comparisons between different methods are fair, and the results show the usefulness of our method.

The recency parameter m in the variable $\mathcal{H}_m$ is unique in SAOR, i.e, the size of the latest clicked item set. The general RoC along with m is displayed in Fig. 1(b). If the size is too small, the performance is not so good as just using one or two latest items cannot cover user's whole preferences. If the size becomes too large, it will also affect the performance as it will include old records in the set and users interests may drift over time. In addition, if the size is too large, it will put much pressure on the calculation as the current item should be compared with each item in the set $\mathcal{H}_m$.

We also do experiments to verify our model on user attention. Here, we choose LinUCB for comparing as it's the baseline and it's fair to verify the assumption. First, the learning speed is analyzed and the recommending accuracy along with recommending times is shown in Fig. 2(a). Though the accuracies of these two algorithms get closer when recommending times get larger, SAOR outperforms LinUCB clearly in the early period. The reason is that SAOR takes record validity into consideration and reduce the bad effect of noisy records. Second, as we make the assumption on the relation between the probability that an item is attractive and its distance with the user's recent clicked items, the effect of the percentage of users that obey our assumption on the algorithm is analyzed. The results are shown in Fig. 2(b). We define the kind of users that obey the assumption as "good users". Both of the two algorithms are fluctuation trends, but the general trends are still straight increased. In general, the more percentage of users obey the assumption, the recommending accuracy is higher and our algorithm performs better throughout.

In perspective of parameter sensitivity, we also analyze the influence of exploration factor α on performance, the results are shown in Fig. 2(c) and (d). The length of recommendation sequences and percentage of users obeying the assumption are fixed respectively and we plot the accuracy along with the value of α. The performance of SAOR is stabler than LinUCB, and the value of α has an obvious effect on LinUCB. It can be inferred that our algorithm is more robust.

6 Related Work

In this section, the developing process of contextual bandit algorithms and their shortcomings will be given. In addition, some existing click models and ways to pick negative samples from unlabeled ones are given, the reasons why they cannot be used to solve the problem in our setting are also listed.

Auer first proposed two contextual bandit algorithms, LinRel and SupLinRel, which assume that users' decisions depend linearly on item feature vectors [3]. Both algorithms use upper confidence bound (UCB) to implement exploration-exploitation tradeoffs [4] and solve linear equations by singular value decomposition. However, both of them require solving SVD of a symmetric matrix during modeling at each round. Then, Li et al. proposed LinUCB for news article recommendation [12], employing ridge regression for solving linear equations to reduce computational cost. Abbasi-Yadkori et al. proposed OFUL [1], which maintains a confidence set to implement exploration-exploitation tradeoffs. Thompson sampling based contextual bandit [2] was proposed, sampling the linear model from Gaussian distribution. All the four algorithms mentioned try to solve the problem of uncertainty and use different solutions respectively. More recently, beyond linear decision models, Li et al. studied generalized linear models, which is a general framework covering linear models, logistical models, etc. [14]. Two algorithms, UCB-GLM and SupCB-GLM, are proposed. Clustering bandits [9,15] are proposed recently which combines clustering technique and exploration-exploitation strategies to solve cold start and dynamic recommendation. All the algorithms above never took the quality of data records and user interaction into consideration.

Click models [6] are commonly used to model user behavior in web-searching scenarios where a search engine displays a list of items in response to a user query. The cascade model is one of the popular click models [8,11]. In this setting, several items are recommended to the user and the user may click one or none of the items, so it's necessary for the system to recognize between negative feedback and no feedback. In the cascade model, the items displayed before the clicked one are treated as negative feedbacks because the user examines them but do not click, the items displayed after the clicked one are considered as no feedbacks as they are unobserved. Those models rely on the relative positioning of items within the list to describe users' behaviors. It's comparatively simple to distinguish negative feedback due to the front and back information of the sequence. However, each time only one item, instead of a list of items, is presented to the user, thus those web-searching click models are not applicable.

7 Conclusions

This paper proposes a novel contextual bandit algorithm SAOR for online recommendation, considering the sparsity of interactions between the users and the recommender system. We have developed techniques to distinguish non-responses and true negative responses, of which are handled differently in our algorithm SAOR. We have provided the regret analysis of our algorithm. Experimental results on both real and synthetic datasets show that SAOR outperforms the existing methods on both effectiveness and efficiency.

Acknowledgments. This work was supported by the National Key R&D Program of China (2017YFB0702600, 2017YFB0702601) and the National Natural Science Foundation of China (61432008, U1435214, 61503178).

References

1. Abbasi-Yadkori, Y., Pál, D., Szepesvári, C.: Improved algorithms for linear stochastic bandits. In: Advances in Neural Information Processing Systems, pp. 2312–2320 (2011)
2. Agrawal, S., Goyal, N.: Thompson sampling for contextual bandits with linear payoffs. In: Proceedings 30th International Conference on Machine Learning (ICML 2013), pp. 127–135 (2013)
3. Auer, P.: Using confidence bounds for exploitation-exploration trade-offs. J. Mach. Learn. Res. **3**(Nov), 397–422 (2002)
4. Auer, P., Cesa-Bianchi, N., Fischer, P.: Finite-time analysis of the multiarmed bandit problem. Mach. Learn. **47**(2–3), 235–256 (2002)
5. Chu, W., Li, L., Reyzin, L., Schapire, R.E.: Contextual bandits with linear payoff functions. In: Proceedings 14th International Conference on Artificial Intelligence and Statistics (AISTTS 2011), pp. 208–214 (2011)
6. Chuklin, A., Markov, I., Rijke, M.d.: Click models for web search. In: Synthesis Lectures on Information Concepts, Retrieval, and Services, vol. 7, no. 3, pp. 1–115 (2015)
7. Cooley, J.W., Tukey, J.W.: An algorithm for the machine calculation of complex Fourier series. Math. Comput. **19**(90), 297–301 (1965)
8. Craswell, N., Zoeter, O., Taylor, M.J., Ramsey, B.: An experimental comparison of click position-bias models. In: Proceedings of International Conference on Web Search Data Min (WSDM 2008), pp. 87–94 (2008)
9. Gentile, C., Li, S., Kar, P., Karatzoglou, A., Zappella, G., Etrue, E.: On context-dependent clustering of bandits. In: Proceedings of 34th International Conference on Machine Learning (ICML 2013), pp. 1253–1262 (2017)

10. Koren, Y.: Collaborative filtering with temporal dynamics. ACM Commun. **53**(4), 89–97 (2010)
11. Kveton, B., Szepesvari, C., Wen, Z., Ashkan, A.: Cascading bandits: learning to rank in the cascade model. In: Proceedings of 32nd International Conference on Machine Learning (ICML 2015), pp. 767–776 (2015)
12. Li, L., Chu, W., Langford, J., Schapire, R.E.: A contextual-bandit approach to personalized news article recommendation. In: Proceedings of 19th International Conference on World Wide Web (WWW 2010), pp. 661–670 (2010)
13. Li, L., Chu, W., Langford, J., Wang, X.: Unbiased offline evaluation of contextual-bandit-based news article recommendation algorithms. In: Proceedings of 4th ACM International Conference on Web Search Data Mining (WSDM 2011), pp. 297–306. ACM (2011)
14. Li, L., Lu, Y., Zhou, D.: Provable optimal algorithms for generalized linear contextual bandits. arXiv preprint arXiv:1703.00048 (2017)
15. Li, S., Karatzoglou, A., Gentile, C.: Collaborative filtering bandits. In: Proceedings of 39th International ACM SIGIR Conference on Research and Development in Information Retrieval, pp. 539–548. ACM (2016)
16. Liu, J., Dolan, P., Pedersen, E.R.: Personalized news recommendation based on click behavior. In: Proceedings of 15th International Conference on Intelligent User Interfaces (IUI 2010), pp. 31–40 (2010)
17. Ren, L., Gu, J., Xia, W.: A temporal item-based collaborative filtering approach. In: Signal Processing, Image Processing and Pattern Recognition (SIP 2011), pp. 414–421 (2011)

Heterogeneous Item Recommendation for the Air Travel Industry

Zhicheng He[1], Jie Liu[2(✉)], Guanghui Xu[1], and Yalou Huang[3]

[1] College of Computer Science, Nankai University, Tianjin, China
{hezhicheng,xugh}@mail.nankai.edu.cn
[2] College of Artificial Intelligence, Nankai University, Tianjin, China
jliu@nankai.edu.cn
[3] College of Software, Nankai University, Tianjin, China
ylhuang@nankai.edu.cn

Abstract. Analyzing the travel behaviors and patterns of air passengers have always been of great significance to the air travel industry. Understanding the demands and interests of passengers behind their behaviors is a crucial and fundamental task for many applications. However, this task is challenging due to the lack of customer information, data sparsity, and the long-tail distribution. In this paper, we investigate the problem of heterogeneous item recommendation by learning representations of items and passengers in a shared latent space. Specifically, we first establish a heterogeneous information network (HIN) through statistical analysis, where the edges represent the interactions between different nodes. Each node also contains some auxiliary attribute information that describes its travel behavior or that of its passenger groups. Then we devise a joint matrix factorization model to learn node representations based on the HIN, where both the heterogeneous edges and the node attributes are incorporated into the learning process. Moreover, a weighting strategy is further utilized to deal with the long-tail distribution of passenger behaviors based on the implicit feedback information. Experimental results conducted on a real-world passenger name record (PNR) dataset demonstrate the effectiveness of the proposed method.

Keywords: Air travel data analysis · Matrix factorization · Air route recommendation · Airline recommendation

1 Introduction

In modern lives, air transportation has always been one of the most important ways for long-distance travels. The huge travel demand promotes the growth and prosperity of the civil aviation industry. According to a bulletin from the Civil Aviation Administration of China (CAAC), the entire industry of China completed a passenger transportation volume of 551.56 million in 2017 achieving an annual growth rate of 13.0% [3]. Among the huge number of passengers, quite a lot of them book tickets through online agents. Thus there is a need for

Q. Yang et al. (Eds.): PAKDD 2019, LNAI 11440, pp. 407–419, 2019.
https://doi.org/10.1007/978-3-030-16145-3_32

recommendation services that can understand passengers' travel demands and give suggestions about flights and airline carriers accordingly.

Inspired by the ubiquitous applications of recommender systems in online retail markets [15] and driven by the potential market and research value, companies and researchers have been devoted to airline customer analysis and service recommendation [2,4,5,14,17]. However, the personalized travel air route and airline predictions still remain challenging. Reasons behind this are threefold. First, there lack enough profiling features that reflect passengers' travel demands or preferences. For security and privacy concerns, detailed customer information like demographics, job titles, or social accounts are strictly confidential to researchers, which creates obstacles for accurate passenger profiling. Second, the passenger behavior data is usually sparse. Due to the high prices of flight tickets, traveling by air is usually to fulfill some specific needs like business or vacations rather than a daily way of traveling for most people. So it is also difficult to fully understand passengers' travel demands from their behavior data. Finally, both the travel frequency and the demands on different routes show a long-tail distribution with respect to the number of passengers, as illustrated in Fig. 1. Therefore, the behavior data of most passengers are submerged by low-frequency travelers and cannot be well modeled.

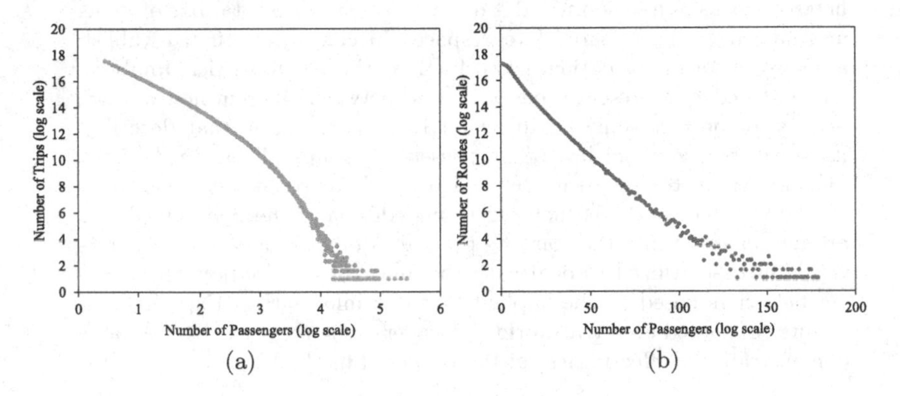

Fig. 1. Travel behavior analysis on a two-year PNR data. (a) The long-tail distribution of travel frequencies; (b) The long-tail distribution of the demands on routes.

In this paper, to deal with the problems mentioned above, we propose a Joint Weighted Non-negative Matrix Factorization (JWNMF) model to learn latent representations of heterogeneous passengers, air routes and airlines in shared semantic space. Specifically, we first establish a heterogeneous information network (HIN) from the Passenger Name Record (PNR) data. Individual nodes can be extracted through statistical analysis, where each of them represents an instance of passengers, routes, or airlines. And the edges between different nodes describe their interactions. For example, a passenger-route edge describes how many times the passenger has taken flights on that route. In the meanwhile, we also extract some auxiliary attributes to depict nodes' characteristics from the perspective of travel behaviors. Attributes of a passenger reflect how he travels, while attributes of a route or airline describe what kind of passenger groups tend to take it. On the basis of the air travel HIN, we further devise a joint matrix factorization framework to learn node representations by integrating the heterogeneous interactions and node attributes, which alleviates the data sparsity problem. Finally, to deal with the long-tail distribution of data, we utilize a weighting strategy based on the analysis of the implicit feedback contained in passenger behavior data. The influence of imbalanced edge weights can be solved, and the performances are improved. Heterogeneous recommendations are conducted in the shared latent space.

To summarize, in this paper, we make the following contributions:

- We analyze the characteristics of PNR data and formulate the air route and airline recommendation problem under a HIN analysis framework that integrates both the interactions and the attributes of different nodes.
- For information integration purpose, a joint factorization model is proposed to simultaneously learn the latent representations of passengers, routes, and airlines based on the HIN.
- Based on the analysis of the implicit feedback information, a weighting strategy is also devised to deal with the imbalanced edge weights caused by the long-tail distribution.
- We conduct experiments to evaluate our proposed framework on the heterogeneous recommendation task with a real-world PNR dataset. Experimental results demonstrate the superiority of our model.

The remainder of this paper is organized as follows. Section 2 highlights related work. In Sect. 3, we formulate the problem and give the technical details of the proposed model. We evaluate the proposed model and analyze the experimental results in Sect. 4. Finally, we conclude our work in Sect. 5.

2 Related Work

With the development of the air travel industry, large quantities of complex and rapidly changing data are being created every second. Air travel data mining has attracted a lot of researchers' attention [1,18], and research results have been achieved on hot issues like security and safety [10,22], intelligent marketing [7,18], customer choice modeling and relation management [16,17], and personalized recommendation [4,5,14], etc. In this paper, we focus on the personalized recommendation problem.

Inspired by the success of recommender systems in online retailing and other industries [2,15], recommendations of flights, air routes, airlines, and auxiliary services are studied to improve the service quality and customer satisfaction. Cao *et al.* proposed a personalized flight recommendation approach based on the maximization of user's choice utility over flight tickets through a paired-choice analysis of historical orders [5]. To overcome the problem of insufficient historical data, air route recommendation is modeled as a cross-domain recommendation problem in which the cross-domain data is integrated [4]. The combinations of user choice models and recommender systems are also explored for airline itinerary suggestion [17]. Other than flights, routes, or airlines, auxiliary services like in-flight music can also be recommended to enhance user experiences [13]. In this paper, we focus on the fundamental air route and airline company recommendation task and propose a matrix factorization framework.

Matrix factorization models are popular in recommender systems [8,12,19, 20], especially the Non-negative Matrix Factorization (NMF) model [11]. Gu *et al.* proposed a weighted NMF model to incorporate the attributes and relations of users and items into the factorization of user-item rating matrix [8]. Lian *et al.* incorporated the spatial clustering information of human mobility behavior into the factorization process for POI recommendation [12]. A deep matrix factorization model is also proposed to make use of both explicit ratings and implicit feedback with the help of deep neural networks [20]. And the joint matrix factorization models are popular as they help incorporate various auxiliary information into the factorization process [19,21].

3 Approach

In this section, we first introduce how the air travel HIN is constructed from PNR datasets, followed by the details of the proposed model. The joint factorization model incorporates both the heterogeneous interactions and attribute information to overcome data sparsity. And the weighting strategy further deals with the imbalanced connection weights caused by the long-tail distribution.

3.1 The Air Travel HIN

The PNR datasets are made up of the flight records of passengers. Each entry usually contains brief passenger information such as ID number, age, and gender, and the flight-specific information such as the air route and the airline company. We focus on learning representations from such PNR datasets. A HIN $\mathcal{G} = \{\mathcal{V}, \mathcal{E}\}$ is first constructed based on the extracted entities and their relations. We focus on the most important three kinds of entities, i.e., passengers, air routes, and airline companies. Thus we have $\mathcal{V} = \mathcal{U} \cup \mathcal{R} \cup \mathcal{C}$, where $\mathcal{U}$, $\mathcal{R}$, and $\mathcal{C}$ denote the set of passengers, routes, and airline companies respectively.

Usually, when a passenger needs to take a flight, he has a departure airport and an arrival airport in mind and just needs to figure out which flight of which airline suits him best. Based on this intuitive understanding, two kinds of relations are extracted, i.e., the passenger-route interaction $\mathcal{E}^{ur} \in \mathcal{U} \times \mathcal{R}$ and the passenger-airline interaction $\mathcal{E}^{uc} \in \mathcal{U} \times \mathcal{C}$, thus we have $\mathcal{E} = \mathcal{E}^{ur} \cup \mathcal{E}^{uc}$.

Apart from the relation information, there also exist some factors that influence passengers' choices over routes and airlines such as passengers' age, gender, total travel mileage, and travel seasons. Therefore, we conduct statistical analysis on how these factors affect passengers' travel behaviors by calculating the percentage of flight records generated by passengers with that attribute or in that season. Results are shown in Fig. 2 where the age and mileage are segmented into groups by maximizing the information gain [6], and the seasons are separated according to the Chinese lunar calendar. It can be observed that passengers have different travel frequency distributions over these factors, which illustrates the necessity to take them into account when modeling passengers' behaviors. Therefore, the passenger attribute matrix $\mathbf{A}^u$ is built where each row describes the corresponding passenger's age group, gender, travel mileage, and travel preference on different seasons. And the route attribute matrix $\mathbf{A}^r$ and airline attribute matrix $\mathbf{A}^c$ are also built by calculating the average attribute values of their passenger groups. In addition, customer loyalty and market share are also considered.

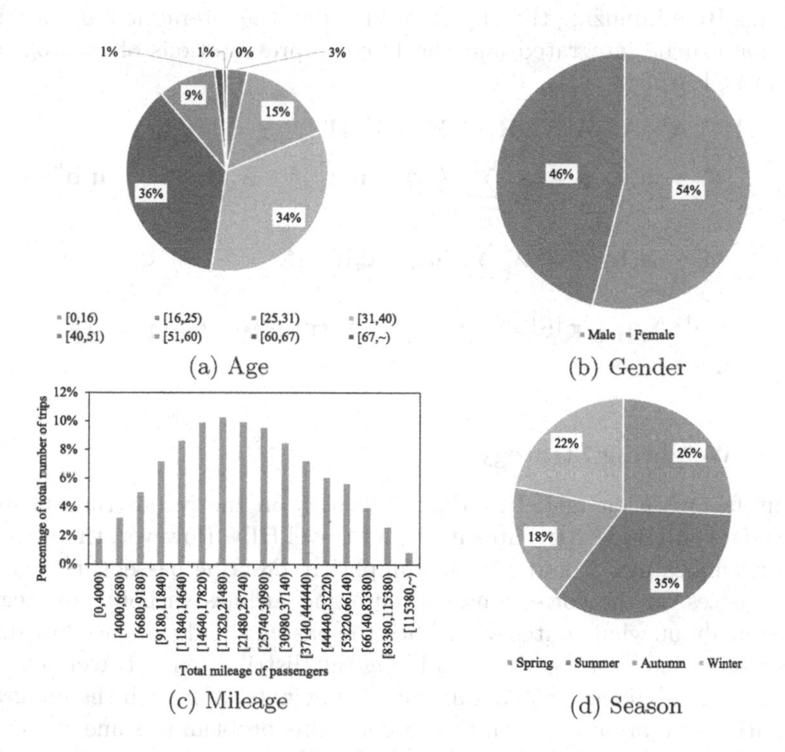

Fig. 2. Distribution of flight records on different factors. (a) Passengers' age; (b) Passengers' gender; (c) Total travel mileage of passengers; (d) Travel season.

3.2 The Joint Factorization Model

Through the above analysis, interactions between nodes and their attributes are extracted. We use matrices to denote them, i.e., the passenger-route interaction matrix $\mathbf{E}^{ur} \in \mathbb{R}_+^{|\mathcal{U}|\times|\mathcal{R}|}$, the passenger-airline interaction matrix $\mathbf{E}^{uc} \in \mathbb{R}_+^{|\mathcal{U}|\times|\mathcal{C}|}$, the passenger attribute matrix $\mathbf{A}^u \in \mathbb{R}_+^{|\mathcal{U}|\times d^u}$, the route attribute matrix $\mathbf{A}^r \in \mathbb{R}_+^{|\mathcal{R}|\times d^r}$, and the airline attribute matrix $\mathbf{A}^c \in \mathbb{R}_+^{|\mathcal{C}|\times d^c}$, where all theses matrices are non-negative and d^u, d^r, and d^c are the dimensions of passenger attributes, route attributes, and airline attributes respectively.

For information integration purpose, we devise a joint non-negative matrix factorization model to learn node representations. Use $\mathbf{U}$, $\mathbf{R}$, $\mathbf{C}$, $\mathbf{H}^u$, $\mathbf{H}^r$, and $\mathbf{H}^c$ to denote the latent representation matrices of passengers, routes, airlines, and their attributes respectively, the model learns them through the reconstruction of the interaction and attribute matrices. Specifically, we aim to minimize the reconstruction loss in Eq. (1), where the conventional squared Euclidean distance [11] is used to measure the reconstruction loss. The non-negative λ_1, λ_2, λ_3, and λ_4 tune the weights of different parts, and K, d^u, d^r, and d^c are the latent dimensions. By minimizing the objective function, the interaction and attribute information can be integrated and the latent representations of heterogeneous nodes can be learned.

$$\begin{aligned}
&\mathcal{D}(\mathbf{E}^{ur},\mathbf{E}^{uc},\mathbf{A}^u,\mathbf{A}^r,\mathbf{A}^c|\mathbf{U},\mathbf{R},\mathbf{C},\mathbf{H}^u,\mathbf{H}^r,\mathbf{H}^c)\\
&=\sum_{e_{ij}^{ur}>0}(e_{ij}^{ur}-\mathbf{u}_i\mathbf{r}_j^\top)^2+\lambda_1\sum_{e_{ij}^{uc}>0}(e_{ij}^{uc}-\mathbf{u}_i\mathbf{c}_j^\top)^2+\lambda_2\sum_{i,j}(a_{ij}^u-\mathbf{u}_i\mathbf{h}_j^{u\top})^2\\
&+\lambda_3\sum_{i,j}(a_{ij}^r-\mathbf{r}_i\mathbf{h}_j^{r\top})^2+\lambda_4\sum_{i,j}(a_{ij}^c-\mathbf{c}_i\mathbf{h}_j^{c\top})^2, \qquad (1)\\
&s.t.\ \mathbf{U}\in\mathbb{R}_+^{|\mathcal{U}|\times K},\mathbf{R}\in\mathbb{R}_+^{|\mathcal{R}|\times K},\mathbf{C}\in\mathbb{R}_+^{|\mathcal{C}|\times K},\mathbf{H}^u\in\mathbb{R}_+^{K\times d^u},\mathbf{H}^r\in\mathbb{R}_+^{K\times d^r},\\
&\mathbf{H}^c\in\mathbb{R}_+^{K\times d^c}.
\end{aligned}$$

3.3 The Weighting Strategy

The joint factorization model in Eq. (1) focuses on the reconstruction loss of positive edges and node attributes in the air travel HIN. However, there are only positive examples in $\mathcal{E}^{ur}$ and $\mathcal{E}^{uc}$ that implicitly describe passengers' demands and preferences on the corresponding route and airlines without any negative information about what routes or airlines the passengers do not need or dislike. What is worse, due to the sparsity and long-tail distribution of travel behaviors, there exists an imbalance problem in the edge weights. Although the integration of attribute information can help to overcome this problem to some extent, it is still difficult to fit the imbalanced edge weights. Therefore, we adopt a weighting strategy that reduces the imbalance of edge weights by taking advantage of the implicit feedback information [9].

First, passengers' preferences on the routes and airlines are extracted from $\mathbf{E}^{ur}$ and $\mathbf{E}^{uc}$ by binarizing the edge weights:

$$p_{ij}^{ur}=\begin{cases}1 & e_{ij}^{ur}>0\\ 0 & e_{ij}^{ur}=0\end{cases},\qquad p_{ij}^{uc}=\begin{cases}1 & e_{ij}^{uc}>0\\ 0 & e_{ij}^{uc}=0\end{cases}. \qquad (2)$$

In other words, if a passenger has taken flights on a route ($e_{ij}^{ur}>0$) or from an airline company ($e_{ij}^{uc}>0$), it implicates his preferences on them. Otherwise, no preferences are assumed. The binary preference strategy effectively reduces the gap between frequent and infrequent interactions and makes the fitting process easier. However, it loses the indication about which routes or airlines attract the passengers more. So a linear weighting strategy is utilized to assign different confidence scores to the preferences according to the edge weights:

$$w_{ij}^{ur} = \alpha e_{ij}^{ur} + 1, \quad w_{ij}^{uc} = \alpha e_{ij}^{uc} + 1, \tag{3}$$

where α is a non-negative hyperparameter that controls the increase rate of the confidence scores. With such a weighting strategy, when there is no interaction between passenger i and route j ($e_{ij}^{ur} = 0$), a minimum confidence score $w_{ij}^{ur} = 1$ is assigned, which means that it is uncertain whether the passenger has interest in the route or not. However, with the growth of e_{ij}^{ur}, there is a larger confidence that the route meets the passenger's needs. And it is the same for passenger-airline pairs.

Together, the binary preferences and the linear confidences solve the imbalance problem caused by data sparsity and the long-tail distributions. So the objective function can be updated as in Eq. (4), where both the observed and unobserved passenger-route and passenger-airline interactions are fitted with different confidences. In this way, there exists no gap between the fitting targets of frequent and infrequent passenger-route or passenger-airline interactions, which makes the learning process easier. However, the valuable frequency information is not abandoned but used to decide how much weight JWNMF should put on each fitting target.

$$\begin{aligned}
&\mathcal{D}(\mathbf{E}^{ur}, \mathbf{E}^{uc}, \mathbf{A}^{u}, \mathbf{A}^{r}, \mathbf{A}^{c} | \mathbf{U}, \mathbf{R}, \mathbf{C}, \mathbf{H}^{u}, \mathbf{H}^{r}, \mathbf{H}^{c}) \\
&= \sum_{i,j} w_{ij}^{ur} (p_{ij}^{ur} - \mathbf{u}_i \mathbf{r}_j^{\top})^2 + \lambda_1 \sum_{i,j} w_{ij}^{uc} (p_{ij}^{uc} - \mathbf{u}_i \mathbf{c}_j^{\top})^2 \\
&+ \lambda_2 \sum_{i,j} (a_{ij}^{u} - \mathbf{u}_i \mathbf{h}_j^{u\top})^2 + \lambda_3 \sum_{i,j} (a_{ij}^{r} - \mathbf{r}_i \mathbf{h}_j^{r\top})^2 \\
&+ \lambda_4 \sum_{i,j} (a_{ij}^{c} - \mathbf{c}_i \mathbf{h}_j^{c\top})^2,
\end{aligned} \tag{4}$$

3.4 Model Optimization

By optimizing the objective function in Eq. (4), the interaction and attribute information can be effectively integrated thus latent representations can be learned. Here we present the details of the optimization process. The derivatives of the objective function $\mathcal{D}$ with respect to the latent variables are:

$$
\begin{aligned}
\frac{\partial \mathcal{D}}{\partial \mathbf{U}} =& -2(\mathbf{W}^{ur} \otimes (\mathbf{P}^{ur} - \mathbf{U}\mathbf{R}^\top))\mathbf{R} - 2\lambda_1(\mathbf{W}^{uc} \otimes (\mathbf{P}^{uc} - \mathbf{U}\mathbf{C}^\top))\mathbf{C} \\
& - 2\lambda_2(\mathbf{A}^u - \mathbf{U}\mathbf{H}^{u\top})\mathbf{H}^u, \\
\frac{\partial \mathcal{D}}{\partial \mathbf{R}} =& -2(\mathbf{W}^{ur} \otimes (\mathbf{P}^{ur} - \mathbf{U}\mathbf{R}^\top))^\top\mathbf{U} - 2\lambda_3(\mathbf{A}^r - \mathbf{R}\mathbf{H}^{r\top})\mathbf{H}^r, \\
\frac{\partial \mathcal{D}}{\partial \mathbf{C}} =& -2\lambda_1(\mathbf{W}^{uc} \otimes (\mathbf{P}^{uc} - \mathbf{U}\mathbf{C}^\top))^\top\mathbf{U} - 2\lambda_4(\mathbf{A}^c - \mathbf{C}\mathbf{H}^{c\top})\mathbf{H}^c, \\
\frac{\partial \mathcal{D}}{\partial \mathbf{H}^u} =& -2\lambda_2(\mathbf{A}^{u\top} - \mathbf{H}^u\mathbf{U}^\top)\mathbf{U}, \quad \frac{\partial \mathcal{D}}{\partial \mathbf{H}^r} = -2\lambda_3(\mathbf{A}^{r\top} - \mathbf{H}^r\mathbf{R}^\top)\mathbf{R}, \\
\frac{\partial \mathcal{D}}{\partial \mathbf{H}^c} =& -2\lambda_4(\mathbf{A}^{c\top} - \mathbf{H}^c\mathbf{C}^\top)\mathbf{C}.
\end{aligned} \tag{5}
$$

With the gradients given in Eq. (5), the objective function can be optimized with any gradient-descent based methods. In this work, we adopt the popular multiplicative update method [11] which is guaranteed to converge to at least a locally optimal solution.

Table 1. Statistics of the datasets.

Datasets	#Passengers	#Records	#Routes	#Airlines	Density of $\mathbf{E}^{ur}/\mathbf{E}^{uc}$
Top100K	100,000	13,074,626	2728	22	0.017/0.414
Rand100K	100,000	3,983,497	2728	22	0.007/0.304

4 Experiments

To investigate the effectiveness of JWNMF in learning latent representations for nodes in the air travel HIN, we evaluate our proposed method on a real-world PNR dataset. The learned representations are evaluated according to the route and airline recommendation performances. And the experimental results prove our points.

4.1 Dataset

We use a two-year anonymized PNR dataset which contains 2,956,088 passengers, 2728 routes, and 22 airline companies. To comprehensively analyze JWNMF's performs on both frequent flyers and normal passengers, two sub-datasets are extracted. The first subset is extracted by selecting the top 100,000 passengers and their records according to the travel frequencies, we denote it as Top100K. While the second subset contains randomly selected 100,000 passengers and their records denoted as Rand100K. Details about the two sub-datasets are demonstrated in Table 1. It can be observed from the number of records and data density that frequent flyers behave differently from normal passengers. And models should perform well on both the valuable frequent flyers and the huge group of normal passengers.

4.2 Baselines

To achieve comprehensive and comparative analysis of our approach, we compare it with three kinds of baselines: the trivial methods, the collaborative filtering methods, and the matrix factorization methods.

- Random. Random guess (Random) is a trivial method in recommender systems. For each passenger, N routes and airlines are randomly selected from the candidate sets and recommended.
- ItemPop. Item popularity (ItemPop) is another trivial method. The routes and airlines are sorted according to the frequencies they appear in the records. And the top N routes and airlines are recommended to all passengers
- UCF. User-based collaborative filtering (UCF) is widely used in a lot of applications. The passenger-item (route or airline) relevance score $\mathrm{UCF(i,j)}$ is calculated as a weighted sum of the passenger's similarity to all passengers that have consumed the item. We adopt the cosine similarity between passengers' flight records and attributes with a parameter $0 \leq \beta_1 \leq 1$ tuning the weight. After that, the top N items are recommended according to $\mathrm{UCF(i,j)}$.
- ICF. Item-based collaborative filtering (ICF) is also widely used in various applications. The passenger-item (route or airline) relevance score $\mathrm{ICF(i,j)}$ is calculated as a weighted sum of the item's cosine similarity to all items that the passenger has consumed. We adopt the cosine similarity between flight records and attributes with a parameter $0 \leq \beta_2 \leq 1$ tuning the weight. After that, the top N items are recommended according to $\mathrm{ICF(i,j)}$.
- NMF. NMF is the most popular matrix factorization method in recommendations and is also the basic model of our JWNMF. Latent representations are learned by independently factorizing the passenger-route matrix $\mathbf{E}^{ur}$ or passenger-airline matrix $\mathbf{E}^{uc}$. And the routes and airlines are recommended according to their similarity to passengers in the latent space.
- JNMF. To evaluate the performances of the integration of the heterogeneous edges and node attributes, the joint NMF (JNMF) model in Eq. (1) is used in the experiments. JNMF simultaneously learns the representations of passengers, routs, and airlines in the shared latent space in which the recommendations are conducted.
- WNMF. The weighting strategy is also independently evaluated by comparison of a weighted NMF (WNMF). The binary preference strategy in Eq. (2) and the linear weights in Eq. (3) are applied to $\mathbf{E}^{ur}$ or $\mathbf{E}^{uc}$. A preference matrix is factorized with the aid of the corresponding weight matrix, and recommendations are conducted accordingly.
- JWNMF. JWNMF combines the advantages of both JNMF and WNMF as shown in Eq. (4). And recommendations are conducted in the shared latent representation space.

4.3 Experimental Settings

For both the Top100K and Rand100K datasets, 10% of the entries in $\mathbf{E}^{ur}$ and $\mathbf{E}^{uc}$ are randomly sampled for test purpose. For both CF models, the hyperparameters β_1 and β_2 are tuned in the range of 0 to 1 with stepsize 0.1 where 0 and

Table 2. Recommendation performances on the Top100K dataset.

Methods	Route recommendation						Airline recommendation					
	P@5	R@5	F1@5	P@10	R@10	F1@10	P@5	R@5	F1@5	P@10	R@10	F1@10
Random	0.002	0.002	0.002	0.002	0.003	0.002	0.042	0.229	0.071	0.042	0.455	0.076
ItemPop	0.077	0.084	0.080	0.057	0.124	0.078	0.165	0.903	0.279	0.090	0.992	0.166
UCF	0.145	0.158	0.151	0.100	0.219	0.138	0.166	0.908	0.280	0.090	0.992	0.166
ICF	0.252	0.274	0.262	0.164	0.356	0.224	0.163	0.892	0.275	0.090	0.985	0.165
NMF	0.181	0.197	0.189	0.123	0.268	0.169	0.137	0.752	0.232	0.087	0.946	0.159
JNMF	0.223	0.243	0.232	0.148	0.322	0.203	0.164	0.901	0.278	**0.091**	0.990	0.166
WNMF	0.261	0.284	0.272	0.177	0.385	0.243	0.161	0.880	0.272	0.090	0.977	0.164
JWNMF	**0.266**	**0.289**	**0.277**	**0.178**	**0.387**	**0.244**	**0.174**	**0.954**	**0.294**	**0.091**	**0.997**	**0.167**

Table 3. Recommendation performances on the Rand100K dataset.

Methods	Route recommendation						Airline recommendation					
	P@5	R@5	F1@5	P@10	R@10	F1@10	P@5	R@5	F1@5	P@10	R@10	F1@10
Random	0.001	0.002	0.001	0.001	0.004	0.001	0.031	0.230	0.054	0.030	0.452	0.057
ItemPop	0.025	0.066	0.036	0.018	0.097	0.031	0.117	0.872	0.206	0.065	0.973	0.122
UCF	0.090	0.238	0.130	0.060	0.318	0.101	0.117	0.876	0.207	0.065	0.979	0.123
ICF	0.122	0.325	0.178	0.072	0.382	0.121	0.117	0.876	0.207	0.065	0.979	0.123
NMF	0.113	0.301	0.165	0.070	0.373	0.118	0.108	0.809	0.191	0.063	0.944	0.118
JNMF	0.134	0.356	0.195	0.084	0.447	0.141	0.123	0.921	0.217	0.066	0.987	0.124
WNMF	0.113	0.301	0.165	0.071	0.377	0.120	0.114	0.853	0.201	0.064	0.961	0.120
JWNMF	**0.150**	**0.397**	**0.217**	**0.097**	**0.514**	**0.163**	**0.127**	**0.950**	**0.224**	**0.067**	**0.996**	**0.125**

1 means the attribute-only and record-only similarity measures respectively. The latent dimension K in all factorization models are tuned in the range of 50 to 500 with stepsize 50, and the maximum iteration number is set to 200. Finally, both the linear parameters α in Eq. (3) and the weight parameters λs are tuned in the range of 10^{-3} to 10^{3}, multiplied by 10 at each step. After the representations are learned, the top $N = 5$ and $N = 10$ routes and airlines are recommended to each passenger according to their relevance scores in the latent space. And the performances are evaluated with the micro-averaged precision (P), recall (R), and F1 scores.

4.4 Experimental Results

Tables 2 and 3 show the results on the Top100K and Rand100K datasets respectively, where the best results are boldfaced. From these results, we have the following observations and analysis:

- JWNMF achieves the best performances on both datasets and all six evaluation measures, which proves the superiority of the proposed model. On both datasets, both JNMF and WNMF perform better than NMF. What is more, by combining both the joint factorization and weighting strategy, JWNMF consistently performs better than all of them. Therefore, both of the proposed modifications are effective and necessary.

- JNMF performs better than WNMF on the Rand100K dataset. The reason is that the Rand100K dataset is more sparse than the Top100K dataset, as demonstrated in Table 1. Because the joint factorization technique is proposed to deal with the data sparsity problem, JNMF achieves more significant improvements than the weighting strategy.
- On the other side, WNMF performs better than JNMF on the route recommendation task on the Top100K dataset. Because frequent flyers often interact frequently with specific routes that differ from each other, the passenger-route interactions are denser and have bigger value differences. Thus WNMF achieves more significant improvements by narrowing the gap between fitting targets while keeping the frequency information.
- Due to the fact that the travel demands and behavior patterns of frequent flyers are more clearly reflected by the dense flight records, performances on the Top100K dataset are generally better than on the Rand100K dataset. However, our JWNMF demonstrates its robustness by achieving the best performances on both datasets.

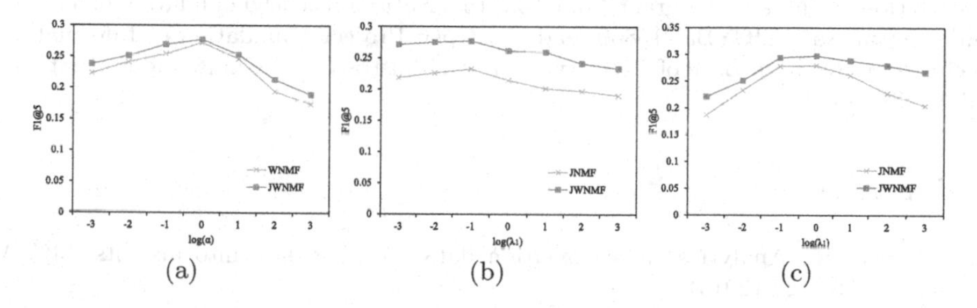

Fig. 3. Analysis of the linear weight parameter α and the balancing parameter λ_1 on the Top100K dataset. (a) The F1@5 scores of route recommendation when α varies; (b) The F1@5 scores of route recommendation when λ_1 varies; (c) The F1@5 scores of airline recommendation when λ_1 varies.

4.5 Parameter Analysis

There are two types of important parameters in JWNMF, the linear weight α in the weighting strategy and the balancing parameters λ_1, λ_2, λ_3, λ_4 in the objective function. Other parameters like the latent dimension K and the iteration number also matter. However, for space limitation, we only analysis how α and λ_1 affect the performances in this subsection as illustrated in Fig. 3.

With the increase of both parameters, all curves rise first and then decline. Because the entries in $\mathbf{E}^{ur}$ and $\mathbf{E}^{uc}$ are integer frequencies, a small α (<1) fails to recognize the relevance information contained in high frequencies, while a large

α (>1) makes the model concentrate too much on high frequencies and overfit. Therefor, we set $\alpha = 1$ in experiments. On the other hand, $\lambda_1 = 0.1$ achieves the best performances on route recommendation, but it is $\lambda_1 = 1$ on airline recommendation, which demonstrates the trade-off between two tasks. Taking full account of the overall performances, we set $\lambda_1 = 0.1$ in the experiments.

5 Conclusion

In this paper, we introduced a heterogeneous item recommendation framework JWNMF which incorporates heterogeneous information from the air travel HIN derived from the PNR dataset. The proposed JWNMF leverages both the interaction information between entities and their attributes, which effectively models passengers' travel demands and behavior patterns. Through a joint weighted factorization framework, representations of multiple kinds of entities are simultaneously learned and mutually enhanced. Experiments conducted on a real-world PNR dataset demonstrated the effectiveness and superiority of JWNMF on air route and airline recommendation tasks.

Acknowledgements. This research is supported by the National Natural Science Foundation of China under grant No. U1633103, Natural Science Foundation of Tianjin under grant No. 18JCYBJC15800, and the Open Project Foundation of Information Technology Research Base of Civil Aviation Administration of China under grant No. CAAC-ITRB-201701.

References

1. Akerkar, R.: Analytics on big aviation data: Turning data into insights. IJCSA **11**(3), 116–127 (2014)
2. Borràs, J., Moreno, A., Valls, A.: Intelligent tourism recommender systems: a survey. Expert Syst. Appl. **41**(16), 7370–7389 (2014)
3. CAAC: Bulletin of the civil aviation industry development statistics in 2017. CAAC Bulletin, pp. 1–19 (2018)
4. Cao, J., Xu, Y., Ou, H., Tan, Y., Xiao, Q.: PFS: a personalized flight recommendation service via cross-domain triadic factorization. In: ICWS, pp. 249–256 (2018)
5. Cao, J., Yang, F., Xu, Y., Tan, Y., Xiao, Q.: Personalized flight recommendations via paired choice modeling. In: BigData, pp. 1265–1270 (2017)
6. Carmel, D., Farchi, E., Petruschka, Y., Soffer, A.: Automatic query refinement using lexical affinities with maximal information gain. In: SIGIR, pp. 283–290 (2002)
7. Chiang, W.: Identifying high-value airlines customers for strategies of online marketing systems: an empirical case in Taiwan. Kybernetes **47**(3), 525–538 (2018)
8. Gu, Q., Zhou, J., Ding, C.H.Q.: Collaborative filtering: weighted nonnegative matrix factorization incorporating user and item graphs. In: SDM, pp. 199–210 (2010)
9. Hu, Y., Koren, Y., Volinsky, C.: Collaborative filtering for implicit feedback datasets. In: ICDM, pp. 263–272 (2008)

10. Lee, A.J., Jacobson, S.H.: Addressing passenger risk uncertainty for aviation security screening. Transp. Sci. **46**(2), 189–203 (2012)
11. Lee, D.D., Seung, H.S.: Algorithms for non-negative matrix factorization. In: NIPS, pp. 556–562 (2000)
12. Lian, D., Zhao, C., Xie, X., Sun, G., Chen, E., Rui, Y.: GeoMF: joint geographical modeling and matrix factorization for point-of-interest recommendation. In: KDD, pp. 831–840 (2014)
13. Liu, H., Hu, J., Rauterberg, M.: iHeartrate: a heart rate controlled in-flight music recommendation system. In: MB, pp. 26:1–26:4 (2010)
14. Liu, J., et al.: Personalized air travel prediction: a multi-factor perspective. ACM TIST **9**(3), 30:1–30:26 (2018)
15. Lu, J., Wu, D., Mao, M., Wang, W., Zhang, G.: Recommender system application developments: a survey. Decis. Support Syst. **74**, 12–32 (2015)
16. Maalouf, L., Mansour, N.: Mining airline data for CRM strategies. Commun. ACS **1**(001) (2008)
17. Mottini, A., Lheritier, A., Acuna-Agost, R., Zuluaga, M.A.: Understanding customer choices to improve recommendations in the air travel industry. In: Workshop on Recommenders in Tourism, pp. 28–32 (2018)
18. Pritscher, L., Feyen, H.: Data mining and strategic marketing in the airline industry. Data Min. Mark. Appl. 39 (2001)
19. Takeuchi, K., Ishiguro, K., Kimura, A., Sawada, H.: Non-negative multiple matrix factorization. In: IJCAI, pp. 1713–1720 (2013)
20. Xue, H., Dai, X., Zhang, J., Huang, S., Chen, J.: Deep matrix factorization models for recommender systems. In: IJCAI, pp. 3203–3209 (2017)
21. Yu, Y., Gao, Y., Wang, H., Wang, R.: Joint user knowledge and matrix factorization for recommender systems. World Wide Web **21**(4), 1141–1163 (2018)
22. Zhao, X., Deng, N., Jing, L.: Application of image recognition in civil aviation security based on tensor learning. J. Intell. Fuzzy Syst. **33**(4), 2145–2157 (2017)

A Minimax Game for Generative and Discriminative Sample Models for Recommendation

Zongwei Wang, Min Gao(✉), Xinyi Wang, Junliang Yu, Junhao Wen, and Qingyu Xiong

School of Big Data and Software Engineering, Chongqing University, Chongqing, China
{zongwei,gaomin,xywang,yu.jl,jhwen,xiong03}@cqu.edu.cn

Abstract. Recommendation systems often fail to live up to expectations in real situations because of the lack of user feedback, known as the data sparsity problem. A large number of existing recommendation methods resort to side information to gain a performance improvement. However, these methods are either too complicated to follow or time-consuming. To alleviate the data sparsity problem, in this paper we propose UGAN, which is a general adversarial framework for recommendation tasks and consists of a generative model and a discriminative model. In UGAN, the generative model, acts as an attacker to cheat the discriminative model to capture the pattern of the original data input and generate similar user profiles, while the counterpart, the discriminative model aims to distinguish the forged samples from the real data. By competing with each other, two model are alternatively updated like playing a minimax game until the generative model has learned the original data distribution. The experimental results on two real-world datasets, Movielens and Douban, show that the user profiles forged by UGAN can be easily integrated into a wide range of recommendation methods and significantly improve their performance, which provides a promising way to mitigate the adverse impact of missing data.

Keywords: Recommendation system · Data sparsity · Generative Adversarial Network · Minimax game · Adversarial training

1 Introduction

Nowadays, personalized recommendation systems become increasingly important due to the problem of information overload. However, it is inevitable for recommendation systems to suffer from data sparsity because of the lack of user feedback. To address this problem, researchers have proposed many approaches. Among them, Kabbur et al. [9] use a structural equation modeling approach to learn the item-item similarity matrix, and He et al. [6] combine methods based on similarity with Markov chain. Breese et al. [3] describe techniques based on correlation coefficients and vector-based similarity calculations to predict additional

Q. Yang et al. (Eds.): PAKDD 2019, LNAI 11440, pp. 420–431, 2019.
https://doi.org/10.1007/978-3-030-16145-3_33

topics or products that a new user might like. Huang et al. [8] apply an associative retrieval framework and related spreading activation algorithms to explore transitive associations among consumers. In addition, Xiong et al. [21] present a reordering model for phrase-based statistical machine translation that uses a maximum entropy model. Besides, there are a lot of studies that [13,16,22–24] pay attention to incorporate other information, such as social information, personal information, and item profiles [7], into recommendation model to mitigate the problem of data sparsity.

The approaches or algorithms above have achieved great success in alleviating data sparsity problem, but to the best of our knowledge, methods that directly add simulated ratings into the input to increase the density of original data have not yet been explored. Existing approaches which simply replicate genuine user profiles are based on heuristic inference, but the input data is generally composed of users with quite different patterns. Directly adding duplicated users may not have a considerable effect on the recommendation quality or even lower the performance. Generative Adversarial Networks (GANs) [1,5,17] give an example that generates a similar sample data distribution from true data distribution and have been applied to many fields. For instance, IRGAN [20] unifies the two types of models in the field of information retrieval to improve recommendation algorithm performance, IDSGAN [14] generates new types of network intrusion and auto-painter [15] deals with image fill problems.

Inspired by GANs in machine learning, in this paper we propose a adversarial framework named UGAN, which takes advantages of both wGAN [1] and cGAN [2,4,17] to forge users that are highly close to genuine users in the input data. In UGAN, the discriminative model uses Wassertein distance to measure the difference between the original distribution and the generated distribution and estimates the probability that samples come from real data rather than generated data, while the generative model make great efforts to fit the original data distribution and generates specified user profiles based on extra information. The two types of models act as two players in a minimax game and each of them strikes to improve itself to 'beat' the other one at every round of this competition. At last, we incorporate the forged user profiles into the input data and conduct recommendation tasks with a wide range of recommendation methods. Extensive experimental results on two real-world datasets, Movielens and Douban, show that the user profiles forged by UGAN can significantly improve the recommendation performance.

The rest of this paper is organized as follow: Sect. 2 introduces the principle of our approach. In Sect. 3, we describe experiments on two real-world datasets and analyze experimental results. In the end, we conclude work in Sect. 4.

2 UGAN Formulation

In this section, we build our framework UGAN by fusing generative and discriminative models in an adversarial setting.

2.1 A Minimax Sample Generation Framework

For recommendation system, we propose a UGAN framework to reduce the sparsity of dataset. In a recommendation system, user profile can be represented by a vector, and each dimension of the vector represents each rating the user gives to the item, as Fig. 1 shows. The generated users with ratings information are named simulated users.

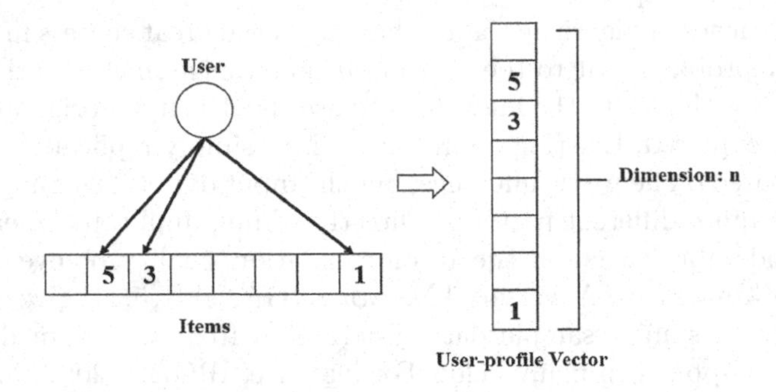

Fig. 1. A user profile can be represented by a user-profile vector.

Our approach are divided into three parts, as illustrated in Fig. 2. Firstly, we put original data that are derived from real datasets into UGAN to generate the simulated users. Secondly, we inject the simulated users into the original datasets to form a mixed datasets. Finally, we use the new datasets to conduct recommendation algorithm. We construct two types of sample models:

Generative Sample Model. This model generates simulated user from given real data distribution, trying to achieve two goals. One is that simulated user data is made to approximate the real data distribution as much as possible. The other is that simulated users need a large number of ratings that can greatly alleviate data sparsity greatly. Thus, we should obtain real users' extra information to guide the generation.

Discriminative Sample Model. On the contrary, this model is a classifier that aims to distinguish between generated data and real data. Concretely, it uses Wassertein distance to measure the difference of two data distribution and meanwhile keeps the process of generation stable without mode collapse.

2.2 Loss Function

Inspired by the idea of cGAN that learns the mapping from the observed input x and random noise vector z to $y : G : \{x, z\} \rightarrow y$. We would like to use the user's activity level information as extra information to guide the generation,

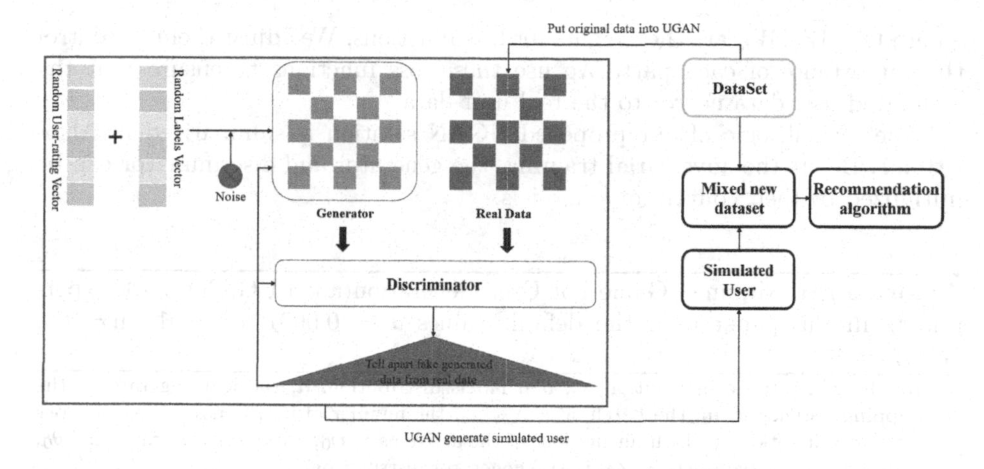

Fig. 2. Our framework.

which makes the simulated user relatively dense. Formally, we have the overall objective (Loss) function:

$$\begin{aligned} \underset{G}{min}\underset{D}{max}V(G,D) &= E_{x,y\sim pdata(x,y)}[D(x,y)] \\ &+E_{x\sim pdata(x),z\sim pdata(z)}[1-D(x,G(x,z))], \end{aligned} \tag{1}$$

where x is the input sketch, y is the label vector and G (x, z) means the simulated user vector. The discriminator D outputs the probability to classify the 'real' and 'fake' input pair by measuring Wassertein distance.

It is hard to simulate the distribution of real user rating data. To solve this problem, we need more constraints for better performance. The generator is trained to minimize the generative loss, and the loss function is given as follows:

$$L_G = E_{x\sim pdata(x),z\sim pdata(z)}\left[1-D\left(x,G\left(x,z\right)\right)\right] \tag{2}$$

Previous studies have shown that the cGAN objective with a traditional loss function is more beneficial for training. Thus, we use the L_1 distance to describe the rating loss L_R, making the generated user data approximates the real user data in an optimized way.

$$L_R = E_{x\sim pdata(x),z\sim pdata(z)}[\|x-G(x,z)\|_1] \tag{3}$$

Then we use the Pearson correlation coefficient to measure the rating correlation between the generated user profiles and the real user profiles.

$$L_P = E_{x\sim pdata(x),z\sim pdata(z)}[P(x,G(x,z))], \tag{4}$$

where P is Pearson coefficient.

Finally, the objective function is defined as follows:

$$L = W_G L_G + W_R L_R + W_P L_P \tag{5}$$

where W_G, W_R, W_P are the weights for loss functions. We adjust them to control the importance of each part. We use these loss functions to ensure that the simulated user data nearer to the real user data.

The overall logic of our proposed UGAN solution is summarized in Algorithm 1. Before the adversarial training, the generator and discriminator can be initialized by their conventional models.

Algorithm 1. Minimax Game For User Generation(a.k.a UGAN). ALL experiments in this paper used the default values $\alpha = 0.0001$, $c = 0.01$, $m = 100$, $N_{critic} = 10$.

Input: R, real ratings distribution, C, real labels distribution. α, the learning rate, c, the clipping parameter. m, the batch size, N_{critic}, the number of iterations of the critic per generator iteration. k, the limit number of the active user. w_0, initial critic parameters. θ_0, initial generator parameters, Z_0, initial generator distribution.

1: **while** not converged **do**
2:     **for** $i = 0, \ldots, N_{critic}$ **do**
3:         Sample$\{R_u\}_{u=1,\ldots,m} \sim R$, $\{C_u\}_{u=1,\ldots,m} \sim C$, a batch from the real rating data.
4:         Sample$\{Z_u\}_{u=1,\ldots,m} \sim Z$, $\{Y_u\}_{u=1,\ldots,m} \sim Y$, a batch of prior sample.
5:         $g_w \leftarrow \nabla_w[\frac{1}{m}\sum_{u=1}^{m} f_w(R_u.concat(C_u)) - \frac{1}{m}\sum_{u=1}^{m} g_\theta(Z_u.concat(Y_u)]$
6:         $w \leftarrow w + \alpha \cdot RMSProp(w, g_w)$
7:         $w \leftarrow clip(w, -c, c)$
8:     **end for**
9:     Sample$\{Z_u\}_{u=1,\ldots,m} \sim Z$, $\{Y_u\}_{u>k} \sim Y$, a batch from the real rating data.
10:     $g_\theta \leftarrow -\nabla_\theta \frac{1}{m}\sum_{u=1}^{m} f_w(g_\theta(Z_u))$
11:     $\theta \leftarrow \theta - \alpha \cdot RMSProp(\theta, g_\theta)$
12: **end while**

Output: Sample $\{Z_u\}_{u=1,\ldots,m} \sim Z$

2.3 Extension to a Specific Case

As Sect. 1 mentioned, the proposed approach can alleviate data sparsity through increasing the average rating number of each user. To achieve this goal, we generate users with more ratings that approximate the original data distribution. As Fig. 3 shows, we use activity level labels to generate a single simulated user with a relatively large number of ratings. For the sake of simplicity, we have only three items and two labels. Besides, users who rate more than two are considered active users, which means the label is one. The generator initially generates random user profiles and labels. Then we put the randomly generated user profiles with labels and the real user profiles with labels into the discriminator at the same time, after which the discriminator returns the discriminant results to the generator, making the generator update to a new version. With continuous generation-discrimination process, user profiles that are similar to real data will be generated. Finally, we use activity level labels to pick out the user profiles which are active.

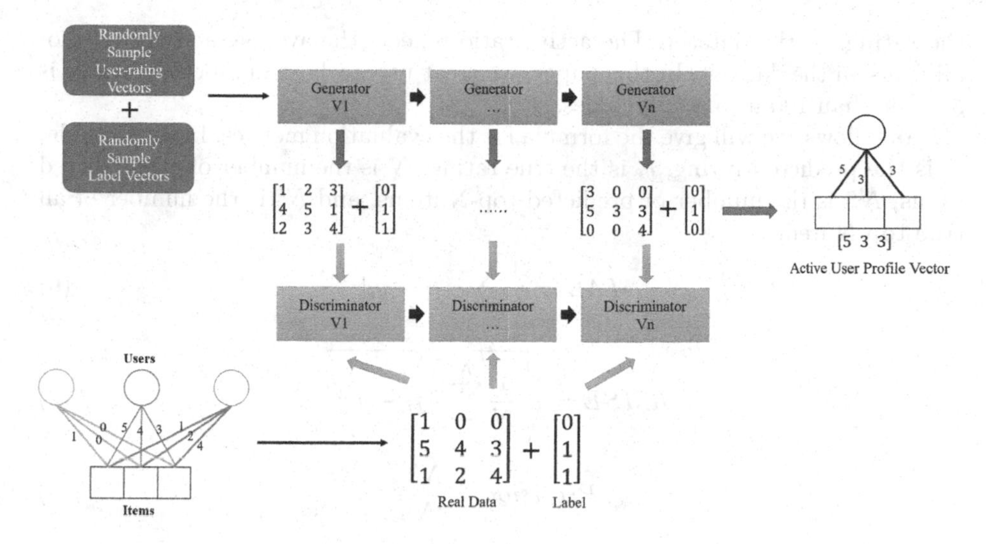

Fig. 3. The process of UGAN

3 Experiments and Analysis

3.1 Datasets and Evaluation Metrics

In the paper, we utilize two real-world datasets, Movielens[1] and Douban[2], in which the average rating numbers are 146 and 27, to evaluate our approach. The details of the two real-world datasets are shown in Table 1. In our experiments, we use 80% of the data for training and the remaining for testing.

Table 1. Basic information of the datasets.

Information	Dataset	
	Movielens	Douban
User number	681	14026
Item number	9125	9066
Rating number	100004	377365

We use MAE and RMSE to evaluate the performance of rating predictions and use Precision, Recall, F1, and MAP to evaluate the recommendation performance of the Top-N rank algorithm.

Furthermore, we use ratings sparsity and active ratio to evaluate the data sparsity of the dataset. The ratings sparsity reflects the average data sparsity of

[1] https://grouplens.org/datasets/movielens/.
[2] https://github.com/CQU-CSE/DatasetCollection.

the ratings in the dataset. The active ratio reflects the average activity level of all users in the dataset. In this paper, we treat users whose number of ratings is greater than 100 as an active ones.

As follows, we will give the formula for the evaluation metrics. In the formula, f_i is the predicted rating, y_i is the true rating, N is the number of all predicted items, N^a is the number of predicted top-N items, and N^t is the number of all true top-N items.

$$MAE = \frac{1}{N}\sum_{i=1}^{N}|f_i - y_i| \tag{6}$$

$$RMSE = \sqrt{\frac{1}{N}\sum_{i=1}^{N}\|f_i - y_i\|^2} \tag{7}$$

$$Precision = \frac{N^a}{N}, \tag{8}$$

$$Recall = \frac{N^a}{N^t}, \tag{9}$$

$$F = \frac{2 * Precision * Recall}{Precision + Recall}. \tag{10}$$

$$MAP = \frac{\sum AveragePrecision}{N} \tag{11}$$

3.2 Experimental Results

In this paper, we operate a number of baseline recommendation algorithms combining with our approach, including UserKNN, BasicMF, SlopeOne [12], SVD [11], all of which has data sparsity problems. The experimental results show that our approach can improve the recommendation performance of these algorithms. At the same time, we also experiment several recommendation algorithms that have positive effects on data sparsity, including BPR [18], PMF [19], EE [10]. From the experimental results, we found that the recommendation performance of these algorithms is also improved when combined with our approach.

We put the preprocessed training data into UGAN for training throughout the experiment. To observe the influence of the dataset on the recommendation algorithm after injecting different portion of simulated users, we generate simulated users whose number is equal to 20%–100% of real users' total number in the original datasets. Besides, simulated user data are respectively injected into the original dataset to run on a variety of recommendation algorithms. The results on the two datasets are shown in Tables 2 and 3.

Because UserKNN, BasicMF, SlopeOne, SVD, PMF and EE are rating prediction algorithms, their evaluation Metrics are MAE and RMSE. BPR is a Top-N rank algorithm, its evaluation Metrics are precision, recall, F1 and MAP. From Tables 2 to 3, we can find that the performance of recommendation algorithms with sparsity problem and the recommendation algorithms that alleviate

Table 2. Results for recommendation algorithm on DouBan.

Algorithm	Evaluation metrics	Original	Add20% simulated users	Add40% simulated users	Add60% simulated users	Add80% simulated users	Add100% simulated users
UserKNN	MAE	0.7056	0.6951	0.6866	0.6779	0.6720	**0.6584**
	RMSE	0.8873	0.8723	0.8599	0.8479	0.8391	**0.8314**
BasicMF	MAE	0.6317	0.6224	0.6056	0.5962	0.5900	**0.5769**
	RMSE	0.8127	0.7993	0.7774	0.7644	0.7553	**0.7465**
SlopeOne	MAE	0.5869	0.5726	0.5643	0.5553	0.5446	**0.5311**
	RMSE	0.7660	0.7567	0.7511	0.7451	0.7379	**0.7287**
SVD	MAE	0.5906	0.5864	0.5839	0.5823	0.5806	**0.5786**
	RMSE	0.7511	0.7408	0.7363	0.7333	0.7290	**0.7130**
PMF	MAE	0.8772	0.8524	0.8300	0.7891	0.7261	**0.6888**
	RMSE	1.1548	1.0980	1.1202	1.0276	0.9679	**0.9342**
EE	MAE	0.6495	0.6493	0.6464	0.6403	0.6319	**0.6177**
	RMSE	0.8079	0.8048	0.9056	0.8009	0.7994	**0.7621**
BPR	Precision	0.1458	0.2262	0.2416	0.2463	0.2663	**0.2693**
	Recall	0.0211	0.0526	0.0594	0.0682	0.0761	**0.0762**
	F1	0.0369	0.0854	0.0954	0.1069	0.1184	**0.1188**
	MAP	0.0795	0.1405	0.1605	0.1644	0.1746	**0.1764**

data sparsity problem on the generated dataset improved compared with the original datasets. As Table 2 shows, the MAE and RMSE values reduce and the precision, recall, F1 and MAP values increase with more simulated user data injecting in DouBan dataset. However, the trend of improvement doesn't continue all the time. As Table 3 shows, we find the improvement of recommendation performance converge after a certain amount of simulated users is injected into the Movielens dataset. For example, the MAE and RMSE on UserKNN algorithm reach the best value when injecting 80% simulated users. Other algorithms also follow the similar disciplines.

Thus, the experimental results show that the performance of the recommendation algorithm can be improved by using our approach. However, we can not improve the performance of the recommendation algorithm all the time with users' number increasing. Though the simulated users generated by UGAN guarantee a relatively large number of ratings and alleviate data sparsity, these simulated users are unable to have fully ratings on every item because such users must conform to the distribution of the original rating data. Therefore, after injecting a certain number of simulated users, data sparsity will not be further reduced. To prove our conclusion, we further study the sparsity change of the dataset with simulated users injecting.

Then, we calculate data sparsity on Movielens dataset. First, we generate simulated users whose number is 20%–200% of original users' total number and inject them into the dataset. As can be seen from Fig. 4, ratings sparsity declines

Table 3. Results for recommendation algorithm on Movielens.

Algorithm	Evaluation metrics	Original	Add20% simulated users	Add40% simulated users	Add60% simulated users	Add80% simulated users	Add100% simulated users
UserKNN	MAE	0.7801	0.7729	0.7669	0.7600	**0.7585**	0.7621
	RMSE	1.0119	0.9979	0.9857	0.9761	**0.9678**	0.9703
BasicMF	MAE	0.7721	0.7402	0.7326	0.7171	0.7123	**0.7008**
	RMSE	1.0173	0.9753	0.9605	0.9421	0.9333	**0.9154**
SlopeOne	MAE	0.6945	0.6850	0.6776	**0.6654**	0.6676	0.6698
	RMSE	0.9086	0.8923	0.8778	**0.8512**	0.8562	0.8631
SVD	MAE	0.6976	0.6872	0.6803	**0.6613**	0.6756	0.6769
	RMSE	0.9128	0.8893	0.8745	**0.8303**	0.8588	0.8666
PMF	MAE	0.8141	0.7646	0.7511	**0.7180**	0.7213	0.7557
	RMSE	1.0402	0.9817	0.9647	0.9473	**0.9336**	0.9673
EE	MAE	0.7535	0.7631	0.7656	0.7636	0.7662	**0.7594**
	RMSE	0.9672	0.9679	0.9576	0.9524	0.9510	**0.9444**
BPR	Precision	0.2192	0.2318	0.2481	**0.2690**	0.2675	0.2637
	Recall	0.0452	0.0561	0.0649	0.0774	**0.0808**	0.0778
	F1	0.0749	0.0904	0.1029	0.1202	**0.1241**	0.1202
	MAP	0.1454	0.1578	0.1631	0.1764	**0.1768**	0.1763

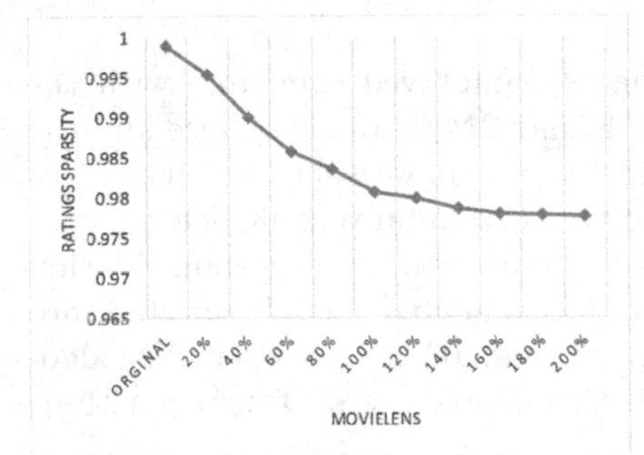

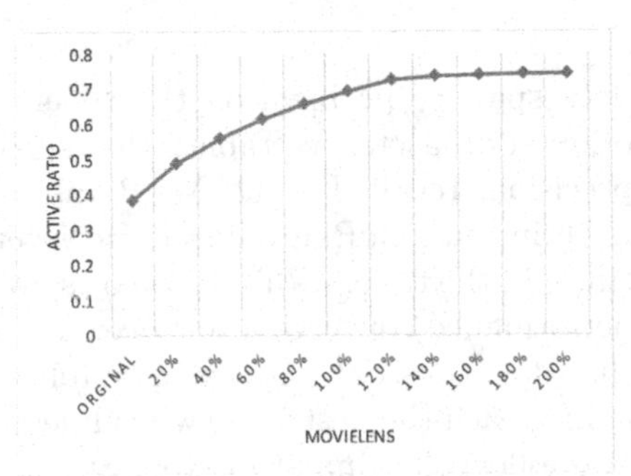

Fig. 4. Data sparsity on Movielens. Ratings sparsity is calculated by the ratio of zero ratings number to total ratings number. Active ratio is calculated by the ratio of active users number to total users number.

and active ratio climes up, which means the sparsity of rating data is significantly reduced and the average activity level of all users is significantly improved. It is shown that our approach plays a great role in alleviating data sparsity. However, we can also find that the curves in both graphs are tended to remain stable after injecting a certain number of users, which means that the sparsity of dataset will finally stop reducing.

Finally, we compare the effect of constraint on the loss function via Eq. (5). We generate simulated users whose number is equal to 100% of original real users' total number without L_P and L_R via Eqs. (3) and (4). Then we generate the same number of simulated users with L_P and L_R and set the weights W_G, W_R, W_P as 0.8, 0.1, and 0.1. Then we respectively inject data into the original dataset. The experimental results on Movielens and DouBan are shown in Fig. 5. The experimental results show that the MAE and RMSE value of datasets that are injected into simulated users without L_P and L_R decline to a certain degree. And the MAE and RMSE value of dataset that are injected into simulated users with L_P and L_R have a further declination. The result show the performance of algorithms with constraints has been improved further.

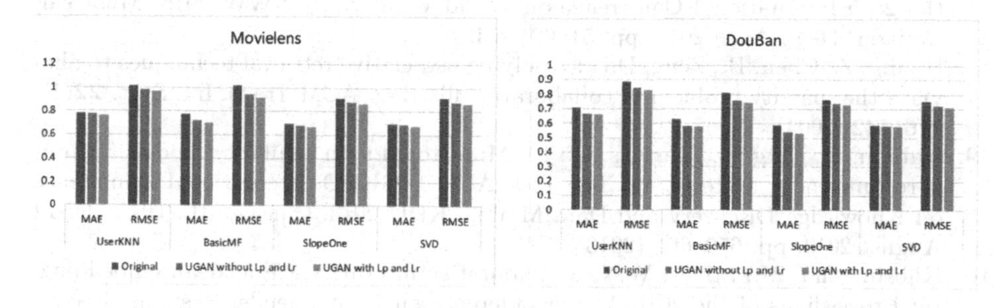

Fig. 5. Results for recommendation algorithm penalty comparison.

4 Conclusion and Future Work

In this paper, we propose a minimax game framework for generative and discriminative sample models to alleviate data sparsity in the recommendation systems. This approach uses UGAN and adds some constraints for a better sample generation process. In addition, whether the algorithm in our experiment has data sparsity problem or not, the experimental results show that the algorithms have obvious improvement with our approach. Furthermore, we analyze the data sparseness of the dataset after using our approach, and find that the degree of data sparsity is significantly reduced. Since the extra information we use for UGAN is only based on the user's activity level labels in the minimax game. In our future works, we plan to use other information to generate more types of sample data, and solve practical issues such as lack of experimental data.

References

1. Arjovsky, M., Chintala, S., Bottou, L.: Wasserstein GAN. CoRR abs/1701.07875 (2017)
2. Bodnar, C.: Text to image synthesis using generative adversarial networks. CoRR abs/1805.00676 (2018)
3. Breese, J.S., Heckerman, D., Kadie, C.M.: Empirical analysis of predictive algorithms for collaborative filtering. CoRR abs/1301.7363 (2013)

4. Eigen, D., Fergus, R.: Predicting depth, surface normals and semantic labels with a common multi-scale convolutional architecture. In: 2015 IEEE International Conference on Computer Vision, ICCV 2015, Santiago, Chile, 7–13 December 2015, pp. 2650–2658 (2015)
5. Goodfellow, I.J., et al.: Generative adversarial networks. CoRR abs/1406.2661 (2014)
6. He, R., McAuley, J.: Fusing similarity models with Markov chains for sparse sequential recommendation. In: IEEE 16th International Conference on Data Mining, ICDM 2016, Barcelona, Spain, 12–15 December 2016, pp. 191–200 (2016)
7. Hsieh, C., Yang, L., Wei, H., Naaman, M., Estrin, D.: Immersive recommendation: news and event recommendations using personal digital traces. In: Proceedings of the 25th International Conference on World Wide Web, WWW 2016, Montreal, Canada, 11–15 April 2016, pp. 51–62 (2016)
8. Huang, Z., Chen, H., Zeng, D.D.: Applying associative retrieval techniques to alleviate the sparsity problem in collaborative filtering. ACM Trans. Inf. Syst. **22**(1), 116–142 (2004)
9. Kabbur, S., Ning, X., Karypis, G.: FISM: factored item similarity models for top-n recommender systems. In: The 19th ACM SIGKDD International Conference on Knowledge Discovery and Data Mining, KDD 2013, Chicago, IL, USA, 11–14 August 2013, pp. 659–667 (2013)
10. Khoshneshin, M., Street, W.N.: Collaborative filtering via Euclidean embedding. In: Proceedings of the 2010 ACM Conference on Recommender Systems, RecSys 2010, Barcelona, Spain, 26–30 September 2010, pp. 87–94 (2010)
11. Koren, Y.: Collaborative filtering with temporal dynamics. Commun. ACM **53**(4), 89–97 (2010)
12. Lemire, D., Maclachlan, A.: Slope one predictors for online rating-based collaborative filtering. CoRR abs/cs/0702144 (2007)
13. Li, W., et al.: Social recommendation using Euclidean embedding. In: 2017 International Joint Conference on Neural Networks, IJCNN 2017, Anchorage, AK, USA, 14–19 May 2017, pp. 589–595 (2017)
14. Lin, Z., Shi, Y., Xue, Z.: IDSGAN: generative adversarial networks for attack generation against intrusion detection. CoRR abs/1809.02077 (2018)
15. Liu, Y., Qin, Z., Wan, T., Luo, Z.: Auto-painter: cartoon image generation from sketch by using conditional wasserstein generative adversarial networks. Neurocomputing **311**, 78–87 (2018)
16. Ma, H., Yang, H., Lyu, M.R., King, I.: SoRec: social recommendation using probabilistic matrix factorization. In: Proceedings of the 17th ACM Conference on Information and Knowledge Management, CIKM 2008, Napa Valley, California, USA, 26–30 October, pp. 931–940 (2008)
17. Mirza, M., Osindero, S.: Conditional generative adversarial nets. CoRR abs/1411.1784 (2014)
18. Rendle, S., Freudenthaler, C., Gantner, Z., Schmidt-Thieme, L.: BPR: Bayesian personalized ranking from implicit feedback. CoRR abs/1205.2618 (2012)
19. Salakhutdinov, R., Mnih, A.: Bayesian probabilistic matrix factorization using Markov chain Monte Carlo. In: Machine Learning, Proceedings of the Twenty-Fifth International Conference (ICML 2008), Helsinki, Finland, 5–9 June 2008, pp. 880–887 (2008)
20. Wang, J., et al.: IRGAN: a minimax game for unifying generative and discriminative information retrieval models. In: Proceedings of the 40th International ACM SIGIR Conference on Research and Development in Information Retrieval, Shinjuku, Tokyo, Japan, 7–11 August 2017, pp. 515–524 (2017)

21. Xiong, D., Liu, Q., Lin, S.: Maximum entropy based phrase reordering model for statistical machine translation. In: ACL 2006, 21st International Conference on Computational Linguistics and 44th Annual Meeting of the Association for Computational Linguistics, Proceedings of the Conference, Sydney, Australia, 17–21 July 2006 (2006)
22. Yu, J., Gao, M., Li, J., Yin, H., Liu, H.: Adaptive implicit friends identification over heterogeneous network for social recommendation. In: Proceedings of the 27th ACM International Conference on Information and Knowledge Management, CIKM 2018, Torino, Italy, 22–26 October 2018, pp. 357–366 (2018)
23. Yu, J., Gao, M., Rong, W., Song, Y., Xiong, Q.: A social recommender based on factorization and distance metric learning. IEEE Access **5**, 21557–21566 (2017)
24. Zhang, C., Yu, L., Wang, Y., Shah, C., Zhang, X.: Collaborative user network embedding for social recommender systems. In: Proceedings of the 2017 SIAM International Conference on Data Mining, Houston, Texas, USA, 27–29 April 2017, pp. 381–389 (2017)

RNE: A Scalable Network Embedding for Billion-Scale Recommendation

Jianbin Lin[1], Daixin Wang[1,2(✉)], Lu Guan[3], Yin Zhao[3], Binqiang Zhao[3], Jun Zhou[1], Xiaolong Li[1], and Yuan Qi[1]

[1] Ant Financial Services Group, Hangzhou, China
daixin.wdx@antfin.com

[2] Computer Science and Technology, Tsinghua University, Beijing, China

[3] Alibaba Group, Hangzhou, China

Abstract. Nowadays designing a real recommendation system has been a critical problem for both academic and industry. However, due to the huge number of users and items, the diversity and dynamic property of the user interest, how to design a scalable recommendation system, which is able to efficiently produce effective and diverse recommendation results on billion-scale scenarios, is still a challenging and open problem for existing methods. In this paper, given the user-item interaction graph, we propose RNE, a data-efficient Recommendation-based Network Embedding method, to give personalized and diverse items to users. Specifically, we propose a diversity- and dynamics-aware neighbor sampling method for network embedding. On the one hand, the method is able to preserve the local structure between the users and items while modeling the diversity and dynamic property of the user interest to boost the recommendation quality. On the other hand the sampling method can reduce the complexity of the whole method theoretically to make it possible for billion-scale recommendation. We also implement the designed algorithm in a distributed way to further improves its scalability. Experimentally, we deploy RNE on a recommendation scenario of Taobao, the largest E-commerce platform in China, and train it on a billion-scale user-item graph. As is shown on several online metrics on A/B testing, RNE is able to achieve both high-quality and diverse results compared with CF-based methods. We also conduct the offline experiments on Pinterest dataset comparing with several state-of-the-art recommendation methods and network embedding methods. The results demonstrate that our method is able to produce a good result while runs much faster than the baseline methods.

1 Introduction

With the exponential growth of data and information on the Internet, recommendation system plays a critical role in reducing information overload. Recommendation systems are widely deployed on many online services, including E-commerce, social networks and online news systems. How to design an effective recommendation system has been a fundamental problem in both academia and industry.

Q. Yang et al. (Eds.): PAKDD 2019, LNAI 11440, pp. 432–445, 2019.
https://doi.org/10.1007/978-3-030-16145-3_34

The key for recommendation system is to model the users' preferences based on their interactions (e.g., clicks and rating) with the items. One of the most popular recommendation methods are known as collaborative filtering (CF) [12]. Its basic idea is to match the users with similar item preferences. Among the various collaborative filtering methods, matrix factorization [10,25] is the mostly used one. However, these matrix factorization based methods are regarded as the linear methods, which are difficult to model the user-item interactions. Then following works use the deep neural networks to model the user-item relationships [11]. Despite of their success, these CF-based methods only aim to model the direct links, i.e. the first-order relationship between the users and items. However, for a graph, only preserving the first-order relationships between the nodes is not enough to characterize the network structure and thus cannot achieve good performance [2,4].

To preserve the second-order local structure in the networks, network embedding is an effective way [4]. Network embedding aims to embed nodes into a low-dimensional vector space with the goal of capturing the low-order and high-order topological characteristics in graphs [9,16,19,20,24]. Although network embedding is able to incorporate local structures, they mainly target on tasks of common link prediction and node classification. Few of them deal with the task of recommendation and thus they seldom consider some specific properties of recommendation, which makes them difficult to get a good performance on recommendation. Last but not least, few of these methods can be applied to the billion-scale networks.

To extend network embedding to recommendation, we meet three challenges. (1) Diversity of user interest. User interest is always diverse and the diverse recommendation can help users explore new items of interest. Therefore, diversity has been a very important measure to evaluate the recommendation system [1]. However, existing network embedding methods seldom consider the diversity. (2) Dynamic changes of user interest. User's preference is dynamic and how to model such a dynamic property is another challenge. (3) Scalability of recommendation system. Existing recommendation scenario often has a huge number of users and items, which is a serious problem with a scale beyond most of existing network embedding methods.

To address these challenges, we propose RNE, a scalable Recommendation-based Network Embedding method. In our method, when discovering the local structure of a user, we will not model all the items the user clicked. Instead, we propose a sampling method, which considers the diversity and dynamics of the user interest, to sample a portion of the items the user has clicked as the user's neighbors. In this way, the sampling method not only can incorporate the important properties of recommendation, i.e. the diversity and dynamics, to improve the recommendation accuracy, but also reduce the computational complexity of the algorithm. Furthermore, we deploy the algorithm on a recommendation system based on the Parameter Server to do distributed and parallel computing, which further facilitates the large-scale training available.

In summary, the contributions of the paper can be listed as follows:

- We propose a network-embedding-based recommendation method, named RNE. When modeling the local structures between the users and items, our method is able to incorporate the dynamics and diversity of the user interest to produce more accurate and diverse recommendation results.
- We implement our recommendation algorithm in a distributed way based on parameter server, which jointly makes the system available for billion-scale recommendation.
- Experimentally, we deploy the whole system on a recommendation scenario of Taobao. Online A/B tests demonstrate that our method is able to achieve more accurate results compared with CF and greatly improve the diversity of the recommendation results. Experiments on offline dataset Pinterest also demonstrate the quality of our method.

Table 1. Multifaceted comparisons between different methods

Method	Local-structure preserving	Diversity	Billion-scale	Complexity
GMF-CF/MLP-CF/NCF	×	×	×	$O(\lvert E \rvert)$
LINE/node2vec	√	×	×	$O(\lvert E \rvert)$
RNE	√	√	√	$O(\lvert V \rvert)$

2 Related Work

2.1 Collaborative Filtering

Recommendation algorithms and systems are well-investigated research fields. In our work, we are only given the user-item interaction data. Therefore, we mainly introduce the CF-based recommendation methods and omit the discussions of content-based recommendation methods and the hybrid recommendation methods.

Collaborative Filtering exploits the interaction graph between the users and items to give the recommendation lists to users. Its basic idea is to match the users which have similar item preferences. Earlier CF methods mainly use the matrix factorization on the user-item matrices to obtain the latent user factors and item factors [5,10,17]. The user factors and item factors together aim to reconstruct the original user-item matrices. However, the matrix factorization is just the linear-based methods, which is difficult to capture the user-item relationships. To overcome such a drawback, following works use the deep neural networks to perform collaborative filtering [18,21]. However, most of the CF-based methods only aim to model the pairwise relationships between the user and item but omit their local structures. And many graph-based works have demonstrate that local structures like second-order relationships are very important for capturing graph structures [19]. In this way, existing CF-based methods are sub-optimal for capturing the relationships between user and items.

2.2 Network Representation Learning

Network embedding has been demonstrated as an effective methods for modeling local and global structures of a graph. It aims to learn a low-dimensional vector-representation for each node. DeepWalk [16] and Node2vec [9] propose to use the random walk and skip-gram to learn the node representations. LINE [19] and SDNE [20] propose explicit objective functions for preserving first- and second-order proximity. Some further works [2,15] use the matrix factorization to factorize high-order relation matrix. Aforementioned methods are designed for homogeneous networks. Then some following embedding methods for heterogeneous networks are proposed, like Metapath2vec [6], HNE [3], BiNE [7] and EOE [23]. Some works further focuses on knowledge graph embedding [22]. Although these network embedding methods are able to preserve the local structures of the vertices, most of them are not specifically designed for the task of recommendation. They do not consider some specific properties of the recommendation tasks like the diversity and dynamic changes of user interest, the scalability issues of large-scale recommendation tasks. Therefore, how to propose an effective network embedding method for billion-scale recommendation is still an open problem.

In summary, we compare our method and the related works in Table 1. Our method is specifically designed for the recommendation scenario and thus consider some specific properties. Furthermore, the proposed method is very scalable and thus can apply to billion-scale recommendations.

3 The Methodology

In our scenario, we have a large number of users and items. Each user may have different ways to interact with the items. For example, the user may view the items, collect the items or buy the items. In this way, we can build the user-item interaction graph, formally formulated as $G = (\mathcal{U}, \mathcal{I}, E)$. Here $\mathcal{U}$ denotes the total of n users and $\mathcal{I}$ denotes the total of T items. $\mathcal{U} \cup \mathcal{I}$ denotes the set of nodes in G. If a user $u \in \mathcal{U}$ views, collects or buys an item $i \in \mathcal{I}$, there is an edge E_{ui} between u and i. We use $E(v), v \in \mathcal{U} \cup \mathcal{I}$ to denote the edges connected to the node v. We assume that G is connected. The recommendation problem is that given a user u, we hope to recommend some personalized items to the user based on his previous behavior.

3.1 Network Embedding for Recommendation

Given the user-item interaction graph $G = (\mathcal{U}, \mathcal{I}, E)$, we aim to map each user and item to a common low-dimensional latent space, where user u can be embedded as $\mathbf{E}_{\mathcal{U}}^{u} \in R^d$ and item i can be embedded as $\mathbf{E}_{\mathcal{I}}^{i} \in R^d$. Then with the embeddings for each user and item, we can retrieve the similar items for the user as his recommendation results.

To achieve this, we propose our method, whose framework can be shown in Fig. 1. It consists of the embedding-lookup layer, embedding layer and softmax

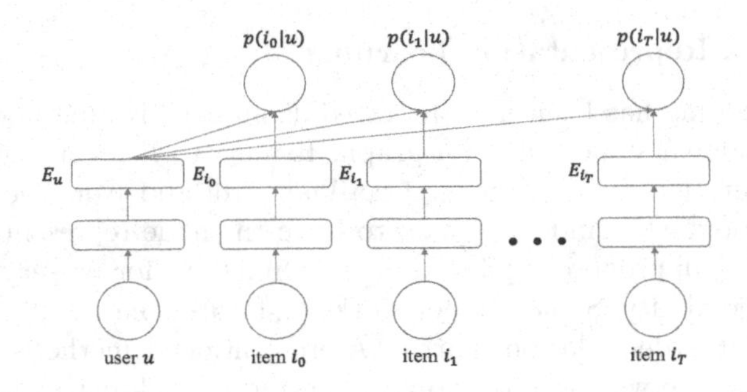

Fig. 1. The framework of RNE.

layer. The embedding-lookup layer helps us obtain the embeddings for the users and items. The embedding layer and softmax layer together model the interactions between the users and items to update the embedding-lookup layer. Then we introduce the designed loss functions to update the embeddings.

We first consider how to model the local structure of a user in the given user-item graph. In the original space, the empirical distributions given a user can be defined as:

$$\hat{p}(i|u) = \frac{w_{ui}}{d_u}, \tag{1}$$

where w_{ui} is the weight between user u and item i and d_u is the degree of user u.

Then we hope to estimate the local structure of a user in the embedding space. Word2vec [14] inspires us to use the inner product between two vertices to model their interactions. Then in our work, given a user u, we define the probability of item i generated by user u as:

$$p(i|u) = \frac{exp(\mathbf{E}_{\mathcal{U}}^{u\,T}\mathbf{E}_{\mathcal{I}}^{i})}{\sum_{j=1}^{|I|} exp(\mathbf{E}_{\mathcal{U}}^{u\,T}\mathbf{E}_{\mathcal{I}}^{j})}, \tag{2}$$

where T means the transpose of a matrix.

Equation 2 is a softmax-like loss function, which defines the conditional distributions $p(\cdot|u)$ of user u over its neighborhoods, i.e. the entire item set, in the embedding space.

With the empirical distributions on the original network and reconstructed distributions on the embedding space, we can learn the embedding by making the defined probability $p(\cdot|u)$ specified by the low-dimensional representations be close to the empirical distributions $\hat{p}(\cdot|u)$. We use the KL-divergence to measure the distance between the distributions. Then the loss functions can be defined as:

$$L = \sum_{u \in \mathcal{U}} \lambda_u KL(\hat{p}(\cdot|u), p(\cdot|u)) \propto - \sum_{(u,i) \in E} w_{ui} log p(i|u), \tag{3}$$

where λ_u denotes the prestige of user u and we set $\lambda_u = d_u$.

Minimizing Eq. 3 will make the vertices with similar neighbors similar to each other. Therefore, it can not only model the observed links on the graph, but also preserve the local structures for each node.

3.2 Recommendation-Based Sub-sampling

However, aforementioned network embedding meets two challenges for large-scale recommendation: (1) Minimizing Eq. 3 is time-consuming since for each edge it needs to run over the entire set of the items when evaluating $p(i|u)$. In this way, the whole complexity is $O(|E||\mathcal{I}|)$, which is unbearable for real recommendation systems. (2) Minimizing Eq. 3 only considers the topology of the graph. It does not consider the diversity and the time decay of the user interest, which are very important properties for recommendation systems.

To reduce the complexity, we first adopt negative sampling as many methods do [19]. For each positive edge (u, i), we will sample some negative edges according to predefined distributions P_{ui}. By performing negative sampling, the objective function for each edge (u, i) can be reformulated as:

$$L_{ui} = log(\sigma(\mathbf{E}_{\mathcal{U}}^{u\,T}\mathbf{E}_{\mathcal{I}}^{i})) + \sum_{j=1}^{k} E_{i_j \sim P_u}(log(\sigma(\mathbf{E}_{\mathcal{U}}^{u\,T} E_{\mathcal{I}}^{i_j}))), \tag{4}$$

where k is the number of negative samples for each user-item pair, $P_{ui} \propto d_i^{3/4}$.

Although negative sampling can reduce the time complexity from $O(|E||\mathcal{I}|)$ to $O(k|E|)$, for billion-scale recommendation, a complexity linear to the number of edges is still a great challenge.

To further reduce the complexity, we only select a portion of the items the user has clicked to obtain his behavior sequence. Then the question comes to how to select the items to effectively represent the user's interest. Here, we mainly consider two properties specified for recommendation. (1) The diversity of user interest: User interest is always diverse. A user will always focus on the items of more than one cluster. (2) The time decay of user interest: User interest is always dynamic. More recent user behavior is more reliable to reflect the recent user interest. Therefore, we should more focus on recent user behavior. Based on these two considerations, we define the selection probability for each user-item pair (u, i) as follows:

$$p(u, i) = 0.999^{t_i} * click(u, c_i)^{\gamma}, \tag{5}$$

where t_i is the hours of the item i from the most recent item, c_i is the cluster index of item i and $click(u, c_i) = \sum_{j \in c_i} w_{uj}$, γ is set to -0.2. Then for each user, we will sample m samples according to the defined probability in Eq. 5 to represent his behavior sequence. Then in this way, the complexity can be reduced from $O(k|E|)$ to $O(km|\mathcal{U}|)$, which is linear to the number of nodes.

In summary, on the one hand, if a user more recently shows the interest to an item, the item should have a larger probability to be sampled. On the other hand, the method is prone to sample the items of the clusters clicked less times by the user. In this way, our method may cover more clusters to ensure the diversity. Therefore, such a sampling strategy can simultaneously model the diversity and time decay of the user interest Furthermore, with the sampling strategy, we do not need to model all the edges in one iteration but instead for each user we only model a portion of its preferred items as the user's behavior sequence. It significantly reduce the time complexity.

3.3 Implementation

In this section, we will introduce the technical implementation of the proposed RNE. The whole process can be divided into two phases: offline model training and online retrieval. This section will describe them in detail.

Off-Line Model Training. To train the proposed RNE, we utilize the Stochastic Gradient Descent (SGD) on the loss function of Eq. 4 to update the node embeddings. In detail, we use E_{pos} to denote all the positive edges sampled by the method we proposed before. Then for each $(u, i) \in E_{pos}$, we can update their embeddings as follows:

$$\mathbf{E}_{\mathcal{U}}^{u} = \mathbf{E}_{\mathcal{U}}^{u} + \lambda\{ \sum_{z \in \{i\} \cup N_{neg}^{k}(u)} [I(z, u) - \sigma({\mathbf{E}_{\mathcal{U}}^{u}}^{T} \mathbf{E}_{\mathcal{U}}^{z})] \cdot \mathbf{E}_{\mathcal{U}}^{z}\}, \quad (6)$$

where $I(a, b)$ is the indicator function that if $a = b$, $I(a, b) = 1$, otherwise $I(a, b) = 0$. $N_{neg}^{k}(i)$ is the negative neighborhoods of vertex i. λ denotes the learning rate. Similarly, we can update embedding $E_{\mathcal{I}}^{i}$ for an item i in a similar way, which we will not discussed more.

From Eq. 6, when given a positive edge, we can update their embeddings. Then we will go over all the pair of positive edges for several iterations to update their embeddings. The whole algorithm can be summarized in Algorithm 1.

Algorithm 1. Training Algorithm for RNE

Input: $G = (\mathcal{U}, \mathcal{I}, E)$
Output: $\mathbf{E}_u$, $\mathbf{E}_I$
1: Initializing E_u and E_I.
2: **while** not converged **do**
3: Construct the positive edge set S_{pos} according to $G = (\mathcal{U}, \mathcal{I}, E)$ and Eq. 5.
4: **for all** $(u, i) \in S_{pos}$ **do**
5: Construct the negative set $N_{neg}^{k}(u)$.
6: Update $\mathbf{E}_u$ and $\mathbf{E}_i$ according to Eq. 6.
7: **end for**
8: **end while**

From Algorithm 1, we find that the learning process from Line 4 to Line 6 is independent for different edges, which inspires us to use some parallelization mechanism to implement it. Then we deploy the whole algorithm on the parameter server, which implements a data-parallelization mechanism. In detail, from Eq. 6 we find that to update a node's embedding, we only need to know the node's previous embeddings, the node's neighborhoods and their embeddings.

Therefore, we can resort to parameter server to implement such a process in a parallelized way. The main workflow of the system is built as follows: (1) In each iteration, the server will assign each worker a subset of the vertices of the graph G. (2) Each worker will pull the assigned vertices from the server and calculate the positive and negative neighborhoods for the assigned vertices. Then with positive and negative sets, each worker can update the embeddings of the assigned vertices according to Eq. 6. (3) After updating, each worker will push his assigned vertices' embddings to the server. Such a training process will be iterated several times.

Online Efficient Nearest Neighbor Search. For online recommendation, we use the nearest neighbor search on the learned embedding space to make recommendations. That is, given a query user u, we can recommend items whose embeddings are the K-nearest-neighbors (K-nn) of the query user's embedding E_u. To achieve the K-nn search, we use the Faiss library [13] which is an efficient implementation for state-of-the-art product-quantization methods. Given that RNE is trained offline and all the user and item embeddings are computed via Parameter Server and saved in database, the efficient K-nn search enables the system to recommend items online.

4 Experiments

The goal of RNE is to produce high-quality and scalable recommendations for real-world systems. Therefore, we conduct comprehensive experiments in two ways: Online A/B tests and Offline experiments.

4.1 Datasets

We use two real-world datasets, i.e. Ali-mobile taobao and Pinterest in this paper.

- Ali-mobile taobao: It is a mobile recommendation scenario deployed on Taobao, the largest E-commerce platform in China. The dataset is extremely large. It has about 1 billion users, tens of million items and a total of about one hundred billion edges. Each edge denotes whether the user has clicked the products. We deploy our algorithm on the service to do online A/B test to evaluate our method.
- Pinterest: The dataset is an image recommendation dataset constructed by [8]. We filter the users which have very few interactions with the items and only retain the users which have more than 20 interactions. After the preprocessing, the dataset consists of 50 thousand users, 10 thousand items and 1.5 million user-item edges. Each edge denotes whether the user has pinned the items.

4.2 Online A/B Tests

The ultimate goal of the recommendation system is to lift the user's interest in the items. Therefore, we perform random A/B experiments on Ali-mobile taobao to demonstrate this, where a random set of users obtain the recommendation results of RNE and another obtain the results of CF-based methods. Any difference in the engagement of the items between the two groups can truly reflect the recommendation quality of two methods. Note that here we only use one baseline because deploying many methods online to do A/B tests will cost a lot of resources. And the reason why we choose CF is that it is well investigated for recommendation and existing network embedding methods cannot scale to billion-scale dataset. For more comparisons with state-of-the-art methods, we do offline experiment, which we will introduce in detail later.

We use the following six metrics to measure the recommendation quality.

- AVD (Averaged View Depth): The metric denotes how deep a user views the page. It measures the recommendation quality.
- ACN (Averaged Click Number): The metric measures the number of clicks on the items for each user in average. It measures the recommendation quality.
- P-CTR (Page Click-through Rate): for a page p, $pctr = \frac{\#click\text{-}throughs(p)}{\#impressions(p)} \times 100\%$. It measures the recommendation quality.
- U-CTR (User Click-through Rate): for a user u, $uctr = \frac{\#click\text{-}throughs(u)}{\#impressions(u)} \times 100\%$. It measures the recommendation quality.
- Re-C (Recommended number of clusters): The averaged number of clusters recommended to users. The clusters are obtained by using our clustering algorithm. The metric measures the recommendation diversity.
- CK-C (Clicked number of clusters): The averaged number of clusters clicked by users. It measures both the recommendation quality and the diversity.

Table 2. Performance of online A/B tests on Ali-mobile taobao

Metrics	AVD	ACN	P-CTR	U-CTR	Re-C	CK-C
Ali-mobile taobao	9.54%	13.21%	4.99%	1.12%	20.49%	16.32%

Table 2 summarizes the lift in engagement of items recommended by RNE compared with CF-based methods in controlled A/B experiments. From Table 2, we have the following observations and analysis:

- We find that RNE can achieve a significant improvement in terms of AVD and ACN over the CF. It indicates by using the results of RNE, users are more willing to go deeper to view more items and click more items, which indirectly demonstrates the ranking quality of RNE.

- In terms of the two CTR metrics, a popular and well-accepted metric to evaluate the recommendation quality, our proposed method also achieves a better result than CF. It further demonstrates that RNE is able to produce personalized items for users. The reason for a better recommendation quality is twofold. (1) Our method is able to capture the local structures between the users and items. (2) Our method considers the dynamic change of the user's interest.
- We find that RNE achieves a higher Re-C compared with CF, which indicates that our recommended results are from more clusters. The reason is that our proposed method incorporates the diversity issue into the model design.
- More importantly, our method achieves a higher CK-C than CF, which demonstrates that not only our method can produce more diverse recommendations, but also the users are willing to click these diverse recommended items. It indicates that our method is able to improve the recommendation quality while improving the recommendation diversity.
- Under the billion-scale scenario, RNE can be deployed online and still obtain good results, which demonstrates the superiority of our method.

4.3 Showcase

In this section, we give some showcase to see some intuitions regarding the embeddings we learn. After the learning process of RNE, all the items will have embeddings. Then in this experiment, given a query item's embedding, we aim to find the most similar 8 items whose embeddings have the smallest distance with the query. Then we display both the query image and the recommended images in Fig. 2.

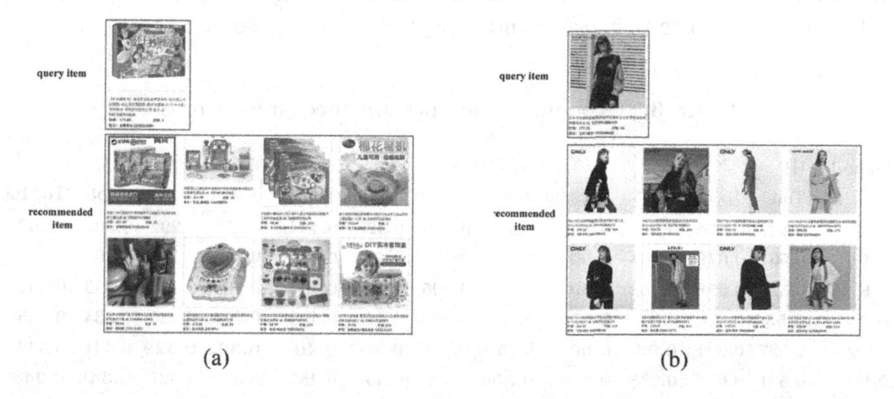

Fig. 2. Real showcase on Ali-mobile taobao: Given an item (in red box), searching for the nearest 8 items (in blue box) using the embeddings learned by RNE. (Color figure online)

In Fig. 2(a), the query item is a princess-style educational toy for girls. When we look at the returned results, these images belong to different categories with the query, like plasticine and origami. But all of them are for fun and a majority of them are also princess-style. It demonstrates that our method is able to find more categories of items but retain the primary style of the item. In Fig. 2(b), the query item is a woman sweatshirt of the brand of Peacebird. The returned images are all coat, sweater or sweatshirt of the brand of Peacebird or Only. Similarly, the returned images and the query image are all casual style but belong to different fine-grained categories. Moreover, actually the brand of Peacebird and Only have very similar styles and our method can learn their inherit relationships.

In summary, in our method, we do not have the item features and the direct relationships between items. We are only given the user-item interaction graph. Although in this case, our method still can model the item relationships by using the user behaviors as the bridge. It demonstrates that by using the network embedding method we propose, the learned embeddings can capture the local relationships between the entities.

4.4 Offline Experiment

To compare more baselines to get comprehensive results, we conduct the offline experiments on the dataset of Pinterest. We randomly sample 90% user-item pairs as the training set and the rest as the testset. For training set, we use 9-fold cross-validation to tune the parameters for all the methods. Note that in this dataset, we do not have the cluster and time information for the item. So we uniformly sample the items to do training. To evaluate the performance, we use the following three metrics: Normalized Discounted Cumulative Gain (NDCG), Mean Reciprocal Rank (MRR) and Hit Rate (HR). NDCG and MRR will consider the rank of the hit and will assign higher scores to hits at top ranks. While HR will only evaluate whether the test items are hit or not. We calculate all the metrics for the test users and report the average score.

Table 3. Recommendation performance on Pinterest.

Method	HR				NDCG				RR			
	Top5	Top10	Top50	Top100	Top5	Top10	Top50	Top100	Top5	Top10	Top50	Top100
GMF	0.501	0.678	0.9	0.97	0.332	0.386	0.425	0.434	0.276	0.292	0.301	0.307
MLP	0.504	0.679	0.908	**0.99**	0.341	0.385	0.426	0.436	0.275	0.295	0.302	0.309
NCF	0.529	0.688	0.912	**0.99**	0.35	0.405	0.443	0.454	0.298	0.32	0.343	0.341
LINE	**0.536**	**0.7**	0.91	**0.99**	0.353	0.409	0.448	0.455	0.297	0.325	0.334	0.335
node2vec	0.527	0.691	**0.93**	**0.99**	0.355	**0.411**	**0.459**	**0.46**	**0.302**	**0.329**	0.341	0.342
RNE	0.531	0.695	0.925	**0.99**	**0.356**	0.41	0.45	0.457	0.3	0.327	**0.345**	**0.348**

We first use the advanced CF-based methods GMF, MLP and NCF [11] as baseline methods. We perform the same process of parameter search as the work [11] did to select the optimal parameters. For network embedding methods, since we only have the graph topology, in this case LINE [19] and node2vec

[9] are state-of-the-art network embedding methods, so we choose them as the baselines. For LINE, we use $\text{LINE}_{1st+2nd}$ with the default parameter settings. For node2vec, we also use the default settings except for the bias parameters p, q, which we conduct the grid search from $\{0.5, 1\}$. The embedding dimension of them is all set as 128.

The results are shown in Table 3. From Table 3, we find that RNE achieves a better performance than all the CF-based methods. The reason is that RNE is able to capture the local structure of each user while CF-based methods only focus on the direct links the user has clicked. It demonstrates that capturing the local structures on the user-item graph is important for recommendation. LINE, node2vec and RNE achieve similar performance in different evaluation metrics and scenarios. But our method runs much faster than node2vec and LINE, which will be discussed later. Therefore, RNE is a better balance between accuracy and efficiency.

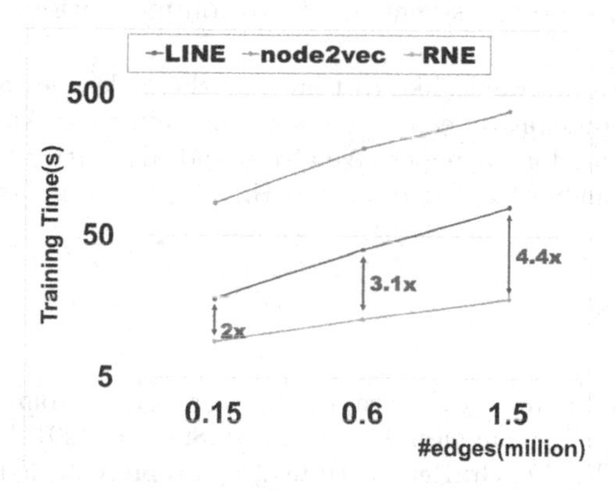

Fig. 3. Time comparisons on Pinterest dataset. We change the number of edges to be trained and report the training time for each network embedding method.

Now we discuss the training time of LINE, node2vec and RNE. For a fair comparison, we do not use the distributed strategy for RNE. From Fig. 3, we find that RNE can boost the running time over LINE and node2vec. Specifically, when the training edges increase from 0.15 million to 1.5 million, the running time improvement of RNE compared with LINE will be larger and larger, from 2x to 4.4x. When the edges continuously increase to the billion-scale dataset like the Ali-mobile taobao dataset, it is difficult for LINE and node2vec to obtain the results. But RNE can still obtain a good result. The reasons why our method can scale to billion-scale dataset are twofold: (1) The proposed sampling method avoids us running over all the edges in the graph. (2) Our method can be deployed on distributed system for parallel computations.

In summary, RNE has a good scalability, which is much more efficiency than baseline methods and can scale to billion-scale recommendation scenario, meanwhile RNE do not sacrifice its recommendation accuracy.

5 Conclusion

In this paper, we propose a novel network embedding method named RNE for scalable recommendation. The proposed network embedding method is able to capture the local structures on the user-item graph to achieve a better recommendation quality. Specifically, to consider the specific properties for recommendation, i.e the diversity and time-decay of user interest, we design a sampling method for embedding process to incorporate these properties. And the sampling method also guarantees the scalability of the proposed method while almost preserving the recommendation quality. We also deploy our algorithm on parameter server to make it available for large-scale recommendation. Experimental results on online A/B tests and offline experiments all demonstrate the superiority of the proposed method.

For the future work, we may consider the user and item features, which can further address the sparsity and cold-start problem. We also want to analyze the role of features and topology structures for recommendation.

Acknowledgement. We would like to thank all the colleagues of our team and all the members of our cooperative team: the search engine team in Alibaba. They provide many helpful comments for the paper. We also would like to thank the support of the Initiative Postdocs Supporting Program and the valuable comments provided by all the reviewers.

References

1. Adomavicius, G., Kwon, Y.O.: Improving aggregate recommendation diversity using ranking-based techniques. TKDE **24**(5), 896–911 (2012)
2. Cao, S., Lu, W., Xu, Q.: GraRep: learning graph representations with global structural information. In: CIKM, pp. 891–900 (2015)
3. Chang, S., Han, W., Tang, J., Qi, G.J., Aggarwal, C.C., Huang, T.S.: Heterogeneous network embedding via deep architectures. In: SIGKDD, pp. 119–128. ACM (2015)
4. Cui, P., Wang, X., Pei, J., Zhu, W.: A survey on network embedding. TKDE (2018)
5. Deshpande, M., Karypis, G.: Item-based top-n recommendation algorithms. ACM Trans. Inf. Syst. **22**(1), 143–177 (2004)
6. Dong, Y., Chawla, N.V., Swami, A.: metapath2vec: scalable representation learning for heterogeneous networks. In: SIGKDD, pp. 135–144. ACM (2017)
7. Gao, M., Chen, L., He, X., Zhou, A.: BiNE: bipartite network embedding (2018)
8. Geng, X., Zhang, H., Bian, J., Chua, T.S.: Learning image and user features for recommendation in social networks. In: ICCV, pp. 4274–4282 (2015)
9. Grover, A., Leskovec, J.: node2vec: scalable feature learning for networks. In: SIGKDD, pp. 855–864 (2016)
10. Harvey, M., Carman, M.J., Ruthven, I., Crestani, F.: Bayesian latent variable models for collaborative item rating prediction. In: CIKM, pp. 699–708 (2011)
11. He, X., Liao, L., Zhang, H., Nie, L., Hu, X., Chua, T.S.: Neural collaborative filtering, pp. 173–182 (2017)
12. Herlocker, J.L., Konstan, J.A., Terveen, L.G., Riedl, J.T.: Evaluating collaborative filtering recommender systems. ACM Trans. Inf. Syst. **22**(1), 5–53 (2004)

13. Johnson, J., Douze, M., Jégou, H.: Billion-scale similarity search with GPUs. arXiv preprint arXiv:1702.08734 (2017)
14. Mikolov, T., Chen, K., Corrado, G., Dean, J.: Efficient estimation of word representations in vector space. arXiv preprint arXiv:1301.3781 (2013)
15. Ou, M., Cui, P., Pei, J., Zhang, Z., Zhu, W.: Asymmetric transitivity preserving graph embedding. In: SIGKDD, pp. 1105–1114 (2016)
16. Perozzi, B., Al-Rfou, R., Skiena, S.: DeepWalk: online learning of social representations. In: SIGKDD, pp. 701–710. ACM (2014)
17. Sarwar, B., Karypis, G., Konstan, J., Riedl, J.: Item-based collaborative filtering recommendation algorithms. In: International Conference on World Wide Web, pp. 285–295 (2001)
18. Strub, F., Mary, J.: Collaborative filtering with stacked denoising autoencoders and sparse inputs. In: NIPS Workshop on Machine Learning for e-Commerce (2015)
19. Tang, J., Qu, M., Wang, M., Zhang, M., Yan, J., Mei, Q.: LINE: large-scale information network embedding. In: WWW, pp. 1067–1077 (2015)
20. Wang, D., Cui, P., Zhu, W.: Structural deep network embedding. In: SIGKDD, pp. 1225–1234 (2016)
21. Wu, Y., Dubois, C., Zheng, A.X., Ester, M.: Collaborative denoising auto-encoders for top-n recommender systems, pp. 153–162 (2016)
22. Xiao, H., Huang, M., Zhu, X.: TransG: a generative model for knowledge graph embedding. In: Proceedings of the 54th Annual Meeting of the Association for Computational Linguistics (Volume 1: Long Papers). vol. 1, pp. 2316–2325 (2016)
23. Xu, L., Wei, X., Cao, J., Yu, P.S.: Embedding of embedding (EOE): joint embedding for coupled heterogeneous networks. In: Proceedings of the Tenth ACM International Conference on Web Search and Data Mining, pp. 741–749. ACM (2017)
24. Zhang, Z., Cui, P., Li, H., Wang, X., Zhu, W.: Billion-scale network embedding with iterative random projection. arXiv preprint arXiv:1805.02396 (2018)
25. Zhou, X., et al.: Enhancing online video recommendation using social user interactions. VLDBJ **26**(5), 637–656 (2017)

Social Network and Graph Mining

Graph Compression with Stars

Faming Li[1], Zhaonian Zou[1](✉), Jianzhong Li[1], and Yingshu Li[2]

[1] Harbin Institute of Technology, Harbin, China
{lifaming2016,znzou,lijzh}@hit.edu.cn
[2] Georgia State University, Atlanta, USA
yili@gsu.edu

Abstract. Making massive graph data easily understandable by people is a demanding task in a variety of real applications. Graph compression is an effective approach to reducing the size of graph data as well as its complexity in structures. This paper proposes a simple yet effective graph compression method called the star-based graph compression. This method compresses a graph by shrinking a collection of disjoint subgraphs called stars. Compressing a graph into the optimal star-based compressed graph with the highest compression ratio is shown to be NP-complete. We propose a greedy compression algorithm called StarZip. We experimentally verify that StarZip achieves compression ratios of 3.8–45.7 and 2.9–241.6 in terms of vertex count and edge count, respectively. Besides, we study the shortest path queries on compressed graphs. On the real graphs, the StarSSSP algorithm for processing shortest path queries on compressed graphs is 4X–20X faster than Dijkstra's algorithm running on original graphs. The average absolute error between the query results of StarSSSP and the exact shortest distances is about 1. On the synthetic graphs, StarSSSP is up to 313X faster than Dijkstra's algorithm, and the average absolute error is also about 1.

Keywords: Graph compression · Star · Shortest path

1 Introduction

In recent years, graphs have been extensively used to model complex relationships between entities in a wide variety of applications. For example, the Web graph represents hyperlinks between Web pages in the World Wide Web. Social networks represent social relationships between people in general or specific domains. So far, massive amount of data represented by graphs, known as *graph data*, has been accumulated in numerous applications. For example, the Web graph consists of at least 4.62 billion vertices (Web pages) in 2017[1]. The total number of monthly active Facebook users has reached 1.754 billion in October 2017[2]. The volume of graph data continues increasing in even faster speed. Currently, graph data has evolved to be a typical class of big data.

[1] http://www.worldwidewebsize.com.
[2] http://www.statisticbrain.com/facebook-statistics/.

Q. Yang et al. (Eds.): PAKDD 2019, LNAI 11440, pp. 449–461, 2019.
https://doi.org/10.1007/978-3-030-16145-3_35

Massive graphs are too large and too complex to be easily understood by people. Recently, numerous studies have been carried out on graph query processing and graph mining. The goal is to develop advanced tools for understanding, querying and mining massive graph data. For a graph analysis problem Q on a large graph G, traditional studies on graph algorithms mostly focus on reducing the time complexity of algorithms to solve Q on G. However, this kind of approaches often do not scale to very large graphs. In recent years, *scale-down approaches* to graph analytics have attracted considerable research attentions. The main idea of scale-down approaches is to reduce the size of the graph G and (approximately) solve the problem Q on the reduced graph of G. Two typical ways to reduce the size of G are *graph sampling* and *graph compression.* Graph sampling [1] randomly selects a subgraph of G that can preserve the characteristics of G. Graph compression merges multiple similar vertices into one vertex, so it reduces the size of G.

We focus on graph compression method in this paper. The concept of *graph compression* is similar to the notion in [2] and [3]. The graph G^*, which is constructed by super vertices and super edges, is a compression of an original graph G and has the following properties:

- G^* is a graph with $|E^*|$ edges, where $|E^*|$ is smaller than the number of edges in G;
- It is computationally easy to convert G into G^*.

The main contributions of this paper are listed as follows.

- We propose a simple yet effective graph compression method called StarZip, which can compress big graph efficiently.
- Besides the impressive compression ratios that the star-based graph compression can achieve, this graph compression method can also support query processing on compressed graphs, such as shortest path queries.
- We conducted comprehensive experiments on the performance of the star-based graph compression and the efficiency and the accuracy of query processing on star-based compressed graphs.

The rest of this paper is organized as follows. Section 2 gives a formal definition of the star-based graph compression and proposes the star-based graph compression algorithm StarZip. Section 3 studies shortest path queries on star-based compressed graphs. Experimental results are reported in Sect. 4. Section 5 reviews the related work. Finally, we conclude the paper in Sect. 6.

2 The Star-Based Graph Compression

This section gives a formal definition of the star-based graph compression and proposes a star-based graph compression algorithm. A *graph* is a pair (V, E), where V is a set of vertices, and E is a set of edges. A graph is *undirected* if (u, v) and (v, u) refer to the same edge. In this paper, we consider undirected graphs. Let $V(G)$ and $E(G)$ denote the vertex set and the edge set of a graph G, respectively.

2.1 Star-Based Compressed Graphs

First, we introduce some basic concepts used in the definition of the star-based graph compression.

Definition 1. *A graph S is a* star *if there is a vertex $s \in V(S)$ such that $E(S) = \{(s, v) | v \in V(S) \setminus \{s\}\}$. The vertex s is called the* center *of the star S. The vertices in $V(S) \setminus \{s\}$ are called the* border vertices *of S.*

In the graph shown Fig. 1(a), the subgraph formed by the edges (v_3, v_0), (v_3, v_1), (v_3, v_2), (v_3, v_4), and (v_3, v_{10}) is a star, where v_3 is the center vertex, and $v_0, v_1, v_2, v_4, v_{10}$ are the border vertices.

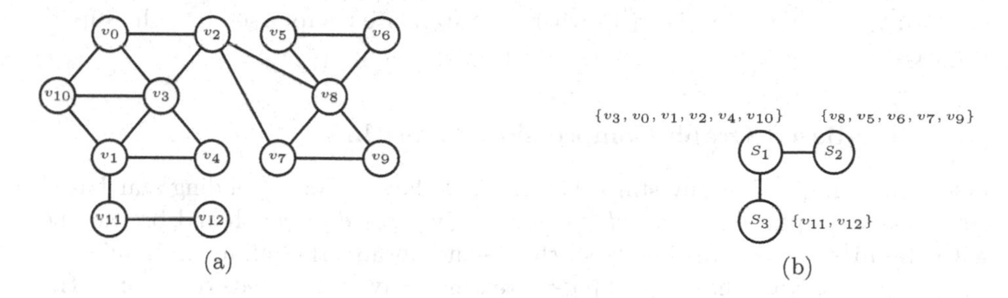

Fig. 1. A sample graph.

Definition 2. *Let G be a graph and $\Phi = \{V_1, V_2, \ldots, V_n\}$ be a partition of $V(G)$, that is, $V(G) = V_1 \cup V_2 \cup \cdots \cup V_n$ and $V_i \cap V_j = \emptyset$ for $i \neq j$. Let H be a graph such that*

- *$V(H) = \Phi$, and*
- *$(V_i, V_j) \in E(H)$ if and only if there exist $u \in V_i$ and $v \in V_j$ such that $(u, v) \in E(G)$.*

We call H the compressed graph *of G with respect to the vertex partition Φ. The vertices in H are called* super-vertices, *and the edges in H are called* super-edges.

Consider the graph G shown in Fig. 1(a). Let

$$\Phi = \{\{v_3, v_0, v_1, v_2, v_4, v_{10}\}, \{v_8, v_5, v_6, v_7, v_9\}, \{v_{11}, v_{12}\}\}.$$

Then, Φ is a partition of $V(G)$. The compressed graph of G with respect to Φ is shown in Fig. 1(b). This compressed graph consists of 3 super-vertices and 2 super-edges.

Graph compression is the process of creating a compressed graph by grouping vertices with similar structural contexts into super-vertices. In this paper, we propose the *star-based graph compression*, which is described as follows.

Definition 3. *Let G be a graph. The* star cover *of G is a set $\{S_1, S_2, \ldots, S_n\}$ of stars in G such that*

- $V(G) = V(S_1) \cup V(S_2) \cup \cdots \cup V(S_n)$, *and*
- $V(S_i) \cap V(S_j) = \emptyset$ *for* $i \neq j$.

Given a star cover $\{S_1, S_2, \ldots, S_n\}$ of a graph G, $\{V(S_1), V(S_2), \ldots, V(S_n)\}$ is a partition of $V(G)$. The compressed graph of G with respect to $\{V(S_1), V(S_2), \ldots, V(S_n)\}$ is called the *star-based compressed graph* of G with respect to the star cover $\{S_1, S_2, \ldots, S_n\}$. In the star-based graph compression, each star in the star cover is compressed into a super-vertex in the compressed graph.

Consider the graph G shown in Fig. 1(a). The stars S_1, S_2 and S_3 constitute a star cover of G. The star-based compressed graph of G with respect to $\{S_1, S_2, S_3\}$ is shown in Fig. 1(b), where the stars are compressed into the super-vertices.

2.2 Star-Based Graph Compression Algorithm

Let G be a graph. For any star cover of G, we have a corresponding star-based compressed graph. The *optimal star-based compressed graph* should be the one with the highest compression ratio, that is, it contains the minimum number of super-vertices. Note that each super-vertex uniquely corresponds to a star in the star cover. The optimal star-based compressed graph is therefore determined by the *minimum star cover*, that is, the star cover with the minimum number of stars.

The minimum star cover of G is closely related to the minimum dominating set of G. The *dominating set* of G is a vertex subset $C \subseteq V(G)$ such that every vertex in $V(G) \setminus C$ is adjacent to at least one vertex in C. The *minimum dominating set* is the one of the minimum cardinality.

Lemma 1. *Let $\{S_1, S_2, \ldots, S_n\}$ be a star cover of a graph G. For $i = 1, 2, \ldots, n$, let s_i be the center of S_i. Then, $\{S_1, S_2, \ldots, S_n\}$ is the minimum star cover of G if and only if $\{s_1, s_2, \ldots, s_n\}$ is the minimum dominating set of G.*

Proof. First, we prove the sufficiency. Assume that $\{S'_1, S'_2, \ldots, S'_m\}$ is the minimum star cover of G, where $m < n$. For $i = 1, 2, \ldots, m$, let s'_i be the center of S'_i. By the definition of stars, s'_i dominates all the border vertices in S'_i. Thus, $\{s_1, s_2, \ldots, s_n\}$ is not the minimum dominating set of G, which is a contradiction. Hence, $\{S_1, S_2, \ldots, S_n\}$ is the minimum star cover of G.

Next, we prove the necessity. Assume that $\{s'_1, s'_2, \ldots, s'_m\}$ is the minimum dominating set of G, where $m < n$. Now, we construct a star cover of G based on $\{s'_1, s'_2, \ldots, s'_m\}$. For $i = 1, 2, \ldots, m$, let s'_i be a center vertex of a star S'_i. For all $v \in V(G) \setminus \{s'_1, s'_2, \ldots, s'_m\}$, we assign v to a star S_i if $(s_i, v) \in E(G)$. Clearly, $\{S'_1, S'_2, \ldots, S'_m\}$ is a star cover of G. Thus, $\{S_1, S_2, \ldots, S_n\}$ is not the minimum star cover of G, which is a contradiction. Hence, $\{s_1, s_2, \ldots, s_n\}$ is the minimum dominating set of G.

Thus, the lemma holds. □

Algorithm 1. MSC

Input: a graph G
Output: a star cover of G
1: $C \leftarrow \emptyset$
2: **while** $V(G) \neq \emptyset$ **do**
3: $s \leftarrow$ the vertex of the maximum degree
4: $S \leftarrow$ the star composed by s and all its neighbors in G, where s is the center
5: $C \leftarrow C \cup \{S\}$
6: delete s and all its neighbors from G
7: **return** C

Algorithm 2. StarZip

Input: a graph G
Output: a compressed graph of G
1: $V \leftarrow$ MSC(G)
2: $E \leftarrow \emptyset$
3: **for all** $S, S' \in V$ and $S \neq S'$ **do**
4: **if** there exist $v \in S$ and $v' \in S'$ such that $(v, v') \in E(G)$ **then**
5: $E \leftarrow E \cup \{(S, S')\}$
6: **return** (V, E)

By Lemma 1, we immediately have the following theorem.

Theorem 1. *Finding the minimum star cover of a graph is NP-hard.*

Proof. The minimum dominating set problem, that is, finding the minimum dominating set of a graph, is NP-hard [4]. By the proof of Lemma 1, the minimum dominating set can be constructed from the minimum star cover in polynomial time. Thus, the theorem holds. □

Since it is infeasible to exactly find the minimum star cover in polynomial time, we propose an approximation algorithm called MSC to find the minimum star cover. The MSC algorithm is developed based on the greedy heuristic minimum dominating set algorithm [5]. The pseudocode of the MSC algorithm is shown in Algorithm 1.

Theorem 2. *The MSC algorithm is* $(\ln \Delta + 2)$*-approximate, where* Δ *is maximal degree of the vertices in* G.

Proof. The minimum dominating set of a graph can be approximated within $\ln \Delta + 2$ [5]. By Lemma 1, the minimum star cover has the same cardinality as the minimum dominating set. Thus, the theorem holds. □

Based on the MSC algorithm, we propose our star-based graph compression algorithm called StarZip. The pseudocode of the StarZip algorithm is shown in Algorithm 2. The StarZip algorithm runs in $O(|V(G)| + |E(G)| \log \Delta)$ time, where Δ is the maximum vertex degree of G.

3 Query Processing on Star-Based Compressed Graphs

In this section, we show that the star-based graph compression is capable of supporting efficient query processing. Particularly, we study single-source shortest path queries on star-based compressed graphs.

3.1 Single-Source Shortest Path Queries

Now, we study how to process *single-source shortest path (SSSP) queries* on star-based compressed graphs. Let G be a graph and G^* be the star-based compressed graph of G computed by the StarZip algorithm. Given a vertex s in G as a source, the single-source shortest path query from s computes the length of the shortest paths from s to all the other vertices in G. Dijkstra's algorithm can process an SSSP query on G in $O(|E(G)| + |V(G)| \log |V(G)|)$ time, where $|V(G)|$ is the number of vertices in G, and $|E(G)|$ is the number of edges in G. Since the star-based compressed graph G^* is significantly smaller than the original graph G, we try to process an SSSP query directly on G^* to save query processing time.

To support SSSP queries on the star-based compressed graph G^*, we associate every super-edge e in G^* with three bits denoted by $b_1(e)$, $b_2(e)$ and $b_3(e)$, respectively. Let u and v be the endpoints of e. The super-vertex u represents a star S_u in G, and the super-vertex v represents a star S_v in G. We assign the bits $b_1(e)$, $b_2(e)$ and $b_3(e)$ as follows.

- $b_1(e) = 1$ if the center vertex of S_u is adjacent to a border vertex in S_v; otherwise, $b_1(e) = 0$;
- $b_2(e) = 1$ if the center vertex of S_v is adjacent to a border vertex in S_u; otherwise, $b_2(e) = 0$;
- $b_3(e) = 1$ if a border vertex in S_u is adjacent to a border vertex in S_v; otherwise, $b_3(e) = 0$.

Notably, it is impossible that the center vertices of S_u and S_v are adjacent because the StarZip algorithm must have identified one of them as a border vertex in the other star.

Given an SSSP query starting from a source vertex s in the original graph G, the SSSP query can be processed on the star-based compressed graph G^* by the StarSSSP algorithm given in Algorithm 3.

For all super-vertices w in G^* that are adjacent to v, we need to update $d[w]$. We propose three strategies to update $d[w]$.

Strategy 1: Update $d[w]$ to $\min(d[w], d[v] + 1)$.
Strategy 2: Update $d[w]$ to $\min(d[w], d[v] + 2)$.
Strategy 3: Let $e = (u, v)$.

- If $b_1(e) = 1$ or $b_2(e) = 1$, update $d[w]$ to $\min(d[w], d[u] + 2)$;
- Otherwise, update $d[w]$ to $\min(d[w], d[u] + 3)$.

The time complexity of the StarSSSP algorithm is $O(|E(G^*)| + |V(G^*)| \log |V(G^*)|)$ since Dijkstra's algorithm runs on the compressed graph G^* in $O(|E(G^*)| + |V(G^*)| \log |V(G^*)|)$ time, and our adaption to Dijkstra's algorithm in StarSSSP only adds $O(1)$ cost to each of the $|E(G^*)|$ iterations.

4 Experiments

In this section, we experimentally evaluate the star-based graph compression as well as the query processing algorithms on star-based compressed graphs.

Algorithm 3. StarSSSP

Input: a star-based compressed graph G^* of a graph G and a source vertex s
Output: the shortest distances $d^*(s, v)$ from s to all the other vertices v in G

```
1: s* ← the super-vertex in G* containing s
2: initialize d[s*]
3: Q ← V(G*)
4: while Q ≠ ∅ do
5:     v ← extract_min(Q)
6:     for all vertices u adjacent to v in G* do
7:         update d[u] by strategy 1, 2 or 3
8: for all w ∈ V(G*) do
9:     for all vertices w' in the super-vertex w do
10:        if w' is center then
11:            d*(s, w') ← d[w]
12:        else
13:            d*(s, w') ← d[w] + 1
14: return d*(s, v) for all v ∈ V(G)
```

Table 1. Statistics of the graph datasets.

Dataset	Type	# vertices	# edges	Average degree	Diameter
Youtube	Social network	1,134,890	2,987,624	5.265	20
DBLP	Collaboration network	317,080	1,049,866	6.622	21
Skitter	Autonomous system	1,696,415	11,095,298	13.081	25
LiveJournal	Social network	3,997,962	34,681,189	17.349	17
Road-PA	Road network	1,088,092	1,541,898	2.834	786
Orkut	Social network	3,072,441	117,185,083	76.281	9
R-MAT-16384	Synthetic	16,384	850,000	103.760	8
R-MAT-65536	Synthetic	65,536	10,000,000	305.176	7
R-MAT-32768	Synthetic	32,768	15,000,000	915.527	5

4.1 Experimental Setting

We implemented the star-based graph compression algorithm StarZip and the query processing algorithm StarSSSP in C++ and compiled them with g++. All the experiments were carried out on a machine with 2 GHz Intel Core 2 CPU and 22 GB of RAM running Ubuntu 14.04.

4.2 Datasets

We carried out the experiments on six real datasets obtained from the Stanford SNAP datasets [6]. In order to control the volume and the density of graphs, we generated some synthetic graph datasets using the R-MAT model [7], a scale-free graph generation model. The characteristics of the real datasets and the synthetic datasets are described in Table 1.

4.3 Performance of the Star-Based Graph Compression

First, we evaluated the performance of the star-based graph compression. Particularly, we examined the compression ratios and the degree distributions of the compressed graphs.

Table 2. Sizes and compression ratios of the star-based compressed graphs.

Dataset	$\lvert V(G^*)\rvert$	$\lvert E(G^*)\rvert$	$\frac{\lvert V(G)\rvert}{\lvert V(G^*)\rvert}$	$\frac{\lvert E(G)\rvert}{\lvert E(G^*)\rvert}$
Youtube	160,660	595,909	7.064	5.014
DBLP	60,191	207,906	5.227	5.050
Skitter	338,713	1,288,202	5.008	8.613
LiveJournal	868,088	9,246,980	4.605	3.751
Road-PA	289,769	531,716	3.755	2.900
Orkut	418,300	24,933,610	7.345	4.700
R-MAT-16384	1,431	77,105	11.449	11.108
R-MAT-65536	3,077	374,996	21.299	28.898
R-MAT-32768	717	61,823	45.701	241.588

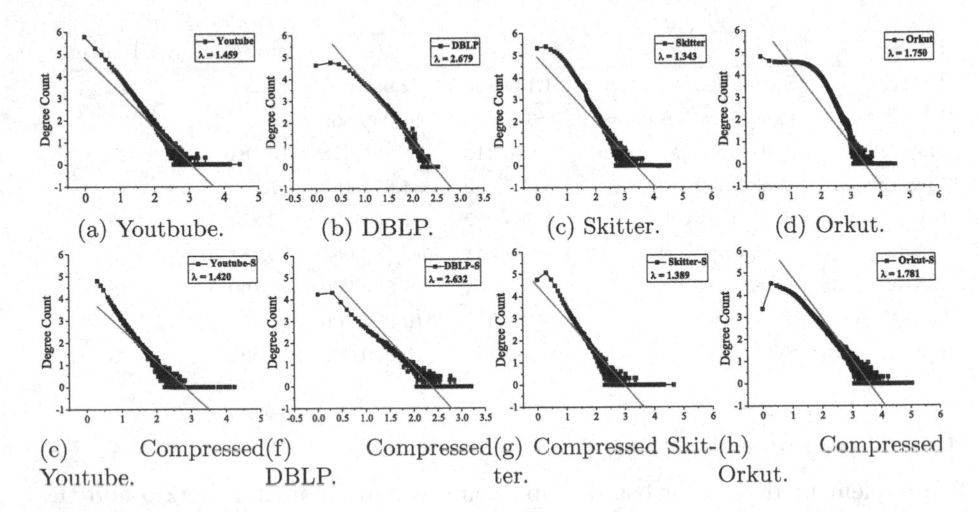

(a) Youtbube. (b) DBLP. (c) Skitter. (d) Orkut.

(c) Compressed Youtube. (f) Compressed DBLP. (g) Compressed Skitter. (h) Compressed Orkut.

Fig. 2. Degree distributions of the real graphs and their star-based compressed graphs.

Compression Ratios. Let G be a graph and G^* be the star-based compressed graph of G produced by the StarZip algorithm. The *compression ratio* is defined as the ratio of the size of G to the size of G^*. Specifically, if the graph size is measured by the number of vertices, we have the *vertex compression ratio*, that is, $|V(G)|/|V(G^*)|$; if the graph size is measured by the number of edges, we have the *edge compression ratio*, that is, $|E(G)|/|E(G^*)|$. Table 2 gives the number of vertices, the number of edges, the vertex compression ratio and the edge compression ratio of each star-based compressed graph returned by the StarZip algorithm. On real graphs, the vertex compression ratio varies from 3.8 to 7.3, and the edge compression ratio varies from 2.9 to 8.6. On synthetic graphs, the vertex compression ratio varies from 11.4 to 45.7, and the edge compression ratio varies from 11.1 to 241.6.

The *correlation coefficient* [8] between the vertex compression ratio and the average vertex degree of the input graph is 0.994, and the correlation coefficient between the edge compression ratio and the average degree is 0.973. Thus, the compression ratio is *positively correlated* with the average degree of the input graph. The denser a graph is, the higher the compression ratio is.

Degree Distributions. A large number of graphs in the real worlds have been shown to be *power-law* graphs, that is, the vertex degrees in a graph follows a power-law distribution. All the real graphs used in our experiments are power-law graphs.

Figure 2 plots the degree distributions of the real graphs before and after compression. The points are plotted in log-log scale. We can see that both the original graph and the star-based compressed graphs follow power-law degree distributions. In Fig. 2, we also give the power law exponents. We can see that the power law exponents of the original graph and the compressed graph are very close.

4.4 Query Processing Performance on Star-Based Compressed Graphs

After examining the performance of the star-based graph compression itself, we evaluated the performance of the query processing algorithms on the star-based compressed graphs.

Efficiency of Shortest Path Query Processing. To evaluate the improvement in query processing efficiency, we use Dijkstra's algorithm running on G as the baseline. For all experimented graphs G, we select $|V(G)|/10$ source vertices uniformly at random and compose $|V(G)|/10$ shortest path queries. For each query, we ran Dijkstra's algorithm on G and ran the StarSSSP algorithm with distance updating strategy 2 on the compressed graph G^*.

Table 3 shows the *speedup ratio*, that is, the ratio of the average execution time of Dijkstra's algorithm to that of the StarSSSP algorithm. We can see that the StarSSSP is 4–20 times faster than Dijkstra's algorithm on the real graphs and is 22–313 times faster on synthetic graphs. It verifies that the StarSSSP algorithm is much more efficient than Dijkstra's algorithm running on the original graphs. Besides, the denser the original graph is, the more efficient StarSSSP is.

Accuracy of Shortest Path Query Processing. The StarSSSP algorithm is an approximate query processing algorithm. Depending on the strategy that the StarSSSP algorithm uses to update distances, the StarSSSP algorithm is able to return lower bounds or upper bounds of the shortest distances from the source vertex to all the other vertices.

To evaluate the accuracy of the StarSSSP algorithm, we measure the *accuracy rate*, the *average absolute error* and the *average relative error* of the query

Table 3. Executing time (s) and Speedup ratios of the StarSSSP algorithm against Dijkstra's algorithm and accuracy rate (A.R.), average absolute error (A.A.E.) and average relative error (A.R.E.) of query results.

Dataset	Dijkstra	StarSSSP	Speedup	A.R.	A.A.E.	A.R.E.
Youtube	2.554	0.124	20.630	0.169	1.424	0.289
DBLP	0.661	0.040	16.408	0.265	1.064	0.164
Skitter	4.495	0.349	12.871	0.290	1.044	0.229
LiveJournal	13.476	3.121	4.318	0.209	1.021	0.207
Road-PA	1.599	0.090	17.774	0.134	18.122	0.081
Orkut	22.097	4.084	5.412	0.302	0.919	0.274
R-MAT-16384	0.104	0.005	22.882	0.365	0.738	0.336
R-MAT-65536	1.248	0.021	58.878	0.238	0.934	0.374
R-MAT-32768	1.601	0.005	313.922	0.540	1.016	0.516

results. Let G be a graph and G^* be the star-based compressed graph of G computed by the StarZip algorithm. For two vertices s and t in G, let $d(s,t)$ be the shortest distance from s to t in G, and let $d^*(s,t)$ be the approximate shortest distance from s to t computed on G^* by the StarSSSP algorithm using distance updating strategy 2. The absolute error between $d(s,t)$ and $d^*(s,t)$ is $|d(s,t) - d^*(s,t)|$, and the relative error between $d(s,t)$ and $d^*(s,t)$ is $|d(s,t) - d^*(s,t)|/d(s,t)$.

Table 3 shows the accuracy rate, the average absolute error and the average relative error of the query results obtained on all experimented graphs. As we can see, the accuracy rate varies from 13.4% to 54%, the average absolute errors are all about 1 except the one on the Road-PA dataset, and the average relative error varies from 8.1% to 49.5%. Note that Road-PA is a road network, which is very sparse. The diameter of Road-PA is 786, and the shortest distances between vertices are generally large. Although the average absolute error on the Road-PA dataset is 18.122, the average relative error is just 8.1%. The experimental results verify that the StarSSSP algorithm is accurate enough in processing shortest path queries.

5 Related Work

Graph compression has been studied for about four decades. Considerable graph compression algorithms have been proposed to compress graphs collected in a variety of applications. Here, we list some related works based on the literal conceptions similar to the **graph compression** in our paper.

Graph Aggregation and Graph Summarization. Graph aggregation and summarization produce small and informative summarization of the original graph to help understand the underlying characteristics of large graphs. *k-SNAP* [9,10] produce summary graph based on the vertex attributes and relationships. It puts some vertices into a vertex with rules that users select or are defined in advance. Navlakha et al. [11] summary unlabelled graphs using Rissanen's Minimum Description Length (MDL) principle. It defines the quality of a graph summary G^* by $cost(G^*)$. This method finds the optimal graph representation by minimizing $cost(G^*)$. But, it becomes difficult when somebody wants to do some queries or operations like the shortest path between two nodes, the cut vertices of graph, etc. In essence, these graph aggregation and summarization algorithms are similar to graph clustering algorithms. They are unable to support queries on compressed graphs without decompression.

Graph Simplification. Ruan et al. [12] simplify a graph by using the concept "gate graph" to preserve the distance of original graph. Then the shortest-path distance between any "non-local" pair can be recovered by consecutive "local" walks through the gate vertices in the gate graph. As we test, the time of compressing a graph with 15,000 edges is more than 2 h, while StarZip only needs 1105 s to compress a graph with 117 million edges. Besides, the accuracy of approximate distance computed through gate graph can not be guaranteed. Bonchi et al. [13] simplify a graph by selecting a subset of arcs in the graphs to maximize the number of nodes reachable in all directed acyclic graphs through some specified root vertices. This method preserves the property of activity of the graph while the graph it gets doesn't support any queries.

Graph Compaction and Graph Partition. The graph compaction here is same to the concept of graph compression in our paper which reduces the scale of graphs and the compact(compressed) graphs can support many operations on graphs. The graph partition is always used in parallel graph processing system [14,15]. It breaks the graph into some small parts to distribute them on different machines to minimize the communication cost between different machines.

Graph Compression. The graph compression method can be applied in several fields. Boldi and Vigna [16] stores the Web graph in adjacency lists. They use multiple lists to copy one list by leveraging locality and similarity. The multiple lists record the list they copy by 0 and 1 sequence. Alder and Mitzenmacher [17] construct the minimum spanning tree to compress the randomly generated Web graph. Different from the studies above, Apostolico and Drovandi [18] make no use of locality and similarity. They compress the Web graph by breath-first search (BFS). To facilitate graph decompression, type labels are used to remember the types of the compressed blocks. In summary, the Web graph compression algorithms in [16] do not apply to graphs in other applications because those graphs usually do not have the locality or the similarity characteristics. Besides, some Web compression algorithms [16,18] just encode the adjacency list to reduce storage space without supporting any queries without decompression. Fan et al. [19] proposes two compression methods on labelled directed graph based on reachability and graph pattern queries. They can get results quickly on compressed graph. But all the methods above aim at weight graph, which are useless when graph is unlabelled.

6 Conclusions

This paper gives a formal definition of the star-based graph compression. We show that finding the optimal star-based compressed graph is an NP-complete problem. The StarZip algorithm uses a greedy compression strategy and achieves an approximation ratio of $\ln \Delta + 2$, where Δ is the maximum vertex degree. In practice, StarZip also achieves impressive compression ratios, which are positively correlated with average vertex degrees. Star-based compressed graphs preserve the distributions of vertex degrees of original graphs. The query results returned by the StarSSSP algorithm on star-based compressed graphs well approximate the exact query results on original graphs. StarSSSP is 4X–313X faster than Dijkstra's algorithm running on original graphs.

Acknowledgements. This work was partially supported by the National Natural Science Foundation of China (No. 61532015, No. 61672189, No. 61732003 and No. 61872106) and the National Science Foundation of USA (No. 1741277 and No. 1829674).

References

1. Leskovec, J., Faloutsos, C.: Sampling from large graphs. In: KDD, pp. 631–636 (2006)
2. Feder, T., Motwani, R.: Clique partitions, graph compression and speeding-up algorithms. J. Comput. Syst. Sci. **51**(2), 261–272 (1995)
3. Toivonen, H., Zhou, F., Hartikainen, A., Hinkka, A.: Compression of weighted graphs. In: KDD, pp. 965–973 (2011)
4. Chvatal, V.: A greedy heuristic for the set-covering problem. Math. Oper. Res. **4**(3), 233–235 (1979)
5. Ruan, L., Du, H., Jia, X., Wu, W., Li, Y., Ko, K.I.: A greedy approximation for minimum connected dominating sets. Theoret. Comput. Sci. **329**(1–3), 325–330 (2004)
6. Leskovec, J., Krevl, A.: SNAP datasets: Stanford large network dataset collection, June 2014. http://snap.stanford.edu/data
7. Chakrabarti, D., Zhan, Y., Faloutsos, C.: R-MAT: a recursive model for graph mining. In: SDM, vol. 4, pp. 442–446 (2004)
8. Li, L.: A concordance correlation coefficient to evaluate reproducibility. Biometrics **45**(1), 255–268 (1989)
9. Tian, Y., Hankins, R.A., Patel, J.M.: Efficient aggregation for graph summarization. In: SIGMOD, pp. 567–580 (2008)
10. Zhang, N., Tian, Y., Patel, J.M.: Discovery-driven graph summarization. In: ICDE, pp. 880–891 (2010)
11. Navlakha, S., Rastogi, R., Shrivastava, N.: Graph summarization with bounded error. In: SIGMOD, pp. 419–432 (2008)
12. Ruan, N., Jin, R., Huang, Y.: Distance preserving graph simplification. In: ICDM, pp. 1200–1205 (2011)

13. Bonchi, F., Morales, G.D.F., Gionis, A., Ukkonen, A.: Activity preserving graph simplification. Data Min. Knowl. Disc. **27**(3), 321–343 (2013)
14. Gonzalez, J.E., Low, Y., Gu, H., Bickson, D., Guestrin, C.: PowerGraph: distributed graph-parallel computation on natural graphs. In: OSDI, pp. 17–30 (2012)
15. Shao, Y., Cui, B., Ma, L.: PAGE: a partition aware engine for parallel graph computation. IEEE Trans. Knowl. Data Eng. **27**(2), 518–530 (2015)
16. Boldi, P., Vigna, S.: The webgraph framework I: compression techniques. In: WWW, pp. 595–601 (2004)
17. Adler, M., Mitzenmacher, M.: Towards compressing web graphs. In: DCC, pp. 203–212 (2001)
18. Apostolico, A., Drovandi, G.: Graph compression by BFS. Algorithms **2**(3), 1031–1044 (2009)
19. Fan, W., Li, J., Wang, X., Wu, Y.: Query preserving graph compression. In: SIGMOD, pp. 157–168 (2012)

Neighbor-Based Link Prediction with Edge Uncertainty

Chi Zhang(✉) and Osmar R. Zaïane

Department of Computing Science, University of Alberta,
Edmonton, AB, Canada
{chi7,zaiane}@ualberta.ca

Abstract. In this work, we are concerned with uncertain networks and focus on the problem of link prediction with *edge uncertainty*. Networks with edge uncertainty are networks where connections between nodes are observed with some probability. We propose the uncertain version of the popular neighbors-based metrics for link prediction. The metrics are developed by considering all possible worlds generated by the uncertain network. We state that by taking all possible worlds of the uncertain network into account, the performance of link prediction can be improved. Since uncertain edges result in a very large number of possible worlds, we propose an efficient divide and conquer algorithm to reduce time complexity and calculate these metrics. Finally, we evaluate our metrics using existing ground truth to show the effectiveness of our proposed approach against other popular link prediction methods.

Keywords: Social network analysis · Link prediction · Uncertain networks

1 Introduction

Link prediction is the problem of determining future or missing associations between entities in networks based on observed links. Because of its broad applications in different domains, link prediction has attracted increasing attention.

In the past decade, many works have been done about link prediction in deterministic graphs, graphs where the network structure is exactly and deterministically known. There are many metrics available for computing the similarity of two nodes. Among all approaches, neighbor-based metrics [1–5] are effective and the simplest way to predict missing links. The other metrics include path-based metrics [6], random-walk-based metrics [7]. Furthermore, there are some learning-based methods [8] and embedding-based methods [9] that have been proposed in recent years.

Most previous studies on link prediction have focused on networks where the structure is exactly known. With the increasing number of applications in which the edges are constructed in the network through uncertain or statistical inference, the problem of link prediction with edge uncertainty has become

Q. Yang et al. (Eds.): PAKDD 2019, LNAI 11440, pp. 462–474, 2019.
https://doi.org/10.1007/978-3-030-16145-3_36

increasingly important. Examples of such networks include protein-protein interaction networks with experimentally inferred links, sensor networks with uncertain connectivity links, or social networks, which are augmented with inferred friendship, similarity, or trust links. However, only few studies take probabilities into consideration. Ahmed et al. [10] proposed the uncertain version of the random walk method for link prediction with edge uncertainty. Mallek et al. [11] put forward an approach combined sampling techniques and information fusion and obtained good results in real-life settings. Up to now, the uncertain version of the popular neighbor-based metrics have not been studied. Murata and Moriyasu [12] proposed weighted similarity indices, including variants of some popular neighbor-based metrics. People may regard probabilities as weights and apply weighted variants of those metrics; however, it may lead to some problems. More details are presented in Sect. 4.

The uncertain scenario will make the problem of link prediction become more complex, and the uncertain version of the most basic neighbor-based methods are not yet studied. Therefore, in this work, we mainly focus on using neighbor-based algorithms to solve the problem of link prediction in the context of uncertain networks. We propose the uncertain version of the popular neighbors-based metrics and efficient algorithms to calculate them. The remainder of this paper is organized as follows. In Sect. 2, we provide the problem definition. In Sect. 3, we review related work. In Sect. 4, we show the limitation of previous work, propose the uncertain version of common-neighbors-based metrics and efficient algorithms to produce them. In Sect. 5, we present the experiment results and our evaluation metric. Finally, we conclude in Sect. 6.

2 Problem Definition

2.1 Uncertain Network

An uncertain graph $\mathcal{G} = (\mathcal{V}, \mathcal{E}, \mathcal{P})$ is defined over a set of nodes $\mathcal{V}$, a set of edges $\mathcal{E}$, and a set of probabilities $\mathcal{P}$ of edge existence. Note the probability over the edge between node $\mathcal{V}_i$ and node $\mathcal{V}_j$ can be represented as $\mathcal{P}_{i,j}$ or $\mathcal{P}_{j,i}$. The multiple links and self-connections are not allowed.

2.2 Link Prediction Problem Definition

The task of link prediction is to discover missing, hidden or future associations between two nodes. Given a network and two unconnected nodes $\mathcal{V}_x$ and $\mathcal{V}_y \in \mathcal{V}$, link prediction is to predict the probability of the existence of a link between the node $\mathcal{V}_x$ and the node $\mathcal{V}_y$. To do this, for each pair of nodes, $\mathcal{V}_x, \mathcal{V}_y \in \mathcal{V}$, which are not directly connected, we assign a score, s_{xy}, according to a given similarity measure. A higher score means nodes $\mathcal{V}_x$ and $\mathcal{V}_y$ are more likely to have an edge. All the nonexistent links are sorted in a descending order according to their scores, and the links at the top are most likely to exist.

Generally, we do not know which links are the missing or future links, otherwise we do not need to do predictions. Therefore, to evaluate algorithms, we use known networks, hide some links, use link prediction algorithms to predict those hidden links and compare the prediction results. Based on the type of network, the observed edges E can be divided into training set E^T and probe set E^P randomly or according to the timestamp. If the known network is time-varying and we know the time each change happens, we can regard the network before a certain time as the training set and the remaining as the probe set. Otherwise, we can just divide the training set and the probe set randomly. To quantify the accuracy of prediction algorithms, we use Precision as our evaluation metric. More experiment details can be found in Sect. 5.

3 Previous Work

As mentioned above, neighbor-based metrics are the simplest yet effective to predict missing links. They assume that two nodes are more likely to be connected if they have more common neighbors. Common neighbors (CN) is one of the most widespread measure used in the link prediction problem mainly due to its simplicity [1]. The Resource Allocation (RA) metric [5] is regarded as one of the best neighbor-based metrics because of its performance. Therefore, in this paper, we concentrate on CN and RA indexes, whose definitions are as follows.

Common Neighbors (CN): Two nodes, $\mathcal{V}_x$ and $\mathcal{V}_y$, are more likely to form a link if they have many common neighbors. Let $\Gamma(x)$ denote the set of neighbors of node $\mathcal{V}_x$. The simplest measure of the neighborhood overlap is the directed count:

$$s_{xy} = |\Gamma(x) \cap \Gamma(y)| \tag{1}$$

Resource Allocation (RA): Considering a pair of nodes, $\mathcal{V}_x$ and $\mathcal{V}_y$, which are not directly connected. The node $\mathcal{V}_x$ can send some resource to $\mathcal{V}_y$, with their common neighbors playing the role of transmitters. In the simplest case, we assume that each transmitter has a unit of resource, and will evenly distribute to all its neighbors. As a results the amount of resource $\mathcal{V}_y$ received is defined as the similarity between $\mathcal{V}_x$ and $\mathcal{V}_y$, which is:

$$s_{xy} = \sum_{z \in \Gamma(x) \cap \Gamma(y)} \frac{1}{k(z)} \tag{2}$$

where $k(z)$ is the degree of node $\mathcal{V}_z$, namely $k(z) = |\Gamma(z)|$.

The above-mentioned similarity metrics, CN and RA, only consider the binary relations among nodes; however, in the real world, links are naturally weighted, which may represent the amount of traffic load along connections in a transportation network or the number of co-authorized papers in a co-authorship network. Murata and Moriyasu [12] proposed weighted similarity metrics as variants of Common Neighbors and Resource Allocation:

$$\text{Weighted Common Neighbors: } s_{xy} = \sum_{z \in \Gamma(x) \cap \Gamma(y)} w(x,z) + w(y,z) \tag{3}$$

$$\text{Weighted Resource Allocation: } s_{xy} = \sum_{z \in \Gamma(x) \cap \Gamma(y)} \frac{w(x,z) + w(y,z)}{s(z)} \quad (4)$$

Here, $w(x, y) = w(y, x)$ denotes the weight of the link between nodes $\mathcal{V}_x$ and $\mathcal{V}_y$, and $s(x) = \sum_{z \in \Gamma(x)} w(x, z)$ is the strength of node $\mathcal{V}_x$.

Besides the aforementioned metrics, we will also use the Local Naïve Bayes model [8] and the Local Random Walk metric [7] to assess our metrics as they are popular. The Local Naïve Bayes model is considered as the state-of-the-art in neighbor-based link prediction algorithms. Due to limited space, we will not cover them in details here.

4 Link Prediction for Uncertain Graphs

To solve the problem of link prediction for uncertain graphs, one very naïve/intuitive way is to regard the probability as a weight and apply weighted similarity metrics. However, there exists some problems. Figure 1 is an example.

Nodes $\mathcal{V}_A$ and $\mathcal{V}_B$ are more likely to be connected than nodes $\mathcal{V}_D$ and $\mathcal{V}_E$ based on Eq. (3) for Weighted Common Neighbors.

$$s_{AB} = 0.2 + 0.9 = 1.1 \quad (5)$$

$$s_{DE} = 0.5 + 0.5 = 1.0 < 1.1 \quad (6)$$

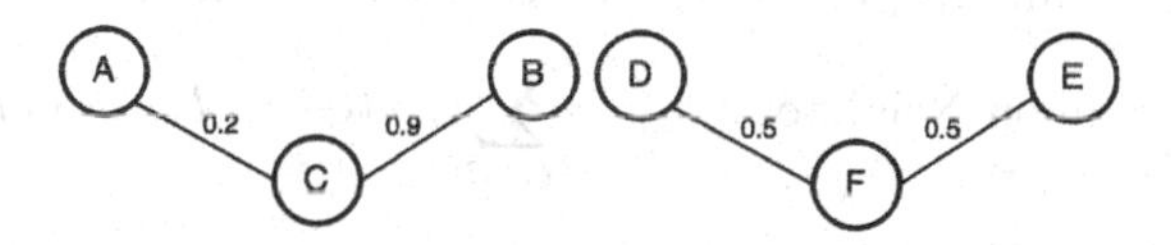

Fig. 1. An example showing the problem when considering the probability as a weight.

However, because each edge may exist or not exist in the real world, both of these two uncertain graphs have four possible worlds, as can be seen in Fig. 2: both links may exist, both may be absent, or either one is present.

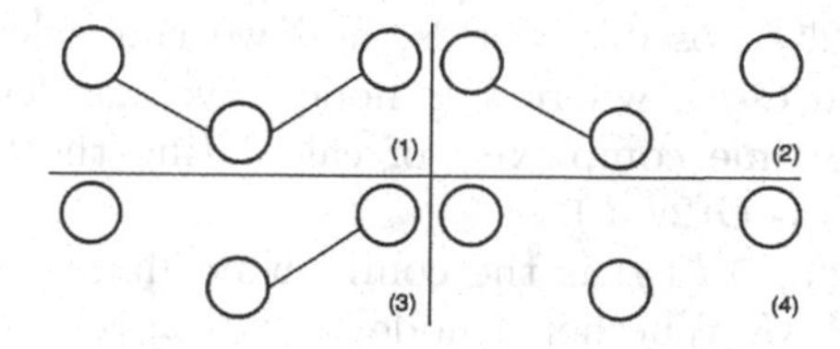

Fig. 2. Possible worlds for two uncertain links between three nodes

Only when both edges $\mathcal{E}_{AC}$ and $\mathcal{E}_{BC}$ exist, node $\mathcal{V}_C$ is the common neighbor of nodes $\mathcal{V}_A$ and $\mathcal{V}_B$, as Fig. 2(1), the probability for this case is $0.2 \times 0.9 = 0.18$ (we assume that the existence of edges are independent with each other). In this case, $s_{AB} = 1$ based on Eq. (1). If node $\mathcal{V}_C$ is not the common neighbor of nodes $\mathcal{V}_A$ and $\mathcal{V}_B$, as Fig. 2(2, 3 and 4), then $s_{AB} = 0$. The probability for this case is 0.82. In comparison, the probability that $s_{DE} = 1$ is $0.5 \times 0.5 = 0.25$, while the probability of $s_{DE} = 0$ is 0.75. Therefore, nodes $\mathcal{V}_D$ and $\mathcal{V}_E$ are more likely to be connected than nodes $\mathcal{V}_A$ and $\mathcal{V}_B$, because the probability of $s_{DE} = 1$ is larger than the probability of $s_{AB} = 1$.

From this example, we can find that each uncertain edge in an uncertain graph may exist or not exist in a real world. If an uncertain graph has $|\mathcal{E}|$ uncertain edges, there will be $2^{|\mathcal{E}|}$ possible worlds in total, since each edge provides us with a binary sampling decision.

Given an uncertain network $\mathcal{G} = (\mathcal{V}, \mathcal{E}, \mathcal{P})$, we can sample each edge in $\mathcal{G}$ according to the probability $\mathcal{P}(e)$ to generate the possible graph $G = (V_G, E_G)$. We have $E_G \in \mathcal{E}$ and $V_G \in \mathcal{V}$. The probability $Pr(G)$ of sampling the possible graph is as follows:

$$Pr(G) = \prod_{e \in E_G} \mathcal{P}(e) \prod_{e \in \mathcal{E}, e \notin E_G} (1 - \mathcal{P}(e)) \tag{7}$$

For each possible world, its corresponding similarity measure may differ. When we calculate its similarity measures, we should take all possible worlds and their possibilities into account. Therefore, Common Neighbor and Resource Allocation in uncertain graphs can be represented as follows.

$$\text{Uncertain Common Neighbors: } s_{xy} = \sum_{G \in \mathcal{G}} (Pr(G) \times |\Gamma_G(x) \cap \Gamma_G(y)|) \tag{8}$$

$$\text{Uncertain Resource Allocation: } s_{xy} = \sum_{G \in \mathcal{G}} (Pr(G) \sum_{z \in \Gamma_G(x) \cap \Gamma_G(y)} \frac{1}{k_G(z)}) \tag{9}$$

Here, $\Gamma_G(x)$ denotes the set of neighbors of node $\mathcal{V}_x$ in the possible world G; $k_G(x)$ is the degree of node $\mathcal{V}_x$ in the possible world G.

4.1 Time Complexity Analysis for the Calculation of Common Neighbors in Uncertain Networks

We have a total of $2^{|\mathcal{E}|}$ possible worlds, and we can calculate CN value for each possible world in $O(k)$, where k is nodes' average degree in the possible world. Therefore, the time complexity of calculating the Common Neighbors value based on Eq. (8) is $O(2^{|\mathcal{E}|}k)$.

Assume $\Gamma_{xy} = \Gamma(x) \cap \Gamma(y)$ is the common neighbors set of nodes $\mathcal{V}_x$ and $\mathcal{V}_y$ in uncertain graph $\mathcal{G}$. Whether a node $\mathcal{V}_z \in \Gamma_{xy}$ is a common neighbor of nodes $\mathcal{V}_x$ and $\mathcal{V}_y$ in a possible world is independent of other nodes because it is determined by the existence of edges $\mathcal{E}_{xz}$ and $\mathcal{E}_{yz}$ in the possible world. Therefore, each node in Γ_{xy} can be considered independently. If the existence probability over uncertain edges $\mathcal{E}_{xz}$ and $\mathcal{E}_{yz}$ are $\mathcal{P}_{x,z}$ and $\mathcal{P}_{y,z}$ respectively, only in $\mathcal{P}_{x,z} \times \mathcal{P}_{y,z}$ of all possible worlds, node $\mathcal{V}_z$ is the common neighbor of nodes $\mathcal{V}_x$ and $\mathcal{V}_y$. Therefore, Eq. (8) can also be represented as:

$$\begin{aligned} s_{xy} &= \sum_{G \in \mathcal{G}} (Pr(G) \times |\Gamma_G(x) \cap \Gamma_G(y)|) \\ &= \sum_{z \in \Gamma(x) \cap \Gamma(y)} \sum_{G \in \mathcal{G}} Pr(G) \times \mathbb{I}_{\Gamma_G(x) \cap \Gamma_G(y)}(z) \\ &= \sum_{z \in \Gamma(x) \cap \Gamma(y)} \mathcal{P}_{x,z} \times \mathcal{P}_{y,z} \end{aligned}$$

When $z \in \Gamma_G(x) \cap \Gamma_G(y)$, $\mathbb{I}_{\Gamma_G(x) \cap \Gamma_G(y)}(z) = 1$, otherwise, $\mathbb{I}_{\Gamma_G(x) \cap \Gamma_G(y)}(z) = 0$.

By doing so, the time complexity for calculating s_{xy} can be reduced to $O(K)$, where K is the nodes' average degree in the uncertain network.

4.2 Time Complexity Analysis for the Calculation of Resource Allocation in Uncertain Networks

We have a total of $2^{|\mathcal{E}|}$ possible worlds, and nodes' average degree in the possible world is k, then we can calculate RA value for each possible world in $O(k)$, so the time complexity of calculating Resource Allocation value based on Eq. (9) is $O(2^{|\mathcal{E}|}k)$.

As mentioned in Sect. 4.1, whether a node $\mathcal{V}_z \in \Gamma_{xy}$ is a common neighbor of nodes $\mathcal{V}_x$ and $\mathcal{V}_y$ in a possible world is independent of other nodes. Besides, the number of edges each common neighbor has is also independent of other nodes. Therefore, each common neighbor can also be considered independently in this case. For the common neighbor node $\mathcal{V}_z$, when we generate possible worlds, we can consider only edges connecting to it, because the existence of other edges will not have an impact on $\mathbb{I}_{\Gamma_G(x) \cap \Gamma_G(y)}(z)$ and $k_G(z)$. The nodes' average degree in the uncertain network is K, so we can consider 2^K possible worlds for the node $\mathcal{V}_z$, and the time complexity can be reduced to $O(2^K t)$, where $t = |\Gamma_G(x) \cap \Gamma_G(y)|$.

$$\begin{aligned} s_{xy} &= \sum_{G \in \mathcal{G}} (Pr(G) \times \sum_{z \in \Gamma_G(x) \cap \Gamma_G(y)} \frac{1}{k_G(z)}) \\ &= \sum_{z \in \Gamma(x) \cap \Gamma(y)} \sum_{G_z \in \mathcal{G}_z} Pr(G_z) \times \mathbb{I}_{\Gamma_{G_z}(x) \cap \Gamma_{G_z}(y)}(z) \times \frac{1}{k_{G_z}(z)} \end{aligned}$$

$\mathcal{G}_z$ here stands for the uncertain sub-graph formed by edges connecting to node $\mathcal{V}_z$, and G_z is the possible world based on the uncertain sub-graph $\mathcal{G}_z$.

4.3 An Efficient Algorithm for the Calculation of Resource Allocation

Only when both edges $\mathcal{E}_{xz}$ and $\mathcal{E}_{yz}$ exist, node $\mathcal{V}_z$ is the common neighbor of node $\mathcal{V}_x$ and node $\mathcal{V}_y$ in the possible world G, which means $\mathbb{I}_{\Gamma_G(x)\cap\Gamma_G(y)}(z) = 1$. When node $\mathcal{V}_z$ is not the common neighbor of node $\mathcal{V}_x$ and node $\mathcal{V}_y$, $\mathbb{I}_{\Gamma_G(x)\cap\Gamma_G(y)}(z) = 0$, it means those possible worlds will not have an impact on the value of s_{xy}. Edges $\mathcal{E}_{xz}$ and $\mathcal{E}_{yz}$ belong to the edge set which connects to node $\mathcal{V}_z$, so for those possible worlds which have an impact on the value of s_{xy}, node $\mathcal{V}_z$ at least has two edges $\mathcal{E}_{xz}$ and $\mathcal{E}_{yz}$.

Assume node $\mathcal{V}_z$ has m extra edges in an uncertain graph except edges $\mathcal{E}_{xz}$ and $\mathcal{E}_{yz}$. Although it will result in 2^m possible worlds, the number of its edges in possible worlds will only range from 0 to m (the number of edges node $\mathcal{V}_z$ has in total ranges from 2 to $m+2$), which means some of the possible worlds share the same number of edges. To calculate s_{xy}, one way is to iterate through all possible worlds, calculate each possible world's possibility based on Eq. (7) and its corresponding count of edges. The other way is to iterate through all the possible number of edges and calculate their corresponding probability, which can be seen as follows:

$$\begin{aligned} s_{xy} &= \sum_{z\in\Gamma(x)\cap\Gamma(y)} \sum_{G_z\in\mathcal{G}_z} Pr(G_z) \times \mathbb{I}_{\Gamma_{G_z}(x)\cap\Gamma_{G_z}(y)}(z) \times \frac{1}{k_{G_z}(z)} \\ &= \sum_{z\in\Gamma(x)\cap\Gamma(y)} \mathcal{P}_{x,z} \times \mathcal{P}_{y,z} \times \sum_{n=0}^{m} (P^n_{1\to m} \times \frac{1}{n+2}) \end{aligned}$$

For the common neighbor $\mathcal{V}_z$, assume there are m edges connecting to it except edges $\mathcal{E}_{xz}$ and $\mathcal{E}_{yz}$, so we can index them from 1 to m. $P^n_{1\to m}$ here stands for from edges e_1 to e_m, the probability that exactly n among them exist in possible worlds. For the node with m edges in the uncertain graph, the number of its edges in possible worlds will range from 0 to m, and in other words, we need to compute $P^0_{1\to m}, P^1_{1\to m}, \ldots, P^m_{1\to m}$.

We propose an efficient way to compute them, which can be regarded as a divide and conquer algorithm. Conceptually, it works as follows:

(1) Divide the probability list into n sublists, each containing 1 element, and compute the probability of having and not having this item respectively.
(2) Repeatedly merge sublists to compute probabilities for sublists with more than 1 element. Here is the equation for merging the left half sublist and the right half sublist.

$$P^n_{1\to m} = \sum_{i=max(0,n-\lceil m/2\rceil)}^{min(n,\lfloor m/2\rfloor)} P^i_{1\to\lfloor m/2\rfloor} P^{n-i}_{\lfloor m/2\rfloor+1\to m} \tag{10}$$

It can be implemented recursively. The result probability list has the length of $m+1$ and $P^0_{1\to m}, P^1_{1\to m}, \ldots, P^m_{1\to m}$ are saved sequentially in the result probability list. The full algorithm description can be found in Algorithm 1.

Algorithm 1. kEdgeProbability

```
   Data: Probability List uncertainEdgeList
   Result: The probability list probList of existing n among m edges,
           n ∈ [0, m]
 1 uncertainEdgeListLength ← len(uncertainEdgeList);
 2 return kEdge(0, uncertainEdgeListLength − 1);
 3 // Inner Function;
 4 Function kEdge(i, j)
 5   length ← j − i + 1;
 6   if length = 1 then
 7     return [1 − uncertainEdgeList[i], uncertainEdgeList[i]]
 8   else
 9     leftLength ← length//2;
10     rightLength ← length − leftLength;
11     left ← kEdge(i, i + leftLength − 1);
12     right ← kEdge(i + leftLength, j);
13     probList ← [0] × (length + 1);
14     for each n ∈ [0, length] do
15       for each k ∈ [0, n] do
16         if k <= leftLength and n − k <= rightLength then
17           probList[n] ← probList[n] + left[k] × right[n − k];
18         end
19       end
20     end
21     return probList;
22   end
```

Based on the description of Algorithm 1, we can find the time complexity of Algorithm 1 is $O(m^2)$. After calculating the probability list, we can easily calculate node $\mathcal{V}_z$'s contribution for s_{xy}. It is reasonable to calculate $\mathcal{V}_z$'s contribution for s_{xy} in $O(m^2)$. However, because the node $\mathcal{V}_z$ has $(m+2)$ neighbors in total, then any two of these neighbors (except those that are already connected, assume u of them are already connected) will regard the node $\mathcal{V}_z$ as a common neighbor when calculating their similarity measures. Then node $\mathcal{V}_z$ will be calculated $(\frac{(m+2)(m+1)}{2} - u)$ times, so the total time complexity will be $O(m^4)$.

This kind of time complexity is still very large. We can use the similar idea as we mentioned in Algorithm 1 to reduce the time complexity. In Algorithm 1, we use the probability lists of the left half sublist and the right half list to compute the probability list of the full list. Actually, Eq. (10) has a more general form, which can be represented as follow:

$$P^n_{1\to m} = \sum_{i=max(0,n+k-m)}^{min(n,k)} P^i_{1\to k} P^{n-i}_{k+1\to m} \tag{11}$$

In Eq. (10), we choose $k = \lfloor m/2 \rfloor$.

When we consider different pairs of unconnected nodes, node $\mathcal{V}_z$'s total edges remain the same, what differs is the set of two edges which connects to the pair of nodes we are considering, and it results in the difference of the remaining edges list which will be used in Algorithm 1. To reduce the time complexity, the idea is to calculate the full edges list's corresponding probability list, which can be represented as $A = [P^0_{1\to m+2}, P^1_{1\to m+2}, \cdots, P^{m+2}_{1\to m+2}]$. For each pair of unconnected nodes, we want to calculate the remaining edges list's corresponding probability list, which can be represented as $B = [P^0_{1\to m}, P^1_{1\to m}, \cdots, P^m_{1\to m}]$. We can firstly find the two edges connecting to the pair of unconnected nodes, and calculate these two edges' corresponding probability list, which can be represented as $C = [P^0_{m+1\to m+2}, P^1_{m+1\to m+2}, P^2_{m+1\to m+2}]$. Then we can use A and C to calculate B based on the Eq. (11). The full equations can be represented as follows:

$$\begin{cases} P^2_{m+1\to m+2}P^m_{1\to m} = P^{m+2}_{1\to m+2} \\ P^1_{m+1\to m+2}P^m_{1\to m} + P^2_{m+1\to m+2}P^{m-1}_{1\to m} = P^{m+1}_{1\to m+2} \\ P^0_{m+1\to m+2}P^m_{1\to m} + P^1_{m+1\to m+2}P^{m-1}_{1\to m} + P^2_{m+1\to m+2}P^{m-2}_{1\to m} = P^m_{1\to m+2} \\ P^0_{m+1\to m+2}P^{m-1}_{1\to m} + P^1_{m+1\to m+2}P^{m-2}_{1\to m} + P^2_{m+1\to m+2}P^{m-3}_{1\to m} = P^{m-1}_{1\to m+2} \\ \cdots \\ P^0_{m+1\to m+2}P^3_{1\to m} + P^1_{m+1\to m+2}P^2_{1\to m} + P^2_{m+1\to m+2}P^1_{1\to m} = P^3_{1\to m+2} \\ P^0_{m+1\to m+2}P^2_{1\to m} + P^1_{m+1\to m+2}P^1_{1\to m} + P^2_{m+1\to m+2}P^0_{1\to m} = P^2_{1\to m+2} \end{cases}$$

These equations are easy to solve. After we get A and C, we can calculate the probability list $[P^0_{1\to m}, P^1_{1\to m}, \cdots, P^m_{1\to m}]$ in $O(m)$. Though it takes $O(m^2)$ time to calculate A, when we consider different pairs of unconnected nodes which have common neighbor $\mathcal{V}_z$, A only needs to be calculated once. To calculate different pairs of unconnected nodes' corresponding probability list B, we can calculate their probability C in constant time, and then use A and C to calculate B in $O(m)$. Because we have $(\frac{(m+2)(m+1)}{2} - u)$ pairs of unconnected nodes, the time complexity of calculating A can be ignored. After we calculate nodes $\mathcal{V}_x$ and $\mathcal{V}_y$'s each common neighbor's contribution for s_{xy}, we can calculate s_{xy} easily.

5 Experiments

5.1 Datasets

Protein-Protein Interaction Network: We used the protein-protein interaction network (PPI) created by Krogan [13]. Two proteins are linked if it is likely that they interact. The core network consists of 2708 proteins and 7123 interactions labeled with probabilities.

Enron Network: The dataset is a subset of Enron employees, comprised of emails sent between employees, resulting in a dataset with 50,572 emails among 151 employees. We used the same method as Pfeiffer and Neville in [14] to assign each edge with a possibility of occurrence.

Synthetic Uncertain Network Based on Deterministic Network: Considering that there are not many publicly available uncertain network datasets on the web, we also generated an uncertain network based on deterministic networks. The dataset we used here is USAir. The US air transportation network contains 332 airports and 2126 airlines. Based on this network, we use an uncertain network generator to generate its corresponding uncertain network. The uncertain network generator used here is adopted from [15]. The percentage of non-existential edges we choose to add in this experiment is 20%.

5.2 Experiments

To test the prediction performance of an algorithm, the observed edges, E, are divided into two separate sets: training set E^T, is regarded as known information; and probe set E^P, is used for testing and no information therein is allowed to be used for prediction. Clearly, we have $E^T \cup E^P = E$ and $E^T \cap E^P = \emptyset$.

For the protein-protein interaction network and the synthetic uncertain network, we only know their connection information, so the training set E^T and the probe set E^P can be randomly divided. In this paper, the training set E^T and the probe set E^P are assumed to contain 90% and 10% of the links respectively. To get more reliable result, each value is obtained by averaging over 100 independent runs of random divisions of the training set and probe set.

Link prediction algorithms should be capable of detecting the dynamic relationships between members in a temporal social network. Because the Enron dataset is time-evolving, the relations among social members change continuously over time. Using link prediction algorithms, we should be able to predict newly added links in future networks. In the experiment, we predict new communications between two employees in Enron Corporation after Jan. 16, 2001, based on historical data. The idea is that, if two employees have email records before Jan. 16, 2001, we generate a potential edge between them. Then we assign these edges with a probability following the method described in [14]. The resulting probabilistic graph consists of 113 nodes and 419 edges, and this graph is regarded as the training set. The testing set is formed by taking in all the edges formed after Jan. 16, 2001. After discarding employees that have not appeared in the list of the 113 employees, as well as the edges that have appeared both before and after Jan. 16, 2001, we obtained 578 ground-truth edges with 113 distinct employees.

To evaluate the performance of prediction algorithms, we apply Precision metric to quantify the accuracy of the prediction, which focuses on top-ranked latent links. It is defined as L_r/L, where among top-L candidate links, L_r is the number of accurate predicted links actually appearing in the testing period.

5.3 Results and Evaluation

As the literature suggested [8], the top L is set to 100 in our experiments. In this section, we compare our metrics (UCN and URA) and other metrics/algorithms using existing ground truth. To evaluate our metrics, we mainly focus on the comparison between the uncertain version of graph proximity measures with weighted and unweighted ones. We also compare our metrics with LNB and SRW. LNB is a local Naïve Bayes model which is based on neighbor-based metrics, and SRW is a local-random-walk based algorithm (we choose $t = 2$ and $t = 3$ in our experiments because they are the optimal choices based on Liu and Lü's experiments in [7]). Since LNB and SRW algorithms are for deterministic networks, in our experiments we ignore the edge probabilities and consider uncertain networks as normal deterministic networks. The prediction accuracies on the three networks are shown in Table 1.

Algorithm name	Description
CN/RA	Pay no attention to probabilities and use the original metrics
WCN/WRA	Regard probability as weight and use weighted metrics
UCN/URA	Use our uncertain version of graph proximity measures
SRW2	Ignore probabilities and run local random walk algorithm [7], choose $t = 2$
SRW3	Ignore probabilities and run local random walk algorithm [7], choose $t = 3$
LNB-CN	Ignore probabilities and use Local Naïve Bayes form of Common Neighbors [8]
LNB-RA	Ignore probabilities and use Local Naïve Bayes form of Resource Allocation [8]

Table 1. Comparative results for different algorithms

Datasets	Common neighbor			Resource allocation			SRW2	SRW3	LNB-CN	LNB-RA
	CN	WCN	UCN	RA	WRA	URA				
PPI	0.472	0.5045	**0.5288**	0.4123	0.45	**0.5728**	0.4136	0.5284	0.4856	0.4992
Enron	0.49	0.52	**0.61**	0.51	0.47	**0.52**	0.43	0.45	0.55	0.46
Synthetic network	0.5812	0.5954	**0.6043**	0.6075	0.6124	**0.6233**	0.5852	0.5992	0.5962	0.5885

From Table 1, we can observe that our uncertain version of the Common Neighbor and Resource Allocation metrics can significantly outperform their original and weighted ones when dealing with uncertain networks. This shows that in the task of link prediction with edge uncertainty, it is worthwhile to take every possible worlds into account.

From Table 1, we can also observe that our metrics (UCN and URA) can outperform the other four baseline methods on PPI and Synthetic datasets. The Enron dataset allows the following observation: the Common Neighbor-based metrics seems to outperform the Resource Allocation-based counterparts on this dataset. It seems that the Resource Allocation metrics are not good choices for Enron dataset.

For run time, based on our experiments, we find UCN to be just a little bit slower than CN, but it has almost the same run time as WCN; and URA is around 2 to 3 times slower than RA and WRA.

6 Conclusion

In this paper, we propose an uncertain version of graph proximity measures for the link prediction problem in uncertain networks. We propose a new algorithm to reduce the time complexity of computing the uncertain version of graph proximity measures. By taking all possible worlds into consideration, the performance of link predictions are improved. In this work, we only focus on the neighbor-based algorithms because they are simple, effective but not yet have been studied, and we have shown the effectiveness of considering all possible worlds when using neighbor-based metrics to do link prediction. When proposing the uncertain version of other link prediction metrics, such as path-based, learning-based metrics and embedding-based algorithms, all possible worlds should also be considered, which would also be very time-consuming. To reduce time complexity, some variants of our algorithm may then be considered.

References

1. Newman, M.E.: Clustering and preferential attachment in growing networks. Phys. Rev. E **64**(2), 025102 (2001)
2. Salton, G., McGill, M.J.: Introduction to Modern Information Retrieval (1986)
3. Jaccard, P.: Étude comparative de la distribution florale dans une portion des alpes et des jura. Bull. Soc. Vaudoise Sci. Nat. **37**, 547–579 (1901)
4. Adamic, L.A., Adar, E.: Friends and neighbors on the web. Soc. Netw. **25**(3), 211–230 (2003)
5. Zhou, T., Lü, L., Zhang, Y.-C.: Predicting missing links via local information. Eur. Phys. J. B-Condens. Matter Complex Syst. **71**(4), 623–630 (2009)
6. Lü, L., Jin, C.-H., Zhou, T.: Similarity index based on local paths for link prediction of complex networks. Phys. Rev. E **80**(4), 046122 (2009)
7. Liu, W., Lü, L.: Link prediction based on local random walk. EPL (Europhys. Lett.) **89**(5), 58007 (2010)
8. Liu, Z., Zhang, Q.-M., Lü, L., Zhou, T.: Link prediction in complex networks: a local naïve Bayes model. EPL (Europhys. Lett.) **96**(4), 48007 (2011)
9. Perozzi, B., Al-Rfou, R., Skiena, S.: DeepWalk: online learning of social representations. In: Proceedings of the 20th ACM SIGKDD International Conference on Knowledge Discovery and Data Mining, pp. 701–710. ACM (2014)
10. Ahmed, N.M., Chen, L.: An efficient algorithm for link prediction in temporal uncertain social networks. Inf. Sci. **331**, 120–136 (2016)
11. Mallek, S., Boukhris, I., Elouedi, Z., Lefevre, E.: Evidential missing link prediction in uncertain social networks. In: Carvalho, J.P., Lesot, M.-J., Kaymak, U., Vieira, S., Bouchon-Meunier, B., Yager, R.R. (eds.) IPMU 2016. CCIS, vol. 610, pp. 274–285. Springer, Cham (2016). https://doi.org/10.1007/978-3-319-40596-4_24
12. Murata, T., Moriyasu, S.: Link prediction of social networks based on weighted proximity measures. In: Proceedings of the IEEE/WIC/ACM International Conference on Web Intelligence, pp. 85–88. IEEE Computer Society (2007)

13. Krogan, N.J., et al.: Global landscape of protein complexes in the yeast Saccharomyces cerevisiae. Nature **440**(7084), 637 (2006)
14. Pfeiffer, J.J., Neville, J.: Probabilistic paths and centrality in time. In: Proceedings of the 4th SNA-KDD Workshop, KDD 2010 (2010)
15. Zhang, C., Zaïane, O.R.: Detecting local communities in networks with edge uncertainty. In: 2018 IEEE/ACM International Conference on Advances in Social Networks Analysis and Mining (ASONAM), pp. 9–16. IEEE (2018)

Inferring Social Bridges that Diffuse Information Across Communities

Pei Zhang, Ke-Jia Chen(✉), and Tong Wu

Jiangsu Key Laboratory of Big Data Security and Intelligent Processing,
Nanjing University of Posts and Telecommunications, Nanjing 210023, Jiangsu, China
zp_njupt@163.com, chenkj@njupt.edu.cn, wutong22@163.com

Abstract. While the accuracy of link prediction has been improved continuously, the utility of the inferred new links is rarely concerned when it comes to information diffusion. This paper defines the utility of links based on average shortest distance and more importantly defines a special type of links named *bridge links* based on community structure (overlapping or not) of the network. In sociology, bridge links are considered to play a more crucial role in information diffusion across communities. Considering that the accuracy of previous link prediction methods are high in predicting strong ties but not much high in predicting weak ties, we propose a new link prediction method named iBridge, which aims to infer new diffusion paths using biased structural metrics in a supervised learning framework. The experimental results in 3 real online social networks show that iBridge outperforms the traditional supervised link prediction method especially in inferring the bridge links and meantime, the overall performance of predicting bridge links and non-bridge links is not compromised, thus verifying its robustness in inferring new links.

Keywords: Bridge link prediction · Information diffusion · Weak ties

1 Introduction

Many complex systems can be described by networks with nodes representing individuals and links denoting the interactions between nodes. As the most widely studied network, social network plays an important role for people to connect with others and to diffuse various types of information. Link prediction [17], one of the important tasks in SNAM (social network analysis and mining), studies the formation of missing links or new links based on current and historical network, with wide application in item recommendation, pre-warning system, biomedical discovery etc. Researchers have been working extensively to study effective link prediction methods for different types of networks and for different application scenarios [1,12,20,21].

Though the accuracy in link prediction continuously improves, it is realized that the new inferred links could be correct, but not particularly novel nor significantly useful to expand new links for users. For example, the users in a social

Q. Yang et al. (Eds.): PAKDD 2019, LNAI 11440, pp. 475–487, 2019.
https://doi.org/10.1007/978-3-030-16145-3_37

network share more common neighbors are more likely to establish friendship link, but this new connection does not help much in getting more new information or having more new friends for both of them.

This phenomenon reminds us of the sociological theory of weak ties proposed by Granovetter [11] and the theory of structural hole proposed by Burt [7]. That is, there has been a large amount of redundant information in the circle of strong ties while the really useful information often comes from *weak ties*. Some researchers [15,16] proposed to measure weak ties using the weights of links. Another work [18] differentiated the strong and weak ties based on the community division of the network. But there is no commonly accepted definition for weak ties till now. Moreover, there is no relevant research on how to accurately infer weak ties, which has strong application background in controlling public opinion and preventing the spread of infectious diseases and computer viruses.

This paper aims to use the bridge link concept to define a type of weak ties, quantify the utility of bridge links and eventually propose an effective method to infer bridge links accurately. The main contributions of this paper are as follows:

- We define a new type of link—bridge link. Some previous work [24] computed the tie strength as the frequency of user interactions. But the tuning of a cutoff threshold has a crucial impact on the correct identification of weak ties. The concept of *bridge* defined by Granovetter [11], which provides the only path between two nodes, is too strict for large-scale networks. The paper redefines bridges as *bridge links* which connect different communities to facilitate the wider diffusion of information.
- We define the utility of links and verify the importance of bridge links on information diffusion. Currently, there is no measure to evaluate the utility of weak ties in information diffusion. The paper designs a *utility function* based on the average shortest path length of nodes, to evaluate the influence of bridge link on information diffusion.
- We propose an efficient inferring method of bridge links named *iBridge* (inferring Bridge links). This method modifies the structural metrics of node pairs according to the statistical characteristics of bridge links, and uses a supervised learning model. The performance of iBridge is higher in predicting bridge links and is not compromised in predicting non-bridge links.

The rest of this paper is organized as followed. Section 2 introduces the related work. Section 3 gives the formalization of related conceptions. Section 4 introduces the measurement of link utility in information diffusion and presents the proposed framework of inferring new bridge links. Section 5 shows the experimental results in details. Section 6 concludes the work.

2 Related Work

In information science, link prediction aims to infer the unknown or missing links from the current or historical networks. The early work is primarily based on Markov chain [20], which proves that the high-level model is more conducive to the prediction accuracy in large-scale networks. Subsequently, some researchers [12] convert link prediction to the binary classification problem in a (semi-)supervised learning framework. Due to the sparsity of the real information networks, the methods using semi-supervised learning [6,13] and active learning [4] are further introduced and achieve good results. Considering that many information networks evolve over time, time-aware techniques [1] are developed, which was proved to be effective for dynamic networks. Recently, the performance of the link prediction method is further improved by leveraging heterogeneous information in the network [21].

The previous link prediction methods are mainly concerned on whether the predicted links are relevant or not, without discussing the quality of these links, such as whether they are useful for information diffusion. Studies [3] show that information propagates more extensively through weak ties in social networks. Some researchers [24] propose the calculation formula of link strength and shows that if links are deleted according to the order of weights, the information coverage in the network will fall sharply. Another work [10] shows that weak ties are able to connect small communities into one large community, helping to reach a wider variety of contacts. In a further study [8], only selected weak ties are helpful for information diffusion.

In link prediction task, weak tie theory was once introduced to solve the problem of information redundancy. Lü et al. [17] studied the role of weak ties in the weighted network link prediction problem. A link is defined as a strong tie if its weight is on the top 50% of all weights; Otherwise, it is a weak tie. In another work [18] weak ties are defined based on community division and then improves the accuracy of link prediction with weak ties.

Different from the above two methods, this paper focuses on how to accurately infer weak ties rather than improving link prediction methods using weak ties. The most related work is the method proposed by Song et al. [22], which aims to find the brokers (a type of nodes) that are critical to information diffusion. But our method is more predictive to find new weak ties.

3 Formalization

This section gives formalization of related conceptions used in this paper.

Definition 1 *Link Prediction.* In a given network $G = \langle V, E \rangle$, where $V = \{v_i\}_{i=1}^{N}$ denotes a set of nodes and $E = \{e_{ij}\}^t$ denotes the set of edges that have been observed at time t, where e_{ij} denotes the link between the node pair $\langle v_i, v_j \rangle$. This task is to predict the possibility of connection between $\langle v_i', v_j' \rangle \notin E$ at time t' $(t' > t)$.

Definition 2 *Community Detection.* Community is an important feature of many networks, especially social networks [19]. Links within the same community are dense while links between different communities are sparse. Given a network $G = \langle V, E \rangle$, the task is to divide all node v_i into different subsets obtaining the collection of communities $Com = \{Com_i\}_{i=1}^{K}$ where $Com_i \subset V$.

Community detection methods can be classified as: non-overlapping methods [19], overlapping methods [2] and hierarchical methods [9]. A non-overlapping method outputs Com, where any $Com_i \bigcap Com_j = \emptyset$ and $i \neq j$. An overlapping method outputs Com, where may exist $Com_i \bigcap Com_j \neq \emptyset$ and $i \neq j$. The hierarchical method is an iteration of non-overlapping method, where each Com_i can be further divided into smaller community set.

Definition 3 *Bridge Link.* In this paper, bridge link is defined as the link across communities. The definition varies depending on different community partitioning methods.

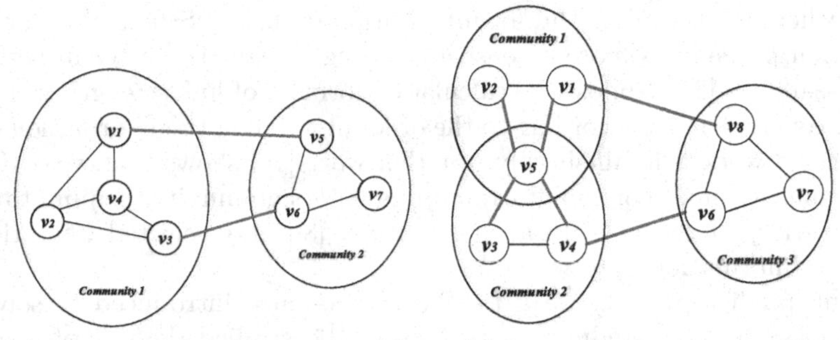

(a) in non-overlapping communities (b) in overlapping communities

Fig. 1. Examples of bridge links across communities.

Bridge Links Across Non-overlapping Communities. Let $C(v_i)$ denotes the community set of node v_i. Since the communities are non-overlapping, for any $v_i \in V$, $|C(v_i)| = 1$. For any given link e_{ij}, if $C(v_i) \neq C(v_j)$, then e_{ij} is a bridge link labeled as $B(e_{ij}) = 1$; otherwise $B(e_{ij}) = 0$. As shown in Fig. 1(a), $e_{1,5}$ and $e_{3,6}$ are bridge links.

Bridge Links Across Overlapping Communities. Sometimes one node may belong to multiple communities. Let $C(v_i)$ denotes the community set of node v_i. For any given link e_{ij}, only if $|C(v_i)| = |C(v_j)| = 1$ and $C(v_i) = C(v_j)$, e_{ij} is labeled as $B(e_{ij}) = 0$; in other cases $B(e_{ij}) = 1$. In Fig. 1(b), two of the communities are overlapping. $e_{1,8}$ and $e_{4,6}$ are bridge links according to the above definition; $e_{1,5}$, $e_{2,5}$, $e_{3,5}$ and $e_{4,5}$ are also bridge links since node v_5 is in the overlapping part of communities.

Bridge Links in Hierarchical Communities. The hierarchical community detection method is based on the iteration of non-overlapping method and there is no overlap between communities in each layer. Therefore, the definition method of bridge link in each layer of the dendrogram is the same as that in non-overlapping communities described above.

Definition 4 *Bridge Link Prediction.* Similar to the general link prediction task, the bridge link prediction aims to predict the bridge links. In a given network $G = \langle V, E \rangle$, where the meaning of V and E are the same as in Definition 1.

$Q = \{e_{ij}|B(e_{ij}) = 1\}$ $(Q \subset E)$ denotes the bridge link set. The definition of bridge link is given in Definition 3. This task is to predict the possibility of connection e'_{ij} between $\langle v'_i, v'_j \rangle$ at time t' where $\langle v'_i, v'_j \rangle \notin E$ but $B(e'_{ij}) = 1$. As shown in Fig. 2, the task is to predict the formation of links like $e_{1,6}$ and $e_{3,5}$ at time t', which are potential bridge links.

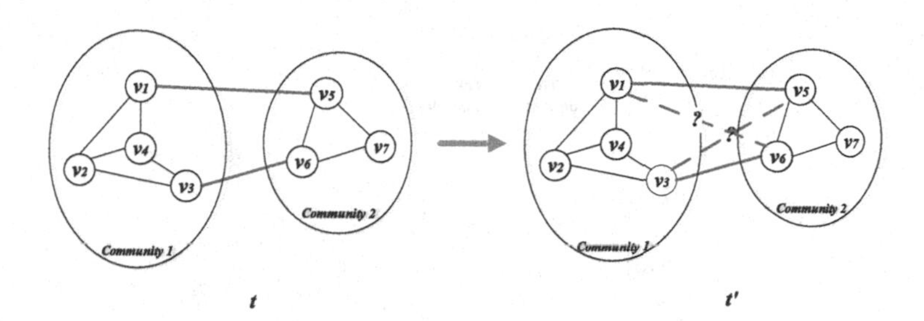

Fig. 2. Description of bridge link prediction.

4 The Proposed Method

4.1 Utility of Bridge Links

To quantify and verify the role of bridge links defined above in information diffusion, we define the utility function based on the average shortest path.

The average shortest distance $\overline{Dist}$ (Eq. 1) is often used to measure the capability of the network to diffuse information. Generally, the smaller the $\overline{Dist}$ value, the more conducive the network is to diffuse information. Here, n denotes the total node number of the network, and $d(v_i, v_j)$ denotes the shortest path length between node v_i and v_j.

$$\overline{Dist} = \sum_{v_i, v_j \in V} \frac{d(v_i, v_j)}{n(n-1)} \tag{1}$$

Based on $\overline{Dist}$, $\Phi(E_k)$ (Eq. 2) is defined to measure the change rate of $\overline{Dist}$ after deleting k edges, where E_k denotes the deleted edges. The higher the value of Φ, the edges are more useful for information diffusion.

$$\Phi(E_k) = \frac{\sum_{e_{ij} \in E - E_k} \frac{d(v_i, v_j)}{n(n-1)} - \sum_{e_{ij} \in E} \frac{d(v_i, v_j)}{n(n-1)}}{\sum_{e_{ij} \in E} \frac{d(v_i, v_j)}{n(n-1)}} \tag{2}$$

Figure 3 shows a comparison of the Φ value in the Facebook dataset, after randomly deleting a certain number of bridge links and non-bridge links (obtained using Louvain method [5]). As expected, the increase rate of the Φ value obtained by deleting bridge links is significantly higher than by deleting non-bridged links.

4.2 Biased Features

Several heuristic structural features like common neighbors (CN), Jaccard coefficient (JC) and Resource Allocation (RA) are often used to describe node pairs due to their low computational complexity and good predictive performance. In bridge link prediction problem, the new features (B-CN, B-JC, B-RA, SBC and SDC) are proposed to describe the node pairs in the bridge position.

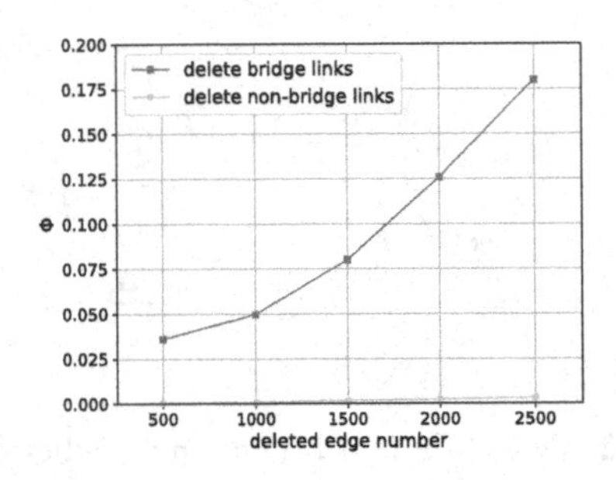

Fig. 3. Comparison of the utility values of bridge links and non-bridge links.

In the following equations, $\tau(v_i)$ denotes neighbor node set of v_i, $d(v_i)$ denotes node degree of v_i, g_i^{st} is the number of geodesic paths from node v_s to node v_t that pass through v_i, n_{st} is the total number of geodesic paths from v_s to v_t, τ_{ij} represents the set of common neighbors of v_i and v_j.

In Eq. 3, $C(v_k)$ denotes the community (communities) where v_k is located. If $|C(v_i) \cap C(v_k)| \geq 1$ or $|C(v_k) \cap C(v_j)| \geq 1$, B-CN will have an additional value. Notice that $|C(v_i) \cap C(v_k)|$ or $|C(v_k) \cap C(v_j)|$ may be larger than 1 in overlapping communities.

$$B\text{-}CN(v_i, v_j) = |\tau_{ij}| + \sum_{v_k \in \tau_{ij}} |C(v_i) \cap C(v_k)| + \sum_{v_k \in \tau_{ij}} |C(v_k) \cap C(v_j)| \tag{3}$$

The definition of B-RA (Eq. 4) is similar to RA but only considers the situation that the $|C(v_i) \cap C(v_k)| \geq 1$ or $|C(v_k) \cap C(v_j)| \geq 1$.

$$B\text{-}RA(v_i, v_j) = \sum_{v_k \in \tau_{ij}} \frac{|(C(v_i) \cap C(v_k)) \cup (C(v_k) \cap C(v_j))|}{d(v_k)} \tag{4}$$

If $|C(v_i) \cap C(v_k)| \geq 1$ or $|C(v_k) \cap C(v_j)| \geq 1$, B-JC$(v_i, v_j)$ will have an additional value.

$$B\text{-}JC(v_i, v_j) = \frac{|\tau_{ij}|}{|\tau(v_i) \cup \tau(v_j)|} + \frac{\sum_{v_k \in \tau_{ij}} (|C(v_i) \cap C(v_k)| + |C(v_k) \cap C(v_j)|)}{|\tau(v_i) \cup \tau(v_j)|} \tag{5}$$

In addition, SBC (Sum of Betweenness Centrality) and SDC (Sum of Degree Centrality) are used because the probability of bridge links connecting to the most influential nodes (top 10% users with highest PageRank scores) in the community is much higher than that of non-bridge links.

$$SBC(v_i, v_j) = \frac{\sum_{s<t} \frac{g_i^{st}}{n_{st}} + \sum_{s<t} \frac{g_j^{st}}{n_{st}}}{\frac{1}{2}n(n-1)} \tag{6}$$

$$SDC(v_i, v_j) = \frac{d(v_i) + d(v_j)}{n-1} \tag{7}$$

The comparison of average SBC and average SDC of bridge links and non-bridge links on three datasets (Facebook, Twitter and NetScience) is shown in Fig. 4, which verifies the effectiveness of SBC and SDC.

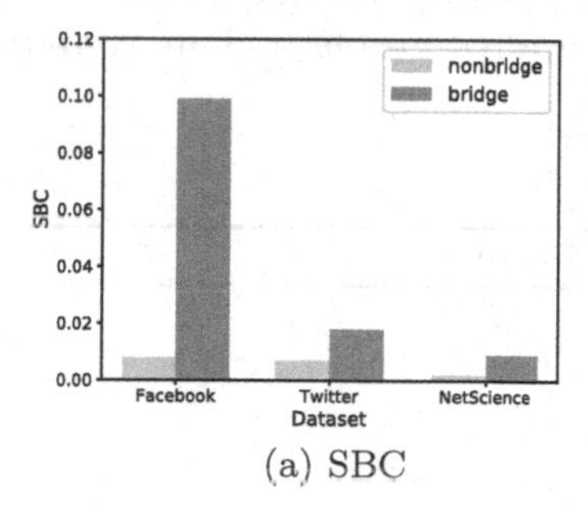

(a) SBC

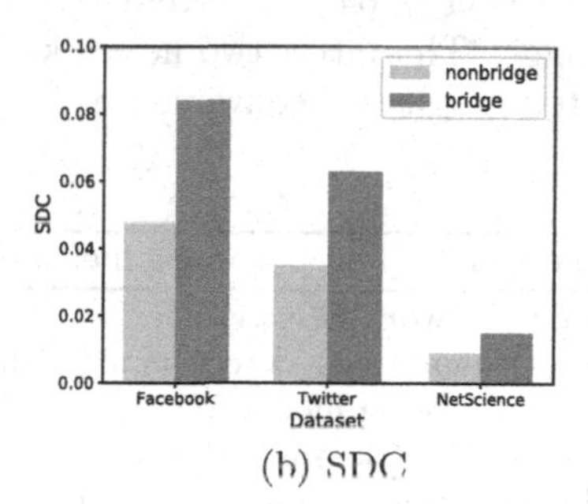

(b) SDC

Fig. 4. Comparison of average SBC and SDC.

4.3 The iBridge Framework

In this paper, inferring bridge links is regarded as a supervised classification problem. The proposed iBridge method is described in Algorithm 1. Firstly, a given network is divided into communities using a community detection method. Secondly, all node pairs which form or may form a bridge link (as defined in Sect. 3) are collected. If there exists an edge between node v_i and v_j, $Label\langle v_i, v_j\rangle = 1$. Otherwise 0. P is a set of positive examples, and N is a set of negative examples. $Fea\langle v_i, v_j\rangle_{n=1}^{k}$ represents the structural features vector of node pair $\langle v_i, v_j\rangle$. Clf_{BL} is a classifier learned by training set, which can infer the label of any node pair $\langle v_i', v_j'\rangle$ in G' at time t', where $Label\langle v_i', v_j'\rangle = 0$.

Here is an example to illustrate how iBridge infers new bridge links (Fig. 5). First, the network is divided into different non-overlapping communities. Take node pair $\langle v_1, v_5\rangle$ as an example, where node v_1 and v_5 belong to different communities. The features of the node pair $\langle v_1, v_5\rangle$ are calculated: $B\text{-}CN(v_1, v_5) = 2$, $B\text{-}RA(v_1, v_5) = 0.25$, $B\text{-}JC(v_1, v_5) = 2$, $SBC(v_1, v_5) = 0.33$, and $SDC(v_1, v_5) = 1.17$. In the illustration, node pair $\langle v_1, v_5\rangle$, $\langle v_4, v_5\rangle$ and $\langle v_3, v_6\rangle$ are positive examples. The classifier Clf_{BL} is then trained based on all node pairs in the training set. Finally, the link probability predicted by Clf_{BL} for the node pair $\langle v_4, v_6 \rangle$ is higher than the threshold and is therefore inferred as a bridge link.

5 Experiment

This section evaluates the performance of iBridge in inferring new bridge links, as well as all new links. All experiments are run on the computer with Windows 10 systems, 2.6 GHz CPU and 12 GB of memory.

5.1 Datasets and Settings

The experiment uses three real-world datasets: Facebook (4,039 nodes and 88,234 edges), Twitter (5,646 nodes and 47,475 edges) and NetScience (1,461 nodes and 2,742 edges). The latter two networks are directed graphs and are converted to undirected graphs for convenience.

Algorithm 1. The proposed method—iBridge.

Input: Network $G = \langle V, E \rangle$;
Community detection algorithm CD;
Learning model M;
Output: The Classifier Clf_{BL};
1 Call CD to find community set $\{Com_i\}$ in G;
2 Collect all node pairs $\{\langle v_i, v_j \rangle\}$ which have or may have bridge links, represented as D;
3 Create $P = \{\langle v_i, v_j \rangle | \langle v_i, v_j \rangle \in D \cap Label\langle v_i, v_j \rangle = 1\}$,
$N = \{\langle v_i, v_j \rangle | \langle v_i, v_j \rangle \in D \cap Label\langle v_i, v_j \rangle = 0\}$;
4 Generate $Trainset$ by sampling from P and N;
5 **for** *each* $\langle v_i, v_j \rangle$ *in* $Trainset$ **do**
6   Calculate $Fea\langle v_i, v_j \rangle_{n=1}^{k}$ with Eq. 3-7;
7   Get $Label\langle v_i, v_j \rangle$;
8 **end**
9 Train M with $Trainset$;
10 Get Clf_{BL}.

The learning settings in all networks are shown in Table 1. The settings are slightly different in non-overlapping communities and overlapping communities. Considering the stability and complexity of the model, the Random Forest classifier is used in both methods. We use the downsampling method to deal with the imbalanced problem, and the experiment uses 10-fold cross validation and outputs the average results.

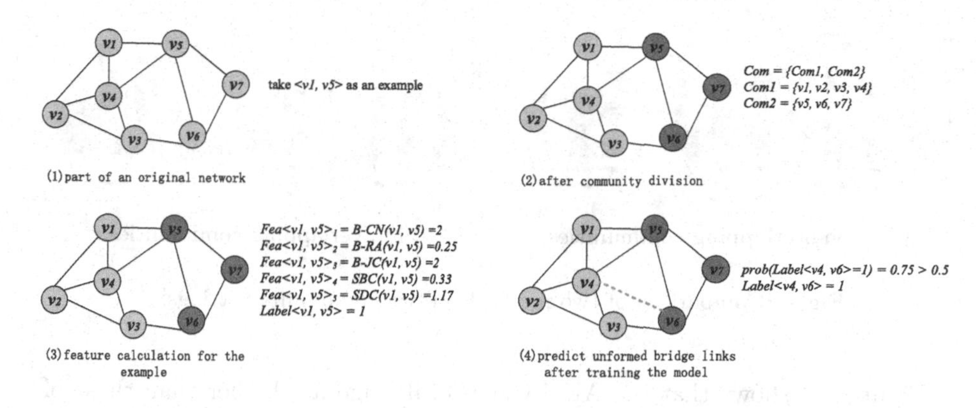

Fig. 5. An illustrative example of iBridge framework.

5.2 Comparative Methods

The method iBridge is compared to baseline method BLiP. Both methods are based on supervised learning framework which first appeared in the work of [12], except that BLiP uses benchmark features while iBridge uses biased features. Both methods can infer bridge links in non-overlapping communities and overlapping communities (BLiP-nc and iBridge-nc for non-overlapping communities, and BLiP-oc and iBridge-oc for overlapping communities).

Table 1. Learning settings in non-overlapping and overlapping community division.

Dataset	Non-overlapping				Overlapping			
	Pos-set	Neg-set	Train-set	Test-set	Pos-set	Neg-set	Train-set	Test-set
Facebook	6,816	335,915	308,457	34,273	8769	7,741,283	6,975,046	775,005
Twitter	1,9778	1,291,729	1,180,356	131,150	640	6,100,969	5,491,448	610,160
NetScience	317	1,062,571	956,599	106,288	337	1,061,901	956,014	106,223

To get non-overlapping communities, the modularity-based Louvain algorithm [5] is used because it is a simple but efficient method for large networks. To get overlapping communities, SLPA algorithm [23] is used because it has good performance in both low density and high density overlapping network without knowing the number of communities.

5.3 Results

AUC (Area under ROC Curve) value, precision and F1-score are used to evaluate the accuracy of iBridge and BLiP in inferring both new bridge links and all types of new links. The comparative results of two methods in inferring bridge links in the non-overlapping and overlapping communities are shown in Figs. 6(a) and 6(b), respectively.

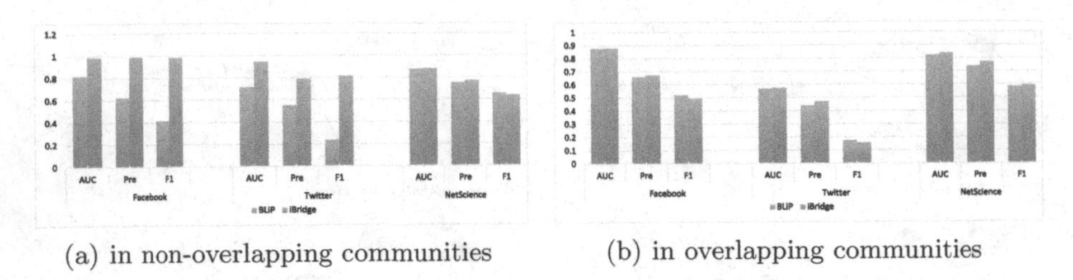

(a) in non-overlapping communities

(b) in overlapping communities

Fig. 6. Comparison of two methods in two community settings.

Figure 6(a) shows that the AUC values of iBridge are higher than those of BLiP in all three datasets. In NetScience, the improvement of AUC is not as significant as in Facebook and Twitter. Moreover, the F1-score has a slight decrease. The possible reason is that in scientific collaboration networks, the ties connecting different clusters are even stronger than ties in densely interconnected local clusters [14]. Figure 6(b) shows that iBridge has higher AUC and precision values in all networks but lower F1-scores in Facebook and Twitter. Moreover, the improvement of AUC and precision values is not as big as that in non-overlapping communities. The possible reason is that in overlapping communities, bridge links are closer to non-bridge links. So the features of these two types of links are not significant different. In addition, the imbalance between positive and negative instances in overlapping communities is more serious than that in non-overlapping communities, which leads to limited improvement of iBridge.

The detailed comparison of values of each indicator is listed in Table 2.

Table 2. Comparison of two methods in non-overlapping and overlapping communities.

	Facebook			Twitter			NetScience		
	AUC	Pre	F1	AUC	Pre	F1	AUC	Pre	F1
BLiP-nc	0.8198	0.624	0.4156	0.7158	0.5514	0.237	0.876	0.7504	**0.6532**
iBridge-nc	**0.9883**	**0.9912**	**0.9856**	**0.9467**	**0.7887**	**0.8122**	**0.8822**	**0.7685**	0.6355
BLiP-oc	0.8783	0.6604	**0.5197**	0.5677	0.4405	**0.174**	0.8219	0.737	0.5833
iBridge-oc	**0.8798**	**0.6735**	0.4945	**0.5738**	**0.4731**	0.1572	**0.8392**	**0.7692**	**0.5921**

The paper also compares the performance of two methods in inferring all new links. The experiment setup and training process are similar to the previous one, except that the dataset contains all node pairs in the network. The learning settings are shown in Table 3. The comparison results of AUC value, precision, and F1-score are shown are listed in Table 4.

Table 3. Learning settings in inferring both bridge links and non-bridge links.

Dataset	Pos-set	Neg-set	Train-set	Test-set
Facebook	88,234	8,066,507	7,339,266	815,474
Twitter	47,475	15,888,367	14,342,257	1,593,584
NetScience	2,742	1,063,788	959,877	106,653

The result shows, in all three networks, the AUC value of iBridge is still higher than that of BLiP in predicting all new links. But the precision and F1-score of iBridge are lower than BLiP in most cases. Considering the structural features in iBridge are biased for inferring new bridge links, the experimental results verify the robustness of iBridge since it is not much worse than BLiP in some indicators, and even slightly better in others.

Table 4. Comparison of two methods in inferring all new links.

	Facebook			Twitter			NetScience		
	AUC	Pre	F1	AUC	Pre	F1	AUC	Pre	F1
BLiP	0.943	**0.7073**	**0.6416**	0.6593	0.5383	**0.2471**	0.9406	**0.8843**	**0.8492**
iBridge	**0.9644**	0.7002	0.6191	**0.676**	**0.5898**	0.1957	**0.9512**	0.8136	0.7838

We also analyze the complexity of all methods. Considering all methods use Random Forest as the base classifier, the time complexity of training and testing model can be ignored when comparing. In a network with m edges and n nodes, the time complexity of BLiP is $O(mn) + O(n^2)$, and the time complexity of iBridge is $O(mn) + O(tm) + O(n^2)$. Here, $O(mn)$ represents the time complexity of calculating the betweenness index, $O(n^2)$ represents the time complexity of generating features for all instances, and $O(tm)$ is the time complexity of detecting communities with Louvain and SLPA where t is the predefined maximum number of iterations of the algorithm.

5.4 Discussion

In this paper, we use BLiP as comparative method because there is currently no other related work on inferring weak ties or bridge links. However, the iBridge framework can be well tuned to other link prediction methods by updating the structural metrics in the network.

The most related work is Song et al.'s [22] method. They developed a heuristic algorithm to find the Top-k brokers based on the weak tie theory. But their method is to mine nodes (brokers)while our method is to mine links (weak ties). In practical applications, deleting the bridges (specific diffusion paths) may be less expensive than deleting brokers, because the latter changes the network structure to a great extent. Another advantage of our method is that it can infer new diffusion paths and therefore play an early warning role.

6 Conclusion

The paper for the first time tries to infer weak ties, aiming to stop the spread of malicious information or to relieve the Matthew effect in sociology. We define weak ties as bridge links based on communities instead of the link weights. The paper proves that bridge links play a more important role in information diffusion than non-bridge links and proposes a utility function for links. New bridge links can be inferred under a proposed supervised learning framework using the biased similarity index. The experiment result shows that the proposed method can effectively infer bridge links as well as non-bridge links.

However, not all bridge links are equally useful, so more sensitive utility function need to be proposed in the future for further differentiation among bridge links. Moreover, weak links and weak ties are not completely equivalent. How to measure the weak relationships between nodes requires further study. Finally, network representation learning can be used to automatically obtain features of weak ties instead of using heuristic biased features.

Acknowledgements. This research was supported by the National Nature Science Foundation of China (No. 61571238, No. 61603197 and No. 61772284).

References

1. Aggarwal, C.C., Xie, Y., Yu, P.S.: A framework for dynamic link prediction in heterogeneous networks. Stat. Anal. Data Min. ASA Data Sci. J. **7**(1), 14–33 (2014)
2. Ahn, Y.Y., Bagrow, J.P., Lehmann, S.: Link communities reveal multiscale complexity in networks. Nature **466**(7307), 761 (2010)
3. Bakshy, E., Rosenn, I., Marlow, C., Adamic, L.: The role of social networks in information diffusion. In: International Conference on World Wide Web, pp. 519–528 (2012)
4. Bilgic, M., Mihalkova, L., Getoor, L.: Active learning for networked data. In: International Conference on Machine Learning, pp. 79–86 (2010)
5. Blondel, V.D., Guillaume, J.L., Lambiotte, R., Lefebvre, E.: Fast unfolding of communities in large networks. J. Stat. Mech. **2008**(10), 155–168 (2008)
6. Brouard, C., D'Alché-Buc, F., Szafranski, M.: Semi-supervised penalized output Kernel regression for link prediction. In: International Conference on Machine Learning, pp. 593–600 (2013)
7. Burt, R.S.: Structural holes and good ideas. Am. J. Soc. **110**(2), 349–399 (2004)
8. Chiu, H.Y., Chen, S.M.: Propagating online social networks: via different kinds of weak ties. In: IEEE/ACM International Conference on Advances in Social Networks Analysis and Mining, pp. 1189–1195 (2013)
9. Clauset, A., Moore, C., Newman, M.E.: Hierarchical structure and the prediction of missing links in networks. Nature **453**(7191), 98 (2008)
10. Ferrara, E., Meo, P.D., Fiumara, G., Provetti, A.: The role of strong and weak ties in Facebook: a community structure perspective. Commun. ACM **57**(11), 78–84 (2012)
11. Granovetter, M.: The strength of weak ties. Am. J. Soc. **78**(6), 1360–1380 (1973)

12. Hasan, M.A., Chaoji, V., Salem, S., Zaki, M.: Link prediction using supervised learning. In: Proceedings of SDM 2006 Workshop on Link Analysis, Counterterrorism and Security (2006)
13. Kashima, H., et al.: Link propagation: a fast semi-supervised learning algorithm for link prediction. In: International Conference on World Wide Web, pp. 1099–1110 (2009)
14. Ke, Q., Ahn, Y.Y.: Tie strength distribution in scientific collaboration networks. Phys. Rev. E Stat. Nonlinear Soft Matter Phys. **90**(3), 032804 (2014)
15. Liu, H., Hu, Z., Haddadi, H., Tian, H.: Hidden link prediction based on node centrality and weak ties. Europhys. Lett. **101**(1), 18004 (2013)
16. Lü, L., Zhou, T.: Link prediction in weighted networks: the role of weak ties. Europhys. Lett. **89**(1), 18001 (2010)
17. Lü, L., Zhou, T.: Link prediction in complex networks: a survey. Phys. A Stat. Mech. Appl. **390**(6), 1150–1170 (2011)
18. Meo, P.D., Ferrara, E., Fiumara, G., Provetti, A.: On facebook, most ties are weak. Commun. ACM **57**(11), 78–84 (2014)
19. Newman, M.E.: Fast algorithm for detecting community structure in networks. Phys. Rev. E Stat. Nonlinear Soft Matter Phys. **69**, 066133 (2004)
20. Sarukkai, R.R.: Link prediction and path analysis using markov chains. Comput. Netw. **33**(1–6), 377–386 (2000)
21. Shi, C., Li, Y., Zhang, J., Sun, Y., Yu, P.S.: A survey of heterogeneous information network analysis. IEEE Trans. Knowl. Data Eng. **29**(1), 17–37 (2017)
22. Song, C., Hsu, W., Lee, M.L.: Mining brokers in dynamic social networks. In: ACM International on Conference on Information and Knowledge Management, pp. 523–532 (2015)
23. Xie, J., Szymanski, B.K.: Towards linear time overlapping community detection in social networks. In: Tan, P.-N., Chawla, S., Ho, C.K., Bailey, J. (eds.) PAKDD 2012. LNCS (LNAI), vol. 7302, pp. 25–36. Springer, Heidelberg (2012). https://doi.org/10.1007/978-3-642-30220-6_3
24. Zhao, J., Wu, J., Xu, K.: Weak ties: subtle role of information diffusion in online social networks. Phys. Rev. E Stat. Nonlinear Soft Matter Phys. **82**(2), 016105 (2010)

Learning Pretopological Spaces to Extract Ego-Centered Communities

Gaëtan Caillaut[1(✉)], Guillaume Cleuziou[1], and Nicolas Dugué[2]

[1] Université d'Orléans, INSA Centre Val de Loire, LIFO EA 4022, Orléans, France
{gaetan.caillaut,guillaume.cleuziou}@univ-orleans.fr
[2] Le Mans Université, LIUM, EA 4023, Le Mans, France
nicolas.dugue@univ-lemans.fr

Abstract. We present a pretopological based approach to extract ego-centered communities. Classical methods often consider only one structural feature of the network, whereas pretopology enables to do multi-criteria analysis. Our approach consists in learning a logical combination of network's descriptors to define a pretopological space. Ego-centered communities are extracted by computing the elementary closure of each node. The quality of such communities is evaluated against the ground truth communities. We show the benefits of our method by comparing it to others on both real and synthetic networks.

Keywords: Community extraction · Pretopology · Ego-centered communities

1 Introduction

Complex network theory highlights the existence of properties shared by many networks modeling real systems. These networks are often structured by communities, that is to say a partition of nodes such that each part is strongly connected toward itself and loosely toward the other nodes [13]. Social networks are a very typical case of such networks, where users gather themselves around various topics. The analysis of these communities is a critical task since it allows to study networks at an intermediate level (mesoscopic) between the local level (neighborhoods) and the global level (the entire network).

A common way to extract network's communities is to find a partition of its nodes that maximizes the modularity [13], that is to say that maximizes the density of internal edges while minimizing the density of external edges inside a community. The entire network's structure must be known in order to compute such a partition. When considering huge networks such as the world wide web or online social networks, this requirement is sometimes hard or impossible to fulfill. Either because the entire network is not known or because it cannot be held in memory. Furthermore a strict partition of the network's nodes is often far from the actual communities, since community overlapping is forbidden [14].

Q. Yang et al. (Eds.): PAKDD 2019, LNAI 11440, pp. 488–500, 2019.
https://doi.org/10.1007/978-3-030-16145-3_38

Hence we focus our work on ego-centered communities which are local communities centered on one (or several) node(s) of interest [3,7]. This approach allows to do the computation on a small/local part of the network, which is a lot less resource demanding. It is even feasible without knowing the entire network since its exploration can be done, if necessary, during the application of the algorithm. Furthermore overlapping communities can be detected by such methods, resulting in communities in accordance with the real ones.

This paper aims to introduce a pretopological approach to uncover ego-centered communities. The pretopology theory is a generalization of the graph theory [6], it is thus a better alternative to study and model complex networks. It allows in particular to model links of various natures between sets of nodes whereas graph theory allows only links between two nodes. A social pretopology can then be made of multiple types of relations between users (friends, co-workers, families, ...).

After an overview of works related to the community extraction task (Sect. 2), we introduce (Sect. 3) some key concepts of the pretopology theory as well as the generic class of logical pretopological spaces. The latest enables to learn a pretopological space adapted to the characteristics of the network. Section 4 sets the framework for our experiments. A description of the predicates that compose our logical formulas is also given. The results of our experiments made on both real world and synthetic networks are shown in Sect. 5. A comparison between classical and pretopological methods is made.

2 Related Works

The task of extracting ego-centered communities has already been tackled by many researchers. At least three different approaches can be distinguished: methods guided by a local modularity inspired measure [13], methods based on propagation algorithms or methods relying on graph embeddings [9].

Modularity based methods extract ego-centered communities by means of a greedy accumulation process. They often start from a single node, called node of interest, which is expanded one node at a time. At each step, the node bringing the greatest gain to the local modularity score is inserted into the community under construction. The algorithm stops when either no node gives an increase to the local modularity or the community is large enough [4]. Some methods have an additional pruning phase that removes the outliers that do not belong anymore to the ego-centered community [3,12].

The carryover-opinion [7] is a proximity measure inspired from opinion or heat propagation. The node of interest can be considered as a heat source which will propagate to other nodes through the network's edges. After a certain amount of time, the temperature of each node will remain stable and a score will be awarded to each node, the hotter the greater the score. An ego-centered community can then be extracted by retaining nodes whose score is greater than a given threshold. This allows to extract communities at different levels of granularity.

Finally, some works apply word embeddings learning methods to learn graph embeddings [8,9]. Let's consider each vertex of a network as a word, then sentences describing paths taken by random walks can be generated. Such sentences can then be fed to any word embedding learning algorithm to learn graph embeddings. Community structures can then be extracted out of these embeddings by various means (e.g. clustering algorithms).

Many approaches have been proposed to solve the community extraction task. There is probably no generic and optimal way to solve this task, nevertheless each method brings a part of the answer which would be a shame to ignore. This motivates the need of our alternative pretopological based method. Indeed, the pretopology allows to model a propagation process defined by various knowledge bases. We propose to leverage this process to extract ego-centered communities.

3 Basics of Pretopology

A pretopological space is defined by a couple (E, a) where E is a finite non-empty set and $a(.)$ a mapping from $\mathcal{P}(E)$ to $\mathcal{P}(E)$ satisfying the following two properties.

$$\forall A \in \mathcal{P}(E), A \subseteq a(A) \tag{1}$$

$$a(\emptyset) = \emptyset \tag{2}$$

The pseudo-closure operator defines an expansion process from a set $A \subseteq E$ to a bigger one. It is commonly defined by a set $\mathcal{V}$ of neighborhoods on E [1] where $V \in \mathcal{V}$ is a reflexive mapping from E to $\mathcal{P}(E)$.

$$\forall A \in \mathcal{P}(E),\ a(A) = \{x \in E \mid \forall V \in \mathcal{V},\ V(x) \cap A \neq \emptyset\} \tag{3}$$

Unlike topological operators, the pretopological pseudo-closure is not necessarily idempotent. It can be applied multiple times on a set $A \in \mathcal{P}(E)$ until a set K such that $A \subseteq K \subseteq E$ and $a(K) = K$ is reached. The set K is called the closure of A and noted $F(A)$. If $|A| = 1$ then $F(A)$ is called an elementary closed set.

A new class of pretopological spaces defined by a logical formula is introduced in [2]. The pseudo-closure operator is defined by a logical formula Q in disjunctive normal form (DNF).

$$\forall A \in \mathcal{P}(E),\ a_Q(A) = \{x \in E \mid Q(A, x)\} \tag{4}$$

This new definition of the pseudo-closure operator is more generic than the previous one, but more importantly, it allows to *learn* a logical combination to define a pretopological space. The LPS (Learning Pretopological Spaces) framework [5] consists in learning a numerical function to define the pseudo-closure operator. This model enforces some limitations, especially the inability to model non-linear combinations. This can be fixed by a logical modelling of the pseudo-closure operator (as proposed in Eq. 4).

The LPS framework proposes to learn a pretopological space based on its expected elementary closed sets. Given a set $S^* = \{F^*(\{x\})\}_{\forall x \in E}$ of elementary closed sets and a list of predicates, LPS learns a positive DNF Q compounded of the predicates fed to the algorithm and such that the elementary closed sets S^* can be retrieved by the learned pseudo-closure operator $a_Q(.)$.

We propose to use the LPS framework to learn a pretopological space such that its elementary closed sets are actually ego-centered communities. To this purpose, we introduce two learning algorithms: LPSMI [2] and LPSFM.

LPSMI allows to learn only pretopological spaces of type V which are defined by a pseudo-closure operator satisfying the isotonic property: $\forall A, B \in \mathcal{P}(E)$, $A \subseteq B \Rightarrow a(A) \subseteq a(B)$. LPSFM does not enforce any restriction on the type of the pretopological space learned.

4 Community Extraction Method

We consider in the following the network in Fig. 1 and we note E its set of nodes. This network is roughly compounded of two communities, namely $\{a, b, c, d\}$ and $\{f, h, g\}$, in addition to the outsider node e. When considering the community centered on d, it is not clear if e belongs to its local community (the same apply for node f). On the contrary it is relatively safe to exclude e from the community centered on a.

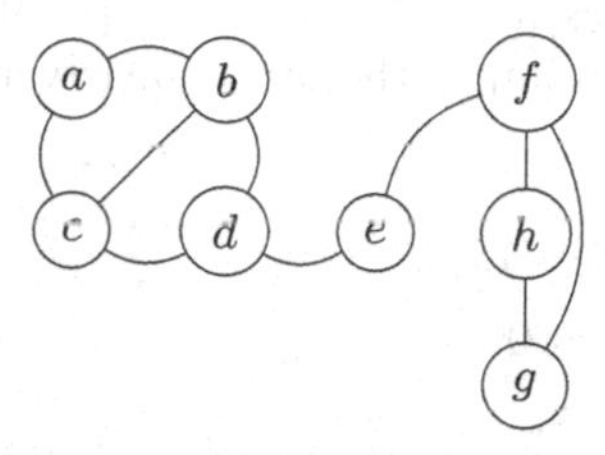

(a) A network with two communities.

	a	b	c	d	e	f	g	h
a	1.00	0.77	0.77	0.59	0.30	0.07	0.00	0.00
b	0.91	1.00	0.86	0.70	0.36	0.07	0.00	0.00
c	0.85	0.79	1.00	0.63	0.30	0.05	0.00	0.00
d	0.68	0.76	0.76	1.00	0.53	0.13	0.00	0.00
e	0.00	0.06	0.06	0.33	1.00	0.44	0.26	0.26
f	0.00	0.04	0.04	0.19	0.58	1.00	1.00	1.00
g	0.00	0.03	0.03	0.14	0.43	0.76	1.00	0.88
h	0.00	0.03	0.03	0.12	0.40	0.73	0.85	1.00

(b) Carryover-opinion proximity matrix.

Fig. 1. Example

4.1 Optimization Functions Targeted at Pretopological Spaces Learning

Given a set of elementary closed sets S^* (the ground truth), LPSMI and LPSFM does not rely on the same criterion to evaluate the quality of a pretopological space in construction and so leading the learning process.

The LPSMI's optimization criterion relies on the structural properties of pretopological spaces of type V to evaluate not only the quality of the learned DNF with regard to the closed sets it engenders, but also its potential relating to its non-elementary closed sets. The pretopological space learned in this fashion will thus be of type V. It is therefore essential that the predicates compounding the learned DNF Q satisfy the same properties as pretopological spaces of type V. That is to say, if a set $A \in \mathcal{P}(E)$ propagates to an element $x \in E$, then any super-set of A must propagates to x too. Let q be a predicate defined on $\mathcal{P}(E) \times E$, it is of type V if it satisfies the following property:

$$\forall A, B \in \mathcal{P}(E), x \in E\,,\, A \subseteq B \Rightarrow [q(A,x) \Rightarrow q(B,x)] \tag{5}$$

The ego-centered communities of the network in Fig. 1 cannot be expressed by the elementary closed sets of this type of pretopological spaces. We assume that the community centered on a is the set $\{a, b, c, d\}$, then the set $\{d\}$ cannot be propagated to e. While it is not absurd in this case, it is probably too restrictive in general. We hence introduce the FM criterion to allow the learning of unconstrained pretopological spaces.

The optimization criterion used by LPSFM is simpler since it does not fit account of the potential of the non-elementary closed sets. Instead it only relies on the raw matching (the F-measure) between the learned elementary closed sets and the ground truth. In practice this optimization criterion is less efficient when the task consists specifically in learning a V-type pretopological space [2]. It is however precious in last resort to lead unconstrained pretopological spaces learning algorithms. Any DNF Q that defines a pseudo-closure operator (satisfying Eqs. 1 and 2) is then allowed. Predicates must therefore satisfy the following properties.

$$\forall A \in \mathcal{P}(E),\, \forall x \in A, q(A,x) = 1 \tag{6}$$

$$\forall x \in E, q(\emptyset, x) = 0 \tag{7}$$

4.2 From Network Descriptors to Predicates

We propose a set of predicates dedicated to the task of ego-centered communities extraction. Each predicate can be seen as a descriptor of a network as it captures one of its features.

We still consider the network in Fig. 1 and we note E its set of nodes, A a subset of E and x an element in E. The whole set of predicates we propose can be organized into three distinct categories described below. The diversity of this set of predicates is a wonderful example showing the multi-criteria analysis ability offered by the pretopology formalism.

Topological Predicates. Let $V(x)$ the neighbors of x in the network and $V(A) = \bigcup_{x \in A} V(x)$ the union of the neighbors of each node in A. We consider only reflexive neighborhoods such that $x \in V(x)$ (hence $A \subseteq V(A)$). In the example, $V(a) = \{a, b, c\}$, $V(d) = \{b, c, d, e\}$ and $V(\{a, d\}) = V(a) \cup V(d) = \{a, b, c, d, e\}$.

A base predicate is defined from the raw adjacency matrix. It is noted $q_{adj}(A,x)$ and is true when an edge between a node in A and x exists.

$$q_{adj}(A,x) = \begin{cases} 1 \text{ if } x \in V(A) \\ 0 \quad \text{otherwise} \end{cases} \tag{8}$$

A good community is commonly described as having strong internal interactions. We propose four predicates defined by the neighborhoods of A and x which aim to capture the intensity of the interactions between A and x.

- $q_{r1}(A,x,k) = \frac{|A \cap V(x)|}{|A|} \geq k$
- $q_{r2}(A,x,k) = \frac{|A \cap V(x)|}{|V(x)|} \geq k$
- $q_{r3}(A,x,k) = \frac{|A \cap V(x)|}{|A \cup V(x)|} \geq k$
- $q_{r4}(A,x,k) = \frac{|V(A) \cap V(x)|}{|V(A) \cup V(x)|} \geq k$

with k a threshold setting in the interval $[0,1]$. Among these four predicates, only q_{r2} is a predicate of type V.

Modularity Based Predicates. Three predicates are defined by the three different local modularity measures given by Clauset [4], Luo [12] and Chen [3]. These predicates are noted $q_X(A,x)$ with $X \in \{clauset, luo, chen\}$. The predicate $q_X(A,x)$ is true if adding x to A makes its local modularity (noted $mod_X(A)$) to increase.

$$\forall A \in \mathcal{P}(E), \forall x \in E\,,\ q_X(A,x) = mod_X(A \cup \{x\}) > mod_X(A) \tag{9}$$

These predicates do not satisfy the properties of pretopological spaces of type V and thus cannot be used by the LPSMI approach.

Proximity Based Predicates. We define the predicate $q_{danisch}(A,x)$ from the carryover-opinion proximity measure [7]. This predicate is true when an element in A is close enough to x. Two elements are said to be close enough when their carryover-opinion proximity is greater than k, with k in the interval $[0,1]$.

$$\forall A \in \mathcal{P}(E), \forall x \in E\,,\ q_{danisch}(A,x,k) = \max_{y \in A}\{carryover(x,y)\} \geq k \tag{10}$$

Many other predicates can be considered, for example predicates based on graph embeddings such as those learned by node2vec. However such approaches require to know the entire network and thus do not suit the local nature of the task. Furthermore experiments shown the inefficiency of this type of predicates for the task of ego-centered community extraction.

4.3 Learning a Pretopological Space

Both LPSMI and LPSFM learn a logical formula in a greedy fashion. The learning process starts from an empty DNF to which conjunctive clauses are appended after each iteration of the algorithm. An iteration of the algorithm consists in finding the best conjunctive clause c knowing a DNF Q already learned during the previous iteration. The clause c is a conjunction of the predicates described above. It is similar to find the clause c maximizing a given optimization criterion for the DNF $Q \vee c$. Since an exhaustive search of the conjunctive clauses space is difficult, a beam search strategy is employed to find a good clause. The algorithm stops when either the maximum number of iterations is reached or the optimization criterion is not improved by the new clause.

The time complexity of these algorithms is mainly governed by the time complexity of the computation of the elementary closed sets. Let (E, a_Q) be a pretopological space, Q a positive DNF and A a subset of E. An application of the pseudo-closure operator $a_Q(A)$ consists in finding all the nodes $x \in E$ such that $Q(A, x)$ is true. Its time complexity is then $O(|E| \cdot O(Q))$ where $O(Q)$ is the time complexity of the evaluation of the DNF Q, which cannot be formally defined since it is different for each iteration. In the context of ego-centered community, A can only propagates to its neighbors, so the time complexity can be reduced to $O(|V(A)| \cdot O(Q))$. The closure operator $F(A)$ needs to apply the pseudo-closure operator until it converges. In the worst case $F(A) = E$ and each pseudo-closure application propagates to one element at a time, so $|E \setminus A|$ steps are required. The worst time complexity of the closure operator is then $O(|E \setminus A| \cdot |V(A)| \cdot O(Q))$. The time complexity of the computation of all the elementary closed sets is hence $O(|E| \cdot |E \setminus A| \cdot |V(A)| \cdot O(Q))$ which can be simplified by its upper bound $O(|E|^3 \cdot O(Q))$. The worst time complexity of the algorithm are then $O(maxiter \cdot |\mathcal{Q}|^2 \cdot beam \cdot |E|^3 \cdot O(Q))$ where *maxiter* is the maximum number of iterations, *beam* is the size of the beam for the beam search and $\mathcal{Q}$ is the set of predicates; $O(|\mathcal{Q}|^2 \cdot beam)$ is the time complexity of the beam search strategy.

As a final remark, we should mention that each elementary closed set can be computed independently of the others. So the computation of all the elementary closed sets can be done parallel which should drastically reduce the overall complexity.

4.4 Community Extraction from a Pretopological Space

In order to illustrate how an ego-centered community is extracted from an elementary closed set, let us again consider the network in Fig. 1. Let (E, a_Q) be a pretopological space with E the set of nodes in the network and Q the DNF defined by $Q = q_{danisch}(A, x, 0.5) \wedge q_{r1}(A, x, 0.5)$. Let the carryover-opinion proximity matrix given in Fig. 1b, the ego-centered community of the node a is obtained by calculating the elementary closed set $F(\{a\})$ in the pretopological space (E, a_Q).

$$\begin{aligned} a_Q(\{a\}) &= \{a, b, c\} \\ a_Q(\{a, b, c\}) &= \{a, b, c, d\} \\ a_Q(\{a, b, c, d\}) &= \{a, b, c, d\} = F_Q(\{a\}) \end{aligned}$$

The elementary closure of element a effectively matches a community in the network. This closed set is obtained by two subsequent applications of the pseudo-closure operator $a_Q(.)$. By construction of Q, the propagation from A to an element x requires that both predicates $q_{danisch}(A, x, 0.5)$ and $q_{r1}(A, x, 0.5)$ are satisfied. The singleton $\{a\}$ is then extended to nodes b and c because:

- on the one hand $carryover(a, b) \geq 0.5$ and $\frac{|\{a\} \cap V(b)|}{|\{a\}|} \geq 0.5$
- on the other hand $carryover(a, c) \geq 0.5$ and $\frac{|\{a\} \cap V(c)|}{|\{a\}|} \geq 0.5$.

The other node d is not reached by the first application of the pseudo-closure since $\frac{|\{a\} \cap V(d)|}{|\{a\}|} = 0$, even if $q_{danisch}(\{a\}, d, 0.5)$ is true. It will be reached throughout the second application thanks to the nodes b and c previously included since $\frac{|\{a,b,c\} \cap V(d)|}{|\{a,b,c\}|} \geq 0.5$.

Next, let's consider the community centered on d.

$$a_Q(\{d\}) = \{b, c, d, e\}$$
$$a_Q(\{b, c, d, e\}) = \{a, b, c, d, e\}$$
$$a_Q(\{a, b, c, d, e\}) = \{a, b, c, d, e\} = F_Q(\{d\})$$

The same community $\{a, b, c, d\}$ is once again retrieved to which node e is added. This community seems consistent with regard to the locality of d.

This example shows how the complex community structure of a network can be extracted from a well-defined pretopological space. The relevance of the DNF defining the pretopological space is the key to this "well-defined" concept. We are tackling this problem with the LPSMI and LPSFM methods.

5 Experiments

In this section we compare the results of the new proposed supervised approaches (LPSFM and LPSMI methods) with existing unsupervised methods devoted to the ego-centred community extraction task. The experiments are performed on two synthetic and two real networks.

5.1 Datasets

The first synthetic network contains 60 nodes distributed across three communities of equal sizes. It is obtained with a simple random model: first, each community is generated according to an Erdős-Rényi model with a probability of 0.2; then, the edges between each pair of nodes from different communities are added with a probability of 0.01.

The second synthetic network arises from the LFR benchmark [10] parameterized in order to obtain a network with 200 nodes. The average degree of the nodes is 15 with at most 30 neighbors per node. The mixing parameter is fixed to 0.3 and 40 nodes belong to three different communities among the 15 communities making up the network. The other parameters are left at their default value.

The first real network used is the famous Zachary's karate club [15]. It models the interactions between the 34 members of a karate club and is composed of two known communities.

The second network from real data describes the interactions between the American football teams of a university championship[1] [3]. The network contains 179 nodes (football teams) and 787 edges (matches); 115 teams are distributed across 11 pre-identified communities and 64 teams are not in any community.

5.2 Experimental Setup and Results

The quality of a community extraction method can be evaluated according to the matching between the obtained vs. expected communities by mean of the F-measure. Let E the set of nodes of the network, an ego-centered community $C(x)$ is extracted from each node $x \in E$ and compared to the expected community noted $C^*(x)$. For the new proposed pretopological methods, the ego-centered community is obtained by the elementary closed set $(C(x) = F(\{x\}))$. The definitions of the precision (P), the recall (R) and the resulting F-measure are reminded in (11).

$$P = \frac{\sum\limits_{x \in E} |C(x) \cap C^*(x)|}{\sum\limits_{x \in E} |C(x)|} \; ; \; R = \frac{\sum\limits_{x \in E} |C(x) \cap C^*(x)|}{\sum\limits_{x \in E} |C^*(x)|} \; ; \; FM = 2 \cdot \frac{P \times R}{P + R} \quad (11)$$

We compare the scores obtained with the new proposed pretopological methods with the ones resulting of the methods from Clauset [4], Luo [12], Chen [3] and Danisch [7]. The methods from Clauset, Luo and Chen provide ego-centered communities by aggregation of nodes maximizing a modularity score. Danisch's approach assumes the proximity of a node with its local community is significantly higher than with nodes from other communities; he shows that, for a given node, the proximity *carryover-opinion* curve reveals successive plateaus separated by sharp decreasing. Different granularity levels are obtained depending of the number of considered plateaus. In the following experiments, three levels of granularity have been used by considering the first two, three or four plateaus; the corresponding methods are noted "Danisch2", "Danisch3" and "Danisch4" respectively.

The LPSMI approach allows to combine only V-type predicates unlike the LPSFM approach that benefits of the whole set of predicates to learn the pretopological space. In the experiments, the threshold is fixed to $k = 0.3$ for the topological predicates, and two predicates are derived from Danisch by considering different parameters : $q_{danisch}(k = 0.15)$ and $q_{danisch}(k = 0.3)$.

LPSMI and LPSFM are evaluated using a 5-fold cross-validation. The results reported in Table 1 are the means of the F-measures obtained by each method

[1] http://www.espn.com/college-football/standings/_/season/2006.

on the (same) five test sets corresponding each to 20% of the communities to extract. Let us notice that the communities serving as references are derived from partitions (except for LFR that contains overlaps) of the network ; actually such references are therefore approximations of the ego-centered communities. The computed scores do not reflect the exact quality of the methods, however they stay the best (and rather good) available indicators at this stage. Finally, the best logical rules learned with LPSFM are given in Table 2.

Table 1. Scores obtained by different ego-centered community extraction methods on four networks. Supervised methods are identifiable by the * symbol.

Method	Erdős-Rényi	Karate	Foot	LFR
Clauset	0.45 ± 0.08	0.68 ± 0.11	0.53 ± 0.07	0.50 ± 0.06
Luo	0.74 ± 0.06	0.82 ± 0.08	0.57 ± 0.07	0.57 ± 0.05
Chen	0.39 ± 0.11	0.37 ± 0.05	0.88 ± 0.04	0.46 ± 0.06
Danisch2	0.70 ± 0.01	0.79 ± 0.05	0.63 ± 0.03	0.43 ± 0.06
Danisch3	0.80 ± 0.03	$\mathbf{0.89 \pm 0.03}$	0.65 ± 0.03	0.51 ± 0.06
Danisch4	0.82 ± 0.04	0.88 ± 0.01	0.52 ± 0.05	0.56 ± 0.06
LPSMI*	0.50 ± 0.00	0.67 ± 0.01	0.31 ± 0.07	0.34 ± 0.09
LPSFM*	$\mathbf{0.85 \pm 0.02}$	0.80 ± 0.07	$\mathbf{0.96 \pm 0.03}$	$\mathbf{0.65 \pm 0.04}$

The scores obtained confirm first that V-type pretopological spaces (generated from LPSMI) are not suitable for modeling local communities. On the other hand, the unconstrained spaces generated with LPSFM are suitable and outperform significantly any other method on most of the datasets. This result reveal in addition the practical interest of the supervision for the community extraction task. Indeed, the supervision allows to take into account the properties of a given network in order to generate a fitting extraction model. Conversely, the performances of the other (unsupervised) methods are highly network dependent. As an example, the communities extracted by the Chen's method on the network *Foot* are good comparatively to the ones from Luo; and it is the opposite on the network *Karate*. The approach from Danisch appears to be more stable across the datasets but remains threshold sensitive.

Table 2. Examples of rules learned with LPSFM. The clauses are presented in the same order they are introduced in the DNF.

Networks	Logical rules
Erdős-Rényi	$(q_{luo} \wedge q_{danisch}(k=0.3)) \vee (q_{r2} \wedge q_{luo}) \vee (q_{adj} \wedge q_{r4} \wedge q_{danisch}(k=0.3))$
Karate	$q_{luo} \wedge q_{danisch}(k=0.15)$
Foot	$(q_{r4}) \vee (q_{r3} \wedge q_{danisch}(k=0.15))$
LFR	$q_{r1} \wedge q_{luo}$

Model Generalization. In order to evaluate its degree of genericity, each pretopological model learned on a given network with LPSFM has been reused to extract ego-centered communities on the other networks. The values reported in Table 3 are the means of the scores obtained by each of the five models learned and tested on the full dataset (the scores on the diagonal are therefore not exactly similar to the ones observed in Table 1).

Table 3. Generalization score obtained by pretopological models learned with LPSFM.

	Erdős-Rény	Karate	Foot	LFR
Erdős-Rény	0.85 ± 0.01	0.62 ± 0.07	0.29 ± 0.21	0.07 ± 0.00
Karate	0.75 ± 0.03	0.80 ± 0.04	0.47 ± 0.14	0.34 ± 0.11
Foot	0.41 ± 0.00	0.59 ± 0.00	0.97 ± 0.00	0.41 ± 0.00
LFR	0.54 ± 0.01	0.74 ± 0.00	0.60 ± 0.00	0.65 ± 0.00

As expected, given that the semantics of the logical rules highly differs from one network to another (Table 2), the pretopological models obtain lower performances when applied on networks unknown a priori. However the apparent robustness of the models derived from the LFR network allows us to consider in perspective a way to lift this lock.

Finally, the new proposed learning strategy opens the way to solutions that exploit all the aspects of a network. The logical formalism used offers a way to combine any type of feature. As an example, the rule learned on the Erdős-Rényi network combines a low-level topological descriptor (q_{r2}) with higher-level one (q_{luo}). On the other hand, this formalism makes the learned model understandable; it's clear that the rule learned on the Erdős-Rényi network is driven by the two predicates q_{luo} and $q_{danisch}(k = 0.3)$. This rule reveals that, on the considered network, the two predicates are (1) additional because they don't appear together in the second and third clauses and (2) too permissive because they must be "controlled" by another predicate.

6 Conclusion

The main contributions in this paper concern:

1. the formalization of the ego-centered community extraction task using the recent techniques for learning pretopological spaces (LPS),
2. the definition of a first collection of local features (as logical predicates), considering the main existing methodologies for the task,
3. the experimental validation on both, real and simulated data, revealing that suitable pretopological spaces exists that are reachable in a supervised learning strategy.

The study shows the suitability of the pretopology for the community extraction task in networks. The use of this theory allows to increase performance with respect to usual methods of ego-centered community detection. The structuring operators offered by pretopolgy have the ability to express various types of interactions between sets of nodes in a natural and elegant way. This ability is crucial when information from different levels have to be capitalized from the network. The present work paves the way to numerous perspectives; some of them are announced hereafter.

Firstly, this work highlighted that a V-type pretopolgy is not suitable for the ego-centered community extraction task by mean of elementary closed sets. However, more investigations have to be led in order to consider other definitions for the concept of community in a V-type pretopological space.

On the other hand, even if the proposed pretopological approach leads to qualitative communities, the underlying models obtain low performances on networks on which they were not trained; but such a training requires in practice a costly recourse to experts making the whole (supervised) strategy crippling[2]. A way to lift the "expert dependent" lock consists in learning pretopological spaces in a totally unsupervised manner. In this perspective, we observed the rather good generalization abilities of the pretopological spaces learned on the LFR network. This result suggests the possibility - recently investigated in [11] - to produce automatically training data from artificial networks satisfying structural properties similar to a target real network.

References

1. Belmandt, Z.: Basics of Pretopology. Hermann, Paris (2011)
2. Caillaut, G., Cleuziou, G.: Learning pretopological spaces to model complex propagation phenomena: a multiple instance learning approach based on a logical modeling. arXiv preprint arXiv:1805.01278 (2018)
3. Chen, J., Zaïane, O.R., Goebel, R.: Local community identification in social networks. In: ASONAM, pp. 237–242 (2009)
4. Clauset, A.: Finding local community structure in networks. Phys. Rev. E **72**(2), 026132 (2005)
5. Cleuziou, G., Dias, G.: Learning pretopological spaces for lexical taxonomy acquisition. In: Appice, A., Rodrigues, P.P., Santos Costa, V., Gama, J., Jorge, A., Soares, C. (eds.) ECML PKDD 2015. LNCS (LNAI), vol. 9285, pp. 493–508. Springer, Cham (2015). https://doi.org/10.1007/978-3-319-23525-7_30
6. Dalud-Vincent, M.: Une autre manière de modéliser les réseaux sociaux. Applications à l'étude de co-publications. Nouvelles perspectives en sciences sociales **12**(2), 41–68 (2017)
7. Danisch, M., Guillaume, J., Grand, B.L.: Towards multi-ego-centred communities: a node similarity approach. IJWBC **9**(3), 299–322 (2013)
8. Figueiredo, D.R., Ribeiro, L.F.R., Saverese, P.H.P.: struc2vec: learning node representations from structural identity. In: KDD (2017)

[2] Let us notice that in practice, few examples are sufficient for driving efficiently the learning process.

9. Grover, A., Leskovec, J.: node2vec: scalable feature learning for networks. In: KDD, pp. 855–864. ACM (2016)
10. Lancichinetti, A., Fortunato, S.: Benchmarks for testing community detection algorithms on directed and weighted graphs with overlapping communities. Phys. Rev. E **80**(1), 016118 (2009)
11. Lu, X., Kuzmin, K., Chen, M., Szymanski, B.K.: Adaptive modularity maximization via edge weighting scheme. Inf. Sci. **424**(C), 55–68 (2018)
12. Luo, F., Wang, J.Z., Promislow, E.: Exploring local community structures in large networks. Web Intell. Agent Syst. **6**(4), 387–400 (2008)
13. Newman, M.E.: Modularity and community structure in networks. Proc. Nat. Acad. Sci. **103**(23), 8577–8582 (2006)
14. Palla, G., Derényi, I., Farkas, I., Vicsek, T.: Uncovering the overlapping community structure of complex networks in nature and society. Nature **435**(7043), 814 (2005)
15. Zachary, W.W.: An information flow model for conflict and fission in small groups. J. Anthropol. Res. **33**(4), 452–473 (1977)

EigenPulse: Detecting Surges in Large Streaming Graphs with Row Augmentation

Jiabao Zhang[1,2], Shenghua Liu[1,2(✉)], Wenjian Yu[3(✉)], Wenjie Feng[1,2], and Xueqi Cheng[1,2]

[1] CAS Key Laboratory of Network Data Science and Technology, Institute of Computing Technology, Chinese Academy of Sciences, Beijing, China
liu.shengh@gmail.com

[2] University of Chinese Academy of Sciences, Beijing, China
zhangjiabao18@mails.ucas.edu.cn

[3] BNRist, Department of Computer Science and Technology, Tsinghua University, Beijing, China
yu-wj@tsinghua.edu.cn

Abstract. How can we spot dense blocks in a large streaming graph efficiently? Anomalies such as fraudulent attacks, spamming, and DDoS attacks, can create dense blocks in a short time window, emerging a surge of density in a streaming graph. However, most existing methods detect dense blocks in a static graph or a snapshot of dynamic graphs, which need to inefficiently rerun the algorithms for a streaming graph. Moreover, some works on streaming graphs are either consuming much time on updating algorithm for every incoming edge, or spotting the whole snapshot of a graph instead of the attacking sub-block.

Therefore, we propose a row-augmented matrix with sliding window to model a streaming graph, and design the *AugSVD* algorithm for computation- and memory-efficient singular decomposition. *EigenPulse* is then proposed to spot the density surges in streaming graphs based on the singular spectrum. We theoretically analyze the robustness of our method. Experiments on real datasets with injections show our performance and efficiency compared with the state-of-the-art baseline.

Keywords: Surge detection · Streaming graphs · Sliding window

1 Introduction

The surges of density in some subgraph are a strong signal to detect anomalies in streaming graphs [2,13]. For example, the controlled user accounts rate fake and high scores to a set of target objects in a short period of time, in Amazon, Yelp, App stores, etc. The spamming phone calls/msgs are sent intensively from a group of phone numbers to another group. And the attacks to a set of servers of target websites from a large pool of IPs. Those cases will result a very dense

Q. Yang et al. (Eds.): PAKDD 2019, LNAI 11440, pp. 501–513, 2019.
https://doi.org/10.1007/978-3-030-16145-3_39

subgraph between some users and objects, phone numbers, and IPs in a short time window. Thus the question is raised:

How can we detect such a dense subgraph, and spot the density surge in a large streaming graph in an efficient and accurate way?

Many existing dense subgraph detection methods, such as M-zoom [11], D-Cube [12], HoloScope [9], have achieved satisfied accuracy in large static graphs. Re-running those methods once a batch of new data comes is very time-consuming and low efficiency, in a streaming graph. The recent work, SpotLight [2], can efficiently detect the sudden changes of a snapshot of the graph in a time period. It was not able to tell which specific part of the snapshot is attacked. DenseAlert [13] detects dense subgraph using an incremental and heuristic algorithm, which updates for every single incoming edge, in order to have a high accuracy. This slows down the algorithm, even though DenseAlert is faster than the batch methods.

Therefore, we reasonably model the streaming graph as a row-augmented matrix, and propose, EigenPulse, to detect surges in large streaming graphs, based on the singular spectrum of the matrix. To get the singular spectrum of a row-augmented matrix, we propose AugSVD for singular decomposition of the streaming graph in a sliding window. Even if attacks may cross windows, we can still detect them since the windows intersect. AugSVD outputs the singular spectrum of every stride, and EigenPulse algorithm calculate the density of a subgraph in first several singular vectors and detect anomalies. The experiments on 5 real data sets show that EigenPulse can detect the suspicious surges of density of some subgraph, achieving high accuracy as the baselines, but consuming much less computation time.

In summary, the main advantages of our algorithms are:

- **Incremental singular value decomposition:** we propose a scalable algorithm, AugSVD, to decompose large streaming graphs, which can output the spectral values of graph nodes at each time window, as long as the graphs can be formulated as matrices augmented in rows.
- **Robust and effective:** we theoretically analyze that the robust approximation of AugSVD to batch SVD. The experiments show that the EigenPulse generated by AugSVD can detect suspicious synchronized activities accurately in real-world graphs.
- **Scalable:** EigenPulse is computation- and memory-efficient. Compared with the state-of-the-art baseline, EigenPulse can be more than 5 times faster.

2 Related Work

2.1 Anomaly Detection in Static Graphs

Anomaly detection in static graphs is well studied in [1]. For example, methods based on spectral decomposition, e.g., EigenSpokes [10], which discovers that the induced sub-graph of the 20 nodes which had the highest magnitude projection along the singular vector almost contains near-cliques. Many existing methods rely on graph's density, e.g., Fraudar [6], several methods even taking into account rating variation and burst of attacks, e.g., CrossSpot [7] and HoloScope which is the only one considers temporal spikes and hyperbolic topology.

2.2 Anomaly Detection in Streaming Graphs

We summarize dense subgraph detection algorithms in streaming graphs. Traditional methods just compare the changes of adjacent graphs via a similarity function based on, e.g., belief propagation [8]. They do not consider evolutionary/periodic trends. Many existing methods, e.g., DenseAlert model dynamic graphs as streaming tensors and aim to approximately identify the top-k densest subblocks, i.e., maintained dense subtensors. DenseAlert divide time into bins and can detect sudden emerging dense subtensors, same with EigenPulse. In contrast, Spotlight detects only the sudden appearance or disappearance of dense blocks in real-time by using randomized sketching-based approach. [15] designs an algorithm MASCOT for counting local triangles to detect anomalies in graph streams. There are some methods based on graph decomposition and partitioning, such as [14] storing a summary of the graph structure based on tensor decomposition and identify change points as anomalies. Some randomized algorithms, i.e., [4] defines a robust random cut data structure that can be used as a sketch or synopsis of the input stream. Some other methods find patterns for anomaly detection, e.g., [3] investigates continuous pattern detection over large evolving graphs with snapshot isolation.

3 Proposed Method

We define a bipartite graph $\mathcal{G}$ to represent the relationships between users and rating objects, callers and callees, attacking IPs and target IPs, etc. The problem that detects surges of density in a large streaming graph $\mathcal{G}$ is described as follows:

Informal Problem 1 (Detecting Density Surges) ***Given:*** *a stream of triplets (user, object, timestamp), where timestamp is the time when a user creates an edge on an object, and a time window of width w,*

- ***find:*** *at each time step t, calculate the density D_t of the subgraph that is the densest one in streaming graph $\mathcal{G}$ within current time window;*
- *detect suspicious surges of density that are above a threshold.*

In our problem, a triplet $(user, object, timestamp)$ is a new edge created in streaming graph $\mathcal{G}$. The triplets come in an order of *timestamp*. A streaming graph $\mathcal{G}$ within a time window indicates that only the triplets coming in the time window are considered as the edges in graph $\mathcal{G}$. The time windows are sliding at each time step.

3.1 Our Model

To develop a fast algorithm for singular decomposition, we model large streaming graph $\mathcal{G}$ as a row-augmented matrix $\mathbf{A}$.

Row-Augmented Matrix: Matrix which is modified in a row augmented manner. For each new piece of data, its corresponding row is incremental or just same with the last row of current matrix.

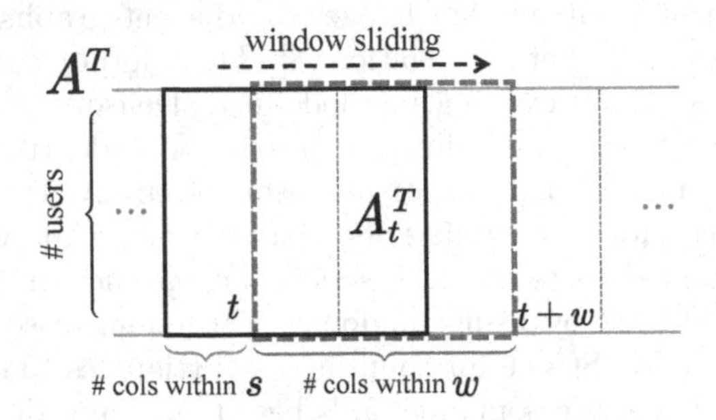

Fig. 1. The sliding window for the row-augmented matrix. w is the window size in a time unit, s is the stride size in a time unit. Note that the number of columns in different time windows may be different.

Figure 1 shows the sliding window for $\mathbf{A}^T$, which $\mathbf{A}$ is the row-augmented matrix. The columns of row-augmented matrix $\mathbf{A}$ represent the *user* ids in streaming graph $\mathcal{G}$. The rows are increasing, and each row is a combination of an *object* id and *timestamp*/s, where s is the stride to a next step. Such ids guarantee the rows coming in the next step are totally different, keeping the property of row-augmented matrix. Note that our model of $\mathbf{A}$ is actually batch-row augmented, and the batch size is decided by the number of new edges between stride s.

We now explain why we can model a streaming graph $\mathcal{G}$ with a row-augmented matrix $\mathbf{A}$. One reason is that since fraudsters and attackers create edges in a relatively short period of time, combining the object ids with binned time can still show a dense block in our matrix $\mathbf{A}$ for anomalies. Besides, an object in different time can mean differently, e.g. the same app may be different versions at different time, the same restaurant and product may have improvement or new generation at later time. With such a combination, one can consider those differences. In another case, a piece of twitter message or news as an object probably no users will read after a short while, which reduces the bias of our model. Finally, the most important of all, such a model, can help us achieve fast algorithm to detect density surges, which is described in the following sections.

Similarly, we can still introduce a sliding window for row-augmented matrix $\mathbf{A}$ as show in Fig. 1. With such a window, we can make algorithm focus on the most recent edges in graph $\mathcal{G}$. When assigning the width of window w as infinity, we can consider all the history at every time step. Or we can have non-overlapped dense subgraphs by setting $w = s$.

Algorithm 1. AugSVD with sliding window

Input: row-augmented matrix $\mathbf{A}$ with sliding window w, column size n, rank parameter k, block size b, two queues *glist* and *hlist*.

1: Choose $l = tb$, where t is an integer, so that l is slightly larger than k
2: $\mathbf{\Omega} = randn(n, l)$; $\mathbf{G} = [\,]$; set $\mathbf{H}$ to an $n \times l$ zero matrix
3: **if** *glist* is not empty **then**
4: *glist.dequeue*(); *hlist.dequeue*()
5: **end if**
6: **repeat**
7: Read rows $\mathbf{a}$ for next stride s
8: $\mathbf{g} = \mathbf{a}\mathbf{\Omega}$; $\mathbf{h} = \mathbf{a}^T\mathbf{g}$
9: *glist.enqueue*($\mathbf{g}$); *hlist.enqueue*($\mathbf{h}$)
10: **until** the elements in *glist* corresponds to a window w
11: **for all** $\mathbf{g}$ in *glist*, $\mathbf{h}$ in *hlist* **do**
12: $\mathbf{G} = [\mathbf{G}, \mathbf{g}]$; $\mathbf{H} = \mathbf{H} + \mathbf{h}$
13: **end for**
14: $\mathbf{Q} = [\,]$; $\mathbf{B} = [\,]$
15: **for** $i = 1, 2, \cdots, t$ **do**
16: $\mathbf{\Omega}_i = \mathbf{\Omega}(:, (i-1)b + 1 : ib)$; $\mathbf{Y}_i = \mathbf{G}(:, (i-1)b + 1 : ib) - \mathbf{Q}(\mathbf{B}\mathbf{\Omega}_i)$
17: $[\mathbf{Q}_i, \mathbf{R}_i] = qr(\mathbf{Y}_i)$
18: $[\mathbf{Q}_i, \widetilde{\mathbf{R}}_i] = qr(\mathbf{Q}_i - \mathbf{Q}(\mathbf{Q}^T\mathbf{Q}_i))$
19: $\mathbf{R}_i = \widetilde{\mathbf{R}}_i\mathbf{R}_i$
20: $\mathbf{B}_i = \mathbf{R}_i^{-T}(\mathbf{H}(:, (i-1)b + 1 : ib)^T - \mathbf{Y}_i^T\mathbf{Q}\mathbf{B} - \mathbf{\Omega}_i^T\mathbf{B}^T\mathbf{B})$
21: $\mathbf{Q} = [\mathbf{Q}, \mathbf{Q}_i]$; $\mathbf{B} = [\mathbf{B}^T, \mathbf{B}_i^T]^T$
22: **end for**
23: $[\widetilde{\mathbf{U}}, \mathbf{S}, \mathbf{V}] = svd(\mathbf{B})$
24: $\mathbf{U} = \mathbf{Q}\widetilde{\mathbf{U}}$
25: $\mathbf{U} = \mathbf{U}(:, 1 : k)$; $\mathbf{V} = \mathbf{V}(:, 1 : k)$; $\mathbf{S} = \mathbf{S}(1 : k, 1 : k)$
26: **return** $\mathbf{U}, \mathbf{S}, \mathbf{V}$

3.2 AugSVD Algorithm

AugSVD is designed for fast singular decomposition of row-augmented matrix $\mathbf{A}$ with sliding windows. It involves only one pass over the data and having accuracy guarantees. The algorithm is described as Algorithm 1.

Initially, for the first window, the queues *glist* and *hlist* are empty. The AugSVD algorithm outputs the first k singular values and vectors for the data observed through the first window. At the second time invoking the algorithm, the window slides one stride forward to form the next window.In such a way, repeatedly calling AugSVD results in the singular vectors of row-augmented matrix $\mathbf{A}$ observed from the sliding windows. Such an algorithm outperforms the standard SVD and other existing randomized algorithms by largely reducing runtime and memory usage.

In Algorithm 1, Steps 3 through 13 prepares matrices $\mathbf{G}$ and $\mathbf{H}$ for the sliding window, while Steps 14 through 25 are just the same as those in the single-pass PCA algorithm [16]. Due to the accumulation of round-off errors, the orthonormality among the columns in $\{\mathbf{Q}_1, \mathbf{Q}_2, \cdots\}$ may lose. To fix this issue, $\mathbf{Q}_i$ is explicitly re-projected away from the span of the previously computed basis vectors (Step 19), just as what is done in [16].

Theorem 1. *The AugSVD algorithm is mathematically equivalent to the basic randomized SVD algorithm in [5] for the row augmented matrix* $\mathbf{A}$.

Proof. As stated before, the AugSVD algorithm is the same as the single-pass PCA algorithm for streaming data in the sliding window, i.e. the row augmented matrix $\mathbf{A}$. It has been proved in [16] that the single-pass PCA algorithm is mathematically equivalent to the basic randomized SVD algorithm in [5]. So, the theorem is proved.

Based on Theorem 1, the AugSVD algorithm inherits the theoretical error bound (if ignoring the round-off error) [5]:

$$\mathbb{E}\|\mathbf{A}-\mathbf{Q}\mathbf{Q}^T\mathbf{A}\| \leq \left(1+\sqrt{\frac{k}{s-1}}\right)\sigma_{k+1} + \frac{\mathrm{e}\sqrt{k+s}}{s}\left(\sum_{j=k+1}^{\min(m,n)}\sigma_j^2\right)^{1/2} \tag{1}$$

where $\mathbb{E}$ denotes expectation, $s = l - k$. If choosing $s = k + 1$, we have

$$\mathbb{E}\|\mathbf{A}-\hat{\mathbf{U}}\hat{\mathbf{\Sigma}}\hat{\mathbf{V}}^T\| \leq 2\sigma_{k+1} + \frac{\mathrm{e}\sqrt{2k+1}}{k}\left(\sum_{j=k+1}^{\min(m,n)}\sigma_j^2\right)^{1/2} \tag{2}$$

Here, we have substituted the computed SVD factors: $\hat{\mathbf{U}}$, $\hat{\mathbf{\Sigma}}$ and $\hat{\mathbf{V}}$ with the single-pass PCA algorithm. Applying a rough analysis, we have

$$\mathbb{E}\max_{i=1,\ldots,k}|\sigma_i-\hat{\sigma}_i| = \mathbb{E}\|\mathbf{\Sigma}-\hat{\mathbf{\Sigma}}\| \leq 2\sigma_{k+1} + \frac{\mathrm{e}\sqrt{2k+1}}{k}\left(\sum_{j=k+1}^{\min(m,n)}\sigma_j^2\right)^{1/2} \tag{3}$$

where σ_i and $\hat{\sigma}_i$ are the accurate and computed the i-th singular value, respectively. Moreover, it can be shown that the likelihood of a substantial deviation from the expectation is extremely small; see Sec. 10.3 of [5] for a proof. This means the expectation symbol in (3) can be removed in an approximate sense. This results in:

$$|\sigma_i-\hat{\sigma}_i| \lesssim 2\sigma_{k+1} + \frac{\mathrm{e}\sqrt{2k+1}}{k}\left(\sum_{j=k+1}^{\min(m,n)}\sigma_j^2\right)^{1/2}, \ i=1,\ldots,k \tag{4}$$

where $\lesssim$ means less than approximately. The right-hand side of (4) means that the error on singular value depends not only on the $(k+1)$-th singular value but also the summation of its subsequent singular values. If $\mathbf{A}$'s singular values do not decay slowly, the second right-hand-side term in (4) would be relatively small, which means the computed singular value has sufficient accuracy.

Algorithm 2. EigenPulse

Input: time t, matrix $\mathbf{A}_t$ within time window $[t-w, t]$, row size m_t and column size n, a pair of left/right singular vectors $\mathbf{u}_t$, $\mathbf{v}_t$.

1: $rowset = [\,]$; $colset = [\,]$
2: $\tau_u = \frac{1}{\sqrt{m_t}}$; $\tau_v = \frac{1}{\sqrt{n}}$
3: **for** $i = 1, \cdots, m_t$ **do**
4:   **if** $abs(u_t[i]) >= \tau_u$ **then**
5:     $rowset.append(i)$
6:   **end if**
7: **end for**
8: **for** $j = 1, \cdots, n$ **do**
9:   **if** $abs(v_t[j]) >= \tau_v$ **then**
10:     $colset.append(j)$
11:   **end if**
12: **end for**
13: [optional] $rowset, colset = dense\_block\_detection(\mathbf{A}_t, rowset, colset)$
14: **return** $D_t(rowset, colset)$

3.3 EigenPulse Algorithm

As we known, the nodes in a dense subgraph probably correspond to larger absolute values in the first several singular vectors. The EigenPulse algorithm is used to detect such subgraph and calculate the density measure based the singular vectors computed with AugSVD. It is described as Algorithm 2.

In Algorithm 2, τ_u and τ_v are two thresholds for left and right singular vectors respectively. The density measure is calculated as:

$$D_t(rowset, colset) = \frac{\sum_{i \in rowset} \sum_{j \in colset} \mathbf{A}_t(i,j)}{|rowset| + |colset|} \tag{5}$$

We calculate this density measure for every time window. If it is obviously larger in a window than that in other windows, the window is very suspicious. Optionally, we can use an existing *dense_block_detection* algorithm, such as Fraudar and HoloScope (HS-α), to further shave *rowset* and *colset* to find a densest subblock, which is efficient for a reduced rows and columns (see step 2–12 in Algorithm 2), and benefit from the existing algorithms. To detect suspiciously surging window, one can simply combine mean value with standard deviation of historical density values as a threshold to take out suspicious windows. By this way, we greatly reduce the data needs to be detected than static methods.

4 Experiments

We design experiments to answer the following questions:

Q1. Efficiency: How fast does EigenPulse analyze the real world data compared with competitor?
Q2. Accuracy: How accurately does EigenPulse detect dense blocks?
Q3. Scalability: Does EigenPulse scale linearly with the size of tensor?

Table 1. Datasets statistic information

Name	Nodes	Edges	Span time
BeerAdvocate	26.5K × 50.8K	1.08M	Jan 2008 – Nov 2011
Yelp	686K × 85.3K	2.68M	Oct 2004 – Jul 2016
Amazon Cellphone	2.26M × 329K	3.45M	Jan 2007 – Jul 2014
Amazon Electronics	4.20M × 476K	7.82M	Dec 1998 – Jul 2014
Amazon Grocery	763K × 165K	1.29M	Jan 2007 – Jul 2014
Sina Weibo	2.74M × 8.08M	50.06M	Sep 2013 – Dec 2013

4.1 Experimental Settings

Machine: We ran all experiments on a machine with 2.7 GHz Intel Xeon E7-8837 CPUs and 512 GB memory.

Data: Table 1 lists the real-world datasets used in our experiments. All of the data are 4-way tensors (users, items, timestamps, ratings) where entry values are the number of reviews. In addition, the AugSVD can only decompose matrices augmented in rows, so we first filter the data with high rating scores, then concatenate these items by the timestamp as row, user as column, and thus the row of modified tensor grows with the forward of time.

Implementations: We chose the state-of-the-art streaming dense-subtensor detection algorithm, DenseAlert, as baseline. In all the experiments, we used sparse tensor format and only considered the first pair left/right singular vector.

4.2 Q1.Efficiency

As we see, EigenPulse chooses suspicious windows based on AugSVD, and then combines other shaving algorithms to obtain smaller dense blocks, finally identifies the fraudulent blocks with density measure. Other streaming methods, however, e.g., DenseAlert needs to update dense subtensor every time when coming a new tensor and SpotLight maintains a streaming tensor about graph sketch information in real-time. So EigenPluse is faster than those algorithms.

We measured the wall-clock time taken by EigenPluse and DenseAlert for analyzing the first 5 datasets and showed the results in the Fig. 2(a). The EigenPluse achieves **2.53× faster** than DenseAlert, or even achieves **12.2×** speed up in BeerAdvocate dataset. According to the performance results of DenseAlert, which is a million times faster than the fastest batch algorithms, e.g., M-Zoom or CP Decomposition. We can draw the conclusion that EigenPluse significantly outperforms most of the state-of-the-art competitors.

In addition, EigenPulse is memory-efficient for only calculating one window's data each time, which up to **2.33 GB** memory consumed.

4.3 Q2.Accuracy

This experiment demonstrates the accuracy of EigenPulse for detecting dense blocks in different datasets.

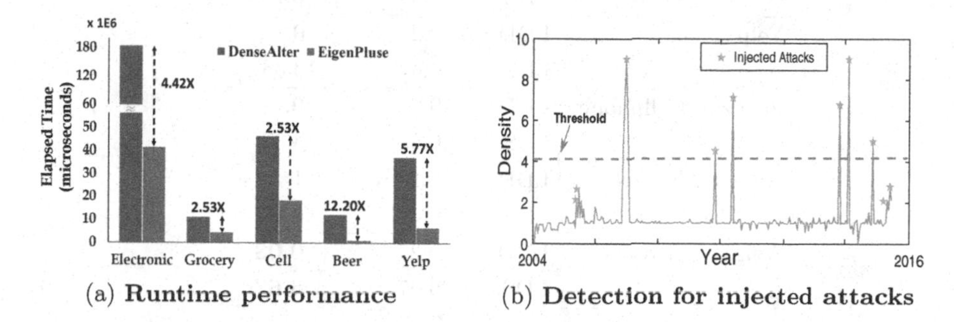

(a) **Runtime performance** (b) **Detection for injected attacks**

Fig. 2. EigenPluse performance: (a). EigenPulse consistently outperforms DenseAlert on 5 datasets, and achieves more than **2.53**$\times$ speed up. ('Beer' denotes BeerAdvocate); (b). EigenPluse successfully detects most of the injected attacks on Yelp dataset.

Detection of Injected Attacks: Here, we set $w = 30$ and $s = 10$ in days.

We injected dense blocks with different densities and different speeds to identify the lowest detection density(LDD) and the lowest detection speed(LDS). The unit of detection speed is $(\#edges/days)$, referring to the maximum number of injected edges in one day which we can detect. To identify the LDS, we keep the injected density unchanged and change the time span until the F-measure value is less than 90%. We randomly choose 20 products whose in-degrees are no more than 100 because they are more likely to buy fake reviews. Since data in windows is part of all the data, so we should inject small blocks into windows. For 0.1 may be a suitable density, we sample out 200 fraudsters as a whole to randomly create 20 edges on each of the 20 products, and also create biased camouflage on other products. Then, we just vary the time span across the data to find out the LDS for each dataset. Having identified the LDS, We choose a proper time span, e.g., 30 days, then inject blocks with different densities until the F-measure value is less than 90% to identify the LDD. We randomly choose 20 products as mentioned above and sample out fraudsters ranges from 20 to 2000 to generate fraudulent blocks with densities ranges from 1.0 to 0.01 for testing to find the lowest fraudulent density. The time was generated for each fraudulent edge: randomly choosing a time in the window range.

In order to give a comparison with DenseAlert, we compare the LDD and LDS on the first 5 datasets in Table 2. As we can see that EigenPulse has the lower LDD than DenseAlert except on Amazon Electronics dataset and has the lower LDS than DenseAlert. In detail, EigenPluse has the LDD which can be as small as 250 on source nodes on Yelp dataset and Amazon Cellphone dataset with the minimum density of 0.08 on sink nodes, which means we can detect fraudsters even if they use 250 accounts to create 20×20 edges across 30 days.

Table 2. Experimental results on real data with injected labels

Data Name	Metrics	DenseAlert	EigenPulse
BeerAdvocate	LDD	0.1	0.1
	LDS	13.33	**6.67**
Yelp	LDD	0.2	0.1
	LDS	26.67	13.33
Amazon Cellphone	LDD	0.2	**0.08**
	LDS	26.67	13.33
Amazon Electronics	LDD	**0.2**	1
	LDS	26.67	**6.67**
Amazon Grocery	LDD	0.2	**0.08**
	LDS	26.67	**6.67**

Besides, we injected 10 dense blocks with density vary from 0.01 to 1 for the Yelp dataset. The Fig. 2(b) shows the densities of all the windows on the EigenPulse. We can see that the injected dense blocks cause significant density surges. By assuming the density follows a Normal distribution, we successfully detect 6 injected blocks after simply set the density detection threshold as $\mu+3\sigma$, where μ and σ are the mean and standard deviation of all windows' density measures.

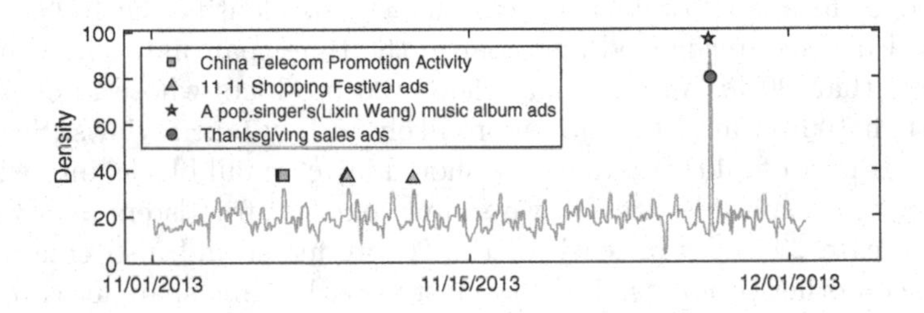

Fig. 3. EigenPulse detects anomalies dense blocks on Sina weibo dataset.

Anomaly Detection on Real Data: For the social network, i.e.,following relationship or message retweets, the dense blocks usually contain anomalous items or correspond to suspicious behaviors, and the sudden surges of density measure can be a significant signal for anomalies. So we applied the EigenPluse to Sina weibo dataset to detect the suspicious dense blocks, and also crawled the detailed content of msgs for verification.

The Fig. 3 illustrates the density change of dense blocks in the sliding windows with $w = 2$ hours, $s = 1$ hour. The Table 3 reports the suspicious features and content of detected blocks. We spot some significant spikes in the Fig. 3, and the message content all gives the tell-tale sign of suspicious behaviors, that is, as the

'Message Topic' shown, most of the messages about advertisements or products promotion information. In particular, We can notice that there are **953** edges for the only 7 users × 8 messages in two hours, which means a user retweeted 20 times for one message in average, and it's very suspicious intuitively. In summary, EigenPluse can detect anomalies dense blocks in real dataset.

Table 3. Dense blocks detected by EigenPulse on Sina weibo dataset

Message topic	Size	Time range	#Edges
China Telecom Promotion Activity	39 × 57	6:00–8:00, Nov 7	2,004
	78 × 58	7:00–9:00, Nov 7	4,051
	151 × 119	8:00–10:00, Nov 7	8,295
11.11 Shopping Festival ads	201 × 139	6:00–8:00, Nov 10	7,012
	196 × 111	7:00–9:00, Nov 10	9,668
	126 × 93	8:00–10:00, Nov 13	638
A pop. singer's (Lixin Wang) music album ads	**7 × 8**	22:00–24:00, Nov 26	**953**
Thanksgiving sale ads	26 × 36	23:00, Nov 26–1:00, Nov 27	629
	43 × 34	1:00–3:00, Nov 27	263

4.4 Q3.Scalability

We demonstrate the linearly scalability with of EigenPluse by measuring how rapidly its update time increases as a tensor grows.

We choose two representative datasets: BeerAdvocate with the highest volume density, and Amazon Electronics with the most edges, and randomly sample sub-tensors with different size of edges. As shown in Fig. 4, the running time of our algorithm increases linearly with the number of the edges.

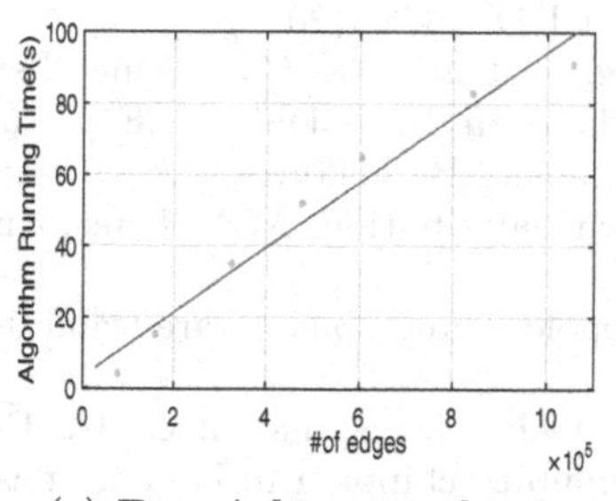

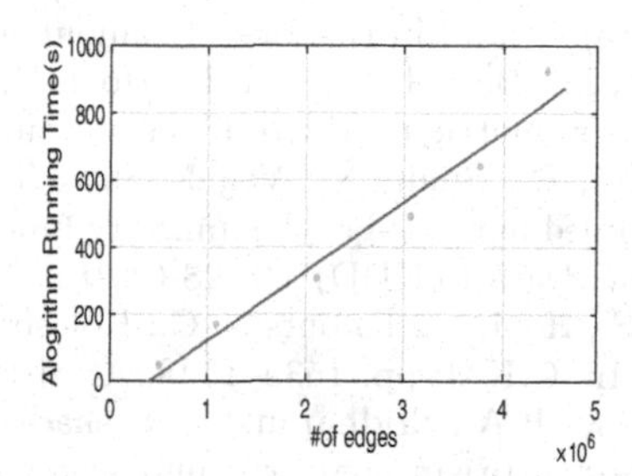

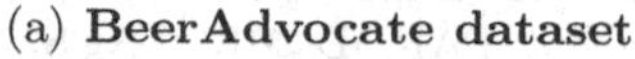

(a) **BeerAdvocate dataset** (b) **Amazon Electronic dataset**

Fig. 4. EigenPulse runs in near-linear time.

5 Conclusion

In this paper, we proposed a surge detection method, EigenPulse, which can spot the density surge in a large streaming graph in a efficient and accurate way. We use row-augmented matrix and Sliding Window to model streaming graph and design the AugSVD algorithm for efficient singular decomposition which is the input of EigenPulse. In conclusion, our algorithm has the following advantages:

- **Incremental singular value decomposition:** we propose a scalable algorithm, AugSVD,which combines Sliding Window to do streaming graph decomposition.
- **Robust and effective:** AugSVD has good robustness and EigenPulse generated by AugSVD can detect suspicious synchronized activities accurately.
- **Scalable:** EigenPulse is near-linear in running time and memory-efficient because it only detects one window's data each time.
- **Reproducibility**: The code and data are available at https://github.com/shenghua-liu/EigenPulse/invitations.

Acknowledgments. This material is based upon work supported by the Strategic Priority Research Program of CAS (XDA19020400), NSF of China (61772498, 61872206, 61425016, 91746301), and the Beijing NSF (4172059).

References

1. Akoglu, L., Tong, H., Koutra, D.: Graph based anomaly detection and description: a survey. Data Min. Knowl. Discov. **29**(3), 626–688 (2015)
2. Eswaran, D., Faloutsos, C., Guha, S., Mishra, N.: Spotlight: detecting anomalies in streaming graphs. In: SIGKDD, pp. 1378–1386. ACM (2018)
3. Gao, J., Zhou, C., Yu, J.X.: Toward continuous pattern detection over evolving large graph with snapshot isolation. In: VLDB (2016)
4. Guha, S., Mishra, N., Roy, G., Schrijvers, O.: Robust random cut forest based anomaly detection on streams. In: ICML (2016)
5. Halko, N., Martinsson, P.G., Tropp, J.A.: Finding structure with randomness: probabilistic algorithms for constructing approximate matrix decompositions. SIAM Rev. **53**, 217–288 (2011)
6. Hooi, B., Song, H.A., Beutel, A., Shah, N., Shin, K., Faloutsos, C.: Fraudar: bounding graph fraud in the face of camouflage. In: KDD. ACM (2016)
7. Jiang, M., Beutel, A., Cui, P., Hooi, B., Yang, S., Faloutsos, C.: A general suspiciousness metric for dense blocks in multimodal data. In: ICDM. IEEE (2015)
8. Koutra, D., Shah, N., Vogelstein, J.T., Gallagher, B., Faloutsos, C.: Deltacon: principled massive-graph similarity function with attribution. ACM Trans. Knowl. Discov. Data (TKDD) **10**, 28 (2016)
9. Liu, S., Hooi, B., Faloutsos, C.: Holoscope: topology-and-spike aware fraud detection. In: CIKM, pp. 1539–1548. ACM (2017)
10. Prakash, B.A., Sridharan, A., Seshadri, M., Machiraju, S., Faloutsos, C.: EigenSpokes: surprising patterns and scalable community chipping in large graphs. In: Zaki, M.J., Yu, J.X., Ravindran, B., Pudi, V. (eds.) PAKDD 2010. LNCS (LNAI), vol. 6119, pp. 435–448. Springer, Heidelberg (2010). https://doi.org/10.1007/978-3-642-13672-6_42

11. Shin, K., Hooi, B., Faloutsos, C.: M-Zoom: fast dense-block detection in tensors with quality guarantees. In: Frasconi, P., Landwehr, N., Manco, G., Vreeken, J. (eds.) ECML PKDD 2016. LNCS (LNAI), vol. 9851, pp. 264–280. Springer, Cham (2016). https://doi.org/10.1007/978-3-319-46128-1_17
12. Shin, K., Hooi, B., Kim, J., Faloutsos., C.: D-cube: dense-block detection in terabyte-scale tensors. In: WSDM (2017)
13. Shin, K., Hooi, B., Kim, J., Faloutsos, C.: Densealert: incremental dense-subtensor detection in tensor streams. In: KDD. ACM (2017)
14. Sun, J., Tao, D., Faloutsos, C.: Beyond streams and graphs: dynamic tensor analysis. In: KDD. ACM (2006)
15. Yongsub Lim, M.J., Kang, U.: Memory-efficient and accurate sampling for counting local triangles in graph streams: from simple to multigraphs. In: TKDD. ACM (2018)
16. Yu, W., Gu, Y., Li, J., Liu, S., Li, Y.: Single-pass PCA of large high-dimensional data. In: IJCAI, pp. 3350–3356 (2017)

TPLP: Two-Phase Selection Link Prediction for Vertex in Graph Streams

Yang Xiao, Hong Huang, Feng Zhao(✉), and Hai Jin

National Engineering Center for Big Data Technology and System,
Services Computing Technology and System Lab, Cluster and Grid Computing Lab,
School of Computer Science and Technology,
Huazhong University of Science and Technology, Wuhan 430074, China
zhaof@hust.edu.cn

Abstract. Currently, data in many applications have naturally been modeled as streams over the massive graph infrastructure, e.g., social networks and electronic business. Graph streams are rapidly changing, enormous and endless networks that are too large to maintain in memory or on disks. An important problem in networks is link prediction, which aims to estimate the likelihood of the existence of a specific link. However, in graph streams, predicting the existence of links connected to one vertex is more common. For example, in social networks, we generally want to recommend several friends to a user rather than determining whether a specific user is your friend. Rapidly and accurately predicting groups of links becomes a formidable challenge because of the tremendous size and rapidly updated information of graph streams. In this paper, we propose the problem of link prediction for vertex in graph streams, which aims to predict the top-k vertices, i.e., the top-k links, that are most likely to connect to the target vertex in graph streams. A two-phase selection framework is proposed to predict top-k links with high efficiency and without loss of accuracy. We also propose a novel method for estimating common neighbor in graph streams, which is a very important measure in link prediction. Extensive experiments show that our algorithms are more efficient and more accurate than state-of-the-art methods.

1 Introduction

Link prediction, which aims to predict unknown links in networks, is a useful and fundamental problem in network science; it has attracted a considerable amount of attention in many fields, such as social networks [5], recommendation systems, [1] and biology [3]. In these fields, many applications can be represented as graph streams, which are rapidly changing, enormous networks with nodes and edges that are received and updated rapidly in a form of a stream. For example, *Twitter*,

F. Zhao—This work was supported in part by National Natural Science Foundation of China under Grants No. 61672256 and Guandong Science and Technology Plan under Grants no. 2017B030305003.

Q. Yang et al. (Eds.): PAKDD 2019, LNAI 11440, pp. 514–525, 2019.
https://doi.org/10.1007/978-3-030-16145-3_40

a type of social network, is a massive, endless graph with nodes and edges that change very quickly over time; thus, they can be represented as graph streams [4].

Traditional link prediction methods generally fail in real-world graph stream scenarios for three reasons. First, these real-world graphs are typically too large to maintain either in memory or even on disks [2,6]. Second, similar to data streams, the edges in graph streams can be processed only once [7]. Therefore, some methods with multipass traversals are no longer feasible in graph streams. Finally, even if we find a way to maintain the full graph in memory or on disk, running algorithms on such a massive graph would be very inefficient and impractical for online link prediction. Therefore, new algorithms and techniques are needed for link prediction in graph streams.

Zhao [6] proposes the problem of link prediction in graph streams, which aims to predict links that may appear in the future. However, in real-world applications, such as recommendation systems, we often encounter scenarios of *link prediction for vertex*, where unknown links connected to one vertex need to be discovered. *Link prediction for vertex* aims to predict the top-k vertices that are most likely to connect to the target vertex. For simplicity, the predicted top-k vertices (or top-k links) are called *k future neighbors* (*KFN*) of the target vertex. Unfortunately, there is no efficient method to solve the *link prediction for vertex* problem in graph streams. State-of-the-art method [6] traverses all of the unlinked vertex pairs and calculates a probability using a type of similarity measure, such as common neighbor or Adamic-Adar [5]. The vertex pairs with the largest probability are predicted to be connected. The time complexity of this method is proportional to $|V|$ because it is necessary to calculate the probability between the target vertex and all the other vertices, where $|V|$ is the number of vertices in the graph streams. Given a graph stream with 10^6 vertices, this method will be time consuming.

In this paper, we propose a *two-phase selection link prediction* (TPLP) framework to solve the problem of *link prediction for vertex in graph streams.* Our main contributions can be briefly summarized as follows:

- We propose the problem of *link prediction for vertex in graph streams* and propose a two-phase selection algorithm to perform link prediction with high efficiency and without loss of accuracy.
- A new method for estimating common neighbor is designed in graph streams, and it is more accurate than state-of-the-art methods. The accuracy of link prediction has been improved by a maximum of 47%.

2 Preliminaries

2.1 Link Prediction for Vertex in Graph Streams

In graph streams, edges are received in the form of a sequence (e_1,e_2...e_t), where t represents the timestamp when edges are received. For convenience, we use $G(t) = (V(t), E(t))$ to indicate the graph received thus far, where $V(t)$ represents the set of vertices in $G(t)$ and $E(t)$ represents the set of edges in $G(t)$. We further denote $\Gamma(u,t)$ as the set of neighbors of vertex u in graph $G(t)$ and $d(u,t)$ as the degree of vertex u in graph $G(t)$. Formally, We define the problem of streaming link prediction for vertex as follows:

Definition 1: (Link prediction for vertex in graph streams). *Given graph streams $G(t)$ and a target vertex u, we aim to find the top-k vertices that are most likely to connect with u in the future. These top-k vertices are called k future neighbors (KFN) of vertex u.*

In our framework, we use two types of similarity-based measures to estimate the likelihood of the existence of a specific link.

Definition 2: (Similarity measures). *Similarity measures are used to estimate the similarity between a given vertex pair. For vertex pair $(u,v) \notin E(t)$, where $u,v \in V(t)$, we define the similarity measures for (u,v) as follows:*

(1) Common Neighbor

$$C(u,v) = |\Gamma(u,t) \cap \Gamma(v,t)| \tag{1}$$

(2) Adamic-Adar

$$AA(u,v) = \sum_{w \in \Gamma(u,t) \cap \Gamma(v,t)} \frac{1}{\log |\Gamma(w,t)|} \tag{2}$$

2.2 Vertex-Biased Sampling

Graph streams are generally too large to maintain in memory or on disk; therefore, sampling is an efficient approach to process graph streams. Zhao [6] proposed vertex-biased sampling to obtain a sampled graph (called *graph sketch*). Each vertex is randomly assigned a hash value, which represents the priority of this vertex. Then they set a threshold L (called the sample size) to limit the number of neighbors for each vertex, i.e., the algorithm only maintains neighbors with the highest L priority for vertices whose degree is larger than L. This sampling algorithm ensures that all of the neighbors of low-degree vertices are retained, while neighbors of high-degree vertices are retained with certain possibilities. This approach yields a reasonable prediction performance. Thus, we also use the vertex-biased sampling algorithm in our work. The graph sketch structure is defined as follows:

Definition 3: (Graph Sketch). *Graph sketch is the graph structure after vertex-biased sampling. Given a vertex u, the graph sketch $S(u)$ is defined as the set of neighbors of u remaining in the sampled graph.*

To measure how much information is retained or lost for the given vertex u in the graph sketch, we use a measure called *sample ratio of graph sketch*. This measure is defined as follow:

Definition 4: (Sample ratio of graph sketch). *Given time t and vertex u, suppose that the size of graph sketch $S(u)$ is $|S(u,t)|$ and that the degree of u in graph $G(t)$ is $d(u,t)$. Then, the sample ratio of graph sketch for vertex u at time t is*

$$\eta(u,t) = \frac{|S(u,t)|}{d(u,t)} \tag{3}$$

For example, if u has a total of 1000 neighbors and only 100 of them remain after sampling, then we say that only 10% of vertex $u's$ information is retained. In fact, we can get the estimation of $\eta(u)$ by using a hash function [6]. In this paper, we take $\eta(u,t)$ as a known variable.

3 TPLP: A Two-Phase Selection Streaming Link Prediction Framework

3.1 Inverted Graph Sketch

To improve the performance of link prediction, we use a structure called inverted graph sketch, an import structure in our work.

Definition 5: (Inverted Graph Sketch). *For each vertex u, inverted graph sketch $I(u)$ is defined as the indices of graph sketch $S(u)$, i.e., $v \in I(u)$ if and only if $u \in S(v)$.*

Similar to the ratio defined in Sect. 2.2, we define sample ratio of inverted graph sketch as follow:

$$\eta'(u,t) = \frac{|I(u,t)|}{d(u,t)} \tag{4}$$

Note that we cannot obtain the true value of $d(u,t)$ because we only have a sampled graph. However, we can estimate $d(u,t)$ based on Eq. (3) as follows:

$$d(u,t) = \frac{|S(u,t)|}{\eta(u,t)} \tag{5}$$

Therefore, we can estimate $\eta'(u,t)$ by

$$\eta'(u,t) = \frac{|I(u,t)| \cdot \eta(u,t)}{|S(u,t)|} \tag{6}$$

3.2 Two-Phase Selection Algorithm

In this section, we discuss our methods for reducing the prediction time. Our main idea is to use the information in the graph sketch and inverted graph sketch to filter those vertices that are unlikely to be the *KFN* of target vertex u. The remaining vertices, called candidates of vertex u, are vertices that can potentially be the *KFN* of vertex u. After this selection step, we only need to calculate the similarity measures of $|Cdd(u)|$ vertex pairs, where $|Cdd(u)|$ is the number of vertex $u's$ candidates. Therefore, for target vertex u, our method is accelerated approximately $|V|/|Cdd(u)|$ times compared to the method without selection, where $|V|$ is the number of vertices in $G(t)$. Here, we only discuss two-phase selection algorithm of Adamic-Adar measure. The selection algorithm for common neighbor measure can be derived in the same way.

Phase-1 Selection. This phase of selection is for all of the vertices. Here, we use inverted graph sketch to estimate the Adamic-Adar measure since we only have sampled graph streams and Eq. (2) is transformed as:

$$AA(u,v) = \sum_{w \in I(u) \cap I(v)} \frac{1}{\log |\Gamma(w,t)|} \tag{7}$$

We do not need to consider vertex v with $I(u) \cap I(v) = \emptyset$ because it has zero similarity to the target vertex u. Therefore, the candidates of vertex u can be formulated as follows:

$$Cdd(u) = \{v | I(u) \cap I(v) \neq \emptyset\} \tag{8}$$

If we traverse all of the vertex $v \in G(t)$ and determine whether $I(u) \cap I(v) = \emptyset$, the time complexity is $O(|V|L)$; the average time complexity to obtain $I(u) \cap I(v)$ is $O(L)$, and we need to perform this step $|V|$ times. This method is highly inefficient. Instead, we use graph sketches $S(u)$ and $S(v)$ to improve computational efficiency. Recall that $v \in I(u)$ if and only if $u \in S(v)$. Assume that $v \in Cdd(u)$; then, we have $I(u) \cap I(v) \neq \emptyset$, which means that there is at least one element in $I(u) \cap I(v)$. Without loss of generality, we assume $q \in I(u) \cap I(v)$. Then, we have

$$q \in I(u)$$
$$q \in I(v) \Rightarrow v \in S(q)$$

Therefore, we have

$$Cdd(u) = \{v | v \in S(q), q \in I(u)\} \tag{9}$$

The number of candidates is now

$$|Cdd(u)| = \sum_{q \in I(u)} |S(q)| \leq L \cdot |I(u)| \tag{10}$$

The link prediction efficiency improves more than $\frac{|V|}{L \cdot |I(u)|}$ times by Phase-1 Selection.

Phase-2 Selection. This phase of selection is for vertices that have been queried previously. In general, the number of queries for different vertices follows a power-law distribution, where some important vertices are frequently queried and other vertices are rarely queried. Suppose that there is a target vertex u that has been queried before. The last query result is denoted as KFN^{old}, and the last query time is denoted as t_1. We now want to query target vertex u again at time t_2, where $t_2 > t_1$; the query result is now denoted as KFN^{now}.

Because graph streams change rapidly over time, the similarity measures and KFN change over time. In other words, some vertices in KFN^{old} may be replaced by new vertices (called KFN^{new}) during time interval (t_1, t_2), although some vertices in KFN^{old} may remain unchanged. Consequently, to obtain KFN^{now}, we only need to focus on KFN^{new}, rather than recalculating the KFN at time t_2.

Based on Eq. (7), the value of Adamic-Adar measure can be changed by two factors. The first is new vertices added to $I(u) \cap I(v)$, and the second is the change in $|S(w)|$ for a vertex $w \in I(u) \cap I(v)$. In fact, we can ignore the influence of the second factor because it is always very small compared to that of the first factor. For example, if $|S(w)|$ increases from 50 to 60, the change in Adamic-Adar measure is $1/log(50) - 1/log(60) \approx 0.011$. However, if a new vertex w with a degree of 10 is added to $I(u) \cap I(v)$, the change in Adamic-Adar measure is $1/log(10) \approx 0.434$. For convenience, we ignore the influence of the change in $|S(w)|$. Therefore, $v \in KFN^{new}$ only if the similarity measure $AA(u, v)$ increases during time interval (t_1, t_2); otherwise, it is unable to replace the vertices in KFN^{old}, which can be formulated as follow:

$$AA(u, v)_{now} > AA(u, v)_{old} \tag{11}$$

where $AA(u, v)_{now}$ is the similarity measure at time t_2, and $AA(u, v)_{old}$ is the similarity measure at time t_1.

Then, based on Eq. (11), we have

$$|I(u) \cap I(v)|^{now} > |I(u) \cap I(v)|^{old} \tag{12}$$

Our candidates now become

$$Cdd(u) = KFN^{old} \cup \{v||I(u) \cap I(v)|^{now} > |I(u) \cap I(v)|^{old}\} \tag{13}$$

For convenience, we divide $S(u)$ into two parts. The first part, denoted as S_u^{old}, is the set of vertices in $S(u)$ that remain unchanged during (t_1, t_2). The second part, denoted as S_u^{new}, is the set of vertices added to $S(u)$ during time interval (t_1, t_2). I_u^{old} and I_u^{new} are denoted in the same way. We can rewrite $I(u) \cap I(v)$ as

$$I(u) \cap I(v) = (I_u^{old} \cup I_u^{new}) \cap (I_v^{old} \cup I_v^{new})$$

Note that $I_u^{old} \cap I_u^{new} = \emptyset$ and $I_v^{old} \cap I_v^{new} = \emptyset$. We have

$$I(u) \cap I(v) = (I_u^{old} \cap I_v^{old}) \cup (I_u^{old} \cap I_v^{new}) \cup (I_u^{new} \cap I_v) \tag{14}$$

Similar to the formula derivation in Phase-1 Selection, the size of the candidates is

$$|Cdd(u)| = \sum_{q \in I_u^{old}} |S(q)^{new}| + \sum_{q \in I_u^{new}} |S(q)| + k \tag{15}$$

It is clear that $|S(q)^{new}| \leq |S(q)|$ because $S(q)^{new} \subset S(q)$. We can ignore k because it is a small value compared to the first two terms. Therefore, we have

$$|Cdd(u)| \leq \sum_{q \in I_u^{old}} |S(q)| + \sum_{q \in I_u^{new}} |S(q)| = \sum_{q \in I(u)} |S(q)| \tag{16}$$

This means that the number of candidates for a queried vertex can be further reduced compared to Phase-1 Selection.

3.3 Estimation of Common Neighbor

Common neighbor is a very important measure in many types of applications, particularly in link prediction. For example, in social networks, the number of mutual friends reflects familiarity between users. However, in graph streams, we are not able to obtain the real value of common neighbor because we only have the sampled graph. Therefore, accurate estimation of the true common neighbor measure is a crucial problem.

Given a vertex pair (u, v), common neighbor estimation is to estimate the number of common neighbor of u and v from the sampled graph streams. Zhao [6] proposed a method to estimate common neighbor in graph streams:

$$C(u, v) = \frac{|S(u) \cap S(v)|}{\max(\eta(u, t), \eta(v, t))} \tag{17}$$

Because the size of reservoir is upper bounded by L, the sample ratio of graph sketch is very low for high-degree vertices. For example, given $L = 50, d(u) = 1000$, the sample ratio of vertex u is 0.05, which means that 95% of the neighbors of u are lost. Consequently, using this method to estimate the common neighbor for high-degree vertices is not accurate. Therefore, we propose a novel method to estimate common neighbor for high-degree vertices. Although the sample ratio of $S(u)$ is very low, $I(u)$, the inverted index structure of $S(u)$, has a relatively high sample ratio, as the size of $I(u)$ is not upper bounded by L. In our experiment, the sample ratio of the inverted graph sketch for a high-degree vertex is larger than 50% in most cases. Therefore, we consider using the inverted graph sketch I for common neighbor estimation.

Suppose that the true value of common neighbor is $C(u, v)$. Since $I(u)$ and $I(v)$ are independent of each other, the number of common neighbor in the sampled graph is $C(u, v) \cdot \eta'(u, t) \cdot \eta'(v, t)$. Therefore, our common neighbor estimation can be formulated as

$$C(u, v) = \frac{|I(u) \cap I(v)|}{\eta'(u, t) \cdot \eta'(v, t)} \tag{18}$$

where $\eta'(u, t)$ and $\eta'(v, t)$ are the sample ratios of the inverted graph sketch for vertices u and v, respectively, which can be obtained by Eq. (6).

4 Experiments

4.1 Datasets

We use four real-world, public datasets that can be formulated as graph streams.

- **Amazon.** This dataset is based on *Customers Who Bought This Item Also Bought* feature of the Amazon website. If a product i is co-purchased with product j, there is an edge between i to j. Note that all of the edges have at a certain timestamp; thus, these networks can be treated as graph streams. There are a total of 262,111 vertices and 1,234,877 edges in the graph streams.
- **DBLP.** In this dataset, vertices represent the distinct authors, and edges correspond to the cooperative relationship between two authors. If author A and author B publish at least one paper together, then there is an edge between A and B. We extracted papers from 1940 to 2015 and discarded vertex pairs without a publication year. There are a total of 1,411,376 vertices and 10,597,380 edges in the graph streams.
- **Wikipedia.** This dataset describes the evolution of the online knowledge base Wikipedia. Vertices represent the articles in Wikipedia, and edges represent the reference relationship between two articles. There are a total of 1,870,709 vertices and 39,953,145 edges in the graph streams.
- **Super-User.** This dataset describes the interactions on the stack exchange website Super User. If user u answers or comments on user $v's$ question at time t, then there will be an edge $<u, v>$ attached with timestamp t. There are a total of 194,085 vertices and 1,443,339 edges in the graph streams.

4.2 Performance of Two-Phase Selection

In this section, we test the average prediction time of three types of algorithms: **Without Selection**, **Phase-1 Selection**, and **Phase-2 Selection**. We design two different query sets for the selection algorithm of different phases. For **Phase-1 Selection**, we randomly choose 1,000 vertices. For **Phase-2 Selection**, we generate a power-law dataset with 10,000 vertices as our query set. In other words, the number of queries for different vertices follows a power-law distribution. The skewness of the power-law distribution is set to 1.

We test three types of similarity measures: common neighbor proposed by Zhao [6] (Base CN), our common neighbor (Our CN) and Adamic-Adar (AA). Table 1 presents the experimental results of **Phase-1 Selection** for the Amazon and DBLP datasets. For the Amazon dataset, the total prediction time of **Phase-1 Selection** for the three algorithms is approximately 0.2 s for a total of 1000 queries, which is almost 500 times faster than the corresponding algorithm **Without Selection**. For the DBLP dataset, the total prediction time for the three algorithms with **Phase-1 Selection** is approximately 0.5 s, while the prediction time (of **Without Selection**) is approximately 15 min. Table 2 presents the results of **Phase-2 Selection**. The results show that **Phase-2 Selection** can reduce the prediction time by approximately 5 times compared to **Phase-1 Selection**.

Table 1. Performance of Phase-1 Selection on the Amazon and DBLP datasets

Methods	Prediction time cost (s)		Methods	Prediction time cost (s)	
	Amazon	DBLP		Amazon	DBLP
Base CN (**Without Selection**)	103.0	835.0	Base CN (**Phase-1 Selection**)	0.20	0.51
Our CN (**Without Selection**)	118.6	1165.0	Our CN (**Phase-1 Selection**)	0.22	0.67
AA (**Without Selection**)	108.9	1019.2	AA (**Phase-1 Selection**)	0.25	0.58

Table 2. Performance of Phase-2 Selection on the Amazon and DBLP datasets

Methods	Prediction time cost (s)		Methods	Prediction time cost (s)	
	Amazon	DBLP		Amazon	DBLP
Base CN (**Phase-1 Selection**)	9.6	16.2	Base CN (**Phase2 Selection**)	1.6	3.7
Our CN (**Phase-1 Selection**)	10.7	21.8	Our CN (**Phase2 Selection**)	1.7	5.5
AA (**Phase-1 Selection**)	9.5	21.1	AA (**Phase-2 Selection**)	1.9	4.2

4.3 Performance of Common Neighbor Estimation

In this section, we test the performance of common neighbor estimation algorithm. Because high-degree vertices have a very low sample ratio, estimation for these vertices is more difficult compared to low-degree vertices. To test the accuracy of common neighbor estimation for vertices with different degrees, we design a series of query sets with different degrees: $V_1 = VSet_{d \geq 20}$, $V_2 = VSet_{d \geq 30} \cdots$ $V_9 = VSet_{d \geq 100}$, where $VSet_{d \geq m}$ means the set of vertices whose degree is not less than m. We define the accuracy of common neighbor estimation as follows.

Suppose that $VSet_{d \geq m} = \{v_1, v_2, \cdots, v_H\}$, where H is the number of vertices in $VSet$. For each vertex $v_i \in VSet$, we find the KFN of v_i based on three different algorithms: **Base CN, Our CN, True CN**, where **True CN** measures the value of common neighbor on the entire graph streams rather than the sampled graph. We define the common neighbor estimation accuracy of a target v_i as

$$r_i = \frac{|KFN^i_{est} \cap KFN^i_{true}|}{k} \tag{19}$$

where KFN^i_{est} is the common neighbor estimation result (i.e., the set of vertices having the largest k common neighbor measure to vertex v_i) for **Base CN** or **Our CN**, while KFN^i_{true} is the predicted result for **True CN**. Then, the average accuracy of common neighbor estimation on the whole query set $VSet_{d \geq m}$ is

$$r = \frac{\sum_{i=1}^{H} r_i}{H} \tag{20}$$

We conduct experiments on the DBLP datasets for different values of k : $k = 5, k = 10, k = 20$. The result is shown in Fig. 1. The result shows that the estimation result of **Our CN** is more accurate than **Base CN** in all query sets regardless of the value of k. It is clear that **our CN** significantly outperforms

Base CN, particularly for high-degree vertices. For $VSet_{d \geq 100}$, our estimation accuracy is approximately 0.8 on the DBLP dataset, which is an improvement of approximately 20% compared to **Base CN**. This result occurs because we utilize the inversed graph sketch structure for high-degree vertices, and the sample ratio is much higher than that in the graph sketch structure.

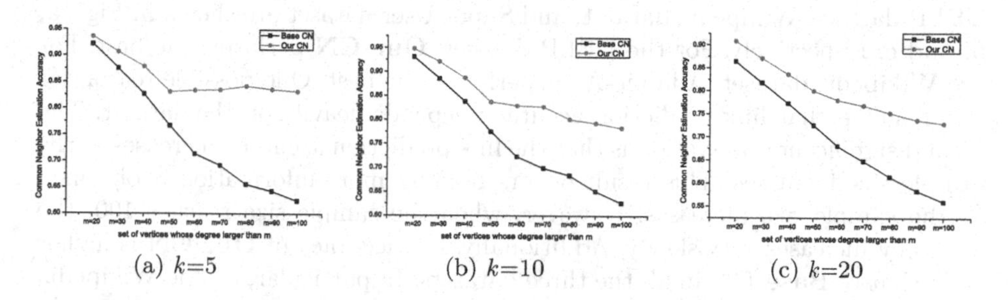

(a) k=5 (b) k=10 (c) k=20

Fig. 1. Estimation accuracy of common neighbor on DBLP for different k

4.4 Performance of Link Prediction Accuracy

In this part, we test the link prediction accuracy of different algorithms. Given graph streams $G(t)$, we consider two intervals $[0, t_1]$ and $[t_1, t_2]$, which are called the training interval and test interval, respectively. The graph in the training interval is called G_{train}, and $G_{train} = (V_{train}, E_{train})$. The graph in the test interval is called G_{test}, and $G_{test} = (V_{test}, E_{test})$. For each dataset, the training interval and the test interval have the same length. We choose a set $Core$ of vertices that arise in both the test interval and training interval for prediction, i.e., $Core = V_{train} \cap V_{test}$. We further denote $E_{new} = \{(u, v)|(u, v) \in E_{train} \backslash E_{test}, u, v \in Core\}$ as the set of edges that arise in the test interval but not in the training interval. Then, we define k_v as the arisen time in E_{new} for vertex v. For example, if $E_{new} = \{(v_1, v_2), (v_1, v_3)\}$, then we say v_1 arises 2 times and v_2, v_3 arise only 1 time. Our query set is all the vertices that arise in E_{new}, i.e., $V_{query} = \{v|k_v \geq 1\}$. For each vertex $v_i \in V_{query}$, we obtain its KFN based on three algorithms: **Base CN**, **Our CN**, and **AA**, where $k = k_{v_i}$. The query result for v_i is called KFN_{v_i}, and the real result is defined as $Res_{v_i} = \{u|(u, v_i) \in E_{new}\}$. We define the link prediction accuracy as follows:

$$acc = \frac{\sum\limits_{v_i \in V_{query}} |KFN_{v_i} \cap Res_{v_i}|}{\sum\limits_{v_i \in V_{query}} k_{v_i}} \tag{21}$$

where the numerator refers to the number of links that are correctly predicted, and the denominator refers to the number of links that need to be predicted. To compare different link prediction algorithms, our link prediction accuracy is measured in terms of relative improvement to a random prediction algorithm on three datasets: DBLP dataset, Wikipedia dataset, and Super-User dataset. Given a vertex v_i, the random prediction algorithm randomly chooses k_{v_i} vertices from *Core* as *KFN* of vertex v_i.

To test the relationship between link prediction accuracy and sample size, we choose different sample sizes L to test prediction accuracy. The results on DBLP dataset, Wikipedia dataset, and Super-User dataset are shown in Fig. 2a, b, and c, respectively. For the DBLP dataset, **Our CN** performs the best. For the Wikipedia dataset, Adamic-Adar performs the best. One possible reason for this result is that link prediction accuracy depends heavily on the dataset. The second significant observation is that the link prediction accuracy increases as the sample size increases. This result occurs because more information is obtained as the sample size increases. However, when the sample size reaches 100, the accuracy increases very slowly. Additionally, the accuracy of **Our CN** is higher than that of **Base CN** in all the three datasets. In particular, on the Wikipedia and Super-User datasets, the accuracy of **Our CN** improves by approximately 28% and 47% compared to **Base CN** when the sample size is equal to 100, respectively. This observation further verifies that proposed common neighbor estimation method has a better estimation of common neighbor compared to state-of-the-art methods.

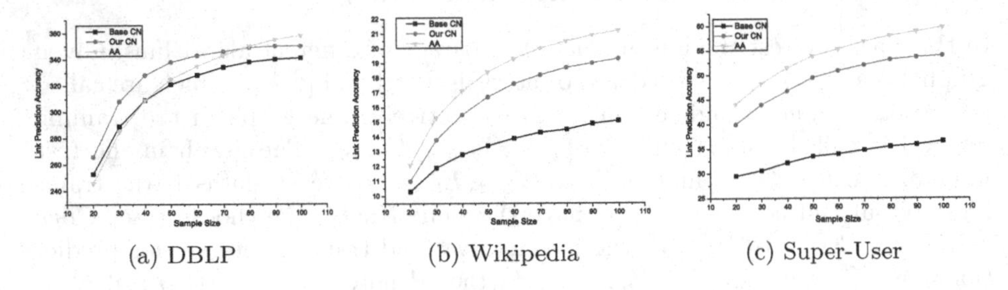

(a) DBLP (b) Wikipedia (c) Super-User

Fig. 2. Link prediction accuracy in terms of sample size on different datasets

4.5 Error Incurred by Sampling

Finally, we evaluate the error incurred by sampling. Specifically, we test link prediction accuracy with and without sampling to obtain the relative error rate. We evaluate the relative error rates of three similarity measures, **Base CN**, **Our CN**, and **AA**, with five different sample sizes: 20, 40, 60, 80, and 100. The experimental results are shown in Table 3. From Table 3, it can be seen that the relative error rate decreases as the sample size increases. Additionally, when the sample size is 100, the relative error rate is very low, even negligible. Particularly for **Our CN**, the relative error rate is only 0.1%. Thus, the error incurred by sampling can be ignored as long as we choose an appropriate sample size (such as 100) such that our proposed algorithm can achieve a particularly good estimation of the similarity measure in graph streams.

Table 3. Link prediction error of different sample size

Algorithm	Sample size				
	L = 20	L = 40	L = 60	L = 80	L = 100
Base CN	10.4%	7.2%	4.0%	3.4%	2.8%
Our CN	9.7%	5.4%	2.1%	0.7%	0.2%
AA	10.9%	7.7%	4.8%	3.6%	3.2%

5 Conclusions

In this paper, we propose the problem of *link prediction for vertex in graph streams*. *Link prediction for vertex* has found many real-world applications, such as recommendation systems and social networks. To support rapid online link prediction, we propose a two-phase selection framework to reduce the prediction time by several orders of magnitude. To improve the link prediction accuracy, we propose a new common neighbor estimation method that performs much better than state-of-the-art algorithms. The experimental results demonstrate that our algorithms are more efficient and accurate than state-of-the-art approaches, and thus can be employed to real-word graph stream applications.

References

1. Li, X., Chen, H.: Recommendation as link prediction in bipartite graphs: a graph kernel-based machine learning approach. Decis. Support Syst. **54**(2), 880–890 (2013)
2. McGregor, A.: Graph stream algorithms: a survey. ACM SIGMOD Rec. **43**(1), 9–20 (2014)
3. Menche, J., et al.: Uncovering disease-disease relationships through the incomplete interactome. Science **347**(6224), 1257601 (2015)
4. Song, C., Ge, T., Chen, C., Wang, J.: Event pattern matching over graph streams. Proc. VLDB Endow. **8**(4), 413–424 (2014)
5. Wang, P., Xu, B., Wu, Y., Zhou, X.: Link prediction in social networks: the state-of-the-art. Sci. China Inf. Sci. **58**(1), 1–38 (2015)
6. Zhao, P., Aggarwal, C., He, G.: Link prediction in graph streams. In: Proceedings of 2016 IEEE 32nd International Conference on Data Engineering (ICDE), pp. 553–564. IEEE (2016)
7. Zhao, P., Aggarwal, C.C., Wang, M.: gSketch: on query estimation in graph streams. Proc. VLDB Endow. **5**(3), 193–204 (2011)

Robust Temporal Graph Clustering for Group Record Linkage

Charini Nanayakkara[(✉)], Peter Christen, and Thilina Ranbaduge

Research School of Computer Science, The Australian National University, Canberra, ACT 2600, Australia
charini.nanayakkara@anu.edu.au

Abstract. Research in the social sciences is increasingly based on large and complex data collections, where individual data sets from different domains need to be linked to allow advanced analytics. A popular type of data used in such a context are historical registries containing birth, death, and marriage certificates. Individually, such data sets however limit the types of studies that can be conducted. Specifically, it is impossible to track individuals, families, or households over time. Once such data sets are linked and family trees are available it is possible to, for example, investigate how education, health, mobility, and employment influence the lives of people over two or even more generations. The linkage of historical records is challenging because of data quality issues and because often there are no ground truth data available. Unsupervised techniques need to be employed, which generally are based on similarity graphs generated by comparing individual records. In this paper we present a novel temporal clustering approach aimed at linking records of the same group (such as all births by the same mother) where temporal constraints (such as intervals between births) need to be enforced. We combine a connected component approach with an iterative merging step which considers temporal constraints to obtain accurate clustering results. Experiments on a real Scottish data set show the superiority of our approach over a previous clustering approach for record linkage.

Keywords: Entity resolution · Star clustering · Vital records · Birth bundling

1 Introduction

Databases that contain personal information, such as censuses or historical civil registries [18], generally contain records that describe a group of individuals, where each individual can occur with different types of roles [6]. A *baby* is born, then recorded as a *daughter* or *son* in a census, and later she or he might marry (as a *bride* or *groom*) and become the *mother* or *father* of her or his own children. Being able to link such records across different databases will allow the reconstruction of whole populations and open a multitude of studies in the health

Q. Yang et al. (Eds.): PAKDD 2019, LNAI 11440, pp. 526–538, 2019.
https://doi.org/10.1007/978-3-030-16145-3_41

and social sciences that currently are not feasible on individual databases [4]. Studying these issues is important to identify how societies evolve over time and discover the changes that influenced and contributed for social evolution [14].

The process of identifying the sets of records that correspond to the same individual is known as *record linkage*, *entity resolution*, or *data matching* [5]. Record linkage involves comparing pairs of records to decide if a pair refers to the same entity (known as a *match*) or to different entities (a *non-match*). In this process, generally, the similarities between the values in selected attributes are compared to decide if two records are similar enough to be classified as a match (if for example all similarities are above a given threshold) [5]. For certain applications such a simple pair-wise linkage does however not provide enough information to identify matching records with high accuracy [6].

In contrast to traditional pair-wise record linkage, *group linkage* [17] has recently received significant attention because of its applicability of linking groups of individuals, such as families or households [6,11]. The identification of relationships between individuals can enrich and improve the quality of data, and thus facilitate more sophisticated analysis of different socio-economic factors (such as health, wealth, occupation, and social structure) of large populations [4].

Historical record linkage [19] involves the linkage of records from historical censuses, as well as birth, death, and marriage certificates, to construct longitudinal data sets about a population. This problem has been studied in the past two decades by researchers working in different domains. In 1996 Dillon investigated an approach to link census records from the US and Canada to generate a longitudinal database to examine changes in household structures [9]. IPUMS is a large project which aims to curate and ultimately link large demographic data collections [19]. The Life-M project is another example of transforming historical records into a multi-generational longitudinal database [3]. The Digitising Scotland project [8], which this work is part of, aims to link civil registration events recorded in Scotland between 1856 and 1973 to create a linked database covering the whole population of Scotland spanning more than a century to allow researchers to conduct studies that currently are not possible.

Here we address one specific step used in historical record linkage as conducted by demographers and historians [18]: the *bundling* (clustering) of birth records by the same mother to identify siblings. Once sibling groups have been identified, they can be linked to census, marriage, and death records using group linkage techniques [6,11]. Linked bundles of siblings allow a variety of studies, for example, about fertility and mortality and how these change over time [18].

Contributions: In this paper we investigate how clustering techniques for record linkage [13,20] can be employed for grouping records where temporal constraints exist between record pairs. We propose and evaluate a novel temporal clustering approach which first creates temporally possible connected components with high precision using only links of high similarities [21], and then employs an iterative refinement step that merges those connected components that are highly similar and temporally possible. We conduct an empirical study on a real historical data set which has been extensively linked semi-manually by domain experts [18] providing us with ground truth data to calculate linkage quality. We show that our temporal clustering approach can outperform a state-of-the-art clustering technique for record linkage in terms of linkage quality.

2 Related Work

Classification techniques for record linkage can be categorised into supervised and unsupervised methods. Unsupervised clustering techniques view record linkage as the problem of how to identify all records that refer to the same entity and to group these records into the same cluster. Hassanzadeh et al. [13] presented a framework to comparatively evaluate different clustering techniques for record linkage. Saeedi et al. [20] proposed a framework to perform clustering for record linkage on a parallel platform using Apache Flink. In their evaluation, star clustering [2] was one of the best performing techniques compared to other clustering methods. In star clustering, records that have high similarities with other records are selected as the centres of possibly overlapping clusters, where the overlapping clusters then need to be split in an iterative second step.

Saeedi et al. [21] recently proposed a novel clustering algorithm based on the strengths of links between records as categorised into strong, normal, and weak (as we discuss in the next section). Connected components are formed based only on strong links initially, which are then refined by adding normal links.

Neither of these clustering approaches, however, has considered temporal constraints. In our recent work [16] we have considered temporal aspects as an improvement to star clustering. While this improved star clustering algorithm was able to achieve better results compared to a greedy temporal clustering approach, it still resulted in low linkage quality due to the requirement of splitting overlapping clusters. Our aim is to improve linkage quality using a novel temporal clustering method which employs the concepts of link strength [21], and integrates them with temporal constraints and an iterative cluster merging step.

3 Overview of Temporal Graph Clustering

Our methodology to conduct temporal graph clustering for group linkage consists of three major phases. In this section, we first describe how we model temporal constraints, and then detail how we generate the initial similarity graph. In Sect. 4 we then propose a connected component generation phase, and in Sect. 5 an iterative refinement phase which merges similar temporally consistent connected components. For notation we use bold letters for clusters, lists, and sets (with upper-case bold letters for sets and lists of sets, lists, and clusters), and normal type letters for numbers and strings. Lists are shown with square and sets with curly brackets, where lists have an order but sets do not.

Modelling Temporal Constraints: One aspect of all three phases of our temporal clustering approach is the consideration of temporal constraints of which pairs of records to consider for linkage. Temporal constrains between records can include that the birth record of a person must be before their death record, a marriage record can only occur once an individual has reached a certain age, or (for our clustering problem) the same mother can only give birth to several babies according to certain biological limitations (at least nine months apart or within a few days for multiple births such as twins) [18].

We model such temporal constraints as a list $\mathbf{T}$ of time intervals where it is *plausible* (or not) for two records to be linked (such as a mother to give birth to two babies). We assume each record $r_i \in \mathbf{R}$ includes a time-stamp, $r_i.t$, such as a date of birth, marriage, or death. Based on these time-stamps we can calculate the temporal difference $\Delta t_{i,j} = r_i.t - r_j.t$ between two records where $\Delta t_{i,j}$ is positive if $r_i.t > r_j.t$ (i.e., r_i refers to a life event that occurred after r_j).

The list $\mathbf{T}$ contains time intervals and *temporal plausibilities*, p, where $p = 1$ means two records are temporally plausible and $p = 0$ means they are not, in the form of tuples $(\Delta t^{start}, p^{start}, \Delta t^{end}, p^{end})$, with $\Delta t^{start} < \Delta t^{end}$. For example, for birth records, $\mathbf{T} = [(0,1,2,1),(3,0,273,0),(274,1,12783,1),(12784,0,99999,0)]$ means that two births by the same mother up-to two days apart are plausible, as are two births nine months to 35 years (274 to 12,783 days) apart, but not two births between three days to nine months or more than 35 years apart.

We calculate the temporal plausibility, $p_{i,j}$, for a pair of records (r_i, r_j), by first identifying the corresponding time difference interval in $\mathbf{T}$ for $\Delta t_{i,j}$, and then calculating the pair's temporal plausibility using linear interpolation as:

$$p_{i,j} = \begin{cases} p^{start}, & \text{if } \Delta t_{i,j} = \Delta t^{start}, \\ p^{end}, & \text{if } \Delta t_{i,j} = \Delta t^{end}, \\ (p^{end} - p^{start}) \cdot \frac{(\Delta t_{i,j} - \Delta t^{start})}{(\Delta t^{end} - \Delta t^{start})} + p^{start}, & \text{if } \Delta t^{start} < \Delta t_{i,j} < \Delta t^{end}. \end{cases}$$

If the calculated $p_{i,j}$ is below a given minimum temporal plausibility threshold p_{min} (provided as input to Algorithm 1), then the corresponding record pair is deemed not to be temporally plausible and it will not be compared. Currently we assume the list $\mathbf{T}$ of temporal constraints is provided by domain experts. As future work, we aim to develop techniques to learn such temporal constraints using true matching record pairs available in ground truth data.

Similarity Graph Generation: In the first phase of our approach, as detailed in Algorithm 1, we calculate pair-wise record similarities. This is a standard record linkage approach [5] using techniques such as approximate string comparisons and a locality sensitive hashing (LSH) [15] based blocking approach.

The main input to the algorithm is a list of records, $\mathbf{R}$, which we aim to link and cluster (in our case we aim to determine which birth records are by the same mother). In order to calculate the pair-wise similarity between record pairs, we use a list of attributes $\mathbf{A}$ and approximate string comparison functions $\mathbf{S}$, such as Jaro-Winkler and edit distance [5], as appropriate to the type of data in an attribute. The calculated attribute similarities may or may not be weighted using the provided list of weights $\mathbf{w}$ (unweighted if all elements of $\mathbf{w}$ are set to 1). In general record linkage, assigning different weights to attributes can increase the quality of the generated links between records [5]. Higher weights can, for example, be assigned to first and last name similarities compared to occupation because names are more likely to help identify matching record pairs.

Algorithm 1. ***Pair-wise similarity graph generation***

Input:
- **R**: List of records to be linked
- **A**: List of attributes from **R** to be compared
- **S**: List of similarity functions to be applied on attributes from **A**
- **w**: List of weights given to attribute similarities, with $|\mathbf{w}| = |\mathbf{S}|$
- **T**: List of temporal constraints
- b, r Number of bands and band size for min-hash based LSH blocking
- p_{min}: Minimum temporal plausibility for record pairs to be compared
- s_{pmin}: Minimum pair-wise similarity for record pairs to be added to the generated graph

Output:
- **G**: Undirected pair-wise similarity graph

```
1:  V = ∅, E = ∅, G = (V, E)                                      // Initialise an empty graph
2:  L = MinHashLSHIndexing(R, b, r)                                // Generate a min-hash index
3:  for l ∈ L do:                                                  // Loop over all min-hash blocks
4:      for (r_i, r_j) : r_i ∈ l, r_j ∈ l, r_i.id < r_j.id do:     // Loop over all record pairs in a block
5:          if IsTempPlausible(r_i, r_j, T, p_min) then:           // Check the temporal plausibility of pair
6:              s_i,j = CompareRecords(r_i, r_j, A, S, w)          // Calculate record pair similarity
7:              s_i,j = NormaliseSim(s_i,j, w)                     // Normalise the similarity
8:              if s_i,j ≥ s_pmin then:
9:                  UpdateGraph(G, (r_i, r_j), s_i,j)   // Add edge and nodes (if they do not exist) to G
10: return G
```

To prevent a full comparison of every possible record pair $(r_i, r_j) : r_i, r_j \in \mathbf{R}$ we employ blocking using min-hash based LSH [15] as parameterised using b (the number of min-hash bands) and r (the band size). Only record pairs (r_i, r_j) in the same LSH block will be compared in detail (line 6 in Algorithm 1). Furthermore, before comparing records, in line 5 we check if a pair of records is temporally plausible with regard to the list **T** of temporal constraints, as described above. The generated undirected similarity graph, **G**, basically contains records as nodes and edges between records if the calculated normalised similarity, $s_{i,j}$ between two compared records r_i and r_j is at least the provided minimum threshold, s_{pmin}, and the two records are also temporal plausible with regard to **T**.

4 Temporal Connected Component Clustering

In the second phase of our approach, based on the ideas of link strength (described below) as proposed by Saeedi et al. [21], we generate a set of connected components (clusters) using the similarity graph **G**, where every pair of records in a cluster must be temporally consistent. The original connected component based clustering approach by Saeedi et al. [21] differs from ours in that it does not consider temporal constraints and also assumes the linkage of records across multiple data sources only. The requirement of incorporating temporal constraints makes the problem much more complex, since simply obtaining the connected components does not ensure temporal consistency between all records within a component, as the example in Fig. 1 shows.

Extending the ideas described by Saeedi et al. [21], and using a minimum cluster similarity threshold, s_{cmin}, with $s_{cmin} \geq s_{pmin}$ (the pair-wise similarity threshold used in Algorithm 1), we categorise the edges in **G** into three types:

- **Strong:** An edge (r_i, r_j) is *strong* if (1) the corresponding similarity $s_{i,j}$ is the highest similarity for both records r_i and r_j with regard to any other edges they have with other records in $\mathbf{G}$, and (2) $s_{i,j} \geq s_{cmin}$.
- **Norm:** An edge (r_i, r_j) is *normal* if (1) the corresponding similarity $s_{i,j}$ is the highest similarity for either record r_i or r_j (but not both) with regard to any other edges they have with other records in $\mathbf{G}$, and (2) $s_{i,j} \geq s_{cmin}$.
- **WeakHigh:** An edge (r_i, r_j) is *weak high* if (1) it is neither strong nor normal, and (2) $s_{i,j} \geq s_{cmin}$.

As detailed in Algorithm 2, one or several of these edge types are used to create the initial connected components (named *base clusters*). Edges (r_i, r_j) with similarity $s_{i,j} < s_{cmin}$ are ignored. The temporal implausible base clusters are then split further until all are temporally consistent.

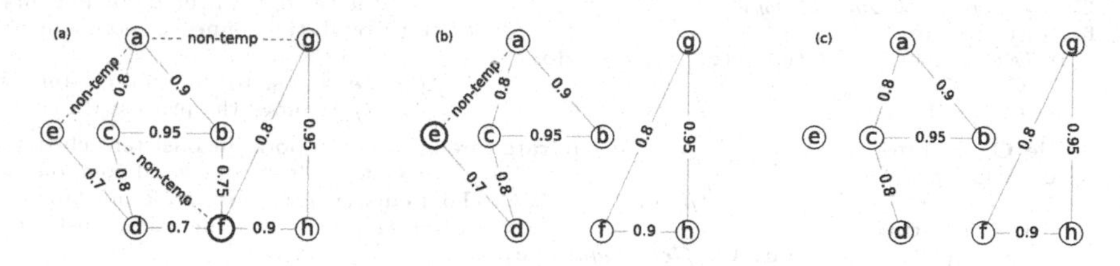

Fig. 1. Example iterative temporal cluster refinement in the base cluster generation phase, as detailed in Algorithm 2, where in each step we identify the best edge(s) to be removed that most improve the temporal consistency of the cluster(s).

First, in lines 1 to 7 of Algorithm 2, we generate the connected components based on the edges in $\mathbf{G}$ of the selected edge type(s) *bt* (one or several of **Strong**, **Norm** and **WeakHigh**, as described above) which we retrieve in the set $\mathbf{E}_b$ in line 2. We then check, in line 5, if all pairs of records in a connected component $\mathbf{c}_i$ are temporal plausible. If they are then $\mathbf{c}_i$ is added to the set of base clusters, $\mathbf{C}_b$, and removed from the set of connected components $\mathbf{C}_{cc}$. At the end of this step the clusters left in $\mathbf{C}_{cc}$ are those that contain record pairs that are temporally implausible (like two birth records five months apart).

We next process the clusters in $\mathbf{C}_{cc}$ (lines 8 to 19) one by one. We pick one $\mathbf{c}_j \in \mathbf{C}_{cc}$ (line 9) and generate a list $\mathbf{N}$ which for each node $v_i \in \mathbf{c}_j$ contains its average similarity with the other nodes in $\mathbf{c}_j$, its neighbours in $\mathbf{c}_j$, and the other nodes in $\mathbf{c}_j$ it is temporally not plausible with. In line 13, using the function $GetNodeToRefineCluster()$ we identify from $\mathbf{N}$ the best $v_i \in \mathbf{c}_j$ to process which reduces by most the number of temporal implausible edges in $\mathbf{c}_j$.

To select the best node v_{ref}, in line 13 we first attempt to find the first node in $\mathbf{N}$ with a non-empty intersection between its set of neighbours $\mathbf{n}_{ref}$ and the set of neighbours of nodes which v_{ref} is temporal inconsistent with, $\mathbf{nnt}_{ref}$. If the intersection is empty for all nodes in $\mathbf{N}$, v_{ref} will be the node with the lowest average similarity in the cluster, the lowest number of neighbours, and that is involved in the highest number of implausible edges in $\mathbf{c}_j$. In the example shown in Fig. 1, assuming the nodes with non-temporal connections are ordered as $\mathbf{N} = [f, e, a, g, c]$, we select node f first since it is the first node in $\mathbf{N}$ with a non-empty intersection ($\mathbf{n}_{ref} \cap \mathbf{nnt}_{ref} = \{b, d\}$). Subsequently (Fig. 1(b)), we check nodes e and a in that order, for non-empty intersection. However, since the intersection is empty for both nodes a and e, node e is selected for removal.

Algorithm 2. *Connected component base cluster generation*

Input:
- $\mathbf{G}$: Undirected pair-wise similarity graph
- $\mathbf{T}$: List of temporal constraints (as discussed in Sect. 3)
- p_{min}: Minimum plausibility threshold for record pairs to be added to a cluster
- s_{cmin}: Minimum similarity for record pairs to be added to a cluster
- bt: Type(s) of edges to use to create base clusters

Output:
- $\mathbf{C}_b$: Set of generated temporal consistent base clusters

```
1:  C_b = {}                                   // Initialise an empty set of clusters
2:  E_b = FindTempEdges(G, bt, s_cmin)         // Get temporal edges of type bt
3:  C_cc = GetConnComp(G, E_b)                 // Get the set of connected components
4:  for c_i ∈ C_cc do:                         // Iterate through the connected components
5:    if IsTempPlausibleCluster(c_i, T, p_min) do:
6:       C_b = C_b ∪ {c_i}                     // Add the cluster to the final cluster set
7:       C_cc = C_cc \ {c_i}                   // Remove the processed cluster
8:  while C_cc ≠ ∅ do:                         // Iterate through the temporal inconsistent clusters
9:    c_j = C_cc.pop()                         // Get the first connected component
10:   N = []                // Initialise a list to hold cluster nodes and node information
11:   for v_i ∈ c_j do:                        // Iterate through the nodes in cluster c_j
12:      N.add((CalcSim(v_i, c_j, G), GetNeigh(v_i, c_j),
              GetTempNotPlausible(c_j, v_i, T, p_min), v_i))
13:   n_ref, nnt_ref, v_ref = GetNodeToRefineCluster(N)     // Select node to refine c_j
14:   C_r = GetTempImproved(c_j, v_ref, n_ref, nnt_ref)     // Partition c_j based on v_ref
15:   for c_i ∈ C_r do:
16:      if IsTempPlausibleCluster(c_i, T, p_min) do:
17:         C_b = C_b ∪ {c_i}                  // Add cluster to the final base cluster set
18:      else:
19:         C_cc = C_cc ∪ {c_i}      // If not temporal, add cluster to C_cc to be refined
20: return C_b
```

Based on the selected node v_{ref}, and sets $\mathbf{n}_{ref}$ and $\mathbf{nnt}_{ref}$, we then partition the cluster $\mathbf{c}_j$ (line 14) using the function *GetTempImproved*() which returns the set $\mathbf{C}_r$ of two or more temporally improved clusters. In lines 15 to 19 we check each cluster $\mathbf{c}_i \in \mathbf{C}_r$ if it is temporally consistent (in which case we add it to the set of base clusters, $\mathbf{C}_b$) or not (in which case we add it to the set of clusters $\mathbf{C}_{cc}$ to be processed further). In Fig. 1, the edges that node f has with its neighbours $\{b, d\}$ are removed first, and then edges of node e are removed next resulting in the three temporally consistent clusters shown in Fig. 1(c).

The algorithm ends once all clusters in $\mathbf{C}_{cc}$ have been processed and the set of temporally consistent base clusters, $\mathbf{C}_b$, that is to be refined in the next phase of our approach, is returned in line 20 of Algorithm 2.

5 Iterative Cluster Merging

In the final phase of our approach we merge base clusters that have high overall similarities between all their individual records. This process is iterative, in that merged clusters will be further compared until no cluster is highly similar with any other cluster. We ensure all merged clusters are temporally consistent.

As detailed in Algorithm 3, we use a priority queue and sets of similar clusters to keep track of cluster pairs that are similar in order to prevent a full pair-wise recalculation of cluster similarities each time a merged cluster is generated. We start the algorithm (lines 1 and 2) by initialising the empty set of final clusters to be generated, $\mathbf{C}_f$, and the empty priority queue, $\mathbf{Q}$, which will hold cluster pairs and their similarities.

Algorithm 3. ***Similar base cluster merging***

Input:
- $\mathbf{G}$: Undirected pair-wise similarity graph
- $\mathbf{C}_b$: Base clusters (as generated in Algo. 2)
- $\mathbf{T}$: List of temporal constraints (as discussed in Sect. 3)
- p_{min}: Minimum plausibility threshold for record pairs to be considered temporal consistent
- s_{mmin}: Minimum similarity threshold for clusters to be merged
- mt: Type of edges to use to merge base clusters
- mm: Method to merge base clusters (cluster similarity calculation method)
- w_{sim}: Weight to assign to cluster similarity versus cluster coverage

Output:
- $\mathbf{C}_f$: Final set of generated clusters

```
1:  C_f = {}                                        // Initialise an empty set of final clusters
2:  Q = []                 // Initialise an empty priority queue which will be sorted by similarity
3:  E_m = FindTempEdges(G, mt, s_mmin)                        // Get temporal edges of type mt
4:  for (c_i, c_j) ∈ C_b, i < j do:                          // Loop over all cluster pairs in C_b
5:     Q.add((CalcSim(c_i, c_j, G, E_m, mm, w_sim), c_i, c_j))  // Add cluster pair and its similarity
6:  while Q ≠ ∅ do:                                       // Iterate through cluster pairs in Q
7:     s_x,y, c_x, c_y = Q.pop()                      // Get the most similar cluster pair from Q
8:     S_x = {c_p : (c_p, c_x) ∈ Q ∧ s_p,x ≥ s_mmin, c_p ≠ c_y}  // Set of clusters highly similar to c_x
9:     S_y = {c_q : (c_q, c_y) ∈ Q ∧ s_q,y ≥ s_mmin, c_q ≠ c_x}  // Set of clusters highly similar to c_y
10:    RemoveAllTuplesWithCluster(Q, c_x)                  // Remove tuples containing c_x from Q
11:    RemoveAllTuplesWithCluster(Q, c_y)                  // Remove tuples containing c_y from Q
12:    if (s_x,y ≥ s_mmin) and IsTempPlausibleClusterPair(c_x, c_y, T, p_min) do:
13:       c_mer = c_x ∪ c_y // Merge clusters if similarity is high enough and if temporally plausible
14:       if S_x ∪ S_y = ∅ do:                      // If no clusters are similar with c_x or c_y
15:          C_f = C_f ∪ {c_mer}                         // Add merged cluster to final clusters
16:       else:
17:          for c_z ∈ S_x ∪ S_y do:     // Add c_mer with clusters similar to c_x or c_y into Q
18:             Q.add((CalcSim(c_z, c_mer, G, E_m, mm, w_sim), c_z, c_mer))
19:    else:
20:       C_f = C_f ∪ {c_x, c_y}   // Add cluster pair c_x and c_y to final clusters if non-mergeable
21: return C_f
```

In lines 3 to 5, we calculate the similarities between every pair of base clusters in $\mathbf{C}_b$, where we consider a set of edge types, mt, different to Algorithm 2, with one or several of **Strong**, **Norm**, and **WeakHigh**, as described before.

In line 5 we calculate the similarity between a cluster pair $\mathbf{c}_i$ and $\mathbf{c}_j$ using the function *CalcSim*() which also takes as input a merge method, mm, and a cluster similarity weight, w_{sim}. mm determines how the overall similarity between clusters is calculated, where it can be one of the aggregation functions minimum, average or maximum. The aggregated similarity between two clusters is assigned a weight of w_{sim} whereas a weight of $1 - w_{sim}$ is assigned for coverage. The coverage is the ratio between the number of edges across $\mathbf{c}_i$ and $\mathbf{c}_j$ in $\mathbf{E}_m$, and the number of edges across $\mathbf{c}_i$ and $\mathbf{c}_j$ in $\mathbf{G}$, which reflects the proportion of edges covered in our similarity calculation. The cluster similarity, $s_{x,y}$, returned by *CalcSim*() is the weighted sum of similarity and coverage.

The main loop starts in line 6 and iterates over each cluster pair tuple in the queue $\mathbf{Q}$. For both clusters in the tuple, $\mathbf{c}_x$ and $\mathbf{c}_y$, we next (lines 8 and 9) identify all other clusters that they are similar with, and we keep these clusters in two sets $\mathbf{S}_x$ and $\mathbf{S}_y$, respectively. We then remove all tuples in $\mathbf{Q}$ that contain $\mathbf{c}_x$ or $\mathbf{c}_y$ since they should not be re-processed. In line 12 we check if the similarity between $\mathbf{c}_x$ and $\mathbf{c}_y$ is at least the minimum cluster merge similarity s_{mmin} and if they are temporally consistent with each other. If this is the case we merge clusters $\mathbf{c}_x$ and $\mathbf{c}_y$ into $\mathbf{c}_{mer}$ in line 13.

If both $\mathbf{c}_x$ and $\mathbf{c}_y$ are not similar with any other clusters (i.e. both $\mathbf{S}_x$ and $\mathbf{S}_y$ are empty; the test in line 14), then based on the triangular inequality we know that the merged cluster $\mathbf{c}_{mer}$ cannot be merged further with any other clusters. Therefore $\mathbf{c}_{mer}$ is added to the set of final clusters, $\mathbf{C}_f$, in line 15. Otherwise, in line 17 we calculate the similarity of the merged cluster, $\mathbf{c}_{mer}$, with each cluster in $\mathbf{S}_x$ and $\mathbf{S}_y$ and add new tuples into the queue in line 18.

If a cluster pair in the queue was not similar enough or not temporally consistent, we do not merge $\mathbf{c}_x$ and $\mathbf{c}_y$ but instead add both into $\mathbf{C}_f$ in line 20. Finally we return the set of merged and temporally consistent clusters, $\mathbf{C}_f$.

6 Experimental Evaluation

We evaluated our temporal clustering approach using a real Scottish data set, as provided by [18], that covers the population of the Isle of Skye over the period from 1861 to 1901. This data set consists of 17,614 birth certificates, where each of these contains personal details about a baby and its parents such as their names, address, marriage date, occupations, and the birth date. As with other historical data [1,11], this data set has a very small number of unique name values (2,055 first and only 547 last names), and the frequency distributions of names are also very skewed. Many records have missing addresses or occupations, and for unmarried women the details of a baby's father are mostly missing.

This data set has been extensively curated and linked semi-manually by demographers who are experts in the domain of linking such historical data [18]. Their approach followed long established rules for family reconstruction, leading to a set of linked birth records. We thus have a set of manually generated links of births that allows us to compare the quality of the different clustering techniques, and to evaluate how temporal constraints can improve linkage quality.

We used three different subsets of attributes to compare record pairs and generate the similarity graph $\mathbf{G}$ discussed in Sect. 3: *All* (parents names, addresses, occupations, and marriage dates), *Parent names and addresses*, and *Parent names* only. To compare attribute values we used approximate string comparison functions such as Jaro-Winkler and edit-distance [5]. We used both unweighted (UW) and weighted (W) similarities, where weights were calculated based on the traditional Fellegi and Sunter record linkage approach [10]. We thus ended up with six different similarity graphs $\mathbf{G}$ where we set $s_{pmin} = 0.7$ to only include pair-wise links with at least this normalised similarity.

As discussed in Sect. 4, we evaluated different combinations of the edge types **Strong**, **Norm**, and **WeakHigh** in our approach. For Algorithm 2 we generated base clusters with only **Strong** edges because these clusters showed much higher precision (95%) in set-up experiments compared to using other edge type combinations. In Algorithm 3 we used '**Norm**', '**Norm** with **WeakHigh**' (where base clusters were merged using both edge types in the same run), as well as '**Norm** and **WeakHigh**' (where base clusters were first merged using **Norm** edges and then the resulting clusters were merged again using **WeakHigh** edges). As discussed in Sect. 5, we used the three cluster merge methods (mm): **Min** (minimum pair-wise record similarity across two clusters), **Avr** (average pair-wise similarity), and **Max** (maximum pair-wise similarity). We also ran experiments where we did not consider any temporal constraints, i.e. we set $\mathbf{T} = [\,]$ in all algorithms.

We calculated linkage quality as precision (the ratio of true links identified against all links within clusters) and recall (the ratio of true links identified against all true links) [5]. We do not present F-measure results given recent work has identified some problematic aspects with this measure when used for record linkage [12]. Instead we present the area under the precision-recall curve (AUC-PR) which has shown to be robust for class imbalance problems [7].

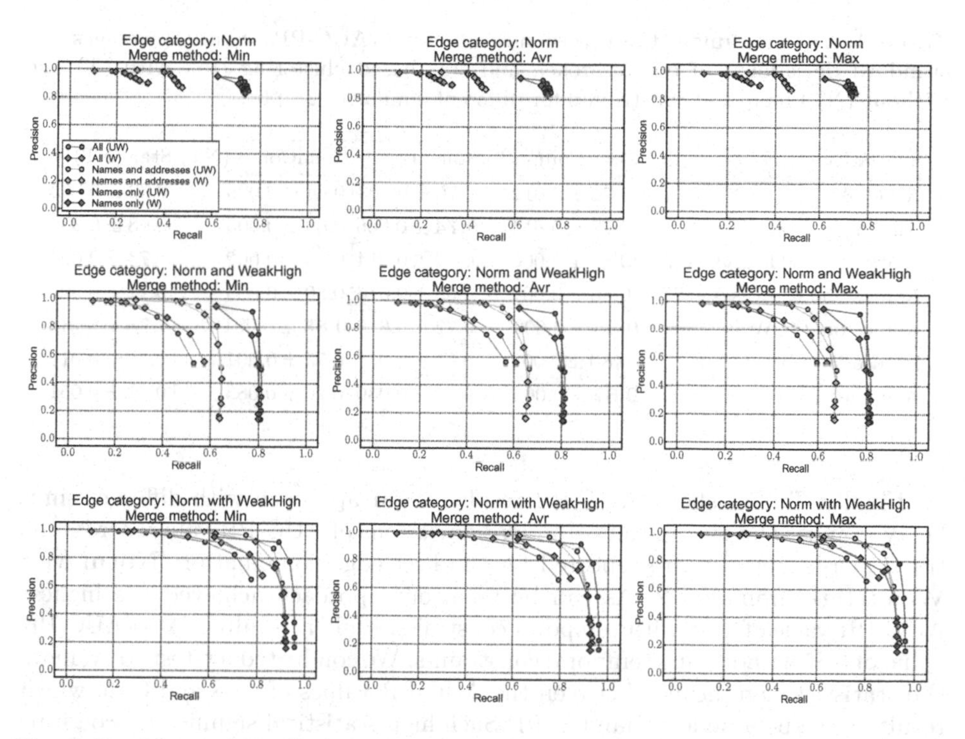

Fig. 2. Precision and recall of our approach with different merge methods (as discussed in Sect. 5) for different similarity graphs as described in Sect. 6.

We compared the proposed approach with our recent star clustering based method [16], as this is the only approach we are aware of that uses temporal aspects for group linkage. We applied LSH [15] (blocking) to limit the number of record pairs being compared, resulting in a recall of 99.7% of true matches for the similarity graph **G**. We set the similarity threshold in Algorithm 2 and 3 from 1.0 to 0.7 in 0.05 steps, and the weight w_{sim} in Algorithm 3 to 0.5. We implemented all techniques using Python 2.7, and the programs and similarity graphs are available from the authors to facilitate repeatability.

Figure 2 shows precision and recall curves of our approach with different edge combinations. As can be seen, the edge combination '**Norm** with **WeakHigh**' provided the best linkage quality compared to other edge combinations. The reason for this is that when using **Norm** edges only, many true links that are in the **WeakHigh** category are ignored. It also appears that merging clusters in a single run using the '**Norm** with **WeakHigh**' method provided better results than conducting the cluster merging in two phases (as done in the '**Norm** and **WeakHigh**' method). We also noted that the quality of clustering does not change much with the merge method mm (**Min**, **Avr**, or **Max**) because most clusters generated only contained between 2 to 4 records.

Table 1. The area under the precision-recall curve (AUC-PR) results (averages and standard deviations) of our approach and star based clustering [16], with (T) and without (NT) temporal constraints for different similarity graphs.

Similarity graph	ConnComp (T)	Star (T)	ConnComp (NT)	Star (NT)
All (UW)	0.72 ± 0.012	0.70 ± 0.003	0.64 ± 0.005	0.63 ± 0.003
All (W)	0.77 ± 0.014	**0.74** ± 0.006	0.69 ± 0.005	0.68 ± 0.004
Names and addresses (UW)	0.87 ± 0.006	0.70 ± 0.014	0.83 ± 0.002	0.73 ± 0.003
Names and addresses (W)	0.86 ± 0.007	0.69 ± 0.016	0.80 ± 0.003	0.72 ± 0.007
Names only (UW)	**0.88** ± 0.002	0.72 ± 0.018	**0.85** ± 0.001	**0.78** ± 0.015
Names only (W)	0.80 ± 0.002	0.65 ± 0.016	0.73 ± 0.001	0.69 ± 0.019
Averages	**0.82** ± 0.064	0.70 ± 0.030	0.76 ± 0.083	0.71 ± 0.051

Finally, Table 1 shows AUC-PR results of our approach with different similarity graphs. The average and standard deviations of AUC-PR values across the three merge methods are reported for the best edge combination '**Norm** with **WeakHigh**' from Fig. 2. As can be seen, our approach achieved the highest AUC-PR value of 0.88 with temporal constrains while it resulted in an AUC-PR value of 0.85 without any temporal constraints. We conducted a t-test to evaluate the statistical significance between the AUC-PR values of 0.88 and 0.85, which resulted in a p-value less than 0.0001. Such high statistical significance confirms that the use of temporal constraints can improve the overall linkage quality in our approach. Further, as this table shows, our approach outperformed both temporal and non-temporal star clustering in terms of linkage quality for all similarity graphs which indicates our approach is suitable to cluster records, such as the births by the same mothers in the context of historical record linkage, with high linkage quality.

7 Conclusions and Future Work

We have presented a temporal clustering approach for group record linkage. Our approach first generates a graph that represents the similarities calculated between individual records, and then generates temporally consistent connected components which are merged to obtain a set of high quality clusters. Our experimental evaluation on a real data set from Scotland has shown that our approach can substantially outperform a previous temporal clustering approach for record linkage. In the future we aim to conduct empirical evaluations for different data sets and parameter settings. We further plan to conduct a complexity analysis on the proposed algorithms and also learn temporal constraints for different time intervals using ground truth data based on true matching record pairs.

Acknowledgements. This work was supported by ESRC grants ES/K00574X/2 *Digitising Scotland* and ES/L007487/1 *ADRC-S*. We like to thank Alice Reid of the University of Cambridge and her colleagues Ros Davies and Eilidh Garrett for their work on the Isle of Skye database, and their helpful advice on historical Scottish demography. This work was partially funded by the Australian Research Council under DP160101934.

References

1. Antonie, L., Inwood, K., Lizotte, D.J., Ross, J.A.: Tracking people over time in 19th century Canada for longitudinal analysis. Mach. Learn. **95**, 129–146 (2014)
2. Aslam, J.A., Pelekhov, E., Rus, D.: The star clustering algorithm for static and dynamic information organization. J. Graph Algorithms Appl. **8**, 95–129 (2004)
3. Bailey, M., Cole, C., et al.: How well do automated methods perform in historical samples? Evidence from new ground truth. Technical report, NBER (2017)
4. Bloothooft, G., Christen, P., Mandemakers, K., Schraagen, M.: Population Reconstruction. Springer, Heidelberg (2015). https://doi.org/10.1007/978-3-319-19884-2
5. Christen, P.: Data Matching. Springer, Heidelberg (2012). https://doi.org/10.1007/978-3-642-31164-2
6. Christen, V., Groß, A., Fisher, J., Wang, Q., Christen, P., Rahm, E.: Temporal group linkage and evolution analysis for census data. In: EDBT, Venice (2017)
7. Davis, J., Goadrich, M.: The relationship between precision-recall and ROC curves. In: ACM ICML, Pittsburgh, pp. 233–240 (2006)
8. Dibben, C., Williamson, L., Huang, Z.: Digitising Scotland (2012). http://gtr.rcuk.ac.uk/projects?ref=ES/K00574X/2
9. Dillon, L.Y.: Integrating nineteenth-century Canadian and American census data sets. Comput. Hum. **30**(5), 381–392 (1996)
10. Fellegi, I.P., Sunter, A.B.: A theory for record linkage. J. Am. Stat. Assoc. **64**(328), 1183–1210 (1969)
11. Fu, Z., Christen, P., Zhou, J.: A graph matching method for historical census household linkage. In: Tseng, V.S., Ho, T.B., Zhou, Z.-H., Chen, A.L.P., Kao, H.-Y. (eds.) PAKDD 2014. LNCS (LNAI), vol. 8443, pp. 485–496. Springer, Cham (2014). https://doi.org/10.1007/978-3-319-06608-0_40
12. Hand, D., Christen, P.: A note on using the f-measure for evaluating record linkage algorithms. Stat. Comput. **28**(3), 539–547 (2018)

13. Hassanzadeh, O., Chiang, F., Lee, H.C., Miller, R.J.: Framework for evaluating clustering algorithms in duplicate detection. PVLDB **2**(1), 1282–1293 (2009)
14. Kum, H.C., Krishnamurthy, A., Machanavajjhala, A., Ahalt, S.C.: Social genome: putting big data to work for population informatics. IEEE Comput. **47**(1), 56–63 (2014)
15. Leskovec, J., Rajaraman, A., Ullman, J.D.: Mining of Massive Datasets. Cambridge University Press, Cambridge (2014)
16. Nanayakkara, C., Christen, P., Ranbaduge, T.: Temporal graph-based clustering for historical record linkage. In: MLG, held at ACM SIGKDD, London (2018)
17. On, B.W., Koudas, N., Lee, D., Srivastava, D.: Group linkage. In: IEEE ICDE, Istanbul (2007)
18. Reid, A., Davies, R., Garrett, E.: Nineteenth-century Scottish demography from linked censuses and civil registers. History Comput. **14**(1–2), 61–86 (2002)
19. Ruggles, S., Fitch, C.A., Roberts, E.: Historical census record linkage. Ann. Rev. Sociol. **44**(1), 19–37 (2018)
20. Saeedi, A., Peukert, E., Rahm, E.: Comparative evaluation of distributed clustering schemes for multi-source entity resolution. In: Kirikova, M., Nørvåg, K., Papadopoulos, G.A. (eds.) ADBIS 2017. LNCS, vol. 10509, pp. 278–293. Springer, Cham (2017). https://doi.org/10.1007/978-3-319-66917-5_19
21. Saeedi, A., Peukert, E., Rahm, E.: Using link features for entity clustering in knowledge graphs. In: Gangemi, A., et al. (eds.) ESWC 2018. LNCS, vol. 10843, pp. 576–592. Springer, Cham (2018). https://doi.org/10.1007/978-3-319-93417-4_37

Data Pre-processing and Feature Selection

Learning Diversified Features for Object Detection via Multi-region Occlusion Example Generating

Junsheng Liang, Zhiqiang Li, and Hongchen Guo(✉)

School of Computer Science and Technology, Beijing Institute of Technology, Beijing, China
guohongchen@bit.edu.cn

Abstract. Object detection refers to the classification and localization of objects within an image by learning their diversified features. However, the existing detection models are usually sensitive to the important features in some local regions of the object. The existing algorithms cannot learn the diversified features regarding to each region effectively, which limit the performance of the model to a certain range. In this paper, we propose a novel and principle method called Multi-region Occlusion Example Generating (MOEG) to guide the detection model in fully learning the features of the various regions of the object. MOEG can generate completely new occlusion examples and it enables our detection model to learn the features of the remaining regions in the object by blocking the important regions in the proposal. It is a general method to generate occlusion examples and it can be implemented to most mainstream region-based detectors very easily such as Fast-RCNN and Faster-RCNN. Our experimental results indicate a 2.4% mAP boost on VOC2007 dataset and a 4.1% mAP boost on VOC2012 dataset compared to the Fast-RCNN pipeline. And as datasets become larger and more challenge, our method MOEG become more effective as demonstrated by the results on the MS COCO dataset.

Keywords: Feature extraction · Data augmentation · Object detection

1 Introduction

In general, we make full use of visual cues from multiple regions of the object to classify an object and localize it. However, it is the fact that the object detection task contains various complex situations, which result in low detection accuracy. For example, some important regions of an object are occluded, deformed or blurred caused by background changes, conditions of illumination or the angle of shooting. Therefore, the detection models may not perform effectively due to the unobvious features extracted from important regions. So, it is a big challenge

Q. Yang et al. (Eds.): PAKDD 2019, LNAI 11440, pp. 541–552, 2019.
https://doi.org/10.1007/978-3-030-16145-3_42

for current detection tasks to fully learn and extract the diversified features in each local region of the object.

In this paper, we propose a novel and principle method called Multi-region Occlusion Example Generating (MOEG) to learn and extract features in different local regions of the object as many as possible, although parts of them are not common features of the object. We utilize these features to classify and localize the object more accurately. MOEG blocks parts of the local regions (which have great impacts on the detection results) in the foreground examples by using masks with different occluded regions (e.g., 30% and 50%). We use these examples to train our model. In order to reduce the influence of such manual occlusion examples on the ConvNet's feature extraction process, we mask the regions in the feature vectors instead of pixels. In order to avoid learning the difference between foreground and background examples caused by these occlusions, we randomly block parts of regions in some background examples. Furthermore, we combine different examples to train our model in each iteration to prevent converging too early and missing some of the important regional features. Examples with different occluded regions can help to learn features in the remaining regions, and make our model able to learn the diversified features in each local region of the object.

In order to show the effectiveness of our method, we conduct substantial experiments over VOC2007 and VOC2012 [18] dataset. We use MOEG to train Fast-RCNN [8] and Faster-RCNN [19]. We compare our model with the standard baseline and the state-of-the-art methods such as A-Fast-RCNN [25] and OHEM [20]. The experimental results indicate that MOEG outperforms the Fast-RCNN pipeline with a 2.4% mAP on VOC2007 and a 4.1% mAP on VOC2012. Furthermore, our accuracies (mAP: 71.8% and 70.5% on VOC2007 and VOC2012) are higher than A-Fast-RCNN (mAP: 71.4% and 69.0% on VOC2007 and VOC2012) and OHEM (mAP: 69.9% and 69.8% on VOC2007 and VOC2012). We integrate MOEG with a stronger baseline Faster-RCNN and we get a boost in VOC2007 (from 70.9%mAP to 72.6%mAP). For MS COCO [17] dataset, we also get a significant promotion which proves that our method can get a good performance even though the dataset itself has provided sufficient examples with objects in different states.

2 Related Work

In recent years, great improvements have been achieved in the field of object detection. Most of these methods come from the promotion of the ability to learn and extract the image features by the detection model. And establishing connections and relationships between features in each region artificially [1,2,7,16,26] can improve the efficiency of information transmission during both feature learning and feature extraction such as Gated bi-directional CNN (GBD-Net) [26]. But such models are too difficult to achieve in engineering due to the complexity. In addition, these models cannot make the best of the self-learning ability of DNN.

In order to make our model learn the diversified features of the object autonomously, one of the most straightforward approaches is to provide large-scale training data which have object in different states. We also can use data augmentation techniques to expand the training data, such as image scaling, cropping, flipping, shifting, rotation and color jittering [12,14,21,23]. There are also some unsupervised data augmentation techniques such as AutoAugment [3] which have got a significant effect in image classification. These techniques can increase both the amount and diversity of training data and have been selectively applied in the training process of object detection networks which can reduce the overfitting phenomenon. Moreover, hard example mining techniques such as OHEM and S-OHEM [15,20] are also used to select the hard examples to train the detection model and improve the performance of the model. OHEM algorithm uses a read-only network for forward pass of all proposals and computing their losses in order to select hard examples. Its network executes forward passes for all proposals, but executes backward passes only for the hard examples selected from the read-only network.

Some novel approaches prove that it is better to generate hard examples instead of searching in the existing dataset [22,25,27]. For example, Hide-and-Seek [22] uses random occlusion image patches in the training process. But it is only applicable for classification and weakly-supervised object localization and random occlusion is not a good strategy. A-Fast-RCNN [25] provides a novel idea that we can generate hard examples using adversarial networks. It uses adversarial networks to generate examples with occlusions and deformations that may be hard for our object detector to classify. The adversarial networks have been used in many fields such as generating images [4,10]. But the current generators cannot achieve good results in generating hard examples in the pixel space for training. A-Fast-RCNN tries to generate occlusions and deformations in feature vectors to avoid this problem. Especially, its Adversarial Spatial Dropout Network can generate occlusion masks for region-based features. However, the adversarial network's prediction result is not accurate enough, the examples generated from it are still not desired hard examples with the most important regions blocked. In order to alleviate this problem, it uses importance sampling to sample the output generated by the adversarial network and jointly train the adversarial network in the detector's training process.

3 Our Method

We introduce our method called Multi-region Occlusion Example Generating (MOEG) in details. Compared with the state-of-the-art methods such as A-Fast-RCNN [25] and Online hard example mining [20] algorithm, MOEG can get a higher performance and it is easier to implement in the mainstream region-based object detectors, such as Fast-RCNN [8] and Faster-RCNN [19].

3.1 Multi-region Occlusion Example Generating

Our goal is to guide our model to learn the diversified features in each region of objects. Firstly, we need to extract the features in each region. SPP-net [11] introduces a spatial pyramid pooling (SPP) layer to generate fixed-length representations (we call it region-based features) for each proposal. These region-based features still keep the spatial message from the proposal. In the Fast-RCNN pipeline [8], we can get the region-based features for each proposal. The size of the features is $d \times d \times c$, where d is the spatial dimension and c is the number of channels (e.g., $c = 512$, $d = 7$ in VGG16). The features in each region of the proposal have been mapped to the original image with size $d \times d$ corresponding to the same position, and has a c-dimensional feature vector.

Secondly, we need to find a good way to distinguish which parts of the regional features are already learned and which parts are not yet. If a part of an object was occluded in an example and this example caused a higher prediction loss compared to others, the features extracted from the occlusion regions of this example must be vital for our current detector to classify the object. That is to say, our current detector has learned these regional features in a good way to some degree and these features have been fully used in detection.

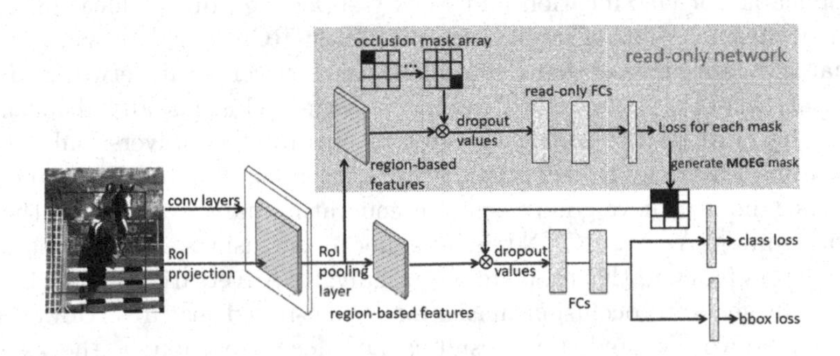

Fig. 1. We implement our method in the standard Fast-RCNN architecture. The network can be divided into two parts: the first part is used to generate occluded examples (on blue background) and the second part is used to accumulate and pass the gradients. (Color figure online)

Since these regional features have been learned, we need to guide our detector to learn features in other regions. So we block these important regions (have higher prediction losses) in the examples and use these examples to train our model. In this way, our model can learn features in the regions that have not been occluded. Instead of generating occlusion masks in the pixel space, we generate masks in the region-based features. It is easier to implement in the network. And it does not affect the feature extraction process of the ConvNet. We can generate a mask M with $d \times d$ values which is either 0 or 1 (0 means block and 1 means maintain). And the element-wise product of mask M and the region-based features X is the final feature map $\tilde{M}$ for the proposal with parts of regions occluded. Using different masks we can get the feature maps corresponding to the examples with different regions occluded. We put these feature maps to the fully connected layer $\mathcal{F}$ and get the class loss of each example. We use exhaustive strategy to generate the final mask $\tilde{M}$. That is, to select the top n% regions which cause highest class loss independently and then combine them. This way can distinguish the most discriminative regions for the current detector and without adding any additional parameters. Our final network loss $\mathcal{L}_{\mathcal{F}}$ can be written down as,

$$\mathcal{L}_{\mathcal{F}} = \mathcal{L}_{\text{softmax}}(\mathcal{F}_c(X{\cdot}\tilde{M}), C) + [C \notin bg]\mathcal{L}_{\text{bbox}}(\mathcal{F}_l(X{\cdot}\tilde{M}), L) \tag{1}$$

We use three kinds of foreground examples to train our model: the original example, the example having occluded 30% of the regions and the example having occluded 50% of the regions. These different kinds of examples can guide our detector in learning diversified features in each region step by step. We only use our method MOEG to generate masks for foreground examples here. We also randomly block values in some regions of background examples to avoid our model learning the difference between foreground and background examples caused by occlusion.

MOEG intends to learn more diversified regional features by masking regions where features have been fully learned and then forcing our model to learn other features in the remaining regions. But the remaining regions in each proposal are not always including the object we interested. Some of them may include many background messages. Because a proposal is counted as true-positive for an object category if the IoU between the proposal and the ground-truth box is greater than a threshold (e.g. 0.5 in our experiments). So, is it necessary to learn such background messages for localizing and classifying the object and will it lead to a decline in the performance of our detector? Many researches [1,6,7,26] have showed that background messages (especially the regions nearby objects) are very useful for object detection. It can be used to improve classification of the proposal when some important object features are missing. And the partially occlusion examples are more difficult to classify for the current detector, so the training will automatically focus more on these examples and these examples can contribute much to the gradient [20,25]. Besides, we use more kinds of foreground examples to train our model which can improve our model's adaptability and prevent the training process falling into the local minimum to some extent.

3.2 Implementation Details

MOEG can be implemented easily by many popular region-based detectors such as RCNN, Fast-RCNN, Faster-RCNN [8,9,14]. In this paper, we use the standard Fast-RCNN detector to achieve MOEG algorithm which can be more convenient for us to compare it with other state-of-the-art methods. We use Python to implement the layers in MOEG and use Caffe [13] to train our model. In Fast-RCNN pipeline, the ConvNet extracts the feature maps from the input image and the RoI pooling layer generates region-based features for each proposal. As shown in Fig. 1, the overall framework can be divided into two parts. The first part of the network is used to generate masks for region-based features of each proposal. In this network, we use exhaustive strategy to generate the final mask $\tilde{M}$ and use this mask to block the region-based features. After that, the processed region-based features are put into the second part of the network to calculate the final loss and execute backward passes.

In the first part, we use a fixed-size window (e.g., the window with size 2×2 is used in our experiments) to slide in a grid with spatial layout $d \times d$ step by step without overlap. We can get $\lceil \frac{d}{2} \rceil \times \lceil \frac{d}{2} \rceil$ masks for each proposal. We use each mask M to drop out the parts of the region-based features in all channels whose corresponding regions in the mask M are occluded. We put these region-based features with partial occlusion to the fully connected layers to get the class losses. We assign those losses to each region where the window is covered and we can eventually get a loss map corresponding to the proposal. Through it, we select the regions that have highest losses and generate the final mask $\tilde{M}$. More specifically, we select the top 30% and top 50% regions with highest losses in our experiments. In order to achieve this process, we set the first part to a read-only network which allocates GPU memory only for forward pass and have not the backward pass process.

In the second part, the element-wise product of the final mask $\tilde{M}$ and region-based features X is the final features for the proposal. We put these features to the fully connected layers $\mathcal{F}$ to get the class loss $\mathcal{L}_{\text{softmax}}$ and the bbox regression loss $\mathcal{L}_{\text{bbox}}$. And this network will execute the forward and backward passes. And then, it will accumulate the gradients and pass them to fully connected layers and ConvNet.

4 Experiments

In order to show the effectiveness of MOEG, we perform most of the experiments on the PASCAL VOC2007 dataset which has a high authority. And we also conduct our experiments on PASCAL VOC2012 [5] and MS COCO [17] dataset. We perform the ablative study to compare our method MOEG with other occlusion methods. And then we compare our method MOEG with A-Fast-RCNN (including ASDN and ASTN) [25] and Online Hard Example Mining (OHEM) [20] algorithm on the PASCAL VOC2007 dataset and PASCAL VOC2012 dataset. In the end, we present result on MS COCO dataset.

4.1 Experimental Settings

Our proposed framework is implemented based on the Fast-RCNN pipeline and we use pre-trained parameters from ImageNet to initialize our ConvNet. We conduct most of our experiments on PASCAL VOC2007 dataset and use the 'trainval' set for training and 'test' set for testing. The standard ConvNet architecture we used is VGG16. We also perform our method MOEG in other architectures such as VGGM, AlexNet, ResNet. We use the Selective Search proposals [24] during training. We train all methods for 40k iterations with an initial learning rate of 0.001 and decay the learning rate to 0.0001 after 30k iterations. Each iteration contains two batches (128 proposals with one image in each batch) and we apply SGD in each iteration.

4.2 Ablative Analysis

To prove that MOEG is better than other occlusion methods. We do an ablative study here. We use VGG16 architecture for these experiments. To make it fair, we use the same Fast-RCNN detector which has been pre-trained for 10k iterations. And then we use different occlusion methods to train this detector with several times and choose the best result for each method. Firstly, we compare different occlusion strategies used in these methods including Random dropout (used in Hide-and-Seek), ASDN (used in A-Fast-RCNN), Exhaustive dropout (used in our method) and Anti-Mask (contrary to Exhaustive dropout). The important thing to note here is we mask region-based features for all experiments, but Hide-and-Seek was originally used to mask pixels. And through our experiments, we found that masking region-based features instead of masking pixels can get better performances. The results have been shown in Table 1.

Table 1. VOC2007 test detection average precision (%). Hide-and-seek mask pixels and random dropout mask region-based features.

method	mAP	aero	bike	bird	boat	bottle	bus	car	cat	chair	cow	table	dog	horse	mbike	persn	plant	sheep	sofa	train	tv
FRCN	69.4	70.4	78.5	70.1	56.8	44.1	77.2	79.6	81.0	50.8	75.1	70.1	81.0	82.9	72.0	73.9	39.7	69.7	67.1	74.8	72.7
Hide-and-Seek	70.2	73.8	80.5	70.4	58.4	46.1	79.7	80.2	83.0	49.2	74.6	69.9	80.0	80.4	75.8	73.6	40.0	69.9	66.1	78.3	75.9
Random dropout	70.5	69.5	79.1	69.7	59.4	44.0	82.2	79.7	82.6	51.5	75.9	69.7	82.8	84.2	76.7	73.9	42.5	67.7	67.4	76.9	73.8
Anti-Mask	69.9	73.9	79.2	69.7	57.9	43.7	79.8	78.7	82.4	50.3	77.1	67.5	79.8	82.2	76.0	73.2	39.2	69.1	69.5	77.4	72.1
ASDN	70.8	73.6	80.5	70.2	59.1	46.2	80.3	79.8	82.4	50.4	77.9	69.4	81.5	81.8	75.6	73.7	40.4	71.9	68.1	78.6	74.1
Exhaustive dropout	71.2	75.0	79.9	70.8	55.9	46.3	81.6	79.6	81.0	50.6	78.1	69.1	83.2	83.4	76.9	74.2	43.0	73.0	70.9	77.9	74.9

We also compare different sampling strategies. The first is used in A-Fast-RCNN which provides proposals in a batch without any occlusions and provides proposals with occlusions (30%) in the other batch. So, it provides the same type of example combination in each iteration. The second is used in our method MOEG which provides three different types of foreground examples (original, 30% and 50%). And in order to verify whether background samples need to be occluded, we also compare different background example combinations. Here, we use Exhaustive dropout to mask foreground examples and use Random dropout to mask background examples. The results have been shown in Table 2. We use exhaustive dropout as our occlusion strategy and combine it with our sampling strategy to train Fast-RCNN. And then we get the best performance. We achieved 71.6% mAP on the PASCAL VOC2007 dataset.

Table 2. VOC2007 test detection mAP(%). Each batch has two ratios representing the ratio of occlusion examples in foreground and background examples respectively.

Batch1	Batch2	Batch3	mAP
0%, 0%	30%, 0%	×	71.2%
0%, 0%	50%, 0%	×	70.7%
30%, 0%	50%, 0%	×	70.3%
0%, 0%	30%, 0%	50%, 0%	71.4%
0%, 30%	30%, 30%	50%, 30%	70.3%
0%, 0%	30%, 30%	50%, 30%	70.9%
0%, 0%	30%, 0%	50%, 30%	71.6%

4.3 Comparisons with A-Fast-RCNN and OHEM

We have compared our method MOEG with ASDN [25] in the previous section. But we used the same pre-trained Fast-RCNN detector for the sake of fairness. That is because ASDN need to be initialized and it also needs feedbacks from the updating detector in the training. But MOEG algorithm does not need these settings. So we directly use pre-trained parameters from ImageNet to initialize our ConvNet and train the Fast-RCNN detector. We increase the total number of iterations to 50k with an initial learning rate of 0.001 and decay the learning rate to 0.0001 after 40k iterations. We compare our method MOEG with A-Fast-RCNN and OHEM in this part. A-Fast-RCNN contains two sub-networks (ASDN and ASTN), which concerns spatial occlusion and spatial deformation respectively but our method MOEG only considers on the aspect of spatial occlusion. OHEM concerns Hard Example Mining which is related to our method MOEG despite we concern Example Generating. We show the results in Table 3. The results show that our accuracy is 1.0% higher than ASDN and 1.9% higher than ASTN. Better yet, our accuracy is still better than A-Fast-RCNN which combines the advantages of two networks ASDN and ASTN. Compared to A-Fast-RCNN, our training time has been reduced by approximately four times (including the time to pre-train ASDN and ASTN). In order to compare MOEG algorithm with OHEM algorithm, we use OHEM algorithm to train Fast-RCNN detector with the training schedule described in [20], because our training schedule is not suitable for it. The result show that our accuracy is 1.9% higher than OHEM. In addition, OHEM algorithm need more GPU memory for forward pass.

MOEG is not just suitable for a single ConvNet architecture and single region-based detector. It has also achieved significant boosts in other mainstream architectures such as VGGM, AlexNet and ResNet. On the PASCAL VOC2007 dataset, our method get a 4.3% mAP boost in VGGM, a 3.5% mAP boost in AlexNet and a 1.9% mAP boost in ResNet. And used in different region-based detectors such as Faster-RCNN [19], our method also can be significant. We train Faster-RCNN in the way of approximate joint training. When we use our

Table 3. VOC2007 test detection average precision (%). Compared with ASDN and OHEM algorithm.

method	arch	mAP	aero	bike	bird	boat	bottle	bus	car	cat	chair	cow	table	dog	horse	mbike	persn	plant	sheep	sofa	train	tv
FRCN	VGG16	69.4	70.4	78.5	70.1	56.8	44.1	77.2	79.6	81.0	50.8	75.1	70.1	81.0	82.9	72.0	73.9	39.7	69.7	67.1	74.8	72.7
OHEM	VGG16	69.9	71.2	78.3	69.2	57.9	46.5	81.8	79.1	83.2	47.9	76.2	68.9	83.2	80.8	75.8	72.7	39.9	67.5	66.2	75.6	75.9
ASDN	VGG16	70.8	73.6	80.5	70.2	59.1	46.2	80.3	79.8	82.4	50.4	77.9	69.4	81.5	81.8	75.6	73.7	40.4	71.9	68.1	78.6	74.1
ASTN	VGG16	69.9	73.7	81.5	66.0	53.1	45.2	82.2	79.3	82.7	53.1	75.8	72.3	81.8	81.6	75.6	72.6	36.6	66.3	69.2	76.6	72.7
A-Fast-RCNN	VGG16	71.4	75.7	83.6	68.4	58.0	44.7	81.9	80.4	86.3	53.7	76.1	72.5	82.6	83.9	77.1	73.1	38.1	70.0	69.7	78.8	73.1
Ours	VGG16	71.8	73.9	80.9	69.5	57.6	47.5	81.8	79.8	84.6	53.3	80.6	71.9	81.6	84.0	78.8	73.8	42.3	71.7	69.5	77.6	73.8
FRCN	VGGM	56.6	66.2	66.7	56.4	40.1	24.8	62.8	71.3	71.0	32.0	58.9	57.2	61.8	70.8	66.4	59.9	30.4	50.2	56.4	65.5	63.4
Ours	VGGM	60.9	68.8	71.0	58.9	46.5	29.3	69.9	73.7	73.5	38.0	64.6	62.8	69.4	73.7	68.5	62.6	29.8	59.1	59.4	72.9	65.6
FRCN	AlexNet	56.4	64.7	66.0	54.3	39.0	28.2	63.0	71.6	67.5	33.3	58.5	59.4	60.2	71.0	63.2	59.8	29.9	54.4	51.7	65.8	65.8
Ours	AlexNet	59.9	66.5	70.3	57.2	41.8	30.8	65.2	72.8	73.6	36.9	64.0	63.9	65.1	75.3	68.5	63.1	30.7	56.5	59.5	70.5	65.1
FRCN	ResNet-101	71.6	77.6	82.5	71.7	55.1	41.9	79.2	80.5	86.9	54.2	81.6	72.2	86.9	84.9	80.3	72.2	35.2	71.2	75.5	78.5	64.0
Ours	ResNet-101	73.5	78.9	84.2	74.9	56.3	42.3	80.8	81.2	85.7	54.6	82.6	74.3	87.2	86.2	79.1	74.5	35.0	73.9	76.2	77.8	65.3
Faster-RCNN	VGG16	70.9	74.1	79.4	68.0	56.2	55.4	76.6	80.4	85.7	52.6	75.7	67.7	77.2	80.5	77.0	78.0	45.5	71.4	67.0	75.7	73.9
Faster-RCNN(Ours)	VGG16	72.6	74.4	81.0	71.7	61.7	54.8	81.8	84.1	82.0	55.3	80.1	70.7	81.9	85.1	78.5	77.8	43.7	71.0	66.3	77.9	71.8

method, we keep all the configs but cut off the backward propagated signal from RPN to ConvNet. The result shows that our method can get a 1.7% mAP boost on the PASCAL VOC2007 dataset.

4.4 PASCAL VOC2012 and MS COCO Results

In order to prove our method MOEG can make effect in more challenging datasets. We also conduct experiments on PASCAL VOC2012 [5] and MS COCO 2014 [17] datasets. On PASCAL VOC2012 dataset, we compare our method MOEG with A-Fast-RCNN and OHEM with VGG16. We show the results in Table 4. From the result of VOC2012 dataset, our Method MOEG has a better performance (mAP: 70.5%) compared to OHEM (mAP: 69.8%) and A-Fast-RCNN (mAP: 69.0%). The result indicates that, when the dataset's diversity increased, A-Fast-RCNN's benefit obviously reduced, but our method MOEG not. It is because our promotion comes from not only the data diversity but also the powerful learning ability of diversified features.

Table 4. VOC2012 test detection average precision (%). Compared with A-Fast-RCNN and OHEM algorithm.

method	mAP	aero	bike	bird	boat	bottle	bus	car	cat	chair	cow	table	dog	horse	mbike	persn	plant	sheep	sofa	train	tv
FRCN	66.4	81.8	74.4	66.5	47.8	39.3	75.9	69.1	87.4	44.3	73.2	54.0	84.9	79.0	78.0	72.2	33.1	68.0	62.4	76.7	60.8
A-Fast-RCNN	69.0	82.2	75.6	69.2	52.0	47.2	76.3	71.2	88.5	46.8	74.0	58.1	85.6	80.3	80.5	74.7	41.5	70.4	62.2	77.4	67.0
OHEM	69.8	81.5	78.9	69.6	52.3	46.5	77.4	72.1	88.2	48.8	73.8	58.3	86.9	79.7	81.4	75.0	43.0	69.5	64.8	78.5	68.9
Ours	70.5	83.3	78.0	70.7	53.9	47.9	77.7	72.6	89.2	47.7	76.7	56.6	87.4	80.9	83.8	77.1	41.6	73.4	62.8	79.7	67.8

To prove our Method MOEG is still significant on the best challenging dataset, we conduct experiments on MS COCO2014 [17]. For the COCO dataset,

Table 5. MS COCO2014 minival detection average precision (%). The architecture is VGG16.

AP@IoU	Area	FRCN	Ours	boost
[0.50 : 0.95]	All	19.2	21.9	2.7
0.50	All	37.0	41.1	4.1
0.75	All	18.2	21.6	3.4
[0.50 : 0.95]	Small	4.0	6.9	2.9
[0.50 : 0.95]	Med.	19.3	22.9	3.6
[0.50 : 0.95]	Large	33.8	35.4	1.6

we use the 'train' set for training and the 'minival' set for testing. During training the Fast-RCNN [8], we apply SGD with 360K iterations. The learning rate starts with 0.001 and decreases to 0.0001 after 320K iterations. For object proposals, we still use Selective Search [24]. We evaluate the mAP averaged for IoU $\in [0.5 : 0.05 : 0.95]$ (COCO's standard metric, simply denoted as mAP@[.5, .95]) and mAP@0.5 (PASCAL VOC's metric). We show the results in Table 5. Using our method MOEG improves the performance of the baseline Fast-RCNN from 37.0% to 41.1% mAP@0.5 on the VOC metric and from 19.2% to 21.9% AP on the standard COCO metric. It proved that our method MOEG can be effective even though datasets become larger and more challenge.

4.5 Visualization

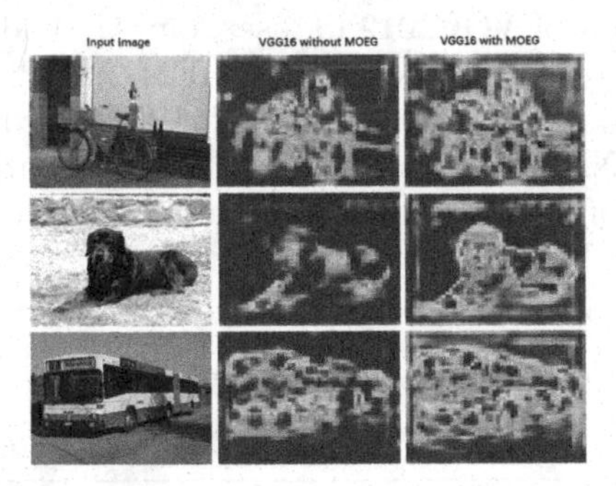

Fig. 2. Visualization of the Fast-RCNN's last convolutional layer output (conv5_3 in VGG16). We compare the result that trained with MOEG and without MOEG.

In order to see whether our method can improve the feature extraction ability of ConvNet or not, we visualize the convolutional layer's outputs, as shown in Fig. 2. We train Fast-RCNN with and without MOEG. The ConvNet architecture we used is VGG16. Compared with the outputs from VGG16 that trained without

MOEG, the outputs that trained with MOEG have more activation regions and larger activation values. Its activation regions not only in the important positions of the object, but also in other less important areas. It proves that our method not only can guide the detector to distinguish the features of different regions, but also can improve the feature extraction ability of the ConvNet even though mask operation is implemented after ROI pooling layer.

5 Conclusions

In this paper, we present a novel and principle method called Multi-region Occlusion Example Generating (MOEG) to learn and make full use of the diversified features in each region of objects. MOEG algorithm can be implemented to most of region-based detectors very easily and it can improve the performance of the detector significantly. MOEG can provide more diverse examples which have a better effect on the learning of diversified features for our detectors. Our experimental results on PASCAL VOC and MS COCO datasets show that using MOEG to train the detector can get a higher detection accuracy and cost less training time compared to most of state-of-the-art methods.

References

1. Bell, S., Lawrence Zitnick, C., Bala, K., Girshick, R.: Inside-outside net: detecting objects in context with skip pooling and recurrent neural networks. In: Proceedings of the IEEE Conference on Computer Vision and Pattern Recognition, pp. 2874–2883 (2016)
2. Chen, X., Gupta, A.: Spatial memory for context reasoning in object detection. arXiv preprint arXiv:1704.04224 (2017)
3. Cubuk, E.D., Zoph, B., Mane, D., Vasudevan, V., Le, Q.V.: Autoaugment: learning augmentation policies from data. arXiv preprint arXiv:1805.09501 (2018)
4. Denton, E.L., Chintala, S., Fergus, R., et al.: Deep generative image models using a laplacian pyramid of adversarial networks. In: Advances in Neural Information Processing Systems, pp. 1486–1494 (2015)
5. Everingham, M., Gool, L., Williams, C.K., Winn, J., Zisserman, A.: The pascal visual object classes (voc) challenge. Int. J. Comput. Vis. **88**(2), 303–338 (2010)
6. Farabet, C., Couprie, C., Najman, L., Lecun, Y.: Learning hierarchical features for scene labeling. IEEE Trans. Pattern Anal. Mach. Intell. **35**(8), 1915–1929 (2013)
7. Gidaris, S., Komodakis, N.: Object detection via a multi-region and semantic segmentation-aware CNN model. In: Proceedings of the IEEE International Conference on Computer Vision, pp. 1134–1142 (2015)
8. Girshick, R.: Fast R-CNN. In: Proceedings of the IEEE International Conference on Computer Vision, pp. 1440–1448 (2015)
9. Girshick, R., Donahue, J., Darrell, T., Malik, J.: Rich feature hierarchies for accurate object detection and semantic segmentation. In: Proceedings of the IEEE Conference on Computer Vision and Pattern Recognition, pp. 580–587 (2014)
10. Goodfellow, I., et al.: Generative adversarial nets. In: Advances in Neural Information Processing Systems, pp. 2672–2680 (2014)

11. He, K., Zhang, X., Ren, S., Sun, J.: Spatial pyramid pooling in deep convolutional networks for visual recognition. In: Fleet, D., Pajdla, T., Schiele, B., Tuytelaars, T. (eds.) ECCV 2014. LNCS, vol. 8691, pp. 346–361. Springer, Cham (2014). https://doi.org/10.1007/978-3-319-10578-9_23
12. He, K., Zhang, X., Ren, S., Sun, J.: Deep residual learning for image recognition. In: Computer Vision and Pattern Recognition, pp. 770–778 (2016)
13. Jia, Y., et al.: Caffe: convolutional architecture for fast feature embedding. In: Proceedings of the 22nd ACM International Conference on Multimedia, pp. 675–678. ACM (2014)
14. Krizhevsky, A., Sutskever, I., Hinton, G.E.: Imagenet classification with deep convolutional neural networks. In: International Conference on Neural Information Processing Systems, pp. 1097–1105 (2012)
15. Li, M., Zhang, Z., Yu, H., Chen, X., Li, D.: S-OHEM: stratified online hard example mining for object detection. In: Yang, J., et al. (eds.) CCCV 2017. CCIS, vol. 773, pp. 166–177. Springer, Singapore (2017). https://doi.org/10.1007/978-981-10-7305-2_15
16. Lin, T.Y., Dollár, P., Girshick, R., He, K., Hariharan, B., Belongie, S.: Feature pyramid networks for object detection. In: CVPR, vol. 1, p. 4 (2017)
17. Lin, T.-Y., et al.: Microsoft COCO: common objects in context. In: Fleet, D., Pajdla, T., Schiele, B., Tuytelaars, T. (eds.) ECCV 2014. LNCS, vol. 8693, pp. 740–755. Springer, Cham (2014). https://doi.org/10.1007/978-3-319-10602-1_48
18. Loshchilov, I., Hutter, F.: Online batch selection for faster training of neural networks. Mathematics (2016)
19. Ren, S., He, K., Girshick, R., Sun, J.: Faster R-CNN: towards real-time object detection with region proposal networks. In: Advances in Neural Information Processing Systems, pp. 91–99 (2015)
20. Shrivastava, A., Gupta, A., Girshick, R.: Training region-based object detectors with online hard example mining. In: Proceedings of the IEEE Conference on Computer Vision and Pattern Recognition, pp. 761–769 (2016)
21. Simonyan, K., Zisserman, A.: Very deep convolutional networks for large-scale image recognition. Computer Science (2014)
22. Singh, K.K., Lee, Y.J.: Hide-and-seek: forcing a network to be meticulous for weakly-supervised object and action localization. In: The IEEE International Conference on Computer Vision (ICCV) (2017)
23. Szegedy, C., et al.: Going deeper with convolutions, pp. 1–9 (2014)
24. Uijlings, J.R., Sande, K.E., Gevers, T., Smeulders, A.W.: Selective search for object recognition. Int. J. Comput. Vis. **104**(2), 154–171 (2013)
25. Wang, X., Shrivastava, A., Gupta, A.: A-fast-RCNN: hard positive generation via adversary for object detection. arXiv preprint arXiv:1704.03414 2 (2017)
26. Zeng, X., Ouyang, W., Yang, B., Yan, J., Wang, X.: Gated bi-directional CNN for object detection. In: Leibe, B., Matas, J., Sebe, N., Welling, M. (eds.) ECCV 2016. LNCS, vol. 9911, pp. 354–369. Springer, Cham (2016). https://doi.org/10.1007/978-3-319-46478-7_22
27. Zhong, Z., Zheng, L., Kang, G., Li, S., Yang, Y.: Random erasing data augmentation. arXiv preprint arXiv:1708.04896 (2017)

HATDC: A Holistic Approach for Time Series Data Repairing

Xiaojie Liu, Guangxuan Song, and Xiaoling Wang(✉)

East China Normal University, Shanghai, China
xiaojie_liu7@126.com, guangxuan_song@163.com, xlwang@sei.ecnu.edu.cn

Abstract. Time series data is prevalent in real life, and time series data mining is also a hot research topic nowadays. However, there may exist lots of anomalous data caused by sensor error in the real data sets, which brings difficulties for data mining. To improve the quality of data mining, it is to repair the data before data analysis. Most of the existing repairing methods use smooth-based or constraint-based techniques, but they only consider a few adjacent points and ignore global holistic information. In this paper, we propose a novel time series data repairing algorithm, named HATDC, that can exploit the holistic information of the time series. First, we use speed constraints and the probability distribution of change rates to detect the dirty data points. After that, the dynamic time warping (DTW) is applied as the distance measure to find similar subsequences in the series, and we estimate the value of these abnormal data points according to the selected similar subsequences from the whole aspect. In addition, we propose an improved algorithm for reducing the time cost based on incremental clustering. Experiments on several real datasets demonstrate that HATDC has a significantly higher repair accuracy and a lower RMS error than other methods.

Keywords: Data repairing · Time series · Anomaly detection · DTW

1 Introduction

With the widespread use of various sensors, time series data is prevalent in our daily life, such as hourly temperatures and GPS trajectories. And the time series data mining is a hot research topic nowadays, because there exists a wealth of implicit information in these data. However, dirty or imprecise values commonly exist in the time series, which can be caused by sensor error [11] or other reasons. The low quality of time series data has a huge impact on data analysis. Ensuring and improving data quality is the two main reasons for data repairing. As shown in the recent study [19], repairing dirty values could improve clustering over spatial data.

In the field of time series data repairing, there are two types of mainstream methods: smooth-based and constraint-based techniques. Smooth-based methods are usually used to eliminate noisy data points in the series, such as simple moving average (SMA) [3] and the exponentially weighted moving average

Q. Yang et al. (Eds.): PAKDD 2019, LNAI 11440, pp. 553–564, 2019.
https://doi.org/10.1007/978-3-030-16145-3_43

(EWMA) [8]. Specifically, SMA just smooths time series data by computing the unweighted mean of the adjacent data points, while EWMA allocates exponentially decreasing weights over time. However, these methods may lead to over-repairing as described in Example 1. Besides, recent work [20] proposed a novel data cleaning approach (SCREEN) under speed constraints, which used the innovative constraints on the trend of value changes to guide the repairing process. The speed constraints are effective in identifying large spike error data points [22]. But the existing constraint-based methods cannot indicate the most likely result among all the valid repairs that satisfy the constraints, and they do not make full use of the data points in the sequence.

Therefore, we propose a novel data repairing approach, named as HATDC, that can exploit the holistic information of the time series: (1) the speed constraints and the probability distribution on change of speeds by statistics on the overall series are used to detect large spike error data points; (2) DTW is used as the distance measure to find the similar subsequences in the whole series, then the value of abnormal data points are estimated by the value of data points at the corresponding position of the selected similar subsequences.

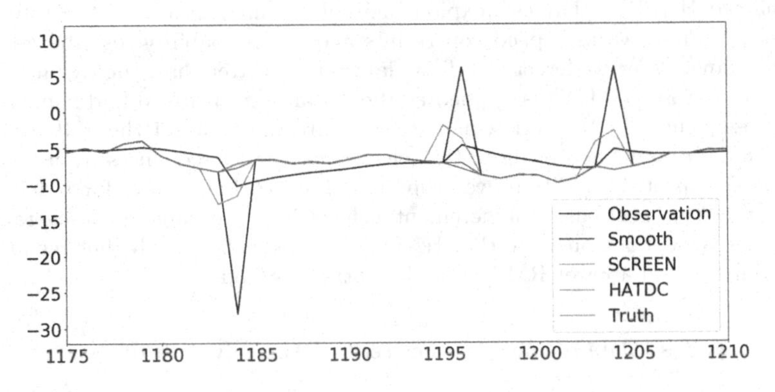

Fig. 1. Example of observations and repairs (Color figure online)

Example 1. Figure 1 presents an example segment of hourly air temperature. Three dirty values appear at time point 1184, 1196 and 1204, respectively, in the observed sequence (in black, the corresponding true values are presented in blue).

The smooth-based cleaning method (in green) modifies almost all the temperature values, most of which are indeed accurate. The speed constraint-based cleaning method (in red), SCREEN [20], *could effectively detect the large spike error in the sequence, but the repair value is not accurate enough. Finally, our proposed HATDC approach, with both effectively error detection and value estimation, obtains repairs closest to the truth.*

Contributions. The main contribution of this paper is to propose HATDC for time series data repairing, which takes full advantage of the holistic information of the series. Specific contributions are as follows:

1. First, we propose a novel anomaly detection method based on the speed (the definition of speed will be described in Sect. 3) constraints and the probability distribution on change of speeds by statistics on the overall series. And it performs well on detecting large spike error data points.
2. Then, we propose a DTW-based time series data repairing algorithm that can estimate the true value of dirty data points according to the selected similar subsequences from the whole aspect. Experiments show our method performces better than the state-of-art methods when the error rate is relatively small.
3. At last, in order to speed up the data repairing algorithm, an improved method (C-HATDC) is proposed based on DTW barycenter averaging (DBA) and incremental clustering.

Organization. The remainder of this paper is organized as follows. We first discuss the related works in Sect. 2. The problem definition and anomaly detection method are introduced in Sect. 3. The dirty data repairing algorithm and some optimization are then presented in Sects. 4 and 5, respectively. The experimental evaluation is demonstrated in Sect. 6. Finally, we conclude our paper in Sect. 7.

2 Related Work

Smooth-based methods are usually used in time series data repairing, such as SWAB [14], SMA [3] and WMA [8]. And the sliding window is used in these methods. Specifically, SWAB splits the time series into many subsequences, which is also applied in our method. And linear interpolation or linear regression is used in SWAB to get the approximating line of a subsequence. With a sliding window, SMA just smooths time series data by computing the unweighted mean of the points in the last window. The data points are weighted equally in SMA, while WMA gives different weights to data points at different positions in the sample window. Moreover, the exponentially weighted moving average (EWMA) [8] allocates exponentially decreasing weights over time. However, these methods may lead to over-repairing as illustrated in Fig. 1. It is obvious that most of the original correct data may be seriously altered, and thus have low repair accuracy.

The constraint-based technique is widely considered in data cleaning, such as holistic data cleaning [5] and sequential dependencies [9]. And the constraint-based data cleaning method identifies and repairs the violations to the given constraints. SCREEN [20] is the first constraint-based stream data cleaning approach, which employs a class of speed constraints. Speed constraint is also adopted in our method to detect dirty data points. AR and ARX [16] indeed have been widely used for anomaly detection [2,4]. And existing anomaly detection techniques could also be adapted to repairing dirty data points by given the labeled truth of some data points [23].

The DTW-based techniques for missing value estimation in gene expression time series have been studied in [10,15,21]. Unlike they use originally clean gene expression time series to impute the series with missing values, we clean a whole time series with erroneous data points in this paper.

3 Anomaly Detection

Consider sequence of n observations $x = \{x_1, x_2, ..., x_n\}$. Each x_i has a timestamp t_i relative to it, and the time intervals between adjacent timestamps are equal. When repairing time series x, we detect the dirty data points in x at first.

In this section, we mainly introduce anomaly detection methods. Instead of just considering speed constraints, we propose a novel anomaly detection method that also considers the probability distribution of speed changes.

Definition 1. By reference to [20], the **speed** from data point $i-1$ to i is defined as $v_{i-1,i} = \frac{x_i - x_{i-1}}{t_i - t_{i-1}}$. And the **speed constraints** $sc = (s_{min}, s_{max})$ are defined as a pair of minimum speed s_{min} and maximum speed s_{max} over the sequence. In the sequence x, the **speed change** before and after the i-th data point is defined as: $u_i = v_{i,i+1} - v_{i-1,i}$.

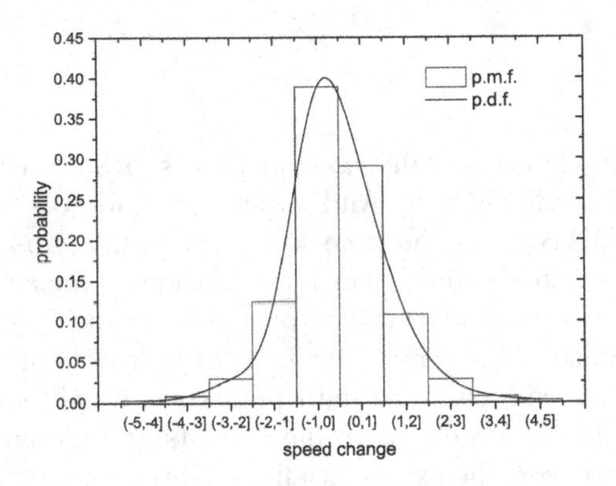

Fig. 2. Probability distributions of speed changes on a real temperature dataset

The probability distribution on change of speeds reflects the holistic information of change rates of the time series, and it can be simply estimated by statistics on the overall series. For each data point in the time series, we calculate its speed change value before and after the point according to Definition 1. And we estimate the probabilities of the speed changes by counting the appearance of speed change values in the series. Figure 2 shows the probability distributions of speed changes on a real temperature dataset presented in Sect. 6. And there is a high similarity between the discrete probability distribution and the approximate continuous probability distribution as demonstrated in [22].

As we know, the value of the i-th data point in the sequence affects the speed change u_i, u_{i-1} and u_{i+1}. Let $p(u_i)$ denote the probability of speed change u_i, let $\overline{u}$ be the mean of the speed changes in the sequence x, and $sc = (s_{min}, s_{max})$ represent the speed constraints. Intuitively, there is a relatively small probability that a speed change $u_i > \overline{u} + s_{max}$ or $u_i < \overline{u} + s_{min}$. And most of the large spike error data points do not satisfy the speed constraints. Therefore, these data points can be easily detected according to the following definition.

Definition 2. In the sequence x, x_i is identified as an **abnormal data point,** if $p(u_{i-1})$, $p(u_i)$ and $p(u_{i+1})$ are smaller than $min\,\{p(\overline{u} + s_{max}), p(\overline{u} + s_{min})\}$.

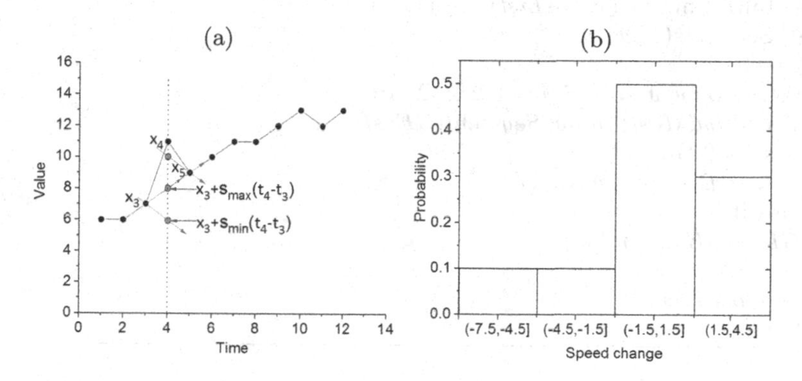

Fig. 3. Speed constraints and probability distribution of speed changes in Example 2.

Example 2 (abnormal data points detection). Consider a sequence $x = \{6, 6, 7, 11, 9, 10, 11, 11, 12, 13, 12, 13\}$. Suppose that $s_{max} = 1$ and $s_{min} = -1$. Figure 3(a) shows the data points and the speed constraints, and Fig. 3(b) shows the corresponding probability distribution of speed changes. After calculation, we get $\overline{u} = 0.1$, $min\,\{p(\overline{u} + s_{max}), p(\overline{u} + s_{min})\} = min\,\{p(1.1), p(-0.9)\} = 0.5$. Because $P(u_3) = P(u_5) < 0.5$ and $P(u_4) < 0.5, x_4$ is identified as an abnormal data point.

4 Dirty Data Repairing

On the one hand, the existing smooth-based methods may seriously alter most of the original correct data in the series, and the existing constraint-based methods cannot indicate the most likely result among all the valid repairs that satisfy the constraints. On the other hand, there are similar subsequences in the time series, but all existing methods do not take advantage of these similar subsequences.

To avoid these problems, we only modify the value of abnormal data points that are identified by the aforementioned method and use the holistic information of similar subsequences to estimate the value of abnormal data points.

In order to find the most similar subsequence in the series, the dynamic time warping (DTW) is applied as the distance measure in our method. And DTW has been commonly used in pattern recognition [7], time series clustering [1] and time series classification (TSC) [6,12,13], because of its good performce in finding similar sequences.

Algorithm 1. HATDC

Input: time series x, speed constraints sc, window size w
Output: repaired time series x'

1: $posList \leftarrow$ AnomalyDetection(x, sc)
2: $x \leftarrow$ InitEstimation(x, $posList$)
3: $S \leftarrow$ Segments(w, x)
4: $GE \leftarrow \emptyset$
5: **for** each segment s_i in S, $i = 1, 2, ..., M$ **do**
6: $\quad s_j \leftarrow FindMostSimilarSegment(GE, s_i)$
7: $\quad$ **if** $s_j! = \emptyset$ **then**
8: $\quad\quad s_i \leftarrow Estimation(s_j, s_i)$
9: $\quad$ **end if**
10: $\quad GE \leftarrow GE.append(s_i)$
11: **end for**
12: $x' \leftarrow Combine(S)$
13: **return** x'

Therefore, we propose a DTW-based data repairing algorithm to repair the dirty data points, named HATDC. The details are shown in Algorithm 1.

Firstly, Algorithm 1 detects the positions of abnormal data points in the time series x (line 1). And we initialize a preliminary estimation of the abnormal data points (line 2). For example, the average of the two nearest non-anomalous neighbors in the series can be used. Then we segment the series x by a window with size w, and x is divided into $M = \lceil n/w \rceil$ subsequences $S = \{s_1, s_2, ..., s_M\}$ (line 3). Since we do not have the originally clean sequences, we initialize the repaired subsequence(GE) as empty (line 4).

Secondly, we traverse $S = s_1, s_2, ..., s_M$ in turn (line 5). If there are n' data points in GE, there should be $n' - w + 1$ subsequences using a sliding window. And we find the most similar subsequence s_j in GE for each subsequence s_i in S (line 6). If the t-th position of s_i is detected as an abnormal data point, it can be estimated by s_j at the corresponding positions to the position t of s_i on the best warping path. Let s'_{jt} denote the mean of all t-aligned values in s_j, which can be an estimated value of the t-th data point of s_i. Therefore, the value of the t-th data point of s_i is estimated as s'_{jt} (line 8). When s_i has been repaired, we append it to GE which can be used in the rest phase (line 10).

Finally, we combine all the subsequences in S to get the output results x'.

5 Clustering-Based Optimization

According to our repairing algorithm, when repairing an abnormal data point, we need to calculate the DTW distance between the subsequence that the data point belongs to and all subsequences that have been repaired to find similar subsequences. As the number of subsequences increases, the number of DTW distances need to calculate also increases, so that the repairing time will increases. Meanwhile, we find that similar candidate subsequences are very similar to each other. For this motivation, we propose an improved algorithm based on time series clustering. In addition, experiments show that there is little difference between using the sliding window and without using the sliding window. In order to reduce the number of comparisons, we only use repaired subsequences instead of the sliding window to do clustering.

Algorithm 2. C-HATDC

Input: time series x, speed constraints sc, window size w, cluster number K
Output: repaired time series x'
1: $posList \leftarrow$ AnomalyDetection(x, sc)
2: $x \leftarrow$ InitEstimation(x, $posList$)
3: $S \leftarrow$ Segments(w, x)
4: $ClusterList \leftarrow InitClusters(S, K)$
5: $CentroidList \leftarrow InitCentroids(ClusterList)$
6: **for** each segment s_i in S, $i = 1, 2, ..., M$ **do**
7: $cid \leftarrow AssignCluster(CentroidList, s_i)$
8: $s_j \leftarrow FindMostSimilarSegments(ClusterList[cid], s_i)$;
9: **if** $s_j! = \emptyset$ **then**
10: $s_i \leftarrow Estimation(s_j, s_i)$
11: **end if**
12: $ClusterList[cid] \leftarrow ClusterList[cid].add(s_i)$
13: $CentroidList[cid] \leftarrow UpdateCentroid(ClusterList[cid])$
14: **end for**
15: $x' \leftarrow Combine(S)$
16: **return** x'

In the traditional K-means algorithm, the centroid computation of a cluster only averages the values of the time series in each position. But for time series averaging, DTW distance is more suitable than Euclidean distance [17]. Hence the centroid of a cluster can be calculated according to average time series for DTW. DTW barycenter averaging (DBA) [18] is the state-of-the-art method to average time series for dynamic time warping, and we will no longer introduce its detail in this paper. K-DBA [6,17] is a clustering algorithm that uses K-means with DTW as distance measure and the DBA method for centroid computation.

In order to reduce the time cost of data repairing, we propose an improved algorithm based on time series clustering, named C-HATDC. The details of C-HATDC is shown in Algorithm 2.

Compared with Algorithms 1 and 2 has the following differences. Firstly, in order to initialize the k clusters, we randomly select a subsequence for each cluster from the subsequences set S (line 4), and we initialize the centroid of each cluster (line 5). Secondly, when repairing the subsequence s_i, a cluster is assigned to s_i by calculating the DTW distance between s_i and k centroids (line 7). And the similar subsequences are selected from the assigned cluster (line 8). Lastly, the repaired subsequence s_i is added into the assigned cluster, and the centroid of this cluster is updated by the DBA algorithm introduced previously.

6 Experiments

The experiments run on two real datasets, STOCK and TEMPERATURE, which can be obtained from YAHOO FINANCE[1] and NOAA[2] respectively. These two datasets are originally clean, so they can be used as ground truth. The positions of dirty data points are selected randomly, and each dirty data point is assigned a value within a range from the minimum to maximum value of the series. The STOCK dataset records the daily prices of stock from August, 2007 to August, 2017. The TEMPERATURE dataset includes the hourly temperature in New York from January 1, 2001 to December 31, 2009.

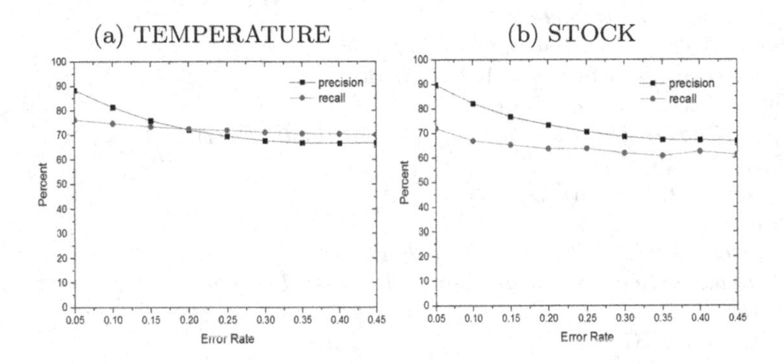

Fig. 4. The accuracy of anomaly detection.

6.1 Accuracy of Anomaly Detection

In the first experiment, we show the accuracy of the anomaly detection algorithm. The speed constraint is natural in most scenarios. Specifically, for STOCK, the price limit in the market declares that the increase or decrease of daily price should not exceed 10%. According to that, we set $s_{max} = 3$ and $s_{min} = -3$. For the particular domain where speed knowledge is not available,

[1] http://finance.yahoo.com.
[2] http://www.noaa.gov.

the speed constraints can be extracted from data by considering the statistical distribution of speed. As noted in [20], s_{max} and s_{min} can be estimated according to a confidence interval (95%). By this way, $s_{max} = 5$ and $s_{min} = -5$ are suggested for TEMPERATURE.

Figure 4 presents the precision and recall of our anomaly detection method, and we consider various error rates from 0.05 to 0.45. As the error rate increases, the detection accuracy decreases.

6.2 Comparison with Existing Approaches

In this experiment, we compare HATDC with WMA, EWMA [8] and SCREEN [20]. Considering various error rates from 0.05 to 0.45 in TEMPERATURE and STOCK respectively. We conduct the experiments for 10 times and evaluate the performce according to the mean value.

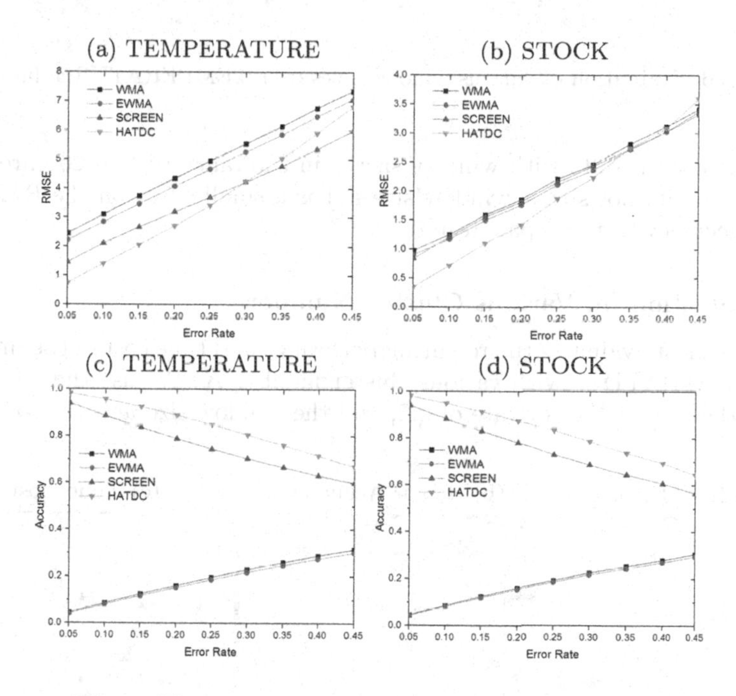

Fig. 5. The repair accuracy under different error rates.

The evaluation criteria include: the root-mean-square(RMS) error between the repair result and truth data; the repair accuracy which is defined in [20].

It is obvious that HATDC perform better than those in other methods. Specifically, Fig. 5(a) and (b) demonstrates that the RMS error of HATDC is lower than other methods when the error rate is relatively small (<30%). Since HATDC is based on similar subsequences, and the higher error rate will affect the similarity calculation, which leads to higher RMS error. Figure 5(c) and (d) presents the repair accuracy of HATDC is the highest.

6.3 Evaluation on Various Window Size

This experiment evaluates the repairing performce of time series divided by various window size. We use TEMPERATURE dataset with error rate of 0.2.

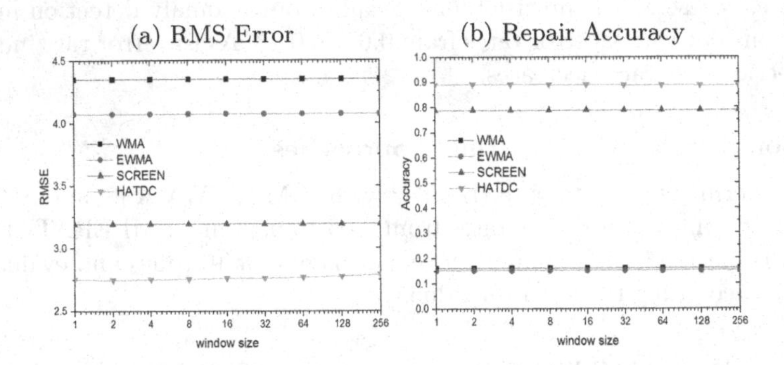

Fig. 6. Evaluation of various window sizes over TEMPERATURE data.

The detailed results with window size w in the range of 1 to 256 are shown in Fig. 6. And it shows that window size w has a small effect on the RMS error and the accuracy of the repair result.

6.4 Evaluation on Various Cluster Number

This experiment evaluates the repairing performce and time cost of the improved algorithm (C-HATDC) with various cluster number K. We use the TEMPERATURE dataset with error rate of 0.2, and the window size w is set to 6.

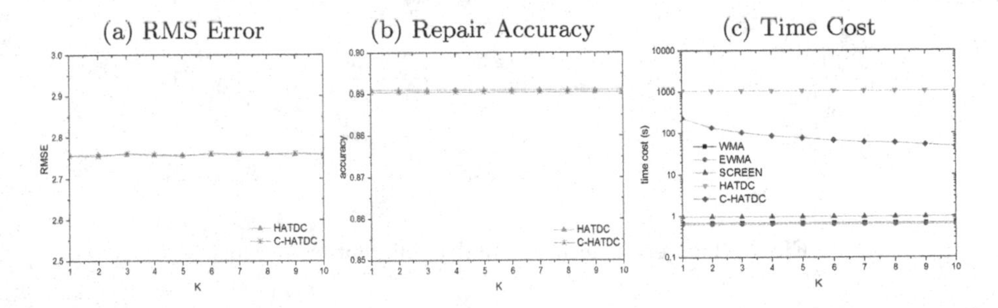

Fig. 7. Evaluation of various cluster number K over TEMPERATURE data.

Figure 7 show the detailed results with cluster number K in the range of 1 to 10. As shown in Fig. 7(a) and (b), the RMS error and repair accuracy barely change with different K. As the number of clusters increases, the time cost decreases as illustrated in Fig. 7(c), and the decline rate of time cost also decreases. It demonstrates the effectiveness of the improved algorithm.

7 Conclusion

In this paper, we study the problem of repairing dirty data in time series. The existing methods don't take full advantage of holistic information or result in over-repairing problems. For this motivation, we propose a novel method, named as HATDC, that can exploit the holistic information of the time series. HATDC consists of anomaly detection and dirty data repairing. First, the speed constraints and the probability distribution of change rates are used to detect abnormal data points. Hence this method cannot work well under the probability distribution of speed change does not conform to the normal distribution. After that, DTW is applied as the distance measure to find similar subsequences in the series, and we estimate the value of these abnormal data points according to the selected similar subsequences from the whole aspect. Experiments on several real datasets show the superiority of our proposal. Moreover, an improved method (C-HATDC) based on K-DBA clustering is proposed by us, it can reduce the time cost effectively and hardly affect the repair results.

Acknowledgments. This work was supported by National Key R&D Program of China (No. 2017YFC 0803700), NSFC grants (No. 61532021 and 61472141), Shanghai Knowledge Service Platform Project (No. ZF1213)and SHEITC.

References

1. Begum, N., Ulanova, L., Wang, J., Keogh, E.: Accelerating dynamic time warping clustering with a novel admissible pruning strategy. In: Proceedings of the 21th ACM SIGKDD International Conference on Knowledge Discovery and Data Mining, pp. 49–58. ACM (2015)
2. Box, G.E., Jenkins, G.M., Reinsel, G.C., Ljung, G.M.: Time Series Analysis: Forecasting and Control. Wiley, Hoboken (2015)
3. Brillinger, D.R.: Time Series: Data Analysis and Theory, vol. 36 (2001). https://doi.org/10.1016/0304-4149(79)90039-5
4. Brockwell, P.J., Davis, R.A.: Introduction to Time Series and Forecasting. STS. Springer, Cham (2016). https://doi.org/10.1007/978-3-319-29854-2
5. Chu, X., Ilyas, I.F., Papotti, P.: Holistic data cleaning: putting violations into context. In: 2013 IEEE 29th International Conference on Data Engineering (ICDE), pp. 458–469. IEEE (2013)
6. Forestier, G., Webb, G.I., Nicholson, A.E., Chen, Y., Keogh, E.: Faster and more accurate classification of time series by exploiting a novel dynamic time warping averaging algorithm. Knowl. Inf. Syst. **47**(1), 1–26 (2016)
7. Furlanello, C., Merler, S., Jurman, G.: Combining feature selection and DTW for time-varying functional genomics. IEEE Trans. Sig. Process. **54**(6 II), 2436–2443 (2006). https://doi.org/10.1109/TSP.2006.873715
8. Gardner, E.: Exponential Smoothing: The State of the Art Part II, vol. 22 (2006). https://doi.org/10.1016/j.ijforecast.2006.03.005
9. Golab, L., Karloff, H., Korn, F., Saha, A., Srivastava, D.: Sequential dependencies. Proc. VLDB Endow. **2**(1), 574–585 (2009)
10. Hsu, H.H., Yang, A.C., Lu, M.D.: KNN-DTW based missing value imputation for microarray time series data. JCP **6**(3), 418–425 (2011)

11. Jeffery, S.R., Berkeley, U.C., Franklin, M.J.: Adaptive cleaning for RFID data streams. In: VLDB, pp. 163–174 (2006)
12. Jeong, Y.S., Jeong, M.K., Omitaomu, O.A.: Weighted dynamic time warping for time series classification. Pattern Recognit. **44**, 2231–2240 (2011). https://doi.org/10.1016/j.patcog.2010.09.022
13. Kate, R.J.: Using dynamic time warping distances as features for improved time series classification. Data Min. Knowl. Discov. **30**(2), 283–312 (2015). https://doi.org/10.1007/s10618-015-0418-x
14. Keogh, E., Chu, S., Hart, D., Pazzani, M.: An online algorithm for segmenting time series. In: Proceedings IEEE International Conference on Data Mining, ICDM 2001, pp. 289–296. IEEE (2001)
15. Kostadinova, E., Boeva, V., Boneva, L., Tsiporkova, E.: An integrative DTW-based imputation method for gene expression time series data. In: Proceedings of 2012 6th IEEE International Conference Intelligent Systems, IS 2012, pp. 258–263 (2012). https://doi.org/10.1109/IS.2012.6335145
16. Park, G., Rutherford, A.C., Sohn, H., Farrar, C.R.: An outlier analysis framework for impedance-based structural health monitoring. J. Sound Vib. **286**(1), 229–250 (2005)
17. Petitjean, F., Forestier, G., Webb, G.I., Nicholson, A.E., Chen, Y., Keogh, E.: Dynamic time warping averaging of time series allows faster and more accurate classification. In: 2014 IEEE International Conference on Data Mining (ICDM), pp. 470–479. IEEE (2014)
18. Petitjean, F., Ketterlin, A., Gancarski, P.: A global averaging method for dynamic time warping, with applications to clustering. Pattern Recognit. **44**(3), 678–693 (2011). https://doi.org/10.1016/j.patcog.2010.09.013
19. Song, S., Li, C., Zhang, X.: Turn waste into wealth: on simultaneous clustering and cleaning over dirty data. In: Proceedings of the 21th ACM SIGKDD International Conference on Knowledge Discovery and Data Mining, pp. 1115–1124. ACM (2015)
20. Song, S., Zhang, A., Wang, J., Yu, P.S.: SCREEN: stream data cleaning under speed constraints. In: Proceedings of the 2015 ACM SIGMOD International Conference on Management of Data, pp. 827–841 (2015). https://doi.org/10.1145/2723372.2723730
21. Tsiporkova, E., Boeva, V.: Two-pass imputation algorithm for missing value estimation in gene expression time series. J. Bioinform. Comput. Biol. **5**(05), 1005–1022 (2007)
22. Zhang, A., Song, S., Wang, J.: Sequential data cleaning: a statistical approach. In: Proceedings of the 2016 International Conference on Management of Data, pp. 909–924 (2016). https://doi.org/10.1145/2882903.2915233
23. Zhang, A., Song, S., Wang, J., Yu, P.S.: Time series data cleaning: from anomaly detection to anomaly repairing. Proc. VLDB Endow. **10**(10), 1046–1057 (2017)

Double Weighted Low-Rank Representation and Its Efficient Implementation

Jianwei Zheng, Kechen Lou, Ping Yang(✉), Wanjun Chen, and Wanliang Wang

Zhejiang University of Technology, Hangzhou 310023, China
ypingpds@163.com

Abstract. To overcome the limitations of existing low-rank representation (LRR) methods, i.e., the error distribution should be known a prior and the leading rank components might be over penalized, this paper proposes a new low-rank representation based model, namely double weighted LRR (DWLRR), using two distinguished properties on the concerned representation matrix. The first characterizes various distributions of the residuals into an adaptively learned weighting matrix for more flexibility of noise resistance. The second employs a parameterized rational penalty as well as a weighting vector $\boldsymbol{s}$ to reveal the importance of different rank components for better approximation to the intrinsic subspace structure. Moreover, we derive a computationally efficient algorithm based on the parallel updating scheme and automatic thresholding operation. Comprehensive experimental results conducted on image clustering demonstrate the robustness and efficiency of DWLRR compared with other state-of-the-art models.

Keywords: Subspace clustering · Low-rank approximation · Nonconvex surrogate function · Proximal gradient method

1 Introduction

Low-rank representation (LRR) [1], as a promising approach to capture the underlying structure of data, has been applied to extensive applications in computer vision and multimedia community. Generally speaking, the success of LRR mainly originates from three merits: a natural hypothesis of underlying multiple low-rank subspaces in observed data, a self-expressive representation with

Supported by organization x.

Electronic supplementary material The online version of this chapter (https://doi.org/10.1007/978-3-030-16145-3_44) contains supplementary material, which is available to authorized users.

Q. Yang et al. (Eds.): PAKDD 2019, LNAI 11440, pp. 565–577, 2019.
https://doi.org/10.1007/978-3-030-16145-3_44

specific noise resistant constraint, and a convex approximation of rank regularization using nuclear norm. However, these characteristics also restrict LRR from wider applications due to the limitations that the structure of errors should be known a prior and the intrinsic rank of data might be loosely approximated.

In order to tackle heterogeneous noise sources and obtain better approximation to the original low-rank assumption, a great variety of clustering methods have been proposed recently to search for a better representation matrix via different choices of constraints, which can be uniformly formulated as the following optimization problem [2]:

$$\min_{\boldsymbol{Z}} \gamma\|\boldsymbol{X} - \boldsymbol{X}\boldsymbol{Z}\|_{\mu} + \Omega(\boldsymbol{X}, \boldsymbol{Z}), s.t.\ \ \boldsymbol{Z} \in C,$$

where $\boldsymbol{X} \in \boldsymbol{R}^{m\times n}$ is the data matrix containing n samples as its columns, $\boldsymbol{Z} \in \boldsymbol{R}^{n\times n}$ is the representation matrix, $\|\cdot\|_{\mu}$ is a specific norm, Ω and C are some regularizer and constraint set on $\boldsymbol{Z}$, respectively, and $\gamma > 0$ is a balance parameter.

For the choice of regularization on the residual term $\boldsymbol{E} = \boldsymbol{X} - \boldsymbol{X}\boldsymbol{Z}$, different norms are exploited to cope with various forms of noise. Especially, Frobenius norm (i.e., $\|\cdot\|_F^2$) is used for modeling the Gaussian noise [3], l_1-norm is adopted for characterizing the Laplacian corruption [2,4], and $l_{2,1}$-norm is introduced for removing sample-specific outliers [1,5]. However, these constraints (or norms) work well only with a correct prior knowledge on error structure, which is often difficult to obtain. Ref. [6] presented another error-removing scheme based on a property named intra-subspace projection dominance (IPD), but the IPD property itself may be disturbed by gross corruption. Ref. [7] integrated feature selection into the residual term for revealing more accurate data relationships on the premise of fixed corruption location, which is a strong assumption that may not be satisfied in most real-world problems. Ref. [8] and [9] employed the maximum likelihood estimation principle to estimate the real distribution of noise, which provides a robust framework to deal with complex errors. However, their optimization algorithm, namely iteratively reweighted inexact augmented Lagrange multiplier (IRIALM) suffers from two limitations. One is that an independent logistic function is suboptimal for the overall optimization. The other is that the model loses the reweights essence by integrating the weights update into the ADMM framework [10].

For the choice of regularization on the representation matrix $\boldsymbol{Z}$, l_1-norm, Frobenius norm, nuclear norm ($\|\cdot\|_*$), and mixtures, e.g., $l_1 + \mathrm{F}$ [2], $l_1 + \|\cdot\|_*$ [11], or replacement [12] of some of them, have been extensively studied for various problems. Despite the success of these convex surrogate functions, recently there have been numerous explorations on employing nonconvex ones for better approximating the intrinsic structure of data. As for the singular values, the key idea is that the larger, and thus more informative, should be less penalized. Such attempts include different nonconvex surrogates [13,14], weighted nuclear norm [15], and their mixtures such as weighted Schatten p-Norm [16]. Empirically, these attempts achieve better performance than the convex counterparts. However, the resultant optimization problem is much more challenging. In addition, most existing noncovex optimization methods take at least $O(mn^2)$ time complexity at each iteration for a $m \times n$ matrix (assuming $m \geq n$), and might be expensive on large matrices.

Motivated by the above works, our goal is to overcome the limitations in the properties of noise resistant constraint and weighted nonconvex regularization. For this purpose, we propose a double weighted LRR (DWLRR) method, which generates clearer block-diagonal representation matrix and facilitates corrupted subspace clustering. Our main contributions lie in the following aspects:

(1) We propose a new scheme to estimate the contributions of different input features, where the tailored residual factors are updated iteratively for revealing the various contributions of the input features.
(2) We employ a parametric nonconvex function to estimate the singular values precisely, while keeping convexity of the cost function via a referenced span of the penalty parameter. An analytical solution of the proposed weighted nonconvex subproblem is also derived in a parallel manner.
(3) By the observation that the singular values obtained from the introduced nonconvex function can be automatically thresholded, we derive an efficient proximal algorithm for model optimization. Moreover, the convergence and the complexity are also discussed.

2 Proposed Double Weighted Model

To begin with, we first introduce LRR, whose goal is to find the underlying subspace structure by

$$\min_{\boldsymbol{Z}} \gamma \|\boldsymbol{X} - \boldsymbol{X}\boldsymbol{Z}\|_{2,1} + \|\boldsymbol{Z}\|_*, \tag{1}$$

where $l_{2,1}$ is for sample specific noise, $\|.\|_*$ is the widely used nuclear norm.

2.1 Weighted Feature Learning for Error Penalizing

As the encountered noise in real scenarios is complex, the distribution of residual $\boldsymbol{E}$ may be far away from the sample specific assumption in (1) or other fixed types of distribution [5]. Recent works on low-rank reconstruction [8], robust regression [9], as well as matrix recovery [10] show that reweighting of the priors significantly promotes the robustness. Similarly, we minimize $\| \boldsymbol{W}^{1/2} \odot (\boldsymbol{X} - \boldsymbol{X}\boldsymbol{Z})\|_{\mu}$, where $\odot$ denotes an element-wise product, by adopting the weighting factor $\boldsymbol{W}$, which aims to use the adaptive weights for flexibly predicting the error distribution.

Considering in large probability that the noisy points are indefinite, thus might be reconstructed with obvious error, whereas the intrinsic features are representable, thus can be reconstructed with subtle error, hence a natural way to determine the weight matrix is solving the following problem:

$$\min_{\boldsymbol{1}^T \boldsymbol{W} \boldsymbol{1}=1, w_{ij} \geq 0} \gamma \| \boldsymbol{W}^{1/2} \odot (\boldsymbol{X} - \boldsymbol{X}\boldsymbol{Z})\|_F^2 + \| \boldsymbol{W}\|_F^2, \tag{2}$$

where the regularization term $\| \boldsymbol{W}\|_F^2$ is used to avoid the trivial solution that all weights are zero except for the smallest e_{ij}, whose weight is 1, the constraints $\boldsymbol{1}^T \boldsymbol{W} \boldsymbol{1} = 1$ and $w_{ij} \geq 0$ are used for numerical stability.

2.2 Weighted Rational Function for Rank Approximation

Although the nuclear norm used in model (1) is the tightest convex approximation to the rank constraint, the obtained solution may seriously deviate from the original one particularly in the presence of noise. Figure 1 illustrates a noisy face image, its rank components, and some penalty functions under same setting of parameters. It can be noted that both the weighted nuclear norm [15], i.e., $\sum_{i=1}^{n} s_i\sigma_i$, and the nonconvex l_p constraint [14], i.e., $\sum_{i=1}^{n} \sigma_i^p$, are tighter approximations than the nuclear norm. We further present a weighted nonconvex constraint, i.e., $\sum_{i=1}^{n} r(s_i, \sigma_i)$, to generalize the nuclear norm into a more flexible model. Figure 1(b) shows that the larger rank components of the noisy image hold a good fit with the original ones, while the smaller singular values deviate far away from the original ones. With this observation, we introduce the parameterized rational function to penalize the larger singular values less than the smaller ones (see Fig. 1(c)) as follows:

$$r(s,\sigma) = \frac{s\sigma}{1 + a\sigma/2}, \tag{3}$$

where s is a weight, σ is some singular value, and a is a tunable parameter.

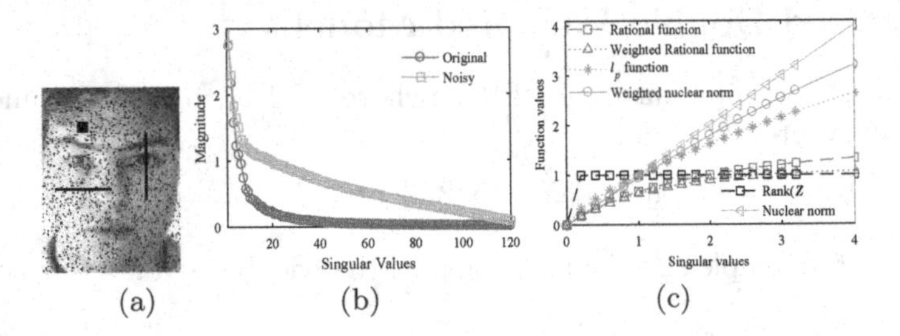

Fig. 1. An instance of (a) corrupted sample and its (b) rank components, as well as some (c) penalty functions.

By enhancing model (1) with above two merits for a better performance, the primitive cost function of DWLRR can be written as

$$J = \min_{\mathbf{1}^T \mathbf{W}\mathbf{1}=1, w_{ij}\geq 0} \gamma \| \mathbf{W}^{1/2} \odot (\mathbf{X} - \mathbf{X}\mathbf{Z})\|_F^2 + \nu \| \mathbf{W} \|_F^2 + \sum_{i=1}^{n} r(s_i, \sigma_i), \tag{4}$$

where ν is a balance parameter, we follow [17] that uses Frobenius norm to measure the fitting residual and denote $\sigma_i = \sigma_i(\mathbf{Z})$ for simplicity.

3 Optimization Algorithm

3.1 Reweighted Framework

We apply the iterative reweighted method (IRM) [10] to solve the optimization problems (4) with two variables $\mathbf{W}$ and $\mathbf{Z}$. The subproblem $\mathbf{W}$ can be minimized as follows:

$$\min_{\boldsymbol{1}^T \boldsymbol{W}\boldsymbol{1}=1, w_{ij}\geq 0} \| \boldsymbol{W}^{1/2} \odot (\boldsymbol{X} - \boldsymbol{X}\boldsymbol{Z})\|_{\mathrm{F}}^2 + \lambda \| \boldsymbol{W} \|_{\mathrm{F}}^2 \tag{5}$$

where $\lambda = \nu/\gamma$. According to the Lagrangian function and KKT condition, we can get the best $\boldsymbol{W}$ as

$$\boldsymbol{W} = (\boldsymbol{\kappa} - \frac{\boldsymbol{E}^2}{2\lambda})_+, \tag{6}$$

where $\boldsymbol{E}^2$ denotes a matrix whose elements are e_{ij}^2, $\boldsymbol{\kappa}$ is the Lagrangian multipliers of $\boldsymbol{1}^T \boldsymbol{W}\boldsymbol{1} = 1$, and $(\cdot)_+$ denotes a nonnegative operator. Without loss of generality, suppose the elements of $\mathrm{vec}(\boldsymbol{E}^2)$ are in nondecreasing order, then $\mathrm{vec}(\boldsymbol{W})$ will be in nonincreasing order. Given parameter l denoting the number of zero elements related to noise, then the $(mn - l + 1)$th element of $\mathrm{vec}(\boldsymbol{W})$ equals 0, where $mn = m \times n$. This together with the constraint $\boldsymbol{1}^T \boldsymbol{W}\boldsymbol{1} = 1$ leads to

$$\kappa = \frac{1}{mn - l} + \sum_{j=1}^{mn-l} \frac{e_j^2}{2\lambda(mn - l)}, \lambda = (mn - l)\frac{e_{mn-l+1}^2}{2} - \frac{1}{2}\sum_{j=1}^{mn-l} e_j^2. \tag{7}$$

With derived parameters κ and λ, we can obtain $\boldsymbol{W}$ analytically by

$$\boldsymbol{W} = \frac{\boldsymbol{M}(e_{mn-l+1}^2) - \boldsymbol{E}}{(mn - l)e_{mn-l+1}^2 - \sum_{j=1}^{mn-l} e_j^2}, \tag{8}$$

where $\boldsymbol{M}(e_{mn-l+1}^2)$ is an $m \times n$ matrix with all elements being e_{mn-l+1}^2.

3.2 Accelerated Proximal Gradient Algorithm

With estimated $\boldsymbol{W}$, the subproblem of $\boldsymbol{Z}$ is as follows

$$F(\boldsymbol{Z}) = \overbrace{\gamma \| \boldsymbol{W}^{1/2} \odot (\boldsymbol{X} - \boldsymbol{X}\boldsymbol{Z})\|_{\mathrm{F}}^2}^{f(\boldsymbol{Z})} + \overbrace{\sum_{i=1}^{n} r(s_i, \sigma_i)}^{r(\boldsymbol{Z})}, \tag{9}$$

We divide (9) into two terms, namely $F(\boldsymbol{Z}) = f(\boldsymbol{Z}) + r(\boldsymbol{Z})$. With the fact that f is L-Lipschitz smooth, i.e., $\|\nabla f(\boldsymbol{Z}_1) - \nabla f(\boldsymbol{Z}_2)\|_{\mathrm{F}} \leq L\|\boldsymbol{Z}_1 - \boldsymbol{Z}_2\|_{\mathrm{F}}$ and $r(\boldsymbol{Z})$ is nonconvex, the accelerated proximal gradient (APG) method [18] can be applied. For our problem (9), APG generates

$$\begin{aligned} \boldsymbol{Z}_{k+1} &= \min \frac{1}{2}\|\boldsymbol{Z} - \boldsymbol{Z}_k + \eta\nabla f(\boldsymbol{Z}_k)\|_{\mathrm{F}}^2 + \eta r(\boldsymbol{Z}) \\ &= \mathrm{prox}_{\eta r}(\boldsymbol{Z}_k - \eta\nabla f(\boldsymbol{Z}_k)) \end{aligned} \tag{10}$$

$$\nabla f(\boldsymbol{Z}_k) = 2\gamma(\boldsymbol{W} \odot \boldsymbol{X})^{\mathrm{T}}(\boldsymbol{W} \odot (\boldsymbol{X}\boldsymbol{Z}_k - \boldsymbol{X})) \tag{11}$$

at iteration k, where $0 < \eta < 1/L$ is the stepsize. Recently, APG methods have been extensively studied. The state-of-the-art is the efficient inexact proximal gradient algorithm [19], within whose framework our LRR subproblem of $\boldsymbol{Z}$ can be solved as in Algorithm 1. Each iteration requires only one proximal step (step 7). Acceleration is performed in step 3 and the objective is then checked to determine whether $\boldsymbol{Y}_{k+1}$ is accepted (steps 5).

Algorithm 1. efficient inexact proximal gradient (EIPG) for problem (8)

Input: Estimated $\boldsymbol{W}$, parameter $\eta \in (0, 1/L)$, stopping criterion $tol>0$, and $k=1$;
Output: $\boldsymbol{Z}$
1. $\boldsymbol{Z}_0=\boldsymbol{0}$, $\boldsymbol{Z}_1 \in R^{n\times n}$ follows $N(0,1)$;
2. **While** $\|F(\boldsymbol{Z}_k) - F(\boldsymbol{Z}_{k-1})\|_F^2 / \|F(\boldsymbol{Z}_k)\|_F^2 > tol$ **do**
3. $k=k+1$; $\boldsymbol{Y}_k = \boldsymbol{Z}_k + \frac{k-1}{k+2}(\boldsymbol{Z}_k - \boldsymbol{Z}_{k-1})$;
4. $\Delta_k = \max_{t=\max(1,k-3),\cdots,k} F(\boldsymbol{Z}_t)$;
5. **if** $F(\boldsymbol{Y}_k) \leq \Delta_k$ **then** $\boldsymbol{G}_k = \boldsymbol{Y}_k$;**else** $\boldsymbol{G}_k = \boldsymbol{Z}_k$; **end if**;
6. $\boldsymbol{V}_k = \boldsymbol{G}_k - \eta \nabla f(\boldsymbol{G}_k)$;
7. $\boldsymbol{Z}_{k+1} = \text{prox}_{\eta r}(\boldsymbol{V}_k)$.
8. **end While**

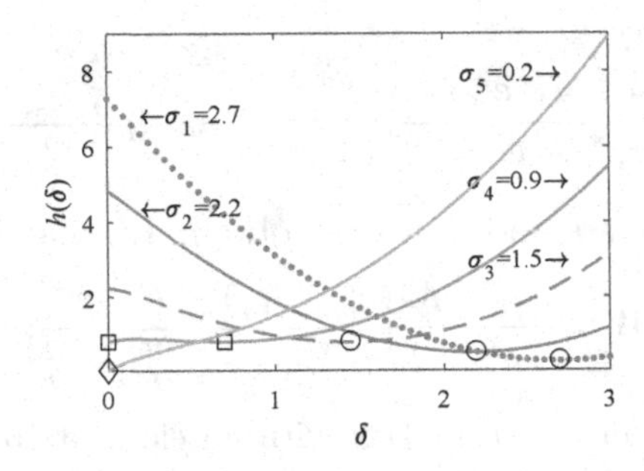

Fig. 2. Illustration of function $h(\delta_i)$ with s_i in nondescending order. The marked points denote the global optimums of $h(\delta_i)$. (Color figure online)

3.3 Automatic Singular Value Thresholding

The main computational complexity of Algorithm 1 lies in the proximal step 7. In this subsection, we show how the solution of this step can be achieved analytically by using the SVT operator. We utilize $\boldsymbol{s} = \nabla r(\sigma(\boldsymbol{Z}))$ for the purpose of preserving the major data components by penalizing the larger singular values less than the smaller ones. Therefore, the weights are in ascending order with the premise that the singular values are descending. Under these conditions, our SVT subproblem $h(\delta_i) = 0.5(\delta_i - \sigma_i(\boldsymbol{Z}))^2 + r(s_i, \sigma_i(\boldsymbol{Z}))$ can be illustrated in Fig. 2, where $a = 4.9$, $\boldsymbol{\sigma} = [2.7, 2.2, 1.5, 0.9, 0.2]^{\mathrm{T}}$, and $\boldsymbol{s} = [0.8, 1.5, 2.5, 2.7, 3.0]^{\mathrm{T}}$ are used. From the figure we can see that the three red lines are convex and their minimal points have the property of $\delta_i^* \geq \delta_j^*$ for $s_i \leq s_i$, $i < j$, the two other lines are nonconvex and their minimal points all lie in $\delta^* = 0$; Fig. 2 guides us to the hypotheses that the SVT function $\text{prox}_{\eta r}(\boldsymbol{V}_k)$ may be strictly convex and parallelly solvable. We present the following lemmas to validate these hypotheses (All the proofs for the Lemmas and Theorems are given in the supplemental material).

Lemma 1. Given the weights as $0 \leq s_1 \leq ... \leq s_n$, $\text{prox}_{\eta r}(\boldsymbol{V}_k)$ can be decoupled into independent subproblems as

$$\min_{\delta_i \geq 0} h(\delta_i) = \frac{1}{2}(\delta_i - \sigma_i(\boldsymbol{V}_k))^2 + \frac{\eta s_i \delta_i}{1 + a\delta_i/2} \tag{12}$$

with their optimal solutions satisfying the order constraint $\delta_1 \geq \delta_2 \geq ... \geq \delta_n$.

Lemma 2. Despite the nonconvexity of rational penalty function, the SVT function $\text{prox}_{\eta r}(\cdot)$ in (10) is strictly convex if $0 < a < 1/(\eta \max(\boldsymbol{s}))$.

From the blue line of Fig. 2, we can further observe that there exists a specific σ for $h(\delta^*) = h(0)$. Thus, the thresholding value τ and the optimal δ^* can be achieved following the generalized iterated shrinkage algorithm (GISA) [16]. For our rational penalty function, we have

$$\begin{aligned} \tau_i &= \frac{2\sqrt{s_i a \eta} - 1}{2}, i = 1, ..., n. \\ \delta_i - \sigma_i + \frac{s_i \eta}{(1 + a\delta_i/2)^2} &= 0, i = 1, ..., n. \end{aligned} \tag{13}$$

Given Lemmas 1 and 2, Theorem 1 ensures a global solution to the problem (10) (Step 7 of Algorithm 1). The solution involves automatic thresholding of singular values in the matrix $\boldsymbol{V}_k$.

Theorem 1. Let $\boldsymbol{V}_k = \boldsymbol{U}\boldsymbol{\Sigma}\boldsymbol{V}^{\mathrm{T}}$ be the SVD of $\boldsymbol{V}_k$. If $0 < a < 1/(\eta \max(\boldsymbol{s}))$, then the global minimizer of step 7 in Algorithm 1 is

$$\boldsymbol{Z}_{k+1} = \boldsymbol{U}\boldsymbol{\Xi}\boldsymbol{V}^{\mathrm{T}}, \tag{14}$$

where $\boldsymbol{\Xi}$ is the threshold function, whose subproblems are defined in (12) with solutions generated by Eq. (13).

3.4 Efficient SVD

The SVD operation is the main burden in Theorem 1, which needs to be repeatedly conducted during each iteration. Proposition 1 shows that $\text{prox}_{\eta r}(\boldsymbol{V}_k)$ can be obtained from the proximal operator on a smaller matrix. To obtain such a $\boldsymbol{Q}$, Ref. [20] resorts to the power method for successfully approximating the SVT in nuclear norm constrained problems. Nevertheless, their algorithm is designed to tackle a fixed rank problem, i.e., the rank of the objective matrix should be given in advance, which may not hold in real-world applications. Motivated by the automatic SVT property from Theorem 1 and the randomized blocked algorithm [21], we attempt to achieve SVD through an adaptive thresholding problem, which does not require any rank parameter defined in advance. A rank shrinkage SVD algorithm is presented, where the required singular values satisfying the shrinkage condition are incrementally estimated by blocked SVD approximation.

Proposition 1. [20]. Assume that $\boldsymbol{Q} \in \mathrm{R}^{n\times q}$, where $q \geq \mathrm{rank}(\boldsymbol{V}_k)$, is orthogonal and $\mathrm{span}(\boldsymbol{U}_q) \subseteq \mathrm{span}(\boldsymbol{Q})$. Then, $\mathrm{prox}_{\eta r}(\boldsymbol{V}_k) = \boldsymbol{Q}\mathrm{prox}_{\eta r}(\boldsymbol{Q}^{\mathrm{T}}\boldsymbol{V}_k)$.

Algorithm 2. $\mathrm{prox}_{\eta r}(\boldsymbol{V}_k)$ with efficient SVD and automatic SVT

Input: $\boldsymbol{V}_k$, block size b;
Output: Estimated left singular vectors $\boldsymbol{U}_Q$, right singular vectors $\boldsymbol{V}_Q$, and thresholded singular values $\boldsymbol{\Sigma}_\delta$.
1. i=1;
2. **While** not converged **do**
3. Generate an $n\times b$ Gaussian random matrix $\boldsymbol{\Omega}_i$;
4. $\boldsymbol{Q}_i$=PowerScheme($\boldsymbol{V}_k$, $\boldsymbol{\Omega}_i$);
5. $\boldsymbol{Q}_i$=orth($\boldsymbol{Q}_i$-$\sum_{j=1}^{i-1}$ $\mathbf{Q}_j\mathbf{Q}_j^{\mathrm{T}}\mathbf{Q}_i$), $\boldsymbol{Q}$=[$\boldsymbol{Q}_1$,...,$\boldsymbol{Q}_i$];
6. $\boldsymbol{B}_i$=$\boldsymbol{Q}_i^{\mathrm{T}}\boldsymbol{V}_k$; $\boldsymbol{B}$=[$\boldsymbol{B}_1$,...,$\boldsymbol{B}_i$];
7. $\boldsymbol{V}_k$=$\boldsymbol{V}_k$-$\boldsymbol{Q}_i\boldsymbol{B}_i$;
8. [$\boldsymbol{Q}_t$, $\boldsymbol{R}_t$]=qr($\boldsymbol{B}^{\mathrm{T}}$,0), [$\boldsymbol{U}_t$,$\boldsymbol{\Sigma}_t$,$\boldsymbol{V}_t$]=svd($\boldsymbol{R}_t$);
9. Obtain $\boldsymbol{\tau}$ given $\boldsymbol{\Sigma}_t$ by (13);
10. if $\max(\boldsymbol{\tau}) \geq \min(\boldsymbol{\Sigma}_t)$, then **break**;
11. **end while**
12. Update $\boldsymbol{\delta}$ given $\boldsymbol{\tau}$ by (13);
13. $\boldsymbol{U}_Q$=$\boldsymbol{Q}\boldsymbol{V}_t$, $\boldsymbol{V}_Q$=$\boldsymbol{Q}_t\boldsymbol{U}_t$.

The entire SVT procedure is shown in Algorithm 2. Step 3–7 use the power method and blocked SVD approximation to efficiently build up an orthogonal matrix $\boldsymbol{Q}$ that approximates $\mathrm{span}(\boldsymbol{U}_Q)$. Step 8–10 perform a small SVD and check the stop criterion. Though SVD operation is still needed, matrix $\boldsymbol{R}_t$ is much smaller than matrix $\boldsymbol{V}_k$. In step 12, the singular values $\boldsymbol{\Sigma}_t$ are thresholded using Theorem 1.

So far, we have presented a new approach to achieve a more appropriate LRR representation matrix $\boldsymbol{Z}$. The learned $\boldsymbol{Z}$ can be used to construct an affinity matrix as $(|\boldsymbol{Z}| + |\boldsymbol{Z}^{\mathrm{T}}|)/2$, which can be further fed into the spectral clustering method for data segmentation.

3.5 Complexity and Convergence

The main computational complexity of DWLRR lies in the SVT operation. With the learned $\boldsymbol{U}_Q$, $\boldsymbol{V}_Q$, and $\boldsymbol{\Sigma}_\delta$, the dot product by $\boldsymbol{W} \in \mathrm{R}^{m\times n}$, the matrix multiplication, and the SVD cost $O(mn)$, $O(mnq)$, and $O(mq^2)$, respectively, where q is the revealed rank of $\boldsymbol{Z}$. In contrast, exact SVT operation takes $O(mn^2)$ time, and is much slower as $n \gg q$.

For the convergence analysis, we first present Theorem 2 to demonstrate the convergence of subproblem $F(\boldsymbol{Z})$, which combining with the closed-form solution (8) for $\boldsymbol{W}$ subproblem leads to $J(\boldsymbol{W}^i, \boldsymbol{Z}^i) \geq J(\boldsymbol{W}^{i+1}, \boldsymbol{Z}^i) \geq J(\boldsymbol{W}^{i+1}, \boldsymbol{Z}^{i+1})$, where i is the iteration number for IRM. Note that the sequence $\{J^i\}_{i=0}^{\infty}$ is bounded from below by zero, hence convergent.

Theorem 2. Given $\eta \in (0, 1/L)$, the sequence $\{\boldsymbol{Z}_k\}$ generated by problem (9) satisfies the following properties:

$$(1) F(\boldsymbol{Z}_k) - F(\boldsymbol{Z}_{k+1}) \geq \frac{1}{2}(\frac{1}{\eta} - L)\|\boldsymbol{Z}_k - \boldsymbol{Z}_{k+1}\|_F^2 \geq 0;$$

$$(2) \lim_{k \to \infty} \|\boldsymbol{Z}_k - \boldsymbol{Z}_{k+1}\|_F^2 = 0.$$

Table 1. Used image datasets. n, m, and c denote the sample size, the feature dimension, and the number of subjects.

Datasets	n	m	c
USPS	2913	256	3
JAFFE	213	676	10
MNIST	2000	784	10
COIL	1440	1024	20
ORL	400	1024	40
AR	1200	4980	100

Table 2. Clustering performance (%) on five different image data sets.

Methods	USPS		JAFFE		MNIST		COIL		ORL		Average	
	AC	NMI	AC	NMI	AC	NMI	AC	NMI	AC	NMI	AC	NMI
LRR	93.2	81.5	94.2	94.4	50.0	51.1	59.0	69.6	51.8	76.4	69.8	74.6
SMR	94.0	81.2	97.3	97.2	62.3	61.8	70.2	79.5	**72.5**	**84.7**	79.3	80.9
FSCNN	93.5	81.8	**99.5**	**99.1**	55.0	54.2	62.0	75.0	65.7	81.8	75.1	78.4
IRIALM	94.0	**82.0**	**99.5**	**99.1**	46.0	44.2	45.3	59.1	66.4	83.2	70.2	73.5
NSGLRR	**94.6**	81.9	98.5	97.8	51.9	52.9	62.0	72.4	65.3	79.8	74.5	77.0
L2Graph	93.2	81.2	97.6	96.5	59.1	54.6	59.7	73.7	69.8	83.8	75.9	78.0
DWLRR	**94.6**	**82.0**	**99.5**	**99.1**	**66.9**	**64.8**	**76.7**	**89.3**	70.2	84.1	**81.6**	**83.9**

4 Experimental Results

We investigate the performance of our DWLRR by conducting comprehensive experiments on image clustering problem. Several recently developed methods including LRR [1], SMR [22], FSCNN [7], IRIALM [8], NSGLRR [11], and L2Graph [6] are used for comparison. The balance parameters of all the competing methods are taken from {1e−3, 1e−2, ..., 1e2} to report the best result. The remaining parameters of all compared algorithms are searched from the suggested candidate sets of the original papers to achieve the best performance. Unless otherwise specified, for DWLRR, the parameters η and a are set as

$0.6\max(\gamma\|\boldsymbol{X}^{\mathrm{T}}\boldsymbol{X}\|_2)$ and $0.9/(\eta\max(\boldsymbol{s}))$, respectively. We utilize accuracy (AC), normalized mutual information (NMI) [11], as well as execution time to evaluate the clustering performance. All experiments are implemented in MATLAB, and are run over a laptop with Intel(R) 2.4-GHz i7 CPU and 8.0-GB RAM.

4.1 Clustering Performance

The first five datasets in Table 1 are used for clustering on nonoccluded images. Table 2 shows the performance of the competing methods, where the last two columns are with the average results of all the evaluated datasets. The first observation is that our method outstandingly outperforms the competitors. For all the datasets, DWLRR achieves the best results except for ORL, where it is second best.

Fig. 3. Some disguised images from the AR dataset.

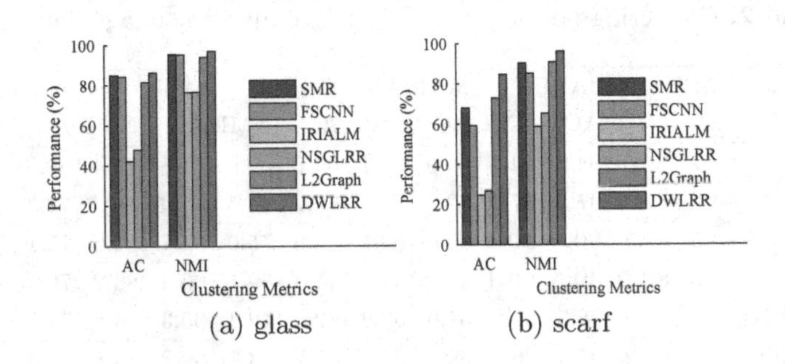

Fig. 4. Clustering performance (%) on the AR dataset.

The disguised images from 100 people of AR dataset are selected to evaluate the robustness of the competing methods. Figure 3 shows some samples. We compare our method with SMR, L2Graph, FSCNN, NSGLRR, and IRIALM. Figure 4 shows the clustering results of AC and NMI. On average, DWLRR achieves an improvement of 1.27%, 1.75%, 32.26%, 29.31%, and 3.82% over SMR, FSCNN, IRIALM, NSGLRR, and L2Graph, respectively. This further verifies the superiority of reweighting scheme and weighted rational penalty.

IRIALM and DWLRR all adopt the idea of weights learning mechanism to eliminate parts of the useless features in input data. Figure 5 illustrates the recovered images and the learned weights maps corresponding to the images with scarf from Fig. 3, where the first and second row are the results from IRIALM and DWLRR, respectively. In Fig. 5, our method clearly learns more accurate face image and weight maps. Moreover, IRIALM assigns close to 0 values (black region) to the deemed occlusion pixels and assigns close to 1 values (white region) to the deemed non-occlusion pixels. However, our method assigns 0 to the occlusion pixels but assigns meaningful values (grey region) to the non-occlusion pixels, which exhibits different contributions of active features and leads to a better clustering results.

4.2 Execution Time

In order to compare the computational complexity of different methods, we measure the execution time of competing algorithms those with relatively better performance, i.e., FSCNN, IRIALM, NSGLRR, and DWLRR. We report the normalized objective values versus execution time of these approaches under their optimal tuned parameters. Figure 6 illustrates the experimental results on MNIST and COIL datasets. The same stopping criterion, i.e., $tol = 1e{-}4$, is used for fair comparison. As can be seen from the figure, the proposed approach is computationally efficient compared to other iterative based clustering methods. In MNIST dataset, NSGLRR converges faster than DWLRR at the initial stage of iterations. However, the decreasing objective values turn to increase at certain point due to the inexactly linearizing approximation to the cost function.

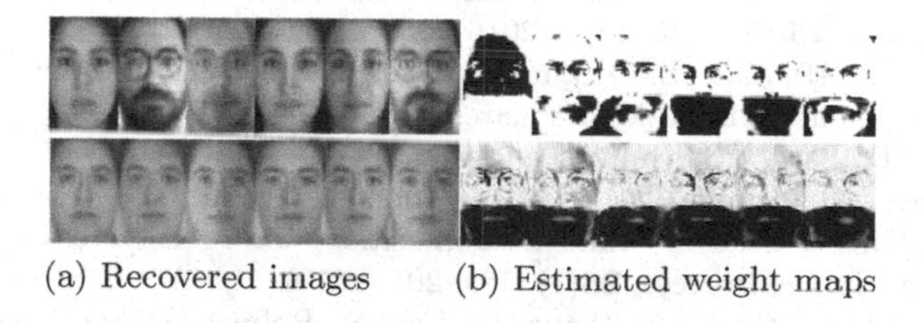

(a) Recovered images (b) Estimated weight maps

Fig. 5. Recovery of AR face images with real disguise.

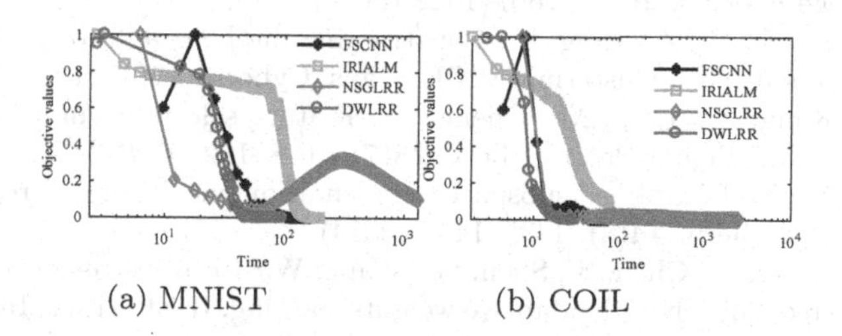

(a) MNIST (b) COIL

Fig. 6. Objective values versus execution time (in seconds).

5 Conclusions

In this paper, we propose a new low-rank representation method, DWLRR, which marries the advantages of feature learning and weighted nonconvex constraint, where the first reveals different contributions of input features in the learning process and the second ensures a closer approximation to the latent low-rank representation matrix. A reweighted APG framework is presented to solve our DWLRR model. Furthermore, based on a key observation that the singular values can be automatically thresholded, we approximate the SVT operator by incrementally blocks stacking and thresholding comparison. Experiments on six image segmentation datasets and four video sequences demonstrate that DWLRR is not only robust to different types of noise corruption, but also more efficient than other state-of-the-art iterative methods.

Acknowledgements. This work is supported by National Natural Science Foundation of China (61602413) and Natural Science Foundation of Zhejiang Province (LY19F030016).

References

1. Liu, G., Lin, Z., Yan, S., Sun, J., Xu, Y., Ma, Y.: Robust recovery of subspace structures by low-rank representation. IEEE Trans. Pattern Anal. Mach. Intell. **35**(1), 171–184 (2013)
2. Kim, E., Lee, M., Oh, S.: Robust elastic-net subspace representation. IEEE Trans. Image Process. **25**(9), 4245–4259 (2016)
3. Lu, C.-Y., Min, H., Zhao, Z.-Q., Zhu, L., Huang, D.-S., Yan, S.: Robust and efficient subspace segmentation via least squares regression. In: Fitzgibbon, A., Lazebnik, S., Perona, P., Sato, Y., Schmid, C. (eds.) ECCV 2012. LNCS, vol. 7578, pp. 347–360. Springer, Heidelberg (2012). https://doi.org/10.1007/978-3-642-33786-4_26
4. Ding, Z., Fu, F.: Dual low-rank decompositions for robust cross-view learning. IEEE Trans. Image Process. **28**(1), 194–204 (2019)
5. Zhang, Z., Li, F., Zhao, M., Zhang, L., Yan, S.: Robust neighborhood preserving projection by nuclear/l2,1-norm regularization for image feature extraction. IEEE Trans. Image Process. **26**(4), 1607–1622 (2017)
6. Peng, X., Yu, Z., Yi, Z., Tang, H.: Constructing the L2-graph for robust subspace learning and subspace clustering. IEEE Trans. Cybern. **47**(4), 1053–1066 (2017)
7. Peng, C., Kang, Z., Yang, M., Cheng, Q.: Feature selection embedded subspace clustering. IEEE Signal Process. Lett. **23**(7), 1018–1022 (2016)
8. Chen, J., Yang, J.: Robust subspace segmentation via low-rank representation. IEEE Trans. Cybern. **44**(8), 1432–1445 (2014)
9. Zheng, J., Yang, P., Chen, S., Shen, G., Wang, W.: Iterative re-constrained group sparse face recognition with adaptive weights learning. IEEE Trans. Image Process. **26**(5), 2408–2423 (2017)
10. Zhang, Z., Li, F., Zhao, M., Zhang, L., Yan, S.: Joint low-rank and sparse principal feature coding for enhanced robust representation and visual classification. IEEE Trans. Image Process. **25**(6), 2429–2443 (2016)
11. Yin, M., Gao, J., Lin, Z.: Laplacian regularized low-rank representation and its applications. IEEE Trans. Pattern Anal. Mach. Intell. **38**(3), 504–517 (2016)

12. Peng, X., Lu, C., Yi, Z., Tang, H.: Connections between nuclear-norm and frobenius-norm-based representations. IEEE Trans. Neural Netw. Learn. Syst. **29**(1), 218–224 (2018)
13. Lanza, A., Morigi, S., Selesnick, I., Sgallari, F.: Nonconvex nonsmooth optimization via convex-nonconvex majorization-minimization. Numer. Math. **136**(2), 343–381 (2017)
14. Lu, C., Tang, J., Yan, S., Lin, Z.: Nonconvex nonsmooth low rank minimization via iteratively reweighted nuclear norm. IEEE Trans. Image Process. **25**(2), 829–839 (2016)
15. Gu, S., Xie, Q., Meng, D., Zuo, W., Feng, X., Zhang, L.: Weighted nuclear norm minimization and its applications to low level vision. Int. J. Comput. Vis. **121**(2), 183–208 (2017)
16. Xie, Y., Gu, S., Liu, Y., Zuo, W., Zhang, W., Zhang, L.: Weighted schatten p-norm minimization for image denoising and background subtraction. IEEE Trans. Image Process. **25**(10), 4842–4857 (2016)
17. Peng, C., Kang, Z., Cheng, Q.: Integrating feature and graph learning with low-rank representation. Neurocomputing **249**, 106–116 (2017)
18. Beck, A., Teboulle, M.: A fast iterative shrinkage-thresholding algorithm for linear inverse problems. SIAM J. Imaging Sci. **2**(1), 183–202 (2009)
19. Yao, Q., Kwok, J., Gao, F., Chen, W., Liu, T.: Efficient inexact proximal gradient algorithm for nonconvex problems. In: Proceedings of the 26th International Joint Conference on Artificial Intelligence, pp. 3308–3314. Melbourne (2017)
20. Yao, Q., Kwok, J., Zhong, W.: Fast low-rank matrix learning with nonconvex regularization. In: International Conference on Data Mining, pp. 539–548. IEEE, Atlantic (2015)
21. Li, Y., Yu, W.: Fast randomized singular value thresholding for low-rank optimization. IEEE Trans. Pattern Anal. Mach. Intell. **40**(2), 376–391 (2018)
22. Hu, H., Lin, Z., Feng, J., Zhou, J.: Smooth representation clustering. In: Conference on Computer Vision and Pattern Recognition, pp. 3834–3841. IEEE, Columbus (2014)

Exploring Dual-Triangular Structure for Efficient R-Initiated Tall-Skinny QR on GPGPU

Nai-Yun Cheng and Ming-Syan Chen(✉)

National Taiwan University, Taipei, Taiwan
nycheng@arbor.ee.ntu.edu.tw, mschen@ntu.edu.tw

Abstract. The QR decomposition is one of the fundamental matrix decompositions in data mining. A particularly challenging case of QR decomposition is to deal with the tall-and-skinny matrix. Tall-skinny QR has lots of applications such as Krylov subspace methods and some subspace projection methods. Furthermore, tall-skinny QR can accelerate the process of principal component analysis (PCA). Although algorithms like TSQR and Cholesky QR have been proposed for computing QR decompositions on tall-and-skinny matrices, none of these algorithms are suitable for being applied to the GPGPU, which has been increasingly used nowadays. In view of the limited memory in GPGPU and also the costly data transmission between CPU and GPGPU, we propose a novel R-initiated TSQR to make the computing of tall-and-skinny QR on the GPGPU efficient. Explicitly, our method is unique in that it utilizes Givens QR to take advantage of the existence of dual-triangular (DT) structure in submatrices in TSQR so as to significantly reduce the computation required. With the R-initiated method, our method can not only meet the memory limitation of GPGPU but also avoid large amounts of data transmission. Theoretical results are derived, showing the merit of the proposed method. The experimental results indicate that our method significantly outperforms the conventional TSQR.

Keywords: TSQR · Tall-and-skinny matrix · QR decomposition

1 Introduction

The QR decomposition is one of the fundamental matrix decompositions with applications throughout data analysis and scientific computing. In a QR decomposition, the target matrix A is factorized into a product of an orthonormal matrix Q and an upper triangular matrix R, i.e.,

$$A = QR.$$

A particularly challenging case of QR decomposition is to deal with a tall-and-skinny matrix. Note that such a kind of matrices exists in several real applications. Specifically, millions of data with only a few attributes can be represented as a tall-and-skinny matrix. The QR decomposition of a tall-and-skinny

Q. Yang et al. (Eds.): PAKDD 2019, LNAI 11440, pp. 578–589, 2019.
https://doi.org/10.1007/978-3-030-16145-3_45

matrix also appears in various numerical methods. For example, tall-skinny QR decomposition can be used in Krylov subspace methods [1,2] and some subspace projection methods [3] for linear systems and eigenvalue problems. Also, the QR decomposition is an efficient way to accelerate principal component analysis (PCA) [4], which is one of the most important methods for dimension reduction.

Given the importance of tall-skinny QR decomposition, several algorithms have been proposed for its efficient execution. However, most of these methods are not applicable to the case of using GPGPU. In Cholesky QR [5], the matrix R can be found by computing the Cholesky decomposition of $A^\top A$. Then the matrix Q can be computed from A and R [6]. Since the Cholesky decomposition cannot be parallelized, Cholesky QR is not applicable when GPGPU is used. CholeskyQR2 [7], an extension of Cholesky QR, has the same restriction.

Since QR decomposition methods currently developed for the GPGPU [8–10] are designed for normal matrices, these methods suffer poor performances on tall-and-skinny matrices. Therefore, a special algorithm, called TSQR, was introduced to compute the QR decomposition of tall-and-skinny matrices in parallel [11,12]. TSQR has been widely used in other algorithm [13,14]. Fig. 1 shows the basic idea of TSQR.

However, the conventional TSQR is designed for distributed computing and is not suitable when GPGPU is used. Note that copying data between the host and global memory is very time-consuming and should be avoided. In Fig. 1, it can be seen that all the orthonormal matrices obtained from submatrices need to be stored during the process. However, due to the limited memory of GPGPU, it is infeasible to store all of the matrices in GPGPU memory.

To remedy this, we utilize the R-initiated approach as in [15,16], meaning that instead of updating the Q matrix along the process, we compute Q after obtaining the final R. However, it is important to point out that our method is unique in utilizing Givens QR, as opposed to the Householder QR previously used, to take advantage of the existence of dual-triangular (DT) structure in submatrices in TSQR so as to significantly reduce the computation required. Moreover, in our design, the computation of Givens QR can be fully parallelized in light of the DT structure. The details will be explained in Sect. 3.

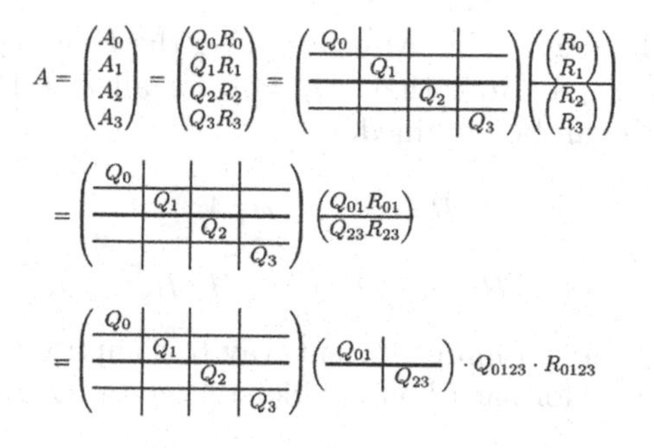

Fig. 1. An example of TSQR

Based on the foregoing, we propose in this paper a scheme to explore dual-triangular structure for efficient R-initiated tall-skinny QR decomposition on GPGPU. The main contributions of our work are listed as follows.

- We employ the R-initiated approach for better efficiency. That is, our method requires much less data transmission between the GPGPU and the host. Also, the R-initiated method remedies the inefficiency of computing Q in Givens QR.
- We prove that Givens QR has the same step complexity, n, as Householder QR for matrices with DT structure in Theorem 1. In Theorems 2 and 3, we prove that using Givens QR for matrices with DT structure can even lead to lower complexity on GPGPU.
- The experiment results indicate that our method not only outperforms others but is also able to accelerate the process of PCA.

This paper is organized as follows. Section 2 briefly summarizes current algorithms for QR factorization. Section 3 presents the proposed R-initiated TSQR and some theoretical results. Section 4 shows the experimental results. Finally, the conclusion is given in Sect. 5.

2 Related Work

In this section, several QR decomposition algorithms are introduced. Householder QR and Givens QR are two fundamental methods for normal QR decomposition while Cholesky QR and TSQR are designed for tall-skinny QR.

2.1 Householder QR

In Householder QR, a sequence of Householder reflections is applied to the target matrix A. Each Householder reflection can be represented by a matrix H_i,

$$H_i = I - 2\frac{v_i v_i^\top}{v_i^\top v_i}$$

where v_i is a Householder vector. After applying the Householder reflection H_1 on A, we have $\tilde{A} = H_1 A$. Note that $\tilde{A}_{1i} = 0$ for all $i > 1$. As $H_1 \dots H_n$ are applied, the R and Q can be obtained.

$$R = H_n \dots H_1 A$$

$$Q = (H_n H_{n-1} \dots H_1)^\top = H_1^\top H_2^\top \dots H_n^\top$$

The WY representation is a more efficient way to compute Q [17]. By updating two matrices, Y and T, for each householder reflection, Q can be obtained.

$$Q = I - YTY^\top.$$

2.2 Givens QR

The Givens QR computes the upper triangular matrix R from the target matrix A by applying a sequence of Givens rotations. Each Givens rotation can replace a selected element with 0. Consequently, each rotation can be represented by a unitary matrix G. Therefore, a sequence of Givens matrices can be used to eliminate all the non-zero elements below the diagonal:

$$R = G_k G_{k-1} \dots G_2 G_1 A.$$

Since all G_i are unitary, it can be inferred from the above equations that

$$Q = (G_k G_{k-1} \dots G_1)^\top = G_1^\top G_2 \top \dots G_k^\top.$$

However, when a Givens rotation is applied, two rows will be rotated at the same time. Hence, the non-zero element a_{ij} can be eliminated only when all $a_{kj} = 0$ where $k < i$. As a result, for dense matrices, the Givens rotation is not suitable for parallel computing.

2.3 Cholesky QR

In Cholesky QR [5], the authors assume $A = QR$, and the following equation can be obtained,

$$A^\top A = R^\top Q^\top QR = R^\top R.$$

This implies that R is the Cholesky factor of $A^\top A$. The following equation is computed to solve the Cholesky decomposition $B = LL^\top$,

$$L_{ii} = \sqrt{B_{ii} - \sum_{j=1}^{i-1} L_{ij}^2} \text{ and } L_{ij} = \frac{B_{ij} - \sum_{k=1}^{j-1} L_{jk} L_{ik}}{L_{jj}}$$

Note that the process of Cholesky decomposition cannot be parallelized, and is hence not able to fully exploit the architecture of GPGPU.

After obtaining matrix R, the Cholesky QR uses the R-initiated method introduced by [6] to obtain the Q factor. Note that Q can be written as $Q = I_{m \times n} - YTY_1^\top$ where Y is a lower triangular unit matrix, and T is an upper triangular matrix [17]. Y_1 represents the top $n \times n$ block of Y. Hence,

$$A = QR = (I_{m \times n} - YTY_1{}^\top)R = \begin{bmatrix} R_{n \times n} \\ 0_{(m-n) \times n} \end{bmatrix} - YTY_1{}^\top R$$
$$\Longrightarrow A - \begin{bmatrix} R_{n \times n} \\ 0_{(m-n) \times n} \end{bmatrix} = -YTY_1{}^\top R$$

Let $V = -TY_1{}^\top R$. Since $T, Y_1{}^\top, R$ are all upper triangular, V is also an upper triangular matrix. Observe that $(A - R) = YV$, which can be a LU decomposition. Therefore, Y and V are obtained, then T can be achieved by solving $TY_1{}^\top R = -V$.

2.4 TSQR

In [12], the authors introduced a parallel algorithm. First, the $m \times n$ matrix A is divided into $M = \lfloor \frac{m}{n} \rfloor$ blocks $\{A_1, A_2, \ldots, A_M\}$. Then QR decomposition is solved for each A_i, so $A_i = Q_{0i}R_{0i}$ is obtained. Additionally, every two R_{0i} are taken as a new submatrix and the QR is solved again. That is,

$$\begin{bmatrix} R_{0,2i} \\ R_{0,2i+1} \end{bmatrix} = Q_{1i}R_{1i} \qquad i = 0, 2, 3, \ldots, \left\lfloor \frac{M-1}{2} \right\rfloor.$$

This is done iteratively until only one R_{k0} is left. The R_{k0} is the desired matrix R, and Q can be computed from $\{Q_{ij}\}$. Figure 1 is a simple example.

3 R-Initiated Tall-Skinny QR on GPGPU

The proposed method is described in this section. The first subsection shows the revision we make to meet the limitation of GPGPU. The second subsection explains how we accelerate the process by the dual-triangular structure. Figure 2 briefly shows the difference between conventional TSQR and our method.

3.1 R-Initiated Method to Meet the Memory Limitation of GPGPU

There are some challenging issues while implementing TSQR on GPGPU. The most severe issue is the limitation of memory. In conventional TSQR, all of the Q_{ij} have to be stored during the process. Thus, the bottleneck becomes the data transmission between GPGPU and the host. To solve this problem, the R-initiated method is used, i.e., the matrix Q is computed from the matrix R. By doing so, large amounts of data transmission can be avoided.

Explicitly, in the R-initiated method, the Q matrix can be computed after the matrix R is obtained. The method mentioned in Sect. 2.3 is used to reconstruct Q.

3.2 Dual-Triangular Structure to Accelerate the Process

Our another design to accelerate the process is to use Givens rotation. Although many modifications for TSQR already exist, most of these methods use the Householder QR. This can be explained by two reasons. First, WY representation from Householder QR can compute the Q matrix more efficiently. Second, Householder QR normally needs less computation. However, these advantages do not exist in this situation of interest. Since the R-initiated method is used, we do not have to compute Q for the submatrices. Also, most of the submatrices in TSQR are not dense and Givens QR usually leads to better performance on sparse matrices.

Explicitly, it can be seen from the conventional TSQR algorithm (Fig. 3) that lots of these submatrices are formed by two upper triangular matrices, which we refer to as dual-triangular(DT) structure. More specifically, there are two kinds of QR decomposition.

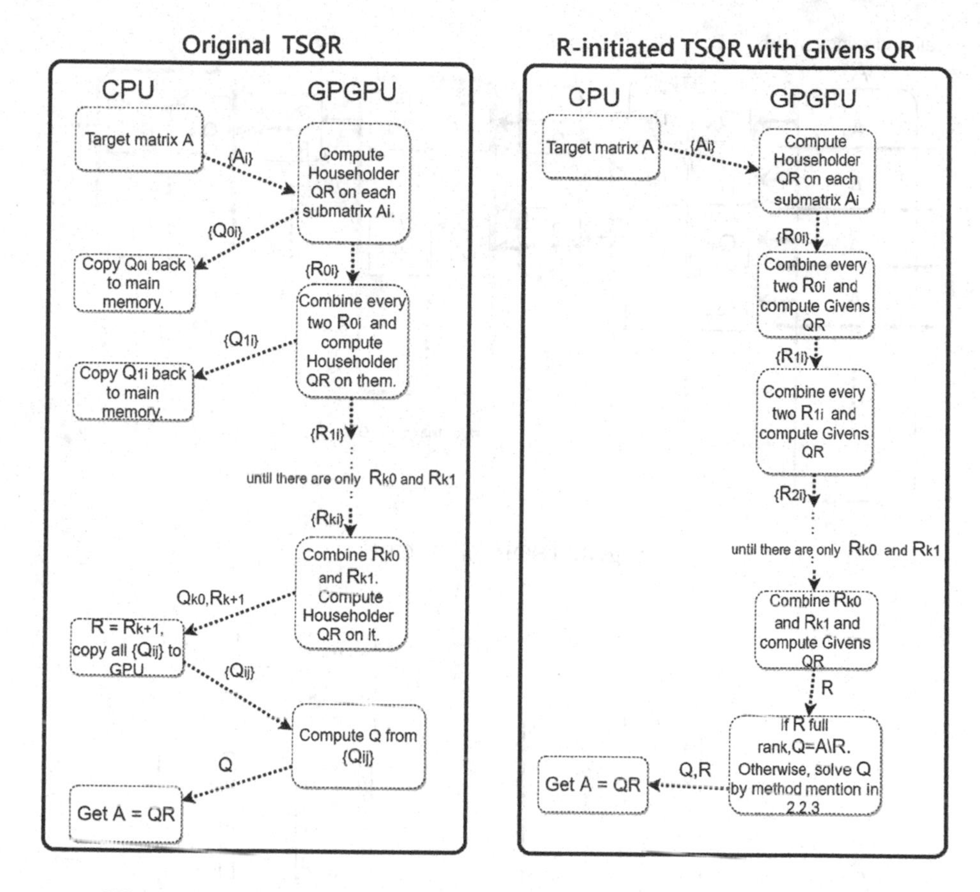

Fig. 2. Comparison between conventional TSQR and R-initiated TSQR

1. **normal QR:** QR decomposition of a normal matrix.
2. **dual-triangular QR:** QR decomposition of a matrix with DT structure.

Despite the conventional TSQR using Householder QR is efficient on CPU, such an approach does not work on GPGPU. For the matrices with DT structure, the Householder vectors tend to have special patterns of zeros as shown in Fig. 4, implying that the computation on CPU can be simply reduced by avoiding the multiplication and addition of zeros. However, this advantage vanishes in the case of using GPGPU since the matrix multiplication can be parallelized. That is, avoiding the operation of zeros on GPGPU will only increase the number of idle threads. In contrast, when Givens QR is used, DT structure can speed up the computing for Givens QR in parallel, which leads to real prominent improvement. We state our theoretical results by the following three theorems, which explain the reason why Givens QR is employed in our method.

Theorem 1. *For submatrices with DT structure ($2n \times n$), the step complexity of Givens QR on GPGPU is n.*

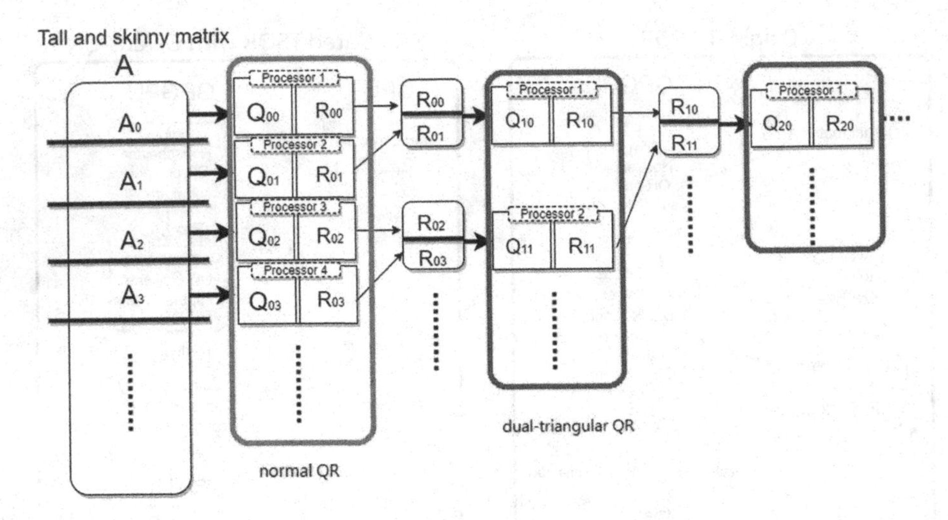

Fig. 3. TSQR on GPGPU

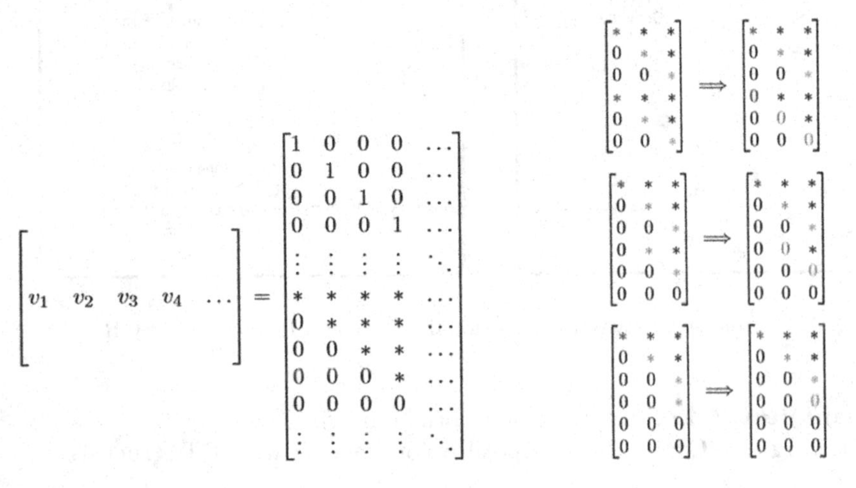

Fig. 4. The special pattern of Householder vector

Fig. 5. Parallel Givens QR by the DT structure

Proof. In the first step, the k-th row is used to eliminate the first non-zero element in $(n+k)$-th row $(k = 1, 2, \ldots, n)$. In the second step, the k-th row is used to eliminate the first non-zero element in $(n+k-1)$-th row $(k = 2, \ldots, n)$. This is done iteratively until no non-zero element remains in the lower half of the matrix. Therefore, the step complexity is n.

An illustrative example of Theorem 1 is given in Fig. 5. From Theorem 1, we learn that the Householder QR and the Givens QR result in the same step complexity n. Next, we compare the amount of computation required for both methods in each step.

Theorem 2. *For matrices with DT structure, the complexity of Householder reflection in each individual step varies. Explicitly, the complexity of the ith step of Householder QR on GPGPU is* $O(\log(i))$.

Proof. For the Householder QR, each step is a Householder reflection. In each Householder reflection, a Householder vector (v_i) needs to be computed and normalized. Then $\tilde{A} = A - v_i v_i^\top A$ is computed. Observe that both v_i and $v_i^\top A$ are vectors. Therefore, the complexity of multiplying v_i and $(v_i^\top A)$ is constant on GPGPU. This leaves the normalization of v_i and the computing of $v_i^\top A$ as bottlenecks. Consider the special pattern in Fig. 4, the number of non-zero elements in (v_i) is $i+1$. Hence, the complexity on GPGPU of the i-th step should be $O(\log i)$.

On the other hand, the merit of the Givens QR is stated below.

Theorem 3. *The complexity of Givens rotation on GPGPU is constant.*

Proof. In Givens rotation, $cos\theta$ and $sin\theta$ are computed and the corresponding two rows are rotated. Notice that, $cos\theta$ and $sin\theta$ can be computed in constant time and all the elements in these two rows can be rotated at the same time. Hence, the complexity of Givens rotation on GPGPU is constant.

For the Givens QR, many Givens rotations may be contained in each step, but all these rotations can be computed in parallel. Therefore, we just have to focus on one Givens rotation. Theorem 3 shows that the complexity of Givens rotation on GPGPU is constant. Hence, the complexity of each step is also constant. From Theorems 2 and 3, it follows that using the Givens QR results in lower complexity for matrices with DT structure on GPGPU. The following table summarizes the complexity (Table 1).

Table 1. Complexity comparison

Compute DT QR on GPGPU	Using Householder	Using Givens
Complexity in a step	$O(\log(i))$	$O(1)$
Step complexity	n	n
Total complexity of a submatrix	$O(\sum_{i=1}^{n} \log(i)) = O(\log(n!))$	$O(n)$
Total complexity of all DT QR	$O(\log(M)\log(n!))$	$O(n\log(M))$

Computing Time for the Whole Matrix

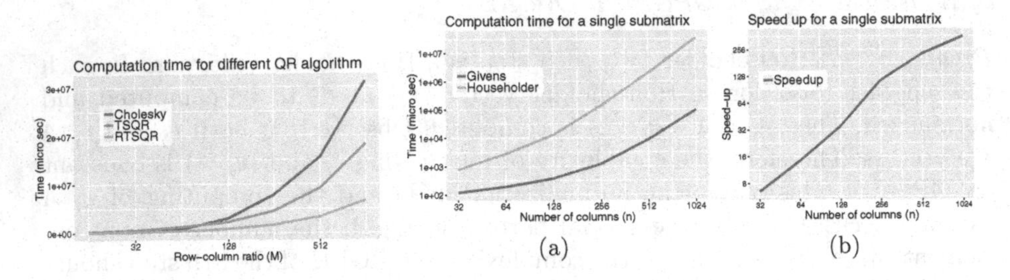

(a) (b)

Fig. 6. Results for different QR algorithms

Fig. 7. Results for a single submatrix

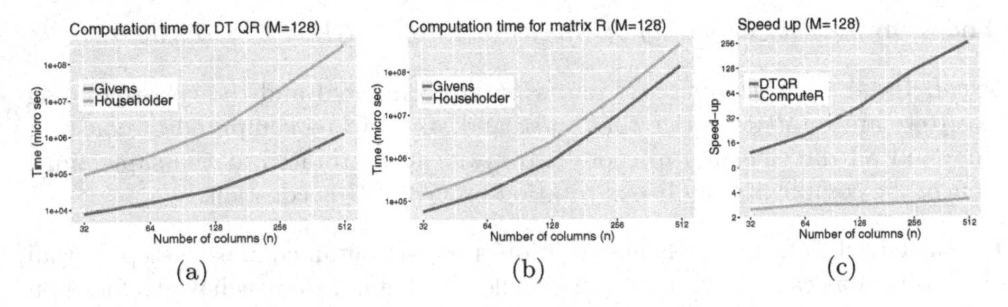

(a) (b) (c)

Fig. 8. Results with different values of n (M = 128)

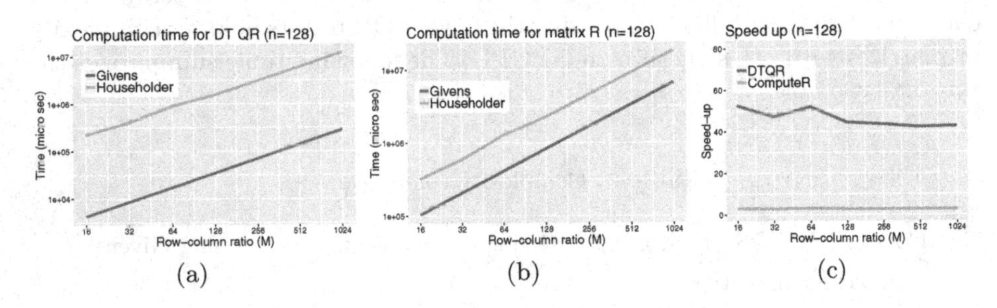

(a) (b) (c)

Fig. 9. Results with different values of M (n = 128)

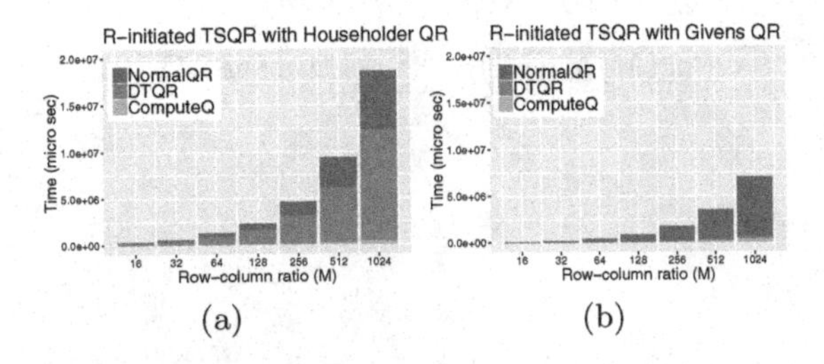

(a) (b)

Fig. 10. Time distribution for R-TSQR (n = 128)

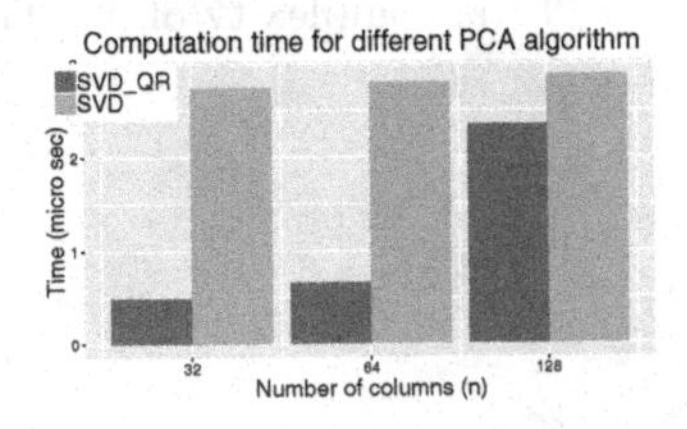

Fig. 11. Computational time of PCA

4 Experimental Results

In this section, we present some experimental results performed on Intel(R) Core(TM) i7-5820K CPU @ 3.30 GHz with 32 GB of main memory and NVIDIA GeForce GTX TITAN X. In Sect. 4.1, our method is compared with other algorithms. Sections 4.2 and 4.3 show how the size of the target matrix A affects the computing time. Section 4.4 presents the time distribution of our method. Finally, Sect. 4.5 shows that proposed method can accelerate the computing of PCA.

4.1 Results for Different Algorithms

Our method is compared with the Cholesky QR (CPU baseline) and the conventional TSQR (GPGPU baseline). The number of columns is fixed ($n = 128$). Figure 6 shows that our method has the best performance. The larger the matrix size becomes, the further our method outperforms the others.

4.2 Sensitivity of Number of Column N

The influence of n is explored in this subsection. First paragraph shows the results on a single submatrix while the second paragraph shows the results for computing the whole matrix.

Computing Time for a Single Submatrix. Figure 7 shows the difference between Givens QR and the Householder QR for a $2n \times n$ submatrix with DT structure on GPGPU. Figure 7(a) is the computing time and Fig. 7(b) shows the speed-up. Figure 7 indicates that our method has better performance and the speed-up is prominent.

Computing Time for the Whole Matrix. In order to explore the influence of n, the row-column ratio is fixed ($M = 128$) in this subsection. Figure 8(a) shows the computation time of DT QR while Fig. 8(b) presents the results for computing the R matrix, which includes both DT QR and normal QR. Figure 8(c) shows the speed-up for computing DT QR and R matrix.

Figure 8(c) indicates that the speed-up increases as n grows. However, compared to the speed-up of DT QR, the speed-up of computing the matrix R is not so significant. This is due to the limited thread numbers in GPGPU, which prevents our method from achieving the theoretical performance. That is, the submatrices in each iteration may not be computed at the same time. Hence, we compare the number of submatrices needed to be factorized in normal QR and DT QR. It can be seen in Fig. 3 that with M submatrices in normal QR, the number of submatrices to be computed in DT QR is $(\frac{M}{2}+\frac{M}{2^2}+\frac{M}{2^3}+\cdots+1) \approx M$. Thus, it can be deduced that the computation time of normal QR and DT QR should be similar in the conventional TSQR. It is recognized that the Givens QR can only accelerate the computing of DT QR. Therefore, the speed-up for computing R is limited even though the speed-up for DT QR is significant.

4.3 Sensitivity of Row-Column Ratio M

In Sect. 4.3, $n = 128$ is fixed in order to show the influence of M. Since M only changes the number of submatrices, the speed-up should be a constant. Figure 9(c) confirms this reasoning and shows that the speed-up is almost a constant.

4.4 The Bottleneck: Computing Normal QR

In this subsection, we fix $n = 128$. Figure 10(a) shows the time distribution of the R-initiated TSQR with Householder QR and Fig. 10(b) is the results of our method. Note that the computation time of computing Q is very little and the DT QR can be significantly accelerated in our method. Therefore, the bottleneck becomes the normal QR.

4.5 PCA with Tall-Skinny QR

An application of PCA is the stationary video background subtraction. Each frame of the video is transformed into a long vector. All these vectors can form a tall-and-skinny matrix in which the number of columns is equal to the number of frames. In order to subtract the background, we apply PCA on the video matrix, and the main computation in PCA is the SVD. In the SVD of the video matrix, the top singular vectors are usually associated with the background. Instead of doing a SVD of matrix A, we use the R-initiated TSQR to accelerate. The SVD of matrix A can be obtained by the SVD of matrix R.

$$A = QR = Q(\tilde{U}\Sigma V^\top) = (Q\tilde{U})\Sigma V^\top = U\Sigma V^\top.$$

In the experiment, there are 16384 pixels in one frame. Therefore, m is fixed and n is the number of frames. Therefore, the row-column ratio ($M = \frac{m}{n}$) decreases when more frames are used. Figure 11 shows the computation time of different values of n. From the blue bars of Fig. 11, we know that the computation time of SVD is not sensitive to n. However, n influences the computation time of the R-initiated TSQR significantly. As a result, our method can speed up the process of PCA prominently as long as n is not too large.

5 Conclusion

In this paper, we propose a scheme to explore dual-triangular structure for efficient R-initiated tall-skinny QR decomposition on GPGPU. First, the R-initiated method is used to avoid data transmission between the GPGPU and the host. Second, we derive that Givens QR has the lower execution complexity for matrices with DT structure on GPGPU. The experimental results validate that our method outperforms others due to its theoretical merit. The speed-up on computation depends on the size of the target matrix A. The bigger the size, the further our method outperforms the others.

References

1. Bai, Z., Demmel, J., Dongarra, J., Ruhe, A., van der Vorst, H.: Templates for the Solution of Algebraic Eigenvalue Problems: A Practical Guide. SIAM, Bangkok (2000)
2. Gutknecht, M.H.: Block Krylov space methods for linear systems with multiple right-hand sides: an introduction (2006)
3. Sakurai, T., Sugiura, H.: A projection method for generalized eigenvalue problems using numerical integration. J. Comput. Appl. Math. **159**(1), 119–128 (2003)
4. Sharma, A., Paliwal, K.K., Imoto, S., Miyano, S.: Principal component analysis using QR decomposition. Int. J. Mach. Learn. Cybern. **4**(6), 679–683 (2013)
5. Nguyen, H.D., Demmel, J.: Reproducible tall-skinny QR. In: 2015 IEEE 22nd Symposium on Computer Arithmetic (ARITH), pp. 152–159. IEEE (2015)
6. Yamamoto, Y.: Aggregation of the compact WY representations generated by the TSQR algorithm. In: Conference Talk Presented in SIAM Applied Linear Algebra (2012)
7. Fukaya, T., Nakatsukasa, Y., Yanagisawa, Y., Yamamoto, Y.: CholeskyQR2: a simple and communication-avoiding algorithm for computing a tall-skinny QR factorization on a large-scale parallel system. In: 2014 5th Workshop on Latest Advances in Scalable Algorithms for Large-Scale Systems (ScalA), pp. 31–38. IEEE (2014)
8. Volkov, V., Demmel, J.: LU, QR and Cholesky factorizations using vector capabilities of GPUS. Technical report, UCB/EECS-2008-49, vol. 49, EECS Department, University of California, Berkeley (2008)
9. Kerr, A., Campbell, D., Richards, M.: QR decomposition on GPUS. In: Proceedings of 2nd Workshop on General Purpose Processing on Graphics Processing Units, pp. 71–78. ACM (2009)
10. Humphrey, J.R., Price, D.K., Spagnoli, K.E., Paolini, A.L., Kelmelis, E.J.: CULA: hybrid GPU accelerated linear algebra routines. In: SPIE Defense, Security, and Sensing, pp. 502–770. International Society for Optics and Photonics (2010)
11. Anderson, M., Ballard, G., Demmel, J., Keutzer, K.: Communication-avoiding QR decomposition for GPUS. In: 2011 IEEE International Parallel & Distributed Processing Symposium (IPDPS), pp. 48–58. IEEE (2011)
12. Demmel, J., Grigori, L., Hoemmen, M., Langou, J.: Communication-optimal parallel and sequential QR and LU factorizations. SIAM J. Sci. Comput. **34**(1), A206–A239 (2012)
13. Constantine, P.G., Gleich, D.F.: Tall and skinny QR factorizations in MapReduce architectures. In: Proceedings of the Second International Workshop on MapReduce and Its Applications, pp. 43–50. ACM (2011)
14. Ballard, G., Demmel, J., Grigori, L., Jacquelin, M., Knight, N., Nguyen, H.D.: Reconstructing householder vectors from tall-skinny QR. J. Parallel Distrib. Comput. **85**, 3–31 (2015)
15. Ballard, G., Demmel, J., Grigori, L., Jacquelin, M., Nguyen, H.D., Solomonik, E.: Reconstructing householder vectors from tall-skinny QR. In: 2014 IEEE 28th International Parallel and Distributed Processing Symposium, pp. 1159–1170. IEEE (2014)
16. Benson, A.R., Gleich, D.F., Demmel, J.: Direct QR factorizations for tall-and-skinny matrices in MapReduce architectures. In: 2013 IEEE International Conference on Big Data, pp. 264–272. IEEE (2013)
17. Schreiber, R., Van Loan, C.: A storage-efficient WY representation for products of householder transformations. SIAM J. Sci. Stat. Comput. **10**(1), 53–57 (1989)

Efficient Autotuning of Hyperparameters in Approximate Nearest Neighbor Search

Elias Jääsaari[1], Ville Hyvönen[2,3(✉)], and Teemu Roos[2,3]

[1] Kvasir Ltd., Cambridge, England
elias.jaasaari@gmail.com
[2] Department of Computer Science, University of Helsinki, Helsinki, Finland
ville.o.hyvonen@gmail.com, teemu.roos@cs.helsinki.fi
[3] Helsinki Institute for Information Technology (HIIT), Helsinki, Finland

Abstract. Approximate nearest neighbor algorithms are used to speed up nearest neighbor search in a wide array of applications. However, current indexing methods feature several hyperparameters that need to be tuned to reach an acceptable accuracy–speed trade-off. A grid search in the parameter space is often impractically slow due to a time-consuming index-building procedure. Therefore, we propose an algorithm for automatically tuning the hyperparameters of indexing methods based on randomized space-partitioning trees. In particular, we present results using randomized k-d trees, random projection trees and randomized PCA trees. The tuning algorithm adds minimal overhead to the index-building process but is able to find the optimal hyperparameters accurately. We demonstrate that the algorithm is significantly faster than existing approaches, and that the indexing methods used are competitive with the state-of-the-art methods in query time while being faster to build.

Keywords: Nearest neighbor search · Approximate nearest neighbors · Randomized space-partitioning trees · Indexing methods · Autotuning

1 Introduction

Nearest neighbor search is a common component of algorithms and pipelines in areas such as machine learning [5,17], computer vision [1] and robotics [11]. In modern applications the search is typically performed in high-dimensional spaces (100–10000 dimensions) over large data sets.

An exhaustive k-nearest neighbor (k-NN) search is often prohibitively slow in applications which either require real-time responses (see e.g. [17]) or run on a resource-constrained device (see e.g. [11]). Hence, *approximate* nearest neighbor (ANN) search is often used instead. ANN algorithms first build an index in an offline phase, after which the index can be used to perform k-NN queries in sublinear time in an online phase. Most of the efficient algorithms fall into one of four categories: product quantization (PQ) [8], locality-sensitive hashing (LSH) [4,7], graph-based methods [10], and tree-based methods [13,14].

Q. Yang et al. (Eds.): PAKDD 2019, LNAI 11440, pp. 590–602, 2019.
https://doi.org/10.1007/978-3-030-16145-3_46

Because ANN algorithms are typically used as an auxiliary component of a pipeline, it can be important for a user that an algorithm requires minimal hand-tuning, especially if the type or size of the data can vary significantly. However, ANN algorithms typically have several hyperparameters which need to be tuned by a time-consuming grid search to achieve a given accuracy level or search time.

This problem is solved by an autotuning algorithm where the user specifies an accuracy level, and the tuning algorithm finds the optimal hyperparameter values. Previously, autotuning methods have been proposed for VP-trees [18], multi-probe LSH [4], k-means trees and RKD trees [13]. In this paper, we propose an autotuning method that is significantly faster than these methods.

Our approach is based on exploiting the structure of randomized space-partitioning trees [3,6,13,14]. ANN algorithms based on randomized space-partitioning trees have been used recently for example in machine translation [5], object detection [1] and recommendation engines [17].

Trees have several advantages: they are fast in high-dimensional spaces (see e.g. experiments in [6,13]); they are simple to implement; they support easy insertion and deletion of points and they are independent, making the parallel implementation trivial. Also of great importance to us is that the structure of a tree-based index can be exploited to speed up the hyperparameter tuning.

Several types of randomized space-partitioning trees have been proposed for ANN search. Randomized k-d (RKD) trees [14] with a priority queue search are used in the popular open-source library FLANN [13]. Random projection (RP) trees [3] with a voting search have a stronger empirical performance than RKD trees with a priority queue search [6]. However, a single principal component (PCA) tree has been found to be more accurate than a single RP tree [16]. The PCA tree has two problems: it is not randomized, and indexing is slow. To solve these problems, we design a randomized variant of the PCA tree.

Typically ANN algorithms are compared in terms of the accuracy–speed trade-off. However, for the algorithm to be useful in practice, the index building procedure must be efficient as well. We test three different types of trees (RKD, RP and randomized PCA) with two search methods (priority queue and voting) considering both the query stage and the index building stage.

More specifically, in this article we:

- Propose an autotuning algorithm to optimize the hyperparameters of tree-based ANN search, and demonstrate that it is faster and more accurate than existing autotuning methods for ANN algorithms.
- Compare experimentally the effect of (a) the randomization strategy and (b) the search method on the efficiency of randomized trees. In particular, we find RP trees combined with voting search to be the best-performing.
- Demonstrate that the best tree-based method is nearly on par with the state-of-the-art ANN algorithms when measured on the accuracy–speed trade-off, and faster when measured on the index building time.

2 Approximate Nearest Neighbor Search

In *k-nn search*, we have a data set $\mathbf{x} = (x_1, \ldots, x_n)$, where each $x_i \in \mathcal{A}$, from which we want to find the indices $f(q)$ of the k nearest neighbors for an arbitrary query point $q \in A$ measured by a dissimilarity measure $\text{dis}(u, v) : \mathcal{A}^2 \mapsto \mathbb{R}$. We assume the dissimilarity measure to be the Euclidean distance $\|u - v\|_2$.

In *approximate nearest neighbor* (ANN) search, it is sufficient that the k points returned by the approximation algorithm are the true nearest neighbors of the query point only with high probability. We denote the returned points by $\hat{f}(q; \boldsymbol{\alpha}, \mathbf{r})$, where $\boldsymbol{\alpha}$ stands for the hyperparameters of the algorithm, and $\mathbf{r}$ stands for the realization of a set of random vectors used by the algorithm.

The accuracy of the approximation is measured by the *error rate* $\text{Err}(q; \boldsymbol{\alpha}, \mathbf{r}) = \frac{1}{k} \sum_{j=1}^{k} \mathbb{1}(f_j(q) \notin \hat{f}(q; \boldsymbol{\alpha}, \mathbf{r}))$, which is the proportion of missed true nearest neighbors; the indices of the true nearest neighbors are denoted by $f(q) = (f_1(q), \ldots, f_k(q))$. Equivalently, we can use *recall*: $\text{Rec}(q; \boldsymbol{\alpha}, \mathbf{r}) = 1 - \text{Err}(q; \boldsymbol{\alpha}, \mathbf{r})$.

In addition to the error rate, we also consider the query time, denoted $\text{Time}(q; \boldsymbol{\alpha}, \mathbf{r})$, when assessing the performance of an ANN algorithm. The hyperparameter optimization problem can be formulated in two ways:

1. Fix the expected error rate $e \in (0, 1)$ and find the hyperparameters $\boldsymbol{\alpha}$ that minimize $E[\text{Time}(Q; \boldsymbol{\alpha}, \mathbf{R})]$ under the constraint $E[\text{Err}(Q; \boldsymbol{\alpha}, \mathbf{R})] \leq e$.
2. Fix the expected query time $t \in (0, \infty)$ and find the hyperparameters $\boldsymbol{\alpha}$ that minimize $E[\text{Err}(Q; \boldsymbol{\alpha}, \mathbf{R})]$ under the constraint $E[\text{Time}(Q; \boldsymbol{\alpha}, \mathbf{R})] \leq t$.

The expectations $E[\cdot]$ are over both the distribution of a query point Q and the random vectors $\mathbf{R}$. These expectations can be estimated using a validation set of query points and a generated sample of random vectors.

3 Randomized Space-Partitioning Trees

3.1 Index Construction

A binary space-partitioning tree recursively divides the data points into different cells with the assumption that nearby points fall into the same cells. At each branch of the recursion, the data set $\mathbf{x}$ is projected onto a chosen direction and assigned into one of the two child nodes by applying a split criterion. In practice we use the median split to ensure balanced trees. This process (Algorithm 1) is continued at the child nodes until the maximum depth ℓ is met.

The type of a space-partitioning tree is determined by its choice of projection direction (see Fig. 1 for an illustration on 2D data). In Algorithm 1, each different type of tree implements its own version of the abstract function GENERATE-DIRECTION which chooses this direction. Its argument ψ represents the tree-type dependent tuning parameters.

In randomized space-partitioning trees, the projection direction is chosen in a non-deterministic fashion. Randomized k-d (RKD) trees [14] choose a coordinate

Algorithm 1

1: **function** GROW-TREE(depth, $\mathbf{x}, \ell, \psi$)
2: **if** depth == ℓ **then**
3: **return** indices of points in $\mathbf{x}$ as a tree node
4: direction ← GENERATE-DIRECTION(ψ)
5: p ← PROJECT($\mathbf{x}$, direction)
6: cut ← SPLIT(p)
7: left ← GROW-TREE(depth + 1, $\mathbf{x}$[p ≤ cut], ℓ)
8: right ← GROW-TREE(depth + 1, $\mathbf{x}$[p > cut], ℓ)
9: **return** (left, right, cut, direction) as a tree node

direction uniformly at random from m directions of the highest variance as the projection direction (we use $m = 5$ as suggested in [14]). Another popular randomized variant is a random projection (RP) tree [3] in which the projection direction is chosen uniformly at random from the d-dimensional unit sphere. We use a sparse version [6], in which only a proportion $a = 1/\sqrt{d}$ of the components of the random vectors are non-zero.

If the first principal component of the data is used as the projection direction, the resulting data structure is a principal component (PCA) tree [16]. However, the original PCA trees are on the one hand deterministic which makes improving accuracy with multiple trees impossible, and on the other hand slow to compute, as computing exact PCA is costly. To speed up the computation, using gradient descent updates to approximate the first principal component of the data at each node of the tree has been suggested [12]. However, index construction still takes $\mathcal{O}\left(nd^2(i+\ell)\right)$ time, where i is the number of gradient descent updates.

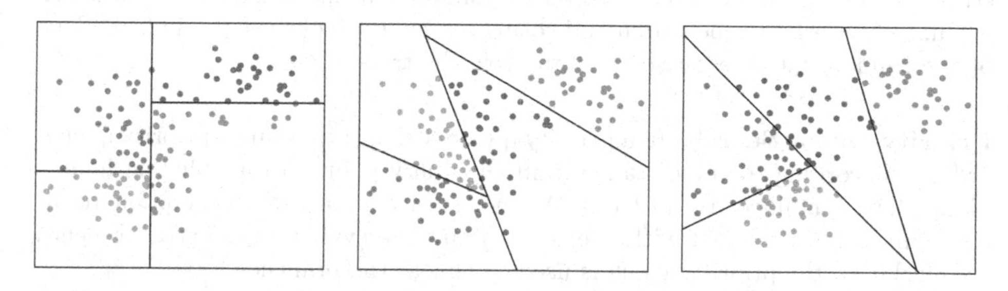

Fig. 1. Different projection directions: k-d (left), RP (middle) and PCA (right).

We make PCA trees more practical for ANN search by modifying the gradient descent update[1] to choose uniformly at random only $a = \sqrt{d}$ dimensions of the data at each node of the tree, and compute the estimated covariance matrix using only these dimensions. Growing a randomized PCA tree is an $\mathcal{O}\left(nd(i+\ell)\right)$

[1] The gradient descent consistently converges with the learning rate $\gamma = 0.01$ in all our experiments; we did not observe further tuning of the learning rate to be necessary.

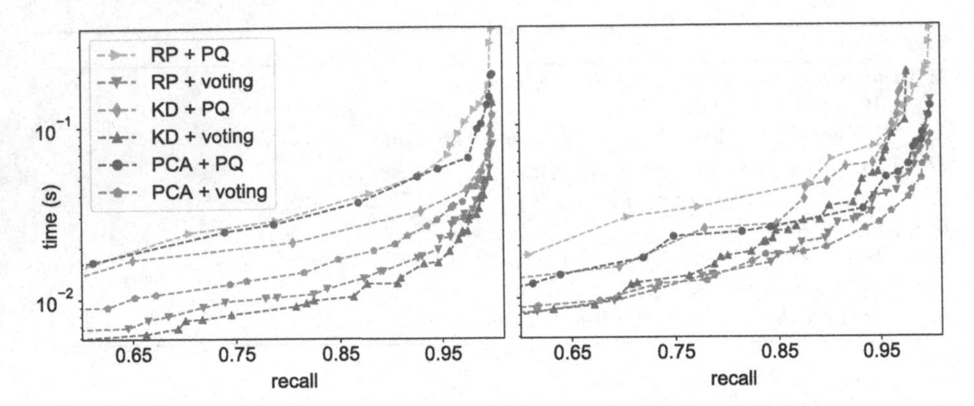

Fig. 2. Recall vs. query time with different trees and search methods for MNIST (left) and Fashion-MNIST (right) for $k = 10$. Towards bottom right is better.

operation since now computing the sample covariance matrix takes only $\mathcal{O}(nd)$ operations. Considering only a sample of dimensions also ensures that the trees are randomized, allowing us to build multiple trees to increase accuracy.

3.2 ANN Search Using Multiple Trees

To use an index consisting of T randomized space-partitioning trees to find k approximate nearest neighbors of a query point q, the query point is first routed down to a leaf at each of the trees: at each level the query point is first projected onto the saved projection direction and then routed into the left or the right child node depending on which side of the split point its projection falls. There are two strategies to choose the candidate set of points for which the true distances are evaluated: priority queue search and voting search. Both of these are independent of the randomization strategy used to grow the trees.

Priority Queue Search. In a priority queue search [14], a single priority queue, ordered according to the distance from the query point to the splitting hyperplanes, is maintained for all trees. When distances from the query point to all the points sharing a leaf with the query point are evaluated, b extra branches are explored; the priority queue is used to choose the branches.

Voting Search. In a voting search [6], distances are computed only to the subset of the points sharing a leaf with the query point. When a data point belongs to the same leaf as a query point in a tree, it gets a vote, and distances are evaluated only to the points that have at least v votes.

3.3 Comparison of Randomization and Search Methods

Figure 2 shows the accuracy–speed trade-off for all combinations of the considered tree types and search methods on two benchmark data sets. For RP trees,

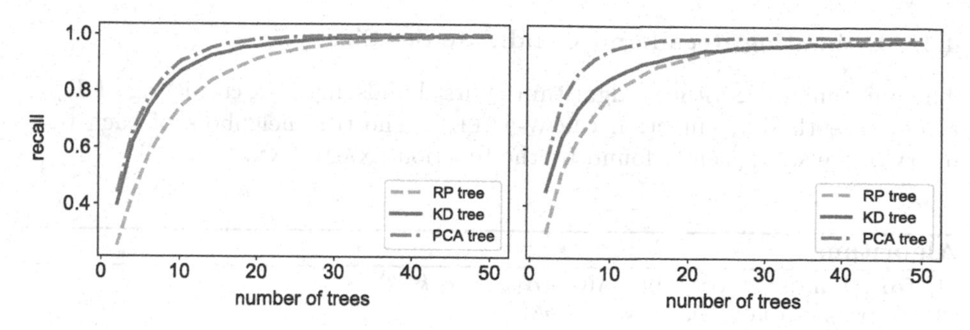

Fig. 3. Recall (for $k = 10$) as a function of the number of trees on MNIST (left) and Fashion-MNIST (right) for RP, RKD and randomized PCA trees. $\ell = 8, v = 1$.

the results are in line with previous experiments [6]. For each type of tree, voting outperforms priority queue (for a given recall level, its query time is faster).

For different tree types, the results vary between the data sets. Note that although both data sets for which results are shown have the same sample size ($n = 60000$) and dimensionality ($d = 784$), the relative order of the trees is different: for MNIST, RKD trees are the fastest, and randomized PCA trees are the slowest; whereas for Fashion-MNIST, randomized PCA trees are the fastest, and RKD trees are the slowest. This means that the relative performance of different randomization strategies depends also on the distribution of the data.

Figure 3 further illustrates the differences between the tree types with fixed parameters: on Fashion-MNIST, PCA trees are noticeably more accurate than RKD trees and RP trees, especially for a small amount of trees. This explains the stronger performance of PCA trees with the optimal parameters; observe that the slightly stronger performance of RKD trees on MNIST is due to their faster projection times (1 vs. $\sqrt{d}$ operations per projection).

However, the differences between tree types are less pronounced than the difference between search methods. Since we can use the same projection vector on each node at the same level of an RP tree, they are the fastest to build (see Table 2). Thus, we present some of the experimental results only for them.

4 An Autotuning Algorithm

Since voting outperforms using a priority queue for all the data sets and the tree types, we present an autotuning algorithm for the voting search. Any of the different tree types can be used. Hence, the tuned hyperparameters $\boldsymbol{\alpha}$ are the number of trees T, the depth of the trees ℓ, and the vote threshold v.

The optimal hyperparameter values are searched from the whole range $\boldsymbol{\alpha}_{\text{lim}} = (1, \ldots, T_{\max}) \times (\ell_1, \ldots, \ell_{\max}) \times (1, \ldots, v_{\max})$; setting a grid interval is not required. Here we use $v_{\max} = T_{\max}$, $\ell_{\max} = \lfloor \log_2 n \rfloor$. Since each individual tree consumes the same amount of memory and takes an equal time to grow, $T_{\max}$ can be chosen as a limit on the building time or the memory consumption.

4.1 Estimating Recall and Candidate Set Size

The autotuning algorithm (Algorithm 2) first builds an index consisting of $T_{\max}$ trees of depth $\ell_{\max}$ (function GROW-TREES). The true neighbors of each test query q_i are subsequently found by the function EXACT-KNN.

Algorithm 2

1: **function** GENERATE-INDEX-AUTO($\boldsymbol{\alpha}_{\text{lim}}, \mathbf{x}, \mathbf{q}, k, \boldsymbol{\psi}$)
2: trees ← GROW-TREES($\mathbf{x}, \boldsymbol{\alpha}_{\text{lim}}, \boldsymbol{\psi}$)
3: **for** $i = 1, \ldots, m$ **do**
4: true-knn ← EXACT-KNN($q_i, k, \mathbf{x}$)
5: A_i ← COUNT-ELECTED($\boldsymbol{\alpha}_{\text{lim}}$, q_i, true-knn)
6: B_i ← COUNT-ELECTED($\boldsymbol{\alpha}_{\text{lim}}$, q_i, $\{1, \ldots, n\}$)
7: recalls ← $\frac{1}{km} \sum_{i=1}^{m} A_i$
8: query-times ← FIT-TIMES($\frac{1}{m} \sum_{i=1}^{m} B_i$, $\mathbf{x}$.dim)
9: **return** recalls, query-times, trees

For each test query, the elected points are counted by COUNT-ELECTED (Algorithm 3) for two sets: the whole data set and the set of true k nearest neighbors. When using an index consisting of the first T trees, all the points that were elected when using an index consisting of the first $T - 1$ trees are also elected for the fixed vote threshold v. This means that we only have to count the points which get their v:th vote at the T:th tree (line 7 of Algorithm 3). Hence, we can count the numbers of elected points for all $1, \ldots, T_{\max}$ number of trees with minimal overhead compared to counting them only for $T_{\max}$ trees.

Algorithm 3

1: **function** COUNT-ELECTED($\boldsymbol{\alpha}_{\text{lim}}$, q, I)
2: initialize three-dimensional tensor A
3: **for** $\ell = \ell_1, \ldots, \ell_{\max}$ **do**
4: initialize votes as zero vector of length n
5: initialize c as zero vector of length $v_{\max}$
6: **for** $T = 1, \ldots, T_{\max}$ **do**
7: $c \leftarrow c +$ COUNT-VOTES(T, ℓ, q, I, votes)
8: write c to A
9: **return** A
10: **function** COUNT-VOTES(T, ℓ, q, I, votes)
11: initialize counts as zero vector of length $v_{\max}$
12: leaf ← node containing q at level ℓ of the T:th tree
13: **for** point in leaf **do**
14: **if** point $\in I$ **then**
15: votes[point] ← votes[point] + 1
16: counts[votes[point]] ← counts[votes[point]] + 1
17: **return** counts

The counting is done by the function COUNT-VOTES (Algorithm 3) which adds a vote for each point of the node, and for each $v = 1, \dots v_{\max}$, counts how many points of this node get their v:th vote.

Finally, the expected recall and candidate set size can be estimated by their sample means for each parameter combination (lines 7 and 8 in Algorithm 2). Since a brute force strategy of performing actual test queries and timing them for each possible hyperparameter combination in the set $\boldsymbol{\alpha}_{\text{lim}}$ is impractically slow, the function FIT-TIMES estimates the expected query time as a function of the candidate set size and data dimension as described in the following section.

4.2 Estimating the Query Time

We exploit linear scaling of the components of a query to build a model which estimates the query time. The query time can be split into the candidate pruning time and the final search time. Further, the candidate pruning phase is dominated by two operations: projecting the points onto the split directions, and vote counting. This suggests that we can estimate each of the three times separately:

$$\text{Time}(q; \boldsymbol{\alpha}, \mathbf{r}) \approx \text{Time}_{\text{proj}}(q; \boldsymbol{\alpha}, \mathbf{r}) + \text{Time}_{\text{vote}}(q; \boldsymbol{\alpha}, \mathbf{r}) + \text{Time}_{\text{dist}}(q; \boldsymbol{\alpha}, \mathbf{r}).$$

Projection Time. The projection time depends on the type of randomization used in the trees. In RKD trees, the projection time is insignificant because coordinate axes are used as split directions. For RP trees and randomized PCA trees, the query point is projected onto a sparse vector at each level of each tree. Hence, the projection time is approximately linear w.r.t. the number of random vectors $z := T\ell$ the query point is projected onto. Thus, we can use a linear model to estimate the projection time for known hyperparameters T and ℓ.

To collect the data for the model, we design an experiment by choosing a representative sample $\mathbf{z} = (z_1, \dots, z_w)$ of sparse random matrices with d columns and $z_1, \dots, z_w$ total components, and measuring the elapsed times to multiply a d-component vector by each of these matrices. The sparsity is fixed as $a = 1/\sqrt{d}$.

When measuring running times, we observed that the random variation is typically small, but sometimes outliers appear, for example due to other processes activating on the background. This is why we use the Theil-Sen estimator [15] to model the dependence between the number of random vectors and projection time. It is a non-parametric estimator for a linear trend, and is much more robust against outliers than ordinary least squares regression.

Now the expected projection time for the hyperparameter values $\boldsymbol{\alpha} = (T, \ell, v)$ can be estimated as $\widehat{\text{Time}}_{\text{proj}}(\mathbf{q}; \boldsymbol{\alpha}, \mathbf{r}) = \hat{\beta}_0 + z\hat{\beta}_1$, where $z = T\ell$, and $\hat{\beta}_0$ and $\hat{\beta}_1$ are the intercept and the slope estimated by the Theil-Sen method.

Voting Time. For one tree, counting the votes means adding a vote for each point of the leaf the query point falls into. For T trees, this means that the whole voting step takes roughly Tn_0 operations, where $n_0 = \lceil n/2^\ell \rceil$ is the maximum leaf size. This means that we can model the voting time as a linear function of $y := Tn_0$, and proceed as in estimating the projection times.

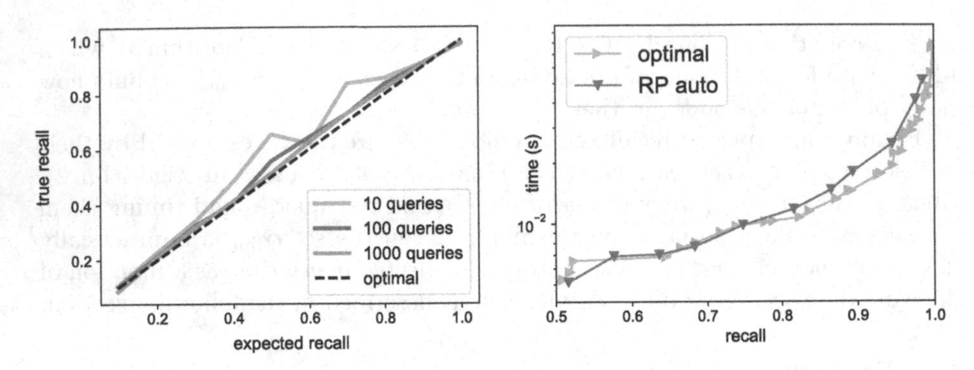

Fig. 4. Left: Recall estimated by autotuning vs. recall on the test set. Right: Recall vs. query time on test set for optimal parameters and auto-tuned parameters. $k = 10$.

Final Search Time. The final search in the candidate set is dominated by computing the distances to all $|S|$ points of the candidate set, which takes $|S|d$ operations; hence it is approximately linear with respect to the candidate set size $|S|$. Thus, we can proceed as before, this time measuring the time it takes to compute the distances from any d-dimensional query point to $|S|$ vectors of dimension d. After fitting the model, the final search time can be estimated as

$$\widehat{\text{Time}}_{\text{dist}}(\mathbf{q}; \boldsymbol{\alpha}, \mathbf{r}) = \hat{\alpha}_0 + |\bar{S}(\mathbf{q}; \boldsymbol{\alpha}, \mathbf{r})|\hat{\alpha}_1,$$

where $\hat{\alpha}_0$ and $\hat{\alpha}_1$ are the coefficients of the Theil-Sen estimator, and $|\bar{S}(\mathbf{q}; \boldsymbol{\alpha}, \mathbf{r})|$ is the observed mean candidate set size for this hyperparameter combination.

4.3 Using the Autotuning Index

After the expected recall levels and the query times have been computed, finding the optimal parameter combination is a matter of a simple table lookup. Since the index has already been built, growing new trees is not required: if the optimal parameter combination is $\hat{\boldsymbol{\alpha}} = (\hat{T}, \hat{\ell}, \hat{v})$, we can just pick the first $\hat{T}$ trees that have already been built, and prune them to depth $\hat{\ell}$.

5 Experimental Results

First, we verify using RP trees that the autotuning algorithm accurately estimates the recall. Figure 4(a) shows estimated recall on a validation set against recall on an independent test set for the MNIST data set. Larger validation sets yield sharper estimates, indicating the consistency of the estimator. The results are similar for other data sets and tree types. Figure 4(b) compares on an independent test set hyperparameters optimized by the autotuning algorithm (RP auto) for the validation set to hyperparameters optimized for the test set (optimal). The parameters found by the algorithm are near-optimal.

Table 1. Comparison of autotuning algorithms. Autotuning times (seconds), query times for 1000 queries (s) and recall (for $k = 10$) measured on a test set (* = did not complete within one hour). For the randomized algorithms (RP and FLANN), average recalls of 10 runs with the corresponding standard deviations are reported. The best result in each case is typeset in boldface.

		Target recall 80%				Target recall 90%			
		RP	LSH	VPtree	FLANN	RP	LSH	VPtree	FLANN
MNIST	tuning	**13.23**	26.84	744.4	102.2	**13.23**	24.61	926.1	113.9
	search	**0.111**	1.164	0.739	0.206	**0.169**	2.513	1.368	0.311
	recall	**0.822**	0.853	0.831	0.654	**0.909**	0.939	0.911	0.790
	stdev	±0.009	–	–	±0.020	±0.004	–	–	±0.017
Fashion	tuning	**13.22**	26.70	396.5	104.8	**13.23**	25.38	427.4	136.4
	search	**0.129**	0.917	0.353	0.310	**0.198**	1.575	0.557	0.216
	recall	**0.798**	0.850	0.813	0.693	0.881	0.927	**0.908**	0.825
	stdev	±0.007	–	–	±0.034	±0.006	–	–	±0.025
Trevi	tuning	**75.89**	156.1	3026	724.9	**76.28**	158.9	*	751.6
	search	**1.730**	14.01	13.58	2.813	**3.371**	25.63	*	4.276
	recall	**0.822**	0.837	0.832	0.566	**0.914**	0.918	*	0.679
	stdev	±0.011	–	–	±0.028	±0.006	–	*	±0.016
Random	tuning	**32.78**	55.48	120.6	134.6	**32.76**	54.56	134.0	149.1
	search	0.074	0.256	0.409	**0.049**	0.095	0.249	0.659	**0.087**
	recall	**0.804**	0.882	0.827	0.602	**0.902**	0.941	0.911	0.728
	stdev	±0.012	–	–	±0.015	±0.007	–	–	±0.015
GIST	tuning	**317.4**	484.1	960.4	*	**318.1**	437.9	1127	*
	search	**9.253**	122.4	41.55	*	**15.51**	205.7	66.54	*
	recall	**0.784**	0.862	0.864	*	**0.881**	0.942	0.940	*
	stdev	±0.011	–	–	*	±0.005	–	–	*

Next, we compare the performance of the presented algorithm with RP trees to other autotuning algorithms for ANN: autotuning for VP-trees [18] and multi-probe LSH [4] implemented in NMSLib [2] and autotuning for RKD trees and hierarchical k-means trees in FLANN [13]. To the best of our knowledge, these are the only available ANN libraries that feature an autotuning method. The compared libraries and our own code are all written in C++.

The data sets used in the experiments are MNIST ($n = 60000$, $d = 784$), Fashion-MNIST ($n = 60000$, $d = 784$), Trevi ($n = 101120$, $d = 4096$), Random ($n = 256000$, $d = 256$) and GIST ($n = 1000000$, $d = 960$). Table 1 shows for two target recall rates (0.8 and 0.9) the autotuning time (including the index-building time), and the query time and recall on a test set which was not used to tune the hyperparameters. The proposed tuning algorithm (with RP trees) is fastest at index building in all cases. Our approach has significantly faster query

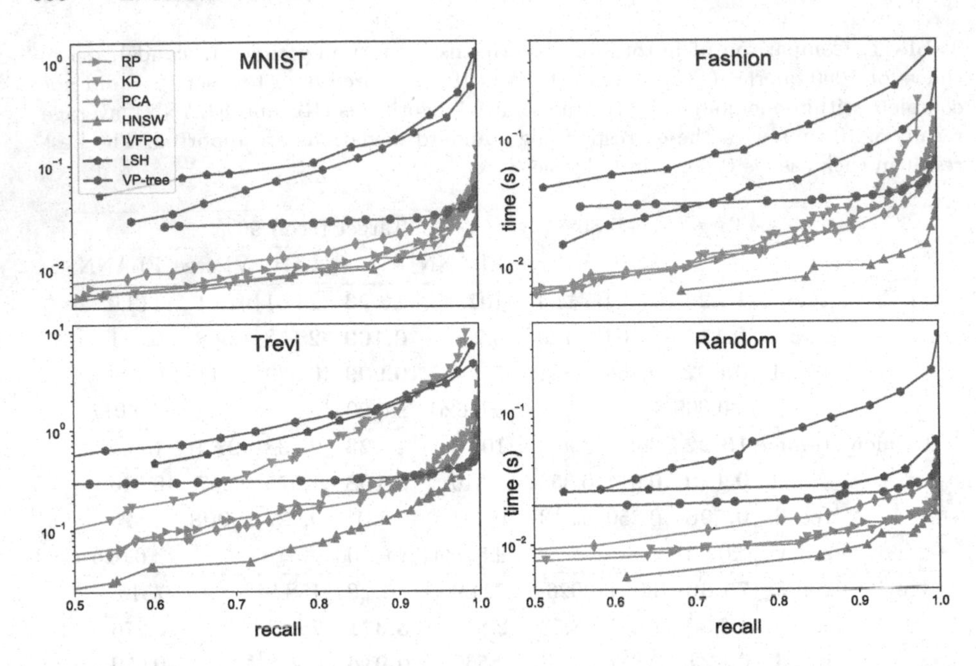

Fig. 5. Recall vs. query time (s) for 100 queries for different ANN algorithms. $k = 10$.

times than VP trees and LSH in all cases, and faster query times than FLANN for all but one data set. Our approach is also the most accurate at estimating the recall in most cases. The other methods systematically over- or underestimate the recall.

We also compare tree-based ANN against state-of-the-art quantization-, and graph-based algorithms: IVFPQ [8] and HNSW [10] implemented in FAISS [9]. As autotuning for these methods is not available, we perform a grid search on the possible parameter values. Figure 5 shows that tree-based methods are faster than PQ (except on highest recall levels) and close to the performance of HNSW. We emphasize that according to an independent benchmarking project[2], HNSW is the fastest ANN algorithm available. Multi-probe LSH and VP tree are also included in the comparison; they are significantly slower than the other methods.

Finally, we compare the index building time (Table 2). Even though HNSW has faster query times, RP trees are significantly faster to build. The whole autotuning takes less time than building a single HNSW index. We emphasize that these results are on a single thread; the differences become more pronounced with multiple threads as the indexing process is embarrassingly parallel for trees. An implementation of the proposed algorithm is available in the MRPT library[3].

[2] https://github.com/erikbern/ann-benchmarks.

[3] https://github.com/vioshyvo/mrpt.

Table 2. Index building times (seconds) for optimal parameters at 90% recall for different ANN algorithms. The best result on each data set is typeset in boldface.

	RP	RKD	PCA	HNSW	IVFPQ
MNIST	**3.62**	7.40	12.63	25.1	27.31
Fashion	**1.86**	14.3	13.12	20.2	30.71
Trevi	102	**43.2**	185	266	262.5
Random	**2.23**	6.83	27.8	90.3	63.59

Acknowledgments. This project was supported by Business Finland (project 3662/31/2018 Advanced Machine Learning for Industrial Applications) and the Academy of Finland (project 311277 TensorML).

References

1. Bilen, H., Pedersoli, M., Tuytelaars, T.: Weakly supervised object detection with convex clustering. In: CVPR, pp. 1081–1089. IEEE (2015)
2. Boytsov, L., Naidan, B.: Engineering efficient and effective non-metric space library. In: Brisaboa, N., Pedreira, O., Zezula, P. (eds.) SISAP 2013. LNCS, vol. 8199, pp. 280–293. Springer, Heidelberg (2013). https://doi.org/10.1007/978-3-642-41062-8_28
3. Dasgupta, S., Sinha, K.: Randomized partition trees for nearest neighbor search. Algorithmica **72**(1), 237–263 (2015)
4. Dong, W., Wang, Z., Josephson, W., Charikar, M., Li, K.: Modeling LSH for performance tuning. In: CIKM, pp. 669–678. ACM (2008)
5. Hassan, H., Elaraby, M., Tawfik, A.Y.: Synthetic data for neural machine translation of spoken-dialects. Small **16**, 17–33 (2017)
6. Hyvönen, V., et al.: Fast nearest neighbor search through sparse random projections and voting. In: IEEE International Conference on Big Data, pp. 881–888 (2016)
7. Indyk, P., Motwani, R.: Approximate nearest neighbors: towards removing the curse of dimensionality. In: STOC, pp. 604–613. ACM (1998)
8. Jégou, H., Douze, M., Schmid, C.: Product quantization for nearest neighbor search. TPAMI **33**(1), 117–128 (2011)
9. Johnson, J., Douze, M., Jégou, H.: Billion-scale similarity search with GPUs. arXiv preprint arXiv:1702.08734 (2017)
10. Malkov, Y.A., Yashunin, D.A.: Efficient and robust approximate nearest neighbor search using hierarchical navigable small world graphs. arXiv:1603.09320 (2016)
11. McBryde, C.R.: Spacecraft visual navigation using appearance matching and multi-spectral sensor fusion. Ph.D. thesis, Georgia Institute of Technology (2018)
12. McCartin-Lim, M., McGregor, A., Wang, R.: Approximate principal direction trees. In: ICML, pp. 1611–1618 (2012)
13. Muja, M., Lowe, D.G.: Scalable nearest neighbor algorithms for high dimensional data. TPAMI **36**(11), 2227–2240 (2014)
14. Silpa-Anan, C., Hartley, R.: Optimised KD-trees for fast image descriptor matching. In: CVPR, pp. 1–8. IEEE (2008)

15. Theil, H.: A rank-invariant method of linear and polynomial regression analysis. In: Henri Theil's Contributions to Economics and Econometrics, pp. 345–381 (1992)
16. Verma, N., Kpotufe, S., Dasgupta, S.: Which spatial partition trees are adaptive to intrinsic dimension? In: UAI, pp. 565–574. AUAI Press (2009)
17. Wang, L., Tasoulis, S., Roos, T., Kangasharju, J.: Kvasir: scalable provision of semantically relevant web content on big data framework. IEEE Trans. Big Data **2**(3), 219–233 (2016)
18. Yianilos, P.N.: Data structures and algorithms for nearest neighbor search in general metric spaces. In: SODA, vol. 93, pp. 311–321 (1993)

An Accelerator of Feature Selection Applying a General Fuzzy Rough Model

Peng Ni[1,2], Suyun Zhao[1,2(✉)], Hong Chen[1,2], and Cuiping Li[1,2]

[1] Key Lab of Data Engineering and Knowledge Engineering of MOE, Renmin University of China, Beijing, People's Republic of China
nipeng@ruc.edu.cn, zhao.suyun@yahoo.com
[2] Information of School, Renmin University of China, Beijing, People's Republic of China

Abstract. Feature selection, also known as variable selection or attribute reduction, is to select a subset relevant features to speedup learning/mining and to improve the learning/mining quality. In the big data era, some feature selection methods have to face the running time problem led by the large-scale data. As a result, in this paper, we try to narrow this gap by proposing a feature selection accelerator. Considering fuzzy rough techniques need no extra expert knowledge, we design the feature selection accelerator based on fuzzy rough reduction techniques. First, we proposed a fuzzy rough accelerator by deleting the learned/discernible instances in the process of feature selection, which decreases the computation and accelerates feature selection. Second, we design a fuzzy rough based feature selection accelerated algorithm. Finally, the numerical experiments demonstrate that the proposed accelerated algorithm could obtain the same reduction results and save much more time, especially on the large-scale datasets.

Keywords: Feature selection · Fuzzy rough techniques · Accelerator · Fuzzy positive region

1 Introduction

Feature selection, also known as variable selection or attribute reduction, is a significant problem in knowledge discovery and data mining. By using an evaluation measure to score every feature and/or different feature subsets, Feature selection selects a subset of relevant features to speedup learning/mining and improve learning/mining quality. The choice of evaluation measure heavily influences the feature selection algorithm, and these evaluation measures which distinguish among the three main categories of feature selection methods: wrappers, filters and embedded algorithms [1]. In this paper, we focus on the filter method, which select features regardless of the learning/mining model and often works as pre-process method.

Till now, there are many known filter feature selection methods. For example, RELIEF [2] is a typical algorithm on the filter method, it could effectively estimate the quality of features in problems with dependencies between features. For another example, mRMR (maximum relevance minimum redundancy) [3] is a mutual

Q. Yang et al. (Eds.): PAKDD 2019, LNAI 11440, pp. 603–614, 2019.
https://doi.org/10.1007/978-3-030-16145-3_47

information based filter feature selection method for finding a set of both relevant and complementary features. What is more, rough/fuzzy rough reduction is also a useful feature selection method [7, 8, 13, 15, 16, 18–25] since rough/fuzzy rough philosophy is human understanding and non-need extra expert knowledge.

In the big data era, the amount of data is large and the dimensionality of data is high, fuzzy rough based feature selection algorithms are rather time consuming. In view of rough set theory [4, 5], some researchers have proposed some heuristic feature selection algorithms [16, 18, 23]. Due to rough set theory can only handle data with categorical values, to deal with continuous values, fuzzy rough set, which combined rough set and fuzzy set, was proposed by Dubois and Prade [6]. Currently, Feature selection algorithms based on fuzzy rough sets mainly include fuzzy positive region based reduction [14, 16], discernibility matrix based reduction [13, 25], fuzzy information entropy based reduction [8, 15]. Whereas it is necessary to point out that all of these algorithms are less effective because of their time and space consumption is too large on large scale datasets.

To reduce intolerable time costs, Qian et al. [9] proposed an accelerator called positive approximation for feature reduction in rough set theory. However, in real applications, data with numerical and symbolic values are ubiquitous. Hence, Qian et al. [10] further developed an extended version of the accelerator called forward approximation for accelerating fuzzy rough feature reduction. Both of them effectively accelerate the algorithm. Whereas it is necessary to point out that forward approximation is based on a kind of cut set of fuzzy relation [8]. The introduction of cut set makes many information hidden in features are omitted and then interferes the reduction results. This motivates us to propose a generalized fuzzy rough based feature selection accelerator which is based on the real fuzzy rough set, not the cut one.

In this paper, we choose Dubois and Prade's fuzzy rough model, a general model without a cut set, to compute the fuzzy positive region, and then we design our accelerated algorithm based on fuzzy positive region reduction. The main contributions in this paper include:

(1) We find that some instances, whose fuzzy positive regions reach maximum, are not useful and meaningful any more for finding the following features. This finding makes our accelerator design possible.
(2) Based on our proposed accelerator, the instances reaching maximum are deleted and then the size of the instances involved in the calculation decreases gradually in the process of feature selection. As a result, feature selection algorithm is accelerated by avoiding redundancy computation on the whole data.
(3) The numerical experiments show that the accelerated algorithm performs more efficiency than the original non-accelerated counterpart.

The remainder of this paper is organized as follows. In Sect. 2, some notations of fuzzy rough set are briefly reviewed. In Sect. 3, we propose a fuzzy rough based feature selection accelerator. In Sect. 4, numerical experiments on eight datasets are given to show our proposed accelerated algorithm outperforms the original one. Finally, we conclude this paper in Sect. 5.

2 Preliminaries

In this section, we briefly review Dubois and Prade's fuzzy rough model and some reduction concepts, such as fuzzy positive region and reducts [12–16].

2.1 Fuzzy Rough Sets

In rough set philosophy, the dataset could be described as one decision table, denoted by $DT = (U, R \cup D)$. Let $U = \{x_1, x_2, \ldots, x_n\}$, called the Universe, be a nonempty set with a finite number of instances. Each instance in U is described by a non-empty finite set of features, denoted by $R \cup D$; R denotes the set of condition features and D denotes the set of decision features, $R \cap D = \emptyset$; When the attribute are crisp, the Universe is split into q equivalence classes $U/D = \{X_1, X_2, \ldots, X_q\}$, where $U = \bigcup_{i=1}^{q} X_i$ and $X_i \cap X_j = \emptyset$ (for any $i \neq j$).

In most practical applications, only the decision features are crisp, the condition features are usually continuous. To handle this type of applications, fuzzy sets [11] is then introduced into rough sets and then fuzzy rough sets are proposed, which is one known generalization of rough sets supporting both continuous and crisp values [12–14].

If the condition attributes are continuous, the decision table is then called a Fuzzy Decision Table, shortly denoted by $FD = (U, R \cup D)$, since each continuous attribute could be transformed into a fuzzy attribute. For example, each attribute value of x_i on the continuous attribute r, denoted by $r(x_i)$, could be normalized into $[0, 1]$. In this paper, we mainly focus on this type of decision tables.

In fuzzy decision table, each continuous attribute subset $P \subseteq R$ corresponds a fuzzy similarity relation. When these is no confusion arise, the fuzzy similarity relation on attribute subset P is denoted by $P(\cdot, \cdot)$. If the distance of x_i and x_j on the attribute subset P is calculated as $d_P(x_i, x_j) = \max_{r \in P}\{|r(x_i) - r(x_j)|\}$, the fuzzy similarity of x_i and x_j on the attribute subset P could be calculated as $P(x_i, x_j) = 1 - d_P(x_i, x_j)$.

Definition 1. Given a fuzzy decision table $FD = (U, R \cup D)$ and $\forall A \subseteq U$, A fuzzy rough set is an order pair $(\underline{R}A, \bar{R}A)$ of A on U such that for every $x \in U$,

$$\underline{R}A(x) = inf_{u \in U} \max\{1 - R(x, u), A(u)\}; \bar{R}A(x) = sup_{u \in U} \min\{R(x, u), A(u)\}$$

This concept is the definition of lower and upper approximations. For the more details the readers are kindly referred to [12–15]. To more illustratively demonstrate this concept, the Proposition 1 is presented as follows.

Proposition 1. Given a fuzzy decision table $FD = (U, R \cup D)$ and $\forall A \subseteq U$. The lower approximation of the instance x to A could be simplified as

$$\underline{R}A(x) = \begin{cases} \min_{\{u \in U, u \notin A\}}\{1 - R(x, u)\}, & x \in A \\ 0, & x \notin A \end{cases}.$$

Proposition 1 gives the topological meaning of the lower approximations. That is, the lower approximation of $x \in A$ is its minimal distance to the instance which does not belong to A, i.e., $u \notin A$. Thus, to find the minimal distance for every x, all the instances in the Universe need be calculated. That is why fuzzy rough sets are slow or even infeasible on the large-scale datasets.

Fuzzy positive region is also an important concept. Based on Propositions 1, 2 is then defined as follows.

Definition 2. Given a fuzzy decision table $FD = (U, R \cup D)$, the fuzzy positive region of D relative to R can be simplified as follows.

$$\forall x \in U, POS_R^U(x) = \underline{R}[x]_D(x) = \min_{\{u \in U, u \notin [x]_D\}}\{1 - R(x,u)\}$$

Definition 2 shows the relation between the fuzzy positive region and the lower approximation.

Definition 3. In a fuzzy decision table $FD = (U, R \cup D)$, the dependency degree of D on R, denoted by γ_R^U, is defined as $\gamma_R^U = \sum_{x \in U} POS_R^U(x)/|U|$.

Definition 4. In a fuzzy decision table $FD = (U, R \cup D)$, $P \subseteq R$ is called a reduct of R w.r.t. D if P satisfies the following two statements: (1) $\gamma_R^U = \gamma_P^U$; (2) for any $r \in P$, $\gamma_R^U \neq \gamma_{P-\{r\}}^U$.

This definition shows that the reduct is the minimal subset of features to keep the dependency degree invariant. To design a feature selection algorithm, we need to know how dependency degree grows with the increasing features. And then Proposition 2 is presented as follows.

Proposition 2. If $P \subseteq R$, then (1) $POS_P^U(x) \leq POS_R^U(x)$; (2) $\gamma_P^U \leq \gamma_R^U$.

Proposition 2 shows that the dependency function is monotonic with the incremental features. This result is the theoretical foundation to design the algorithm of feature selection.

Based on Definition 4 and Proposition 2, the feature selection/attribute reduction algorithm based on dependency function, shortened by DAR, is described in Algorithm 1 [16].

Algorithm 1. DAR

INPUT: $FD = (U, R \cup D)$

OUTPUT: red

1. $P \leftarrow \emptyset$;
2. $lef \leftarrow R - P$;
3. **Calculate** γ_R^U;
4. **While** $\gamma_P^U < \gamma_R^U$ **do**
   - $a^* = arg(max\{\gamma_{P\cup\{a\}}^U | a \in lef\})$;
   - $P \leftarrow P \cup \{a^*\}$;
   - $lef \leftarrow lef - \{a^*\}$;
5. **Let** $red \leftarrow P$, $i = 0$;
6. **While** $i < |P|$ **do**
   - Take the ith feature p_i in P
   - **if** $\gamma_{red-\{p_i\}}^U = \gamma_R^U$ **then** $red \leftarrow red - \{p_i\}$;
   - i++;
7. **Output** red;

Proposition 3. In a fuzzy decision table $FD = (U, R \cup D)$, $P \subseteq R$ is called a reduct of R w.r.t. D if P satisfies the following two statements: (1) $\forall x \in U, POS_P^U(x) = POS_R^U(x)$; (2) for any $a \in P, \exists x \in U$, $POS_{P-\{a\}}^U(x) \neq POS_R^U(x)$.

This proposition shows that the reduct could also be seen as a minimal subset of features to keep all the positive regions invariant.

3 Fuzzy Rough Based Feature Selection Accelerator

In this section, we would like to propose an accelerator. And by using this accelerator, an accelerated algorithm is then designed.

3.1 Some Theorems

Definition 5. Given a fuzzy decision table $FD = (U, R \cup D)$, $\Delta U = \{x \in U | POS_P^U(x) < POS_R^U(x)\}$ is called a Key Instance Set of P on FD.

Theorem 1. Given a fuzzy decision table $FD = (U, R \cup D)$. If $P \subseteq R$ is a reduct of R w.r.t. D, then $\forall x \in U - \Delta U$, $POS_P^U(x) = POS_R^U(x)$.

Proof. By Proposition 2, we get $\forall x \in U$, $POS_P^U(x) \leq POS_R^U(x)$. By Definition 5, we get $\forall x \in \Delta U$, $POS_P^U(x) < POS_R^U(x)$. Based on the above two results, we get that if $x \notin \Delta U$, then $POS_P^U(x) = POS_R^U(x)$, i.e., $\forall x \in U - \Delta U$, $POS_P^U(x) = POS_R^U(x)$. ■

By Proposition 3, we find that P is the minimal subset of R to keep all positive region values reaching the maximum. However, to effectively handle the noise in the real applications, it is necessary to propose Parameterized Key Instance Set based on fuzzy rough sets to handle this kind of problem.

Definition 6. Given a fuzzy decision table $FD = (U, R \cup D)$ and a threshold $\alpha \in [0, 1)$, $\Delta U^\alpha = \{x \in U | POS_P^U(x) + \alpha < POS_R^U(x)\}$ is called a Parameterized Key Instance Set of P on FD^U.

Theorem 2. Given a fuzzy decision table $FD = (U, R \cup D)$ and a threshold $\alpha \in [0, 1)$. If $P \subseteq R$ is a reduct of R w.r.t. D, then $\forall x \in U - \Delta U^\alpha$, $POS_P^U(x) + \alpha \geq POS_R^U(x)$.

Based on the Parameterized Key Instance Set, we could measure the relative significance of the features with respect to the obtained reduct in a fuzzy decision table in a new way.

Definition 7. Let $FD = (U, R \cup D)$, $P \subseteq R$, $\forall r \in R - P$ and a threshold $\alpha \in [0, 1)$.

The significance of r in P is defined as

$$Sig^\alpha(r, P, D, U) = \left|\Delta U^\alpha_{P \cup \{r\}}\right|,$$

where $|\cdot|$ denotes the cardinality of a set and $\Delta U^\alpha_{P \cup \{r\}} = \{x \in U | POS^U_{P \cup \{r\}}(x) + \alpha \geq POS_R^U(x)\}$.

If we choose an important feature r into the candidate reduct, that is to say, the feature $r \in R - P$ could distinguish the most instances.

Theorem 3. Let $FD = (U, R \cup D)$, $P = \{r_1, r_2, \ldots, r_n\} \subseteq R$ and a threshold $\alpha \in [0, 1)$. For $\forall a \in R - P$, we have $U_{i+1}^{\alpha} = U_i^{\alpha} - \Delta U_{P \cup \{r\}}^{\alpha}$, where $U_1^{\alpha} = U$ and $U_{n+1}^{\alpha} = \emptyset$.

Theorem 3 shows that when we choose the first feature into the candidate reduct, we need to calculate all instances' fuzzy positive regions. With the increment of the size of P, the size of instances involved in the calculation is decreasing gradually.

3.2 Fuzzy Rough Based Feature Selection Accelerator

Based on Proposition 3, we design a feature selection/attribute reduction method based on positive region, shortened by PAR. It is easy to get that the stop criterion of DAR and PAR in step 4 is essentially consistent, both of which is to keep the information invariant before and after reduction. The time complexity of DAR and PAR is same. The algorithm is described in Algorithm 2.

Algorithm 2. PAR

INPUT: $FD = (U, R \cup D)$
OUTPUT: red

1: $P \leftarrow \emptyset$;
2: $lef \leftarrow R - P$;
3: **Calculate** $POS_R^U(x)$;
4: **While** $\exists x \in U, POS_P^U(x) + \alpha < POS_R^U(x)$ **do**
  For each $a \in lef$, calculate $\Delta U_{P \cup \{a\}}^{\alpha} = \{x \in U \mid POS_{P \cup \{a\}}^U(x) + \alpha \geq POS_R^U(x)\}$,
  Select $a^* \in lef$ satisfying $a^* = \arg(max_{a \in lef} \{Sig^{\alpha}(a, P, D, U)\})$;
  $P \leftarrow P \cup \{a^*\}$;
  $lef \leftarrow lef - \{a^*\}$;
5: **Let** $red \leftarrow P$;
6: $i = 0$;
7: **While** $i < |P|$ **do**
  Take the ith feature p_i in P
  if $POS_{red-\{p_i\}}^U + \alpha \geq POS_R^U$ **then** $red \leftarrow red - \{p_i\}$;
  $i{+}{+}$;
8: **Output** red;

The stop criteria of PAR need check every instance in the whole dataset. It is really time consuming. However, Definition 6 and Theorem 3 already show that with the incremental of the features, some instances, whose fuzzy positive regions reach maximum, are not meaningful selecting the following feature. This finding motivates us to accelerate the PAR by using Theorem 3.

The accelerated algorithm is presented in Algorithm 3.

Algorithm 3. PARA (Positive Region based Selection Accelerator)

INPUT: $FD = (U, R \cup D)$
OUTPUT: red

1. $P \leftarrow \emptyset$;
2. $lef \leftarrow R - P$;
3. **Calculate** POS_R^U;
4. **Let** $i = 1, U_i^\alpha = U$;
5. **While** $|U_i^\alpha| \neq 0$ **do**
   For each $a \in lef$, calculate $\Delta U_{P\cup\{a\}}^\alpha = \{x \in U_i \, |POS_{P\cup\{a\}}^U(x) + \alpha \geq POS_R^U(x)\}$;
   Select $a^* \in lef$ satisfying $a^* = \arg(max_{a\in lef} \{Sig^\alpha(a, P, D, U_i)\})$;
   $U_{i+1}^\alpha = U_i^\alpha - \Delta U_{P\cup\{a^*\}}^\alpha$;
   $P = P \cup \{a^*\}$;
   $lef = lef - \{a^*\}$;
   i++;
6. **Let** $red = P, i = 0$;
7. **While** $i < |P|$ **do**
   Take the ith feature p_i in P
   if $\forall x \in U, POS_{red-\{p_i\}}^U(x) + \alpha \geq POS_R^U(x)$,
   then $red = red - \{p_i\}$;
   i++;
8. **Output** red;

In PARA, Step 5's time complexity is $O\left(\sum_{i=1}^{|C|}(|C|+1-i)i|U_i||U|\right)$. Whereas Step 4's time complexity is $O\left(\sum_{i=1}^{|C|}(|C|+1-i)i|U|^2\right)$ in PAR, it is obvious that PARA could save much time than PAR.

4 Experimental Analysis

In this section, we conduct several numerical experiments to compare three feature selection algorithms mentioned in Sect. 3.

4.1 Experimental Setup

(1) The data used in the experiment are showed in Table 1. In addition to 'stock 1', 'stock 2', 'stock 3', which are stock data and not public, the rest of datasets in Table 1 can be downloaded from [26].
(2) We normalize the feature value into the interval [0, 1] with MinMaxScaler.
(3) We choose the K-nearest neighbor (K is set as 7 in this paper) [17] as the classifier to measure the performance of reduct. And 5-fold cross validation is used to guarantee the stability and fairness of classification results.

(4) All experiments are conducted on a computer with Ubuntu 16.04.4 LTS, Intel(R) Xeon(R) W-2145 CPU @ 3.70 GHz and 32 GB memory. The programming language is C++.

Table 1. Data description.

Datasets	Features	Instances	Classes
waveform	21	5000	3
letter	16	20000	26
credit	23	30000	2
stock 1	619	938	3
stock 2	1350	2018	3
stock 3	1350	2018	3
optdigits	64	5620	10
coil2000	85	9822	2

4.2 Compare DAR and PAR

In this subsection, to demonstrate that DAR and PAR has the similar reduction results, we compare DAR and PAR in Table 2.

Table 2 shows the execution time and selected features number of DAR and PAR on eight datasets, we can find that both of them can obtain nearly same feature numbers of selection and the execution time is comparable. For example, their average running time ratio is almost equal to 1.0 when they obtain the similar size of reducts. What is more, their running time is also close. All these shows that PAR and DAR are comparable reduction methods no matter on the running time or the reduction results.

Table 2. The time and selection of DAR and PAR.

		DAR		PAR		Ratio: DAR/PAR	
Datasets	Attr. size	Reduct size	Time (s)	Reduct size	Time (s)	Reduct	Time
waveform	21	14	1060	14	937	1	1.1
letter	16	9	9227	9	8821	1	1.0
credit	23	9	16387	9	13541	1	1.2
stock 1	619	9	659	9	835	1	0.8
stock 2	1350	17	20559	17	21165	1	1.0
stock 3	1350	18	26090	18	31442	1	0.8
optdigits	64	24	16923	24	17220	1	1.0
coil2000	85	33	19899	33	17593	1	1.1
average	441	17	13850	17	13944	1	1.0

4.3 Compare PAR and PARA

In this subsection compare our proposed accelerator with PAR in Table 3 and Fig. 1.

Table 3 demonstrates that PARA spends less time than PAR. For example, on the dataset 'stock 3', PARA save almost 28540 s (7.9 h) compared with PAR. It is also easy to find that the time ratio between PAR and PARA are 7.2 in average. Or even, sometimes, the time ratio is more than 10, for example, on the cases of 'stock 3' and 'coil2000'. All these show that accelerator is effective and efficiency.

Table 3. The time and selection of PAR and PARA.

		PAR		PARA		Ratio: PAR/PARA	
Datasets	Attr. size	Reduct size	Time (s)	Reduct size	Time (s)	Reduct	Time
waveform	21	14	937	14	345	1	2.7
letter	16	9	8821	9	1296	1	6.8
credit	23	9	13541	9	1499	1	9.0
stock 1	619	9	835	9	264	1	3.2
stock 2	1350	17	21165	17	2558	1	8.3
stock 3	1350	18	31442	18	2902	1	10.8
optdigits	64	24	17220	24	7469	1	2.3
coil2000	85	33	17593	33	1187	1	14.8
average	441	17	13944	17	2190	1	7.2

To furtherly show the performance of our accelerator, we graph the time trendlines of PAR and PARA on gradually incremental datasets. That is, we divide all datasets into 10 groups equally. Firstly, one group is used as the first dataset, secondly, one more group is added to the first one as the second dataset. And so on, we get ten gradually incremental dataset. The time trendlines on those ten gradually incremental datasets are graphed in Fig. 1.

The time trendlines of two algorithms PAR and PARA are graphed in Fig. 1. Figure 1 demonstrates that the accelerator works obviously or even significantly faster and faster. It is easy to see from Fig. 1 that the blue trend (non-accelerated one, PAR) grows dramatically. This demonstrates that PAR spends more and more time with the increment of the size of datasets. However, the red trend (the accelerated one, PARA) is always significantly lower than the blue one, which shows the accelerated algorithm works significantly faster than the non-accelerated one. With the increase of the size of data, the accelerated algorithm runs faster and faster.

4.4 The Classification Performance Comparison of Three Algorithms

Finally, we compare the classification performance of these three algorithms in Table 4. Table 4 mainly shows that DAR, PAR and PARA have the comparable classification performance. As a result, it is easy to get that this accelerator PARA is effective.

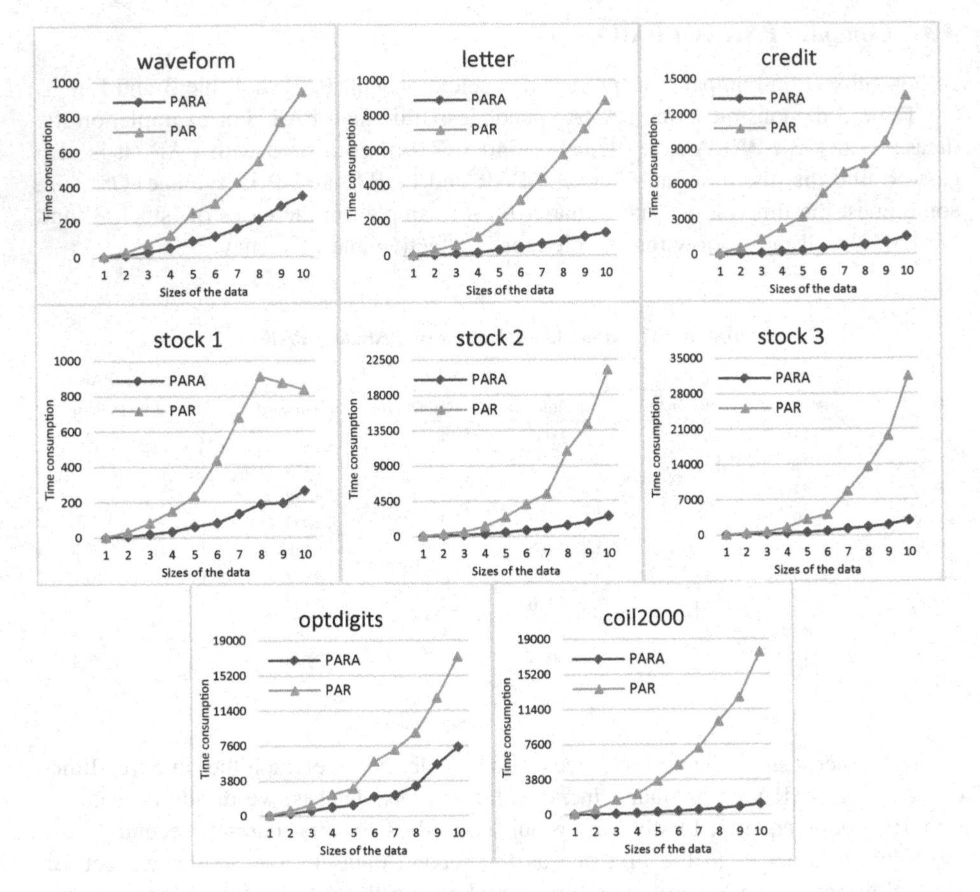

Fig. 1. Times of PAR and PARA versus the size of data.

Table 4. The classification performance of PARA, PAR and DAR.

Datasets	Accuracy			
	PARA	PAR	DAR	All features
waveform	**0.81 ± 0.013**	**0.81 ± 0.013**	0.81 ± 0.012	0.83 ± 0.009
letter	0.90 ± 0.007	0.90 ± 0.007	**0.91 ± 0.004**	0.95 ± 0.001
credit	0.81 ± 0.007	0.81 ± 0.007	**0.81 ± 0.010**	0.80 ± 0.008
stock 1	**0.72 ± 0.025**	**0.72 ± 0.025**	0.71 ± 0.032	0.52 ± 0.104
stock 2	**0.49 ± 0.019**	**0.49 ± 0.019**	0.48 ± 0.023	0.52 ± 0.020
stock 3	**0.39 ± 0.028**	**0.39 ± 0.028**	0.38 ± 0.021	0.36 ± 0.026
optdigits	0.97 ± 0.007	0.97 ± 0.007	**0.97 ± 0.005**	0.98 ± 0.005
coil2000	0.94 ± 0.002	0.94 ± 0.002	**0.94 ± 0.002**	0.94 ± 0.001
average	0.75 ± 0.014	0.75 ± 0.014	0.75 ± 0.014	0.74 ± 0.022

5 Conclusions

In this paper, we proposed an accelerator based on fuzzy rough sets for feature selection. Based on the theoretical analysis and experimental results, it is easy to draw a conclusion that the accelerator proposed in this paper can vastly decrease the execution time, especially on large-scale datasets, without classification performance loss.

Acknowledgements. This work is supported by National Key Research & Develop Plan (No. 2016YFB1000702), National Key R&D Program of China (2017YFB1400700), and NSFC under the grant No. 61732006, 61532021, 61772536, 61772537, 61702522 and NSSFC (No. 12 \&ZD220), and the Fundamental Research Funds for the Central Universities, and the Research Funds of Renmin University of China (15XNLQ06). It was partially done when the authors worked in SA Center for Big Data Research in RUC. This Center is funded by a Chinese National 111 Project Attracting. This work is also supported by the Macao Science and Technology Development Fund (081/2015/A3).

References

1. Guyon, I., Elisseeff, A.: An introduction to variable and feature selection. JMLR. **3**, 1157–1182 (2003)
2. Kira, K., Rendell, L.A.: A practical approach to feature selection. In: Proceedings of the 9th International Conference on Machine Learning, pp. 249–256. Morgan Kaufmann, Los Altos (1992)
3. Peng, H.C., Long, F., Ding, C.: Feature selection based on mutual information: criteria of max-dependency, max-relevance, and min-redundancy. IEEE Trans. Pattern Anal. Mach. Intell. **27**(8), 1226–1238 (2005)
4. Pawlak, Z.: Rough Sets: Theoretical Aspects of Reasoning About Data. Kluwer Academic Publishers, Boston (1991)
5. Pawlak, Z., Grzymala-Busse, J.W., Slowiski, R., Ziako, W.: Rough sets. Commun. ACM **38** (11), 89–95 (1995)
6. Dubois, D., Prade, H.: Rough fuzzy sets and fuzzy rough sets. Int. J. Gen. Syst. **17**, 109–137 (1990)
7. Wang, X.Z., Tang, E.C.C., Zhao, S.Y., Chen, D.G.: Learning fuzzy rules from fuzzy samples based on rough set techniques. Inf. Sci. **177**, 4493–4514 (2007)
8. Hu, Q., Yu, D.R., Xie, Z.X.: Information-preserving hybrid data reduction based on fuzzy-rough techniques. Pattern Recogn. Lett. **27**(5), 414–423 (2006)
9. Qian, Y.H., Liang, J.Y., Pedrycz, W., Dang, C.Y.: Positive approximation: an accelerator for feature reduction in rough set theory. Artif. Intell. **174**(9), 597–618 (2010)
10. Qian, Y.H., Wang, Q., Cheng, H.H., Liang, J.Y., Dang, C.Y.: Fuzzy-rough feature selection accelerator. Fuzzy Sets Syst. **258**(C), 1–78 (2015)
11. Zadeh, L.A.: Fuzzy sets. Inf. Control **8**(3), 338–353 (1965)
12. Dubois, D., Prade, H.: Rough fuzzy sets and fuzzy rough sets. Int. J. Gen. Syst. **17**(2–3), 191–209 (1990)
13. Tsang, E.C.C., Chen, D.G., Yeung, D.S., Wang, X.Z., Lee, J.W.T.: Attributes reduction using fuzzy rough sets. IEEE Trans. Fuzzy Syst. **16**(5), 1130–1141 (2008)
14. Yeung, D.S., Chen, D.G., Tsang, E.C.C., Lee, J.W.T., Wang, X.Z.: On the generalization of fuzzy rough sets. IEEE Trans. Fuzzy Syst. **13**, 343–361 (2005)

15. Hu, Q.H., Zhang, L., An, S., Zhang, D., Yu, D.R.: On robust fuzzy rough set models. IEEE Trans. Fuzzy Syst. **20**(4), 636–651 (2012)
16. Yao, Y.Y., Zhao, Y., Wang, J.: On reduct construction algorithms. Trans. Comput. Sci. **2**, 100–117 (2008)
17. Coomans, D., Massart, D.L.: Alternative k-nearest neighbour rules in supervised pattern recognition: part 1. K-Nearest neighbour classification by using alternative voting rules. Analytica Chimica Acta **136**, 15–27 (1982)
18. Kryszkiewicz, M., Lasek, P.: FUN: fast discovery of minimal sets of attributes functionally determining a decision attribute. Trans. Rough Sets **9**, 76–95 (2008)
19. Zhao, S.Y., Chen, H., Li, C.P., Zhai, M.Y., Du, X.Y.: RFRR: robust fuzzy rough reduction. IEEE Trans. Fuzzy Syst. **21**(5), 825–841 (2013)
20. Bhatt, R.B., Gopal, M.: On fuzzy rough sets approach to feature selection. Pattern Recogn. Lett. **26**(7), 965–975 (2005)
21. Chen, D.G., Tsang, E.C.C., Zhao, S.Y.: Attributes reduction with fuzzy rough sets. In: IEEE International Conference on Systems, Man, and Cybernetics, Vol. 1, pp. 486–491 (2007)
22. Zhao, S.Y., Wang, X.Z., Chen, D.G., Tsang, E.C.C.: Nested structure in parameterized rough reduction. Inf. Sci. **248**, 130–150 (2013)
23. Swiniarski, R.W., Skowron, A.: Rough set methods in feature selection and recognition. Pattern Recogn. Lett. **24**(6), 833–849 (2003)
24. Chen, D.G., Yang, Y.Y.: Attribute reduction for heterogeneous data based on the combination of classical and fuzzy rough set models. IEEE Trans. Fuzzy Syst. **22**(5), 1325–1334 (2014)
25. Chen, D.G., Zhao, S.Y.: Local reduction of decision system with fuzzy rough sets. Fuzzy Sets Syst. **161**(13), 1871–1883 (2010)
26. http://archive.ics.uci.edu/ml/datasets.html

Text Feature Extraction and Selection Based on Attention Mechanism

Longxuan Ma[(✉)] and Lei Zhang

Beijing University of Posts and Telecommunications, Beijing, China
{malongxuan,zlei}@bupt.edu.cn

Abstract. Selecting features that represent a particular corpus is important to the success of many machine learning and text mining applications. However, the previous attention-based work only focused on feature augmentation in the lexical level, lacking the exploration of feature enhancement in the sentence level. In this paper, we exploit a novel feature extraction and selection model for information retrieval, denoted by Dynamic Feature Generation Network (DFGN). In sentence dimension, features are firstly extracted by a variety of different attention mechanisms, then dynamically filtered by thresholds automatically learned. Different kinds of characteristics are distilled according to specific tasks, enhancing the practicability and robustness of the model. DFGN relies solely on the text itself, requires no external feature engineering. Our approach outperforms previous work on multiple well-known answer selection datasets. Through the analysis of the experiments, we prove that DFGN provides excellent retrieval and interpretative abilities.

Keywords: Feature extraction and selection · Machine learning · Question answering

1 Introduction

Modeling textual relevance between document query pairs lives at the heart of information retrieval (IR) research. Comprehending logical and semantic relationship between two sentences is the core challenge and a fundamental technology in natural language processing (NLP). Meanwhile, acquiring the ability to rank is versatile and essential for many information retrieval tasks, and serves as a core function to more complex and sophisticated systems.

In recent years, the attention mechanism, allocating different weight to words or sub-phrases in sentences [1,14], is widely employed in natural language processing. Searching appropriate semantic information for ranking answers naturally conforms to the characteristics of attention mechanism. By computing word level similarity matrix to extract matching patterns from different text granularity that is useful for prediction, attention significantly improve the performance of convolutional [7,26] or recurrent [23,24] networks. Compare-aggregate models with soft-attention [11,22] provide a new way of comparing attention

Q. Yang et al. (Eds.): PAKDD 2019, LNAI 11440, pp. 615–627, 2019.
https://doi.org/10.1007/978-3-030-16145-3_48

results and obtain great achievement. Soft-attention considers all the words in a sentence when assigning weight, also known as the word by word attention. Intra-attention based models [9,10] distills relationship between words in a sentence. Co-attention based models [5,8,19] learn joint information with respect to two sentences then distribute weight to both sides, which embrace a great success in matching natural language sentences. Attention mechanism comes in different forms and in company with extractive max and mean pooling [15,27] or alignment pooling [3,16]. Extractive max-pooling selects each word based on its maximum importance of all words in the other text. Extractive mean-pooling is a more wholesome comparison, paying attention to a word based on its overall influence on the other text. Alignment-pooling aligns semantically similar sub-phrases together, extracting only the most relevant information. Although attention mechanism is typically applied as weight allocation strategy, recent approaches [17,18] utilize attention as feature augmentation tool. They utilize co-attention and intra-attention in conjunction with extractive max, mean, alignment pooling to compute weighted representations, then compress each representation into a scalar, attaching these scalars as extra features to original embedding, proven to be efficient in retrieval-based question answering tasks.

However, previous approaches have two disadvantages. The first is that they only focus on the augmentation of lexical level [3,17]. Research on feature enhancement of sentence level is not enough. Therefore, in this paper, we first enhance features at the sentence level and then perform information retrieval process. The second deficiency is that they usually use fixed features and fail to select features of different quantities based on different tasks. Alignment results are usually directly employed [16] or compress into scalars [18]. In this work, we apply parametric co-attention to perform thresholds in filtering operation. Different kinds of characteristics are distilled according to specific tasks.

The main contributions of our work are as follows:

- We propose a sentence level feature enhancement method, extend feature augmentation method that used to focus on the word level. In addition to the traditional attention algorithms, we propose a new feature extraction algorithm and a new dynamic threshold feature selection algorithm.
- We design Dynamic Feature Generation Network for answer selection, DFGN acquires the ability to automatically abandon useless and inefficient data, reducing the cost of adjusting parameters. Our code is publicly available[1].
- Experimental results show that DFGN outperforms current work on WikiQA, TREC-QA and InsuranceQA datasets. We give an in-depth analysis to illustrate why DFGN owns the excellent retrieval and interpretative abilities.

2 Our Proposed Model

The overall structure of DFGN is shown in Fig. 1. We apply pre-trained 300 dimensional Glove [12] as word embedding. The inputs of our model are query

[1] https://github.com/malongxuan/QAselection.

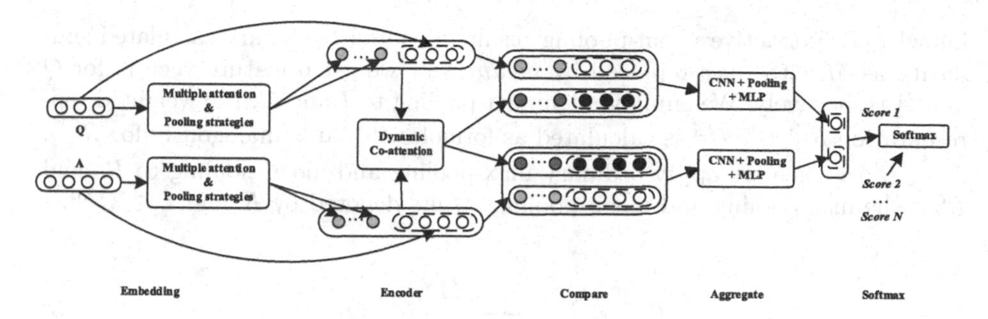

Fig. 1. The over structure of DFGN.

$Q \in \mathbb{R}^{q \times d}$ and and answer $A \in \mathbb{R}^{a \times d}$, d represents the 300 embedding size, q and a represent the length of Q and A respectively.

2.1 Multiple Attention and Pooling Strategies

Co-attention and intra-attention are the most commonly used attention mechanisms. In our model, we use both parametric and non-parametric computations to extract intrinsic features while learning extended features with parameters, we define self-attention and intra-attention respectively for the sake of distinction. Self-attention matrices $S^Q \in \mathbb{R}^{q \times d}$ and $S^A \in \mathbb{R}^{a \times d}$, intra-attention affinity matrices $I^Q \in \mathbb{R}^{q \times q}$ and $I^A \in \mathbb{R}^{a \times a}$, co-attention affinity matrices $C^Q \in \mathbb{R}^{q \times a}$ and $C^A \in \mathbb{R}^{a \times q}$ are calculated by fomulas (1), (2), and (3), where δ means $Tanh$, $\otimes$ means matmul product, W^{S^Q} and $W^{S^A} \in \mathbb{R}^{d \times d}$ are parameters.

$$S^Q = \delta(Q \otimes W^{S^Q}) \,\&\, S^A = \delta(A \otimes W^{S^A}) \tag{1}$$

$$I^Q = Q \otimes Q^T \,\&\, I^A = A \otimes A^T \tag{2}$$

$$C^Q = Q \otimes A^T \,\&\, C^A = A \otimes Q^T \tag{3}$$

We extract feature vectors in a variety of ways then concatenate them to sentence representations. The features are extracted from three levels: word interaction between sentences, word interaction within a sentence, and the hidden layer interaction of a sentence. They are corresponding to co-attention, intra-attention, self-attention respectively. We use formula (4) to obtain M^{S^Q}. Where $M^{S^Q} \in \mathbb{R}^{1 \times d}$, and analogously for M^{S^A}, M^{I^Q}, M^{I^A}, M^{C^Q} and M^{C^A}.

$$M^{S^Q} = \sum_{i=1}^{q} \frac{exp(\max_{j=1}^{d} S^Q_{i,j})}{\sum_{t=1}^{q} exp(\max_{j=1}^{d} S^Q_{t,j})} Q_i \tag{4}$$

It is worth noting that M^I and M^C extract the sentence level dimensional features, while M^S extracts the hidden dimensional features. Employing extractive pooling to hidden dimension in answer selection is the first time to our knowledge. Extractive mean-pooling results, denoted by N, are calculated similarity as M with max replaced by $mean$. Now we get 6 feature vectors for Q and A respectively. We employ alignment pooling to I and C, the weighted representation $R^{I^Q} \in \mathbb{R}^{q \times d}$ is calculated as formulas (5), and analogously for R^{C^Q}, R^{I^A}, R^{C^A}. Then we apply ordinary max-pooling and mean-pooling to R^I and R^C. The max-pooling and mean-pooling results denoted by K and $L \in \mathbb{R}^{1 \times d}$.

$$R_j^{I^Q} = \sum_{i=1}^{q} \frac{exp(I_{i,j}^Q)}{\sum_{t=1}^{q} exp(I_{t,j}^Q)} Q_i \tag{5}$$

By attaching 10 feature vectors to Q, we obtain final question representation $X^Q = [Q; M^{S^Q}; M^{I^Q}; M^{C^Q}; N^{S^Q}; N^{I^Q}; N^{C^Q}; K^{I^Q}; K^{C^Q}; L^{I^Q}; L^{C^Q}] \in \mathbb{R}^{(q+10) \times d}$, and analogously for answer representation $X^A \in \mathbb{R}^{(a+10) \times d}$.

2.2 Encoder Layers

After multiple attention, we use nonlinear functions as encoder to compute the input of the second co-attention layer separately, denoted by $E^Q \in \mathbb{R}^{(q+10) \times d}$, $E^A \in \mathbb{R}^{(a+10) \times d}$. Where σ represent $Sigmoid$ and δ means $Tanh$. Experiments show that this encoding method not only achieves equal level accuracy as other complex structures like CNN and LSTM but also has fewer parameters and saves training time. Equation (6) computes E^Q, and analogously for E^A. $W_{en1}^{X^Q}$ and $W_{en2}^{X^Q}$ are parameters to be learned, $*$ is the element-wise product.

$$E^Q = \sigma(X^Q \otimes W_{en1}^{X^Q}) * \delta(X^Q \otimes W_{en2}^{X^Q}) \tag{6}$$

2.3 Second Attention Layer

The second co-attention layer learns joint information between features to perform interactive confirmation. We use the formula (7) to calculate the affinity matrices G^{E^Q} and $\widetilde{G}^{E^Q} \in \mathbb{R}^{(q+10) \times (a+10)}$. $\widetilde{G}^{E^Q}$ performs threshold. $\widetilde{W}^Q \in \mathbb{R}^{d \times d}$ is a parameter matrix to be learned. We employ Eq. (8) to get weighted representation $H^{E^A} \in \mathbb{R}^{(a+10) \times d}$, $\phi(x, y)$ is an operation that if $x < y$ then x set to 0, otherwise x keeps original value. The advantage of this method is that it can remove redundant information dynamically instead of using empirical values. For different corpora, the number of retained features are different. The calculations are analogously for G^{E^A}, $\widetilde{G}^{E^A}$ and H^{E^Q}.

$$G^{E^Q} = (E^Q) \otimes (E^A)^T \quad \& \quad \widetilde{G}^{E^Q} = (E^Q) \otimes \widetilde{W}^Q \otimes (E^A)^T \tag{7}$$

$$H_j^{E^A} = \sum_{i=1}^{q+10} \phi\left(\frac{exp(G_{i,j}^{E^Q})}{\sum_{t=1}^{q+10} exp(G_{t,j}^{E^Q})}, \frac{exp(\widetilde{G}_{i,j}^{E^Q})}{\sum_{t=1}^{q+10} exp(\widetilde{G}_{t,j}^{E^Q})}\right) E_i^Q \tag{8}$$

2.4 Compare, Aggregate, Softmax Layers

There are several different forms of compare function [22]. We choose element-wise multiplication. The goal of the comparison layer is to match each feature with its weighted version. The comparison result, $Y^E = H^E * E$, is fed to CNN with both max and mean pooling to get aggregate features, then we employ multilayer perceptron to get the final scores of Q and A, then two branches are concatenated and compressed to a scalar for the final softmax layers, as shown in Fig. 1. We feed related answer set $A\{A_1, A_2, \ldots, A_N\}$, target label set $Y\{y_1, y_2, \ldots, y_N\}$ along with Q into the model. We select all positive answers to this question, denoted by p, then randomly select $N - p$ negative answers from the answer pool. We train our listwise model with KL-divergence loss to optimize the ranking results. Please consult our code to see the implementation.

3 Experiments

3.1 Datasets and Experimental Protocol

Statistical information of experimental datasets is shown in Table 1. WikiQA [25] is constructed by crowd-sourcing through sentences extraction from Wikipedia and Bing search logs. TREC-QA [21] is from the TREC Question Answering tracks. Recent work [2,16,18] use clean version by removing questions that have only positive/negative answers or no answers. InsuranceQA [4] is collected from a community question answering website which contains two versions ($V1$ and $V2$). We use the $V1$ version which has two test sets (Test1 and Test2).

Table 1. Statistics of WikiQA, TREC-QA and InsuranceQA datasets.

Dataset(train/test/dev)	WikiQA	TrecQA(clean)	InsuranceQA V1
Questions	873/243/126	1162/68/65	12887/1800 * 2/1000
Sentences	20360/6165/2733	5919/1442/1117	24981(ALL)
Average length of questions	7.16/7.26/7.23	11.39/8.63/8.00	7.16
Average length of sentences	25.29/24.59/24.59	30.39/25.61/24.9	49.5
Question answer pairs	5.9k/1.4k/1.1k	53.4k/1.4k/1.1k	1.29m/1.8m/1m
Average candidate answers	9	38	100/500/500

To train our model in mini-batch, we truncate the question to 12 words, the answer to 50 words, candidate answers to 15 and batch size to 6. We add 0 at the end of the sentence if it is shorter than the specified length. We remove all the symbols, keep only the words and fix the word representations during training. We set a dropout rate as 0.1 at encoder layer. The CNN windows are [1, 2, 3, 4, 5]. We resort to Adam algorithm as the optimization method and update the parameters with the learning rate as 0.001, $beta_1$ as 0.9, $beta_2$ as 0.999. We add $L2$ penalty with the coefficient parameter λ as 10^{-5}. We resort to the gradient global norm clipping method to avoid the gradient exploding problem and set the clip norm as 5. We design the model with Keras[2] and Theano library[3]. All experiments are carried out on Ubuntu 16.04, a single GPU (GeForce GTX 1080). We resort to Mean Average Precision (MAP), Mean Reciprocal Rank (MRR) and accuracy (Precision@1) to measure the experimental results.

3.2 Experimental Results

Table 2 reports our experimental results on all datasets. We select 11 models for comparison and use the original papers data. DFGN(reduce) represents a model without enhanced sentence features, and DFGN(full) means full model. On WikiQA, we observe that DFGN(full) outperforms a myriad of complex neural architectures. Notably, we obtain a performance gain of 1.6% in terms of MRR against strong models such as MULT and DCA. Table 2 also reports our results on the clean version of TrecQA. On the MRR index, DFGN(full) model outperforms latest MCAN by 2.4%. On InsuranceQA V1 dataset, our models outperform the strongest model SUBMULT+NN [22]. DFGN(full) gains accuracy promotion of 2.3% both in Test1 and Test2. In all experiments, the DFGN(full) model achieve better performance than the DFGN(reduce) model.

Table 2. Performance on WikiQA, TREC-QA and InsuranceQA datasets.

Models	WikiQA		TrecQA(clean)		InsuranceQA V1
	MAP	MRR	MAP	MRR	Top1(Test1/Test2)
AP-BiLSTM([15])	0.671	0.684	0.713	0.803	0.717/0.664
MP-CNN([6])	0.693	0.709	0.777	0.836	-/-
MPCNN+NCE([13])	0.701	0.718	0.801	0.877	-/-
PWIM([7])	0.709	0.723	-	-	-/-
BiMPM([24])	0.718	0.731	0.802	0.875	-/-
MS-LSTM([19])	0.722	0.738	0.813	0.893	0.705/0.669
IWAN([16])	0.733	0.750	0.822	0.889	-/-
IABRNN([20])	0.734	0.742	-	-	0.701/0.651
MULT([22])	0.743	0.754	-	-	**0.752/0.734**
DCA([2])	**0.756**	**0.764**	0.821	0.899	-/-
MCAN-FM([18])	-	-	**0.838**	**0.904**	-/-
DFGN(reduce)	0.745	0.753	0.828	0.905	0.762/0.744
DFGN(full)	**0.766**	**0.780**	**0.848**	**0.928**	**0.775/0.757**

[2] https://github.com/keras-team/keras.

[3] http://www.deeplearning.net/software/theano/.

4 Discussion and Analysis

4.1 Question Type Analysis

In this paragraph, we use DFGN(reduce) and DFGN(full) to analyze two problems, first is the efficiency of the model itself for different types of questions, second is the impact of dynamic features on different questions. Figure 2 demonstrates all five types of questions in WikiQA test set. The histograms represent the MRR metric and the proportion of each type of questions in dataset respectively. Both models own better results for 'Where' and 'Who' questions because locations and characters are easier to retrieve. In DFGN(reduce), the MRR value of 87.2% and 83.8% are respectively achieved for 'Where' and 'Who' questions. While due to the proportion of 55.1% and 15.2%, the effect on 'What' and 'How' questions decide the overall performance of the model. In DFGN(reduce), the MRR value 75.1% and 65.1% are respectively achieved for 'What' and 'How' questions. After adding dynamic sentence features, we find that DFGN(full) improves the MRR value by 2.3% on 'What' questions, polish up the MRR value of 'How' question by 5.7%. 'What' and 'How' problems increase more than 'Where' and 'Who' problems, which shows that the dynamic feature generation successfully improves the comprehension ability of complex semantic relations.

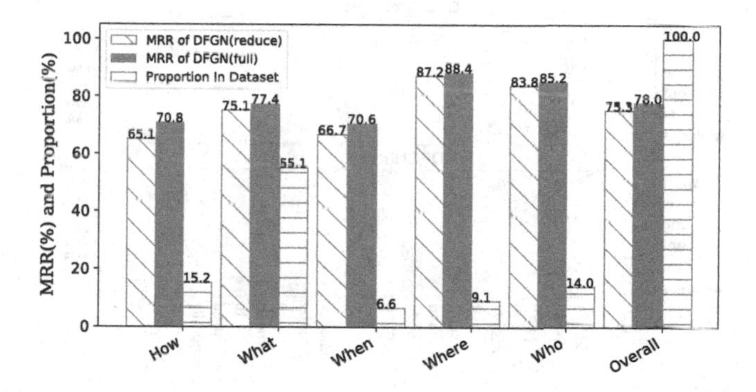

Fig. 2. Comparison between DFGN(reduce) and DFGN(full) on type of questions.

4.2 In-Depth Analysis

We resort to a question-answer pair of WikiQA dataset to perform in-depth analysis. Figure 3 illustrates how sentence features affect the weight allocation of the second co-attention matrix in DFGN(reduce) and DFGN(full). Since the second co-attention matrix in DFGN(full) owns 10 additional features both in rows and columns, we only compared the common parts. The heat map above is from DFGN(reduce), and the heat map below is from DFGN(full). The question is "what is the concept of wellness" and the answer is "wellness is generally used to mean a healthy balance of the mind body and spirit that results in an overall feeling of well being". In DFGN(reduce), we can see that the weighted matrix fails to find the words related to the question. In DFGN(full), there are three obvious differences. Firstly, the distribution of weight is more centralized. Secondly, the number of features filtered out by different rows and columns is different. Thirdly, the weighted matrix with dynamically generated features not only filters out useless alignment information but also better understands the meaning of the question. It focuses on a more appropriate sub-phrases such as "wellness is", "healthy balance", "mind body and spirit", "overall feeling" and "well being". This is because after adding additional features, the weight distribution range changes from the original sentence length to the sentence length plus 10. Unimportant information will be allocated less weight. Then the dynamic selection mechanism filters out irrelevant information and highlights important information. Thus the retrieval result is more precise. This also explains why the "what" and "how" questions improve more in Sect. 4.1. The difference between this method and former methods is that the number of features filtered out is neither empirical nor fixed but automatically learned by parameters. Dynamic feature generation mechanism corrections the deficiencies of the original algorithm.

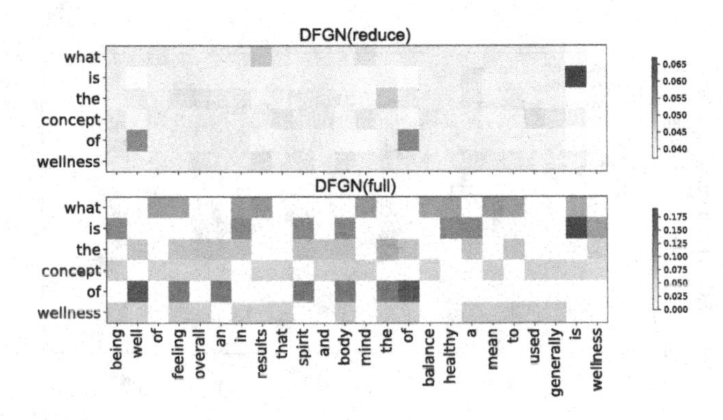

Fig. 3. Comparison between DFGN(reduce) and DFGN(full) in second co-attention.

We also present the co/intra/self extractive pooling attention weight in Fig. 4. The left and right graphic are the weight allocation on the question and answer respectively. We can observe that different attention mechanism focuses on different positions. In this example, two self-extractive poolings assign weight more evenly. In the question part, co-extractive-mean pooling pays more attention to interrogative word such as "what", while intra-extractive-max pooling put all weight at word "concept". In the answer part, co-extractive-mean pooling puts more weight at sub-phrases such as "results in" and "well being", while intra-extractive-max pooling assigns most weight at "wellness" and "healthy". When all six extractive pooling strategies act synthetically, different features

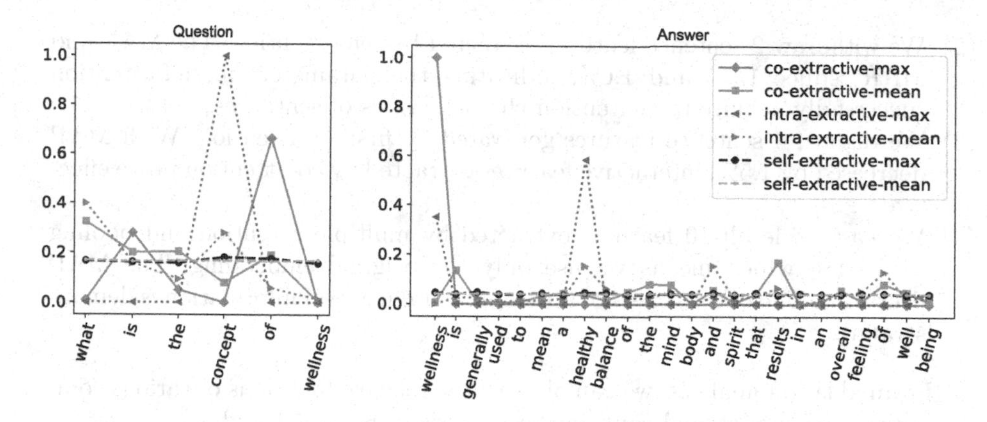

Fig. 4. Comparison between extractive pooling with c o/intra/self attention.

are extracted. Six extractive pooling features, four alignment pooling features, and the original features together construct a feature set. Dynamic selection mechanism sets unimportant features to zero and retains useful ones according to different texts. In practice, the longer the sentence is, the more words are related to the overall semantics, the more obvious the difference heat map and line graph between DFGN(reduce) and DFGN(full) will be. The DFGN(full) model extracts more appropriate semantic information for the ranking task.

4.3 Ablation Analysis

This section shows the relative validity of the different components of our DFGN(full) model. Table 3 presents the 4 different structures on the TrecQA(clean) test set.

Table 3. Ablation analysis On TrecQA(clean) test set.

Setting	MAP	MRR
Full model	0.848	0.928
(1) without Intra-attention features	0.831	0.908
(2) without Self-attention features	0.835	0.912
(3) without Co-attention feature	0.833	0.908
(4) without All sentence features	0.828	0.905

(1) We take away 4 sentence features acquired by intra-attention. With MAP decreased by 1.7% and MRR drop by 2.0%, intra-attention proves that it extracts the internal information in a sentence.

(2) We withdraw 2 sentence features extracted by self-attention. The MAP and MRR reduce 1.3% and 1.6%, indicating the parameters in self-attention successfully acquire the extension characteristics of sentences.
(3) We discard 4 sentence features generated by first co-attention. With MAP decreased by 1.5%, interactive features extracted by co-attention prove effective.
(4) We cast aside all 10 features extracted by multiple attention and pooling strategies, which means we use only the original embedding. The MAP drops 2%. The effectiveness of attaching sentence level information is demonstrated.

From ablation analysis, we can observe the relative functions of various components to our model, and confirm the analysis of Sects. 4.1 and 4.2.

5 Related Work

Learning to rank candidate answers is a long-standing problem in NLP and IR. The state-of-the-art models today are mostly neural attention based approaches. These models focus on different aspects. MPCNN [13] proposes a new ranking method in answer selection. PWIM [7] combines LSTM and deep CNN to investigate the similarity matrix. Gated attention in RNN [14,23] explore the internal semantic relations of sentences, obtain remarkable improvement in natural language inference. IABRNN [20] further develop gated attention and achieve success in answer selection. Inner-attention [10] and self-attention [9] are introduced to LSTM, extracting an interpretable sentence embedding.

Recent advances in neural matching models go beyond independent expression learning. Major architectural paradigms that invoke interaction between document pairs directly improve performance because matching has deeper and finer granularity. Multi-Perspective CNN [6] and BiMPM [24] match sentences with multiple views and perspectives. AP-BiLSTM [15] utilize extractive max-pooling to learn the relative importance of a word based on its maximum effect to all words in the other document. IWAN [16] extract features from a constructed word-by-word alignment matrix with self-attention. We use all previous extraction strategies and design a novel fixed attention features with parameterized self-attention. Meanwhile, Compare-aggregate framework [11,22] are proven to be effective in answer selection tasks. DCA [2] further improves former work by dynamic clipping useless information, which inspires our feature filtering method.

However, the approaches using deeper layers lead to more progress in performance. MAN [19] applies multihop-sequential-LSTM to achieve step by step learning. Most recently, Compare-propagate model CAFE [17] and multi-cast approach MCAN-FM [18] compress attention vectors into scalar valued features, which are used to augment the word representations. Inspired by them, we excavate the approach which uses multiple attentions and extractive pooling strategies to enhance representations power for answer selection task. But unlike all previous work, we design dynamic feature generation methods at sentence levels.

6 Conclusion

We propose a novel architecture Dynamic Feature Generation Network (DFGN) for retrieval-based question answering. Unlike previous work which only focused on feature augmentation in the lexical level, we study dynamic feature extraction and selection in the sentence level. Features are extracted by a variety of attention mechanisms, attached to the sentence level, and dynamically filtered. DFGN acquire the ability to extract and select features according to different tasks dynamically. Different kinds of characteristics are distilled according to specific tasks, enhancing the practicability and robustness of the model. Our model needs no external resources and feature engineering, relies solely on the semantic information of the text itself. The experimental results outperform current work on multiple well-known datasets, which illustrates our approach effectively improves information retrieval efficiency. Moreover, we give an in-depth analysis of our model which enables us to comprehend its inner working principle further. In the future, we plan to validate the efficiency of our model on more sentence matching tasks, such as natural language inference and paraphrase identification.

References

1. Bahdanau, D., Cho, K., Bengio, Y.: Neural machine translation by jointly learning to align and translate. CoRR abs/1409.0473 (2014)
2. Bian, W., Li, S., Yang, Z., Chen, G., Lin, Z.: A compare-aggregate model with dynamic-clip attention for answer selection. In: Proceedings of the 2017 ACM on Conference on Information and Knowledge Management, CIKM 2017, Singapore, 06–10 November 2017, pp. 1987–1990. ACM (2017)
3. Chen, Q., Zhu, X., Ling, Z., Wei, S., Jiang, H., Inkpen, D.: Enhanced LSTM for natural language inference. In: Proceedings of the 55th Annual Meeting of the Association for Computational Linguistics, ACL 2017, Long Papers, Vancouver, Canada, 30 July–4 August, vol. 1, pp. 1657–1668. Association for Computational Linguistics (2017)
4. Feng, M., Xiang, B., Glass, M.R., Wang, L., Zhou, B.: Applying deep learning to answer selection: a study and an open task. In: 2015 IEEE Workshop on Automatic Speech Recognition and Understanding, ASRU 2015, Scottsdale, AZ, USA, 13–17 December 2015, pp. 813–820. IEEE (2015)
5. Gong, Y., Luo, H., Zhang, J.: Natural language inference over interaction space. CoRR abs/1709.04348 (2017)
6. He, H., Gimpel, K., Lin, J.J.: Multi-perspective sentence similarity modeling with convolutional neural networks. In: Proceedings of the 2015 Conference on Empirical Methods in Natural Language Processing, EMNLP 2015, Lisbon, Portugal, 17–21 September 2015, pp. 1576–1586. The Association for Computational Linguistics (2015)
7. He, H., Lin, J.J.: Pairwise word interaction modeling with deep neural networks for semantic similarity measurement. In: NAACL HLT 2016, The 2016 Conference of the North American Chapter of the Association for Computational Linguistics: Human Language Technologies, San Diego California, USA, 12–17 June 2016, pp. 937–948. The Association for Computational Linguistics (2016)

8. Kim, S., Hong, J., Kang, I., Kwak, N.: Semantic sentence matching with densely-connected recurrent and co-attentive information. CoRR abs/1805.11360 (2018)
9. Lin, Z., et al.: A structured self-attentive sentence embedding. CoRR abs/1703.03130 (2017)
10. Liu, Y., Sun, C., Lin, L., Wang, X.: Learning natural language inference using bidirectional LSTM model and inner-attention. CoRR abs/1605.09090 (2016)
11. Parikh, A.P., Täckström, O., Das, D., Uszkoreit, J.: A decomposable attention model for natural language inference. In: Proceedings of the 2016 Conference on Empirical Methods in Natural Language Processing, EMNLP 2016, Austin, Texas, USA, 1–4 November 2016, pp. 2249–2255. The Association for Computational Linguistics (2016)
12. Pennington, J., Socher, R., Manning, C.D.: Glove: global vectors for word representation. In: Proceedings of the 2014 Conference on Empirical Methods in Natural Language Processing, EMNLP 2014, A Meeting of SIGDAT, a Special Interest Group of the ACL, Doha, Qatar, 25–29 October 2014, pp. 1532–1543. ACL (2014)
13. Rao, J., He, H., Lin, J.J.: Noise-contrastive estimation for answer selection with deep neural networks. In: Proceedings of the 25th ACM International Conference on Information and Knowledge Management, CIKM 2016, Indianapolis, IN, USA, 24–28 October 2016, pp. 1913–1916. ACM (2016)
14. Rocktäschel, T., Grefenstette, E., Hermann, K.M., Kociský, T., Blunsom, P.: Reasoning about entailment with neural attention. CoRR abs/1509.06664 (2015)
15. dos Santos, C.N., Tan, M., Xiang, B., Zhou, B.: Attentive pooling networks. CoRR abs/1602.03609 (2016)
16. Shen, G., Yang, Y., Deng, Z.: Inter-weighted alignment network for sentence pair modeling. In: Proceedings of the 2017 Conference on Empirical Methods in Natural Language Processing, EMNLP 2017, Copenhagen, Denmark, 9–11 September 2017, pp. 1179–1189. Association for Computational Linguistics (2017)
17. Tay, Y., Tuan, L.A., Hui, S.C.: A compare-propagate architecture with alignment factorization for natural language inference. CoRR abs/1801.00102 (2018)
18. Tay, Y., Tuan, L.A., Hui, S.C.: Multi-cast attention networks for retrieval-based question answering and response prediction. CoRR abs/1806.00778 (2018)
19. Tran, N.K., Niederée, C.: Multihop attention networks for question answer matching. In: The 41st International ACM SIGIR Conference on Research and Development in Information Retrieval, SIGIR 2018, Ann Arbor, MI, USA, 08–12 July 2018, pp. 325–334. ACM (2018)
20. Wang, B., Liu, K., Zhao, J.: Inner attention based recurrent neural networks for answer selection. In: Proceedings of the 54th Annual Meeting of the Association for Computational Linguistics, ACL 2016, Long Papers, Berlin, Germany, 7–12 August 2016, vol. 1. The Association for Computer Linguistics (2016)
21. Wang, M., Smith, N.A., Mitamura, T.: What is the jeopardy model? A quasi-synchronous grammar for QA. In: EMNLP-CoNLL 2007, Proceedings of the 2007 Joint Conference on Empirical Methods in Natural Language Processing and Computational Natural Language Learning, Prague, Czech Republic, 28–30 June 2007, pp. 22–32. ACL (2007)
22. Wang, S., Jiang, J.: A compare-aggregate model for matching text sequences. CoRR abs/1611.01747 (2016)
23. Wang, S., Jiang, J.: Learning natural language inference with LSTM. In: NAACL HLT 2016, The 2016 Conference of the North American Chapter of the Association for Computational Linguistics: Human Language Technologies, San Diego California, USA, 12–17 June 2016, pp. 1442–1451. The Association for Computational Linguistics (2016)

24. Wang, Z., Hamza, W., Florian, R.: Bilateral multi-perspective matching for natural language sentences. In: Proceedings of the Twenty-Sixth International Joint Conference on Artificial Intelligence, IJCAI 2017, Melbourne, Australia, 19–25 August 2017, pp. 4144–4150. ijcai.org (2017)
25. Yang, Y., Yih, W., Meek, C.: WikiQA: a challenge dataset for open-domain question answering. In: Proceedings of the 2015 Conference on Empirical Methods in Natural Language Processing, EMNLP 2015, Lisbon, Portugal, 17–21 September 2015, pp. 2013–2018. The Association for Computational Linguistics (2015)
26. Yin, W., Schütze, H., Xiang, B., Zhou, B.: ABCNN: attention-based convolutional neural network for modeling sentence pairs. TACL **4**, 259–272 (2016)
27. Zhang, X., Li, S., Sha, L., Wang, H.: Attentive interactive neural networks for answer selection in community question answering. In: Proceedings of the Thirty-First AAAI Conference on Artificial Intelligence, San Francisco, California, USA, 4–9 February 2017, pp. 3525–3531. AAAI Press (2017)

[illegible] matching [illegible] [illegible] Australia [illegible]
[illegible] [illegible] [illegible] Portugal [illegible] [illegible]
[illegible] [illegible]
[illegible] [illegible] of the [illegible]

Author Index

United States
rs